25th Edition of IManEE 2021 International Conference (IMANEE 2021)

IOP Conference Series: Materials Science and Engineering
Volume 1235

Online
21 - 22 October 2021

Editors:

Laurentiu Slatineanu
Margareta Coteata
Andrei Mihalache
Nicanor Cimpoesu
Carmen Bujoreanu
Vasile Merticaru
Adriana Munteanu
Felicia Banciu
Mihai Boca
Marius Ionut Ripanu
Dragos Chitariu

ISBN: 978-1-7138-4692-5
ISSN: 1757-8981

Printed from e-media with permission by:

Curran Associates, Inc.
57 Morehouse Lane
Red Hook, NY 12571

Some format issues inherent in the e-media version may also appear in this print version.

Institute of Physics
Dirac House, Temple Back
Bristol BS1 6BE UK

Phone: 44 1 17 929 7481
Fax: 44 1 17 920 0979

techtracking@iop.org

Additional copies of this publication are available from:

Curran Associates, Inc.
57 Morehouse Lane
Red Hook, NY 12571 USA
Phone: 845-758-0400
Fax: 845-758-2633
Email: curran@proceedings.com
Web: www.proceedings.com

TABLE OF CONTENTS

IOP Conf. Series: Materials Science and Engineering 1235 (2022) 011001 doi:10.1088/1757-899X/1235/1/011001

INNOVATIVE MANUFACTURING ENGINEERING & ENERGY INTERNATIONAL CONFERENCE - IManEE 2021

Preface

This volume of "IOP Conference Series: Material Science and Engineering" contains the selected papers presented in the ***25th International Conference on Innovative Manufacturing Engineering & Energy - IManEE 2021 (http://www.2021.imane.ro/)***, edition organized by Department of Machine Manufacturing Technology, Faculty of Machines Manufacturing and Industrial Management, from the "Gheorghe Asachi" Technical University of Iasi, Romania.

The conference was held as planned during ***21st- 22nd of October 2021*** and, due to the pandemic, all the sessions were held virtually, on Google Meet. We acknowledge that the pandemic had a huge impact on research and mobilities and we were happy that despite these difficulties, the communication and the exchange of ideas and knowledge were experienced by the international participants from Greece, Israel, Ukraine, Poland, Hungary, France, Germany, Serbia, Portugal, Bulgaria, Russia, Moldova Republic, Iraq, and Romania. The parallel sessions on the main topic ensured 15 minutes of presentation and discussion for each presented paper. One must know that all submitted papers were single-blind reviewed by at least two reviewers.

Topics

The papers of this volume are related to the following topics, as the IManEE 2021 conference sections were held:

Additive manufacturing
Advanced Processes and surface engineering
Advanced materials
Automotive & transport engineering
Mechanical and manufacturing equipment
CAD/CAM/CAE/ CAx & Automation
Process quality and evaluation
Design, analysis and modelling
Green energy & energy management
Industrial and product management.

Keynote Speakers of IManEE 2021 were:

Elizabeth Bauer Grosse, University of Lorraine, Institut Jean Lamour, Nancy, (FR): *Search for novel carbides in transition metal -carbon systems*

Arik Sadeh, Industrial Engineering and Technology Management Faculty - HIT Holon Institute of Technology, Israel - *Measuring Efficiency Using Data Envelopment Analysis DEA: Developing and Manufacturing Cellular Modem*

Tamás Bányai, Institute of Logistics, University of Miskolc, Hungary: *Industry 4.0-enabled transformation of conventional production and logistics into cyber physical systems*

Miloš Madić, Faculty of Mechanical Engineering in Niš, University of Niš, Serbia: *Status and development trends of laser processing technology*

Răzvan Păcurar, Technical University of Cluj-Napoca, Faculty of Machine Manufacturing, Cluj: Napoca, (RO) - *Boosting the scientific excellence and innovation capacity of 3D printing methods in pandemic period*

Evangelos Bellos, George Chatzistelios, Angeliki Deligianni & Konstantinos Kirytopoulos, National Technical University of Athens, School of Mechanical Engineering, (GR): *Stakeholder analysis for the development of LNG bunkering facilities*

Members of Organizing Committee from TU of Iasi:

Eugen Axinte; Mihai Boca; Constantin Cărauşu; Viorel Cohal; Oana Dodun, Petru Duşa; Laurenţiu Ghenghea; Octavian Lupescu; Vasile Merticaru; Gavril Muscă; Florin Negoescu; Gheorghe Nagit; Marius Rîpanu; Lucian Tăbăcaru, Gheorghe Creţu – Conference Chair, and program chairs Andrei Mihalache and Margareta Coteaţă.

Department of Manufacturing Engineering, Dunarea de Jos University of Galati, (RO):

Cătălin Fetecău, Gabriel-Radu Frumuşanu, Viorel Păunoiu, Cătălina Maier, Virgil-Gabriel Teodor, Cezarina Afteni.

Scientific Committee

Ovidiu-Dorin ALUPEI-COJOCARIU (RO), Daniel-Constantin ANGHEL (RO), Eugen AXINTE (RO), Qingshun BAI (CN) Nicolae BALC (RO), Felicia BANCIU (RO), Alexandra BANU (RO), Irina BESLIU BANCESCU (RO), Katarzyna BIAŁAS (PL), Florin-Sandu BLAGA (RO), Mihai BOCA (RO), Ioan BONDREA (RO), Toufik BOUDOUH (FR), Paul Dan BRÎNDAŞU (RO), Mihaela BUCIUMEANU (RO), Carmen BUJOREANU (RO), Leandru-Gheorghe BUJOREANU (RO), Constantin BUNGAU (RO), Miguel CAVIQUE (PT), Kai CHENG (UK), Francisco CHINESTA (FR), Neculai Eugen SEGHEDIN (RO), Dragos CHITARIU (RO), Lucian-Ionel CIOCA (RO), Margareta COTEATA (RO), Vasilis DIMITRIOU, (GR), Oana DODUN (RO), Cristian-Vasile DOICIN (RO), George DRAGHICI (RO), Daniel DRAGOMIR-STANCIU (RO), Catalin DUMITRAS (RO), Petru DUSA (RO), Andrzej DZIERWA (PL), Nicoleta ENE (USA), Cristina FENISER (RO), Catalin FETECAU (RO), Alexandru Cătălin FILIP (RO), Lorelei GHERMAN (UK), Daniel GHICULESCU (RO), António GONÇALVES-COELHO (PT), Richard Ioan Emil HERMAN (RO), Radu IBANESCU (RO), Romeo IONESCU (RO), Daniela-Monica IORDACHE (RO), Juliana JAVOROVA (BG), John KECHAGIAS (GR), Apostolos KORLOS (GR), Irene KORONAKI (GR), Panagiotis KYRATSIS (GR), Ionut LAMBRESCU (RO), Stanisław LEGUTKO (PL), Dorin LUCA (RO), Miloš MADIĆ (RS), Catalina MAIER (RO), Angelos P. MARKOPOULOS (GR), Sergiu MAZURU (MD), Vasile MERTICARU (RO), Eugen MERTICARU (RO), Andrei Marius MIHALACHE (RO), Alina Adriana MINEA (RO), António MOURÃO (PT), Adriana MUNTEANU (RO), Ilie MUSCA (RO), Irina NEAGA (UK), Florin NEGOESCU (RO), Pantelis NIKOLAKOPOULOS (GR), Eduard Laurentiu NITU (RO), Anna NOCIVIN (RO), Gheorghe OANCEA (RO), Dumitru OLARU (RO), Ancuta PACURAR (RO), Razvan PACURAR (RO), Viorel PALEU (RO), Carmen Ema PANAITE (RO), Viorel PAUNOIU (RO), Marek PŁACZEK (PL), Marcela POKUSOVA (SK), Marcel Sabin POPA (RO), Daniela POPESCU (RO), Fabrizio QUADRINI (IT), Sever-Gabriel RACZ (RO), Miroslav RADOVANOVIC (RS), Victorita RADULESCU (RO), Marius Ionut RIPANU (RO), Alin Daniel RIZEA (RO), Tetiana ROIK (UA), Arik SADEH (IL), Loredana SANTO (IT), Francisco SANTOS MARTIN (ES), Christophe SAUVEY (FR), Carol SCHNAKOVSZKY (RO), Elena SCUTELNICU (RO), Ioan SEBESAN (RO), Lorenzo SEVILLA HURTADO (ES), Alexandru SOVER (DE), Vasilios SPITAS (GR), Marius-Adrian SPIROIU (RO), Felicia STAN (RO), Nicolae-Doru STĂNESCU (RO), Gabriela STRNAD (RO), Mirela SUCHEA (GR), Petr SVOBODA (CZ), Roberto TETI (IT), Irina TIȚA (RO), Aurel Mihail TITU (RO), Alexei TOCA (MD), Pavlina TOTEVA (BG), Cristian ULIANOV (UK), Gyula VARGA (HU), Aleksandar VENCL (RS), Andrzej WRÓBEL (PL).

IOP Conf. Series: Materials Science and Engineering 1235 (2022) 011001 doi:10.1088/1757-899X/1235/1/011001

List of Editors of IManEE 2021

Laurentiu Slătineanu, PhD
Margareta Coteata, PhD
Andrei Mihalache, PhD
Nicanor Cimpoesu, PhD
Carmen Bujoreanu, PhD
Vasile Merticaru, PhD
Adriana Munteanu, PhD
Felicia Banciu, PhD
Mihai Boca, PhD
Marius Ionut Ripanu, PhD
Dragos Chitariu, PhD

Remember

Since 1996 the Department of Manufacturing Technology from Gheorghe Asachi Technical University of Iasi, Romania, had successfully organized the conference annually by alternating the coordination in partnership with the Department of Manufacturing Technology from the Technical University of Moldova, Republic of Moldova. In the recent years some other partners from Romania – Pitesti and from Greece – Athens and Kozani were main organisers.

This 25th edition of the conference is dedicated to the memory of Professor Emeritus - Doctor Honoris Causa Octavian Vasile Pruteanu (1933 -2021), the initiator of the international conference organized by our department.

IOP Conf. Series: Materials Science and Engineering 1235 (2022) 011002 doi:10.1088/1757-899X/1235/1/011002

Peer Review Statement

All papers published in this volume have been reviewed through processes administered by the Editors. Reviews were conducted by expert referees to the professional and scientific standards expected of a proceedings journal published by IOP Publishing.

- **Type of peer review:** Single Anonymous
- **Conference submission management system:** Morressier
- **Number of submissions received:** 105
- **Number of submissions sent for review:** 97
- **Number of submissions accepted:** 79
- **Acceptance Rate (Submissions Accepted / Submissions Received × 100):** 76.2
- **Average number of reviews per paper:** 2.6
- **Total number of reviewers involved:** 85
- **Contact person for queries:**
  Name: Margareta Coteata
  Email: mcoteata@tcm.tuiasi.ro
  Affiliation: Gheorghe Asachi Technical University of Iasi, Romania

IOP Conf. Series: Materials Science and Engineering 1235 (2022) 012001 doi:10.1088/1757-899X/1235/1/012001

Study behaviour of ratchet mechanism manufactured with 3D printing technology

C I Rizescu, D E Baciu and D Rizescu

University POLITEHNICA of Bucharest, Mechatronics and Precision Mechanics Department, Splaiul Independentei 313, sector 6, 060042, Bucharest, Romania

danarizescu@netscape.net

Abstract. In the present paper, there was developed research for the behavior of ratchet mechanism and ratchet wheel considering several materials and technologies. A ratchet wheel and pawl allow rotation of the shaft in only one direction. Considering 3D printing technologies, the paper approaches the designing and machining of a classic ratchet mechanism, generally well known, but there were highlighted some special particularities. That is known that the ratchet mechanisms are used mainly to low-speed applications, as well as in those applications where noise and vibration are of low importance. Using materials like PLA (Polylactic Acid) or ABS (Acrylonitrile Butadiene Styrene) for example, noise and vibrations can be considerably reduced. Another advantage is that the entire mechanism has lightweight. In this work, there were considered different infill percentages, 20% and 30%, for the 3D printed parts. The designed experimental setup allows rapid changing of the mechanical elements. There were developed several mechanisms, considering different numbers of teeth for the ratchet wheel and two materials. This paper aims the study of the behavior of machine elements obtained as 3D Printed parts like gears, bellows, bushes, sliders, brakes, diaphragms, leaf, and helical springs, bearings, clutches, etc., presented by authors in the previous papers.

1. Introduction

The last five years have brought incredible achievements to these still new additive technologies, which makes it possible in our days to find a large variety of products manufactured by these technologies and some in development, making them more and more competitive in different areas.

AM technologies applications in this area are obtaining more importance. Finding and approving new materials are the real challenges related to that can be put into desirable physical forms with the most suitable mechanical characteristics, enhancing dimensional precision and surface quality, making elements from various materials with a operational role.

Ratchet mechanisms are used as control and actuation mechanisms or as mechanisms for blocking movement. In the first situation, they are used in combination with other mechanisms (cams, levers) resulting in an intermittent movement for the driven element.

Operating conditions. The determination of the operating conditions is based on the relative position of the pawl with respect to the ratchet wheel, described by the position of the oscillation axis of the pawl with respect to the normal n-n, considered at half of the working height of the ratchet wheel tooth, figure 1.

Three different ratchet positions are identified. In the case of variant I, the O_2 oscillation center of the ratchet is above the normal *n-n*, the condition is set that the ratchet can be inserted in the gear. For this, the moments of the active forces must overcome the moment due to the friction force from the point of contact. Thus, neglecting the friction in bearing O_2, the relation can be written [1]:

$$F \cdot l \cdot tg\alpha + Q \cdot a \geq \mu \cdot F \cdot l, \tag{1}$$

where l represents the distance from the oscillation centre of the ratchet O_2 to the contact point B;
a - the distance from the force Q to the ratchet axis,
Q - the force Q in the spring,
F – active force, $F = \frac{M_t}{R}$, M_t – ratchet wheel torque, R – ratchet wheel radius (R=28.5 mm);
μ – friction coefficient between ratchet and wheel teeth;
α - ratchet axis elevation angle (above the normal n-n)

$$Q \geq \frac{F \cdot l \cdot (\mu - tg\alpha)}{a}, \tag{2}$$

In the situation where there is no spring (Q = 0) the necessary condition to achieve the gearing is:

$$tg\alpha > tg\varphi = \mu$$

or α > ϕ where ϕ is the friction angle.

Considering, for example, μ ≈ 0.2 results α > 12°.

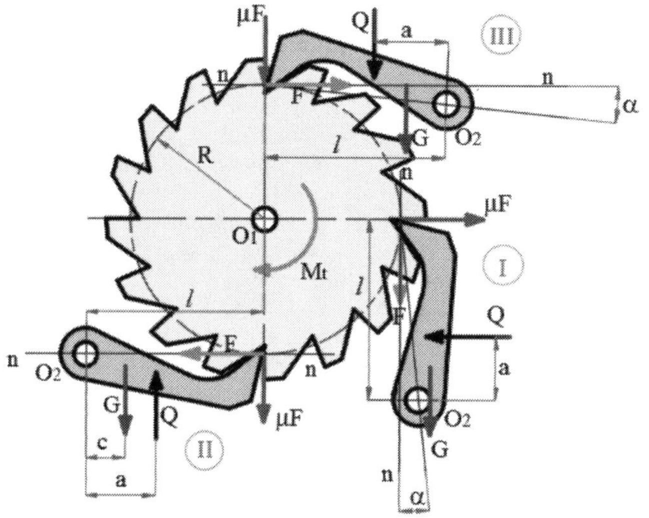

Figure 1. Ratchet mechanism [1].

In variant II (the centre O_2 is on the normal n-n-), condition (1) becomes [1]:

$$Q \cdot a > \mu \cdot l \cdot F \tag{3}$$

In this case, if a spring is not used, the ratchet cannot work and then it will be mounted in variant I, the condition being:

$$G \cdot c \geq \mu \cdot l \cdot F \tag{4}$$

where G represents the ratchet weight, c - the distance from the centre of gravity of the ratchet to the oscillation point O_2.

The operation in variant III (the centre of oscillation located below the normal *n-n*) is the most disadvantageous because there is a tendency to expel the ratchet from the gear position. To avoid this, the condition must be met [1]:

$$\mu \cdot F \cdot l + Q \cdot a \geq F \cdot l \cdot tg\alpha \tag{5}$$

If no spring is used, then relation (5) becomes:

$$tg\varphi \geq tg\alpha \quad \text{or} \quad \varphi \geq \alpha$$

Ratchet mechanisms can be found in many applications such as: tachometer construction, clock mechanisms, musical and mechanical toys, time relays, feed mechanisms [2]. The specialized literature offers numerous solutions for achieving intermittent movement. Their comparison is most often limited to questioning the size of the angular step, the number of moving elements and coupling positions, spatial arrangement or other issues related to the execution, adjustment, and accuracy of obtaining the angular step. The comparison of the kinematic parameters is possible only after the complete dimensioning of the mechanism. The mechanism was designed using Inventor 2018, in two variants: with ratchet wheel provide with 12 teeth and with 24 teeth. These mechanisms were printed using two materials: PLA (Polylactic Acid) and ABS (Acrylonitrile Butadiene Styrene) [3].

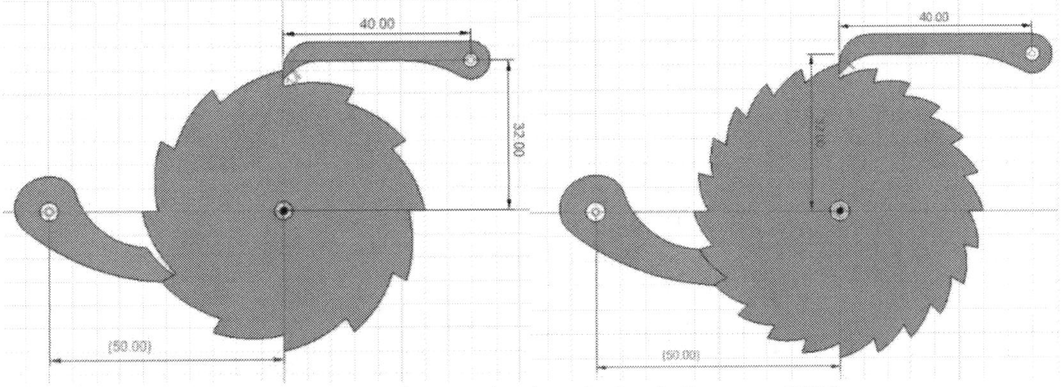

Figure 2. Ratchet mechanism design in Inventor 2018.

2. Experimental setup

The ratchet mechanism consists of the subsequent elements, see figure 3: 1 – electric motor for driver ratchet actuation, 2 – driver ratchet, 3 – driver ratchet tension helical spring, 4 – ratchet wheel, 5 – compression helical spring, 6 – locking ratchet. There were developed four ratchet mechanisms with respect to the materials (PLA and ABS) and ratchet wheels teeth numbers (12 and 24).

Figure 3. Experimental ratchet mechanism setup.

In order to determine the force Q, from the two helical springs, there was used the Hans-Schmidt HV 500N vertical stand, as it is shown in figure 4, [4]. The stand consists of 1 - base table, 2 - displacement measuring system for spring deflection determination, 3 - Imada force transducer, and 4 – helical spring. Both springs were tested, and their characteristics are presented in figure 5.

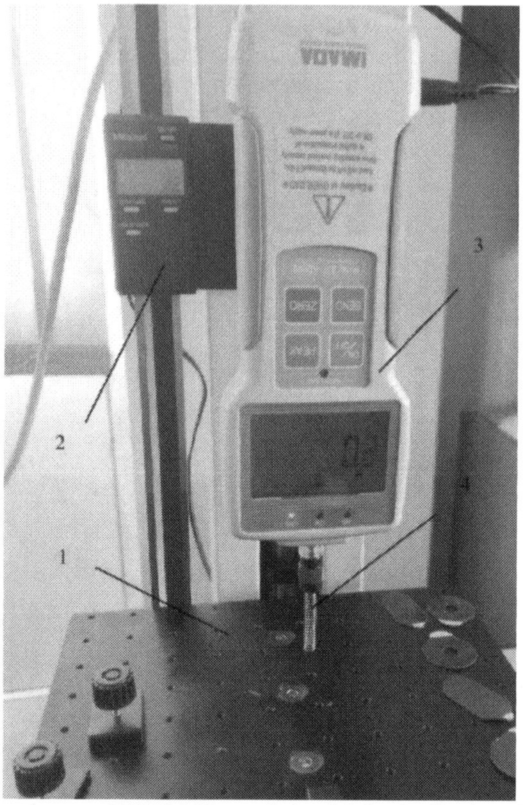

Figure 4. Experimental setup HV500 Hans-Schmidt for helical spring characteristic.

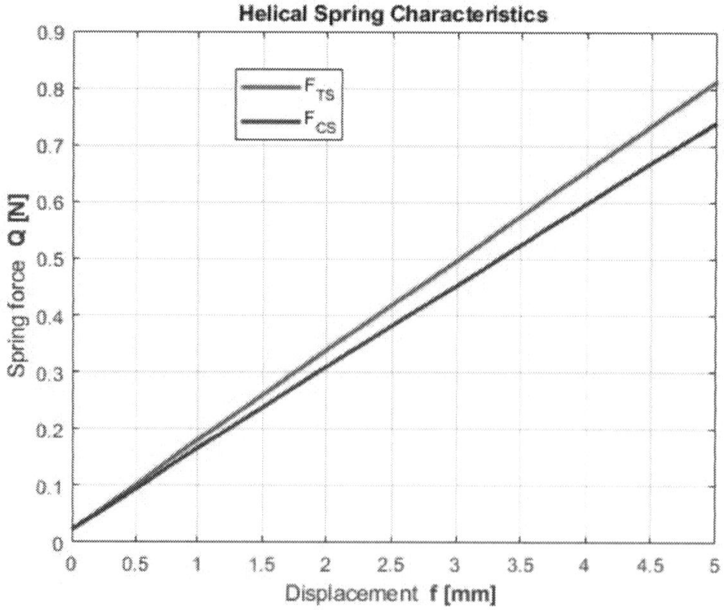

Figure 5. Experimental setup HV500 Hans-Schmidt for helical spring characteristic.

The force for tension helical spring, operating on driver ratchet is labelled with F_{TS}, while F_{CS} is the force for compression helical spring, operating on locking ratchet. The maximum values for spring forces are 0.8 N.

3. Mechanism components manufactured using 3D Printing

The process of additive manufacturing by FDM has a remarkable development in the last 6-10 years, its main advantage being the achievement of robust parts with a relatively low cost [5]. However, an important disbenefit of this technology is still identified by the accuracy and the roughness of the parts obtained.

It is known that, due to the advantages of additive manufacturing FDM, the development of the specific control systems could be strong support for new open-source for software applications. Considering this point of view, the assessment of the quality of the parts obtained by FDM is the subject of several studies and researches on the mutual software solutions and the properties of the materials.

In the specialty literature, there is a number of researches which had as main objective the optimization of part surfaces obtained by FDM, addressing solutions like application of metallic paint on the surfaces of the printed, the application of chemical treatments or post-processing technologies of adjustment controls [6].

These solutions have the main disadvantage of entering a new phase in the process of obtaining the parts, a phase that increases production times and costs.

Further production processes are required to obtain complex components in a short time, at reduced production costs. To meet these requirements, the Fused Deposition Modelling (FDM) process parameters must be established for every application. The obtained quality is determined by the material properties, process parameters and characteristics of the similar Prusa MK3S + machine used to print these components. PLA and ABS are the two most common FDM desktop printing materials [7,8]. The mentioned materials are thermoplastics, meaning they enter a soft and mouldable state when heated and then return to a solid when cooled. Through the FDM process, both are melted and then extruded through a nozzle to build up the layers that create a final part. In figure 6 there are shown all four ratchet wheels with respect to the materials (PLA and ABS) and ratchet wheels teeth numbers (12 and 24). The infill percentage, defined as the density of the pattern of interior surfaces, was 20%. The thickness of the lateral wall of the wheel is 1.5 mm.

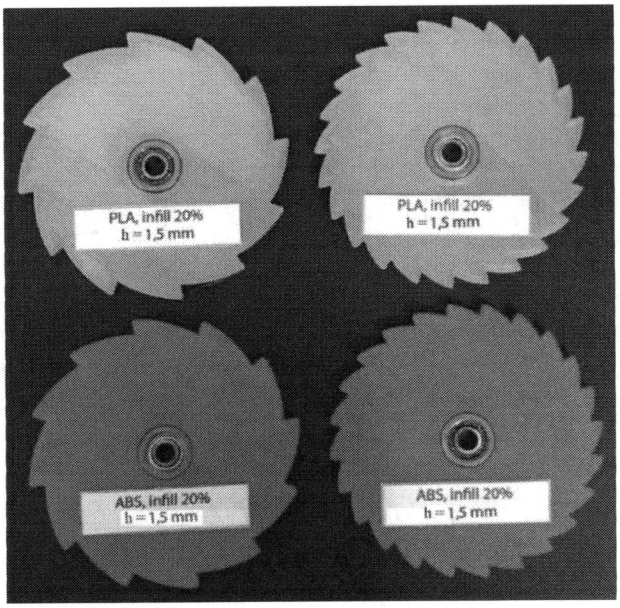

Figure 6. Ratchet wheels printed from PLA and ABS with 12 and 24 teeth.

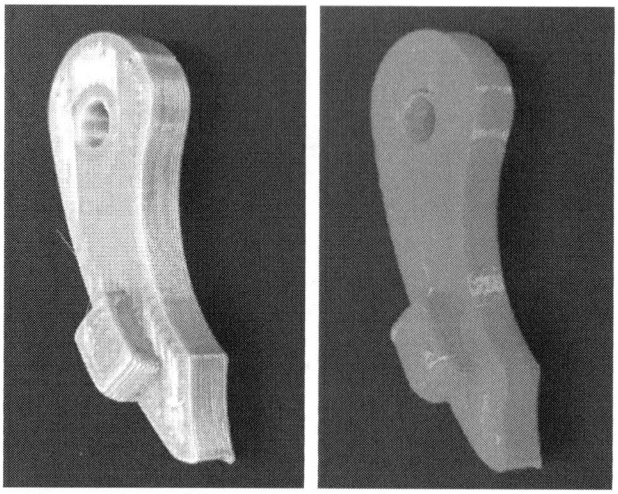

Figure 7. Locking rachet: PLA left and ABS right.

Acrylonitrile Butadiene Styrene is a 3D printing plastic and was practiced for the locking ratchet as shown in Figure 7, on the right side. It is an adaptable petroleum-based substance that belongs to a class of thermoplastic polymers. There are different trademarks and grades of ABS filament, which makes them unchanging in terms of use and design. PLA is the most widely used material for filament. It is a biodegradable, printable and also durable material and was used for the ratchet mechanism. Consequently, it is a good option for printing objects with relatively big dimensions due to its poor thermal expansion (little or without warping) and for printing weeny elements due to its reduced fusing temperature. In figure 7, on the left side there is the locking ratchet [9,10].

4. Experimental results obtained on the experimental setups

The experimental results include several angular rotation speeds for actuating motor for driver ratchet. There were considered values of the angular rotation speed as: $\omega_1 = 24$ rpm and $\omega_1^{`} = 66$ rpm. Considering these speeds, the experimental setup allows studying the cinematic and dynamic behaviour of ratchet mechanisms [6]. There was developed a program (software) for actuating DC motor for each experimental stand, corresponding to the number of teeth of the ratchet wheel. The motor used to drive the ratchet is an SG90 servo motor with 180 degrees rotation (90 degrees in each direction). The rotation angle of the electric motor is 45 degrees for the ratchet wheel with 24 teeth and 65 degrees for the wheel with 12 teeth. It is controlled by an Arduino UNO, and power is supplied directly from the control board.

Thermoplastic extrusion technology is based, like other additive manufacturing processes, on the design of an apparent image of the real object in a 3D editor in Computer-Aided Design program. The authors have studied other researches from literature dedicated to design and printing parts, mechanisms and entire precision mechanics instruments [11-13]. In the case of this study, the three-dimensional model of a ratchet mechanism was developed.

Table 1. Mass of ratchet assembly

Material	Mass [g]
Polylactic Acid	32
Acrylonitrile Butadiene Styrene	26

After 3D modelling is completed, the obtained file is exported to *.stl* format, which is recognized by most modern 3D printers [14,15]. Then the *.stl* file is processed by a special slicing program, which transforms it into a control code G for the FDM printer [7].

The processed file was sent to our Prusa MK3S+ printer. The dimensions of the printer printing table are 200 x 250 mm, the maximum height of a printed object being 210 mm. Other important printing parameters were: layer height was set to 0.15 mm, heat bed temperature was 110 °C, the environmental temperature was 22 °C.

Figure 8. Four experimental stands to test ratchet wheel mechanisms.

The reason for which the authors developed such experimental setups was to determine cinematic parameters of ratchet mechanisms. Consequently, there were designed and manufactured four ratchet mechanisms. Considering only the weight of ratchet wheel and locking ratchet and neglecting other parts like shafts and bearings, the weights of the printed mechanisms are presented in Table 1.

Using equation (2) the friction coefficient between the locking ratchet and the ratchet wheel was determined. There were considered $M_t = 0.08$ Nm, $R = 28.5$ mm, $l = 40$ mm, $a = 20$ mm, $Q = 1$ N, and angle $\alpha = 12°$. The friction coefficient results $\mu = 0.39$ for both materials.

5. Conclusions

The work performs a theoretical and experimental study which has as its notable objective the estimation of the integration capacity of the ratchet mechanisms acquired by FDM in high accuracy mechanisms. The ratchet mechanism has been studied and flourishing tested in operation with the help of additive technologies from two different printable materials. The set-ups made are used as demonstrative stands in the laboratory, for educational applications. Considering both materials, ABS and PLA, it turned out that their properties and characteristics are recommended for mechanical engineering applications. The advantage of ABS is its reduced weight, which makes it proper for applications where reduced mass is necessary. For the work purpose, there were printed using additive manufacturing technology four ratchet mechanisms from two different materials. In order to analyse the behaviour of the printed ratchet mechanisms, they were operated at different drive angular speeds, 24 rpm, and 66 rpm respectively. Because the mechanisms weights are nearly equal, there are no significant differences in the cinematic behaviour due to the different materials, PLA and ABS. In order to prevent the ratchet from disengaging, all mechanisms were provided with tension/compression helical springs. The friction coefficient between the locking ratchet and the ratchet wheel was determined for both materials. Even the friction coefficient was relatively high, its value can be reduced by using different pairs of materials. This approach will be developed in another work. Furthermore, a future research target is represented by the fatigue and resistance tests of the ABS ratchet wheel. The research team developed multiple previous researches to evaluate the mechanical behaviour of stamped machine parts such as different types of bearings, springs, clutches, gears, bearings, etc. The experimental setups can be used for teaching purposes for laboratory works in disciplines such as mechanisms for precision mechanics or precision instruments.

ORCID iDs
C I Rizescu https://orcid.org/0000-0002-2764-520X
D E Baciu https://orcid.org/0000-0001-5722-3027
D Rizescu https://orcid.org/0000-0002-7738-2540

References
[1] Demian T, Tudor D and Grecu E 1982 Mechanisms for precision mechanics (Bucharest: Editura Didactica si Pedagogica) in Romanian
[2] Rizescu C 2013 Mechanisms and elements for precision mechanics Part I (in Romanian) (Bucharest: Printech) p 41-55 in Romanian
[3] Budzik G 2011 *Advanced Applications of Rapid Prototyping Technology in Modern Engineering* 339-364 https://doi.org/10.5772/22848
[4] https://www.hans-schmidt.com/en/produkt-details/test-stand-hv-500n/
[5] Voicu A, Gheorghe G and Bădiţa L 2014 *Studies and Research on the measurement and 3D integrated control of complex components from automotive industry by laser scanning* Int. Conf. ICNAAM 2014 https://doi.org/10.1063/1.4912852
[6] Bickford J H 1972 *Geneva Mechanisms, Mechanism for Intermittent Motion* New York: Industrial Press Inc. p 264 https://hdl.handle.net/1813/57662
[7] Mohamed O A, Masood S H, and Bhowmik J L 2015 *Adv. Manuf.* **3** 42–53 https://doi.org/10.1007/s40436-014-0097-7
[8] Ravi P, Shiakolas P S and Thorat A D 2017 *Journal of Manufacturing Science and Engineering*, **139** (7) 071007 https://doi.org/10.1115/1.4035963
[9] Pearce J M, Blair M, Laciak C, Andrews K J, Nosrat R, Zelenika-Zovko A 2010 *Journal of sustainable development* **3**(4) 17-29 https://doi.org/10.5539/jsd.v3n4p17
[10] Khosravani M R, Schüürmann J, Berto F and Reinicke T 2021 *Polymers* **13** 1559 https://doi.org/10.3390/polym13101559
[11] Mishra P K, Senthil P 2020 *Mater. Today Commun.* **23** 100955 https://doi.org/10.1016/j.mtcomm.2020.100955
[12] Zhang X, Fan W, Liu T 2020 *Compos. Commun.* **21** 100413 https://doi.org/10.1016/j.coco.2020.100413
[13] Akhouri D, Banerjee D, Mishra S.B. 2020 *Mater. Today Proc.* **26** 2140–2 https://doi.org/10.1016/j.matpr.2020.02.461
[14] Novakova-Marcincinova L, Kuric I 2012 *Manuf. and Ind. Eng.* **11** (1) https://doi.org/10.1109/INES.2012.6249805
[15] Sharma V and Sing S 2016 *International Journal of Computational Intelligence Research* **12** (1) p 55-61

Experimental characterisation of physical properties of different PLA filaments used in FFF 3D printing technology

A Lang

Ansbach University of Applied Sciences, Faculty of Technology, Residenzstr. 8, Ansbach, Germany

anette.lang@hs-ansbach.de

Abstract. In the field of 3D printing, fused filament fabrication (FFF), many PLA filaments are offered by different providers. The physical properties of the filaments from various manufacturers differ significantly from each other. These differences have corresponding effects on the processing of the filaments, as well as on the quality of the final product. This article compares PLA filaments from different manufacturers and additionally gives an overview of their physical properties. The investigated properties were namely the melting point, cold crystallization, glass transition temperature, melt flow index and residual content. The results show the differences between the various manufacturers' specifications in the data sheets and point out others. The melt flow index in particular stands out here, which can have a strong influence on the printing process. The effect especially on the mechanical properties has to be determined in further investigations.

1. Introduction

The use of bio-based polymers such as polylactide (PLA) has become an integral part of everyday life due to its wide application spectrum, e.g. packaging material, medical products and 3D printing filament. The biodegradability of PLA should also be highlighted [1]. In the field of 3D printing, it is often used as a filament in the FDM method. Depending on the manufacturer, there are major price differences, which do not have a significant effect on the printing application. The comparison of the filaments from three different manufacturers was made with regard to melt flow behaviour, melting and crystallisation point and fillers. These tests should filter out differences between the individual characteristics of PLA polymers. A white filament from each manufacturer was selected for the tests.

Already when comparing the datasheets, differences in the melting temperature T_m, glass transition temperature T_g, melt flow rate MFR and the printing temperature of the PLA filament can be noticed (see Table 1). Manufacturer B has no information about the MFR and T_m on the datasheet or online.

Table 1. Comparison of PLA data sheets of different manufacturers

	Manufacturer A	*Manufacturer B*	*Manufacturer C*
Density	1.24 g/cm³	1.25 g/cm³	1.26 g/cm³
MFR (210°C, 2.16 kg)	9.56 g/10 min		6.09 g/10 min
T_g	57 °C	~ 50 °C	~ 60 °C
T_m	77 – 146 °C		145 – 160 °C
Printing temperature	180 – 210 °C	190 – 230 °C	200 – 220 °C
Printing speed			30 – 80 mm/s

The differences in the data sheets can have various causes

- Different crystalline forms (such as α-, α´-, β- or γ-PLA) [2]-[8]
- Additives
- Filler materials
- Pretreatment
- Etc.

In order to investigate the characteristics of these PLA grades, different analytical tests were carried out and the results were compared. The performed analyses are Thermo Gravimetry Analyses (TGA), Differential Scanning Calorimetry (DSC) and Melt Flow Rate (MFR).

2. Tests

All measurements were taken in the laboratory of Ansbach University of Applied Sciences. The TGA measurements were carried out with TGA/SDTA851 from Mettler Toledo to determine the proportion of fillers. The test was perfomed in two stages: Stage 1 in the temperature range 25 - 600 °C with a heating rate of 10 K/min in a nitrogen atmosphere; Stage 2 from 600 - 800 °C also with a heating rate of 10 K/min in an oxygen atmosphere. The results are shown in Table 2. The PLA from the company Manufacturer B has the lowest residual with 0.8%. The other two companies have a similar residual of 2.7% and 2.9% respectively. It is noticeable that Manufacturer B has a degradation stage at approx. 300 °C. This degradation stage does not occur with the other two manufacturers (see Figure 1).

Table 2. TGA measurement to detect the residuals

	Manufacturer A	*Manufacturer B*	*Manufacturer C*
Residual in N_2 atmosphere up to 300 °C	--	98.7%	--
Residual in N_2 atmosphere up to 600 °C	2.8%	1.5%	3.2%
Residual in O_2 atmosphere up to 800 °C	2.7%	0.8%	2.9%

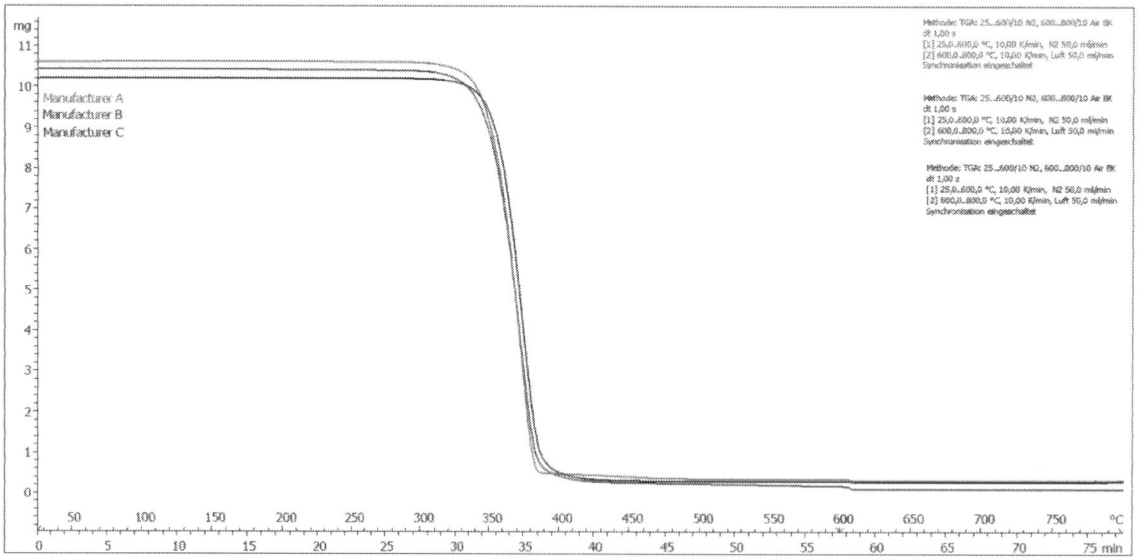

Figure 1. TGA measurements of the PLA material of the three manufacturers

To compare the glass transition temperature and the melting temperature, a DSC measurement was done with the DSC822 from Mettler Toledo. The DSC measurement was carried out in a temperature range of 0 – 200 °C at a heating rate of 10 K/min. In order to distinguish the thermal properties, two heating runs were performed (H1 and H2). In between, a cooling run (C) with a cooling rate of 10 K/min was performed. The results are shown in Table 3. The most important information obtained from the DSC measurements is:

Manufacturer A
A glass transition temperature of 55 ° and approx. 60 °C was measured which corresponds with the range of the manufacturer's specifications. The manufacturer specifies a melting range of 77 – 146 °C. Using DSC, the melting peak was measured to be between 160 – 163 °C. Here, the manufacturer's specifications differ considerably from the measured values, see table 1. Cold crystallisation occurs in both heating runs. This is typical behaviour of PLA, as it requires very low cooling rates or corresponding nucleating agents to crystallise PLA.

Manufacturer B
A glass transition temperature of 61 – 62 °C was determined for the filament from Manufacturer B. This is approx. 10 K higher than in the manufacturer's specifications, see table 1. The melting peak is at 152 - 153 °C. The manufacturer does not provide any information on the melting temperature. Similar to Manufacturer A, cold crystallization also occurs with Manufacturer B in both heating runs. This indicates that PLA is partially crystalline.

Manufacturer C
For Manufacturer C PLA filament, a glass transition temperature of 60 °C is specified. This was confirmed in the DSC measurement. The manufacturer's specifications also match the DSC measurement for the melting temperature. This PLA has the lowest melting point (151 °C) compared to the other two manufacturers. The cold crystallisation peak occurs at 114 – 119 °C.

With regard to the melting point, the three manufacturers' specifications already lie in a wide temperature range, extending from 77 – 160 °C.

Table 3. DSC results

	Manufacturer A			***Manufacturer B***			***Manufacturer C***		
	H 1	C	H 2	H 1	C	H 2	H 1	C	H 2
T_g [°C]	55	56	60	62	56	61	59	56	61
With relaxation			x	x		x	x		x
Cold crystallisation [°C]	98		125	122		124	114		120
T_m [°C]	163		160	153		152	151		151

The MFR was determined with the Meltfixer LT from ThermoFisher. A temperature of 190 °C and a load of 2.16 kg were selected for the measurements. Table 4 shows the results.
The measuring temperature had to be reduced from 210 °C to 190 °C in contrast to the manufacturer's specifications, as otherwise no qualified measurement could have been carried out. Therefore, a direct comparison with the manufacturer's data is not possible. However, the trend can be recognised.

Table 4. MFR results

	Manufacturer A	***Manufacturer B***	***Manufacturer C***
MFR (190 °C, 2.16 kg) *(measurements)*	13.1 g/10 min	7.2 g/10min	4.8 g/10min
Printing temperature *(Data sheet)*	180 – 210 °C	190 – 230 °C	200 – 220 °C

3. Results

The PLA filament from Manufacturer B has the lowest residue in the TGA measurement with 0.8%. A white filament was chosen by all three manufacturers. From this it can either be concluded that the other two manufacturers use additional fillers. Or that the Manufacturer B filament has a lower proportion of anorganic colour pigment. This difference must be investigated by further investigations - both in terms of colour intensity and fillers.

For all three manufacturers, the PLA filament was partially crystalline (see melting peak and glass transition temperature in DSC). The proportion of crystalline areas has a major impact on the mechanical properties of PLA.

Likewise, cold crystallisation is measured for all three filaments. No crystallisation peak appears in the defined cooling during the DSC measurement. Therefore, cold crystallisation is also measured in the second heating run of the DSC. It can be assumed that printed components with these filaments are partially crystalline. In order to support this thesis, detailed analyses must be carried out on 3D-printed components.

For better comparability, the values from the DSC measurement and the measured MFR value were shown in Figure 2 for all three manufacturers.

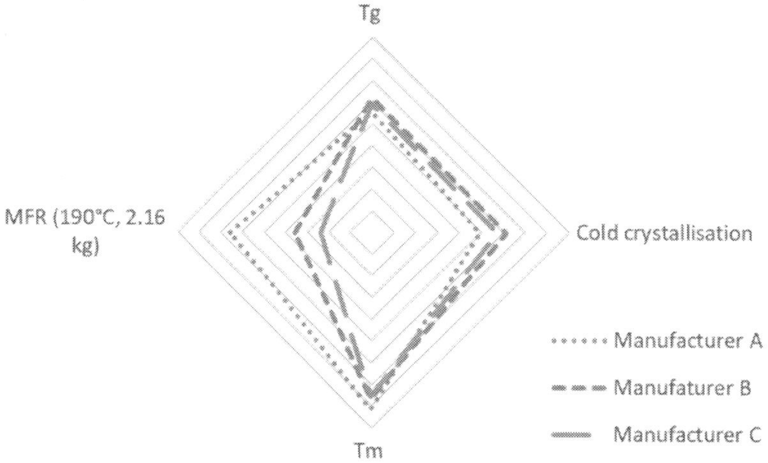

Figure 2. Comparative presentation of DSC and MFR results

The glass transition temperatures and the melting temperatures are in the same order for Manufacturer B and C. The melting temperature of Manufacturer A is up to 10 K higher. The following factors can lead to this:

- Different structures
- Crystallinity
- Molar mass distribution

With regard to the melt flow behaviour, the following relationship can be established between MFR and the printing temperature: The lower the printing temperature, the higher the MFR (see Figure 3).

The product from Manufacturer A has the highest MFR value of 13.1 g/10min and Manufacturer C the lowest with 4.8 g/10min.

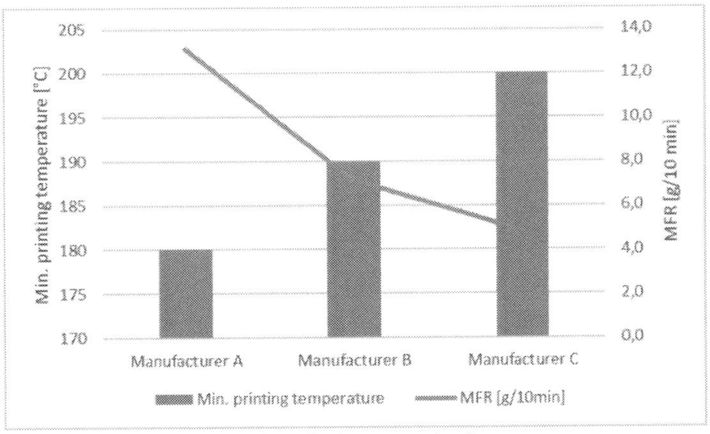

Figure 3. MFR in relation to printing temperature

4. Conclusions

The available datasheets for PLA filament already show some differences between the various manufacturers (density, MFR, T_m, etc.) This is also evident in the investigations using TGA, DSC and MFR. By means of the DSC measurements, the manufacturers' specifications for T_g and T_m could be verified. Only for manufacturer A, the measured T_m value is higher than the values from the datasheet. The TGA measurement did not show any noticeable differences between the manufacturers. Manufacturer B had the lowest residue and consequently the lowest filler content.

The MFR values stand out in particular. The values could not be verified in comparison to the datasheet. For example, the MFR value measured for Manufacturer A was twice as high, although the test temperature had been lowered by 20 K in order to receive reliable results. This may have a strong influence on the printing application. Further investigations concerning the printing application are necessary here. Moreover, the effect especially on the mechanical properties should be examined in further tests.

Acknowledgements

A special gratitude I give to the technical staff of the Ansbach University of Applied Sciences.

ORCID iDs

A Lang https://orcid.org/0000-0001-7820-663X
A Sover https://orcid.org/0000-0003-3408-898X

References

[1] Södergård A and Stolt M 2002 *Progress in Polymer Science* **27** 1123-63 https://doi.org/10.1016/S0079-6700(02)00012-6
[2] De Santis P and Kovacs J 1968 *Biopolymers* **6** 299-306 https://doi.org/10.1002/bip.1968.360060305
[3] Hoogsteen W, Postema A R, Pennings A J, ten Brinke G and Zugenmaier P 1990 *Macromolecules* **23** 634-42 https://doi.org/10.1021/ma00204a041
[4] Kalb B and Pennings A J 1980 *Polymers* **21** 607-12 https://doi.org/10.1016/0032-3861(80)90315-8
[5] Puiggali J, Ikada Y, Tsuji H, Cartier L, Okinara T and Lotz B 2000 *Polymer* **41** 8921-30 https://doi.org/10.1016/S0032-3861(00)00235-4
[6] Cartier L, Okihara T, Ikada Y, Tsuji H, Puiggalli J and Lotz B 2000 *Polymer* **41** 8909-19 http://dx.doi.org/10.1016/S0032-3861(00)00234-2
[7] Zhang J M, Duan Y X, Sato H, Tsuji H, Noda I, Yan S and Ozaki Y 2005 *Macromolecules* **38** 8012-21 https://doi.org/10.1021/ma051232r
[8] Cho T Y and Strobl G 2006 *Polymer* **47** 1036-43 https://doi.org/10.1016/j.polymer.2005.12.027

IOP Conf. Series: Materials Science and Engineering 1235 (2022) 012003 doi:10.1088/1757-899X/1235/1/012003

Design and physical validation of a non-planar interlocking element for tubular structures

V Ermolai[1,2] A Sover[2] and G Nagîț[1]

[1] "Gheorghe Asachi" Technical University of Iasi, Department of Machine Manufacturing Technology, Blvd. D. Mangeron, 59A, 700050, Iasi, Romania
[2] Ansbach University of Applied Sciences, Faculty of Technology, Residenzstraße 8, 91522, Ansbach, Germany

ermolai.vasile@gmail.com

Abstract. An interlocking topological structure represents an arrangement of parts designed so that the degrees of freedom of each component are takeover by the surrounding neighbours. For the planar arrangements, there are multiple possibilities to design such assemblies via platonic solids such as tetrahedron, cube, octahedron and osteomorphic geometries. This paper describes the geometrical modelling process of a non-planar interlocking element that can be used to produce tubular structures. The geometric element was defined by two curved surfaces identified with the inner and outer shell of the tubular structure and joined by six curved surfaces, providing through those the interlocking design. The virtual geometry was produced using an additive manufacturing process generating the physical model of a tubular construction assembled from non-planar interlocking elements.

1. Introduction

A Topology Interlocking (TI) assembly represents a configuration of elements of a specific shape held together by adjacent elements through contact. Theoretically, those elements mutually interlock due to their arrangement and shape, disregarding the use of binders to keep the pieces together [1]. However, in practice, the contact between elements is not enough to maintain the integrity of a TI assembly; therefore, a surrounding frame is necessary to maintain the constant pressure between element's contact surfaces [1,2].

The simplest TI configurations can be achieved using any of the five platonic solids in a planar arrangement. Those geometries are tetrahedron, cube, octahedron, dodecahedron and icosahedron [2]. As an example, for TI configurations with tetrahedral elements (figure 1a), the interlocking is achieved if on each row of elements there are two sections normal to the assembly plane, so that one section provides the kinematic constraints from one direction, and the other section ensures the same response from the opposite direction as can be observed from figure 1a.

If we imagine multiple cross-sections of the tetrahedra TI configuration, either starting from up or bottom, the first, the mating between the tetrahedral shells begins with lines, followed by rectangles and squares at the level of the middle plane (figure 1b). Due to this fact, contact between elements is maintained in each cross-section of the assembly.

In 2007 Molotnicov *et al.* [3] proposed a new concept of interlocking element known as osteomorphic bricks, similar to the platonic elements regarding the interlocking aspect, but different from those through more complex contact geometry. Their shape is characterized by two flat surfaces and pairs of concave and concave geometries that provide the interlocking behaviour. Since then, many researchers investigated the behaviour of these elements and newly developed elements in various scenarios: load-bearing capacity, bending stiffness, crack propagation, energy absorption and more, proving their capacity to absorb more load compared with the monolithic blocks [4,5]. Due to their already fragmented nature, when an element yields, the other elements maintain the integrity of the configuration. Moreover, it is possible to obtain hybrid configurations by combining elements with different physical properties with many possible applications in civil engineering [6].

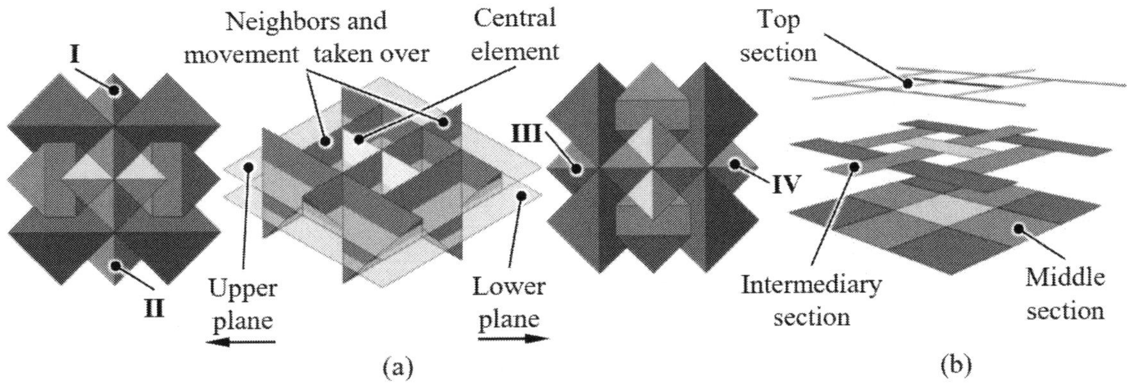

Figure 1. TI assembly configuration with regular tetrahedral elements. (a) The central frame represents an isometric view of a nine elements configuration with two cutting planes. Elements **I** and **II** are blocking the upward movement, while elements **III** and **IV** are restraining the downward movement; (b) Contact geometry for multiple sections: top (line), a quarter (rectangle) and middle (square) of the tetrahedra TI configuration.

The concept of TI configuration was further extended in structural and architectural design related research by adding the third dimension to the existing planar configurations. This way, several papers investigated the possibilities of designing and producing non-planar TI assemblies. Wang *et al.*, 2019 developed an algorithm capable of generating convex interlocking elements that can approximate various double-curved surfaces for architectural design and validated their findings through functional mock-ups [7]. Bejarano & Hoffman, 2019, investigated the possibility of non-planar tessellation using the angle-choice approach for the tetrahedron element, but the method needs further improvement to reduce elements overlap [1]. Xu *et al.* 2020 designed a novel Non-planar Interlocking Element (NpIE) to construct tubular structures and compared the design with a monolithic tube through finite element analysis, the new configuration providing better impact resistance, crack propagation behaviour and higher energy absorption capacity [8].

Based on Xu *et al.* findings, this paper presents the design methodology and CAD structure of adapted non-planar interlocking element for tubular structures with a single symmetry plane, capable of auto-adapting element shape following the input parameters. In addition, samples from the simulated configuration were physically validated through Rapid Prototyping.

2. Design method

Below, the design methodology is described for a parametric non-planar element for tubular structures using the CATIA V5 CAD environment. The design methodology can be used in any CAD software with parametric design features.

The first step consists of defining the input data for the tubular structure, which are described by:

- *Number of elements:* the value of this parameter defines the tessellation on each row of the tubular structure;
- *Value of the inner diameter*: has two important aspects. First, as a constructive parameter for the inner shape of the interlocking element. Secondly it is essential for the practical use aspect, as a functional requirement for the tubular structure;
- *Element thickness:* similar to the previous parameter, it also has constructive importance because of its value depends on the outer shape of the element and practical relevance being as well a functional parameter for the tubular structure;
- *Assembly tolerance:* the value of this parameter (measured in degrees) defines the constructive gap (radial) between the interlocking elements. For the first step of the design process, a fixed value was

used (i.e., 0.5 degree), and later to be changed depending on the other considered parameters. A higher number of elements (e.g., 15 elements) will require a tighter tolerance, while less of them (e.g., six elements) needs a greater value.

The second step consists of defining all the parameters and relations. Those were used to define and automatically update the tubular structure based on the input data described above. All used parameters and equations between the construction and assembly parameters are described in table 1. The presented values were used to define the construction method and changed depending on functional requirements (i.e., number of elements, inner diameter and wall thickness). All used equations implied geometrical aspects and trigonometric functions associated with the used shapes: circle, the arc of circle and hexagon.

Table 1. Constructive and assembly parameters for the NpIE for tubular structures.

	Parameters and equation description	Notation	Values ranges and equations	Step value
Constructive parameters & radial assembly	No. of interlocking elements	NIE	3 to 12	1
	Inner diameter (mm)	ID	40 to 100	10
	Element thickness (mm)	ET	1 to 10	1
	Outer diameter (mm)	OD	*ID+2ET*	
	Radial tolerance (°)	RT	0.05 to 2	0.05
	Assembly angle (°)	AA	*360/NIE*	
	Construction angle (°)	CA	*AA-RT*	
Vertical assembly	External diameter arc chord (mm)	EDAC	*(ID+2•ET)•sin(AA/2)*	
	Translation tolerance (mm)	TT	0.05 to 0.2	0.05
	Translation length (mm)	TL	*EDAC•tan(49.1)•3/4-TT*	
	Rotation angle (°)	RA	*AA/2*	
	No. of rows	NR	2 to n	1

Based on the construction parameters (upper half of table 1) and the build principles of the circular segments figure 2, I), the base sketches were defined for the inner (red) and outer (blue) shell of the NpIE based on the values of the inner and outer diameter, respectively the value of construction angle. In the next step, the planes (figure 2, II) for the construction sketches were defined as the inner and outer shells of the NpIE. Those planes are parallel with the YZ plane and situated on a midpoint referred to Arc Chord, AC (red dot highlighted in figure 2, I).

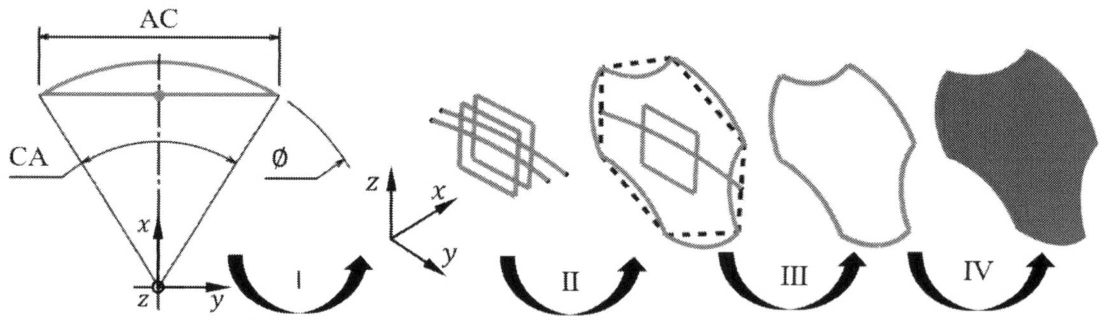

Figure 2. Construction steps of the outer shell of the NpIE.

Further, the construction frame was designed for the outer shape, represented by a regular hexagon with the Width Across Flats (WAF) coincident with the outer diameter arc chord (figure 2, II). Then the two groups of arcs are reflected (concave and convex). The external arcs must be coincident with the circumscribed circle of the hexagon, and the internal arcs must be symmetrical to the same circle having the hexagon edges as reference axis. The resulting contour (figure 2, III) was filled and warped on a circle coincident with the outer diameter of the tubular structure (figure 2, IV). The same build principle was applied for constructing the inner shell but rotated by 60° to ensure the interlocking requirement, with the resulting shape warped on the circle coincident with the inner diameter. Before completing the element shell, first, the assembly quality of the NpIE was checked. It was observed that in the case of the outer shape, the surfaces matched correctly according to the imposed tolerances (i.e. radial tolerance and vertical tolerance), but in the case of the inner shape, the joining was correct just for radial connections. For the vertical assembly direction, the gap was visibly more significant than the radial one (see figure 3 right side). However, it was possible to obtain an acceptable value of the gap for much smaller element thickness (see figure 3 left side). It was identified that the value of the translation length (associated with the outer shell) was responsible for this issue. Thus, it was established that the height of the inner shell must match with the outer shell.

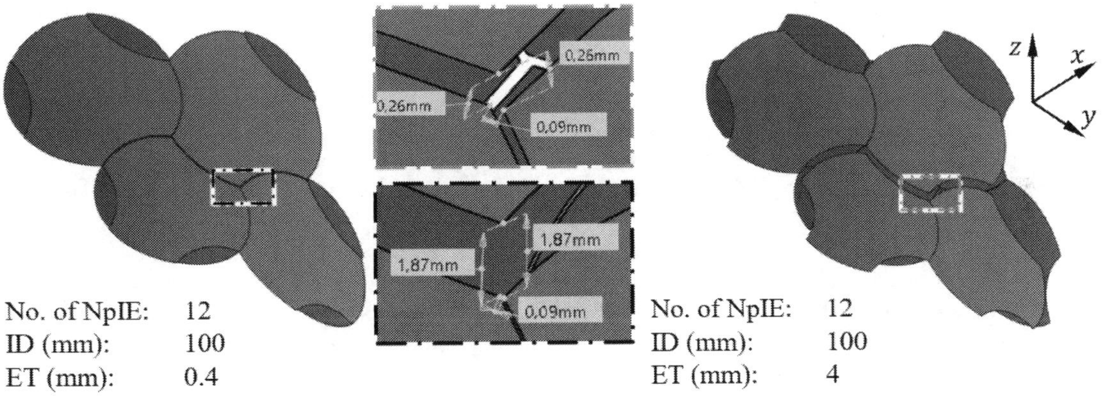

Figure 3. Simulation of the assembly process for NpIE.

In the first stage, the best possible solution was determined for the inner profile of the shell using dimension constraints. Afterwards, the profile was constrained through coincidence between construction lines and angles associated with the base hexagons of the inner respectively outer limits. This way, it was ensured that the inner shape of the NpIE is capable of auto-adjusting when input parameters are changed. A preview of the difference between the regular and the adapted shape is highlighted in figure 4a.

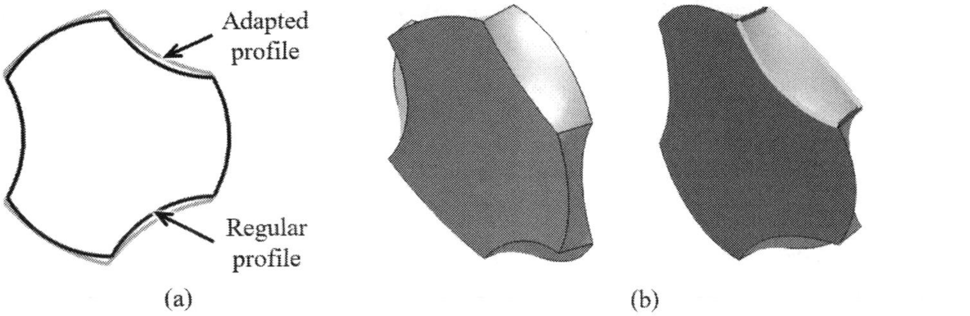

Figure 4. (a) The difference between the regular and adapted profile; (b) Views of the NpIE (left side inner shape and left side outer shape)

The last step of the design process consists of defining the side surfaces that ensure the interlocking aspect of the non-planar element. First, the guidelines were created based on the arches of the inner

IOP Conf. Series: Materials Science and Engineering 1235 (2022) 012003 doi:10.1088/1757-899X/1235/1/012003

and outer warped surfaces, as shown in figure 4b (green lines), followed by the profiles obtained by connecting the extremity points of the arches (brown lines). Then, the profile was lofted, obtaining the side surface, and the process was repeated for all six sides of the NpIE.

3. Results

In the end, based on the assembly parameters (see table 1), the first two rows of the tubular structure were defined and tested in various configurations of number of interlocking elements, inner diameter and element thickness. It was observed that the best configurations for the current inner profile construction imply the use of a higher number of elements, with a wall thickness of 2-3 mm, thickens arranged on a larger diameter. A part of the simulated configurations is presented in figure 5. When the number of elements is increased, the value of the radial tolerance must be reduced; otherwise, the gap value between the NpIE increases.

It can be observed that the adapted profile for the NpIE presented in figure 4 has certain limitations regarding the minimum number of elements that a row may contain. As an example, in the case of six elements tessellation, the smallest gap value was obtained with 100 mm for the inner diameter, 2 mm for the element thickness and a radial tolerance of 0.3 mm. Thus, a smaller tessellation was possible, but the result could not be physically validated.

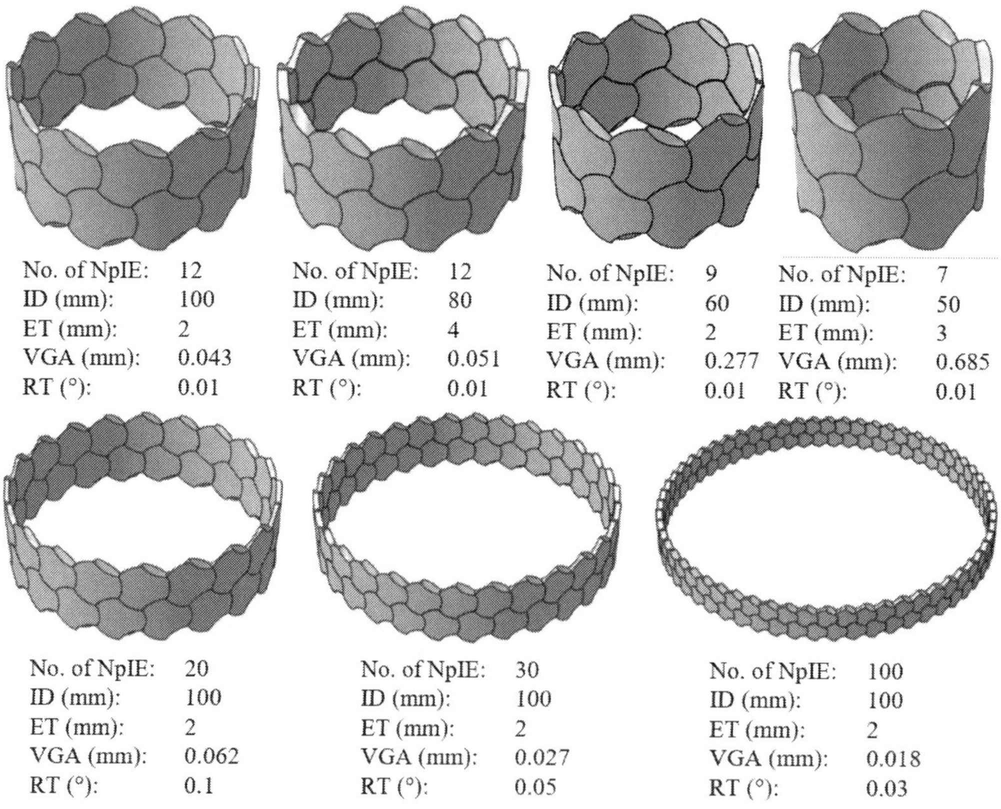

Figure 5. Simulation of the NpIE assembly for tubular structures (VGA is Vertical Gap Average).

The physical validation of NpIE for tubular structures was made through Fused Filament Fabrication (FFF) technology. All parts were printed using polylactic acid (PLA) material for the 12 elements configuration. Besides CAD configuration precision, the assembly quality was influenced by the NpIE orientation during printing (figure 6). In addition, it was observed that the prototypes could not sustain their structural integrity; external pressure is required to hold the side surfaces in contact.

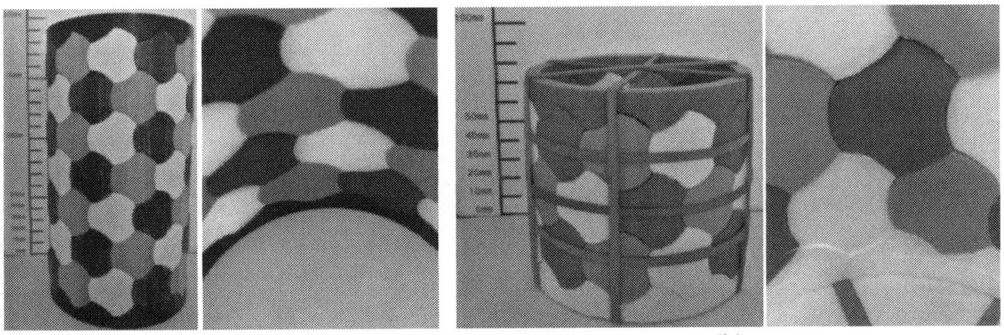

(a) (b)

Figure 6. Prototypes for 12 NpIE configuration, ID 80 mm: (a) ET 3 mm, seven rows, on external surface build orientation (glued elements); (b) ET 6 mm, three rows, on side build orientation.

4. Conclusions

In this paper, a construction methodology of a non-planar interlocking element for tubular structures was proposed, aiming to improve the geometry of the existing solutions to produce interlocking elements. The resulting CAD configuration can generate an assembly geometry for no less than six elements and up to 100 elements per row tessellation on multiple inner diameters and wall thicknesses.

Two tubular structures were validated via Rapid Prototyping to demonstrate the capability of the CAD setup. However, there are some limitations of the current form, and further research must be done in this regard. First, the assembly quality must be improved for elements thicker than 3 mm, followed by an improvement regarding the minimum number of elements.

Furthermore, the adapted profile for the inner shape is not malleable enough to fit all possible combinations of the input parameters. Thus, a correlation between profile constraints and input parameters must be investigated. Another possible solution to the current limitation of the inner profile could be to have multiple profiles divided into classes, into pairs of the number of elements, inner diameter and element thickness. Further investigation will be performed in this regard.

Acknowledgments
This work was funded by "Gheorghe Asachi" Technical University of Iaşi through CNFIS-FDI-2020-0354 grant.

ORCID iDs
V Ermolai https://orcid.org/0000-0001-5967-5748
A Sover https://orcid.org/0000-0003-3408-898X
G Nagîţ https://orcid.org/0000-0001-9649-8525

References

[1] Bejarano A and Hoffmann C 2019 *Int. J. of Architectural Computing* **17** (1) 53-73 https://doi.org/10.1177%2F1478077119827187
[2] Dyskin A V, Estrin Y, Belov-Kanel A J and Pasternak E 2003 *Philosophical Mag. Letters* **83** (3) 197-203 https://doi.org/10.1080/0950083031000065226
[3] Estrin Y, Dyskin A V and Pasternak E 2011 *Mater. Sci. and Eng. C* **31** (6) 1189-94 https://doi.org/10.1016/j.msec.2010.11.011
[4] Molotnikov A, Estrin Y, Dyskin A V, Pasternak E and Kanel-Belov A J 2007 *Eng.Fracture Mec.* **74** (8) 1222-32 https://doi.org/10.1016/j.engfracmech.2006.07.012
[5] Razaee Javan A, Seifi H, Xu S, Lin X and Xie Y M 2018 *Int. J. of Impact Eng.* **116** 79-93 https://doi.org/10.1016/j.ijimpeng.2018.02.008
[6] Sover A 2004 Erweiterung der Fertigungsmethoden für Systeme aus topologisch selbstverzahnten Elementen *IMW - Institutsmitteilung* **24** 57-8
[7] Wang Z, Song P, Isvoranu F and Pauly M 2019 *ACM Transactions on Graphics* **38** (6) 193 https://doi.org/10.1145/3355089.3356489
[8] Xu W Lin X and Xie Y M 2020 *Tunnelling and Underground Space Technology* **103** 103503 https://doi.org/10.1016/j.tust.2020.103503

Influence of contact geometry over the filament bond of polylactic acid blends

V Ermolai[1,2], **A Sover**[2] **and G Nagîț**[1]

1 "Gheorghe Asachi" Technical University of Iasi, Department of Machine Manufacturing Technology, Blvd. D. Mangeron, 59A, 700050 Iasi, Romania
2 Ansbach University of Applied Sciences, Faculty of Technology, Residenzstraße 8, 91522 Ansbach, Germany

ermolai.vasile@gmail.com

Abstract. Fused Filament Fabrication is one of the most common Additive Manufacturing technologies among industrial and domestic users because of the satisfying quality of parts and the great variety of thermoplastic polymers available on the market. New possibilities were provided along with the equipment with multiple extrusion systems, referring to multi-colour and multi-material components. Besides part orientation during the manufacturing process, filament bond influences the mechanical properties of parts produced through Fused Filament Fabrication technology. This paper aims to investigate if the bond performance for polylactic acid-based polymers can be enhanced by designing new contact geometries, different from the current surface-to-surface approach. To comprehensively understand the influence of the designed bonding geometries over the tensile strength, all specimens were produced according to a factorial design of experiment method and contact geometries constrained with function related to the nozzle size.

1. Introduction

Fused Filament Fabrication (FFF) is an extrusion-based technique part of the Additive Manufacturing (AM) technology, which uses molten thermoplastic materials to create physical components. Similar to the other AM technologies, a product is obtained as follows. Firstly, CAD data is exported as a mesh file (e.g., STL, OBJ, 3MF). Next, the mesh is imported into the machine's software, orientated and positioned, followed by process parameters adjustment (a step that depends on the model's application and used material). The third step is represented by the actual manufacturing and, depending on the used material, may require pre-manufacturing preparation (e.g., build surface gluing, environment preheat). Finally, after parts removal, a post-processing operation is required [1] (e.g., removing construction elements like support structures or build plate adhesion elements).

In FFF technology, all parts are made layer by layer with a defined step value (e.g., 0.1, 0.3 mm). Each layer is defined by multiple passes of the extrusion system on defined paths creating the outer shell, consisting of a sum of perimeters, bottom and top solid layers, and the internal geometry, infill, characterised by pattern and density (figure 1).

One of the advantages of FFF technology is material sourcing [2], with many polymer options starting from conventional ones (e.g., PLA, PET) up to engineering-grade (e.g., ABS, PC) and high-performance blends (e.g., PEEK, PEI), provided in the form of wire in various colours. Due to the additive way of producing products, depending on model geometry or orientation in the build volume, support structures may be needed [3]. Those structures are made from the same material as the part for single extrusion systems, extending post-processing time because of layer adhesion between supports interface and part.

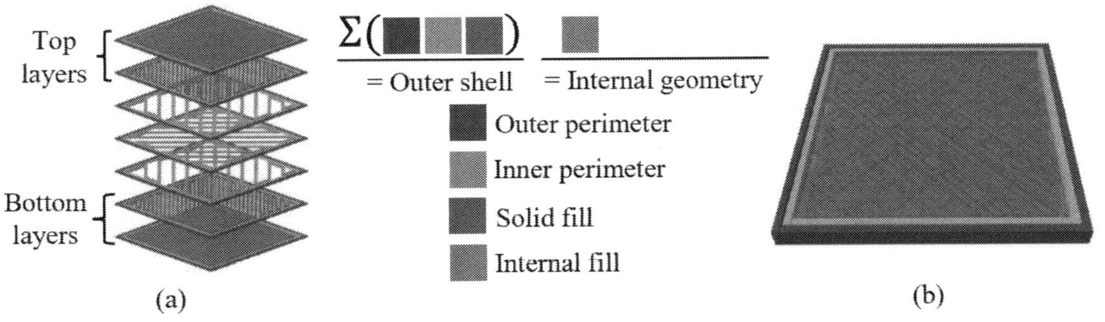

Figure 1. The structure of a model produced through FFF: a. Exploded view of a part's layers; b. Assembled view of the part's layers.

Multiple extrusion systems provided new possibilities regarding support use by introducing soluble materials (e.g., PVA, HIPS [4]) or non-compatible materials and offered new opportunities to manufacture multi-colour and multi-material components. The deposition way remains the same, but in the desired locations, another material is deposited. A distinct mesh is required for each material, and their assembly is done in the machine's software. A disadvantage of a multiple extrusion system is the purge block, a sacrificial part made from all materials used in the process, which is used to ensure the proper transition between materials and increases the production time. However, this can be overlooked considering the advantages: good surface quality of the resulting parts, easy to remove supports, multi-material and multi-colour components.

Lopes *et al.*, 2019 evaluated the mechanical properties of contours adhesion for PLA, PET and TPU. Thus, head-to-head joining between zones of "hourglass type" specimens was investigated for both the same material blends boundary adhesion and PLA with PET and TPU. The results demonstrated that mechanical properties are affected in the contact areas regardless of the material and concluded that a proper boundary interface design must be performed for functional exploitation [5].

Ando *et al.*, 2021 investigated the tensile strength of multi-colour PLA "hourglass type" specimens with flat and on edge orientation having contact surfaces with different incline. It concluded that horizontal adhesion is much stronger than vertical adhesion, suggesting that for stronger parts, an overlap degree is necessary to ensure horizontal adhesion for contours and solid fills [6].

Regarding the aforesaid, this paper aims to investigate if the mechanical properties of multi-material specimens can be enhanced by using contact geometry patterns associated with nozzle size.

2. Methods

2.1 Default multi-material bond formation

Regarding layer adhesion, based on the information provided by the slicing software (figure 2a), for the process preview simulation, it can be assumed that materials adhesion will occur only in the vertical plane, forming a head-to-head joining between layers (preview made through BCN Cura and Prusa slicer). In practice, besides the vertical adhesion, a horizontal one was identified (figure 2b) made by default during the printing process (part made through BCN Cura).

During printing, a zigzagging between geometries boundaries was observed (perimeters) referring to the specimen midplane. Staring with the first layer, the left boundary (blue material) is located before the midplane, and the right one crosses the mid reference (yellow material). The second layer is created on top of the previous similarly, but this time the left boundary exceeds the midplane, and the right side is printed before the mid reference. As shown in the section of figure 2b (right side), the horizontal adhesion is done (by default) over a slightly longer area than the used nozzle diameter of 0.4 mm.

This observation means that by using a simple head-to-head merging between meshes, the slicing tool will create by default a zig-zag welding pattern composed of zones vertically and horizontally bonding.

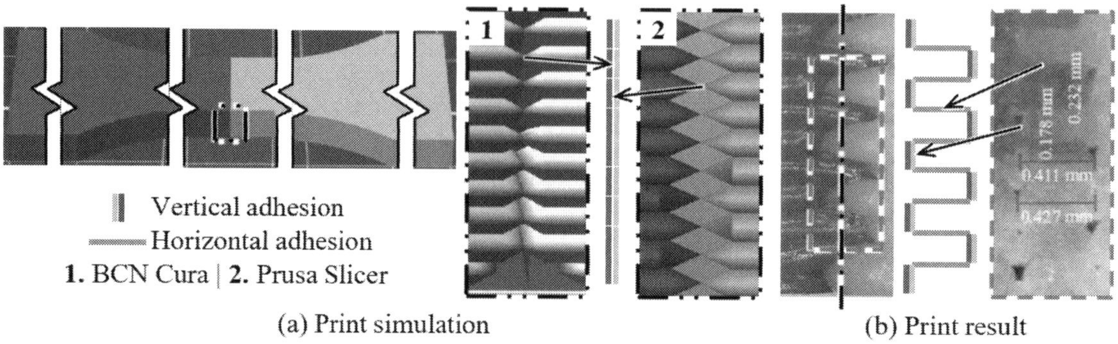

Figure 2. (a) Multi-material joining simulation vs (b) Actual print result (side view) with broken section of the multi-material joining formation (Slicing tools: BCN Cura 3.4.3; Prusa slicer 2.3.0).

2.2 Specimens contact design

Two patterns were considered for the "hourglass type" (specimen 1A of ISO 527:2) specimens, zig-zag and sinusoidal, to increase the contact surface between materials. Each of those profiles is characterised by pitch and height and was constrained as a functions of nozzle size (e.g., 0.4, 0.6 mm), as shown in figure 3. Those patterns were created across two directions, X and Z. Therefore, using the patterns across X direction, vertical contact area was extended and using the patterns across Z, more horizontal contact area was provided.

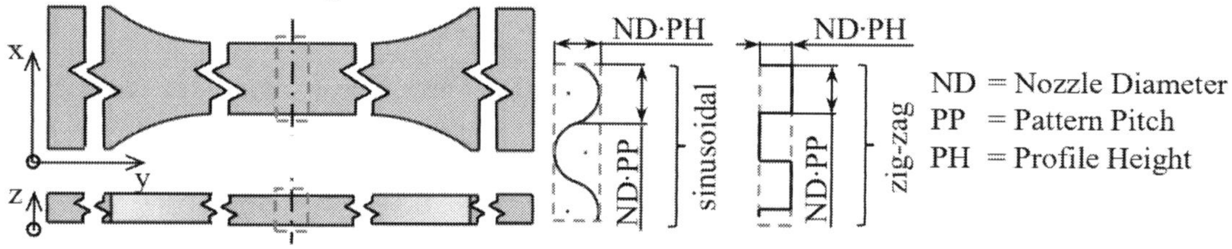

Figure 3. Contact geometry construction for multilateral 1A specimens.

2.3 Design of experiment

Two Polylactic Acid (PLA, AprintaPro PrintaMent) materials were used to fabricate the specimens with yellow and blue colours. The values of process parameters were kept constant, and the most relevant ones are presented in table 1.

As variables, parameters related to the contact geometry and profiles interactions were considered: *Pattern profile (Patt. Prof.), Pattern direction (Patt. Dir.), Pattern pitch (Patt. Pich)* and *Profile height (Patt. Hgt)*. Besides those, the *Overlap degree (Ovp.)* between the two halves of the specimens was considered to verify if a small value of the overlap (25% from the nozzle diameter) can improve the vertical adhesion by forming an interpenetration between bodies's external perimeters.

As the design of the experiment method, a half factorial plan was considered with the variables mentioned above on two levels. Details regarding all considered parameters are presented in table 1. The Minitab 19 tool was used to determine the factors with the most significant influence over the bond strength. In order to provide a better understanding of the tensile properties, the results were compared with standard bond (SB) formation provided by the slicing software and with the tensile strength of the conventional probes made from a single material, PLA Blue (PB), respectively PLA Yellow (PY). In total, 16 experimental runs were performed with five replicates for each specimen configuration. Regarding infill degree, the chosen value was 30% because, in practice, higher densities are avoided due to increased manufacturing time and material consumption. Therefore, the experiment must also consider the economic aspects of the part production and not just the ideal ones.

IOP Conf. Series: Materials Science and Engineering 1235 (2022) 012004 doi:10.1088/1757-899X/1235/1/012004

Table 1. Process and contact geometry parameters for 1A specimens.

Process parameters (constant)				Contact geometry parameters (variable)		
					Levels	
					1st	2nd
Nozzle diameter (mm)	0.4	Perimeters (no.)	4			
Extrusion temp. (°C)	210	Infill pattern	Grid			
Bed temp. (°C)	50	Infill degree (%)	30	Pattern profile	zig-zag	sin
Printing speed (mm/s)	40	Fan speed (%)	100	Pattern direction	X	Z
Layer thickness (mm)	0.2	Retraction distance (mm)	8	Pattern pitch	2·ND	4·ND
Extrusion width (mm)	0.4	Retraction speed (mm/s)	30	Profile height	1·ND	1.5·ND
Raster angle (°)	45	No. top layers	5	Overlap degree	0·ND	0.25·ND
Raster width (mm)	0.4	No. bottom layers	5			

3. Results

All specimens were manufactured according to Minitab's randomised order using a BCN Sigmax R19 3D Printer. The results of bond fomation by runs can be observed in figure 4. For specimens R3, R8, R10, R11 and R14, with the pattern orientated across x, bumps can be observed most probably due to the increased value of the overlap between specimens' bodies. The same observation for y-direction profile with an increased welding zone between walls, represented by the green material in the join area.

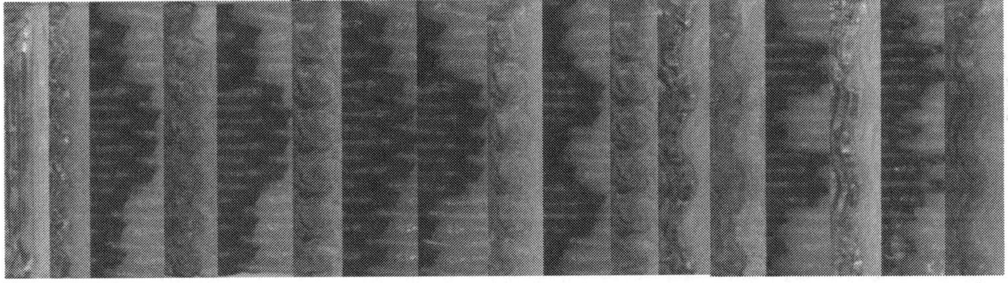

SB R1 R2 R3 R4 R5 R6 R7 R8 R9 R10 R11 R12 R13 R14 R15 R16

R1-sin_x_0.8_0.4_0
R2-zig-zag_z_1.6_0.4_0.1
R3-sin_x_0.8_0.6_0.1
R4-zig-zag_z_1.6_0.6_0
R5-zig-zag_x_0.8_0.6_0
R6-sin_z_1.6_0.6_0.1
R7-sin_z_0.8_0.4_0
R8-zig-zag_x_0.8_0.4_0.1
R9-sin_z_0.8_0.6_0
R10-sin_x_0.8_0.4_0.1
R11-zig-zag_x_1.6_0.4_0
R12-zig-zag_x_1.6_0.6_0.1
R13-zig-zag_z_0.8_0.6_0.1
R14-sin_x_1.6_0.4_0.1
R15-zig-zag_z_0.8_0.4_0
R16-sin_x_1.6_0.6_0

,with: Run number → R5 - zig-zag _ x _ 0.8 _ 0.6 _ 0 ← Overlap degree

Pattern profile (zig-zag); Pattern direction (x); Pattern pitch (0.8); Pattern height (0.6)

Figure 4. Layer bond formation for the experimental run contact geometries.

Specimens were tested for tensile strength using an Instron 4411 universal testing machine in an environment having 23 °C and 55% moisture level at 1mm/min speed. The resulted tensile strength for each specimen configuration is presented in figure 6. As expected, the two-material configuration is weaker compared to single material specimens (25.57 MPa for PLA Blue and 27.21 MPa for PLA Yellow). Although, an improvement in the mechanical properties of the specimens was observed compared to the standard bond formation provided by the slicing software (only 19.54 MPa). Even a slight change in the contact geometry between the specimen's bodies positively influences the tensile strength. It was observed that single material specimens entered the plastic domain of deformation and yield before the break. In the standard boding formation case, the break was observed precisely in the middle plane where the specimens' bodies mate. The break region was situated between the end of shell perimeters and mating surfaces for the specimens included in the factorial design. It was also observed that specimens failed juts in the PLA Blue domain, which was expected due to weaker tensile properties than PLA Yellow (see figure 6). All runs were characterised by rapid failure after reaching the maximum stress value (see figure 5).

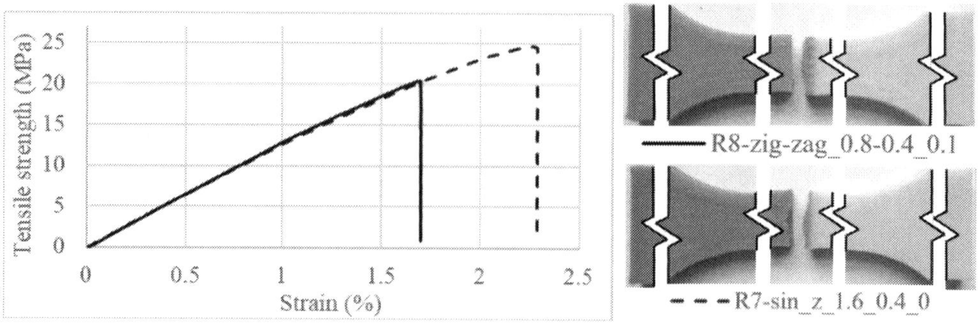

Figure 5. Tensile strength of R7&R8 specimens.

After determining the average tensile strength for each group of specimens, the resulting values were used as a response for Minitab to analyse. Initially, as terms for the factorial design analysis, only the main factors (table 1) and first-level interactions were considered. After examining the variance of the considered terms, the factorial analysis was performed again only with the statistically relevant factors.

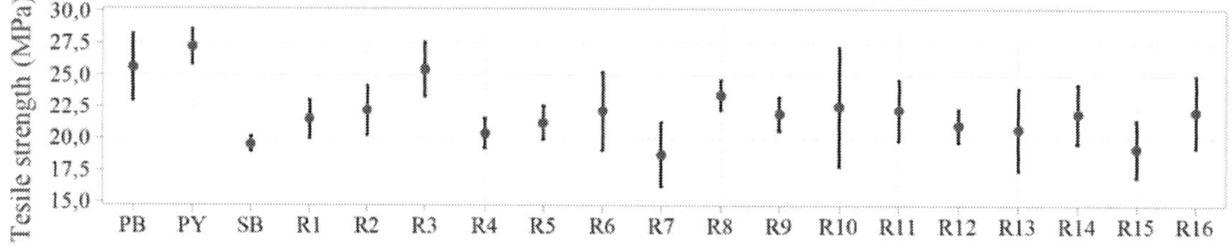

Figure 6. Tensile strength of all tested specimens.

The final Pareto chart as presented in figure 7a shows that only two of the five considered factors, B (Pattern direction) and E (Overlap degree), are significantly influencing the tensile properties of the "hourglass type" specimens, and their influence is shown in the main effect plot (figure 7b). However, even if factors A (Pattern profile) and D (Pattern height) separately are not statically significant at a level of 0.05, their interaction has a significant influence over the mechanical properties of the samples, the peak being reached at *sin·0.6* configuration (see figure 7c).

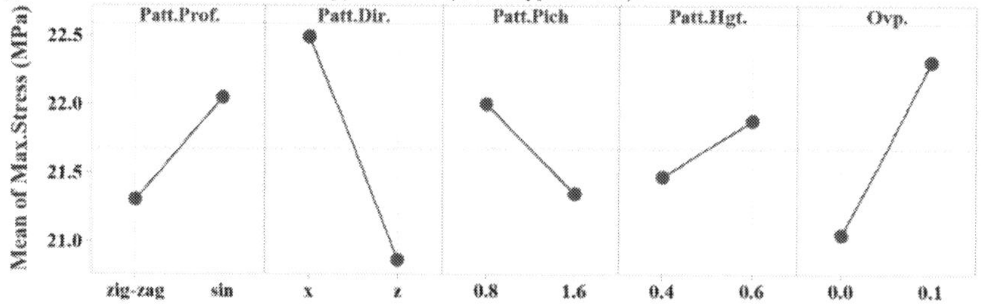

(a) Main effect plot for Max. stress

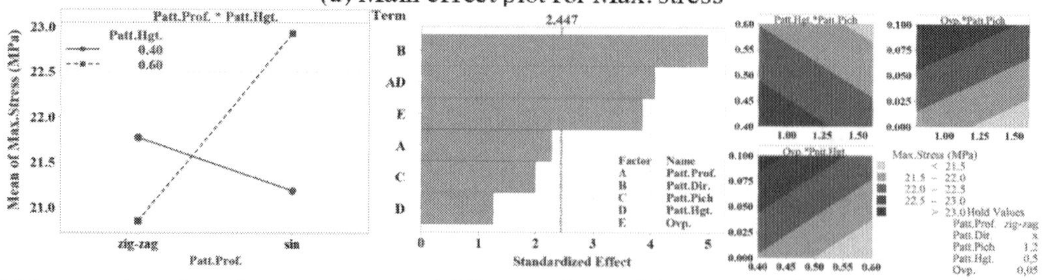

(b) Interaction plot of pattern profile and pattern height (c) Pareto chart of the standardised effects (d) Contour plot of others factors' interaction

Figure 7. Factorial design analysis results of contact geometries over filament bond strength.

By analysing the main effects and interaction plot, the strongest configuration is a combination of a sinusoidal profile having a height of 1.5·ND (0.6 mm) and an overlap degree of *0.25·ND* (0.1 mm). Of the performed runs, the strongest configuration was *R3_sin_x_0.8_0.6_0.1*, with an average tensile strength of 25.43 (MPa) weaker by 6.99% than PLA Yellow and by 0.55% than PLA Blue, but stronger than standard bond formation by 30.14%. Furthermore, other factors interaction also positively influences the specimens' mechanical properties (see figure 7d).

4. Conclusions

Multi-material printing using Fused Filament Fabrication introduced new possibilities regarding components production. However, standard bond formation provided by the slicing software is ≈35% weaker on average compared with single material specimens. By modifying the contact geometries between "hourglass type" specimen's bodies with patterns constrained by nozzle diameter-based relations, the capacity to bear mechanical loads was increased by approximately 30% compared to the regular bond formation.

Regarding the pattern profile orientation, only the X or Y direction should be considered in the design process. This statement is a consequence of the default zigzagging between parts layers produced by the slicing software.

When producing parts with multi-materials, more attention must be paid regarding perimeters. Although, in single material printing, the perimeters form a continuous shell that holds the part together, when using multiple materials, the external perimeter ensures the vertical adhesion between components. For this reason, a certain overlap degree must be considered to provide proper bonding.

Acknowledgments

This work was funded by "Gheorghe Asachi" Technical University of Iași through CNFIS-FDI-2020-0354 grant.

ORCID iDs

V Ermolai https://orcid.org/0000-0001-5967-5748
A Sover https://orcid.org/0000-0003-3408-898X
G Nagîț https://orcid.org/0000-0001-9649-8525

References

[1] Gibson I, Rosen D and Stucker B 2015 Additive Manufacturing Technologies (New York: Springer) p 1-6 https://doi.org/10.1007/978-1-4939-2113-3
[2] Oztan C, Ballikaya S, Ozgun U, Karkkainen R and Celik E 2019 Applied Materials Today **15** 77-82 https://doi.org/10.1016/j.apmt.2019.01.001
[3] Gordelier T J, Thies P R, Turner L and Johanning L 2019 Rapid Prototyp. J. **25** 953-971 https://doi.org/10.1108/RPJ-07-2018-0183
[4] Biggs M 2019 Essentials of Autopsy Practice ed Rutty G (Cham: Springer) p 19-49 https://doi.org/10.1007/978-1-84628-835-7
[5] Lopes L R, Silva A F and Carneiro O S 2018 Addit. Manuf. **29** 45-52 http://dx.doi.org/10.1016/j.addma.2018.06.027
[6] Ando M, Birosz M, Jeganmohan S 2021 Measurement **169** 108583 https://doi.org/10.1016/j.measurement.2020.108583

IOP Conf. Series: Materials Science and Engineering 1235 (2022) 012005 doi:10.1088/1757-899X/1235/1/012005

Rapid prototyping method for replacing non-circular gears

D Rizescu, D Besnea, C Rizescu, E Moraru and I Panait

Politehnica University of Bucharest, Splaiul Independentei Street No.313, Romania

ciprian.rizescu@upb.ro

Abstract. The paper presents a modern method of obtaining non-circular gearwheels from by means of additive technologies. Such a method is useful in case of wear or failure of a mechanical gear which has in its composition such non-circular metal gearwheels in the case of technical systems that must ensure continuous operation. With the help of the FDM technology (Fused Deposition Modeling) it will be created a model of the non-circular gearwheel that will constitute casting model for future gear, or they can even be used as functional gears if a suitable material and a precise execution technology with optimal printing conditions are considered.

1. Introduction

Generally, the gear is considered to have a cylindrical shape. But there are models of a non-circular shape and their main feature consist in a variable gear ratio, depending on the angle of rotation of the part. Several technical applications can be found for these non-circular mechanisms and for this, triangular, square models and elliptical gears can be developed. With constant drive shaft rotation, these models provide an uneven output speed. The high complexity of their technological obtaining did not give such designs widespread use, but with the advent and dynamic development of additive technologies, the problem of manufacturing can be solved.

Sometimes it becomes necessary to change the speed of a actuating element within one rotation. For this purpose, elliptical gears can convert constant speed rotary motion into variable speed rotary motion within one cycle and in this case the value of change in speed depends on the distance between the focuses of the ellipse.

By using specially selected systems of several gears, a variety of changeable speed domains can be achieved that are ideally suited to the needs of specific automation devices operating in a wide variety of applications. Elliptical gears can be manufactured from different materials according to the application, additive technologies making it possible to realize components from different materials, including metal. The most important applications of non-circular gears can be considered to be polygraphic and textile field. Their use can also be highlighted in the gearboxes of some pumps and in specific measuring instrument systems. If the elliptical gears do not meet the operational needs and application area, other constructive forms can be used, for example triangular, hexagonal and other shapes. It should be noted that the shapes of the initial diameters of the two gears must match perfectly in order to transmit regular movement [1,2].

The article will discuss some different cases from geometrical point of view of gearing using non-circular gears. Certain important parameters will be analytically determined using the Matlab environment for all cases. Also, some non-circular gears were manufactured using additive methods of fabrication, and namely fused deposition modeling.

IOP Conf. Series: Materials Science and Engineering 1235 (2022) 012005 doi:10.1088/1757-899X/1235/1/012005

2. Material and methods

If the problem of obtaining a variable transmission ratio according to a given or imposed law is raised, then gears or smooth surfaces wheels whose shape is not circular can be used. The contour of the two wheels represents two curves wrapped around each other. The authors took into consideration some cases: a) the case of two elliptical wheels rotating around their focal points, b) the case of two elliptical wheels rotating around geometric centres, c) the case of an eccentric circular wheel which rotates together with its conjugate wheel, d) the case of two wheels having the contour after a logarithmic spiral, e) the case of two wheels having the contour according to a trigonometric function. When such gear is damaged it must be replaced and a new gear could be executed with the help of additive technologies.

The contour of the two gears represents two mutually wound curves. The transmission ratio is defined as:

$$i_{12} = \frac{\omega_1}{\omega_2} \tag{1}$$

where the angular speed of the drive element, ω_1, este is constant, and the angular speed of the driven element, ω_2, is variabile. It should also be added that the distance between the axles of the two gears is constant, $a_0 = r_{w1} + r_{w2}$, where r_{w1} is the vector radius of the drive gear, and r_{w2} is the vector radius of the driven gear. The vector radii are also the radii of the rolling circles. In practice, several laws of variation of the transmission ratio are used. Below are some examples more commonly used in the construction of apparatus [1].

a) the case of two elliptical gears that rotate around their foci, figure 1 – a.

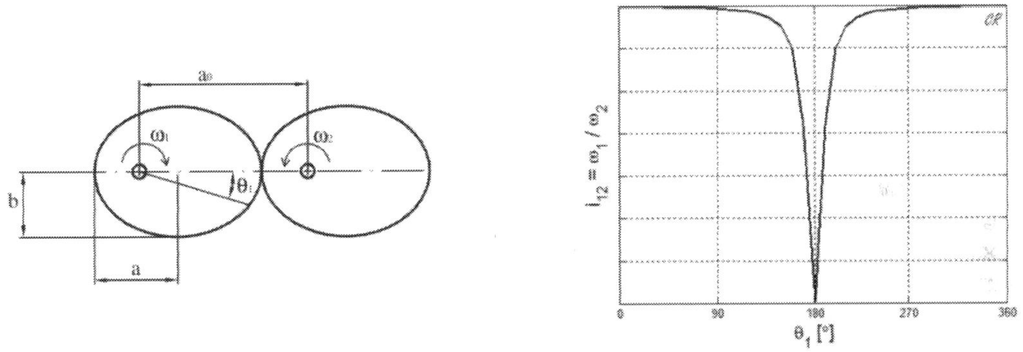

a **b**

Figure 1. a - Elliptical gears with centers of rotation in the foci of ellipses;
b - Variation of transmission ratio i_{12} with the angle of the drive gear, θ_1.

The transmission ratio i_{12} is:

$$i_{12} = \frac{\omega_1}{\omega_2} = \frac{-2\cdot r}{r^2+1+(r^2-1)\cdot cos\theta_1} \tag{2}$$

Where $r=R_{max}/R_{min}$, for which $R = \frac{b^2}{a(1+e\cdot cos\theta_1)}$, θ_1 is the angle of rotation of the drive gear, a_0 the distance between axis, e – eccentricity of the ellipse, $e = \sqrt{1-(\frac{b}{a})^2}$ and a and b are the semiaxes of the ellipse. In this case the wheels are identical and that is why it is an advantage [2]. They are used in the case of mechanisms where a fast return stroke is required, printing presses (automatic), automatic mechanisms. The variation of the transmission ratio depending on the rotation angle θ_1 is presented in the figure 1 – b [1].

b) the case of two elliptical gears rotating around geometric centers, figure 2 – a.
The transmission ratio i_{12} is::

$$i_{12} = \frac{\omega_1}{\omega_2} = \frac{-2r}{r^2+1+(r^2-1)\cdot cos2\theta_1} \tag{3}$$

Where r=a/b, θ_1 is the angle of rotation of the drive gear, a_0 distance between axis, a and b are the semiaxes of the ellipse. It is observed that the distance between the axes, a_0=a+b.

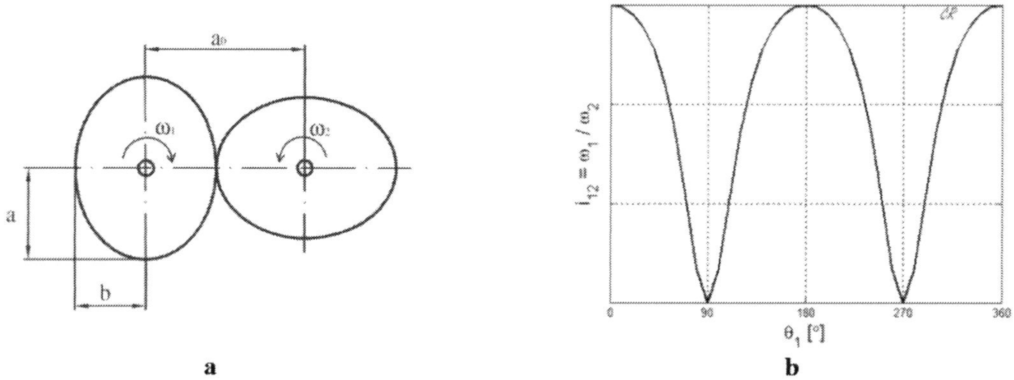

Figure 2. a - Elliptical gears identical with the centers of rotation in the geometric centers;
b - Variation of transmission ratio i_{12} with the angle of the drive gear, θ_1.

In this case the wheels are identical, and the geometric properties are the known ones. They are used in cases where two full speed cycles are required for one rotation. The variation of the transmission ratio depending on the rotation angle θ_1 is presented in the figure 2 – b [1].

c) the case of an eccentric circular gear that rotates together with its conjugate gear – figure 3 – a. The transmission ratio i_{12} is:

$$i_{12} = \frac{\omega_1}{\omega_2} = \frac{a_0-\sqrt{a^2+e^2+2aecos\theta_1}}{\sqrt{a^2+e^2+2aecos\theta_1}} \tag{4}$$

Where e - circular gear eccentricity, a – circular gear radius, a_0 distance between axis, and θ_1 is the angle of rotation of the drive gear.

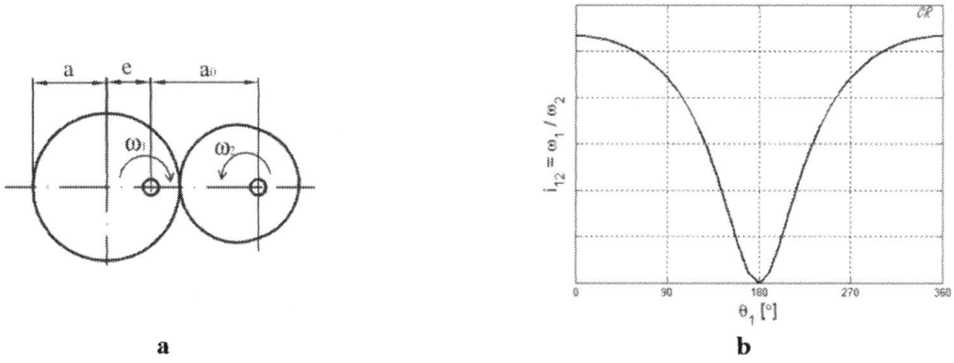

Figure 3. a - Circular gear, which rotates together with its conjugate;
b - Variation of transmission ratio i_{12} with the angle of the drive gear, θ_1.

The variation of the transmission ratio depending on the rotation angle θ_1 is presented in the figure 3-b [1].

d) the case of two gears having the contour after a logarithmic spiral, figure 4 – a. The transmission ratio i_{12} is:

$$i_{12} = \frac{\omega_1}{\omega_2} = \frac{a_0 - Ae^{k\theta_1}}{Ae^{k\theta_1}} \tag{5}$$

Where a_0 – distance between axis, A and k are two coefficients that define the logarithmic spiral, e is the basis of the natural logarithm.

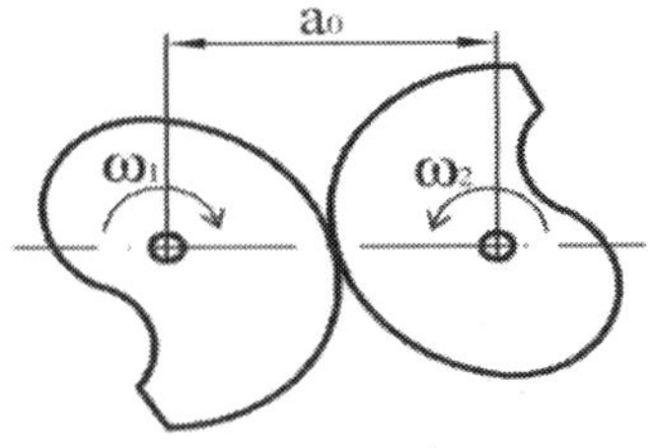

a

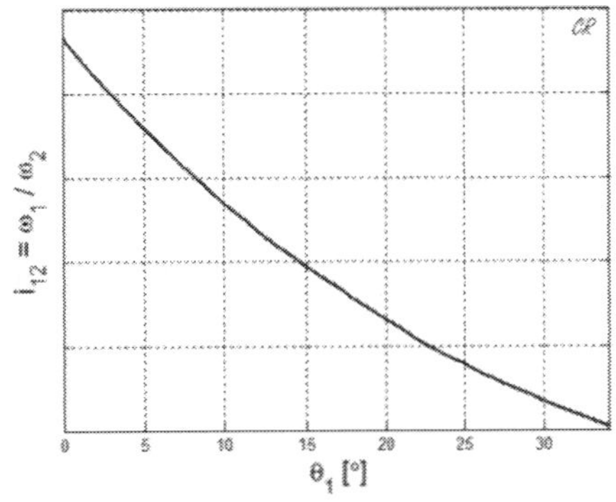

b

Figure 4. a – Gears with a contour by a logarithmic spiral;
b - Variation of transmission ratio i_{12} with the angle of the drive gear, θ_1.

The variation of the transmission ratio depending on the rotation angle θ_1 is presented in the figure 4-b [1].

e) the case of two gears having the contour after a trigonometric function (sine), figure 5 - a. The transmission ratio i_{12} is:

$$i_{12} = \frac{\omega_1}{\omega_2} = \frac{1}{k\cos\theta_1} \tag{6}$$

Where k is a constant, a_0 is distance between axis, and θ_1 is the angle of rotation of the drive gear.

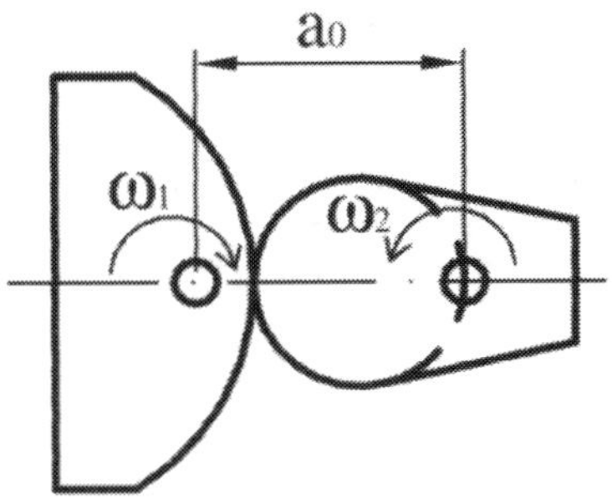

a

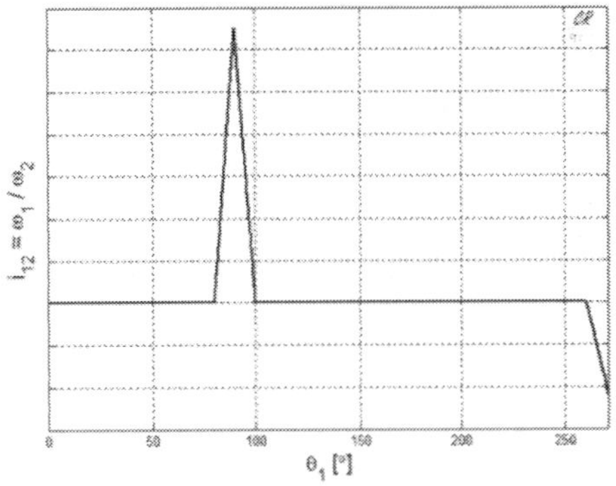

b

Figure 5. a - Gears having the contour after a trigonometric function (sine);
b - Variation of transmission ratio i_{12} with the angle of the drive gear, θ1.

The variation of the transmission ratio depending on the rotation angle θ_1 is presented in the figure 5-b [1].

3. Experimental tests

Some non-circular gears have been designed and executed by means of additive technologies, namely the FDM method of thermoplastic extrusion using PLA (polylactic acid) as material, and the virtual and realized models are represented in figure 6 [6-8].

The printing conditions were as follows: the fill factor 20%, the extrusion temperature of 200 °C, 200 µm thickness layer and about 40 mm/s printing speed. The printer used was Makerbot, which also has software (slicer software) with the same name to manage the STL digital file or to change the printing conditions.

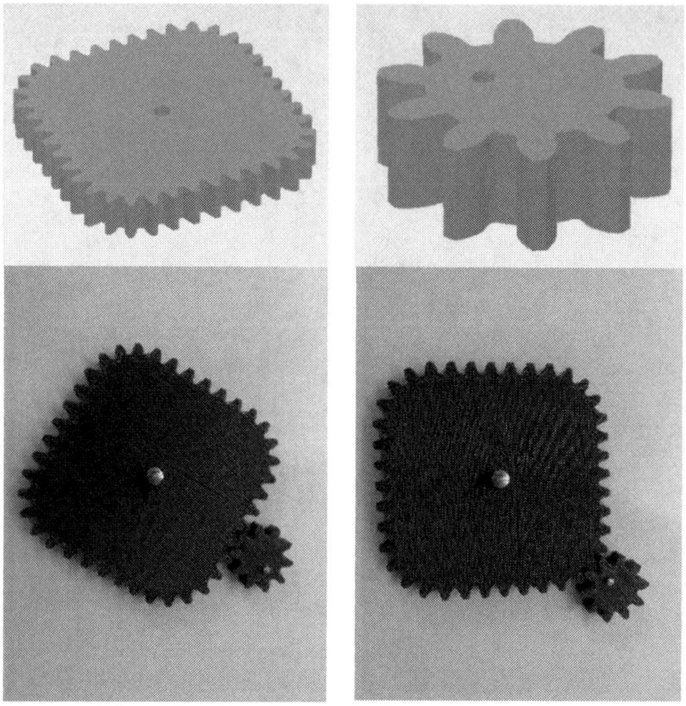

Figure 6. CAD models and realized through FDM additive technolgy non-circular gears

The experimental results were obtained on a special gear inspection machine, by the examination of the gears manufactured using FDM method [3,9]. The inspection equipment software allows the control of the gearing, as well as the evaluation of some important parameters that characterize the gears engaging.

The acquired data are highlighted in the software interface with different colors: the compliant results are highlighted in green, and the non-compliant ones in red. Basically, the software determines the following values: total radial composite deviation Fi", tooth-to-tooth radial composite deviation, fi", runout deviation by composite test, Fr", short-wave component, fk" [4,10].

Figure 7 shows some results obtained from the performed tests. If the DIN 3963 standard is taken into account and a certain quality level, then it can be concluded that the measured gear doesn't pass the requirements. All above mentioned parametres have been exceeded: Fi", fi", Fr", fk", as one can observe. In figure 7, both coordinate systems, carthesian and polar were considered for the gear. This dual representation of the gears allows to determine the vulnarable zones of tested gears. A new gear design could be considered in order to improve the gear quality level [5].

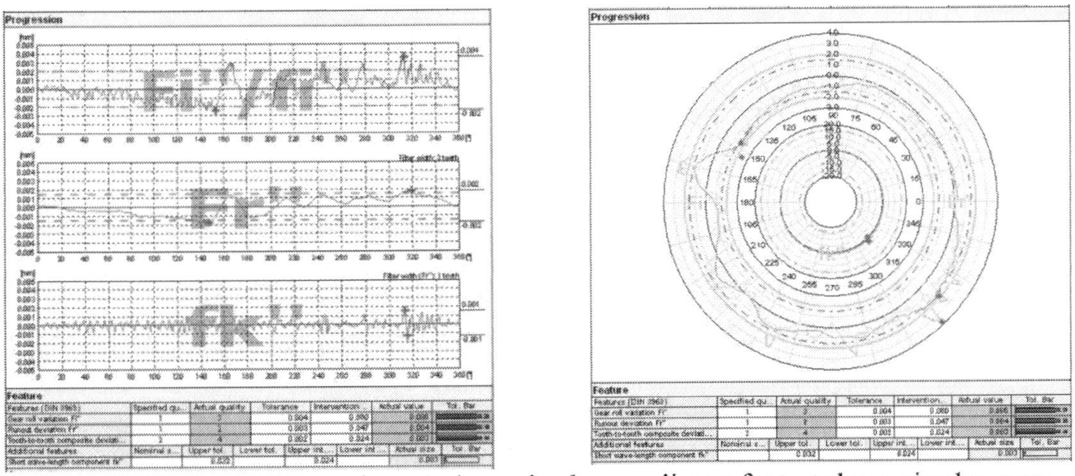

Figure 7. Determined errors in cartesian and polar coordinates for tested non-circular gears

4. Conclusions

Rapid prototyping manufacturing methods are gaining increasing interest among multiple applications due to its indisputable advantages. In some specific situations and applications, it is recommended to use non-circular gears, and the additive manufacturing method is a promising candidate in terms of their execution technology. These technologies can be used both for the realization of non-circular gears provided that an optimal material and equipment is used, as well as for the realization of casting models for the future gears. The authors presented several laws of variation of the transmission ratio for non-circular gears, also presenting some experimental data obtained regarding the testing of non-circular gears executed by FDM thermoplastic extrusion technology. In the future work we will study the influence of fill factor on the gears precison parameters and the influence of materials on the gears mechanical behaviour.

ORCID iDs

D Rizescu: https://orcid.org/0000-0002-7738-2540
D Besnea: https://orcid.org/0000-0002-8513-810X
C I Rizescu: https://orcid.org/0000-0002-2764-520X
E Moraru: https://orcid.org/0000-0002-5745-8165
I Panait: https://orcid.org/0000-0003-4744-5296

References

[1] Rizescu C I 2013 Mechanism and Precision Instruments – Part I *Printech Publisher*
[2] Voicu A, Gheorghe G and Bădiţa L L 2014 Studies and Research on the measurement and 3D integrated control of complex components from automotive industry by laser scanning *International conference ICNAAM* 2014 https://doi.org/10.1063/1.4912852
[3] https://www.frenco.de/en/double-flank-gear-inspection-machines/
[4] Pawar P B and Utpat A A 2015 *Materials Today: Proceedings* **2** 2968-74 https://doi.org/10.1016/j.matpr.2015.07.278
[5] Van Melick H G H Tooth-Bending effects in plastic Spur Gears 2007 *Gear technology*
[6] Mao K, Langlois P, Hu Z, Alharbi K, Xu X, Li M, Hooke C J and Chetwynd D 2015 *Wear* **332-33** 822-6 https://doi.org/10.1016/j.wear.2015.01.084
[7] Senthilvelan S and Gnanamoorthy R 2007 *Materials & Design* **28** (3) 765-72 https://doi.org/10.1016/j.matdes.2005.12.002
[8] Mao K, Li W, Hooke C J and Walton D 2010 *Tribology International* **43** 433-9 https://doi.org/10.1016/j.triboint.2009.07.006
[9] Ashish N. Taywade, Dr. V.G. Arajpure 2014 *IJERT* **3** (9)
[10] Yousuef S, Osman T A, Abdalla A H and Zohdy G A *IEEE Transactions on Nanotechnology* **12** (4) 616-620 https://doi.org/10.1109/TNANO.2013.2264902

IOP Conf. Series: Materials Science and Engineering 1235 (2022) 012006 doi:10.1088/1757-899X/1235/1/012006

Effect of the infill patterns on the mechanical properties of the carbon fiber 3D printed parts

M A Pop[1], S M Zaharia[2], L A Chicos[2], C Lancea[2], V M Stamate[2], G R Buican[2] and I S Pascariu[2]

[1] Transilvania University of Brasov, Materials Science Department, Brasov, 29 Eroilor Ave., Romania;
[2] Transilvania University of Brasov, Manufacturing Engineering Department, Faculty of Technological Engineering and Industrial Management, Brasov, 29 Eroilor Ave., Romania;

mihai.pop@unitbv.ro

Abstract. The future in the aeronautical industry is in our conception to obtain 3D printed parts from composite materials of the polymer type reinforced with various materials (fibreglass, carbon, Kevlar, etc.), with very good mechanical properties. In this paper authors analyse the mechanical properties of 3D printed parts obtained by material extrusion of Polyamide based filament reinforced with 15% chopped carbon fibres by three types of infill pattern: grid, triangular and lines and with an infill density of 50%. After testing the specimens, the grid infill pattern showed the lowest average value in the two types of mechanical tests, but the values are higher than those reported in the literature by other researchers. Another fact, it can be concluded that the core configuration does not have a significant influence on the mechanical property's performance. After microstructures analysis with an optical microscope was observed the defects specific to 3D printing (pores, gaps and agglomerations) were and we determined that the share of defects decreases from grid pattern to lines pattern and for the triangular pattern they are minor even insignificant.

1. Introduction

Additive manufacturing is a technological process and are transform a (CAD) model by building of 3D objects from materials, layer by layer, using different materials, from natural resources (Polylactic acid, also known as poly (lactic acid) or polylactide and the abbreviation PLA), (Acrylonitrile butadiene styrene (ABS), nylon and so on) or composite materials made in diverse combination (natural or artificial) [1-4].

In this moment are two main directions in this researches field: obtaining new materials and optimizing the printing geometry of walls. Because, 3D printing process have a low environmental impact and it is possible to obtain complex parts with materials such as liquid, powder or filament it is in a continuous development and intensely researched process. Various properties can be obtained in the printed parts depending on the materials used. From this point of view powder bed fusion can be mentioned, including selective laser sintering and laser cladding [5-8].

In this paper the authors aim to find out the influence of three types of infill pattern: grid, triangular and lines, on the mechanical properties namely bending and tensile strength. The samples were printed

IOP Conf. Series: Materials Science and Engineering 1235 (2022) 012006 doi:10.1088/1757-899X/1235/1/012006

3D with Filament Fused Fabrication (FFF) process using Polyamide based filament reinforced with 15% carbon fibres. The infill value was kept constant for all samples, namely 50%. The build defects influence the mechanical behaviour together with the types of filaments used and have already been extensively reported in the literature. The properties can be improved by optimization of the printing parameters [9-13].

2. Materials and Methods

2.1. Samples design

Using the Solid Works 2016 software was designed all the specimen types for mechanical tests. We started printing test specimens with three types of infill patterns, namely: grid (Figure 1.a), lines (Figure 1.b) and triangles (Figure 1.c).

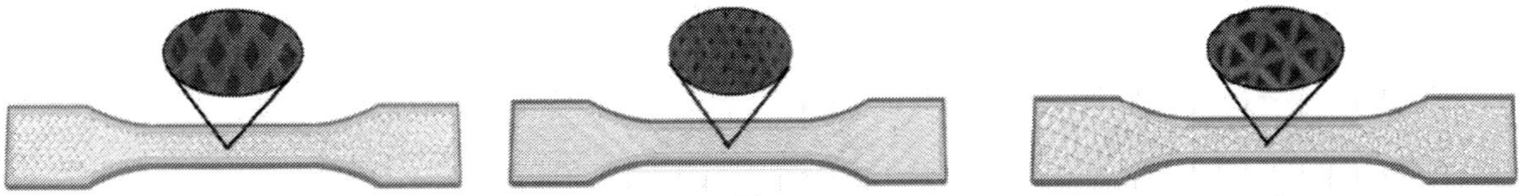

a) b) c)

Figure 1. Specimens for tensile tests with infill pattern: a) grid; b) lines; c) triangles.

2.2. Sample manufacturing

The BCN3D Epsilon printer was used to obtain the composite specimens were manufactured using the UltraFuse PAHT CF15 filament.

The PAHT CF15 filament properties (mechanical and thermal) provided by the Ultrafuse manufacturer [14], were presented in Table 1. PAHT CF15 filament contains 15% of carbon fiber chopped reinforcement and thus allowing the printed parts to withstand higher thermal and mechanical loads [15].

Table 1. Mechanical and thermal properties of the material used in 3D printing [12].

Mechanical and thermal properties	PAHT CF15	Standard
Tensile strength [MPa]	103.2	ISO 527
Elongation at break	1.8%	ISO 527
Modulus of elasticity [MPa]	8386	ISO 527
Flexural stress [MPa]	160.7	ISO 178
Flexural modulus [MPa]	8258	ISO 178
Impact strength [kJ/m^2]	20.6	ISO 179
Melting temperature [°C]	234	ISO 11357

This printer machine is a high quality one, characterized by dual extrusion possibility, with 420 mm x 300 mm x 400 mm. construction volume. The printing parameters of the 3D printed composite specimens were described in Table 2.

Table 2. FFF process parameters of the PAHT CF15 specimens.

Parameter	Value
Layer height	0.2 [mm]
Infill	50 [%]
Print speed	40 [mm/sec]
Printing temperature	260 [°C]
Building plate temperature	90 [°C]
Hotend	0.6 [mm]

2.3. Mechanical tests

The mechanical tests (three point bending and tensile) were done on the WDW - 150S universal testing machine. It is possible to perform static and dynamic tests even alternating fatigue, between 1 and 150 kN test force [16].

2.4. Optical Microscopy

The most representative samples including filaments and prints were cross-sectioned, embedded into an acrylic resin, and levelled using an automatic metallographic polishing device (Al_2O_3 suspension 0.5 μm and 1 μm grit) for the structure and morphology analysis.

By using an OMNIMET-BUEHLER microscope made by Nikon (Tokyo, Japan) two different micrographs of the cross-section on x-y view and on the z view, at 200x magnification, at each sample was made.

3. Results and Discussions

For the mechanical tests (three-point bending and tensile), were manufactured five specimens by the FFF process in order to obtain each configuration of the structure (grid, lines, triangles).

The mechanical tests of the specimens, manufactured by the thermoplastic extrusion process, had as main purpose the determination of the mechanical characteristics (tensile strength, three-point bending strength), in accordance with ASTM D638 and ASTM C393 standards.

A number of five tests were performed for each configuration of infill density studied (grid, lines, triangles), and the test results were shown in Figure 2. Specimens with infill configuration triangles showed the best results in the tests at three-point bending. Instead, at the tensile test, the specimens with infill lines configuration showed the best results. The specimens with grid infill pattern showed the lowest average value in the two types of mechanical tests, but the values are higher than those reported in the literature [17].

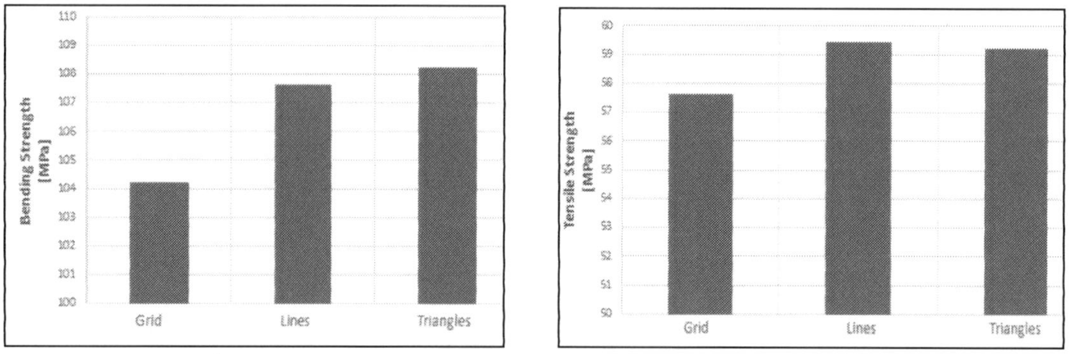

a) b)

Figure 2. The results of the mechanical tests: a) three-point bending; b) tensile.

The statistical indicators, for the results obtained from the tensile and three-point bending tests (tensile strength/ three point bending strength), for the infill pattern specimens (grid, lines and triangles), were presented in Table 3. Following the analysis of experimental data obtained we concluded that those are homogeneous and the average is representative. If the coefficient of variation is less than 30%, for all types of infill pattern, the data are homogeneous.

For all six sets of the experimental data, the coefficient of variation, was between 2.30% and 6.79% and the mean it can be estimated that is representative, as can be seen in Table 3.

After microstructures analysis with a Nikon Omnimet-Buehler microscope was observed from figure 3a, b, c and d, the chopped carbon fibres both in section and longitudinally as well as their arrangement.

IOP Conf. Series: Materials Science and Engineering 1235 (2022) 012006 doi:10.1088/1757-899X/1235/1/012006

Table 3. Tensile and three point bending statistical parameters determined for the composite specimens.

Infill pattern \ Statistical parameters	Mean (m)	Standard deviation (s)	Coefficient of Variation (CV)%
Grid – Bending strength [MPa]	104.9	2.40	2.30
Lines – Bending strength [MPa]	107.6	3.29	3.05
Triangles – Bending strength [MPa]	108.2	3.90	3.60
Grid – Tensile strength [MPa]	57.6	3.91	6.79
Lines – Tensile strength [MPa]	59.4	3.21	5.40
Triangles – Tensile strength [MPa]	59.2	3.90	6.59

Also, the length of the carbon fibres can be observed, defects of the filament before 3D printing (gaps, pores and small agglomerations) as well as defects specific to 3D printing (pores, gaps and agglomerations).

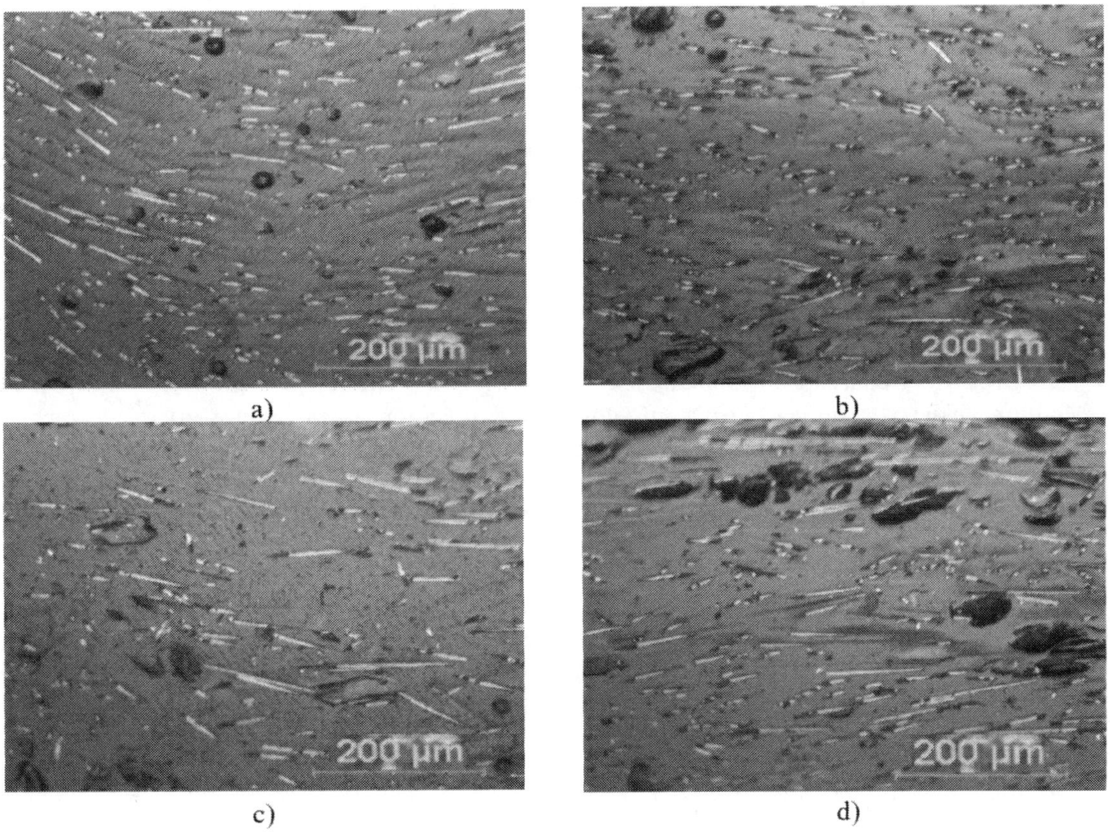

Figure 3. Analysis of defects in the cross section on x-y view composite specimens: a) PAHT CF15 filament; b) Infill pattern grid; c) Infill pattern lines; d) Infill pattern triangular

IOP Conf. Series: Materials Science and Engineering 1235 (2022) 012006 doi:10.1088/1757-899X/1235/1/012006

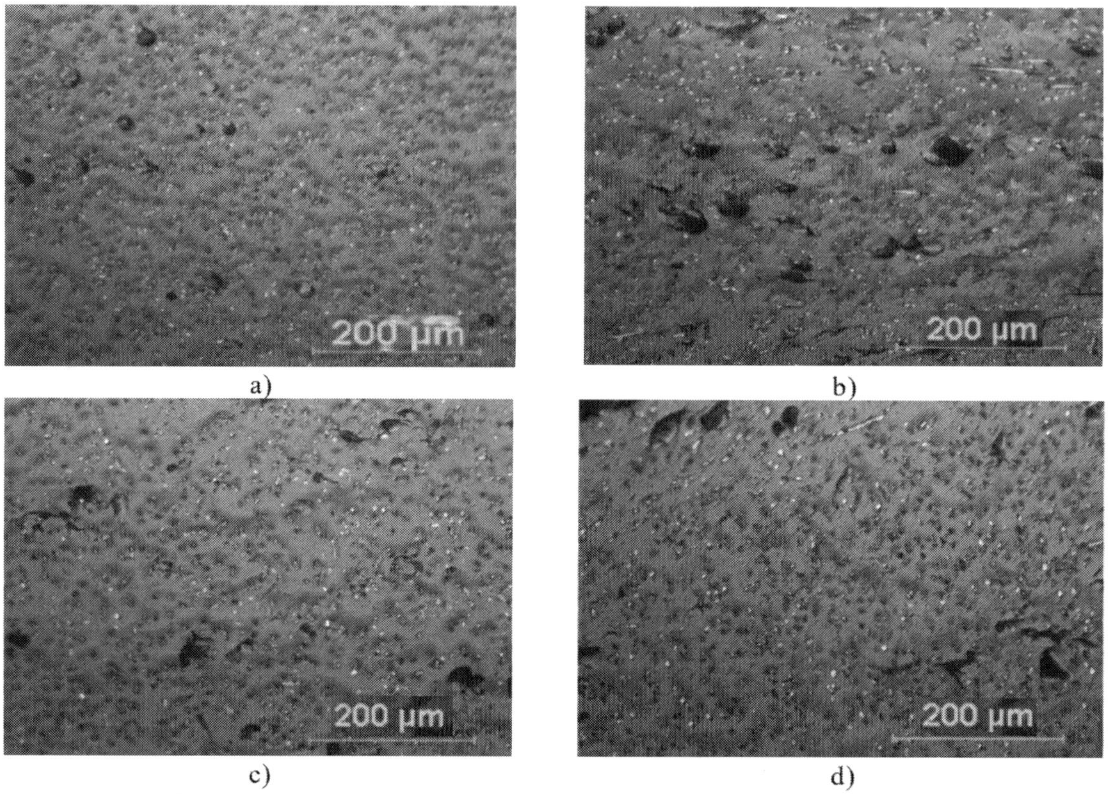

Figure 4. Analysis of defects in the cross section on z view composite specimens: a) PAHT CF15 filament; b) Infill pattern grid; c) Infill pattern triangular; d) Infill pattern lines

Were not observed any macrostructural major defects and a relatively uniform fibre reinforcement distribution can be seen in the filament and a compact mass (Figure 3a and Figure 4a).

The defects observed in the filament (dimensional unevenness, small agglomerations, gaps) were implicitly transferred to the printed part but in a smaller proportion. In addition to these defects, the triangular defects formed at the intersection of the walls as well as when changing the printing direction were also observed. Possible solutions for decreasing the proportion of these defects: we recommend decreasing the printing speed as well as increasing the infill value.

As can be seen in the micrographs shown in Figure 3 (b, c and d) and Figure 4 (b, c and d), and compared with mechanical properties obtained for them, the share of defects decreases from grid pattern to lines pattern and for the triangular pattern they are minor even insignificant.

4. Conclusions

By the present paper, the authors try developing and after characterizing a grid, line and triangles model structure with 50% infill density. Those patterns type is built architectures for 3D-printed parts. The following conclusions are drawn from our experiments:

- It can be stated that in the case of mechanical tests (three-point bending and tensile), for the three infill patterns analysed, the performances are similar. Thus, it can be concluded that the core configuration does not significantly influence the three-point bending and tensile performance.
- The results obtained are quite close in value, better in the case of the tensile test for lines pattern (59.4 MPa) and in the case of the triangles pattern (108.2 MPa) for three points bending test. The composite filaments with 15% chopped carbon fibre from a mechanical test point of view, showed superior values compared to standard filaments (ABS, PLA, PETG), materials currently used in 3D printing.

- Composite filaments have a high degree of difficulty in 3D printing, and the result of the process showed specific defects (cracking between the extruded carbon filament layers, rectangular and triangular voids).
- Taking into account the above affirmation, due to the good values obtained from mechanical testing, it is recommended to use these materials / printing configurations in the field of aeronautics and not only.
- A general conclusion is that this material (PAHT CF15) having the three printing configurations has safety in operation and can be used in obtaining printed parts usable in engineering, but only after a documented study and preliminary strength calculations.
- In order to eliminate the deficiencies that appeared during the printing, defect and to improve the mechanical properties of the parts printed with composite materials with carbon chopped, new researches are necessary, which must include the modification of the printing parameters and infill value.

Acknowledgements

This work was supported by a grant of the Ministry of Research, Innovation and Digitization, CNCS/CCCDI – UEFISCDI, project number PN-III-P2-2.1-PED-2019-0739, within PNCDI III. We also acknowledge to PRO-DD Structural Founds Project (POS-CCE, O.2.2.1., ID 123, SMIS 2637, ctr. no 11/2009) for providing the infrastructure used in this work.

ORCID iDs

M Pop: https://orcid.org/0000-0003-2753-2907
S Zaharia: https://orcid.org/0000-0002-8636-5558
G Buican: https://orcid.org/0000-0002-5838-0056
L Chicos: https://orcid.org/0000-0002-7674-2874
C Lancea: https://orcid.org/0000-0002-4746-6241
V Stamate: https://orcid.org/0000-0003-0275-4263
I Pascariu: https://orcid.org/0000-0002-4844-4132

References

[1] Ngo T D, Kashani A, Imbalzano G, Nguyen K T Q and Hui D 2018 *Compos. Part B Eng.* **143** 172–96 https://doi.org/10.1016/j.compositesb.2018.02.012
[2] Symes M D, Kitson P J, Yan J, Richmond C J, Cooper G J T, Bowman R W, Vilbrandt T and Cronin L 2012 *Nat. Chem.* **4** 349–54 https://doi.org/10.1038/nchem.1313
[3] Szuman K, Krucińska I, Boguń M and Draczyński Z 2016 *Autex Res. J.* **16** 119–27 https://dx.doi.org/10.1515/aut-2015-0047
[4] Pop M A, Croitoru C, Bedő T, Geamăn V, Radomir I, Coşniţă M, Zaharia S M, Chicoş A L and Miloşan I 2019 *J. Appl. Polym. Sci.* **136** 47382 http://dx.doi.org/10.1002/app.47382
[5] Wang K, Li S, Rao Y, Wu Y, Peng Y, Yao S, Zhang H and Ahzi S 2019 *Polymers* **11** 1878 https://doi.org/10.3390/polym11111878
[6] Guessasma S, Belhabib S and Nouri H 2019 *Polymers* **11** 1778 https://doi.org/10.3390/polym11111778
[7] Liao Y, Liu C, Coppola B, Barra G, Di Maio L, Incarnato L and Lafdi K 2019 *Polymers* **11** 1487 https://doi.org/10.3390/polym11091487
[8] Vinyas M, Athul S J, Harursampath D and Thoi T N 2019 *Mater. Res. Express* **6** 115301 https://doi.org/10.1088/2053-1591/ab43ab
[9] Liu Z, Lei Q and Xing S 2019 *J. Mater. Res. Technol.* **8** 3741–51 http://dx.doi.org/10.1016/j.jmrt.2019.06.034
[10] Brites F, Malça C, Gaspar F, Horta J F, Franco M C, Biscaia S and Mateus A 2017 *Procedia Manuf.* **12** 156–65 https://dx.doi.org/10.1016/j.promfg.2017.08.020

[11] Le Duigou A, Castro M, Bevan R and Martin N 2016 *Materials & Design* **96** 106-14 https://doi.org/10.1016/j.matdes.2016.02.018
[12] Liao Y, Liu C, Coppola B, Barra G, Di Maio L, Incarnato L and Lafdi K 2019 *Polymers* **11** (9) 1487 http://dx.doi.org/10.3390/polym11091487
[13] Vinyas M, Athul S J, Harursampath D and Thoi T N 2019 *Materials Research Express* **6** (11) 115301 https://doi.org/10.1088/2053-1591/ab43ab
[14] https://s.cdnmpro.com/490505258/custom/prod/1_fisa_tehnica_1671.pdf
[15] https://www.ultrafusefff.com/wp-content/uploads/2019/06/Innofil3D_PAHT_PET_CF15_Factsheet-online.pdf
[16] https://icdt.unitbv.ro/en/research-centres/advanced-metal,-ceramic-and-mmc-composite-materials-and-technologies/infrastructure.html, accessed at 20.07.2021
[17] Pop M A, Croitoru C, Bedo T, Geamăn V, Radomir I, Zaharia S M and Chicoş L A 2020 *Polymers* **12** (5) 1129 https://doi.org/10.3390/polym12051129

IOP Conf. Series: Materials Science and Engineering 1235 (2022) 012007 doi:10.1088/1757-899X/1235/1/012007

A study of Fused Filament Fabrication process efficiency: ABS vs PLA materials

K Ninikas[1], J Kechagias[1], N A Fountas[2,3] and N M Vaxevanidis[2]

[1] Design & Manufacturing Lab (DML), Department of FWSD, University of Thessaly, Karditsa, 43100, Greece.
[2] Department of Mechanical Engineering, School of Pedagogical and Technological Education (ASPETE), GR 151 22, Amarousion, Greece.
[3] Department of Mechanical Engineering, University of West Attica (UNIWA), 250 Thivon & P. Ralli Str., Egaleo, GR 122 41, Greece.

vaxev@aspete.gr

Abstract. A study of the shape accuracy of pieces created using the Fused Filament Fabrication (FFF) process is presented. Accuracy of the parts coming from 3D Printing (3DP) technologies is an essential criterion, primarily when used in assembly configurations. Two polymers; PLA and ABS, three infill percentages; 20, 50, and 70 (%), three numbers for shells; 1, 2 and 3 and three different layer heights; 0.1, 0.2 and 0.3 (mm) were the variables and their corresponding levels. A benchmark specimen was designed whilst nine experiments were performed according to the Taguchi methodology. Dimensional accuracy readings then followed, and the results were analysed using descriptive statistical tools such as analysis of means (ANOM), analysis of variance (ANOVA) and contour plots. The presented results indicate that 3D printed parts of PLA material exhibited better dimensional accuracy than ABS ones.

1. Introduction

Fused Deposition Modeling (FDM), also termed Fused Filament Fabrication (FFF), is a 3-dimensional printing (3DP) method that commonly refers to systems that create components by gradual deposition or addition of solid material [1,2]. Initially the term "Rapid Prototyping (RP)" was used with reference to the creation of 3D items.

Items produced by RP methods are used with an increasing trend in a number of industrial sectors, such as automotive, fashion, aerospace, medicine etc. Consequently, a great number of 3DP alternatives have been developed over the last few years. These techniques, which are termed generally as Rapid Prototyping and Manufacturing, include FFF, stereolithography (SLA), laminated object manufacturing (LOM), selective laser sintering (SLS), direct metal deposition, etc. [2]. FFF is considered to be a dominant technique among the 3DP technologies since the majority of the equipment used in the relative market belongs to this category [3].

The FFF method is user friendly since the operator can select most of the input parameters. On the other hand, the output (quality) is determined by a number of indices, the most important of which is the product's shape accuracy. Hence, the shape accuracy of an item produced by the FFF process and the control factors influencing it to constitute significant research issues in the bibliography. The most significant parameters controlling shape accuracy are layer height, infill rate, number of shells, printing and movement speed [4-8].

This manuscript investigates the influence of various controlling factors on the shape accuracy of parts fabricated by the FFF technique. These factors include the deposited material; the infill rate which indicates the amount of material that occupies the internal part of the item; the number of shells which stands for the number of outlines printed on each layer of the model and indicating the density of the outside walls and the layer height, which determines the print quality by setting the height of each layer that is being deposited

2. Experimental set-up

A study of the effect of the layer height (LH), infill rate (IR), printing material (PM) and a number of shells (NS) on the shape accuracy of items fabricated by a low-cost FFF printer (Duplicator® 4X) was realised by performing nine printing operations.

The selected part shape was prepared in neutral surface (STL) format. The CAD model of the experimental part is depicted in figure 1. The printer used as well as an indicative printed experimental part are shown in figure 2.

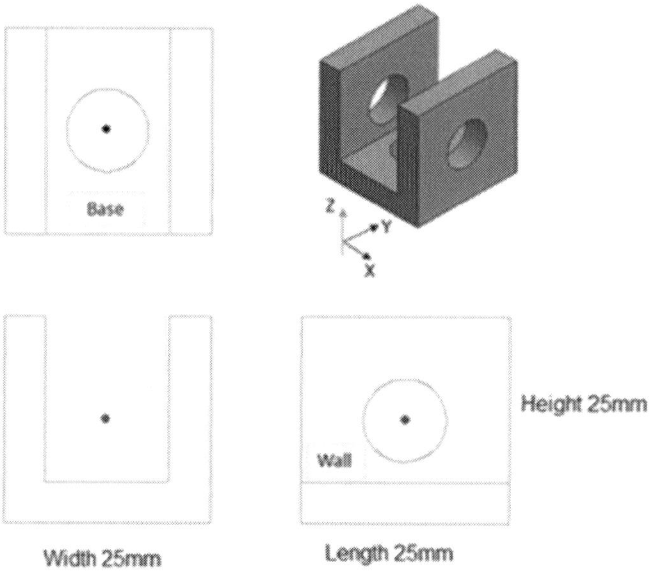

Figure 1. Drawing details of the CAD model.

Nine tests were conducted employing the orthogonal array L9(3^4) proposed by Taguchi [9,10]. In this methodology, the main factors considered to affect process results are placed at different rows in a designed orthogonal array. In the case under consideration, the output characteristic investigated is the dimensional accuracy; its most desirable value is the nominal dimensions (nominal-the-best attribute).

The summary statistic for this procedure, called signal-to-noise ratio, is defined for this case by the objective function described in equations 1, 2 and 3:

$$\eta_i = 10\,log_{10}\frac{\mu_i^2}{\sigma_i^2} \tag{1}$$

where,

$$\mu_i = \frac{1}{n}\sum_i y_i \tag{2}$$

and

$$\sigma_i^2 = \frac{1}{n-1}\sum_i (y_i - \mu_i)^2 \tag{3}$$

are the perceived data average and variance for the *i-th* test respectively, y_i represents the observed data for the *i-th* experiment and n is the number of tests [9-11].

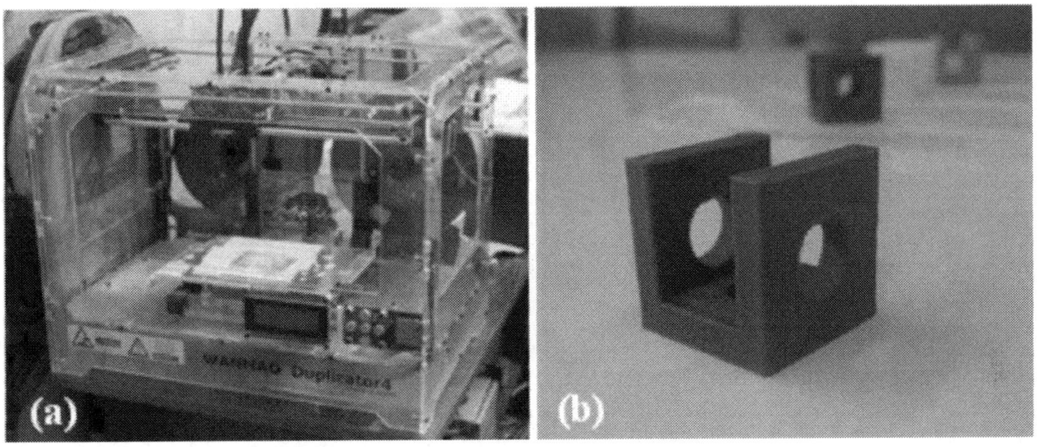

Figure 2. (a) Wanhao Duplicator® 4X 3D Printer, (b) Indicative printed part.

The goal of the scenario examined is to maximize the objective function η.

The input factors employed were the following: layer height (LH); number of shells (NS); printing material (PM) and infill Rate (IR). The PM is determined by the control factors Acrylonitrile Butadiene Styrene-ABS with "1" and Polylactic Acid-PLA with "2".

For IR, three levels equal to 20%, 50% and 70% of the solid material were selected. NS represents the number of outlines deposited on each object's layer and is defined by three values, namely 1, 2 and 3. Finally, for LH, three values for the height of each layer were selected, i.e., 0.1, 0.2 and 0.3 mm.

The process parameters and their corresponding levels are tabulated in Table 1.

The standard L9(3^4) orthogonal array has been selected based on a robust design to conduct the matrix experiment (Table 2).

All the other processing parameters were kept constant as proposed by the materials and 3D printing machines manufacturer due to the fact that the core intention of this study is to monitor and compare each of the two different material dimensional accuracy performances. Moreover, in this study, the bonding quality and the surface roughness of the printed part were not studied; these topics will be studied and reported in future work.

Table 1. Parameter design using the *L9* orthogonal array.

No. Exp.	Variable parameters	Level 1	Level 2	Level 3
1	PM (ABS or PLA)	ABS	PLA	-
2	IR (%)	20	50	70
3	NS (number)	1	2	3
4	LH (mm)	0.1	0.2	0.3

Table 2. Series of experiments by applying the *L9* Taguchi's orthogonal array.

No Exp.	PM	IR	NS	LH	DA XY (avg) mm	DA Z (avg) mm
1	1	20	1	0.1	0.234	-0.085
2	1	50	2	0.2	0.188	-0.145
3	1	70	3	0.3	0.053	0.013
4	2	50	3	0.1	0.147	-0.190
5	2	70	1	0.2	0.226	-0.200
6	2	20	2	0.3	0.054	-0.080
7	1	70	2	0.1	0.245	-0.120
8	1	20	3	0.2	0.102	-0.075
9	1	50	1	0.3	0.333	0.500

The parts shape is evaluated by measuring the linear dimensions in XY plane and Z direction for each test part.

A digital caliper with a 0.01 mm precision was utilized. The effect of each factor was monitored by applying Analysis of Means (ANOM) and Analysis of Variance (ANOVA). The best factor values were identified; the optimized combination was then constructed and evaluated. The nine test parts are illustrated in figure 3.

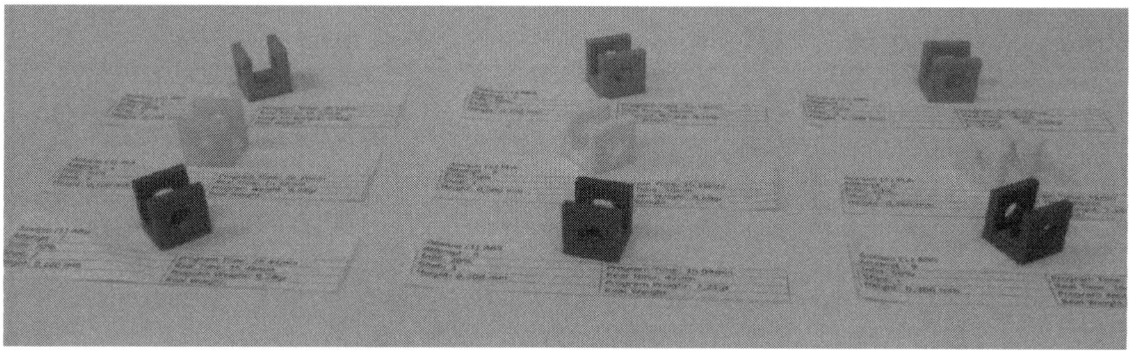

Figure 3. 3D printed experimental parts.

3. Results and discussion

Based on the measurements received from the ninth test item for the dimensions examined, ANOM and ANOVA analyses have been performed in order to determine the parameters affecting shape accuracy.

The best factor level based on the analysis of means was chosen as the one that leads to the higher value of the objective function η within the experimental region.

This means that the highest dimensional accuracy is required. The results pertaining to ANOM are illustrated in figures 4 and 5, whilst the corresponding ones for ANOVA are shown in Tables 3 and 4.

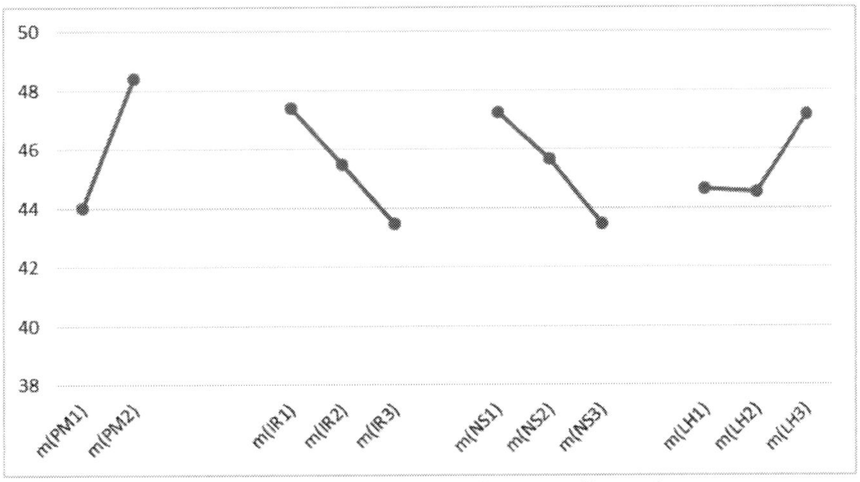

Figure 4. ANOM diagrams for X-Y dimensions.

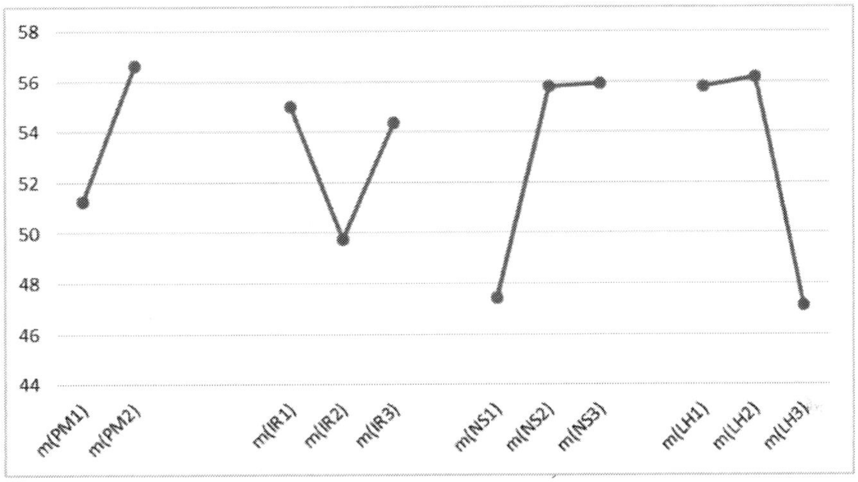

Figure 5. ANOM diagrams for Z dimension.

From these results it is evident that the shape accuracy in the X-Y plane follows a different pattern from the one valid for the Z-axis as far as the influence of the input parameters is concerned. Note that from figures 4 and 5 the optimum level for each parameter can be defined; moreover, from Tables 3 and 4 the corresponding contribution of each parameter in the objective of shape accuracy for the X-Y plane and Z-axis is observed.

Table 3. Analysis of variances for X-Y dimensions (ANOVA).

Variable	Deg. Fr.	SSq	Mean Sq	%
PM	1	38.59	38.59	39.9
IR	2	23.10	11.55	23.9
NS	2	21.76	10.88	22.5
LH	2	13.29	6.64	13.7
Total	7	96.750	67.66	100

Table 4. Analysis of variances for Z direction (ANOVA).

Variable	Deg. Fr.	SSq	Mean Sq	%
PM	1	58.01	58.01	14.3
IR	2	48.86	24.43	12.0
NS	2	142.66	71.33	35.2
LH	2	156.17	78.08	38.5
Total	7	405.71	231.85	100

It is concluded that the shape accuracy referring to the XY plane is mainly affected by PM, followed by ID and the NS. On the other hand, the results pertaining to the shape accuracy in the Z-axis indicate that all parameters are important for shape accuracy, with LH and NS being the dominant factors. The results referring to the optimum levels for control (input) parameters are summarized in Table 5.

Table 5. Optimum levels for control parameters.

characteristic	PM (type)	IR (%)	NS (number)	LH (mm)
XY plane	PLA	20	1	0.3
Z direction	PLA	20	3	0.2

Considering the above observations, the combination of factors resulting in the optimum shape accuracy reads as follows: PLA for PM and IR of 20% both for XY plane and Z-axis. Furthermore, the optimum number of NS is "1" for the XY plane and "3" for the Z-axis.

Moreover, a value for LH equal to 0.3 mm seems advantageous for shape accuracy in the XY axes. On the contrary, the optimum for the Z axis is achieved with a layer height value equal to 0.2 mm.

Further examination of the effects of FDM parameters on dimensional accuracy is provided by the contour plots referring to XY plane and to Z-axis.

However, contour plots suggest that control parameters are studied in continuous form by considering the interactions among parameters that have not been examined by applying Taguchi's L9 orthogonal array. Figure 6 and figure 7 illustrate the effect of control parameters referring to FDM for dimensional accuracy in XY plane and Z-axis, respectively.

Worth mentioning that these objectives are optimized in their minimum values since they minimize the deviation from the nominal dimension, corresponding to maximization of accuracy.

By examining the plots depicted in both figures, it can be verified that PLA (level "2" for printing material, PM) is the most advantageous material for 3D printing when the objective is to maximize dimensional accuracy, i.e., minimizing the deviation from nominal dimensions of the part's features.

In all contour plots, the regions under interest where beneficial values for control parameters can be selected are anywhere but the dark-green ones. By carefully examining the regions, it is observed that experimental results referring to control parameters agree with those obtained by the analysis of means to a significant extent.

Besides, it is observed by the analysis of means that effects do not always follow a straight trajectory. This is a strong indication of variations among the levels of process parameters. The latter encourages further research with more experimental runs to obtain a larger experimental sample to be examined.

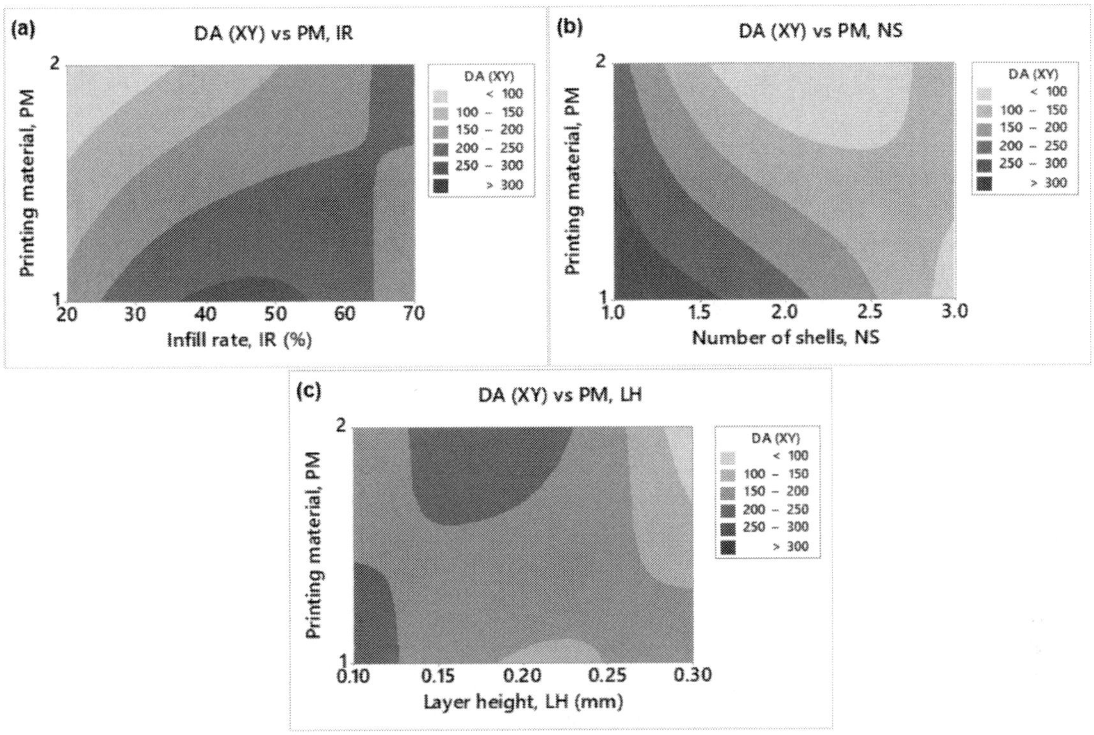

Figure 6. Contour plots for XY plane.

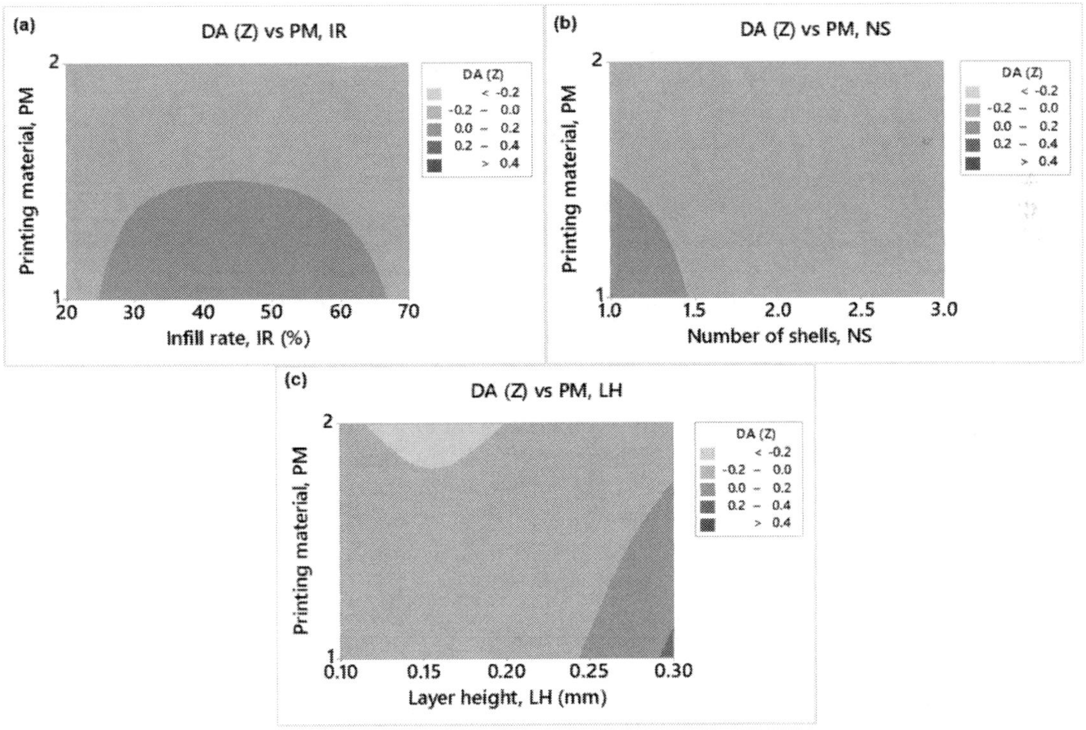

Figure 7. Contour plots for Z height.

4. Conclusions and future perspectives

This work deals with a study concerning the accuracy of 3D printed parts produced by the FFF process. Two printing polymeric materials; ABS and PLA, three infill percentages; 20, 50, and 70 (%), three numbers for shells; 1, 2, and 3 and three different layer heights; 0.1, 0.2 and 0.3 (mm) were the variables and their corresponding levels. A benchmark specimen was designed whilst nine experiments were performed according to the Taguchi methodology. It was concluded that when it comes to dimensional accuracy, PLA material is a promising consumable in comparison to ABS. By considering the combination of parameters resulting in the most desirable dimensional accuracy, XY plane and Z direction are optimized by an infill rate of 20%. The optimum number of shells is "1" in the XY plane and "3" in the Z direction. A value for layer height equal to 0.3 mm is the most beneficial for dimensional accuracy referring to the XY plane, whilst the best value for Z direction is obtained with a layer height value equal to 0.2 mm. A preliminary analysis using contour plots for examining dimensional accuracy referring to XY plane and Z direction, has shown that, when process parameters are investigated in a continuous form, significant interactions occur and they should be carefully considered when conducting parameter optimization research. The authors plan to look further ahead for performing additional series of experiments and processing the new results. Artificial neural networks or genetic algorithms are currently considered to be applied to single or multi-objective optimization problems with reference to this work, with the aim to provide a solution domain from which one may select advantageous values for optimally controlling the FFF process.

Acknowledgments

The authors acknowledge the financial support for the dissemination of this work from the Special Account for Research of ASPETE (ELKE) through the funding program "*Strengthening ASPETE's research*".

ORCID iDs

K Ninikas https://orcid.org/0000-0002-1304-3363
J Kechagias https://orcid.org/0000-0002-5768-4285
N A Fountas https://orcid.org/0000-0001-5859-6503
N M Vaxevanidis https://orcid.org/0000-0002-7641-5267

References

[1] Yan X and Gu P 1996 *Comput. Aided Des.* **28** 307-18 https://doi.org/10.1016/0010-4485(95)00035-6
[2] Kruth J-P, Leu M-C and Nakagawa T 1998 *CIRP An. Manuf. Technol.* **47** 525-40 https://doi.org/10.1016/S0007-8506(07)63240-5
[3] Chaidas D, Kitsakis K, Kechagias J and Maropoulos S 2016 *IOP conference series: materials science and engineering* **161** (1) 012033 http://dx.doi.org/10.1088/1757-899X/161/1/012033
[4] Ambrogio G, Costantino I, De Napoli L, Filice L, Fratini L and Muzzupappa M 2004 *J. Mater. Process. Technol* **153–154** 501-7 https://doi.org/10.1016/j.jmatprotec.2004.04.139
[5] Dimitrov D, Wijck WV, Schreve K and Beer ND 2006 *Rapid Prototyp. J.* **12** 42-52 https://doi.org/10.1108/13552540610637264
[6] Kechagias J 2007 *Rapid Prototyp. J.* **13** 316-23 https://doi.org/10.1108/13552540710824823
[7] Kechagias J, Stavropoulos P, Koutsomichalis A, Ntintakis I and Vaxevanidis N 2014 *Proc. of the 18th Int. Conf. on Circuits, Systems, Communications and Computers* 17-21
[8] Zhou J G, Herscovici D and Chen CC 2000 *Int. J. Mach. Tools Manuf* **40** 363-79 https://doi.org/10.1016/S0890-6955(99)00068-1
[9] Phadke M S 1995 *Quality engineering using robust design* Prentice Hall PTR
[10] Tsui K-L 1992 *IIE Trans.* **24** 44-57 https://doi.org/10.1080/07408179208964244
[11] Anitha R, Arunachalam S and Radhakrishnan P 2001 *J. Mater. Process. Technol.* **118** 385-8 http://dx.doi.org/10.1016/S0924-0136(01)00980-3

Studies on the influence of FFF parameters on the tensile properties of samples made of ABS

A Portoaca, I Nae, D G Zisopol and I Ramadan

Mechanical Engineering Department, Petroleum – Gas University of Ploiesti, Blvd. Bucharest, No. 39, 100680, Ploiesti, ROMANIA

ing_ramadan@yahoo.com

Abstract. FFF (Fused Filament Fabrication) prototyping method is mostly used in the fabrication of 3D complex geometric shapes. Layer by layer deposition technology offers a wide range of shapes for plastic parts. The mechanical characteristics of the manufactured parts depends on the plastic material used and the FFF printing parameters. The studies conducted in this paper show the influence of the FFF parameters on the tensile properties of ABS (Acrylonitrile Butadiene Styrene). The results indicate a considerable effect of infill percentage on tensile properties.

1. Introduction

With the increasing complexity of industrial products, requirements related to environmental protection and resource conservation, new manufacturing technologies are needed. Fusion deposition modelling is the most significant technique in additive manufacturing (AM) that refers to the process in which successive layers of material are stored in a computer-controlled environment to create a three-dimensional object [1]. Since 2009, the demand for FFF has grown steadily from year to year and many experts believe that this technology has the potential to revolutionize production in many sectors [2].

However, there are still some limitations and disadvantages that have been identified, especially in terms of the lower mechanical properties of FFF parts compared to products by conventional methods, such as injection and compression techniques [2,3]. Since mechanical characteristics are extremely important for functional parts, it is necessary to investigate the influence of process parameters on mechanical performance and geometrical qualification [3,4]. Further research is also needed to determine the printing parameters, such as: deposition orientation, layer thickness, deposition speed, mainly because the information in the literature is diverse in terms of the values obtained by the mechanical characteristics of parts made with 3D printers. This applies to PLA material, which unlike ABS has not been extensively analyzed [5-9]. The mechanical behavior of parts made by additive technologies is also evaluated in other scientific research. However, the FFF process is not widely studied. Printing variable parameters like layer thickness, nozzle temperature, building orientation and raster orientation, have a significant influence on AM 3D printed parts. The influence of layer orientation on tensile strength, impact resistance and rupture modulus of rapid prototyped specimens was studied by other researchers.

Regarding to 3D Printing technologies, if the beginnings consisted in obtaining prototypes for esthetic and geometrical verifications, now functional parts, mold and tools can be obtained, including metal parts made by additive manufacturing. The difference between these techniques is reduced to the application for which the additive manufacturing technology is used [10].

This means that additive technologies can be considered a manufacturing alternative for rapidly obtaining functional elements such as devices and molds. One question can be answered: The manufacture of a motion transmission with gears, fully functional in its housing, without any assembly actions, would be possible? Although the answer can be yes, there are some issues that must be addressed, like the advantageousness of its price considering a mass production.

A characteristic common to the different additive manufacturing techniques is that of the need for a minimum number of phases in the manufacturing process starting with the development of the "idea" by the designer through to obtaining the finished product [15].

2. Comparison of raw materials used for 3D printing in FFF technology

The principal advantage of FFF technology is the possibility of using a diversity of specific materials. Physical and mechanical characteristics of the most filaments used in FFF technology are shown in Table 1 shows. The data was collected from manufacturers data sheets specifications. These indicative values depend very much on the printing conditions and filament providers.

Table 1. Physical-mechanical characteristics of the most used filaments in FFF technology [20-22]

Materials	Extrusion temperature (°C)	Bed temperature (°C)	Density (g/cm^3)	Tensile Strength (MPa)	Specific deformation (%)	Charpy impact strength (kJ/m^2)
PLA	210 ± 10	25-60	1.31±0.02	15.5-72	34.5 ± 8.1	5.7 ± 0.4
PP	240 ± 10	80	0.89	14	>200	10
BVOH	210 ± 10	60	1.14	45	9	21
PETG	240	80	1.25	31.9 ± 1.1	6.8 ± 0.9	5.1 ± 0.3
316 L	240 ± 10	105 ± 15	5.6 ±0.2	561	53	1110
ULTEM	350 ± 10	150 ± 10	1.27	81	3.3	6
ABS	210 + 40	110 ± 10	1.10	33.9	4.8	10.5

The mechanical properties considered in Table 1 refer to the orientation of the horizontal test specimen (XY orientation). Under these conditions, this paper analyzes the influence of technological conditions of deposition on the mechanical characteristics of the deposited material. Economical environmental and safety challenges have caused scientists and economic agents to partially replace petrochemical polymers with other materials [11,12], as may be the ABS material used in this study.

3. Experimental methodology

Physical and mechanical characteristics of 3D printed samples are important indicators for the quality of printing. The experiment planning includes tests performed to determine how the technological parameters of the printing influence the mechanical characteristics of the deposited material. In these conditions, the working parameters of the printer used must be taken into account.

The initial data targeted in the experimental determinations were grouped into two categories (as shown in Table 2). The first category refers to process parameters considered as constant quantities: deposition direction, printing speed, filling pattern, deposition temperature. The second category refers to process parameters considered variable such as: layer thickness (g), infill percentage (g_u), flow rate (F_r). Test specimens were made, grouped in sets of samples consisting in 5 specimens, belonging to a group of process parameters, highlighted in Table 2. The geometrical form and dimensions of the specimen are shown in Figure 1. The specimens (Figure 2) used in this study to assess dimensional accuracy, repeatability and mechanical properties are modeled based on ASTM D638, Type IV standards for tensile testing of plastic [13, 14].

Table 2. Constant and variable printing parameters

Layer thickness (mm) and infill percentage	Printing parameters				
	Flow rate (mm^3/s)	Temperature (°C)	Print speed (mm/s)	Build orientation	Infill model-lines
0.2 (100%)	6.4	240	40	X-Y	45°
0.2 (75%)	6.4	240	40	X-Y	45°
0.2 (50%)	6.4	240	40	X-Y	45°
0.15 (100%)	4.8	240	40	X-Y	45°
0.15 (75%)	4.8	240	40	X-Y	45°
0.15 (50%)	4.8	240	40	X-Y	45°
0.1 (100%)	3.2	240	40	X-Y	45°
0.1 (75%)	3.2	240	40	X-Y	45°
0.1 (50%)	3.2	240	40	X-Y	45°

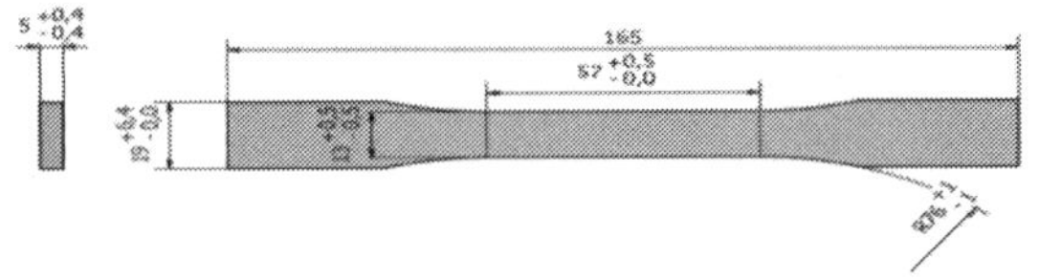

Figure 1. Tensile test specimen- Dimesional details [14,19]

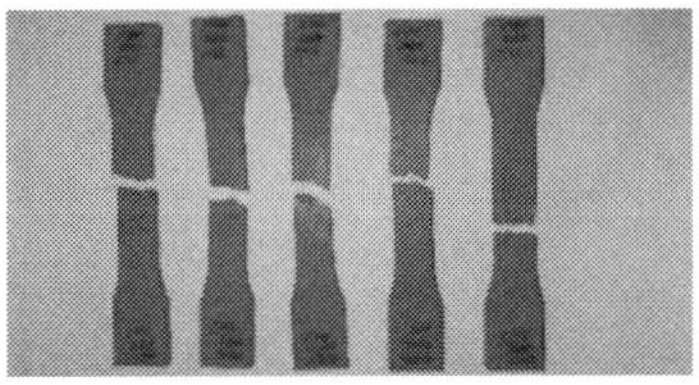

Figure 2. Fracture samples

The specimen model was designed using Autodesk Inventor software, transformed into STL format, and the G-code was made using the 3D printer software: Idea Maker. The software offers the possibility to adjust printing parameters.

In total, a number of 45 specimens were manufactured by 3D printing on a Raise E2 printer. The specimens were subjected to mechanical tests to determine the main material characteristics: ultimate tensile strength (R_m), elongation at break (ε), tensile Young's modulus (E_1). Tensile tests were performed on an electro-mechanical machine with a force cell of 2.5 kN, at a speed of 5 mm/min and elongation at break was measured using an axial extensometer.

4. Results and discussion

This paper studies the influence of layer thickness and infill percentage on Young's modulus, linear specific deformation and ultimate tensile strength of additively manufactured ABS were explored.

The test methodology was performed in the context were the variation parameters are: the thickness of the deposited material layer (g = 0.1 mm; 0.15 mm; 0.20 mm) and the infill percentage (g_u = 50%; 75%; 100%). Graphs were drawn to express the dependence of the mechanical characteristics (CM) of the tested specimens according to the infill percentage CM = f(g_u), for different thicknesses of the deposited layer (g = 0.1 mm; 0.15 mm; 0.20 mm) – Figure 3-11.

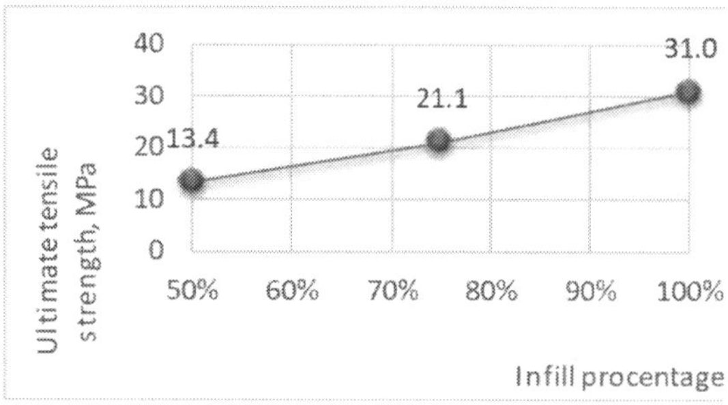

Figure 3. Dependence $R_m = f(g_u)$, for (g = 0.1 mm) and (g_u=50%; 75%; 100%)

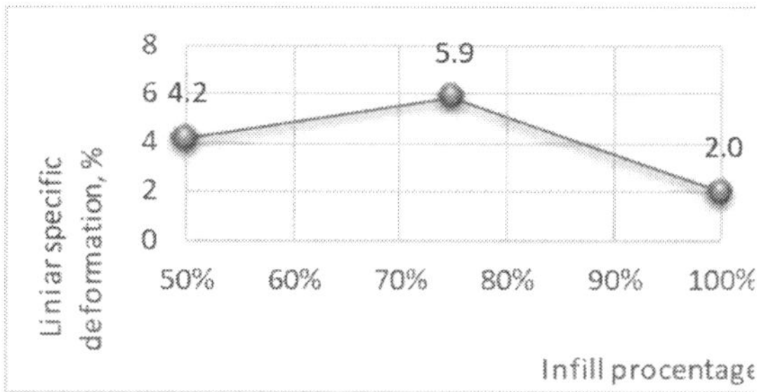

Figure 4. Dependence $\varepsilon = f(g_u)$, for (g = 0.1 mm) and (g_u=50%; 75%; 100%)

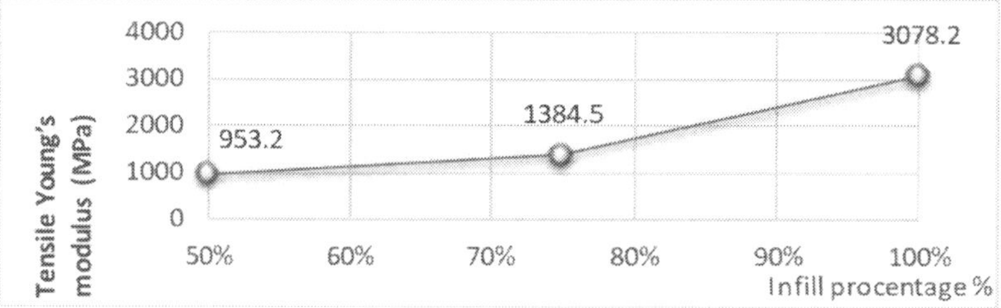

Figure 5. Dependence $E_l = f(g_u)$, for (g=0.1 mm) and (g_u= 50%; 75%; 100%).

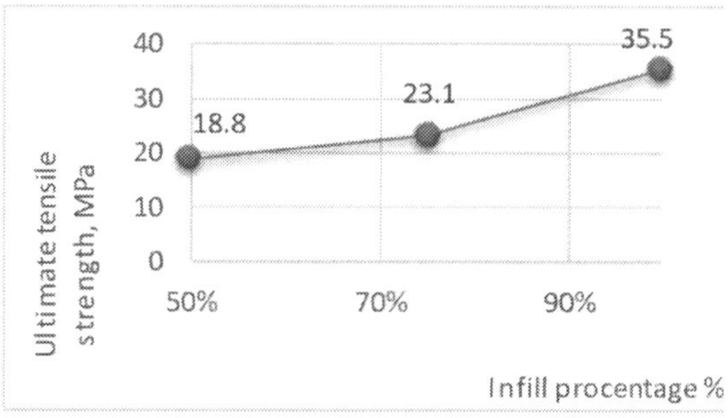

Figure 6. Dependence $R_m = f(g_u)$, for (g= 0.15 mm) and (g_u=50%; 75%; 100%)

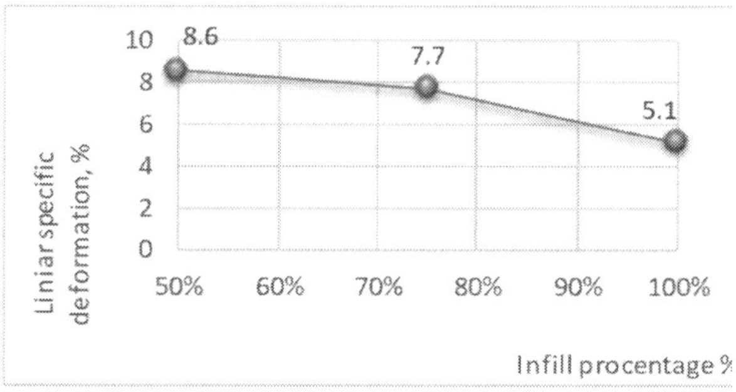

Figure 7. Dependence $\varepsilon = f(g_u)$, for (g = 0.1 mm) and (g_u=50%; 75%; 100%)

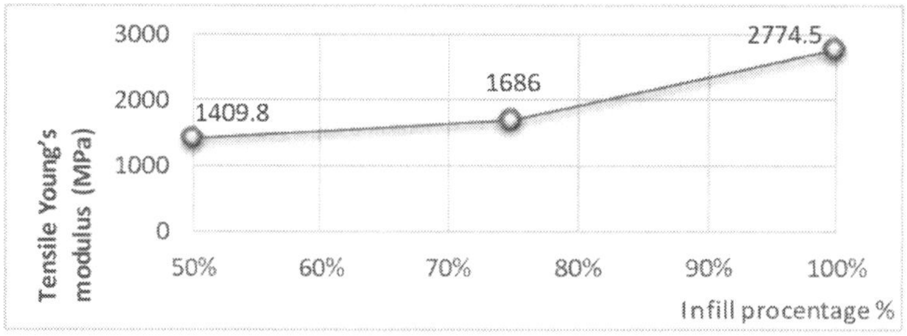

Figure 8. Dependence $E_l = f(g_u)$, for (g=0.15 mm) and (g_u=50%; 75%;100%).

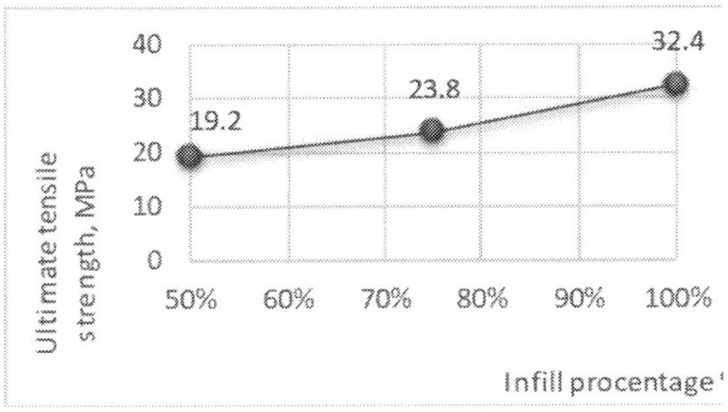

Figure 9. Dependence $R_m = f(g_u)$, for (g= 0.20 mm) and (g_u =50%; 75%; 100%)

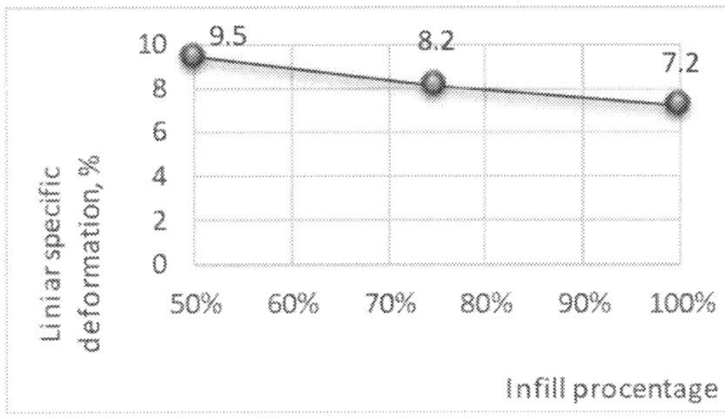

Figure 10. Dependence $E_l = f(g_u)$, for (g= 0.20 mm) and (g_u=50%; 75%; 100%).

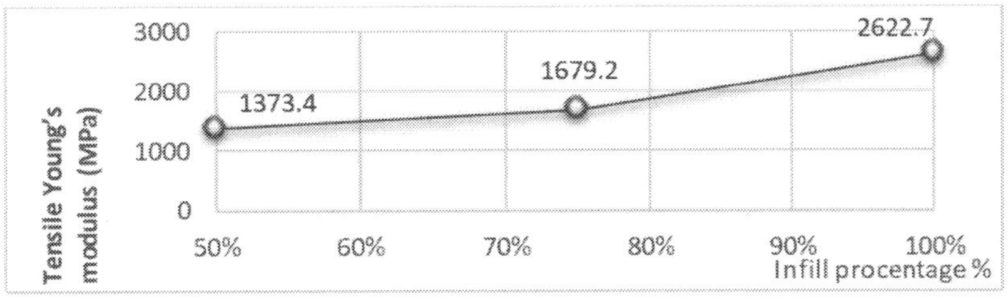

Figure 11. Dependence $E_l = f(g_u)$, for (g_u=0.20 mm) and (g_u=50%; 75%; 100%).

5. Conclusions

The objective of the paper was to establish the variation of the mechanical characteristics of the ABS material, deposited by 3D printing, depending on the parameters of the technological process.

Analyzing the dependence between ultimate tensile strength (R_m) and infill percentage it can be seen that higher values of Ultimate tensile strength are obtained for 100% infill percentage. The low adhesion of tested sample layers is directly influenced by the deposition temperature gradient. In Table 2 a direct dependence between the flow rate of the ABS filament and the thickness of the deposited layer can be observed. All graphs reveal better performance with the increase of infill percentage and less influence of the layer thickness (similar results).

References [5,13,16-18] reveal that the lower thickness of the deposited layers will improve the ultimate tensile strength of samples. In this study, although the maximum value was highlighted for the layer thickness of 0.15 mm, higher elongation at break is observed in the specimens with a layer thickness of 0.2 mm. Taking into account the variation of linear specific deformation it can be seen in Figures 4, 7, 10 that lower deformation values correspond to samples with higher infill percentages.

Young's modulus may be thought of as a substance's resistance to elastic deformity; the stiffer the material, the higher its elastic modulus. So, the set of samples with a layer thickness of 0.1 mm and 100% infill tends to be more brittle than the samples with a lower layer thickness (0.15 mm, 0.2 mm) as shown in Figures 5, 8, 12.

In this context, further experiments will be developed in order to evaluate the effect of other input factors of the manufacturing process on the tensile stress behaviour of material parts.

ORCID IDs
I Nae: https://orcid.org/0000-0001-8866-217X
D G Zisopol: https://orcid.org/0000-0002-0824-0514
I N Ramadan: https://orcid.org/0000-0002-0920-5710
A I Portoaca: https://orcid.org/0000-0003-1986-8419

References

[1] Mohan N, Senthil P, Vinodh S and Jayanth N 2017 *Virtual Phys Prototyp* **12** (1) 47-59 https://doi.org/10.1080/17452759.2016.1274490
[2] Rahim T N A T, Abdullah A M and Akil H M 2018 *Polym Rev* **59** (4) 1-16 https://doi.org/10.1080/15583724.2019.1597883
[3] Yilmaz H, Turan M K and Fahamec V 2017 *ETASR* **7** (6) 2160-66 https://doi.org/10.48084/etasr.1464
[4] Belarbi B, Ghernaout M and Benabdallah T 2019 *ETASR* **9** (3) 4182-87 https://doi.org/10.48084/etasr.2689
[5] Vaezi M and Chua C K 2011 *J. Adv. Manuf. Technol* **53** 275–84 http://dx.doi.org/10.1007/s00170-010-2821-1
[6] Tymrak B M, Kreiger M and Pearce J M 2014 *Mater Design* **58** 242–46 http://dx.doi.org/10.1016/j.matdes.2014.02.038
[7] Vălean C, Marşavina L, Mărghitaşa M, Linula E, Razavi J and Berto F 2020 *Procedia Struct. Integr.* **26** 313–20 https://doi.org/10.1016/j.prostr.2020.06.040
[8] Lanzotti A, Grasso M, Staiano G and Martorelli M 2015 *Rapid Prototyp. J.* **21** 604–17 http://dx.doi.org/10.1108/RPJ-09-2014-0135
[9] Majid F, Zekeriti N, Rhanim R, Lahlou M, Rhanim H and Mrania B 2020 *Procedia Struct. Integr.* **28** 1719-26 https://doi.org/10.1016/j.prostr.2020.10.147
[10] Saenz F, Otarola C, Valladares K and Rojas J 2021 *Addit. Manuf.* **39** https://doi.org/10.1016/j.addma.2021.101841
[11] Al-Ghamdi K A 2019 *Int. J. Lightweight Mater. Manuf.* **2** 338-45 http://dx.doi.org/10.1016/j.ijlmm.2019.05.004
[12] Zhang Z, Ortiz O, Goyal R and Kohn J 2014 Biodegradable Polymers *Principles of Tissue Engineering* (Fourth Edition) chapter 23 p 441-73 https://doi.org/10.1016/B978-0-12-398358-9.00023-9
[13] Weake N, Pant M, Sheroan A, Haleem A and Kumar H 2020 *Procedia Manuf.* **51** 704–9 https://doi.org/10.1016/j.promfg.2020.10.099
[14] ASTM Standard test method for tensile properties of plastics D638 2010
[15] Jimenez J, Luis R, Dominguez A, Espinosa M and Dominguez M 2019 *Hindawi Complexity* **2019** 9656938 https://doi.org/10.1155/2019/9656938
[16] Nomani J, Wilson D, Paulino M and Mohammed M I 2020 *Mater. Today Commun.* **22** 100626 https://doi.org/10.1016/j.mtcomm.2019.100626
[17] Chacóna J M, Caminerob M A, García-Plazab E and Núñezb P J 2017 *J. Mater. Des.* **124** 143–57 https://doi.org/10.1016/j.matdes.2017.03.065
[18] Chevrychkina A, Volkov G. A and Estifeev A D 2017 *Procedia Struct. Integr.* **6** 283-5 https://doi.org/10.1016/j.prostr.2017.11.043
[19] Zisopol D G, Nae I, Portoaca A I, and Ramadan I 2021 *ETASR* **4** 7458-63 https://doi.org/10.48084/etasr.4311
[20] https://eu.polymaker.com/ accessed at 14.05.2021
[21] https://www.verbatim.ro/ accessed at 24.06.2021
[22] https://www.basf.com/ accessed at 10.05.2021

IOP Conf. Series: Materials Science and Engineering 1235 (2022) 012009 doi:10.1088/1757-899X/1235/1/012009

Innovative process for material separation of printed circuit boards (PCB) with laser technology

M Zink, A Sover and T Vogel

Ansbach University of Applied Sciences, Faculty of Technology, Residenzstr. 8, Ansbach, Germany

zink.markus@hs-ansbach.de

Abstract. In the last years, the worldwide amount of electronic waste has increased, according to a report of the United Nations University. Our society urgently needs new solutions for the recycling of electronic waste, which is often not recycled in an environmentally friendly way. However, the electronic waste contains a large quantity of noble metals. They are excellent for reuse and can be melted and reshaped several times. For this purpose, they must be separated into their components before recycling. The laser technology has a great potential for an environmentally friendly, cost-effective and at the same time material-appropriate separation of the individual materials from different hybrid components. The possibility of using a laser to mechanically and physically separate the individual layers of a printed circuit board was studied for this paper. The first tests have shown that it is possible to separate the individual PCB materials to facilitate optimal reuse of them.

1. Introduction

According to a 2017 report by the United Nations University, the amount of e-waste generated worldwide each year has increased over the past years, and this increase is predicted to continue in the following years. This development should be viewed critically, considering that only 20% of e-waste is recycled worldwide. [1]

Printed circuit boards (PCBs) are an important component of all electronic devices and are made of valuable metals such as copper, gold, palladium and silver. Various processes are already used today for the recycling of these materials. The recycling processes of printed circuit boards applied up to now are classified into three categories. [2]

- In the thermal recycling process, the metals are recovered through burning and metallurgy.
- Mechanical processes are applied to shred the PCBs and then separate the metallic from the non-metallic materials.
- In addition, PCBs can be separated and recycled using chemical processes.

All these processes have specific disadvantages in terms of the environment. [2]

The material separation of rigid printed circuit boards using a fibre laser is the object of research in this paper. Laser technology is already showing good results in separating composite materials such as painted components. [3]

The aim is to study a new environmentally friendly method for material separation. The focus here is on the recovery of used metals.

2. Materials and basics of the printed circuit board construction

Printed circuit boards can be very different in terms of their materials and construction methods, depending on their use and individual requirements. [4] The main difference is the number of copper layers; these can be single-layer, double-layer or multilayer. [5] Figure 1 shows two cross-sections of printed circuit boards with six copper layers, which have a thickness of approximately 33 to 38 μm.

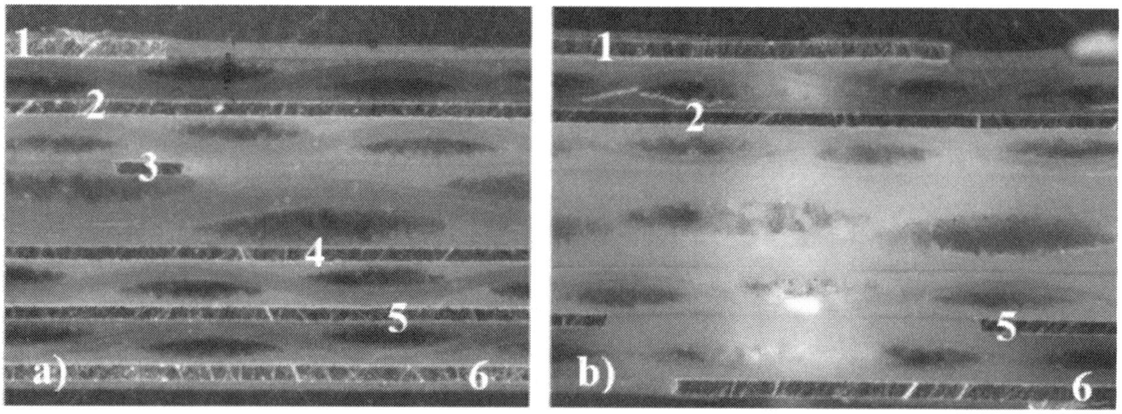

Figure 1. Microscope images of the embedding.

The total thickness of the PCB is approximately 801 μm. As shown in Figure 1a), there are six copper layers in total, but in some places, there are only two to five copper layers on top of each other, which can be seen in Figure 1b). Figure 2 shows the top side of the PCB used. The PCB is covered with a green solder resist. The first copper layer is under the darker areas.

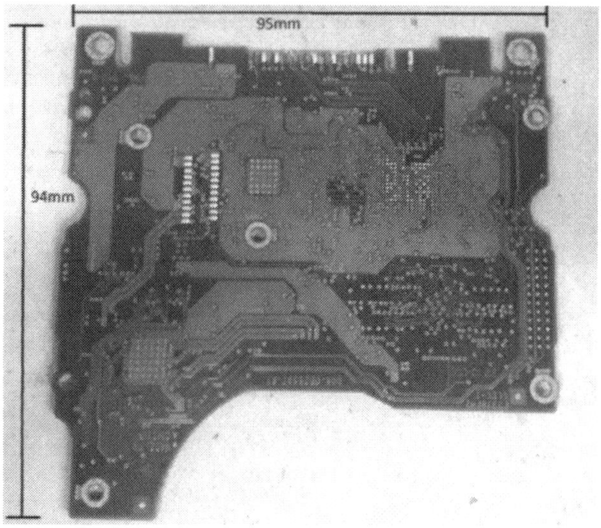

Figure 2. Topside of the used printed circuit board.

3. A new technique for separating printed circuit boards

Laser technology has a great potential for the environmentally friendly, cost-effective and at the same time material-appropriate separation of the individual layers as well as the separating of the materials used in a printed circuit board.

In the future, this technology will play a decisive role in the recycling industry for the sustainable recycling of printed circuit boards to manage large quantities of e-waste.

The laser technology can be used for separating the different layers of a PCB. In the process of the separation of layers, the material is vaporised or solid particles are detached because of tensions in the different layers of the PCBs. The layer that is to be removed must lie on the outside in this process and must not be covered by other layers. In addition, the material to be removed should have a high absorbance to the wavelength of the laser so that conversion into thermal energy is possible at the surface of the material and the material can be vaporised.

The tests were carried out with the Powerline F 50 fibre laser from Rofin-Coherent. The laser transmits a laser beam with a wavelength of 1064 nm, has a power of 50 W and offers a pulse frequency of 50 to 200 kHz and a pulse width of 1 to 120 ns. [6]

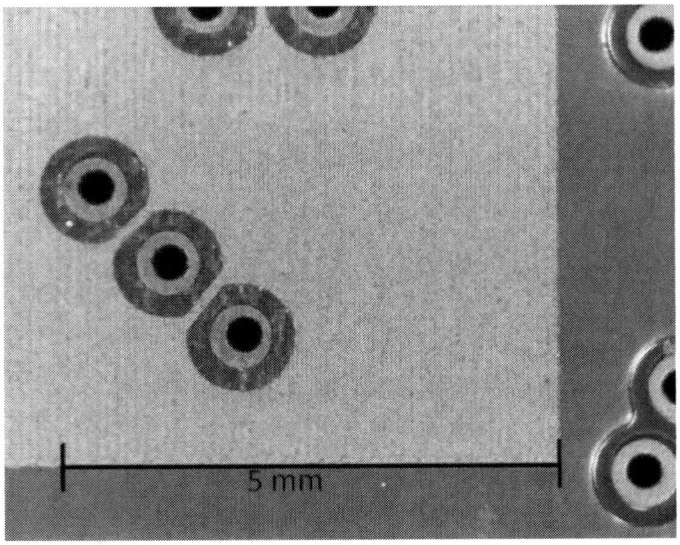

Figure 3. Exposing the first copper layer.

The settings of the laser beam can be precisely adjusted to optimise the process, its speed and the load on the surface. [7] Hence, high quality of the recycled material can be achieved. For an optimal separation of the individual layers, the laser parameters need to be determined.

The following parameters can be set for the laser system: frequency, speed, point distance, pulse width, line width.

Depending on these parameters, the system calculates the required laser power. All the other parameters remained unchanged and are the standard settings of the laser. Identical printed circuit boards were used to find these laser parameters.

During the tests, an attempt was made to expose the first copper layer of the PCB with a size of 10 mm x 10 mm. Error! Reference source not found. shows a microscope image of the surface that was processed.

The exposed copper layer (light-coloured area) is visible in contrast to the non-exposed area, which was not processed and is still covered with solder resist.

A larger area was then processed with the determined laser parameters, see Error! Reference source not found..

After the tests were completed, only slight residues of the solder resist were visible on the processed surfaces.

The following parameters were used for processing the printed circuit board:

- Frequency: 100 kHz
- Speed: 500 mm/s
- Point distance: 0.005 mm
- Pulse width: 50 ns
- Line width: 0.1 mm
- Laser power: 33 W

Processing of this area takes about 2.68 seconds, for the complete used PCB about 200 seconds.

During the working process, smoke is produced when the first layer is separated. The smoke was evacuated through a filter system. The processing of PCB can be done also underwater, to avoid smoke propagation. In this case, the laser parameters were changed.

The frequency was increased from 100 kHz to 200 kHz and the pulse width from 50 ns to 120 ns. As a result, the energy for the system automatically increases from 33 W to 50 W. No smoke was observed.

This research work proves that it is possible to remove the copper layer from printed circuit boards.

For different material combinations, individual laser parameters (energy, pulse duration, speed) must be applied in each case.

Furthermore, in addition to the fibre laser used here, which is a solid-state laser, other laser quantities types could also be considered, such as gas lasers or semiconductor lasers.

The first copper layer was easily separable after the laser processing. After removing this layer, a second laser processing was applied to detach the second copper layer.

Figure 4. The detached copper layer of the printed circuit board.

4. Conclusion

Laser technology shows good results in separating the different layers of printed circuit boards. Compared to other methods, this technology offers great advantages: it is easy to operate and quiet, does not require any consumables or environmentally hazardous media such as solvents and the operating costs are low. Furthermore, apart from radiation protection, no special environmental protection measures are required.

The results of this research project have shown that copper can be completely separated from the remaining materials in single-layer printed circuit boards. For multilayer printed circuit boards more work should be done to avoid the carbonisation of thermoset materials. To manage the enormous amount of e-waste and at the same time separate the materials that are used for recycling, laser technology could be used in the recycling industry in the future. The knowledge acquired in this project will be used in a follow-up project dealing with the assembly of laser equipment suitable for separating PCB and offering an efficient solution for recycling e-waste on an industrial level.

All information collected regarding the necessary pre- or post-processing will be used to ensure a high degree of automation of the process. However, to optimise this process with regard to industrialisation, further investigations of the materials are necessary. Considering the opportunities provided by the research of printed circuit board materials separation by laser, it is worth being continued. Moreover, the recycling process of printed circuit boards can be supported by a previous partial separation of the materials.

ORCID iDs
A Sover https://orcid.org/0000-0003-3408-898X
M Zink https://orcid.org/0000-0002-6883-5595

References

[1] Baldé C P, Forti V, Gray V, Kuehr R and Stegmann P 2017 *The Global E-waste Monitor 2017: Quantities, Flows, and Resources* https://www.itu.int/en/ITU-D/Climate-Change/Documents/GEM%202017/Global-E-waste%20Monitor%202017%20.pdf Accessed 10.05.2021

[2] Xiang D, Mou P, Wang J, Duan G and Zhang H C 2007 Printed circuit board recycling process and its environmental impact assessment *The Int. J. of Advanced Manufacturing Technology* 34 1030-6 https://doi.org/10.1007/s00170-006-0656-6

[3] Sover A 2019 Research on the Process of Paint Removal from Thermoplastic Materials by Laser *IOP Conference Series: Materials Science and Engineering* **564** 012028

https://doi.org/10.1088/1757-899X/564/1/012028

[4] Db electronic *PCB base materials and their properties: Base material types classified according to their use* [Online] https://db-electronic.com/en/pcb-manufacturing_s35.htm. Accessed 11.01.2021

[5] Pcbart *Printed Circuit Board Introduction & PCB Types* [Online] https://www.pcbcart.com/article/content/PCB-introduction.html Accessed 12.04.2021

[6] Coherent *PowerLine F Series: Cost-Effective Laser Markers* [Online] https://www.coherent.com/assets/pdf/COHR_PowerLine_F_Series_DS_1020_4.pdf Accessed 27.04.2021

[7] Zink M and Sover A 2020 *Study on the Process of Paint Removal from Thermoplastic Materials by Laser Technology* Proc. of the 2nd Int. Conf. of the University of Applied Sciences Ansbach 3rd - 4th of July 2020

IOP Conf. Series: Materials Science and Engineering 1235 (2022) 012010 doi:10.1088/1757-899X/1235/1/012010

Superalloy manufacturing application on CNC machine tool

Adrian But

Politehnica University of Timişoara, Mechanical Faculty, Department of Materials and Manufacturing Engineering, Blvd Mihai Viteazu 1, Timişoara 300222, Romania

adi.but@gmail.com

Abstract. The paper presents one real application to manufacturing a superalloy material, taking into consideration their properties (creep-rupture) that recommend them for high-temperature applications dedicated to the aerospace manufacturing area. Aspects related to milling operations on the CNC machine tool are underlined. Thus, several milling strategies are detailed, depending on the type of cutting tool and the material of the removable inserts, but also aspects related to the milling process itself, which involves information's about the characteristics of the material of the part, optimal ways of chip formation depending on the type of inserts. Three different milling strategies were analysed with the target to obtain the best solution to manufacturing the superalloy materials to keep is possible, more the time live of the cutting tools, but at the same time, to reduce the time of manufacturing. It was obtained a significant reducing time of manufacturing and was implemented different milling CAM solutions with useful conclusions for future applications.

1. Introduction

Superalloys are composed of alloys based on Fe, Ni, Co and Cr, with various amounts of W, V, Ta, Nb, Ti, Al, to which can be added small amounts of B, Zr, and Hf, which increase the strength to creep and ductility by finishing the granulation.

In the case of superalloys based on Fe and Ni, carbon is present in relatively small quantities (below 0.03%), and slightly higher amounts in the case of superalloys based on Co.

The most important physical properties of superalloys are density, specific heat, electrical conductivity, electrical resistivity, liquid curve temperature, solid curve temperature, magnetic permeability, Curie temperature.

In 1950 only 10% of the composition of a jet engine was made of superalloys. This percentage reached 50% in 1985, and 60% in 1993. At the same time, the technology of creating the components was based on powder metallurgy, gradually eliminating the classic casting methods.

Although the advantages offered by powder metallurgy recommend this technology, there are also problems related to this type of processing:

- the presence of metallic and non-metallic inclusions reduces fatigue resistance
- the size, shape and distribution of the powders are difficult to control because there is no model for simulating the atomization of molten metal,
- there is no model to evaluate the effect of cracks on fatigue strength
- there are no effective methods for determining pore size and number.

IOP Conf. Series: Materials Science and Engineering 1235 (2022) 012010 doi:10.1088/1757-899X/1235/1/012010

2. Milling of superalloys

Superalloys are divided into three groups: nickel-based, iron-based or cobalt-based. The physical properties and the manufacturing of each superalloy family depend on the chemical composition and the manufacturing process used. The application of annealing or ageing treatments will influence the further processing properties of the superalloys.

Nickel-based superalloys are the most common, and now, over 50% of an aircraft's engine is made from these types of alloys. The development department says this percentage will increase over time. The most commonly used nickel-based superalloys are Inconel 718, Waspaloy, Udimet 720 (with secondary phase precipitation), Inconel 625 (solid solution alloy). Unlike other materials, in the case of superalloys, the coolant must be used in all applications made with carbide cutters, except for milling with ceramic inserts. The shape of the piece and its fixing is very important. Aerospace components often have thin walls and many complex surfaces, which can be easily deformed or vibrated. For this reason, it is necessary to develop complex clamps, dedicated to certain components, to reduce vibrations and to take over the cutting forces.

The last years have brought a significant development in the field of cutting tools, offering processors a variety of possibilities. Removable carbide and ceramic tiles have seen an improvement in productivity, reliability, and quality. The target is to use a processing strategy appropriate to both the cutting tool used and the type of manufacturing (planning, pocket milling, shoulder milling, etc.) that is to be achieved, in order to find the optimal point between the amount of material removed and the life of the cutting tool. In manufacturing the superalloys, the cutting speed and the hardness of the material are the factors that most influence the life of the cutting tool. The temperatures developed during processing amount to 750-1000°C. These temperatures allow the hardened surfaces to harden, which contributes to the wear of the cutting tool. The results of some studies show that by increasing even by 5 m/min the cutting speed can decrease the life of the cutting tool by 50%.

2.1. Milling of superalloys with ceramic plates

An advantage is that the characteristics of the coolant must not be taken into account because the ceramic cutting tools work on dry land without coolant.

The cutting speed is between 700-1200 m/min. The heat is a part of the cutting mechanism - this heat comes from high cutting speeds and dry processing.

The heat developed is almost certainly destructive to carbide cutters. Compared to carbide cutting tools, ceramic cutting tools have a much higher heat resistance, which makes them a very good choice for machining superalloys.

The main applications in which ceramic tool cutters are used are machining of motor housings and oil drilling equipment, due to the large amounts of material that must be removed.

Figure 1 shows the effects of the coolant on the ceramic tiles. In the case of both ceramic tiles, the same cutting parameters were used, and the processing time was equal.

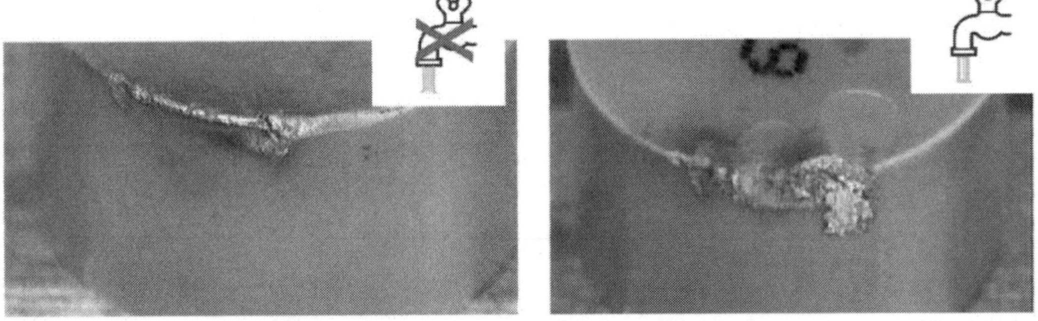

Figure 1. The effect of coolant on ceramic tiles.

The cutting speed must be chosen in such a way as to generate sufficient heat in the cutting area. If the cutting speed is too low, it can wear the cutting edge, and if it is too high it can destroy the plate. Figure 2 shows the effects of cutting speed on ceramic tiles. In the case of both ceramic inserts, they were used to manufacture the same type of material, the same cutting parameters were used, except for the cutting speed:

- Figure 2.a.) Vc = 700 m/min;
- Figure 2.b.) Vc = 1000 m/min.

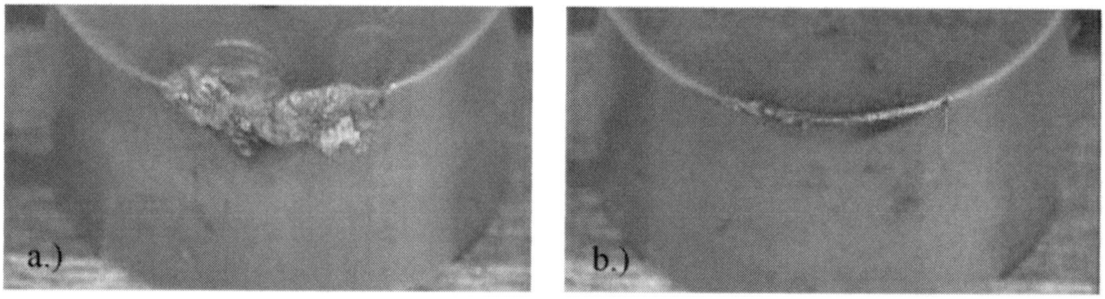

Figure 2. The effect of cutting speed on ceramic tiles: a.) Vc = 700m / min; b.) Vc = 1000m/min.

3. Manufacturing Inconel 625

According to those presented, the processing of superalloys can be done with different types of cutting tools. In the following, the machining program will be performed on a machine tool with numerical control using different milling methods.

The roughing operation of the part will be done by 3 different methods, the finishing operations being the same.

The roughing operations will be performed by the following methods:

- milling with carbide inserts;
- milling with ceramic tiles;
- HSM milling with monobloc milling cutters.

Inconel 625 can also be found under other names, such as: Haynes 625, Nickelvac 625, Nicrofer 6020, Altemp 625 or Chronic 625.

3.1. Face milling

The face milling operation is performed with a milling cutter with round carbide inserts, with 7 teeth, and with ϕ 63mm diameter. The plate radius is 6mm. The cutting depth ap is 1mm and ae is 27mm. The trajectory of the cutting tool is represented in Figure 3. The cutting parameters are: cutting speed Vc = 35 m / min, rotation speed = 177 rpm, feed per tooth fz = 0.1 mm / tooth, working feed F = 123 mm / min. According to the simulation performed in the CAM program, the manufacturing time was 5 minutes and 29 seconds.

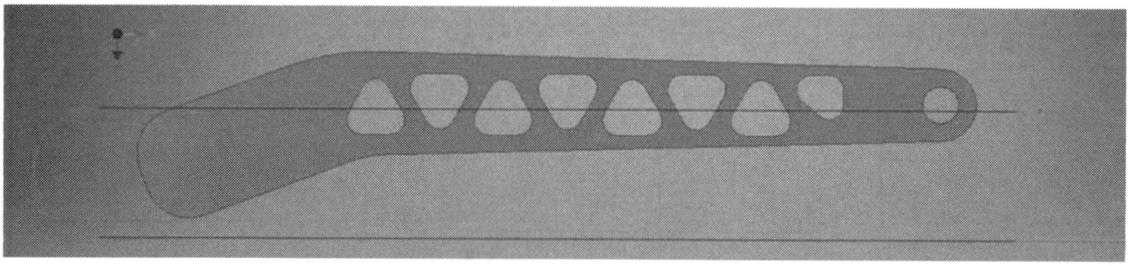

Figure 3. The face mill operation.

3.2. Contour roughing

Taking in consideration that the roughing contour involves the removal of a large amount of material, it is best to make a comparison between the processing strategies mentioned above. Contouring with milling cutter with round carbide plates. The first option is with a ϕ50mm diameter milling cutter with round carbide inserts with 6 teeth. The manufacturing was made by successive passes of 1 mm depth, with a finishing spring by 0.2 mm. The trajectory of the cutting tool is represented in Figure 4. The cutting parameters are: cutting speed Vc = 40 m / min, rotational speed S = 254 rpm, feed per tooth fz = 0.1 mm / tooth, working feed F = 133 mm / min.

The manufacturing time was 2 hours, 6 minutes and 29 seconds.

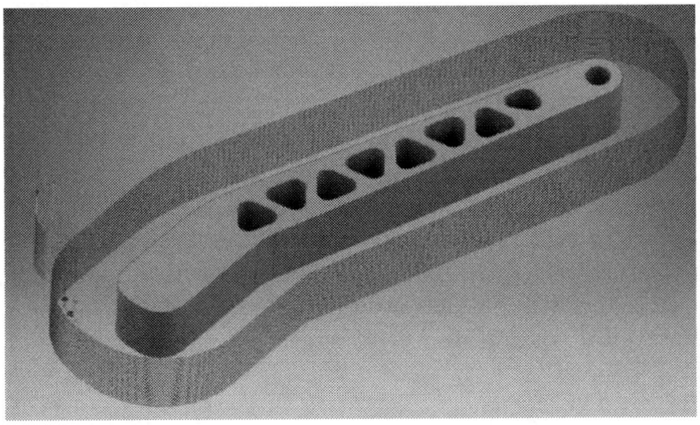

Figure 4. Contouring with round plates.

3.3. Contouring with milling cutter with round ceramic tiles

We will use the same face mill cutting tool ϕ50 mm, with 6 teeth, but the material of the inserts was changed in ceramic. The trajectory is the same as in the previous operation (Figure 4), with successive passes of 1 mm each, but there are significant changes in the cutting parameters. As mentioned above, milling with ceramic plates requires high cutting speeds and a slightly lower feed per tooth than milling with carbide plates. These parameters of the cutting process are: cutting speed Vc = 800 m/min, rotation speed S = 5092 rpm, feed per tooth fz = 0.06 mm / tooth, working feed F = 1833 mm/min.

The processing time was 10 minutes and 37 seconds.

3.4. HSM milling with monobloc end milling cutting tool

The third way of manufacturing applied to make the contour is HSM (High-Speed Milling) milling, with a monobloc milling cutter of ϕ12mm in 4 teeth, with a corner radius R 0.5mm, and with the slot length 30mm.

This strategy involves a large axial cutting depth, in this case, 26mm (1mm below the bottom of the part), and small radial penetrations, of about 6% of the diameter of the cutting tool. Given that the processed material is a superalloy, Inconel 625, the cutting parameters are small: cutting speed Vc = 51 m / min, rotation speed S = 1352 rpm, feed per tooth fz = 0.05 mm / tooth, working feed F = 270 mm / min. The trajectory of the cutting tool is represented in (Figure 5).

The processing time using this strategy is 43 minutes and 5 seconds.

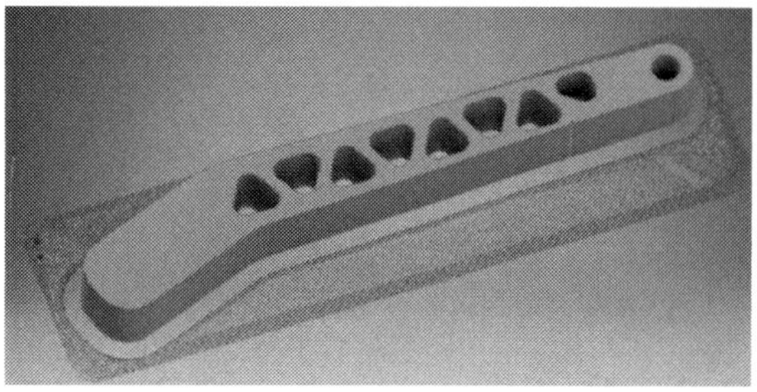

Figure 5. The cutting HSM trajectory.

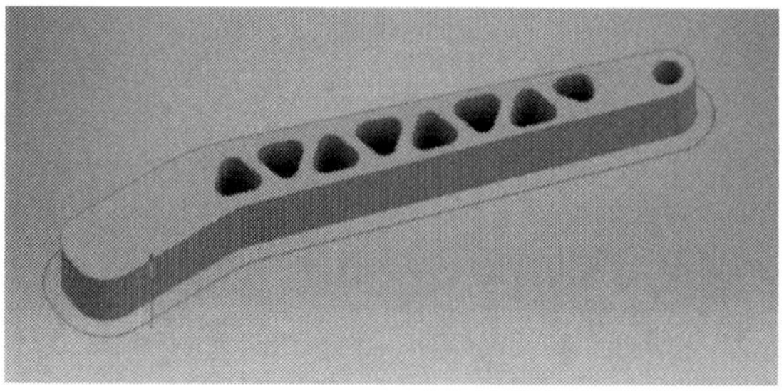

Figure 6. Finishing trajectory.

3.5. Contour finishing

After the roughing, the next operation is the finishing one, in order to obtain a good quality and roughness of the surface. It is made with a diameter 12mm 4-tooth, end mill cutter. Being a finishing operation, where only the addition of 0.2 mm left by the roughing cutter needs to be removed, this operation can be approached with a higher cutting speed, and with a lower advance, in order to obtain the desired surface quality. The finishing operation is done in a single pass, the cutting depth is 26mm, cutting speed Vc = 65 m/min, rotation speed S = 1724 rpm, feed per tooth fz = 0.02 mm/tooth, working feed F = 140 mm/min. Figure 6 shows the trajectory of the cutting tool for the finishing operation.

Taking in consideration that the row material is a superalloy and that the finishing operation is performed in a single pass, it is possible to apply for a pass with addition 0 (Spring pass), which has the target of removing the remaining material due to deflection of the cutting tool.

The time of the contour finishing operation is 4 minutes and 10 seconds.

3.6. Pocket milling

The triangular-shaped pockets are machined with a carbide end mill cutter, monobloc diameter ϕ8mm, with 4 teeth having the slot length 26 mm, and corner radius R 0.5 mm. The manufacturing strategy used is HSM speeds, the total depth of the pockets being achieved in 3 passes. The axial cutting depth ap is 8.66 mm, and the radial ae is 5% of the diameter of the cutting tool : 0.4 mm. Figure 7a) shows the processing from top view, and Figure 7.b isometric view, with the 3 levels of cutting.

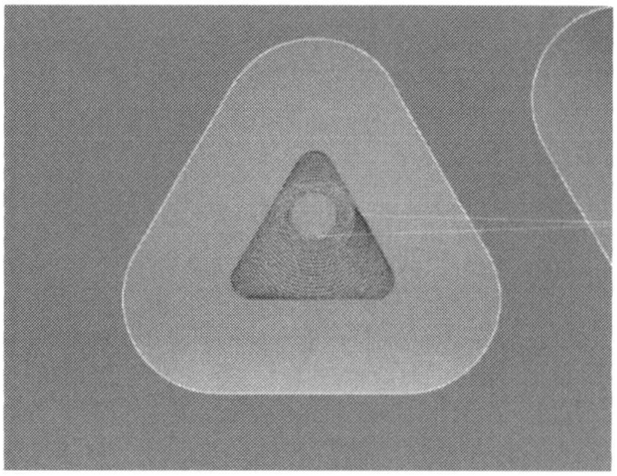

Figure 7.a) Pocket milling.

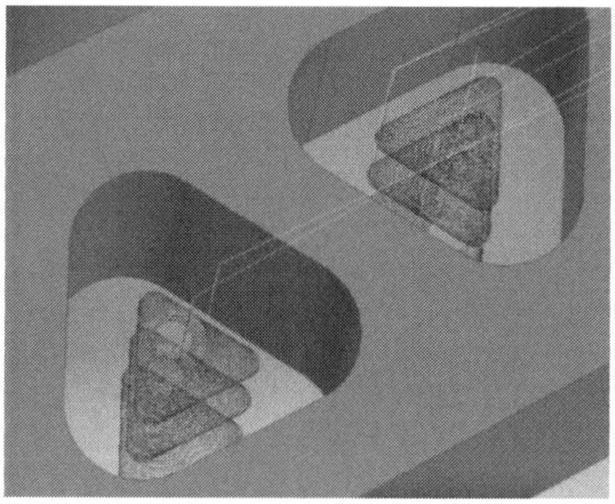

Figure 7.b) Pocket milling -isometric view.

4. Conclusions

The histogram below show the machining times of the part depending on the type of cutting tools used, and the machining strategy applied. It can be seen that the most productive method of processing a superalloy is with ceramic inserts.

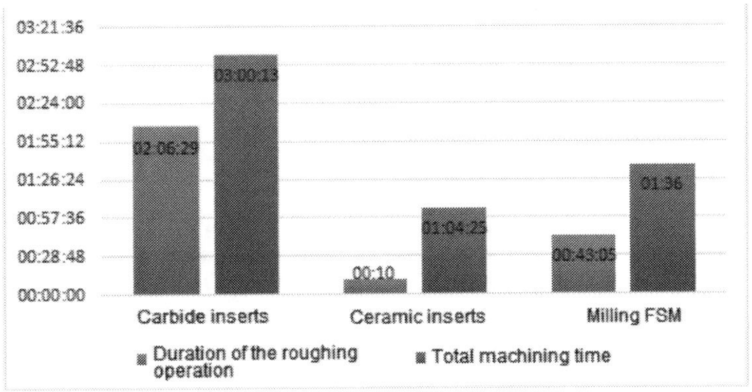

Figure 8. Histogram with the manufacturing time – clamping I.

The optimization depends on several factors:

Carbide inserts are 4-5 times cheaper than ceramic inserts, and the amount of material that can be removed with them is higher:

- it could be more advantageous to process single or small series parts, which do not need to remove a lot of material, and processing times do not represent an important role in the final price calculation.

- If will be mass production to be processed, and the goal is productivity, it is recommended solution with ceramic inserts, and the money invested in ceramic inserts will be recovered in reducing processing times.

Take into consideration the last version of MASTERCAM (the CAM software that was used in this application) have more and very useful milling strategies that are dedicated to optimizing the milling applications and obtaining more and accurate parts.

ORCID iDs

A But https://orcid.org/0000-0002-4032-1381

5. References

[1] Hazarika M and Dixit U S 2015 *Setup planning for machining* (Springer International Publishing) https://doi.org/10.1007/978-3-319-13320-1

[2] Metal Cutting Technology Technical Guide 2010

[3] SPINWORKS -Pokolm cutting tools catalogue- 2016

[4] Kumar C and Deb S 2012 Generation of optimal sequence of machining operations in setup planning by genetic algorithms *Int J Adv Manuf. Syst* **11** (1) 67-80 https://doi.org/10.1142/S0219686712500059

[5] Sormaz D N, Arumugam J and Ganduri C 2007 Integration of rule-based process selection with virtual machining for distributed manufacturing planning *Process planning and scheduling for distributed manufacturing* (Springer London) chapter 3 https://doi.org/10.1007/978-1-84628-752-7_3

Digital service application for the recalibration of a machine tool through Augmented Reality-supported data acquisition

F Arnold[1], G Lemme[2], M Hess[3], S Witt[4] and S Ihlenfeldt[1,2]

[1] Institute of Mechatronic Engineering, Chair of Machine Tools Development and Adaptive Controls, TU Dresden, Germany
[2] Fraunhofer Institute for Machine Tools and Forming Technology (IWU), Dresden, Germany
[3] Software AG, Darmstadt, Germany
[4] METROM Mechatronische Maschinen GmbH, Hartmannsdorf, Germany

frank.arnold@tu-dresden.de

Abstract. The effort for checking and correcting the spatial movement accuracy of a processing machine with 5 axes is very high. Calibration and recalibration must be carried out directly on the machine by the machine supplier or a company specializing in this. This is associated with high personnel and long machine downtimes. These constraints can be improved by using modern methods. A consistent approach of an AR-supported measurement procedure for the preparation and execution of a required measurement run and the execution of the calibration itself as a digital service is presented. This enables rapid execution by the machine operator himself using a DoubleBallBar measurement system without the need of the presence of the machine manufacturer. The determination of the calibration parameters on the basis of the measurement data is then carried out by the machine manufacturer and provided as a service. The basis for such a secure and auditable service is a digital service platform. This serves as an intermediary between the user and the machine manufacturer and uses distributed ledger technology. The approach presented is the subject of current development work.

1. Introduction

The manufacturing accuracy of machine tools plays a decisive role in securing and increasing the quality of products. Due to the production-related wear of the machine components as well as thermal influences during machining, inaccuracies and deviations occur which have a decisive influence on the product quality. To counteract these inaccuracies, a machine operator must have his machine tool calibrated by a machine manufacturer at regular intervals. These intervals depend on the production orders, the number of operating hours or to make sure that the machine can be used for production of high-precision components. Other occasions can be the replacement of machine components or the occurrence of an unexpected event, such as a collision in the work area. This requires close coordination between the machine operator and the machine manufacturer with regard to medium-term scheduling for implementation, free employee capacities and appropriate business travel planning so that a service employee of the machine manufacturer can perform the work at the machine location. This coordination effort inevitably leads to a time delay for the machine operator, during which he cannot use the machine.

In this paper, tools are presented, an IT architecture is outlined, and a workflow is presented on the basis of which this work can be carried out event-based and, thanks to smart services, without business trips. This can simplify and significantly accelerate the recalibration service presented here, as well as other services in the area of machine tools and production technology for the machine operator.

2. Machine tool and measuring device for the new approach

First, in addition to the machine tool used in the test environment, the selected measuring device for the new approach of calibration as a digital service is described.

2.1. Machine tool

Parallel kinematic structures underwent spectacular development in the 1990s due to the considerable advantages associated with kinematics [1]. These include, for example, the low masses to be moved, which thus promise excellent dynamic behaviour. However, in the field of machine tools, the lack of mass has the disadvantage of lower inertia for thermal changes, a generally smaller working space due to the design, and lower positioning and repeatability accuracies compared to serial precision machines. The latter aspect is the subject of extensive study in order to obtain stable and repeatable accuracy results through both design and software compensation techniques [2]. Using the example of a commercially available machine tool from Metrom (model P800), which is designed as a structure with driven constrained chains in degree of freedom 5, studies will be carried out to increase accuracy. The machine system is suitable for 5-sided complete machining and is determined by the following technical parameters (Table 1). The machine is shown in Figure 1.

Table 1. Technical properties of the machine tool Metrom P800 [1]

Description	Unit	Value
Working space	mm^3	800x800x450
Max. Speed v_f	m/min	60
Max. acceleration	m/s^2	12
Spindle power P	kW	14/30
Max. spindle speed	1/min	30000

Figure 1. machine tool Metrom P800.

2.2. Extended DBB as measuring device

In addition to the IT architecture, the implementation of the processes requires a measuring device necessary for a calibration measurement to be carried out by the machine operator himself. This must be sufficiently accurate, robust for use by a machine operator in a production environment and allow to perform a fast test. DoubleBallBar's (DBB) are used as measuring devices for various tasks. In the first place, they are used for circular tests according to ISO 230-4 **[3]**. For use within the new approach of calibration as a digital service, a DBB with an extended measuring range is necessary. This enables calibration in the entire working space of the machine without violating the small measuring range of currently available DBB [4]. Such a measuring device with strategies for the fast evaluation of the volumetric motion accuracy of a multi-axis kinematic system has been developed as part of previous research projects [2] [5] [6]. This extended DBB (Figure 2) fulfils the required suitability of a measuring system within the scope of the new approach of calibration as a digital service and can easily be mounted in the machine tool (Figure 3).

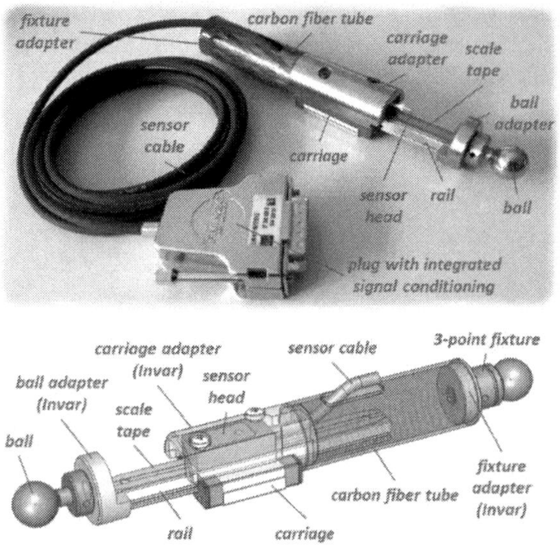

Figure 2. DBB measuring device with extended measuring range [2].

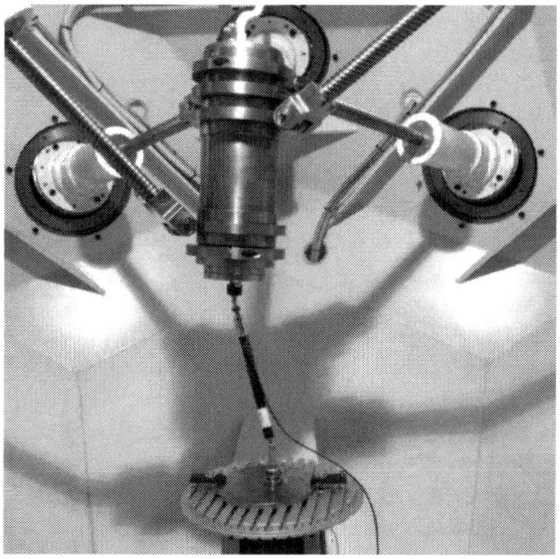

Figure 3. DBB measuring device in Metrom P800.

3. IT System Architecture

3.1. Digital Service Platform

The machine recalibration service is provided through a digital service platform (Figure 4). It acts as an intermediary between the user and the machine manufacturer and comprises the following features:

1. Digital marketplace for offering and selling the machine recalibration service;
2. Integrated web user interface to access the machine recalibration service;
3. Auditable documentation and logging of purchases and service executions through Distributed Ledger Technology (DLT).

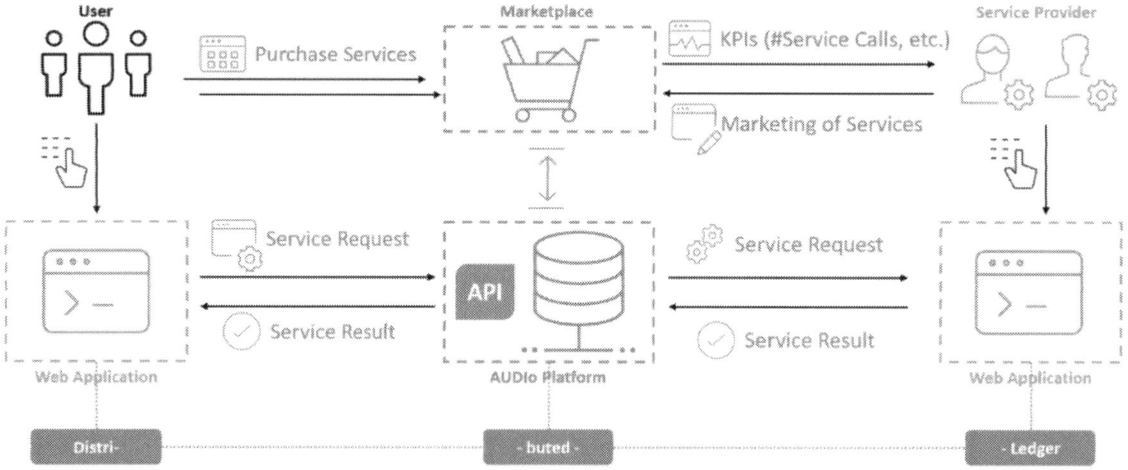

Figure 4. Digital Service Platform concept.

IOP Conf. Series: Materials Science and Engineering 1235 (2022) 012011 doi:10.1088/1757-899X/1235/1/012011

3.1.1. Auditable documentation

The lack of trust between a customer and a service provider and the lack of legal certainty are common problems when dealing with services of any kind, especially on digital platforms. A seamless documentation of the performed steps can provide clarifications to mitigate these issues but there is no guarantee that the documentation has not been manipulated.

Our digital service platform addresses these issues by implementing an auditable documentation mechanism based on Distributed Ledger Technology. All crucial steps such as making purchases or issuing service orders are immutably documented by the platform's web client and backend in a Distributed Ledger. For our platform prototype, we selected the Distributed Ledger IOTA [7] in combination with IOTA's Masked Authenticated Messaging technology (MAM) [8]. Through this, all data is stored in encrypted form and can only be read by the entities involved in the corresponding service contract, i.e., the customer, the service provider and the platform.

3.1.2. Service execution

As a first step, the machine operator decides to perform a recalibration. To do this, he uses the extended DBB and, supported by an Augmented Reality (AR) assistance system, carries out a measurement run. The needed program to perform this run is purchased from the machine manufacturer, because the necessary traverse movements to determine the calibrating parameters are machine-specific. The assistance system guides the machine operator through each individual step of preparing and performing the measurement. Each individual step must be acknowledged for its correct execution. This ensures the data quality and plausibility of the measurement data necessary for the service provider to perform the service.

Second, the machine operator purchases the recalibration service on the marketplace and requests the new recalibration parameters from the machine vendor, in this example the company Metrom, using the platform's web client. Figure 5 illustrates this workflow and the steps that are documented in the Distributed Ledger (colored diamonds).

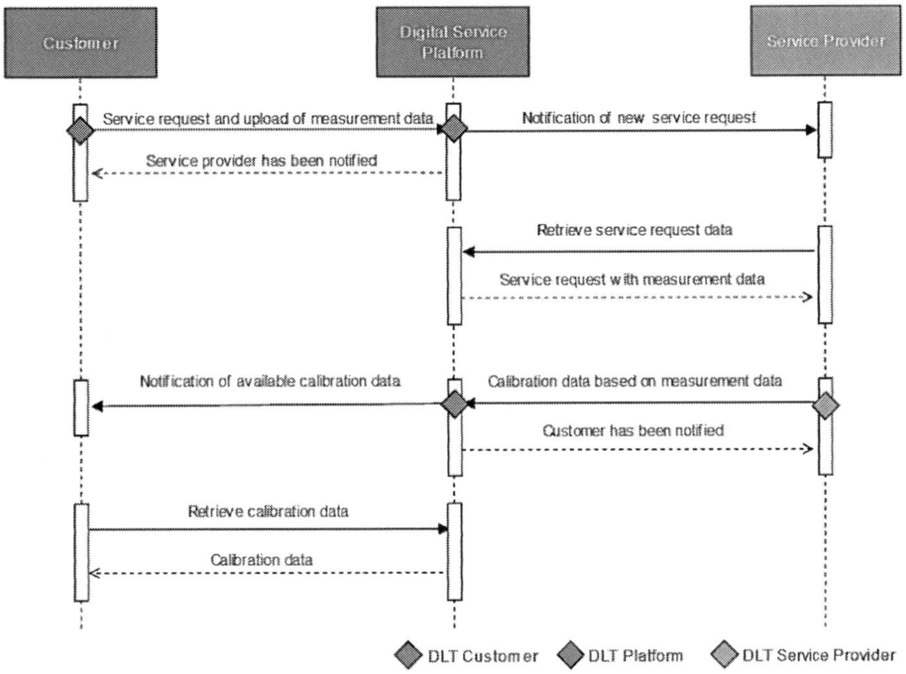

Figure 5. Illustration of the Recalibration Service workflow. The colored diamonds depict the steps that are documented in the distributed ledger and their colours indicate the entity storing the information.

Therefore, the customer uploads the measurement data obtained for the machine to the platform. The platform's web client encrypts the data on the client-side before uploading it to the platform (end-to-end encryption) and stores the file hash and additional metadata in the Distributed Ledger. Once finished, the platform notifies the service provider of the new service request and records the receipt of the service request in the Distributed Ledger.

Third, the service provider (Metrom) downloads and decrypts the service request data using the platform's web client, computes the calibration parameters from the measurement data offline and subsequently uploads the set of new calibration data to the platform. As in the second step, the web client applies end-to-end encryption to the data and stores the file hash and additional metadata in the Distributed Ledger. Upon completion, the platform records the receipt of the service request in the Distributed Ledger and notifies the customer that the calibration parameters are now available for download.

In the last step, the customer downloads the set of new calibration parameters and updates his machine with this data.

3.2. Augmented Reality - Assistance System

As described, precision and repeatability represent the greatest main requirements for modern machine tools. To achieve these goals in machine tools, standard measurement methods (e.g., ISO 230-4:2005[1]) and costly measurement systems exist to perform the necessary calibrations. We present an augmented reality (AR) based measurement approach to support the calibration process and the handling of the DBB measuring system of machine tools with live data. The AR support enables the machine operator to assemble the DBB, place it in the machine and perform measurements independently of specially trained service personnel from the machine manufacturer. By using suitable AR glasses (Vuzix Blade Security Certified), the machine operator is shown step-by-step instructions, receives information from the machine tool control system and can trigger actions on the AR glasses using the integrated touchpad. At the same time, he has both hands free for the assembly or disassembly of the DBB during the entire installation process and is not restricted by an oversized AR system.

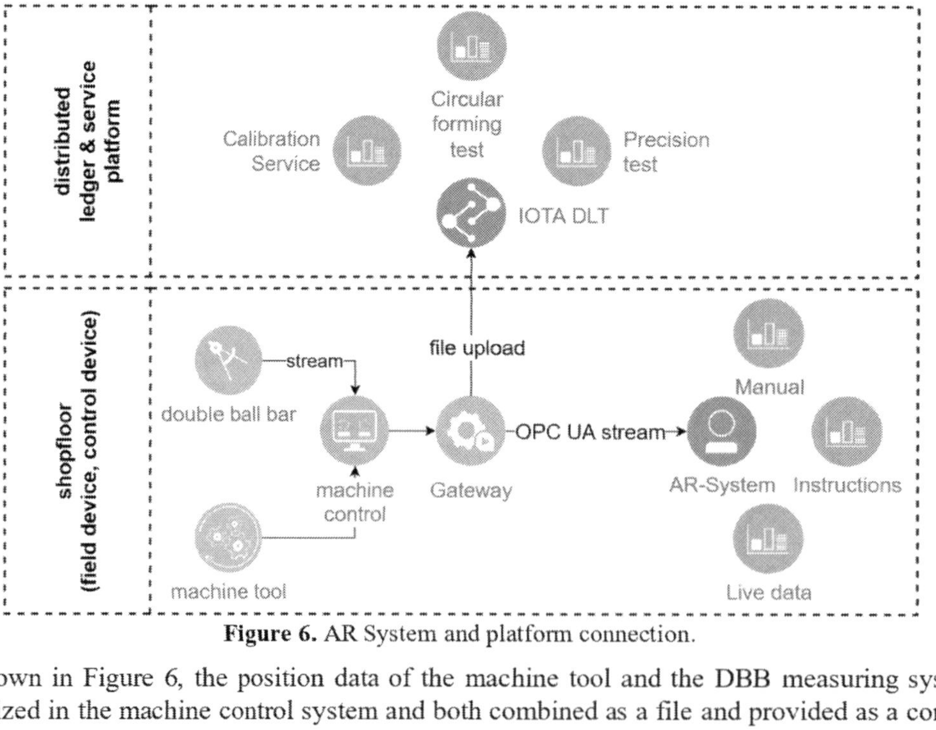

Figure 6. AR System and platform connection.

As shown in Figure 6, the position data of the machine tool and the DBB measuring system are synchronized in the machine control system and both combined as a file and provided as a continuous

[1] ISO 230-4, Test code for machine tools - Part 4: Circular tests for numerically controlled machine tools, 2005.

OPC UA stream via a gateway. The machine operator can then use the OPC UA stream to view the live data, call up the operating manual or receive instructions on how to use the measuring system correctly. With the help of the recorded measurement file, the machine operator can use further services (e.g. precision test) on the distributed ledger-based service platform in addition to recalibration (see 3.1.2 Service execution).

4. Conclusion

Increasing networking, digitized machine tools and available measuring systems allow smart services to be developed to increase machine quality. In combination with augmented reality (AR) systems and the use of Distributed Ledger Technology (DLT), service ecosystems are emerging that relieve maintenance personnel and strengthen the role of the machine operator. The necessary measurement run with a DoubleBallBar (DBB) for data acquisition for recalibration as a digital service from the machine manufacturer was carried out on a Metrom P800 machine by the machine operator himself. The processing of the calibration as a service was handled via a digital service platform. The data was exchanged in a tamper-proof manner according to the DLT. The service personnel of the machine manufacturer no longer need to record the data for data-driven services themselves on the machine tool. This eliminates the need for planning, business trips and waiting times at the machine operator's premises, allowing him to use a service on an event-based basis and at short notice.

Acknowledgements

This paper has been financed by BMWi Program “Smarte Datenwirtschaft” – Project “AUDlo-Auditlösung für ML-basierte, datengetriebene Dienstleistungen” Reg.-Nr.: 01MD19005, Germany. It was supported by the Federal Ministry for Economic Affairs and Energy (BMWi) based on a decision taken by the German Bundestag.

ORCID iDs

F Arnold https://orcid.org/0000-0003-4974-8581
G Lemme https://orcid.org/0000-0002-8781-9179
M Hess https://orcid.org/0000-0001-5827-2736
S Witt https://orcid.org/0000-0001-8622-4846
S Ihlenfeldt https://orcid.org/0000-0002-9258-5178

5. References

[1] Weck M 2006 Werkzeugmaschinen 5 (Berlin Heidelberg: Springer Vieweg) https://doi.org/10.1007/978-3-540-32951-0
[2] Kauschinger B, Friedrich C, Zhou R and Ihlenfeldt S 2020 Fast evaluation of the volumetric motion accuracy of multi-axis machine tools using a Double-Ballbar *Journal of Machine Engineering* **20** 44-62 https://doi.org/10.36897/jme/119678
[3] ISO 230-4 2005 Circular tests for numerically controlled machine tools
[4] Renishaw https://www.renishaw.de/de/qc20-w-kreisformmessgeraet--11075 Accessed 20.05.2021
[5] Zhou R, Kauschinger B and Ihlenfeldt S 2021 Data Synchronisation by Continous Spatial Measurement with Double Ballbar *Measurement* **174** 108909 https://doi.org/10.1016/j.measurement.2020.108909
[6] Zhou R, Kauschinger B and Ihlenfeldt S 2020 Path generation and optimization for DBB measurement with continuous data capture *Measurement* **155** 107550 https://doi.org/10.1016/j.measurement.2020.107550
[7] IOTA Foundation https://www.iota.org Accessed 16.07.2021
[8] IOTA Foundation 2017 https://blog.iota.org/introducing-masked-authenticated-messaging-e55c1822d50e Accessed 16.07.2021

IOP Conf. Series: Materials Science and Engineering 1235 (2022) 012012 doi:10.1088/1757-899X/1235/1/012012

Considerations on implementing Industry 4.0 in a manufacturing process

G Udvuleanu and V Nicolae

University of Pitesti, Faculty of Mechanical and Technology, Department of Automotive Engineering, Romania

georgian.u@gmail.com

Abstract. An industrial manufacturing company must adapt to the market all the time. If not, there is a high risk of losing money and the business can fail (for example in 2020, due to COVID 19 situation, more than 40% of businesses have temporary close their business and 60% of them have remained permanently closed). Wouldn't be wonderful if the manufacturing process can be flexible and adapted in function of the customer demand? Industry 4.0 does that – in the function of orders, software can tell us with how many people or machines can work, what will be the rooting/flow, how much energy is needed for that manufacturing process, etc. and most important – when the goods will be on the market. Industry 4.0 is playing a role more and more important in all the fields, as everybody is looking for performance and efficiency. This article is showing a study on an assembly line for the gear shifter of a manual gearbox. People might be taking wrong decisions when there is a tight situation regarding breakdowns or capacity crisis, but a good system in place can guide the humans to apply the best solution to avoid losses and increase productivity. In this article you will find an example of how Industry 4.0 can influence an assembly line, for a simple product, by increasing the productivity at 133% versus a traditional line who lost 40% in productivity due to a machinery breakdown.

1. Introduction

Industry 4.0 is a wording associated with the 4th wave of the Industrial Revolution. It means ongoing automation of traditional manufacturing and industrial practices, using modern smart technology. Industry 4.0 is based on machine-to-machine communication (M2M), GPS, mobile devices, advanced human-machine interfaces, 3D printing, authentication and fraud detection, smart sensors using the internet for taking actions without human intervention, self-monitoring and diagnose, virtual reality, cloud computing, etc.

First Industrial Revolution (1760 – 1840) has done the transition from production with human force to machines using steam and waterpower.

Second Industrial Revolution (1871 – 1914) also known as the Technological Revolution was based on electricity discovery and involve installations of railroad and telegraph networks, which allowed for faster transfer of people and ideas.

Third Industrial Revolution, also known as the Digital Revolution, occurred in the late 20th century, after the end of the two world wars. This third wave includes the discovery of the computers and internet. Forth Industrial Revolution started in 2013 and is ongoing today, having more and more applications that doesn't need the human involvement. The wording "Fourth Industrial Revolution" was first

introduced by a team of scientists developing a high-tech strategy for the German government in 2013 and it was introduced worldwide by the World Economic Forum Annual Meeting in 2016. [1]
An industrial manufacturing company must adapt to the market all the time. If not, there is a high risk of losing money and the business can close (for example in 2020, due to COVID 19 situation, more than 40% of businesses have temporary close their business and 60% of them have to remain permanently closed).[2]

Wouldn't be wonderful if the manufacturing process can be flexible and adapted in function of the customer demand? Industry 4.0 does that – in the function of orders, software can tell us with how many people or machines can work, what will be the rooting/flow, how much energy is needed for that manufacturing process, etc. and most important – when the goods will be on the market. We can see already worldwide examples with Industry 4.0 systems that are working properly: Rotterdam harbour where a big part of cranes and trucks that carry containers have been replaced by autonomous machines [3]; Built Robotics that have developed in San Francisco autonomous machines for flattening the land in constructions [3]; Audi smart factory where AGV's are transporting components or car bodies [3]. All the industries can use Industry 4.0 systems in function of their needs. Why do we need this digitalization? To improve work, to make it more efficient and sometimes more fun and easier.

2. Case study. Evaluation

In order to present practically the advantages of Industry 4.0, we consider a manufacturing line in automotive industry engineering, gearshift mechanism. The entry data for this manufacturing line are:

- Parts/year to be produced 3 50 000 pcs.
- Production line must be modular.
- Assembly line to fit in the available area.
- To have as many sensors as possible and poka-yoke systems in order to be able to use same manufacturing line for similar products.

For this job, first the process engineering team build the synoptic of the assembly operations, then the technical specifications for each assembly machine and in the end (after receiving the dimensions of the machines from the tooling supplier) the layout of the process.

Figure 1 presents the components of the gear shifter.

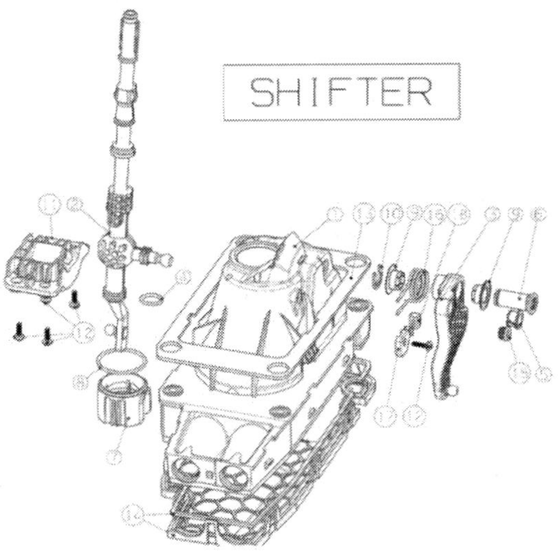

Figure 1. Gear shifter components.

Component number	Component name	How many are needed
1	Lever housing	1
2	Main lever assembly with over moulded geometry	1
3	Select lever	1
4	O-ring	1
5	Select Bush	1
6	Select Pin	1
7	Spherical Bearing	1
8	O-ring of Spherical Bearing	1
9	Bush	2
10	Safety Clip for Shaft	1
11	Spherical Bearing Cover	1
12	Screw	5
13	Foam Seal	1
14	Lever Housing Cover	1
15	Obturator Cap	1
16	Select Spring	1
17	Spring Holder	1
18	Spring Damper	1

The succession of the manufacturing operations is presented in figure 2.

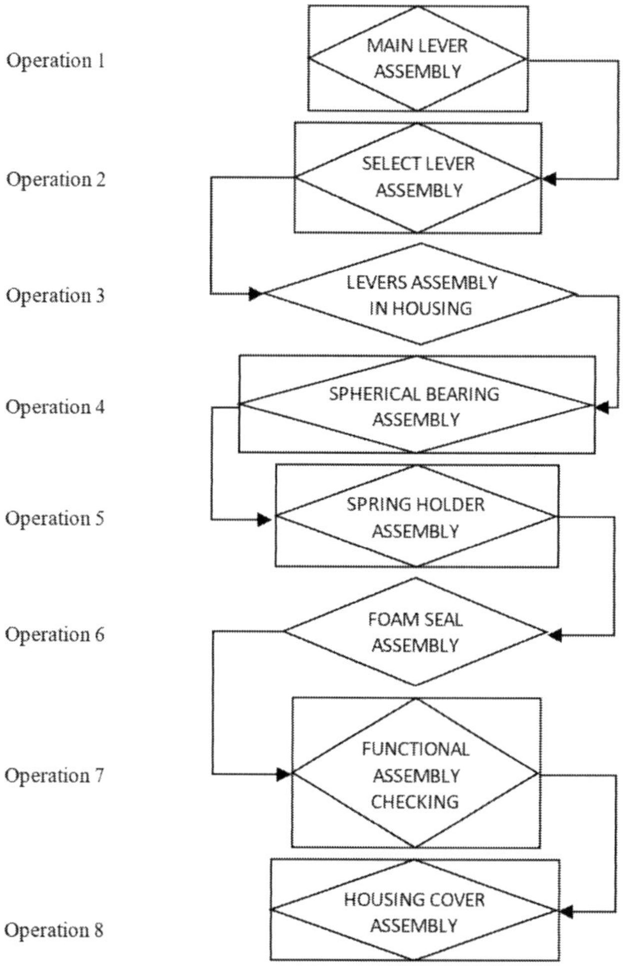

Figure 2. Synoptic of the manufacturing process.

For operations 3 and 6 there is no need to have an assembly machine, because the operator will do the job manually. For all other assembly operations, the manufacturing process need a machine.

As **technical specifications** for the assembly, machinery must comply with the legal rules, ergonomics and checking of a correct assembly operation. Also, the production capacity of the machine is very important, as usually the process engineers ask for a minimum of 20% capacity more than what is needed, in order to be safe in case of extra customer demand. In our case, all the machinery are specific and mainly represent clipping machines with sensors for detection of the components and counters for the good or bad parts. Only the operation 4, for fixing the Spherical Bearing cover and Spring Holder, has a machinery that contains an electrical tightening machine with poka-yoke for torque, angle and counter for number of screws assembled on one semifinished part. The functional assembly checking is done with a robotic hand that contains force and distance measurement sensors. These sensors are measuring the friction in the bearing and also movement distance for the two levers while the robotic hand is simulating all the gears passing.

Layout of the manufacturing process can be in the "U" shape in order to facilitate the loading of the line with components and unloading of the finished goods.

Figure 3. Layout of the manufacturing process.

The time allocated for each operation splited by movements in the function of machines, operators and synoptic are presented in Table 1.

Table 1. Time allocated for each process step.

	Steps of assembly process	Time (s)	Percent (%)
1	Assembling of main lever (2+4+7)	12	5.8%
2	Assembling of select lever (3+9+16+5)	20	9.9%
3	Spring pretensioning	6	3%
4	Assembling of O-ring (1+8)	6	3%
5	Insert levers assembly into housing (1+2+3)	12	5.8%
6	Assembling of select pin (6+1+3+10)	6	3%
7	Assembling of spherical bearing cover (11+12+1)	19	9.4%
8	Assembling of spring holder (17+18)	7	3.4%
9	Assembling of spring holder in housing (17+12+1)	17	8.4%
10	Assembling of obturator cap (1+15)	5	2.5%
11	Assembling of foam seal (1+13)	12	5.8%
12	Functional assembling of mechanism control	41	20.2%
13	Assembling of housing cover (1+14)	20	9.9%
14	Packaging	20	9.9%

In Table 1 could be noticed that the Functional Test is the bottleneck of the process. So, any Process Engineer can group these steps as follows:

Table 2. Steps of process allocated by operator.

Operator	Time (s)	Percent (%)
Operator 1: Step 1+2+3	38	18.7%
Operator 2: Step 4+5+6+7	43	21.2%
Operator 3: Step 8+9+10+11	41	20.2%
Operator 4: Step 12	41	20.2%
Operator 5: Step 13+14	40	19.7%

As Table 2 shows, the new process bottleneck will be Operator 2, with a time of 43 s/part.

Table 3. Capacity of production.

Takt time (s)	Parts/hour	Parts/shift	Parts/week	Parts/year
43	84	627	9.405	442.035

It could be noticed that the yearly capacity of the line is assured and there is a production reserve of 26% (simulation has been done in 5 working days per week and 3 shifts per day). This is the ideal situation when efficiency of the line is 100%. By efficiency it is considered the number of good parts produced, divided by the target parts to be produced.

In a traditional manufacturing process, these parameters will be:

Table 4. Capacity vs. efficiency.

Efficiency	Takt Time (s)	pcs/h	pcs/shift	pcs/week	pcs/year	Operators
100%	43	84	627	9.405	442.035	5
90%	47,3	76	570	8.550	401.850	5
85%	49,5	73	546	8.190	384.930	5
80%	51,6	70	523	7.845	368.715	5
75%	53,8	67	502	7.530	353.910	5
70%	55,9	64	483	7.245	340.515	5

It is obvious that anytime our production is running under 75% efficiency, there is a risk of capacity (remember, the target is 350 000 pcs/year). In order to achieve the capacity, if the manufacturing line has a low efficiency, the manufacturer must work overtime or to quickly improve the process somehow. It means extra costs for labour, for transportation of people and energy for running the working place.

3. Proposal for improvement

Considering Industry 4.0, the situation for this manufacturing line is totally different: the loss is minimum and the risk of failing the capacity is close to zero.

Each step of the manufacturing process is monitored with sensors, PLC's (Programmable Logic Controllers) and counters for good and bad parts. The engineers have the chance to see virtually any kind of scenario: what happen when a machine breakdown, when the operators are missing, when there will be an extra operator, when the power breakdown etc.

Table 5. Simulation of efficiency loss.

Manufacturing	Present situation					Proposal			
	Efficiency	Time (s)	Pcs/ hour	Pcs/hour needed	Parts lost	Operators	New time (s)	Pcs/ hour	Recovery time (min)
Operator 1: Steps 1+2+3	100%	38	94	67	0	1	38	94	0
Operator 2: Steps 4+5+6+7	60%	60	59	67	5	2	35	102	19
Operator 3: Steps 8+9+10+11	100%	41	87	67	0	1	41	87	0
Operator 4: Step 12	100%	40	90	67	0	1	40	90	0
Operator 5: Steps 13+14	100%	41	87	67	0	1	41	87	0

The system can be designed to be capable to manage all the maintenance operations schedule and adapt the manufacturing process based on the efficiency of each manufacturing step, or on the operator's availability. System has a theoretical manufacturing time setup by the process engineer for each step of the assembly and is continuously proposing solutions to recover the lost parts due to different reasons. For example, in Table 5 the efficiency decreased at 60% for the second operator due to the fact that the machine had a malfunctioning.

After 10 minutes of malfunction, the system shows that five parts are lost and propose the solution to add for 19 minutes a second operator in that working station (this is in case the machine-tool is not repaired). As could be noticed, if the line is supplemented with two operators, the time it is not dividing by two because the system already configurate the manufacturing steps 4 and 5 (18 s) for operator 2 and steps 6 and 7 (25 s) for the additional operator. The additional operator will still depend on the malfunctioning machine (60% efficiency), so his time is adjusted from 25 s to 35 s.

To compare a traditional manufacturing system with the one where it was integrated the Industry 4.0, in this example could be considered only on the costs of labour. So, the values presented in Table 5 show that the capacity has been recovered in 19 minutes with one additional operator. In traditional manufacturing process, most probably after a full shift running with five operators there will be a loss of the parts due to efficiency of only one machine-tool and the recovery of parts can be done only with a full team and one shift overtime (5 people * 8 hours vs. 1 man * 0.3h).

4. Conclusions and future research

The new era of industrialization gives us the opportunity to:

- create optimum processes by using virtual reality,
- maintain properly the machines once the line is ready for production,
- adapt the process in function of all the variables that can appear during the manufacturing.

Virtual reality can help us to build a strong containment plan that can eliminate 90% of manufacturing breakdowns.

Maintenance Management of the Machines reduce downtime (Time to repair) and increase Time between Failure, two important KPI's (Key Performance Indicators) for production. In the case presented above we saw that Productivity has decreased 40% due to a machinery problem. If the maintenance is done properly and in time, this kind of issues are very rare.

Adaptability of the Manufacturing Process is crucial for a Plant because it directly affect the Productivity and in the end the Profitability of the Company. In our exercise above we can see an improvement of Productivity with 133%, because the system has avoided an overtime with five people. Industry 4.0 can bring more benefits if we are thinking to use management of the energy in the manufacturing line. Let's take into consideration the support machinery for the manufacturing process, like compressors, steam generators, chillers, electrical transformers, etc. In the traditional manufacturing plants, all these machineries are working 24/7, but in a plant with Industry 4.0 implementation they are working only when it is necessary and only at capacity needed to support the manufacturing process. Therefore, I believe that the future research must be done in the field of energy and utilities management to reduce the production costs and to be more friendly with the environment.

The amortization of costs to transform a manufacturing plant from traditional running to Industry 4.0 could be obtain in a period of five years to fifty years, depending on technology and plant capacity. This is the reason why the companies are not jumping so easy to a technology like that, even if they are aware of the benefits.

ORCID IDs

G Udvuleanu https://orcid.org/0000-0001-8566-3214
V Nicolae https://orcid.org/0000-0002-2536-3372

IOP Conf. Series: Materials Science and Engineering 1235 (2022) 012012 doi:10.1088/1757-899X/1235/1/012012

5. References

[1] Schwab K 2021 *The fourth industrial revolution* Encyclopedia Britannica, https://www.britannica.com/topic/The-Fourth-Industrial-Revolution-2119734

[2] Sundaram A 2020 *Yelp data shows 60% of business closures due to the coronavirus pandemic are now permanent* https://www.cnbc.com/.html

[3] *A Few Examples of Industry 4.0* https://www.rewo.io/real-world-examples-of-industry-4-0/

[4] Pfeiffer S 2017 The Vision of Industry 4.0 in the Making – A Case of Future Told, Tamed and Traded *Nanoethics* **11** 107-21 https://doi.org/10.1007/s11569-016-0280-3

[5] Abele E, Metternich J, Tisch M 2019 *Best-Practice Examples.* Learning Factories (Darmstadt: Springer Nature Switzerland AG) pp 335-452 https://doi.org/10.1007/978-3-319-92261-4

[6] Mourtzis D, Zogopoulos V, Vlachou E 2018 Augmented Reality supported Product Design towards Industry 4.0: a Teaching Factory paradigm *Procedia Manufacturing* **23** 207-12 https://doi.org/10.1016/j.promfg.2018.04.018

Research on the solid-state joining of AlMg3

A I Feier and F V Banciu

Materials and Manufacturing Department, Faculty of Mechanics, University Politehnica Timisoara, Bd. Mihai Viteazu No.1, 300222, Timişoara, România

ana-maria.feier@upt.ro

Abstract. Nowadays, new products are wanted to reach the final consumer as soon as possible, at the lowest possible costs and in the shortest possible time. This time limitation puts its mark on all the stages necessary for product development, implicitly on the stage of their conception. Given the need for products that require assembly, of major importance, in the design phase, becomes the choice of materials and ways of joining them, the parameters of the processes to be used, equipment, tools, devices with which it is made and then joint control. Having this knowledge inventoried, for the types of processes and materials possible to be used, the time in which the choice of the way of joining is made can decrease substantially. In this context, the present paper presents experimental tests on the solid-state joining of 3 mm sheets of AlMg3 alloy, a very well weldable material with WIG process, corrosion-resistant especially in the marine environment, very good formability at low temperatures. The solid-state joining procedure of the test specimens will be presented with the help of a modified milling cutter, and the conclusions of this type of joint highlighting also the destructive examination part of the tested specimens constitute a set of knowledge that can be inventoried, stored, and used, when necessary, in the products development process. The paper thus demonstrates the potential ease of applying and reproducing Friction Stir Welding (FSW) with simple machinery (CNC milling cutter), while contributing to an understanding of the mechanical behaviour (initialization of failure) of joints. This, the behaviour and the defects are very friendly explained in the macroscopic images from the paper.

1. Introduction

In the early 1990s, more precisely in 1991, a new joining technology called Friction Stir Welding was invented and patented by a team from TWI Cambridge, led by W.M. Thomas. The process was first applied industrially in Sweden in 1995. Due to the qualities of this process, it aroused the interest of economically strong countries, such as the USA and Japan [1-4].

The principle of the welding process is explained in figure 1 [6,7]. Friction stir welding uses the heat released by the friction between the active element and the material that will be joined, heat that produces the plasticization of the marginal areas of the materials to be joined. By the advanced movement of the tool along the joint, the plasticized material is homogenized, and by cooling, a joint is made between the welding parts, in solid phase.

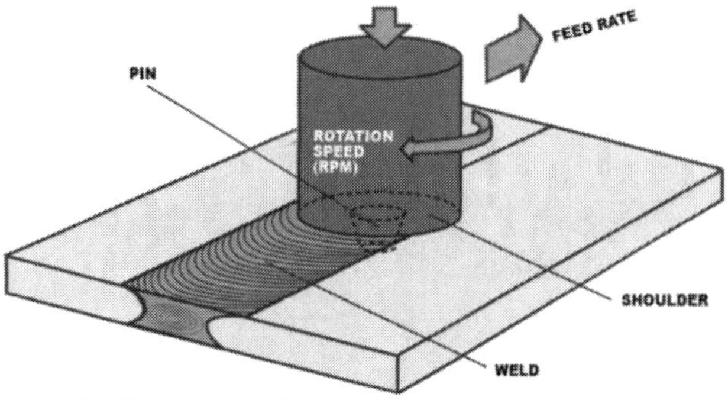

Figure 1. The process description of Friction Stir Welding.

2. Materials and infrastructure used for joining

The material used in the experimental tests is the AlMg3 alloy (EN AW 5754). This material is recommended for welded structures in the nuclear environment, production of boilers, applications in the marine environment (boat production), and road signs. It is an alloy with medium hardness. It can be easily anodized. In table 1 is the alloy's chemical composition an in table 2 the mechanical characteristics.

Table 1. Chemical composition of the alloy.

Chemical composition %												
Melt	Si	Fe	Cu	Mn	Mg	Cr	Ni	Zn	Ti	Ga	V	Al
S37106411	0.25	0.33	0.059	0.26	2.8	0.035	0.01	0.04	0.016	0.011	0.015	96.174
Min. Max.	0.4	0.4	0.1	0.5	2.6 3.6	0.3	0.05	0.2	0.15	0.05	0.05	
Obs: 0.10-0.06 Mn+Cr						Each max.: 0.05			All max. 0.15			

Table 2. Mechanical characteristics of the alloy.

Mechanical characteristics (MPa)					
Melt	Values from Standard (EN 485)				
S37106411	Rm (MPa)		Rp0.2 (MPa)		Elong. % 50 mm
	Min.	Max.	Min.	Max.	Min.
	190	240	80		16
	Measured Values				
	Rm (MPa)		Rp0.2 (MPa)		Elong. % 50 mm
	212		128		26.5

The infrastructure used for joining materials in the solid state was CNC milling cutter, the material was AlMg3 which was joined with a steel pin made from 56SiCr7. Al the infrastructures are part of Materials Engineering and Manufacturing laboratory equipment (MFP) of the Faculty of Mechanical Timisoara. In table 3 is the Chemical composition of pine alloy% (56SiCr7) according to SR EN 10089: 2003.

Table 3. Chemical composition of pin alloy (56SiCr7) according to SR EN 10089: 2003.

C	Si	Mn	P	S	Cr	-
0.52-0.6	1.6-2.0	0.7-1	Max. 0.025	Max. 0.025	0.2-0.45	Cu + 10 Sn <0.6

(a) (b) (c)

Figure 2. FSW infrastructure (a), (b) and FSW joint (c).

3. Experimental tests -the specimens in solid phase

The research program started during the project Diaspora - PN-III project P11.1-MCT-2018-0032 [6] in which was tried to join an Aluminium alloy with a milling machine, the study was conducted in the Materials and Manufacturing Engineering Department. In the incipient phases of the research, defects in the joint resulted, but by optimizing the technology and the shape of the pin, it was obtained the results presented in this paper. After optimizing the welding technology, 3x250x350 mm pieces were made and from the pieces, some specimens were extracted to perform the tensile test.

3.1. The welding technology

- Milling speed - 800 rotations/min
- Feed rate - 40 mm/min
- Tilt angle of the pin - 2 degrees
- Pin length - 2.62 mm

By optimizing the welding technology and observing the non-penetration area, the decision was made to join the specimens on both sides, i.e., after the joint on one side; the specimen was turned on the opposite side using the same joining technology. By joining the specimens on both sides, the imperfection (the area of impenetrability) disappeared. There were two stages of the research:

- The first step was the joining on one side of the pieces of AlMg3 alloy
- The second step was the joining on both sides of the pieces of AlMg3 alloy

4. Examination of test specimens

The specimen's examination was performed destructively, i.e. they were made tensile tests, and then a microscopic analysis was performed. As the lack of penetration area is visible only under a microscope, it was decided to test both types of joints, the specimens joined on one side and on both sides to see where the failure occurs and at what force compared to the base material.

The usual tensile test was performed, the specimens were made according to the tensile standard, the tensile test parameters were observed: speed, time, grips of the specimens, etc.

3 specimens were taken from each type of joint (on one side as well as on both sides) and the dimensions of the test pieces for the tensile test are 3x66x300 mm with a radius of R5 [6,7].

Tensile tests were performed within the CMMC-UPT Department on the TESTWELL / UTS machine.

Machine used on tests was TESTWELL / UTS traction test machine with:

- Hydraulic tanks
- Tensile tests and compression tests can be performed
- Maximum force: 250 kN
- Computer aided machine control, acquisition, and post-processing of results

In the first stage, the tensile tests were performed on the joint only on one side and it can be seen in figure 3 that only one specimen out of 3 yielded close to the joint. The failure of the joint was close to to the base material yield, namely 200.9 MPa.

Table 4. Tensile tests results.

	Specimen ID	R_{eH}	R_m	A_t
		MPa	MPa	%
1 A-I	1 A-I	112.6513	221.1408	21.67622
2 A-I	2 A-I	111.6845	200.9386	8.610308
3 A-I	3 A-I	111.9555	221.5049	20.88182

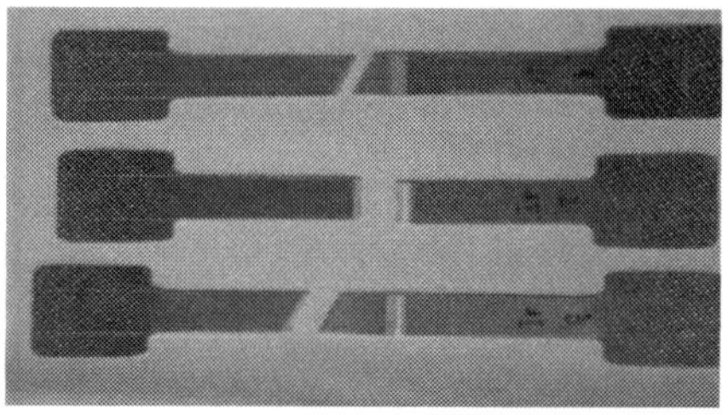

Figure 3. The tensile tests of each test piece joined on one side only

The second stage of the tensile test was the tests on the specimens obtained by joining on both sides (the zone of lack penetration was removed by joining on the both slides).In the case of these specimens, the failure occurred in the base material at recommended forces within the intervals received from the manufacturer. In addition to the tensile tests, several micro-macro analyses were also performed to observe if the area of non-penetration / imperfection from the first specimens (joined on one side) was maintained in the case of those joined on both sides figure 4 [8].

Table 5. Tensile test results.

	Specimen ID	R_{eH}	R_m	A_{gt}
		MPa	MPa	%
1 A-II	1 A-II	122.0289	221.6105	17.0021
2 A-II	2 A-II	111.5833	222.2222	18.38929
3 A-II	3 A-II	129.997	224.1296	19.61197

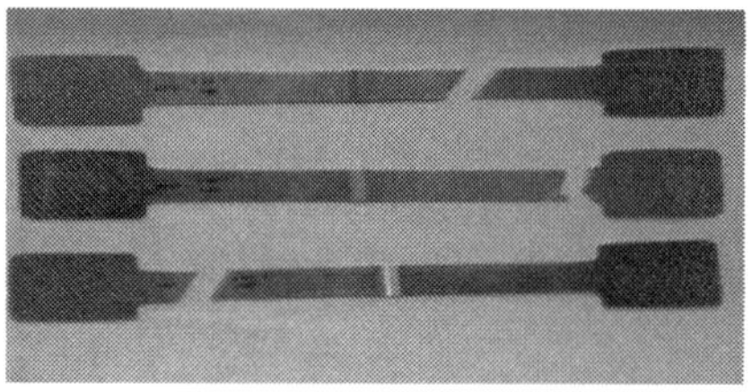

Figure 4. The tensile test on each test piece joined on both sides

Figure 5. Microscope specimens

The following figures show the shape of the nugget obtained at the AlMg3 pieces joined on one side. The nugget is the central part that can be seen as a nucleus of a lighter color. In the picture on the right, the joint is enlarged so that the imperfection (non-penetration) can be observed. It was observed under the microscope that the non-penetration area started right in the end area of the pin, therefore the length of the pin can be one of the factors that can generate that imperfection [9,10].

Specimen No.1

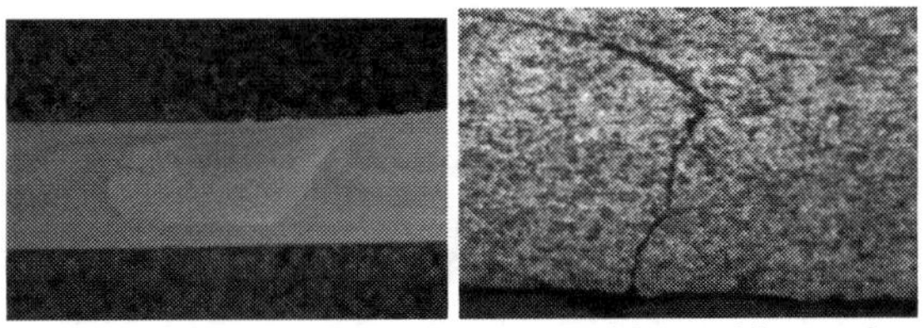

Figure 6. Specimen 1 joined on one side and imperfection b

It can be seen in figure 7a the shape of the nugget under a microscope and in figure 7b the area of non-penetration / imperfection that starts from the area of the end of the pin and develops towards the lower part of the joint.

Specimen No.2

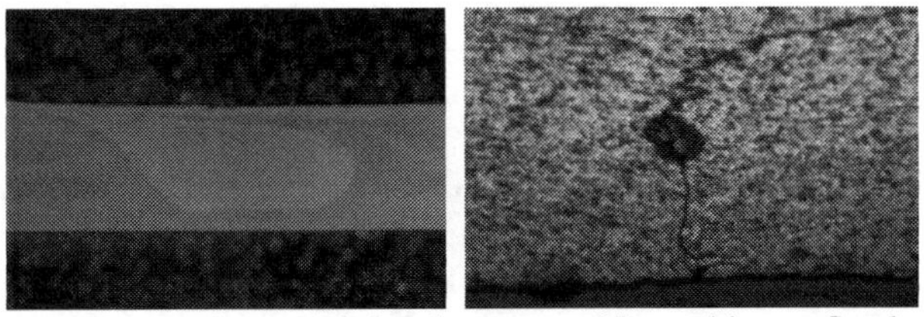

Figure 7. Specimen 2 joined on one side and imperfection b

In the case of test piece no. 2, the imperfection is slightly higher, therefore even when the tensile tests the failure took place near the joint area.

Specimen No.3

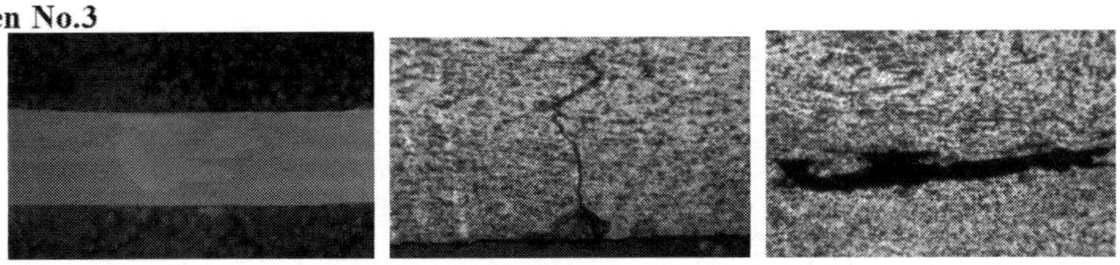

Figure 8. Specimen 3 joined on one side and imperfection b, c

In the case of test no. 3, although the defect existed and in addition there was an area of lack of penetration somewhere in the middle of the joint, the tensile test no. 3 yielded outside the joint area.

The following figures 9 presents the shape of the resulting joint welded on both sides. In this case the imperfection no longer appears. The imperfection disappeared when the pieces were joined also on the opposite side.

Specimen No.1

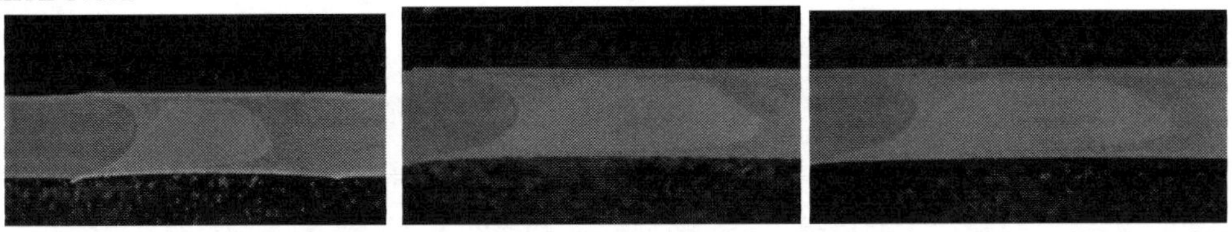

Figure 9. Shape of the resulting joint welded on both sides for Specimen No.1, No.2 and No. 3

5. Conclusions

Research on this type of joint shows that, although it was a non-penetrating area at the bottom of the joint, two out of three tensile specimens did not fail in the joint, and the specimen that failed almost in the joint has the strength close to base material (200.9 MPa).

Joining the specimens on both sides, the lack of penetration zone disappeared and at the tensile test, the failure was far outside the joint. Because the number of specimens made was quite small, research will continue in this direction because the joint can be obtained relatively quickly, without too many preparation operations and with the help of infrastructures that are quite low compared to machines designed exclusively for welding with rotating active element (FSW).

In the near future, the research is intended to be continued both by the constructive modification of the pin (its length) and of the working parameters of the milling machine, because both factors can generate the non-penetration area of the joint.

The novelty of this paper is the realization of a solid-state joint, a primitive FSW using a CNC milling machine, to which a small upgrade was made and to which a pin/tool specific to the FSW process was attached. The pin/tool was made in the laboratory so the tests were made only on the length of 2.62 cm, research will continue with other lengths. The costs of the research have been reduced considering that the joining was done on a CNC milling cutter of about 1000 Euro (the infrastructure is from 1960), which can be compared with the joints obtained on specific FSW process infrastructures, infrastructures that cost from 200000 Euro upwards.

ORCID iDs

A I Feier https://orcid.org/0000-0003-4174-8608
F V Banciu https://orcid.org/0000-0001-5246-9829

6. References

[1] Blaga L, Amancio ST, dos Santos JF and Bancila R 2012 Fricriveting of Civil Engineering Composite Laminates for Bridge Construction *Proc. Annual Conf. of the Society of Plastics Engineers* ANTEC 2012 (Orlando, Florida:) *Society of Plastics Engineers* pp 1691-7

[2] Blaga L, Dos Santos J F, Bancila R and Amancio-Filho S T 2015 Friction Riveting (FricRiveting) as a new joining technique in GFRP lightweight bridge construction *Construction and Building Materials* **80** 167–79 https://doi.org/10.1016/j.conbuildmat.2015.01.001

[3] Cipriano G P, Blaga L A, dos Santos J F, Vilaça P and Amancio-Filho S T 2018 Fundamentals of force-controlled friction riveting: Part I-joint formation and heat development *Materials* **11** (11) 2294 https://doi.org/10.3390/MA11112294

[4] Cipriano G P, Blaga L A, Dos Santos J F, Vilaça P and Amancio-Filho S T 2018 Fundamentals of Force-Controlled Friction Riveting: Part II—Joint Global Mechanical Performance and Energy Efficiency *Materials* **11** (12) 2489 https://doi.org/10.3390/MA11122489

[5] Panaitescu Ş 2009 *Stir friction welding with rotation active element* (in Romanian, Timisoara: Ed. Sudura)

[6] Feier A 2018 Timişoara Research Project Report Diaspora - PN-III- P11.1-MCT-2018-0032

[7] Atanasiu C *et al.* 1982 Materials Testing (in Romanian, Bucureşti: Editura Tehnica)

[8] Borba N Z, Blaga L, dos Santos J F and Amancio-Filho S T 2018 Direct-Friction Riveting of polymer composite laminates for aircraft applications *Materials Letters* **215** 31–4 https://doi.org/10.1016/J.MATLET.2017.12.033

[9] Feier A, Becheru A, Brînduşoiu M and Blaga L 2021 Process Transferability of Friction Riveting of AA2024-T351/Polyetherimide (PEI) Joints Using Hand-Driven, Low-Cost Drilling Equipment. *Processes* **9** 1376 https://doi.org/10.3390/pr9081376

[10] Lambiase F, Balle F, Blaga L A, Liu F and Amancio-Filho S T 2021 Friction-based processes for hybrid multi-material joining *Composite Structures* **266** 113828 https://doi.org/10.1016/J.COMPSTRUCT.2021.113828

IOP Conf. Series: Materials Science and Engineering 1235 (2022) 012014 doi:10.1088/1757-899X/1235/1/012014

Obtaining one-sheeted revolution hyperboloid surface by cutting

L Slătineanu[1], A Hriţuc[1], O Dodun[1], G Nagîţ[1] and I Beşliu-Băncescu[2]

[1] Department of Machine Manufacturing Technology, "Gheorghe Asachi" Technical University of Iaşi, Blvd. D. Mangeron 59A, 700050 Iaşi, Romania
[2] Department of Mechanics and Technologies, "Ştefan cel Mare" University of Suceava, Universităţii Street 13, Suceava 720229, Romania

hrituc.adelina3295@yahoo.com

Abstract. There are situations when obtaining surfaces that correspond to one-sheeted revolution hyperboloid surface is formulated in industrial practice. Solving such a problem may require the use of expensive procedures and equipment. The analysis of some solutions from the specialized literature led to the acceptance of some possibilities to obtain certain surfaces specific to the one-sheeted revolution hyperboloids by turning and milling. A technological solution that involves using a milling device adaptable to a universal lathe has been proposed.

1. Introduction

One of the categories of flat curves is the so-called *conics*. From a mathematical point of view, a conic is a curve that can be obtained by intersecting a plane with the surface of a straight, circular cone. Among the types of conics studied are the circle, the ellipse, the parabola, and the hyperbola. In principle, *the hyperbola* is considered the geometric place of the points for which the difference in distances from two fixed points, called foci, is constant.

A hyperboloidal surface is obtained by rotating a hyperbola around an axis in its plane [1-3]. Hyperboloidal surfaces can be found in the practice of civil constructions, in the case of cooling towers and roof categories, where they confer specific characteristics to those constructions.

In machine manufacturing, hyperbolic surfaces must be provided to the drive discs used for centerless grinding with longitudinal feed movement performed by the cylindrical workpiece. Another practical situation that necessitates hyperboloidal surfaces is that of the rollers for driving cylindrical workpieces that must perform a rotational movement and a feed movement along their axis, in the case of straightening machines. Such a machine is used, for example, to straighten the bars from which the workpieces for the cylindrical rollers in the bearings are cut.

It was necessary, as such, to identify and study both specific aspects of hyperboloidal surfaces and procedures to obtain such surfaces by removing material from the workpiece.

Thus, Li *et al.* proposed an analytical mathematical model for determining roller profiles in pipe straightening equipment [4]. In the envelope theory, they used coordinate transformation matrices and elements of differential geometry. They suggested that using a spherical-end mill cutter, the active surface of the roller can be obtained. In Romania, large-scale research on getting hyperboloidal surfaces by wire electrical discharge machining was initiated and developed by C. Deneş [5-7]. Numerical modeling of the cross-roll straightening process using a hyperboloidal roller was proposed by Mutrux *et al.* [8]. The generation of a hyperbolic settlement surface when sharpening helical drills was approached by Baroiu *et al.* [9]. They found that such an area has advantages both in terms of the drills' behavior and the kinematics of the process of machining that area. Abadjieva *et al.* have developed research on the possibilities of synthesizing hyperboloid gear drives to transmit motion between two shafts whose axes are not parallel or intersect [10]. Rods of the hyperbolic form were still considered by Benešovský *et al.* when they addressed the issue of verifying an operative algorithm usable for cross-roll straightening [11]. Zhang used rolls of shapes close to those of a rotating hyperboloid when modeling cross-sectional ovalization of thin pipe walls in the case of a straightening process [12].

From the examination of the specialty literature, it is found that the approaches for obtaining by cutting the surfaces corresponding to one-sheet revolution hyperboloids were less considered and materialized. For this reason, the objective of the research whose results are presented in this paper was to identify a solution for machining hyperboloidal surfaces of known dimensions on universal machine tools, including by using specialized devices adapted to such machine tools.

2. Theoretical considerations regarding the obtaining of one-sheeted revolution hyperboloidal surfaces

The mathematical relation used to define a one-sheeted hyperboloidal surface (figure 1) has the general form:

$$\frac{x^2}{a^2}+\frac{y^2}{b^2}-\frac{z^2}{c^2}-1=0, \quad (1)$$

where a, b and c are constants. If $a = b$, a revolution hyperboloidal surface of rotation is reached.

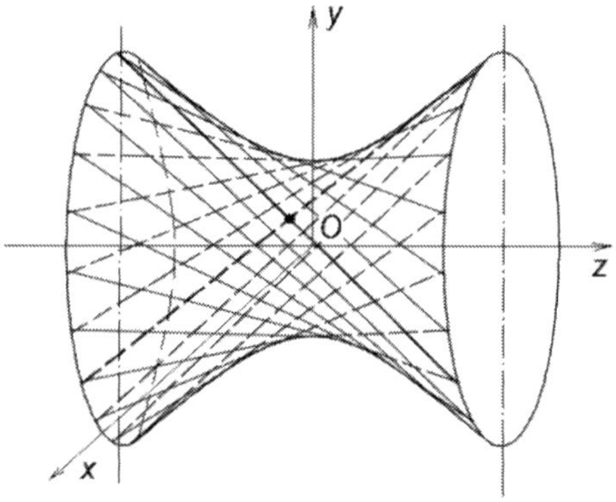

Figure 1. The one-sheeted hyperboloid as a double-ruled surface.

There are several ways to obtain one-sheeted hyperboloidal surfaces and with a circular cross-section. Thus, a variant generates the hyperboloid's surface by rotating a hyperbola around an axis (figure 2). In the *xOyz* coordinate system, the equations that define such a hyperbola are the following:

$$\begin{cases}\frac{x^2}{a^2}+\frac{z^2}{c^2}-1=0 & (2)\\ y=0, & (3)\end{cases}$$

where a and c have constant values.

On the other hand, the equations of a so-called *generator circle* are the following:

$$\begin{cases}x^2+y^2+z^2=\lambda^2 & (4)\\ x+y+z=\mu, & (5)\end{cases}$$

where λ and μ are not zero real numbers. Considering all the four equations (2)-(5), it can generate a one-sheeted revolution hyperboloid, obtained by rotating the hyperbola defined by the system of equations (2) and (3) around the *Oz* axis. Such a way of generating a one-sheeted hyperboloidal surface (by rotating a hyperbola around an axis of rotation) can be translated into practice by moving the peak of a lathe tool in the plane of the rotation axis of the main shaft of a numerically controlled lathe (figure 2). In this case, the moving of the peak lathe tool will define the hyperbola whose rotation around the axis of rotation of the lathe main shaft will generate the surface of the revolution hyperboloid (with circular cross-section).

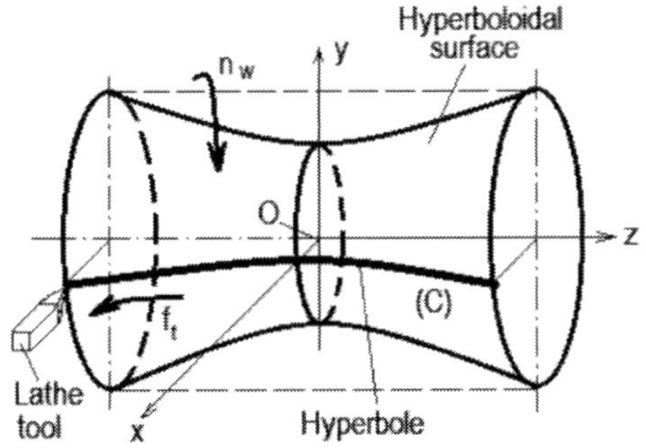

Figure 2. Obtaining a one-sheeted hyperbolic surface by moving the peak of the lathe tool along a trajectory in the shape of a hyperbolic arc.

A second variant of generating the surface corresponding to a revolution hyperboloid is when a line that is not included in the plane of the rotation axis rotates around this axis (figure 3). The mathematical model valid in such a situation will first consider the definition of the line using equations of the form [3]:

$$\begin{cases} \alpha\left(\frac{x}{d}-\frac{z}{e}\right)=\beta\left(1-\frac{y}{f}\right) & (6) \\ \beta\left(\frac{x}{d}+\frac{z}{e}\right)=\alpha\left(1+\frac{y}{f}\right), & (7) \end{cases}$$

where d, e, and f are constants, and α and β are not zero real numbers.

Since each of the two equations of the previous system defines a plane, their intersection will correspond to the line to be rotated around the axis of rotation Oz. The circle of variable radius defined by the system of equations (2) will move.

Examination of such a surface shows that in addition to the line defined by the system of equations (6) and (7), there is another line whose equations of definition are similar to those in the system of equations (6) and (7), but in which, instead of the constants α and β, will be used other constants, γ and δ, having values in correlation with those of the constants α and β. For this reason, it is stated that the hyperboloid defined above is a double-ruled surface.

If we wanted to generate the surface of a one sheeted revolution hyperboloid following this second way of defining and we intend to use a turning process, we would find that, of course, it is possible to move the peak of the lathe tool along an inclined line, in the vertical plane, when using an ordinary universal lathe. The difficulty that we meet in such a situation is the continuous variation of the functional values of the side relief angle and side rake angle. The possible device solution able to ensure the continuous change of the position of the lathe tool (by rotating it around an axis that coincides with or it is parallel to the Oz axis) so that the values of the side relief angle and the side rake angle of the lathe tool are approximately constant seems quite complicated and, as such, less efficient. For this reason, the solution that involves the use of an end cutter with a hemispherical active area becomes more interesting to approach (figure 3).

Such a milling tool could be used on a milling machine or a universal lathe. In the case of a vertical milling machine, the workpiece could be located and clamped in a universal chuck and a live center, the latter placed on a rotating table of the vertical milling machine or on a plate that can be rotated around a vertical axis and which is a component of a device placed on the table of the vertical milling machine. In its simplest form, the feed movement could be done manually.

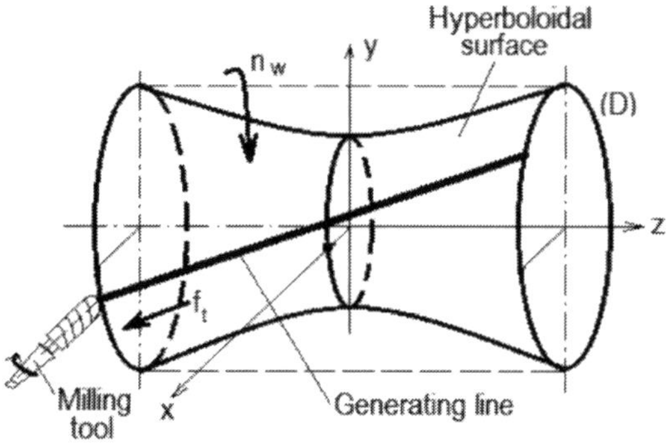

Figure 3. Obtaining a one-sheeted hyperbolic surface by moving an end mill with the hemispherical active area along a straight line inclined to the axis of rotation of the workpiece on a universal lathe.

3. Device for milling a surface corresponding to a one-sheeted revolution hyperboloid

The materialization of a surface milling process corresponding to a one-sheeted revolution hyperboloid on a universal lathe made needs to design a device used for this purpose and adaptable on a universal lathe. A constructive solution was considered as simple as possible but which would allow the milling of the one-sheeted hyperboloidal surfaces with dimensions variable between certain limits and would not require significant changes in the machine tool used.

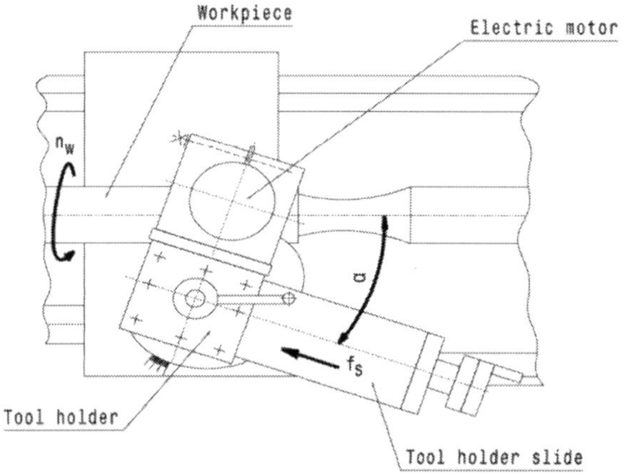

Figure 4. Top view of the device for milling a hyperboloidal surface on a universal lathe.

We started from the premise of the possibility of driving the milling tool in rotating motion with the help of a separate electric motor and possibly allowing a change in the speed of the milling tool. The rotation of the workpiece corresponds to the feed movement from the milling processes of the flat surfaces. Since this feed movement takes place at a rather low speed, it is obvious that the lowest rotation speeds of the main lathe shaft will have to be used.

As a machine tool adaptable for the materialization of a surface milling process corresponding to a one-sheeted revolution hyperboloid, a lathe of the type SNA 500x1000 (made in Romania) was considered. This lathe ensures the main shaft rotational speeds in the range of 16-1,600 rev/min. For the rotary motion of the end mill, an ISEL UFM 1050 device (made in Germany) [13] was used. This ensures a speed of 11,000 - 25,000 rev/min. Using an end mill with two teeth and an outer diameter of 8 mm, the achievable speed for a tool rotation speed n_T = 11,000 rev/min has a value of 276 m/min.

Note that the above aspects were made by considering an end mill with a flat front surface. When using an end mill with a hemispherical active area, more recommended for use when it is necessary to obtain profiled surfaces, if milling tools with low radius values corresponding to the hemispherical active area are used, the values of the feed motion speed will be lower. This fact will allow obtaining lower values of the roughness parameters corresponding to the machined surface.

The device incorporating the electric motor with the output shaft having mounted a tool holder chuck has been assembled in a plate provided with a central hole and with an elastic slit which allows the secure clamping of the rotating tool motor (figures 4 and 5).

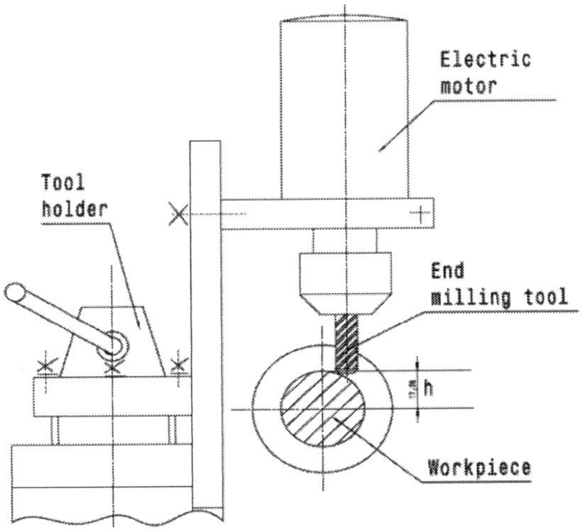

Figure 5. Side view of the positioning of the device for milling one-sheeted hyperboloidal surface in the tool holder of the universal lathe.

In its turn, the electric motor support plate was placed on a vertical plate. To ensure the possibility of changing the position of the peak of the milling tool in the vertical plane and to machine, in this way, hyperboloidal surfaces with different minimum diameters, two slots have been provided in the vertical plate to allow the movement of the clamping screws of the electric motor support plate. The actual mounting of the electric motor support plate on the vertical plate takes place with the help of nuts.

For a precise adjustment of the position of the milling tool peak to the rotation axis of the lathe main shaft, an initial positioning in relation to the cone peak corresponding to a live center mounted in the movable tailstock was taken into account. The value of the subsequent vertical displacement of the electric motor support plate could be tracked, for example, using a dial gauge mounted in a device with magnetic clamping on the vertical plate. A parallelepiped-shaped part was assembled to the vertical plate using screws, allowing the device to be mounted in the tool holder of the universal lathe. A feed movement of the milling tool peak along an inclined path relative to the rotation axis of the lathe main shaft can be achieved by rotating the tool holder slide guide with the recommended value of the

Figure 6. Image of the milling device placed on a universal lathe and using an end mill.

inclination angle. As the apron cannot normally be moved with a mechanical feed on the universal lathes, the handwheel of the tool holder slide will be rotated, and a feed movement will be performed manually. A mechanical feed movement could be obtained using the templates used for turning by tapping conical surfaces. In the latter situation, it is necessary to disengage the transverse slide of the lathe from the screw to achieve the cross-feed movement and, respectively, to materialize a longitudinal feed by the lathe apron. Under the conditions mentioned above, the device whose image is shown in figure 6 was made. Preliminary experimentation of the device proved its proper functioning to the conditions taken into account in its design.

4. Conclusions

Different processes can be used to obtain one-sheeted revolution hyperboloidal surfaces by cutting. The analysis of these processes showed that a relatively simple solution could be conceived starting from the use of an end milling process but on a universal lathe. It is thus possible to materialize the generation of a one-sheeted hyperboloidal surface by rotating around an axis a straight line inclined relative to the rotation axis of the main shaft and located above the rotation axis horizontal plane. For this purpose, a milling device for machining one-sheeted hyperboloidal surfaces adaptable to a universal lathe has been designed. The device provides possibilities for adjusting the position of the milling tool in the vertical direction and has its own motor for driving the rotating milling tool. The movement of the milling tool peak along an inclined line to the rotation axis of the lathe main shaft takes place by manually feeding the lathe tool slide after rotating the slide guide with an angle of appropriate value. Preliminary experimentation of the device has shown its adequate functioning. In the future, it is envisaged to design an improved variant of the device, which will allow the controlled positioning along a vertical direction of the milling tool peak, using slide-guide and screw-nut mechanisms. It is also intended to extend the conditions for testing the possibilities of using the device.

ORCID iDs

L Slătineanu https://orcid.org/0000-0002-5976-9813
A Hriţuc https://orcid.org/0000-0003-3871-7800
O Dodun https://orcid.org/0000-0003-4047-3550
G Nagîţ https://orcid.org/0000-0001-9649-8525
I Beşliu-Băncescu https://orcid.org/0000-0003-3132-7818

5. References

[1] Murgulescu E, Flexi S, Kreindler O, Sacter O and Tîrnoveanu M 1965 *Analytical and differential geometry* (in Romanian) (Bucharest: Editura Didactică şi Pedagogică)
[2] Weisstein E W 1999 https://mathworld.wolfram.com/One-SheetedHyperboloid.html Accessed 12.05.2021
[3] Rotaru L G 2019 *Technical-scientific conference of students, masters and doctoral students* vol 2 (Chişinău, Republic of Moldova), pp 77-80 http://repository.utm.md/handle/5014/3076
[4] Li K Y, Chen, C K and Yang S C 1999 *J. Mater. Process. Technol.* **94** (2-3) 157-66 https://doi.org/10.1016/S0924-0136(99)00089-8
[5] Deneş C 2018 *Nonconv. Technol. Rev.* **22** (3) 15-20
[6] Deneş C 2018 *Nonconv. Technol. Rev.* **22** (4) 3-9
[7] Deneş C 2002 Contributions on the processing by electrical erosion with wire electrode (in Romanian). Doctoral thesis. (Sibiu, România, Universitatea "Lucian Blaga")
[8] Mutrux A, Berisha B, Hochholdinger B and Hora P 2008 *Proc. of the 7th LS-Dyna Anwederforum* (Bamberg, Germany) C-I-34 https://www.researchgate.net/publication/264848051_Numerical_Modelling_of_Cross_Roll_Straightening
[9] Baroiu N, Teodor V G, Berbinschi S and Oancea N 2011 *Proc. Int. Conf. Innovative Technologies, In-Tech* (Bratislava, Slovacia) https://www.researchgate.net/publication/237076194_Multi-Flute_Helical_Drills_-_The_Modeling_of_a_Hyperboloid_Sharpening_Method_of_the_Back_Faces
[10] Abadjieva W, Abadjiev V, Kawasaki H and Mouri T 2013 *Proc. ASME 2013 Int Design Eng Technical Conf and Computers and Information in Engineering Conference IDETC/CIE* (Portland, U.S.A) DETC2013-12234 https://doi.org/10.1115/DETC2013-12234
[11] Benešovský M, Návrat T and Petruška J 2017 *Proc. of 23rd Int. Conf. Engineering Mechanics* (Svratka, Czech Republic) *p 134 https://www.engmech.cz/improc/2017/0134.pdf*
[12] *Zhang Z* 2019 *Int. J. Mech. Sci.* **153** 83-102 https://doi.org/10.1016/j.ijmecsci.2019.01.021
[13] Engraving spindle UFM 1050 https://www.isel.com/en/engraving-spindle-ufm-1050.html Accessed 8.10.2021

IOP Conf. Series: Materials Science and Engineering 1235 (2022) 012015 doi:10.1088/1757-899X/1235/1/012015

Analysis of specific cutting energy in laser fusion cutting of an aluminium alloy

M Madić[1], M Coteaţă[2], P Janković[1] and M Trifunović[1]

[1] Faculty of Mechanical Engineering in Niš, University of Niš, Aleksandra Medvedeva 14, 18000 Niš, Serbia
[2] "Gheorghe Asachi" Technical University of Iaşi, Blvd. D. Mangeron, 39A, 700050 Iaşi, Romania

madic@masfak.ni.ac.rs

Abstract. Laser cutting is a non-contact, non-conventional method for contour cutting which is characterized by several process performances. The present paper aims to analyze the specific cutting energy in CO_2 laser cutting of an aluminium alloy using nitrogen as an assist gas. An experimental investigation, based on the use of central composite design, was conducted to define mathematical models for the prediction of the kerf width, surface roughness, and specific cutting energy. Thus, an analysis of the effects of the laser power, cutting speed, and nitrogen pressure on the specific cutting energy is enabled. Moreover, the analysis of experimental data included the correlation between the specific cutting energy and surface roughness. Pareto optimization, considering specific cutting energy and surface roughness, was conducted to propose alternative laser cutting conditions concerning conflicting criteria.

1. Introduction

Among different contour cutting technologies, laser cutting is one of the most widely used technologies in today's industry because of its numerous advantages including the ability to cut almost all engineering materials at high cutting rates, high levels of precision and accuracy, and overall cut quality.

From the technology point of view, the laser cutting process is a complex and diffuse process governed by several process-related parameters and associated multi-physical phenomena including heating, melting, evaporation, heat transfer, melt removal by assisting gas stream, etc. Theoretical modelling and analysis of the cutting process based on the heat conduction, mass-and energy-balance equations are very complicated and with limited practical applicability for technological purposes [1]. In regard to this, several researchers and industrial practitioners apply less general, but more accurate empirical mathematical models for analysis and optimization of considered experimental hyperspace to get the most out of laser cutting technology for a given application. Parthiban *et al.* [2] applied response surface methodology and genetic algorithm to model and optimize the kerf width by determining the optimized CO_2 laser cutting parameter values for gas pressure, cutting speed, and laser power. By using the full factorial design, Nassar *et al.* [3] optimized cutting speed, laser power, and gas pressure for maximizing the surface quality in CO_2 laser cutting of AISI 307 stainless steel. Girdu *et al.* [4] studied the effect of laser power and cutting speed on different efficiency indicators of the laser cutting process such as linear specific energy, surface-specific energy, and volume-specific energy. Hardox 400 and Hardox 450 steel with different thicknesses were used as workpiece materials and were cut with different combinations of process parameters as defined by the adopted experimental matrix. Anghel *et al.* [5] investigated the effects of the process parameters on the material removal rate (MRR) and dimensional deviation in CO_2 laser cutting of miniature spur gears of stainless steel. Recently, Genna *et al.* [6] applied factorial designs to analyze the effect of material type (AlMg3 aluminium alloy, carbon steel, and stainless steel), workpiece material thickness, cutting speed, and pressure of assist gas on the kerf geometry characteristics, surface roughness and cut edge quality in high power CO_2 laser cutting. A hybrid machine learning approach, consisting of clustering and classification methods, aimed at enhancing the planning process of the laser cutting operations was proposed by Tercan *et al.* [7].

The goal of the present study is to analyze the effects of the laser power, cutting speed, and nitrogen pressure on the specific cutting energy in CO_2 laser cutting of an aluminium alloy. Moreover, the Pareto optimization problem, considering specific cutting energy and surface roughness as objective functions and kerf width as functional constraints, was formulated and solved.

2. Experimental setup details

In the present study, the standard face centred central composite design (CCD) design was employed for experimental investigation of CO_2 laser fusion cutting of aluminium alloy (AlMg3). A plate with a thickness of 3 mm was used to cut specimens with dimensions 20 mm × 40 mm. Application of the design of experiments methodology allowed for systematical analysis of the effects of selected laser cutting parameters (laser power, cutting speed, and nitrogen pressure) on the selected process performances (kerf width, mean roughness depth, and specific cutting energy) by performing 15 experimental trials with unique combinations of parameter levels, as given in table 1.

Table 1. Laser cutting parameter levels used in CCD.

Laser cutting parameter	Unit	Coded level		
		-1	0	1
Laser power, *P*	kW	3	3.5	4
Assist gas pressure, *p*	bar	6	8	10
Cutting speed, *v*	m/min	3	3.25	3.5

The cuts were conducted using TEM_{00} on Prima Industry CO_2 laser cutting machine while suppling coaxially with the laser beam nitrogen 5.0 using the conical shape nozzle. The lens with a focal length of 127 mm was used to focus the laser beam to the lower part of the plate. The stand-off distance was kept constant at 0.8 mm.

For the given laser cutting regime, i.e., the combination of the particular set of laser cutting parameter values, kerf geometry was measured using the 3D coordinate measuring machine (Mitutoyo, QSL-200Z) upon which averaged kerf width values were obtained. Also, cut surface roughness was assessed in terms of the mean roughness depth (R_z) with the help of the digital measuring instrument MahrSurf-XR.

Taking into account the measured average kerf width one can estimate the corresponding material removal rate (MRR). Then by dividing the irradiated laser power and the estimated MRR one can derive the specific cutting energy for each combination of the cutting parameter values. Specific cutting energy (E_{cs}) characterizes the actual cutting process from an energy consumption point of view. In other words, it represents the cutting energy consumed in removing a unit volume of the workpiece material.

3. Empirical mathematical models

Upon realization of the CCD plan and measurement of cut quality characteristics mathematical relationships between laser cutting parameters and average kerf width (K_w), mean roughness depth (R_z) and specific cutting energy (E_{cs}) were defined. The unknown model coefficients were obtained using the least squared method. The mathematical models were obtained as:

$$K_w = 0.46 - 0.07P + 0.031p - 0.013v - 0.11P^2 + 0.021p^2 + 0.053v^2 + 0.003Pp - 0.022Pv - 0.023pv \quad (1)$$

$$R_z = 24.3 - 0.87P + 3.717p + 2.777v - 1.002P^2 + 0.403p^2 + 2.176v^2 + 3.558Pp - 2.311Pv - 1.227pv \quad (2)$$

$$E_{cs} = 48.23 + 8.87P - 3.51p - 2.242v + 14.1P^2 - 3.795p^2 - 6.458v^2 - 0.711Pp + 2.171Pv + 2.996pv \quad (3)$$

The statistical validity of the developed models was assessed in terms of coefficient of determination (R^2). R^2 values of 97.8 %, 98.6 %, and 97.3 % were obtained for kerf width, mean roughness depth, and specific cutting energy mathematical models, respectively. The obtained results indicate that developed models are able to explain significant variability of the considered responses. Moreover, p values from ANOVA confirm the significance of the developed mathematical models.

4. Results and discussion

The main aim of the present paper was to analyze the effects of the process parameters on the specific cutting energy. Therefore, three 3D surface plots were developed by considering the two-parameter interaction effects of the process parameters (figure 1).

As could be observed from figure 1 there exists a direct positive, almost linear correlation between the laser power and specific cutting energy. As could be expected an increase in the cutting speed results in a specific cutting energy decrease, as a consequence of the increased MRR at higher cutting speeds. Orishich *et al.* [8] also observed non-linear inverse dependence of the specific cutting energy with respect to cutting speed in CO_2 laser cutting of AISI 321 austenitic stainless steel.

Although nitrogen does not chemically react in contact with the molten material and does not contribute to melting in the kerf zone, it is apparent that an increase in the nitrogen pressure reduces the specific cutting energy. This phenomenon may be explained considering the change in kerf geometry. Namely, it has been previously observed for CO_2 laser cutting of aluminium alloy [9] that for given laser power and cutting speed, an increase in nitrogen pressure may result in a kerf width increase. Moreover, it has been observed that the nitrogen pressure has the maximum influence on the kerf width followed by the cutting speed and laser power. As noted by Abdel Ghany and Newishy [10], although using high nitrogen pressure is not economical, providing high nitrogen flow is necessary as non-reactive gas is required to act as a mechanical force to clean the low viscosity oxides formed during melting the stainless steel. The significant main effect of the nitrogen pressure on the kerf width as well as the two-factorial interaction effect of the nitrogen pressure and cutting speed on the kerf width was also reported in CO_2 laser cutting of AISI 316L austenitic stainless steel [11].

From figure 1 one can also observe that the specific cutting energy ranges from about 40 J/mm³ up to approximately 70 J/mm³, with the mean value of 50.5 J/mm³. By optimization of the specific cutting energy model (equation 3), one can determine the minimal specific cutting energy and corresponding combination of the process parameters. However, the focus of the present study was to extend the optimization model to determine the correlation between the specific cutting energy and corresponding surface roughness. Moreover, as it is clear that MRR increases with an increase in kerf width which ultimately decreases specific cutting energy and that smaller kerfs are generally favourable in laser cutting, a functional constraint related to the maximal allowable kerf width was added. Thus, the following laser cutting constrained optimization problem was formulated:

$$\begin{aligned}&\text{Determine}: P, p, v\\&\text{to minimize } E_{cs} \text{ and } R_z\\&\text{Subject to}: 3 \le P \le 4\,(\text{kW}), 6 \le p \le 10\,(\text{bar}), 3 \le v \le 3.5\,(\text{m/min}), K_w \le K_{wmax} = 0.4\,(\text{mm})\end{aligned} \quad (4)$$

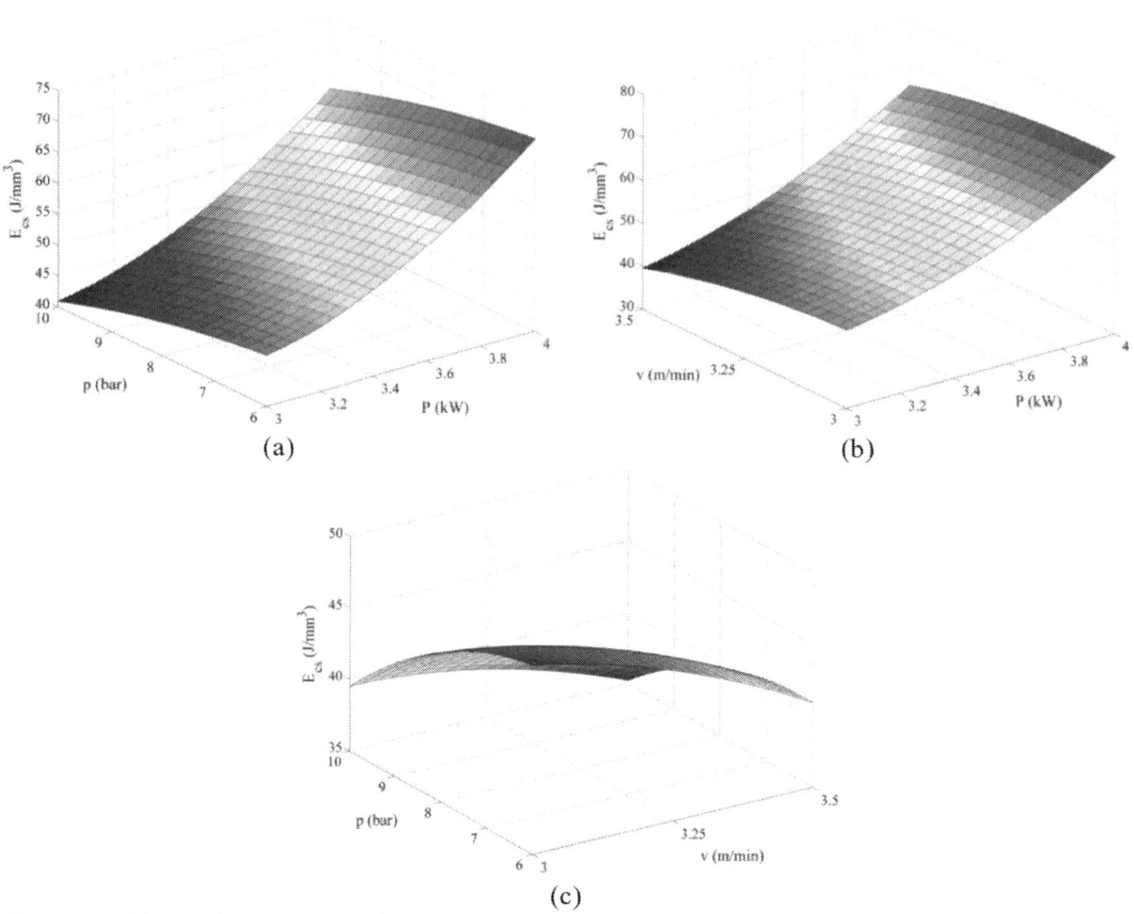

Figure 1. Change in specific cutting energy with respect to (a) laser power and nitrogen pressure, (b) laser power and cutting speed, (c) cutting speed and nitrogen pressure.

The goal of the optimization process is to determine the optimal laser cutting parameter values so as to minimize the specific cutting energy and mean roughness depth at the same time while ensuring that the average kerf width is less or equal to 0.4 mm. The formulated Pareto constrained optimization problem is solved by the application of the exhaustive iterative search algorithm combined with the epsilon constraint method using the developed software prototype [12]. This method was selected as it represents a parameter-free approach that only requires performing a large number of objective function calculations that are, however, executed very fast. Moreover, this approach, unlike metaheuristic optimization algorithms, guarantees the optimality of the determined solutions for the given discrete optimization hyper-space. As a result of optimization, a single Pareto optimal set, consisting of 14 optimal solutions, was obtained based on the best objective function values (E_{cs} and R_z) as given in the determined discontinuous Pareto front (figure 2). The entire optimization process lasted only 54 seconds.

The optimal alternative laser cutting conditions are summarized in table 2. It should be noted that each Pareto solution is an acceptable optimization solution and the choice of a particular one can be guided by the specific production requirements and/or process planner's findings. For example, if one considers the Trumpf criterion for maximal R_z values for laser cutting of aluminium alloys with a thickness of 3 mm (R_{zmax}=22 μm), the Pareto solutions from 4 to 12 could be considered as acceptable.

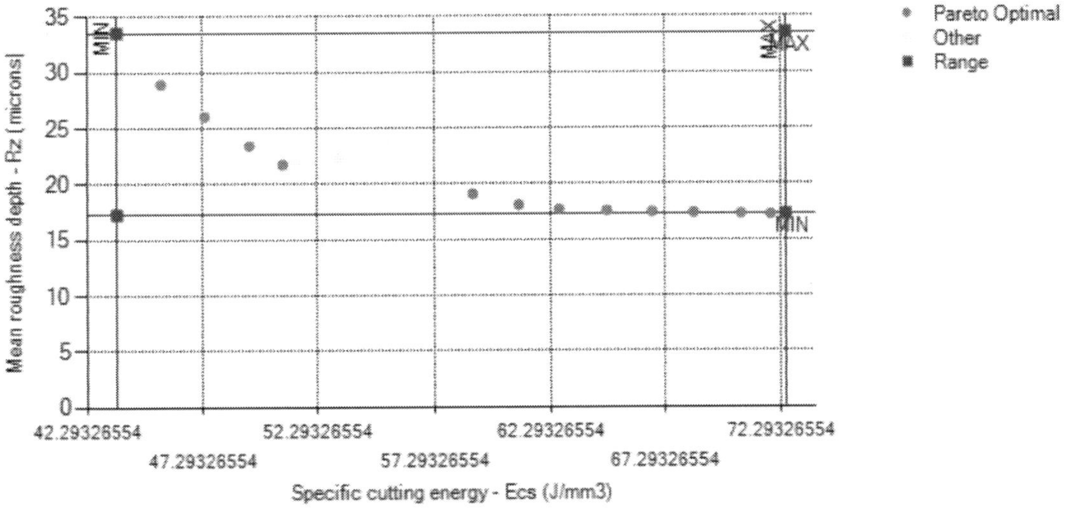

Figure 2. Determined Pareto front for specific cutting energy and mean roughness depth (screenshot from the used software prototype).

Table 2. Determined Pareto set of optimization solutions.

Pareto solution	P (kW)	p (bar)	v (m/min)	R_z (μm)	E_{cs} (J/mm^3)
1	3.07	6	3.37	28.93	45.51
2	3.12	6.02	3.30	26.07	47.41
3	3.15	6	3.19	23.43	49.36
4	3.08	6	3.01	21.76	50.79
5	3.82	6	3.30	19.11	58.95
6	3.83	6	3.20	18.10	60.93
7	3.87	6	3.13	17.69	62.70
8	3.90	6	3.13	17.57	64.78
9	3.93	6	3.13	17.47	66.72
10	3.96	6	3.14	17.39	68.53
11	3.99	6	3.13	17.30	70.56
12	4.00	6	3.15	17.25	71.85

The analysis of the determined Pareto set reveals that the combination of laser power and cutting speed ultimately condition producing as low as possible surface roughness or as low as possible specific cutting energy. Thereby, the combination of high laser power levels and low cutting speeds minimizes surface roughness while the combination of low laser power levels and intermediate cutting speeds leads to minimal specific cutting energy. Also, it is clear that for all cutting conditions nitrogen pressure is approximately 6 bar, which confirms the previous statement that increasing the nitrogen pressure increases the kerf width, and therefore the kerf width constraint was included in the optimization problem formulation.

5. Conclusion

In the present study, an experimental investigation was performed to model and analyze the effects of laser cutting parameters i.e., laser power, cutting speed, and nitrogen pressure on the specific cutting energy in CO_2 laser cutting of aluminium alloy. To this aim, a face-centred central composite design was realized, wherein average kerf width and mean roughness depth were measured and modelled using second-order polynomials. In addition, the Pareto optimization problem considering mean roughness depth and specific cutting energy as objective functions and kerf width as functional constraints were formulated and solved by the application of the exhaustive iterative search algorithm combined with the epsilon constraint method. Analyzing the obtained results lead to the following ensuing conclusions:

- For the covered experimental hyperspace, the laser power has the maximum influence on the specific cutting energy followed by the nitrogen pressure and cutting speed.
- There are no significant interaction effects on the change in specific cutting energy. In other words, the effect of a given parameter on the specific cutting energy does not qualitatively change when the given parameter changes its level.
- The specific cutting energy and surface roughness are contradictory objectives, wherein higher specific cutting energy levels are needed to ensure better surface quality.
- Multi-objective optimization revealed that for the minimization of the specific cutting energy and surface roughness the balance between the cutting speed and laser power is essential. It was also seen that the mean roughness depth decreases linearly as the specific cutting energy increases up to 50 J/mm^3.
- The observed discontinuity of the determined Pareto front may arise due to the inclusion of the kerf width functional constraint.

Acknowledgements
This research was financially supported by the Ministry of Education, Science and Technological Development of the Republic of Serbia (Contract No. 451-03-9/2021-14/200109)

ORCID IDs
M Madić https://orcid.org/0000-0002-2310-2590
M Coteata https://orcid.org/0000-0003-0443-1441
P Janković https://orcid.org/0000-0000-9951-4500
M Trifunović https://orcid.org/0000-0002-4653-6957

6. References

[1] Cicala E F, Zsivanov D and Nichici A 1998 *Fifth Conf. on Optics* International Society for Optics and Photonics **3405** 293-8
[2] Parthiban A, Chandrasekaran M, Muthuraman V and Sathish S 2018 *Mater. Today Proc.* **5** 14531–38 https://doi.org/10.1016/j.matpr.2018.03.042
[3] Nassar A, Nassar E and Younis MA 2016 *Leonardo Electronic J. Pract. Technol.* **29** 127–36
[4] Girdu C C, Mihail L A and Dragoi M V 2019 *IOP Conf. Series: Materials Science and Engineering* **659** 012045 https://doi.org/10.1088/1757-899X/659/1/012045
[5] Anghel C, Gupta K and Jen TC 2020 IOP Conf. Series: Materials Science and Engineering **971** 022081 https://doi.org/10.1088/1757-899X/971/2/022081
[6] Genna S, Menna E, Rubino G and Tagliaferri V 2020 *Appl. Sci.* **10** 4956 https://doi.org/10.3390/app10144956
[7] Tercan H, Al Khawli T, Eppelt U, Büscher C, Meisen T and Jeschke S 2017 *Prod. Eng. Res. Devel.* **11** 195–203 https://doi.org/10.1007/s11740-017-0718-7
[8] Orishich A M, Malikov A G, Shulyatyev V B and Golyshev A A 2014 *Phys. Procedia* **56** 875–84 https://doi.org/10.1016/j.phpro.2014.08.106
[9] Madić M, Radovanović M and Kovačević M 2015 *J. Prod. Eng.* **18** 39–42
[10] Ghany K A and Newishy M 2005 *J. Mater. Process. Technol.* **168** 438–47 https://doi.org/10.1016/j.jmatprotec.2005.02.251
[11] Eltawahni H A, Hagino M, Benyounis K Y, Inoue T and Olabi A G 2012 *Opt. Laser Technol.* **44** 1068–82 https://doi.org/10.1016/j.optlastec.2011.10.008
[12] Madić M, Kovačević M, Radovanović M and Blagojević V 2016 *Facta Universitatis, Series: Mech. Eng.* **14** 135–45 https://doi.org/10.22190/FUME1602135M

IOP Conf. Series: Materials Science and Engineering 1235 (2022) 012016 doi:10.1088/1757-899X/1235/1/012016

Influence of cutting parameters on surface topography in high-speed milling of slots in titanium alloy TC4

S Kurpiel, J Krupa, K Zagórski and J Cieślik

Faculty of Mechanical Engineering and Robotics, Department of Manufacturing Systems, AGH University of Science and Technology, al. Mickiewicza 30, 30-059 Cracow, Poland

szkurpiel@agh.edu.pl

Abstract. This paper summarizes the experimental investigation of influence of process parameters on surface quality and dimensional accuracy of slots milled in titanium alloy TC4. Samples were milled using a 10 mm HXT coated monolith end mill for HSM milling of superalloys and titanium alloys. Process parameters, such as rotational speed, feed rate, and depth of cut were selected based on the Taguchi L9 method. After machining samples were measured, including surface quality parameters such as roughness, waviness and dimensional deviations. Measurement results indicate that within inspected parameter range rotational speed to feed ratio has a greater influence on surface quality than the depth of cut while having less influence on dimensional accuracy.

1. Introduction

Titanium alloys represent a group of materials increasing in importance in various branches of industry, including aviation, aerospace and power industries [1]. Of this group, one of the most commonly used is Ti6Al4V, an alloy combining low density with exceptionally high mechanical properties at high temperatures. Due to this fact it is a challenging material to machine [2].

Demand for efficient manufacturing processes with the capability to create high-quality parts from titanium alloys has driven the development of specialized tools, process parameters and machining strategies [3]. General guidelines for appropriate parameter selection in manufacturing processes are being widely researched including tool life prediction [4], influence of coatings on tool life and product quality [2,5]. Constant development in the cutting tools sector drives the necessity to update those results.

The analysis of literature shows, that most of the studies on high-speed machining of Ti6Al4V concerns turning processes. Similar investigation [6] of influence of turning parameters on surface quality of Ti6Al4V samples has shown, that feed rate has greatest influence on surface quality – an increase in feed rate and decrease in rotating speed led to significant deterioration of surface roughness. Authors of the study [7] on hole drilling in TC4 alloy investigated influence of process parameters on surface quality and dimensional accuracy. Results of their study indicate, that there is a significant dependence of those parameters.

This study is focused on investigating general relationship between cutting parameters, surface quality and geometric accuracy in HSM milling using *JS754100E2C.0Z4A-HXT* end mill. 15 slots were machined with different cutting parameters and measured for roughness, waviness, dimensional and

shape accuracy which led to determining relationship between those factors. This paper summarizes experimental research of dependency of product quality on milling parameters.
Results indicate, that cutting speed to feed ratio has greater influence on surface quality than the depth of cut, while having less influence on dimensional accuracy.

2. Materials and methods

In this study series of samples was prepared under controlled conditions, with varying process parameters such as depth of cut a_p, rotating speed n, and feed rate V_f. Values of those parameters for each sample are included in table 1. Selection of those values was based on manufacturer's recommendations [8] - minimal, maximal, and recommended values of those parameters were chosen. Selected parameters are within the range of HSM processing [9,10].

Table 1. Process parameters.

No	Depth of cut a_p [mm]	Rotating speed n [rpm]	Feed Rate V_f $\left[\frac{mm}{min}\right]$
1	2	1592	267
2	2	1592	374
3	2	1592	481
4	2	2228	267
5	2	2228	374
6	2	2228	481
7	2	2865	267
8	2	2865	374
9	2	2865	481
10	1	2228	267
11	1	2228	374
12	1	2228	481
13	3	2228	267
14	3	2228	374
15	3	2228	481

Preparation of the samples was accomplished using *Mikron HPM 600 HD* milling center manufactured by *GF Machining Solutions, Switzerland* with iTNC 530 control software developed by *Heidenhain*. The tool was mounted in a hydraulic holder *HSK 63-12-85*. All samples were machined with emulsion cooling. Figure 1 shows the machining setup.

The material used in this study was titanium alloy TC4. The chemical composition of investigated material is shown in table 2. This alloy is the equivalent of the American Ti6Al4V [11,12].

All samples were machined using *JS754100E2C.0Z4A-HXT* coated monolith end mill intended for machining superalloys and titanium alloys. The tool was manufactured by *SECO Tools* as a part of the *JS754* series. The outer diameter of the mill is 10 mm and the flute length equals 20 mm. The end mill has 7 teeth with a flute helix angle of 48°. Material used for cutting samples was a blank, sized 16x100x100 mm. Each of the slots has a length of 22 mm and a width of 10 mm. Figure 2 shows the tool and the sample with markings.

The samples were then measured, including surface topography parameters such as surface roughness (R_a, R_z), waviness (W_a, W_z) and dimensional deviations from the reference model. Surface quality parameters were measured using *TOPO 01P* contact profilometer in accordance with ISO 25178-2. 31 roughness and waviness profiles covering a length of 1.25 mm offset 0.1 μm from each other were measured at the center of each slot.

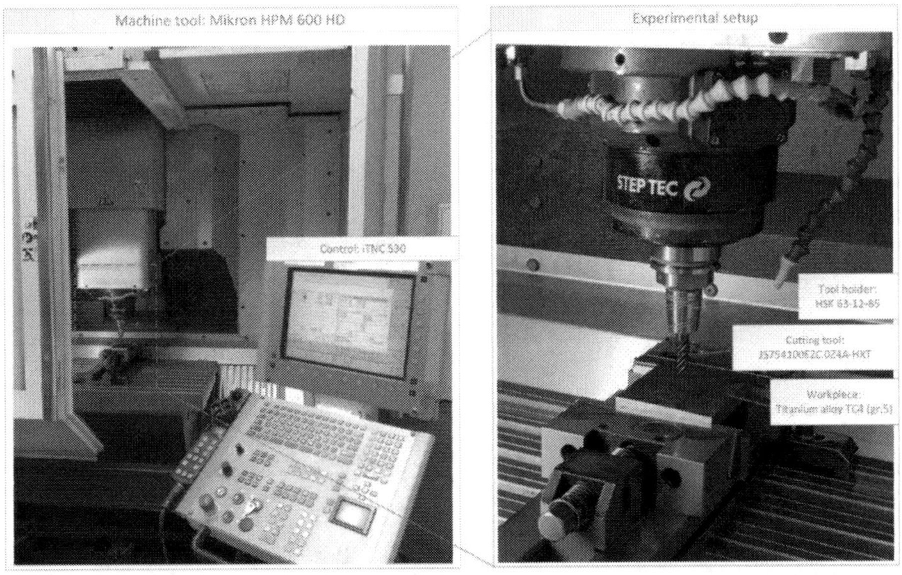

Figure 1. The machine tool and experimental setup used for the experiments.

Table 2. Chemical composition of titanium alloy TC4.

Ti	Al	V	Fe	N	C	H	O
The rest	5.8-6.0	3.2-3.5	0.21-0.23	0.02-0.04	0.02-0.05	0.008-0.01	0.13-0.17

a) b)

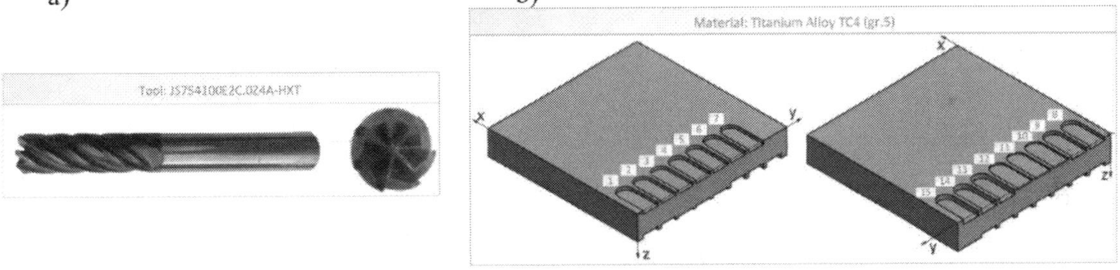

Figure 2. (a) Solid end mill used for the experiments and (b) local coordinate system and description of the slots.

The dimensional accuracy was measured using optical scanner GOM model *ATOS ScanBox 6130*, which allowed us to create a 3D map of dimensional deviations. The reference point for GOM measurement matched the reference point used in the manufacturing process (figure 2). The results of this measurement allowed us to evaluate the dimensional and shape accuracy of each sample.

3. Result and discussions

3.1. Surface topography analysis

As a result of topography measurement average values of roughness and waviness for each slot was determined. Values of surface roughness R_a ranged from 0.466 μm to 1.440 μm, R_z from 3.154 μm to 10.569 μm. Values of surface waviness W_a varied in the range of 0.229 μm to 0.694 μm and W_z from 0.816 μm to 3.248 μm. Results of measurements are graphically presented in figure 3.

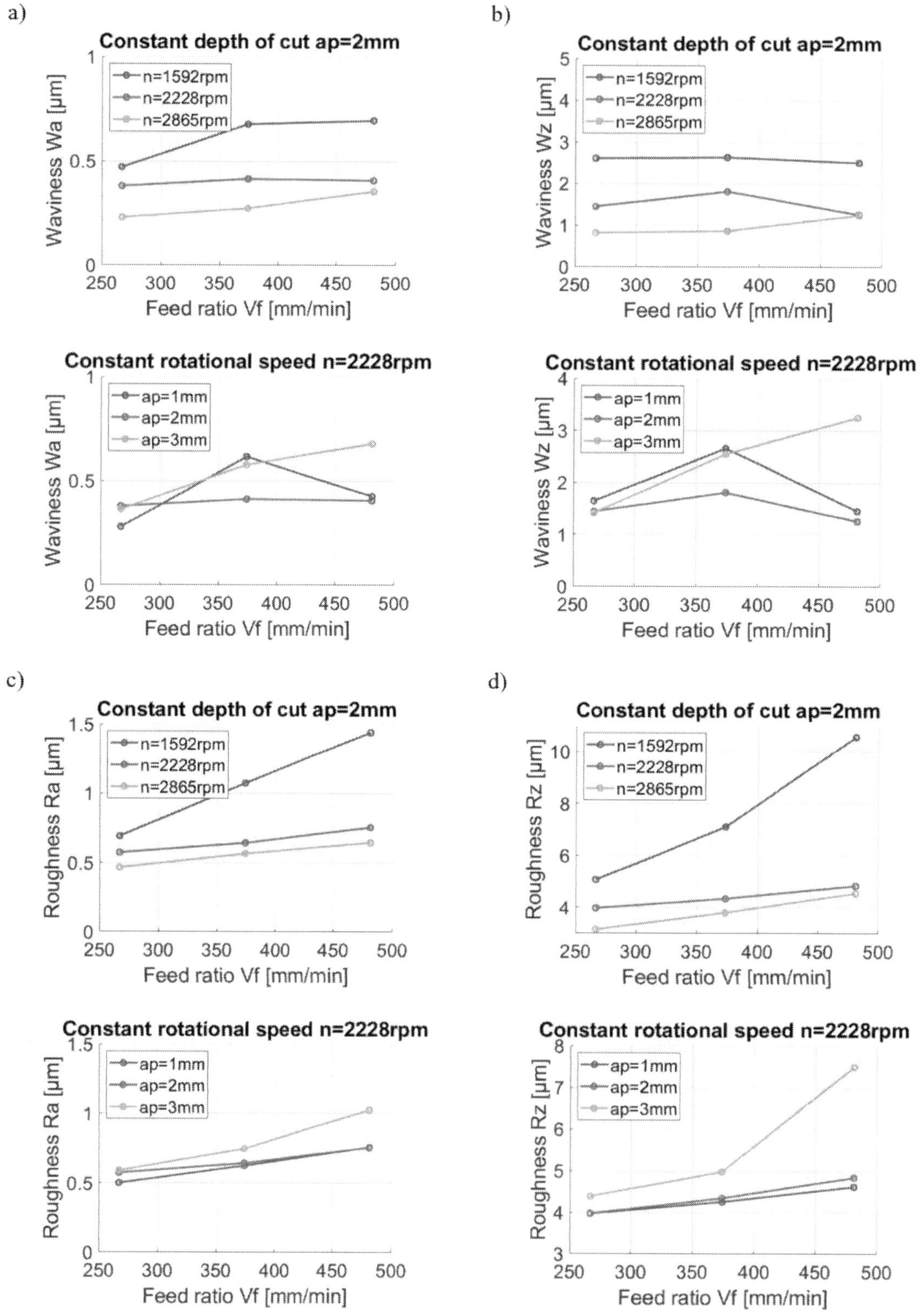

Figure 3. Comparison of experimental results: (a) W_a, (b) W_z, (c) R_a and (d) R_z.

IOP Conf. Series: Materials Science and Engineering 1235 (2022) 012016 doi:10.1088/1757-899X/1235/1/012016

3.2. Dimensional and shape accuracy analysis

The result of measurement of dimensional and shape accuracy conducted using GOM was used to plot a color map presenting deviations from the nominal shape of each sample (figure 4). Each slot was labeled with a number corresponding to a set of process parameters in table 1. The center of local coordinate system was positioned in accordance with figure 2 for purposes of the measurement. The color map shows deviations ranging from -0.1 mm to +0.03 mm for constant depth of cut of 2 mm, and from -0.17 mm to +0.11 mm for constant rotating speed (2228 rpm), where negative value indicates, that more material was removed, than necessary. The deviations near the bottom edge result from chamfer at the end of the end mill 0.125x45°. Asymmetry observed at the end of each groove result from the geometry of the tool.

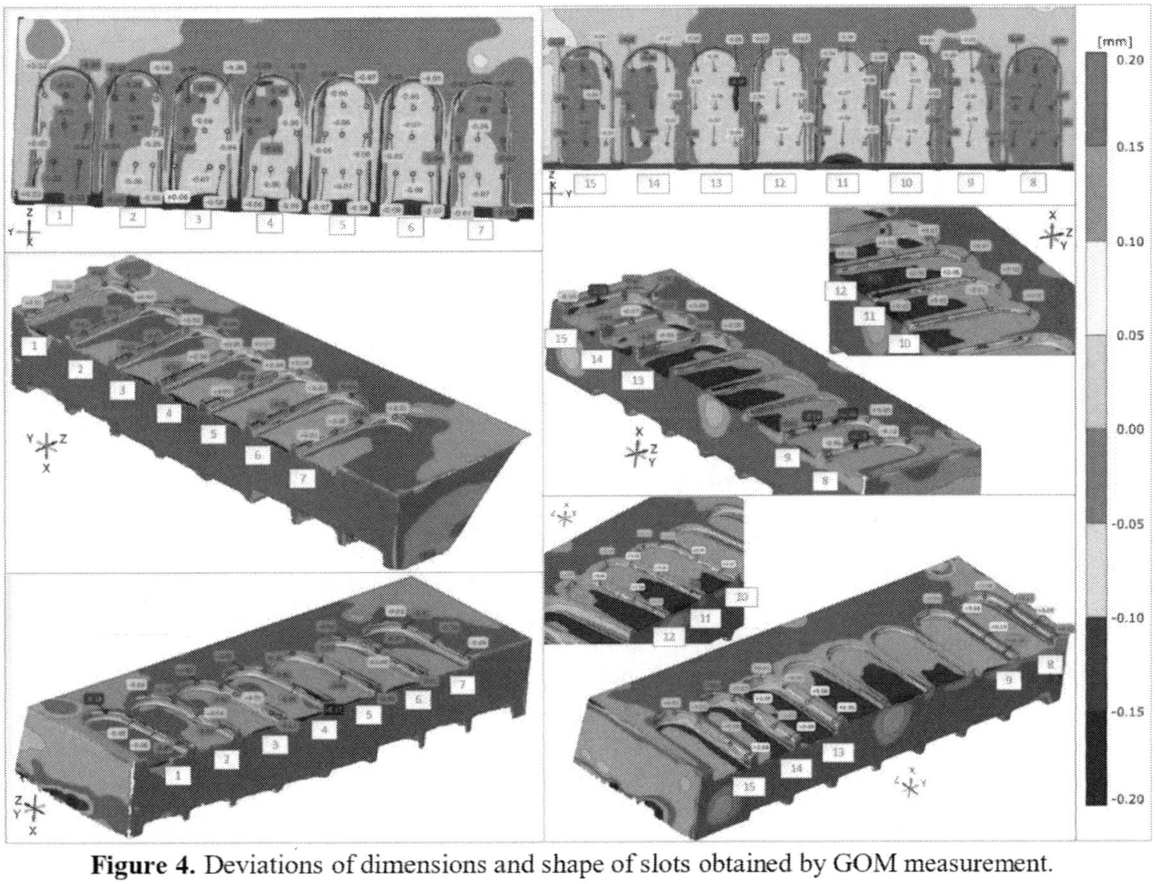

Figure 4. Deviations of dimensions and shape of slots obtained by GOM measurement.

4. Conclusions

During this study a series of samples - slots machined in Titanium TC4 with varying process parameters were prepared and measured to investigate the influence of cutting parameters on surface topography and dimensional accuracy. Measurement included determining the surface topography described by values of surface roughness (R_a and R_z) and waviness (W_a and W_z), and measuring dimensional deviations from a reference model. Conducting the experiment allowed us to draw several conclusions.

Based on surface quality results for measured samples following conclusions were drawn:

- increase of feed rate (V_f) and depth of cut (a_p) with a constant rotational speed (n) result in deterioration of surface quality,
- increase of rotational speed (n) and decrease of feed rate (V_f) with a constant depth of cut (a_p) result in improvement of surface quality,
- a ratio of rotational speed (n) and feed rate (V_f) has significantly greater influence on surface quality, than the depth of cut (a_p).

Those conclusions reflect the expected dependencies - an increase of feed rate with constant rotational speed caused an increase in the distance between grooves created by flutes, thus deteriorating the surface quality.

GOM measurement showed no significant dependence of dimensional accuracy on cutting parameters.

During preparation of samples no significant wear of the end mill has been noticed. The influence of the tool wear as well as broader range of process parameters should be taken into account in further studies.

ORCID iDs

S Kurpiel https://orcid.org/0000-0003-4787-9801
J Krupa https://orcid.org/0000-0002-1820-8905
K Zagórski https://orcid.org/0000-0002-1898-8307
J Cieślik https://orcid.org/0000-0002-0203-5149

References

[1] Dudek Ł, Hryniewicz T and Rokosz K 2016 (in Polish) Buses: technology, operation, transport systems **R17** (8)
[2] Zaleski K and Matuszak J 2017 (in Polish) RUT Mechanika **89** (4) 563–572 https://dx.doi.org/10.7862/rm.2017.53
[3] Karkalos N E, Karmiris-Obratański P, Kurpiel S, ZagÅłrski K and Markopoulos A P 2021 *Machines* **9** (4) 75 https://doi.org/10.3390/machines9040075
[4] Kuczmaszewski J, Zaleski K, Matuszak J, Pałka T and Garwacki R 2017 *Mechanik* **90** (3) 198–200 https://dx.doi.org/10.17814/mechanik.2017.3.40
[5] Haja Syeddu Masooth P, Jayakumar V and Bharathiraja G 2020 *Materials Today* **22** (3) 726–736 https://doi.org/10.1016/j.matpr.2019.10.036
[6] Abishekraj N, Gowtham T, Krishnaraj V, Jahaziel R B and Priyadarshini B G 2020 *IOP Conference Series: Materials Science and Engineering* **912** 032076 https://doi.org/10.1088/1757-899X/912/3/032076
[7] Waqar S, Asad S, Ahmad S, Abbas C A and Elahi H 2016 *4th Asia Conference on Mechanical and Materials Engineering* (*Materials Science Forum* vol 880) (Trans Tech Publications Ltd) pp 33–6 https://doi.org/10.4028/www.scientific.net/msf.880.33
[8] SECO Tools 2020 Solid end mills URL https://ipaper.ipapercms.dk/SecoTools/seco-news/20201/solid-end-mills/pl-solid-20201/?page=1
[9] Grzesik W 2017 *Advanced Machining Processes of Metallic Materials (Second Edition)* ed Grzesik W (Elsevier) pp 285–397 https://doi.org/10.1016/B978-0-444-63711-6.00015-6
[10] Ezugwu E 2004 *Journal of the Brazilian Society of Mechanical Sciences and Engineering* **26** (1) 1-11 https://doi.org/10.1590/S1678-58782004000100001
[11] Wei D B, Zhang P Z, Yao Z J, Liang W P, Miao Q and Xu Z 2013 *Corrosion Science* **66** 43–50 https://doi.org/10.1016/j.corsci.2012.08.063
[12] Wu Z R, Hu X T and Song Y D 2014 *International Journal of Fatigue* **59** 170–75 https://doi.org/10.1016/j.ijfatigue.2013.08.028

IOP Conf. Series: Materials Science and Engineering 1235 (2022) 012017 doi:10.1088/1757-899X/1235/1/012017

On the laser beam absorption efficiency in laser welding of aluminium thin sheet with copper pipe

P Karmiris-Obratański[1,2], E L Papazoglou[2], N E Karkalos[2], E Hontzopoulos[3] and A P Markopoulos[2]

[1] School of Mechanical Engineering and Robotics, AGH University of Science and Technology, Cracow, Poland
[2] School of Mechanical Engineering, National Technical University of Athens, Athens, Greece
[3] Prime Laser Technology SA, Keratea, Greece

amark@mail.ntua.gr

Abstract. Welding is considered the most versatile joining method, with extensive applicability in every modern industrial field. Especially, laser welding technology is widely utilized due to its numerous advantages, and its capability to process a variety of different materials. Among the several process parameters that impact the welding results, the absorbed energy and the power density are of major importance. These parameters mainly depend on the utilized materials and their physical properties, as well as the laser beam inherent characteristics. Moreover, the welding of more complex shapes and geometries is even more challenging, since multiple scatterings are taking place as well. The current paper presents a simulation model of the laser beam path and its absorption in the case of welding an aluminium thin plate and a copper pipe. The Ray Tracing method has been employed, while, the model's geometrical parameters, are dimensionless, reduced to the pipe's radius. The materials' absorption coefficient is calculated as a function of the incident angle, while the multiple scatterings are taken into consideration. From the simulation, the total absorption, as well as the relevant power density are calculated for a series of different initial process control parameters, i.e., spot size, incident angle, relevant laser spot position.

1. Introduction

In the modern industrial environment, lasers are widely employed in numerous applications, including laser welding processes. By utilizing laser technology in welding, deep and narrow welds can be resulted, with reduced Heat Affected Zones (HAZ) and low thermal distortions. Moreover, high production rates can be achieved, increased versatility and enhanced design capabilities [1,2]. Nevertheless, laser welding is a complex thermal process that involves the interaction of a high-power density beam with a solid surface, resulting in material phase change. During heating, the material undergoes solid heating, melting, and evaporation, while complex physical phenomena are taking place. Since most of the metals have high reflectivity, at first, only a small amount of the laser beam energy is initially absorbed by the workpiece. However, this energy is sufficient to rapidly heat the material surface, resulting in the generation of energy ionized metal vapors, which subsequently accelerate the absorption process. Laser welding can produce a wide variety of joints, ranging from thin sheets of a couple of hundred micrometers thickness, up to thick plates. Although almost the same range of materials as other fusing welding techniques can be processed, aluminum and copper alloys do face significant difficulties, mainly due to their high degree of reflectivity [1,3].

The absorbed power density is of major importance and a key factor, related to the welding quality and the process' efficiency. The absorbed power directly depends on various parameters like the materials' optical properties, the laser beam characteristics and the set up geometrical features. All these parameters interact in a non-linear way, resulting to the overall final absorbed power density. In literature, the power density has been studied as the result of utilizing different laser power and beam's focal diameters. Namely, in the work of Faye *et al.* [4] the effect of welding parameters on the joining of AA6061 aluminum was experimentally studied, while Kubiak *et al.* [5], based on experimental results, presented analytical and numerical models concerning the laser power distribution. As it was aforementioned, aluminum and copper alloys face difficulties in laser treatment due to their inherent optical properties, thus, extensive research has been conducted regarding these materials' laser welding. Case in point, Hijazi *et al.* [6] presented a statistical based methodology for parameters choice in laser welding of copper materials, while Lee *et al.* [7] investigated the keyhole welding of copper by utilizing femtosecond laser in order to enhance the coupling efficiency. Similar studies have been carried out for aluminum alloys as well. Song *et al.* [8], by employing the Response Surface Methodology, presented an experimental investigation concerning the correlation between the process parameters and the porosity in laser welding of 7A52 aluminum alloy, while Sathish *et al.* [9], also experimentally investigated the butt joint laser welding of AA7068 aluminum alloy. Apart of studying the laser welding of a single material, it is also extremely interesting to investigate bi-material welding. For instance, Chen *et al.* [10], studied the manner the processing parameters affect the welding of stainless steel with copper, while Mathivanan *et al.* [11] in their work investigated the laser welding of copper-aluminum sheets.

As it was priorly mentioned, the geometrical set-up of the workpiece-laser beam also significantly affects the laser absorbance. In the simplest cases, where plates or other flat surfaces are welded, the laser beam is emitted vertically on the surface and beam secondary scatterings occur only due to the keyhole formation [12]. On the other hand, when more complex geometries have to be welded, multiple reflections are taking place, which mainly depend on the welding geometry, the laser beam's incident angle and its focus spot. Ray Tracing method has been successfully adopted in studying the multiple reflections of the beam after the keyhole formation [13]. Additionally, Ray Tracing algorithms have been also utilized in estimation of absorption coefficient in Powder Bed Fusion processes [14], where the laser beam interacts with powder material and not with a solid flat surface, indicating that this method is suitable to determine the path and the interactions of a laser beam with surfaces of complex shape.

In the current study, the Ray Tracing method has been utilized in order for the multiple reflections and scatterings of the laser beam that take place during the longitudinal plate-pipe welding, to be calculated. By employing the Fresnel equations and according to the beam's path, the absorption of the laser beam materials is estimated, along with other significant parameters like the maximum power density and the total irradiated area. The model and the simulations pertain to the welding of a copper pipe on an aluminum plate, while the model has become dimensionless in relation to the pipe's radius and has been solved for different process parameters combinations.

2. Materials and methods

In the current paper, the longitudinal laser welding of a copper pipe on an aluminum plate has been studied. By employing the Ray Tracing method, for different parameters, i.e., beam's angle and focus spot, the successive reflections were calculated, and based on the Fresnel absorbance theory the absorbed power was estimated as well. More specific, the laser beam path, the absorbed power by each element (pipe and plate), the maximum power density, and the total irradiated area for different beam's angle and focus spots have been calculated. In Figure 1a, the model's geometrical characteristics are graphically presented. Aiming to the generality of the model, the focus spot, i.e., the distance of the laser beam focus spot from the pipe-plate contact point, introduced into the simulation dimensionless, namely, reduced to the pipe's radius. For the current simulation, a pipe with radius of 5 mm and thickness of 0.5 mm was considered, while the aluminum plate's thickness was also 0.5 mm. Although these dimensions do not affect the model's results since the focus spot distance is reduced to the pipe's radius, they are listed for the sake of completeness. The calculations were conducted for focus spot values from 0.2 to 1 with 0.2 step, meaning that the actual focus spot distance varied from 1mm up to 5mm with a step of 1mm. The laser beam angle values ranged from 10° up to 50° with 10° step, while all the geometrically achievable combinations were calculated. As a result, in total, 45 simulations were carried out for the different parameters combinations. Finally, the laser beam was considered uniform (top-hat), with spot size diameter of 100 μm and nominal power of 100 W. Again, the absorbed power was calculated as the percentage of the initial beam's nominal power.

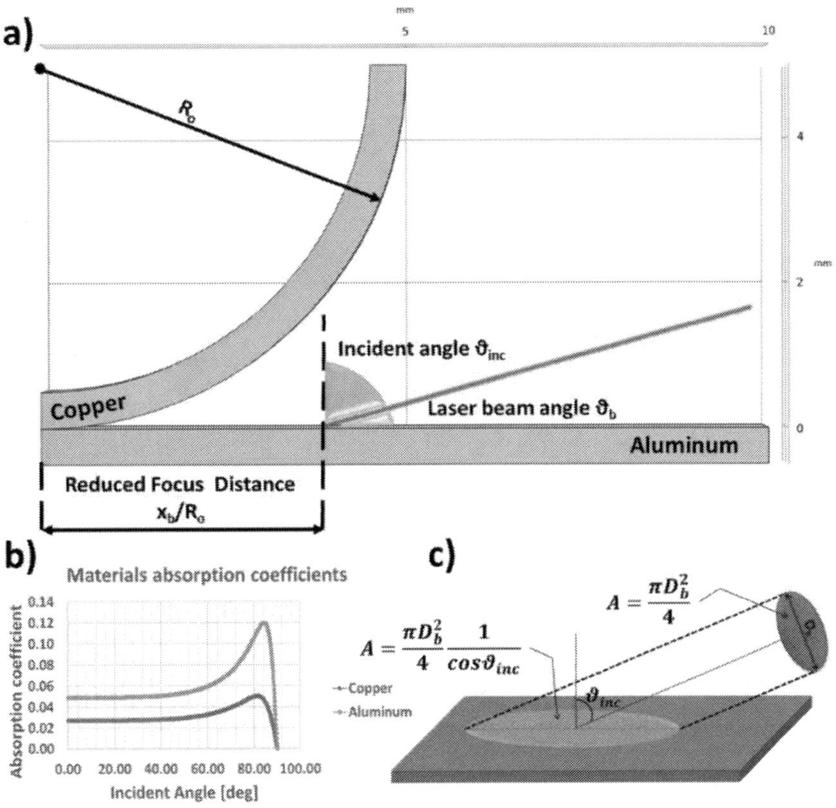

Figure 1. a) Model's basic geometrical characteristics, b) copper and aluminium absorption coefficients in respect of the incident angle, c) irradiated area in respect of the incident angle.

The absorption coefficients of copper and aluminum were calculated based on the Fresnel equations:

$$a_s(\theta) = 1 - \left|\frac{\cos\theta - (n^2 - \sin^2\theta)^{1/2}}{\cos\theta + (n^2 - \sin^2\theta)^{1/2}}\right|^2 \quad and \quad a_p(\theta) = 1 - \left|\frac{n^2\cos\theta - (n^2 - \sin^2\theta)^{1/2}}{n^2\cos\theta + (n^2 - \sin^2\theta)^{1/2}}\right|^2 \tag{1}$$

where, α_s and α_p refer to the absorption to S-polarized and P-polarized light, respectively and n is the relative refractive index of the material. The refractive indexes of copper and aluminum for 1.07 μm wavelength were n_{cop}=0.35308, k_{cop}=7.0792 and n_{al}=1.3717, k_{al}=10.314 [15,16]. Based on Fresnel equations, the absorption coefficients for copper and aluminum in respect of the incident angle were calculated (see Fig. 1b). According to Fresnel equations, for greater incident angles (between 60° and 85°) the absorption coefficients become higher. On the other hand, the irradiated area increases as function of the incident angle. Thus, two inverse and competitive mechanisms are simultaneously taking place, which are expected to significantly affect the results.

IOP Conf. Series: Materials Science and Engineering 1235 (2022) 012017 doi:10.1088/1757-899X/1235/1/012017

3. Results and discussion

In Figure 2 the percentage of the laser beam nominal power, which is absorbed by the aluminum plate and the copper pipe are presented for different focus spots and laser beam angles. It is deduced that the absorbance increases as the focus spot gets closer to the plate-pipe contact point, and for lower beam angles. The total absorbance, from the plate and the pipe, may reach over 70% for focus spot 0.2 and laser beam angle 10°, when the materials' absorption coefficients are approximately 2.5 and 5% for the copper and aluminum respectively. The order of magnitude of higher absorbance percentage can be attributed to the beam's multiple scatterings, as it is "trapped" between the plate and the pipe. Finally, as it is reasonable and expected, the aluminum absorbance is almost double that of copper, as a result to the difference in the materials' absorption coefficients.

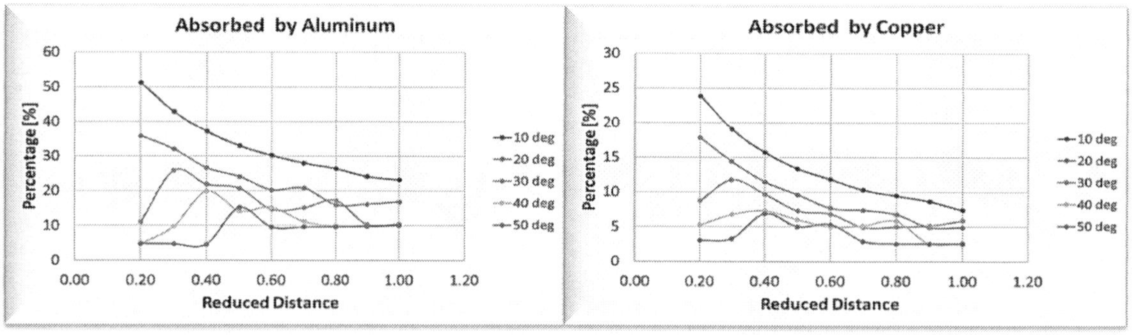

Figure 2. Absorbance percentage of the laser beam from the Aluminium plate and the Copper pipe

The power density is an important parameter since it is directly related to the material melt and ablation mechanisms, and mostly in the first stages of the process. In Figure 3 the maximum power density in aluminium and copper are presented. Again, the maximum power density is increased as the focus spot gets closer to the contact point, while, the beam angle does not significantly affect the maximum power density. Especially in copper, the maximum power density is almost irrespective to the beam's angle, with dependence only from the focus spot. This result can be justified and interpretated by considering the total irradiated area (Figure 4). Although for lower beam angles, higher power absorption is calculated, the total irradiated area, due to the multiple scatterings, and the different beam paths, changes irregularly, thus, the higher power absorption does not compulsorily result to higher power density.

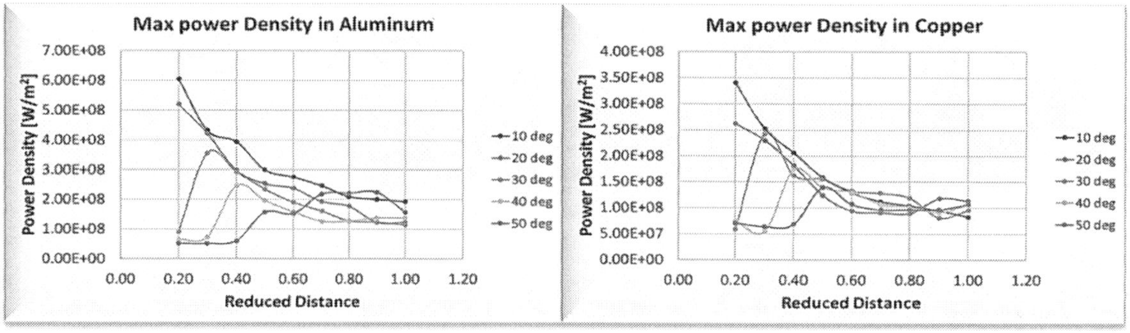

Figure 3. Max power density in Aluminium plate and the Copper pipe

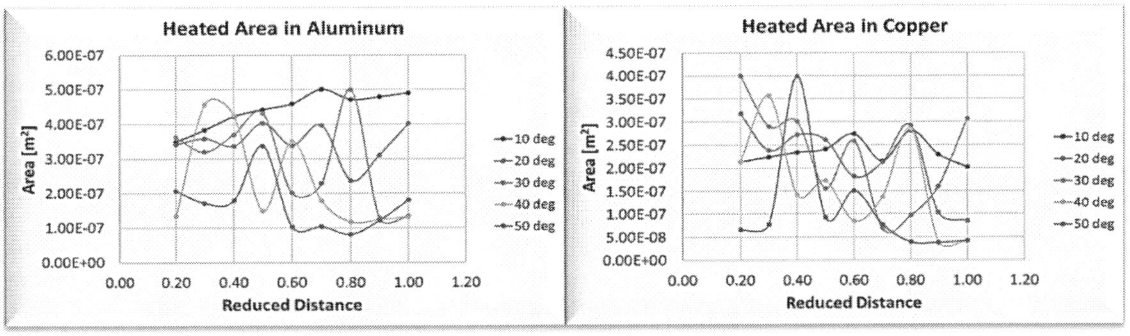

Figure 4. Irradiated area in Aluminium plate and the Copper pipe

Finally, a notable conclusion pertains to the point of the maximum power density. In Figure 5, the power densities for focus spot 0.2 and beam angle 10° are presented, when the laser beam incident the aluminium plate (Figure 5a) and after multiple scattering (Figure 5b). It is deduced that the point of maximum power density is not the initial laser beam incident point, rather one closer to the contact point as the result of the multiple reflections. This is of major importance since higher power density leads in increased material phase transformations, thus, the predicted higher material melt and ablation will occur not at the beam initial incident point but closer to the plate-pipe contact point.

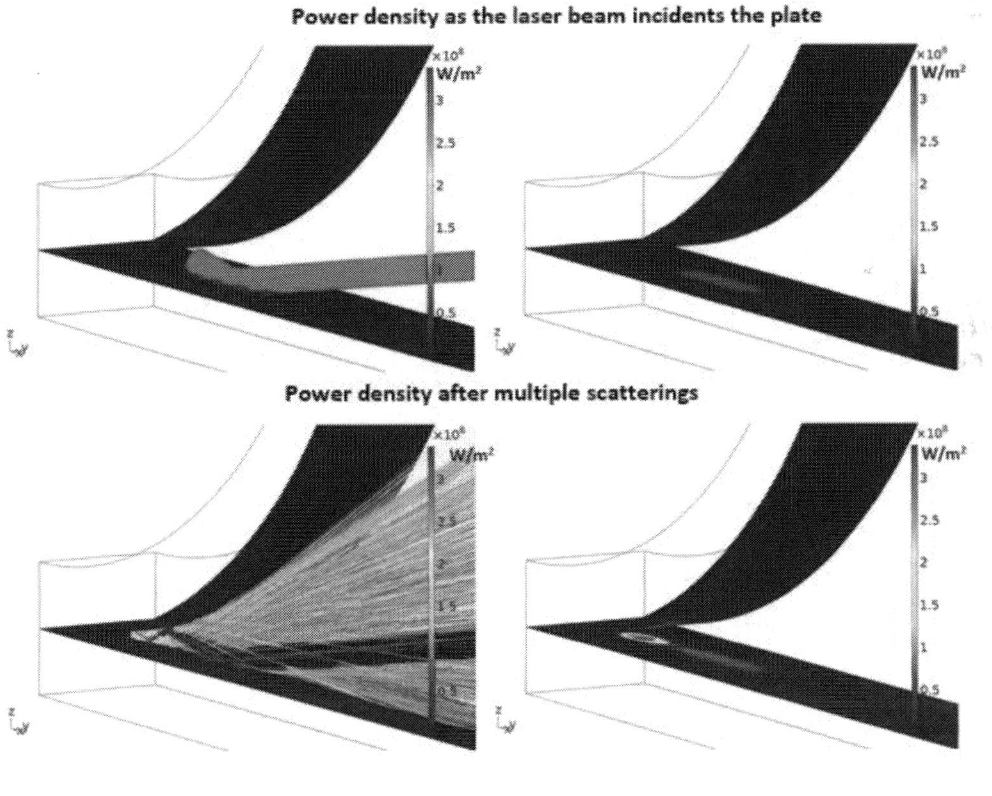

Figure 5. Power density for 0.2 focus spot and 10° beam angle
a) when the beam incident the aluminium plate b) after multiple scattering

4. Conclusions

In the current study, the Ray Tracing method was adopted in order for the multiple scatterings and reflections of the laser beam during the longitudinal plate-pipe welding to be calculated. The absorption coefficients were defined based on the Fresnel equations, while, the absorbed power, the power density and the total irradiated area were estimated. The main conclusions are:

- the absorbed power is increased as the focal point gets closer to the plate-pipe contact point, and for low beam angles. Moreover, the aluminum plate absorbs almost double the power compared to the copper pipe.
- the power density is also increased as the focal point gets closer to the plate-pipe contact point, while, the beam angle does not seem to have a significant impact.
- the point of the maximum power density is not necessarily the laser beam initial incident point, but one closer to the plate-pipe contact point, as the result of the multiple reflections that take place.

ORCID IDs

P Karmiris-Obratański https://orcid.org/0000-0002-2507-0764
E L Papazoglou https://orcid.org/0000-0002-2220-8350
N E Karkalos https://orcid.org/0000-0001-6913-825X
E Hontzopoulos https://orcid.org/0000-0002-6082-0399
A P Markopoulos https://orcid.org/0000-0002-3754-5202

References

[1] Dawes C 2008 *Laser Welding-A practical guide* (Woodhead Publishing Limited)
[2] Yilbas B S, Akhtar S and Shuja S Z 2013 *Laser Forming and Welding Processes* (Springer) https://doi.org/10.1007/978-3-319-00981-0
[3] Katayama S 2013 *Handbook of laser welding technologies* (Woodhead Publishing Limited)
[4] Faye A, Balcaen Y, Lacroix L and Alexis J 2021 *J. Adv. Join. Process.* **3** 100047 https://doi.org/10.1016/j.jajp.2021.100047
[5] Kubiak M, Piekarska W and Stano S 2015 *Int. J. Heat Mass Transf.* **83** 679–89 https://doi.org/10.1016/j.ijheatmasstransfer.2014.12.052
[6] Hijazi L, Kaiser E and Altarazi S 2020 *Procedia Manuf.* **51** 890–6 https://doi.org/10.1016/j.promfg.2020.10.125
[7] Lee K and Ki H 2021 *Opt. Laser Technol.* **139** 106943 https://doi.org/10.1016/j.optlastec.2021.106943
[8] Song C, Dong S, He P, Yan S and Zhao X 2019 *Procedia Manuf.* **37** 294–8 https://doi.org/10.1016/j.promfg.2019.12.050
[9] Sathish T, Sevvel P, Sudharsan P and Vijayan V 2020 *Mater. Today Proc.* **37** (2) 1672–7 https://doi.org/10.1016/j.matpr.2020.07.196
[10] Chen S, Huang J, Xia J, Zhao X and Lin S 2015 *J. Mater. Process. Technol.* **222** 43–51 https://doi.org/10.1016/j.jmatprotec.2015.03.003
[11] Mathivanan K and Plapper P 2019 *Procedia Manuf.* **36** 154–62 https://doi.org/10.1016/j.promfg.2019.08.021
[12] Volpp J and Vollertsen F 2013 *Phys. Procedia* **41** 460–8 https://doi.org/10.1016/j.phpro.2013.03.102
[13] Cho J H and Na S J 2006 *J. Phys. D. Appl. Phys.* **39** 5372–8 https://doi.org/10.1088/0022-3727/39/24/039
[14] Yang Y, Gu D, Dai D and Ma C 2018 *Mater. Des.* **143** 12–9 https://doi.org/10.1016/j.matdes.2018.01.043
[15] B. Johnson P and W. Christy R 1974 *Phys. Rev. B* **9** 5056–70 https://doi.org/10.1103/PhysRevB.9.5056
[16] Rakić A D 1995 *Appl. Opt.* **34** (22) 4755-67 https://doi.org/10.1364/AO.34.004755

IOP Conf. Series: Materials Science and Engineering 1235 (2022) 012018 doi:10.1088/1757-899X/1235/1/012018

Using the finite element analysis to study the cutting speed influence on temperature and internal stresses in the turning process

Aurel Mihail Ţîţu[1,2,3] and Alina Bianca Pop[4]

[1] "Lucian Blaga" University of Sibiu, 10, Victoriei Street, Sibiu, România
[2] The Academy of Romanian Scientists, 54, Splaiul Independenței, Sector 5, Bucharest, Romania
[3] Romanian Association for Alternative Technologies Sibiu - A.R.T.A. Sibiu, 10 Victoriei Street, Sibiu, Romania
[4] Technical University of Cluj-Napoca, North University Center of Baia Mare, 62A, Victor Babeş Street, Baia Mare, Romania

mihail.titu@ulbsibiu.ro

Abstract. The tendency to use the approach ing of finite element method in the turning simulation is becoming increasingly pronounced. The cutting process study based on this type of method requires the identification and knowledge of the geometric tool parameters that define the whole process. This simulation method aims to establish and describe the entire development process, characterized by trying to identify and present the optimal solution or close to an optimal realization and practical development of the entire cutting process. This study aims to draw up tabular analysis reports based on performing a series of simulations on the C45 material using the finite element approaching on DEFORM 2D. During the performed simulations, the same input values of the cutting conditions will be kept, except in the case of the cutting speed, whose modification is the object of the study. In this context, the cutting speed change influences the effective stresses distribution, the temperatures, and the cutting force variation in the cutting process. These issues were followed, and a package of own conclusions and points of view could be operationalized immediately.

1. Introduction

Since the last decades, the tendency to use the finite element analysis method to simulate cutting processes is becoming more and more pronounced [1,2].

The study of the cutting process based on this type of method requires the identification and knowledge of the geometric parameters of the tool that define the whole process [3,4].

This simulation method aims to establish and describe the development of the entire process, characterized by trying to identify and present the optimal or near-optimal solution for the realization and practical development of the entire cutting process. These issues are addressed by authors such as Jia [5], List [6], Klocke [7], and Vavruska [8].

As it is known, two components are always involved in the cutting process, namely: the workpiece and the tool [9]. In the literature, the cutting process is defined as the deformation process - cutting, detachment, and removal of excess material in the form of chips [10,11].

The cutting tool is the tool with which the material can be processed to the desired size and shape [12,13]. The geometry of a cutting tool involves the following elements: the face or clearing surface, the flank, the edges, and the tool's tip. Based on the consultation of the specialized bibliography, it was found that, in the simulation with finite elements of the solid's deformation processes, two numerical formulations were widely used: the Lagrange formulation and the Euler formulation. Individually, both Lagrange's nor Euler's approaches can be adapted to model deformation processes involving large deformations and unconstrained material flows. Thus, the advantages of one method are disadvantages for the other method, and in this respect, the two approaches complement each other.

The approach that combines the advantages of the two methods mentioned above is represented by the Arbitrary Formulation - Lagrange - Euler. Each of these techniques has its own advantages and disadvantages, making them suitable for some modeling cases and unsuitable for others. The concept of the Arbitration-Lagrange-Euler (ALE) approach in finite element analysis was initially created for application in fluid mechanics problems and was later adopted for solid mechanics applications. The ALE technique was developed mainly enjoy the advantages of the two traditional techniques and avoid their disadvantages, using each technique where it is appropriate. However, correct modeling and interpretation of the physical problem must be made, a numerical challenge. In the literature that has been identified, a series of studies address the techniques mentioned in the finite element simulation of cutting processes. Among these studies, Laghari [14], Wang [15], Laghari and Liu [16,17] as well as Attanasio [18] are mentioned. The latter explicitly analyzed tool wear using finite element analysis (FEA), along with Cui [19], Xiong [20], Rasaee [21], and Setti [22]. In terms of residual stresses, these are some of the most important parameters that characterize almost any surface layer of any mechanical component, which plays a very important role in controlling its performance. Residual stresses are those that exist in part, without any other stresses acting on it.

Consequently, they must be in a state of equilibrium. In general, the residual stresses are induced by uneven plastic deformations due to mechanical and/or thermal loads and transformation phases.

2. Research methodology

In carrying out this research, the aim was to prepare tabular reports based on performing a series of simulations with the finite approaching in DEFORM 2D application.

During the simulations performed, the same input values of the cutting conditions will be kept, except for the cutting speed, the modification of which is the research objective.

Therefore, next, we will follow the influence of the speed change on the distribution of effective stresses, temperatures, and the variation of forces within the cutting process by orthogonal turning.

The first necessary steps in FE simulation using the FEM of the orthogonal turning process consist in defining and modeling the geometry of the lathe cutting tool (figure 1) and also of the workpiece (figure 2), in this case - using the tools provided by the FEA application.

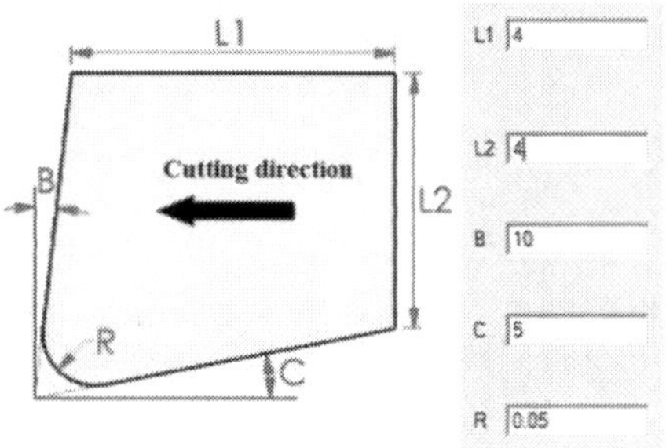

Figure 1. Defining tool geometry.

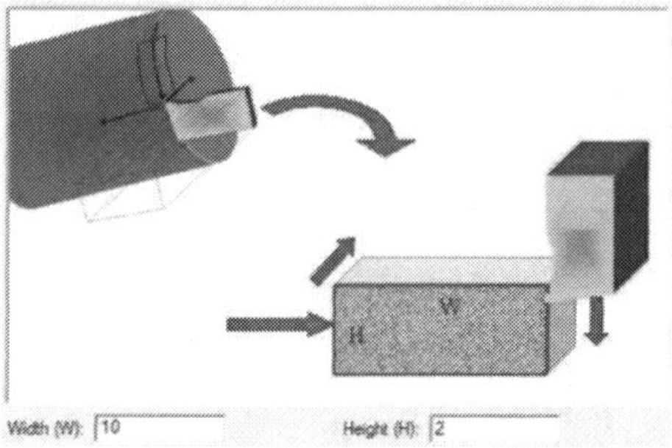

Figure 2. Defining tool geometry in DEFORM 2D application.

Also, the choice of materials and the discretization of previously created geometric patterns should not be neglected.

Using the calculation algorithm and the material models used by the FEA application, in the same processing conditions, 3 simulations were performed in which the cutting speed is the only variable parameter of the turning process.

Table 1. Cutting regimes established in the simulations.

Simulation conditions		
The diameter of the workpiece	[mm]	50
Cutting edge radius	[mm]	0.05
Cutting speed	[m/min]	125 200 470
Depth of cut	[mm]	1
Feed	[mm/rot]	0.1
Rake angle	[deg]	10
Clearance angle	[deg]	5
Ambient temperature	[°C]	20
Tool material		Coated TiC
Workpiece material		C 45

Thus, in table 1 are shown all the conditions based on which the simulations were performed.

Using the tool profile model and the processing conditions given in Table 1, the distribution of stresses and temperatures in the contact area between the tool and the workpiece and the variation of forces in the direction of the chip formation were followed.

To be able to graphically represent the influence of the speed change in the cutting process, the results obtained from the simulations will be used.

Table 2. Distribution of stresses and temperatures during simulations.

V=125 [m/min]	v=200 [m/min]	v=470 [m/min]

Effective stress distribution in workpiece [Mpa]

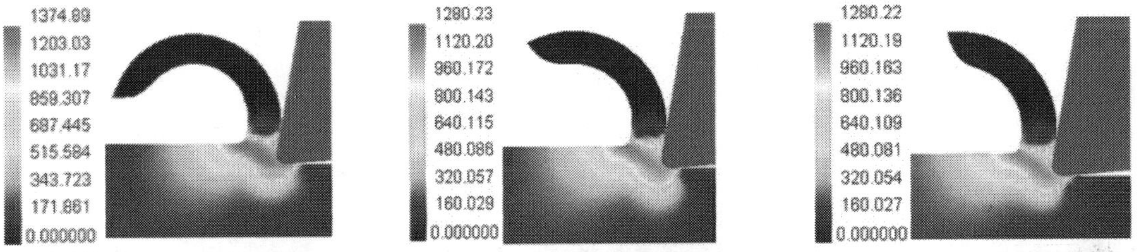

Temperatures distribution in tool [°C]

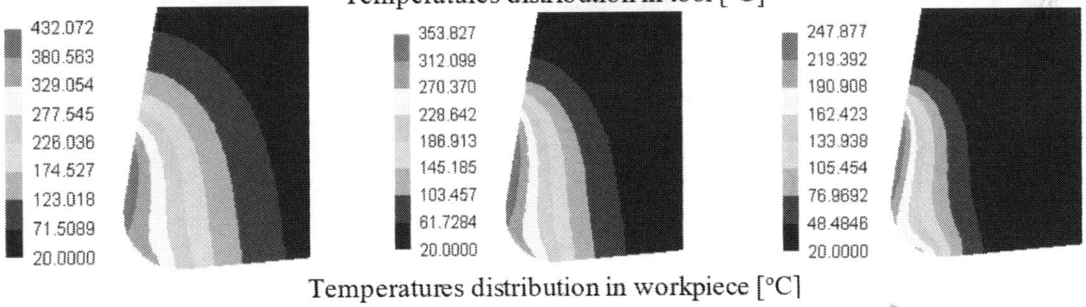

Temperatures distribution in workpiece [°C]

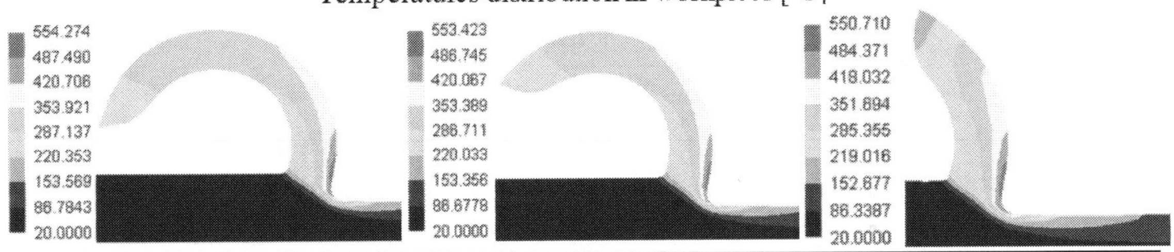

IOP Conf. Series: Materials Science and Engineering 1235 (2022) 012018 doi:10.1088/1757-899X/1235/1/012018

Table 3. Graphical representation of stresses and temperature distribution.

V=125 [m/min]	v=200 [m/min]	v=470 [m/min]
Distribution of effective stresses in the part [Mpa]		
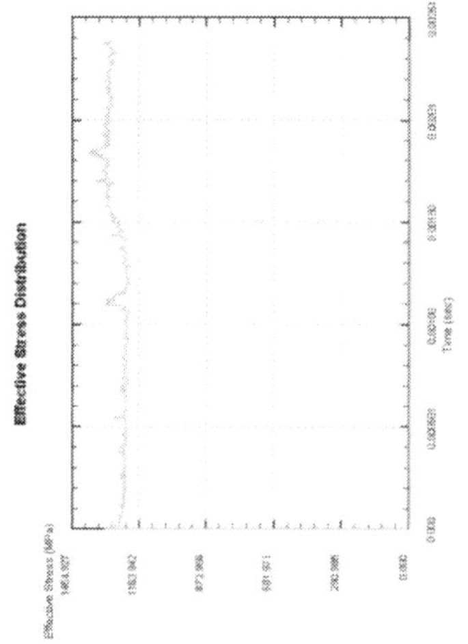	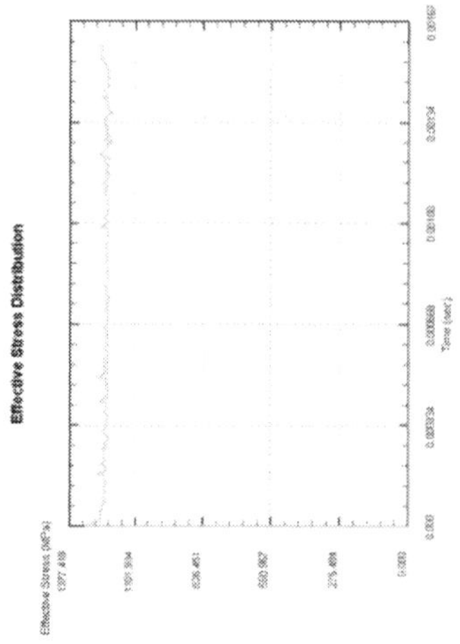	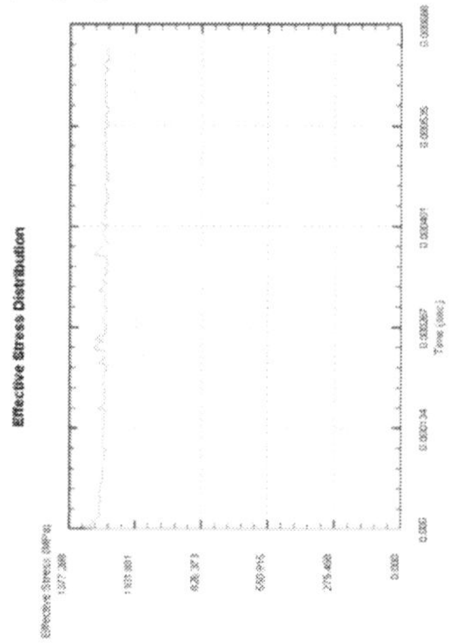
Distribution of temperature in the tool [oC]		
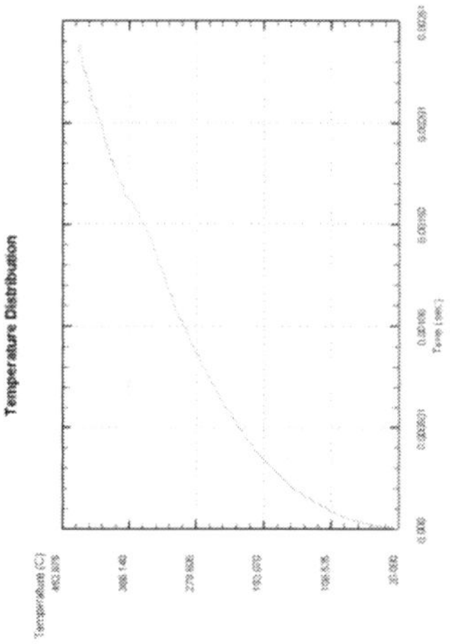	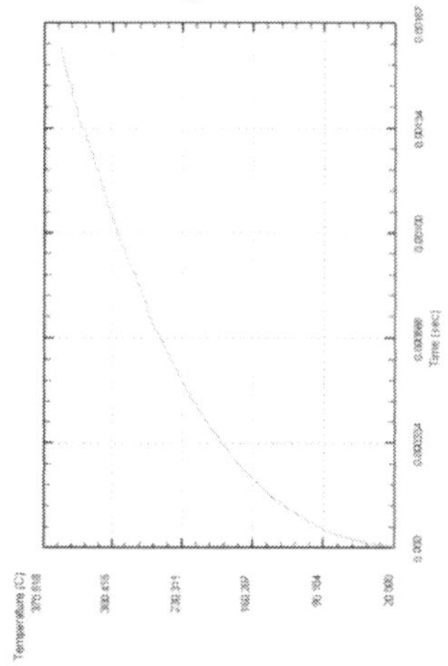	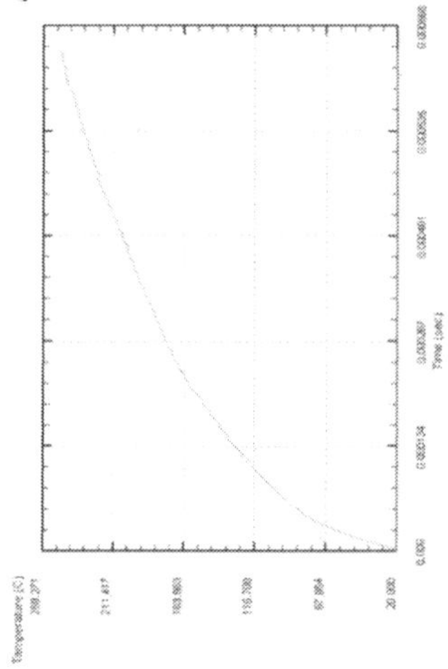
Temperatures distribution in the part [°C]		
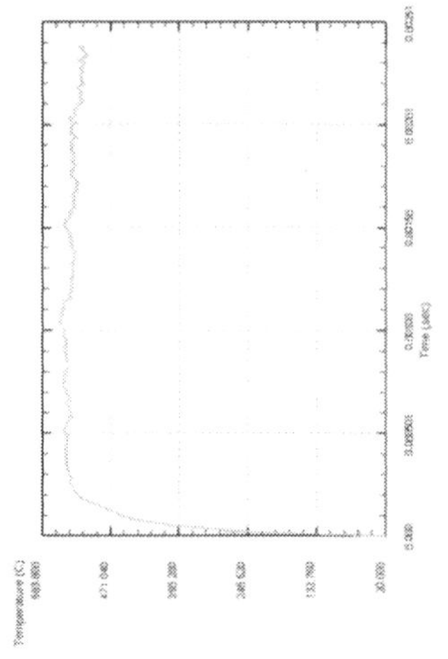	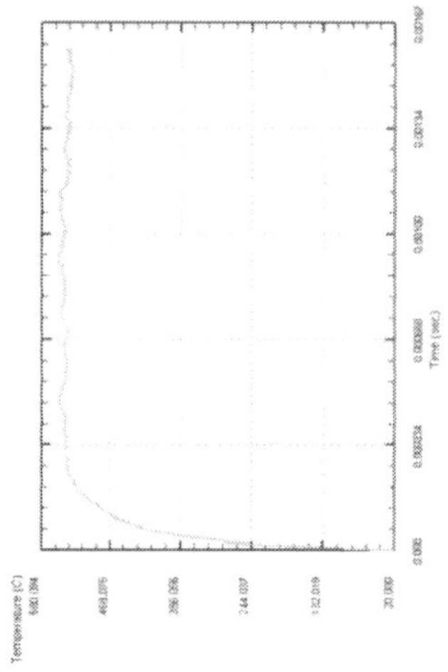	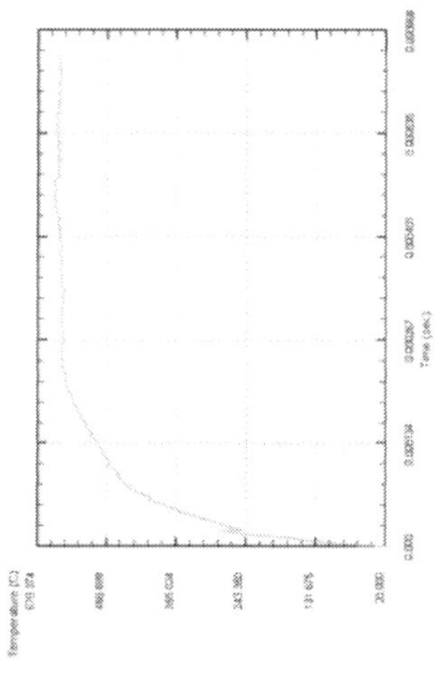

Table 2 shows the distributions and maximum values reached during the temperatures and stresses simulation in the part and tool in the contact area.

A decrease in the maximum values of both temperatures and stresses can also be observed as the speed increases. This phenomenon is also presented in the literature.

For a better understanding of the phenomenon of decreasing the maximum values with increasing cutting speed, the graphs in table 3 show the variations of stresses and temperatures occured during the turning process over 10 mm machined.

In accordance to the graphs presented in table 3, it is also observed that both the stresses and the temperatures, in part, at high speeds, once reaching the maximum, tend to remain constant without oscillating.

Table 4. Variation of forces in the X direction of the XY plane of chip formation.

v=125 [m/min]	v=200 [m/min]	v=470 [m/min]
Variation of forces in the X direction		
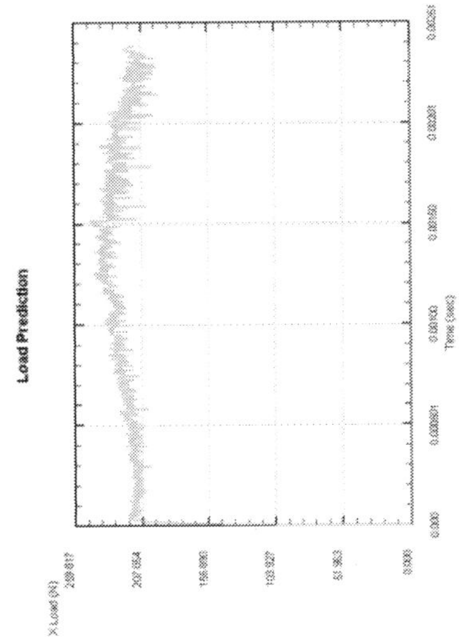	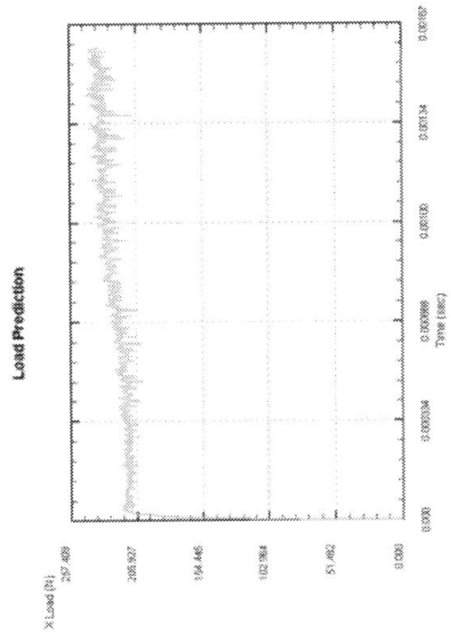	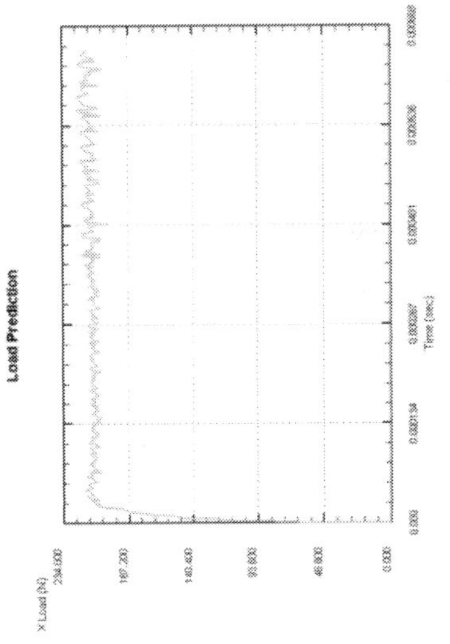

Regarding the graphs represented in table 4, hence the aspect confirmed by the theories, namely - it can be seen that, with the increase of the speed, the forces in the X direction - from the XY plane of chip formation decrease.

3. Conclusions

The research highlights the usefulness and advantages of simulating cutting processes based on modern simulation procedures using the finite element analysis method.

The mathematical model that is the starting point in parametric modeling is basic in the preprocessing phase which is a stage of finite element analysis. Therefore, if we have a properly defined geometric pattern, it can be easily discretized through any finite element application.

The Lagrange method is most common in cutting machining simulations since this approach can determine as accurately as possible the deformations and how the temperature is distributed both at the tip of the tool and in the workpiece, as well as the distribution of the stresses in the workpiece, the forces oscillation, and the formation of chipp in the turning process.

The study and comparison of scientific works identified at the national and international level have led to the enrichment or clarification of some already existing fields that present some ambiguities. Some of these identified problems are of real interest to conduct detailed research.

The finite element analysis method software provides various ways of modelling and simulation of the turning processes, providing information of real interest on the material flow and heat during the process, useful in order to facilitate the design of the product and the tools necessary to perform the processing operations.

ORCID iDs
A M Ţîţu https://orcid.org/0000-0002-0054-6535
A B Pop https://orcid.org/0000-0002-4784-8485

4. References

[1] Meng X, Lin Y and Mi S 2020 *Metals* vol **10** (8) 1038 https://doi.org/10.3390/met10081038
[2] Luo H, Fu J, Wu T, Chen N and Li H 2021 *Materials* **14** (3) 553 https://doi.org/10.3390/ma14030553
[3] Ma W, Yu J, Yang Y and Wang Y 2021 *J Manuf Mater Process* **5** (1) 2 https://doi.org/10.3390/jmmp5010002
[4] Laghari RA, Li J and Wu Y 2020 *Materials* **13** (23) 5524 https://doi.org/10.3390/ma13235524
[5] Jia B, Rusinek A, Pesci R, Bahi S and Bernier R 2020 *Int J Mech Sci* **170** 105356 https://doi.org/10.1016/j.ijmecsci.2019.105356
[6] List G, Sutter G, Bi X F and Bouthiche A 2013 *J Mater Process Technol* **213** 693–9 https://doi.org/10.1016/j.jmatprotec.2012.11.021
[7] Klocke F, Brockmann M, Gierlings S, Veselovac D, Kever D, Roidl B, Schmidt G and Semmler U 2015 *Procedia CIRP* **31** 352–6 https://doi.org/10.1016/j.procir.2015.03.067
[8] Vavruska P, Zeman P and Stejskal M 2018 *Procedia CIRP* **77** 578–81 https://doi.org/10.1016/j.procir.2018.08.216
[9] Yang J, Aslan D and Altintas Y 2018 *CIRP J Manuf Sci Technol* **23** 78–90 https://doi.org/10.1016/j.cirpj.2018.08.005
[10] Pelayo G U, Olvera-Trejo D, Luo M, De Lacalle L L and Elías-Zuñiga A 2020 *Measurement* **173** 108670 https://doi.org/10.3390/app10249135
[11] Liu H Z and Zong W 2019 *Int J Adv Manuf Technol* **100** 2689–700 https://doi.org/10.1007/s00170-018-2853-5
[12] Lotfi M 2018 *Ultrasonics* **88** 106–14 https://doi.org/10.1016/j.ultras.2018.03.013
[13] Schmitz T L 2000 *CIRP Ann Manuf Technol* **49** 303–8 https://doi.org/10.1016/S0007-8506(07)62951-5
[14] Laghari R A, Li J and Mia M 2020 *Metals* **10** (6) 840 https://doi.org/10.3390/met10060840
[15] Wang J, Zuo J, Shang Z and Fan X 2019 *Appl Math Model* **70** 1–17 https://doi.org/10.1007/s42452-021-04689-z
[16] Laghari R A, Li J, Xie Z and Wang S-Q 2018 *3D Res* **9** 46 https://doi.org/10.1007/s13319-018-0199-2
[17] Laghari R A, Li J, Laghari A A and Wang S-Q 2020 *Arch Comput Methods Eng* **27** 1363–77 https://doi.org/10.1007/s11831-019-09340-0
[18] Attanasio A, Faini F and Outeiro J 2017 *Proc CIRP* **58** 440–4 https://doi.org/10.1016/j.procir.2017.03.249
[19] Cui D, Zhang D, Wu B and Luo M 2017 *Int J Mech Sci* **131** 613–24 https://doi.org/10.1016/j.ijmecsci.2017.07.027
[20] Xiong Y, Wang W, Jiang R and Kunyang L 2018 *Int J Mech Sci* **149** 258–97 https://doi.org/10.3390/met7060197
[21] Rasaee S and Mirzaei A H 2019 *Trans Indian Inst Metals* **72** 1023–30 https://doi.org/10.1007/s12666-019-01576-5
[22] Setti D, Arrabiyeh P A, Kirsch B, Heintz M and Aurich J C 2020 *Int J Mach Tools Manuf* **149** 103489 https://doi.org/10.1016/j.ijmachtools.2019.103489

IOP Conf. Series: Materials Science and Engineering 1235 (2022) 012019 doi:10.1088/1757-899X/1235/1/012019

Study about the chip formation in the turning process using the finite element analysis

Alina Bianca Pop[1] and Aurel Mihail Țîțu[2,3]

[1] Technical University of Cluj-Napoca, 62A, Victor Babeş Street, Baia Mare, Romania,
[2] "Lucian Blaga" University of Sibiu, 10, Victoriei Street, Sibiu, România,
[3] The Academy of Romanian Scientists, 54, Splaiul Independenței, Sector 5, Bucharest, Romania

mihail.titu@ulbsibiu.ro

Abstract. At this time, there is a significant interest in the chip shape and formation at the international level, an issue studied based on mathematical calculations, experiments, and finite element analysis. The analysis of the chip formation process showed that the thickness of the undetached chip also varies according to the rake angle of the cutting tool, and using 3D models to analyze the chip formation process demonstrates this fact. Following the research, a study was conducted in which an analysis of the chip morphology is presented. Using Deform 2D application, a finite element simulation - a comparative analysis of the chip formation was performed when the value of the clearance angle varies. The increasing variety of cutting radius values has an influence on the way the chips are formed in terms of thickness, which generate differences in its shape. Following the study, a series of conclusions and own findings emerged that with the increase of the cutting-edge radius, the stress distribution is radial over a greater area of the part, reason for which the deformation of the chip is smaller. Arguments have also been made on distribution of the temperature, which is directly influenced by the shape of the tool, influencing the temperature of the generated chip. The results also show that with the increase of the cutting-edge radius of the tool, there is a tendency for the detached chip to settle.

1. Introduction

The study of cutting processes requires identifying and knowing the geometric parameters that define it [1,2]. At the international level, there is a significant interest in the shape and method of chip formation studied based on mathematical calculations, experiments, and finite element analysis. In this context, authors such as Liu [3] studied tool wear using the finite element method in the milling process, Papanikolaou [4] followed the problem of temperatures using dynamic simulations, He [5] followed the formation of chips in high-speed machining. Vandana [6] researched metal cutting simulation, and Vavruska [7] aimed at reducing processing time by changing spindle speed and feed rate using different milling strategies. Wibowo also studied [8] the optimal combination of cutting parameters during the pocket mill process of plastic mold steels material, and Arruda [9] focused on studying the optimization of the finishing milling process using ball nose end mill. Other research belongs to Rathod [10], Liu [11], Binder [12], Lofti [13], Attanaio [14], and Yue [15], who addressed the issue of tool wear and chip formation using FEA. Also, Lofti and Farid, in their study [16], address chip breaker geometry on chip shape and the forces that appeared in the cutting process. Cui [17] and De-Buruaga [18] focused on how the chip formed and the problem of end milling temperatures. Other relevant research belongs to Möhring [19], Karaguzel [20], and Xie [21], who studied the effects of changing the geometric parameters of the tool on the formation of the chip and on the whole cutting process. The analysis of the chip formation process showed that the thickness of the non-detached chip also varies depending on the angle of inclination of the cutting tool, and the use of 3D models to analyze the chip formation process shows this fact. Regarding the area of the contact area for different angles of clearance of the cutting tool and also depending on the cutting-edge radius of the tool, the thickness and shape of the non-detached chip registers a variation for each change of one of the two parameters that establish the position of the edge at some point. This paper presents a series of studies on chip morphology in different situations. The first study compares the chip creation in the situation of variation of the clearance angle.

2. Research methodology

2.1. Case 1. How the chip is formed in case of variation of the clearance angle

The input data adopted in this study are:

Software applications: CAD application - Autodesk INVENTOR; finite element analysis software - Deform 2D; Cutting process: orthogonal turning; cutting tool - coated TiC; cutting speed - 800 rpm; cutting depth -0.05 mm; cutting feed - 0.1 rot/min; cutting edge radius of the tool: 0.05 mm; rake angle γ [°] - variable: -5, 0, 5, 10, 15, 20; clearance angle α [°]: 20; workpiece material - C45; workpiece diameter: 50 mm. The 3D model of the tool and the part was defined in the CAD application and subsequently imported into the finite element analysis application. Starting with the combinations of geometric factors of the tool, a series of simulations was performed using FEM. Table 1 presents this analysis together with the shape indication of the chip and its deformation related to each simulation.

Table 1. Simulation package and the representation of the chip morphology when the clearance angle is 20°, and the rake angle varies.

γ [°]	Chip shape	Effective deformation of the chip [mm/mm]
-5 Y X	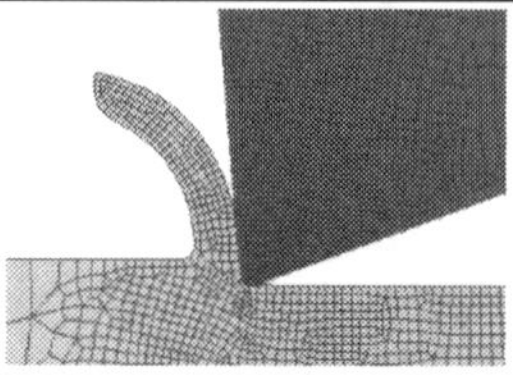	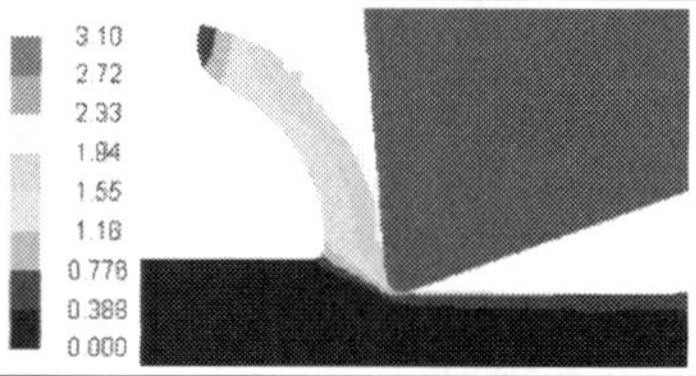
0 Y X	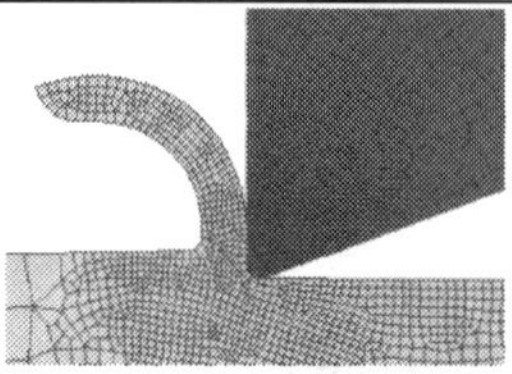	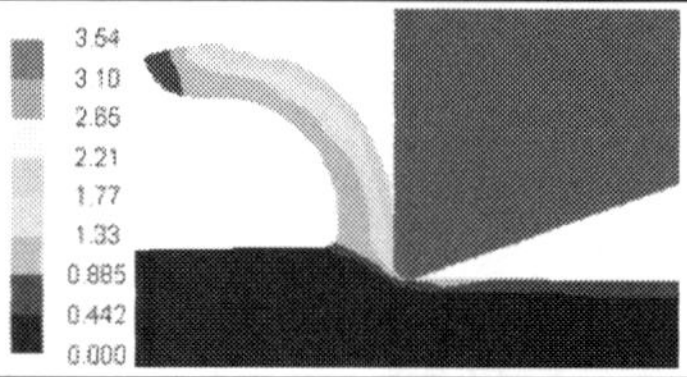
5 Y X	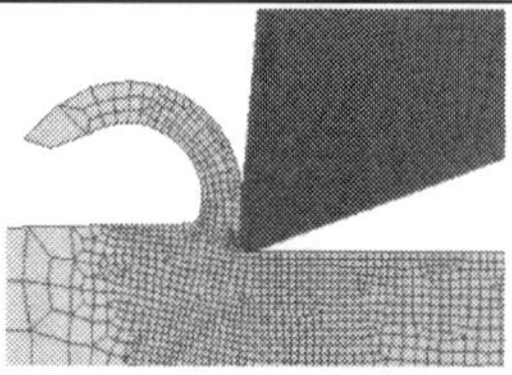	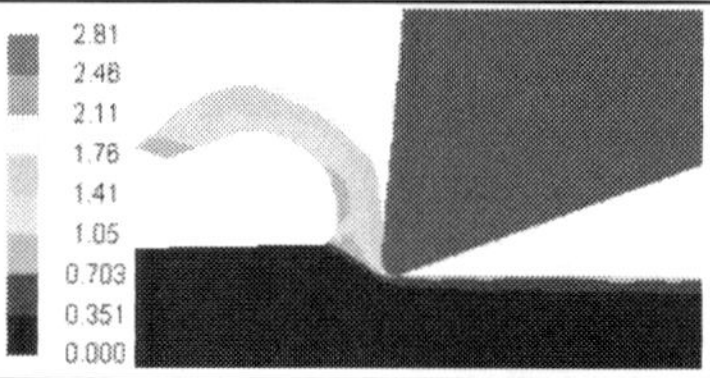
10 Y X	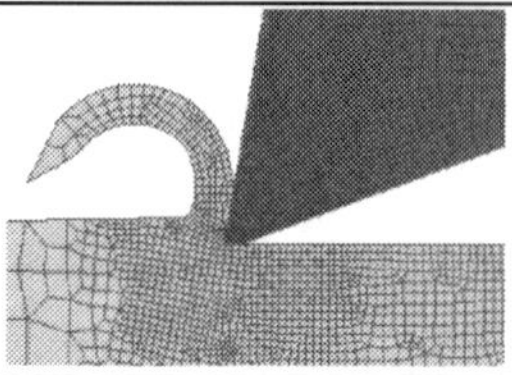	

15 Y X	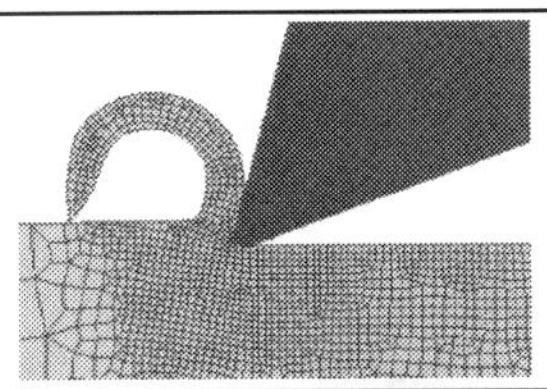	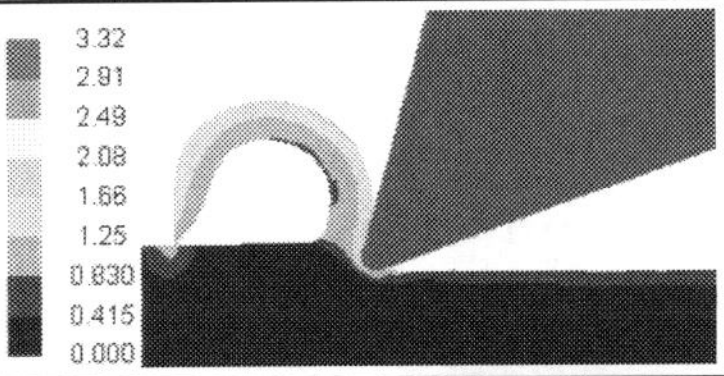
20 Y X	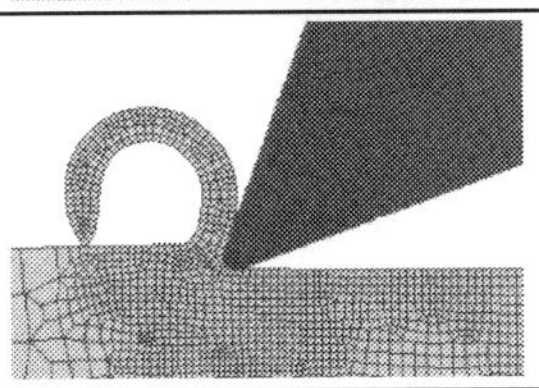	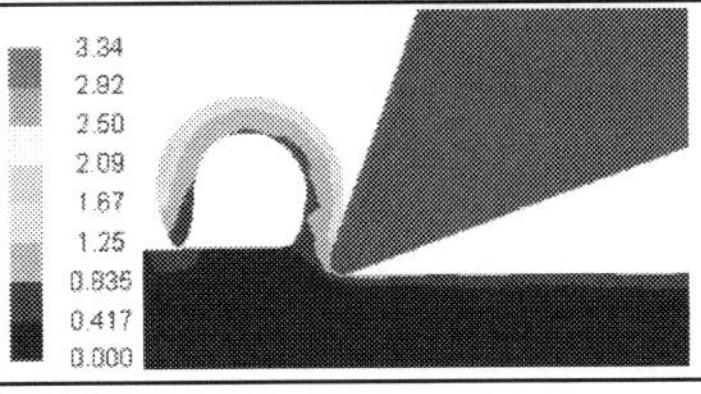

The chip formation plays an essential role in the cutting process. This mode of formation is directly influenced by the values assigned to the rake angle, as shown in the illustrations in Table 1. As the rake angle increases from the negative angle, the shear angle and forces increase, and the length of the shear plane decreases. At the same time, the resistance of the tool edge decreases, but the chip no longer tends to compression. The material is chipped in a supervised manner in each of the six situations indicated so that the existing geometry can be modified while gradually obtaining a new chip shape. This aspect is also of particular importance for what happens when the cutting tool enters the material in ordinary cutting feed and cutting speeds and, according to the manufacturing catalogs, is known.

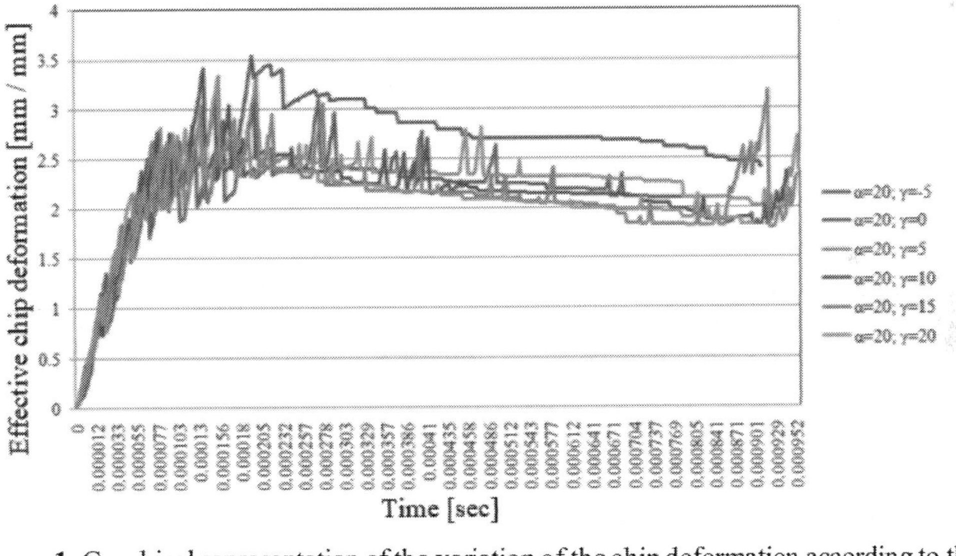

Figure 1. Graphical representation of the variation of the chip deformation according to the variation of the rake angle.

The graph in figure 1 shows the variation of the chip deformation in the six simulation variants. From this graph, it can be seen that the variation is not a constant over time, it shows oscillations in all simulations. The situation of higher temperatures in the area in front of the tool leads to increased plastic deformation during the chip formation process. As a result of the plastic deformations that occurred when the chip layer was transformed into a chip, the chip dimensions: length L_c, width b_c and thickness a_c differs in size from the corresponding dimensions of the cutting layer from which the chip was formed. The dimensions of the cutting layer involve L - the length of the tool edge, in the direction of the main cutting motion, in contact with the machined material; b – the width of the cutting layer, dimension considered on the main cutting surface and measured perpendicular to the direction of the main movement; a – thickness of the cutting layer, dimension considered in the plane perpendicular to the direction of the main movement and measured perpendicular to the main cutting surface. Under normal working conditions, when processing plastics, the chip shortens $L_c < L$, thickens $a_c > a$ and widens $b_c < b$. Therefore:

The longitudinal contraction coefficient represents:

$$k_l = \frac{L}{L_c} \tag{1}$$

The thickening coefficient of the chip represents:

$$k_a = \frac{a_c}{a} \tag{2}$$

The chip widening coefficient represents:

$$k_b = \frac{b_c}{b} \tag{3}$$

For usual working conditions (angle of inclination of the main edge, λ<30°), can be considered that $k_b = 1$. Consequently, given that the transformation of the material into the cutting process takes place without a change in volume, it appears that:

$$L \cdot a \cdot b = L_c \cdot a_c \cdot b_c \tag{4}$$

Therefore, the coefficient of plastic compression in the situation where the resulting chip is continuous is:

$$k_a = \frac{a_c}{a} = \frac{L}{L_c} \tag{5}$$

The chip's plastic compression coefficient can qualitatively characterize the degree of deformation of the chip material layer. For normal working conditions, when processing steels, the compression coefficient according varies between$k$=1,5...6. In this situation, the compression coefficient according to finite element analysis is 3. Indeed, the increase in the plastic compression coefficient indicates an increase in the degree of deformation of the chipped material.

2.2. Case 2. How the chip is formed in case of variation of the cutting tool edge radius

Input data that differ from the previous study are: cutting tool - uncoated metal carbides; cutting speed 300 rot/min; cutting depth 0.3 mm; cutting feed 0.1 rev/min; cutting edge radius of the tool varies: 0.05, 0.1, 0.2, 0.3, 0.4 mm; rake angle γ: 10°; clearance angle α: 5°; ambient temperature 20 °C. The second study is based on a comparison of how the chip is formed in the case of the variation of the sharpening radius of the tool edge.

Table 2. Simulation package and the representation of the chip morphology.

Cutting edge radius [mm]	Chip shape	Effective deformation of the chip [mm/mm]
0.05 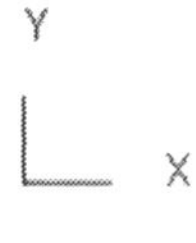		

IOP Conf. Series: Materials Science and Engineering 1235 (2022) 012019 doi:10.1088/1757-899X/1235/1/012019

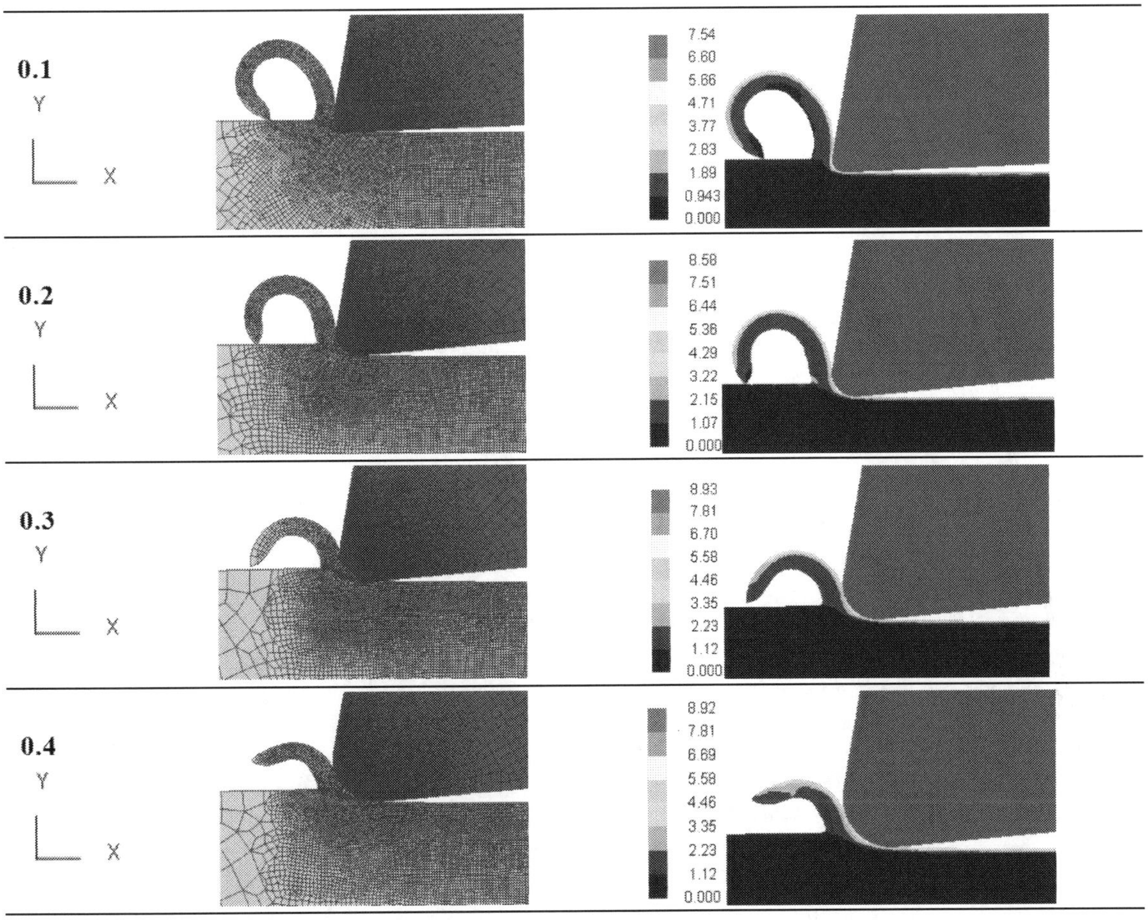

Similar to the study performed previously in table 2, the variant of the series of finite element simulations is presented in the situation of variation of the sharpening radius and the resulting chip shape and its deformation related to each simulation.

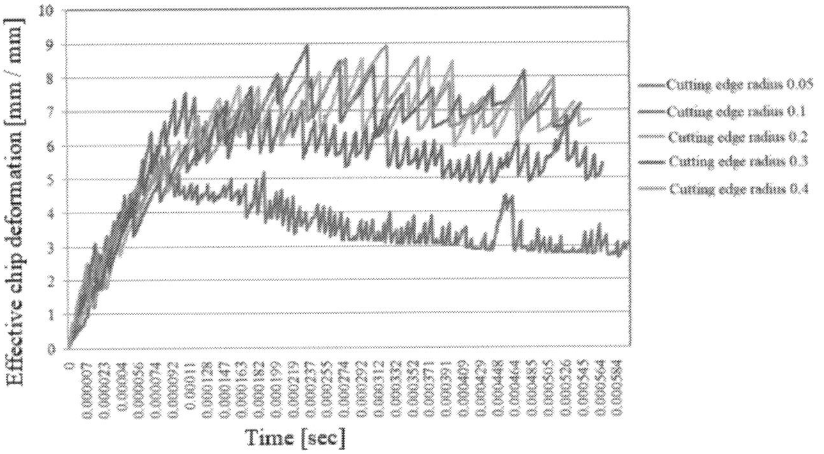

Figure 2. Graphical representation of the variation of the chip deformation according to the variation of the cutting-edge radius.

The variation with a trend to increase the sharpening radius affects the chip formation, specifically its thickness, which significantly influences its shape. In this context, if the sharpening radius changes in the upward direction, the stress distribution will be radially occupying a larger area of the workpiece. For this reason, the deformation of the chip will be smaller. Regarding the temperature distribution, that is automatically depending on the shape of the tool tip. In turn, it influences the temperature of the generated chip. If the aim is to obtain a temperature distribution on a smaller surface of the chip, it is recommended to use a smaller radius. But in this way, we obtain a more accentuated deformation of the chip.

For a greater radius, we obtain distribution of temperature over a biger surface of the chip. With the sharpening radius increasing, it can be seen how the chip deforms over time. As a result of the obtained results, it also appears that with the increase of the sharpening radius of the tool edge, the detached chip tends to settle. Figure 2 shows very clearly how the actual deformation of the chip appears - very pronounced at the beginning of the chip, after which the growths are oscillating.

3. Conclusions

The chips are formed by deforming the part material. Once the tool starts machining in the contact zone with the part, high temperatures and stresses are generated. This is much weaker at the edge of the tool. These generated stresses and temperatures they are already sufficient during cutting for the plastic deformation of the material to take place. Therefore, as it appears from the analysis of the first case, the deformations of the material appear quickly inside the cutting area - an aspect that supports the confirmation of the specialized theory. As the material reaches the cutting point, the chip leaves the material sliding into the cutting plane, pushing it out of the tool. Once the chip slides in front of the tool, the friction produced causes temperature rises in that area. The edge of the tool enters the material deforming it to obtain the chip, after the tool, the material returns due to the elastic deformation. The most suitable variant regarding the contact between the tool and the workpiece material is given by the moment of deformation, and based on this moment, the required cutting force can be determined in real-time. At the same time, it is worth mentioning the importance of determining the amount of heat dissipated in the tool. Concerning the second case, it is found that the deformation of the chip increases suddenly and tends to normalize or even decrease if the radius is 0.05 mm, a variation different from that in which the radius is 0.1 - in the sense that the latter shows a visible differentiation from the first variant. As the radius values increase from 0.2 to 0.4 mm, it is concluded that there are differences in deformations, but these are lower than those generated by processing with radii of 0.05 and 0.1 mm. At the same time, it was found that these variations - in the same time interval, are framed approximately in the same range of values of the effective deformation.

ORCID iDs

A B Pop https://orcid.org/0000-0002-4784-8485
A M Țîțu https://orcid.org/0000-0002-0054-6535

4. References

[1] Zhang K, Wang K, Liu Z and Xu X 2020 *Appl Sci* **10** (9) 3057 https://doi.org/10.3390/app10093057
[2] Qin P, Wang M and Sun L 2020 *Appl Sci* **10** (24) 9135 https://doi.org/10.3390/app10249135
[3] Liu Z, Yue C, Li X, Liu X, Liang S Y and Wang L 2020 *J Manuf Mater Proces* **4** (4) 121 https://doi.org/10.3390/jmmp4040121
[4] Papanikolaou M, Hernandez F R and Salonitis K 2020 *Metals* **10** (9) 1220 https://doi.org/10.3390/met10091220
[5] He N, Lee T C 2002 *J Mater Process Technol* **129** 101–4 https://doi.org/10.1016/S0924-0136(02)00583-6
[6] Vandana A S and Sundaram N K 2018 *Tribol Lett* **66** 94–7 https://doi.org/10.1007/s11249-018-1047-5
[7] Vavruska P, Zeman P and Stejskal M 2018 *Procedia CIRP 2018* **77** 578–81 https://doi.org/10.1016/j.procir.2018.08.216

[8] Wibowo Y T, Baskoro S Y and Manurung V A T 2018 *IOP Conf. Series Mater. Sci. Eng.* **306** 012137 https://doi.org/10.1088/1757-899X/306/1/012137
[9] Arruda T M, De Paiva A P, Brandão L C and Ferreira J R 2019 *Precis Eng* **60** 194–214 https://doi.org/10.1016/j.precisioneng.2019.07.013
[10] Rathod K B and Lalwani D I 2018 *Mater. Today Proc* **5** 6692–701 https://doi.org/10.1016/j.matpr.2017.11.326
[11] Liu H Z and Zong W 2019 *Int J Adv Manuf Technol* **100** 2689–700 https://doi.org/10.1007/s00170-018-2853-5
[12] Binder M and Klocke F 2017 *Simul Model Pract Theory* **70** 65–82 https://doi.org/10.1016/j.simpat.2016.09.001
[13] Lotfi M 2018 *Ultrasonics* 88 106–14 https://doi.org/10.1016/j.ultras.2018.03.013
[14] Attanasio A, Faini F and Outeiro J C 2017 *Proc CIRP 2017* **58** 440–4 https://doi.org/10.1016/j.procir.2017.03.249
[15] Yue C, Liu X and Liu Z 2019 *Chin J Mech Eng* **52** 199–207 https://doi:10.3390/jmmp4040121
[16] Lotfi M, Farid A A and Soleimanimehr H 2014 *Int J Adv Manuf Technol* **78** 917–25 https://doi.org/10.1007/s00170-014-6676-8
[17] Cui D, Zhang D, Wu B and Luo M 2017 *Int J Mech Sci* **131** 613–24 https://doi.org/10.1016/j.ijmecsci.2017.07.027
[18] de-Buruaga M S, Soler D, Aristimuño P X, Esnaola J A and Arrazola P J 2018 *Appl Therm Eng* **145** 305–14 https://doi.org/10.1016/j.applthermaleng.2018.09.051
[19] Möhring H C, Kushner V, Storchak M and Stehle T 2018 *CIRP Annals* **67** 61–4 https://doi.org/10.1016/j.cirp.2018.03.009
[20] Karaguzel U and Budak E 2018 *J Materi Process Technol* **262** 532–40 https://doi.org/10.1016/j.jmatprotec.2018.07.024
[21] Xie W and Fang F 2020 *Appl Surf Sci* **504** 144451 https://doi.org/10.1016/j.apsusc.2019.144451

IOP Conf. Series: Materials Science and Engineering 1235 (2022) 012020 doi:10.1088/1757-899X/1235/1/012020

Corrective profiling method for tools that generate by enwrapping

G A Moroşanu, V G Teodor and R Crăciun

"Dunărea de Jos" University of Galaţi, Faculty of Engineering, Department of Manufacturing Engineering, Galaţi, Romania

virgil.teodor@ugal.ro

Abstract. There are known methods of corrective profiling of tools, which can predict the quality of parts generating obtained with cutting tools that process by enwrapping. The purpose of this profiling is to avoid errors caused by improper profiles of generating tools, using the so-called method of "reverse generation", which is based on the reproduction of the relative movement between the tool and the part, but in which the tool profile is known, and the goal is determining the part's profile. The method presented in this paper uses the "virtual pole" method to profile the tools generating by enwrapping, by the rolling method and avoids the need to explicitly write the enveloping condition. This paper demonstrates the possibility of applying this method to the profiling of gear-shaped tools and presents the application of the proposed algorithm for a tool whose profile was determined directly by measurement. The numerical application developed demonstrates the ability of the method to simply solve the problems related to the corrective profiling of the tools that generate by rolling.

1. Introduction

In industrial practice, techniques are often used to correct the profiles of the parts obtained by cutting, especially if their generating is done by rolling [1,2].

These techniques are fundament on the so-called "reverse generation", which is based on the reproduction of the relative movement between the tool and the part, but in which the goal is determining the part's profile, starting from the knowns tool's profile. For this purpose, the enwrapping condition is determined either using fundamental theorems such as Gohman or Willis [3,4], or using complementary theorems, of which we can mention: the method of minimum distance, the method of trajectories or the method of the family of substitutive circles [5].

In this paper it is proposed to demonstrate the possibility of using the virtual pole method for the purpose of corrective profiling of gear shape cutter tools. The idea of corrective profiling starts from the possibility of predicting the quality of the generating of profiles obtained by enwrapping. In this case, the quality of the generating was considered to be the accuracy of the overlap between the theoretical profile (the one to be obtained) and the actual profile, generated by the cutting tool.

In order to be able to perform this corrective profiling, it is necessary to know the effective generating profile of the tool. This is possible if the tool is investigated by scanning with a video scanner or projector profile. In this way, the coordinates of the points belonging to the cutting tool profile are determined [6-9]. Subsequently, the reverse generating method is applied, which analytically simulates the process of generating the desired profile by rolling. Unlike the classic profiling algorithm, in reverse generating, the profile of the cutting tool is considered known and the profile of the part is determined, the real profile that will be determined by the tool.

Next, the generated real profile is compared with the theoretical one and it's fit between the tolerance limits is verified. If it is necessary to modify the profile obtained by generating, the coordinates of any points outside the permissible domain are recalculated so that those points are brought within the tolerance field.

The last step is to change the positions of the conjugate points belonging to the generating profile accordingly, to ensure its desired shape.

2. The principle of the virtual pole method

The virtual pole method [6-10] is a variant of the Willis theorem application, which eliminates the need to write the equations of relative movements between the part and the tool. The method is based on the well-known observation that, at the contact point, the profiles in reciprocally enwrapping admit a common normal which passes through the gearing pole [3,9,10].

As with other analytical profiling methods, four reference systems are used in this case: $x_o y_o z_o$; xyz - fixed reference systems; XYZ - the local reference system, solidary with the part; $\xi\eta\zeta$ - the local reference system, solidary with the tool. The two locals reference systems are mobiles.

For gear shape cutter tool generating, the positioning of the 4 reference systems is the one shown in figure 1.

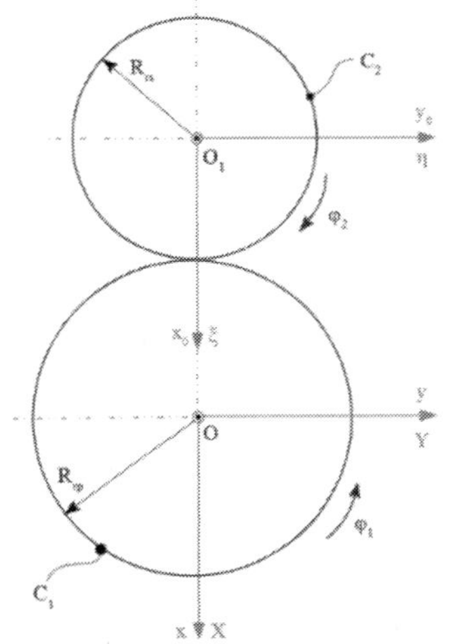

Figure 1. Generating with gear shaped cutter tool. Reference systems.

Figure 1 shows the systems of coordinates and the two centrodes: C_1, associated with the workpiece and C_2 associated with the cutter. The radii of the centrodes are: R_{rp} - the rolling radius of the part and R_{rs} - the rolling radius of the tool.

According to the method, considering a current point M whose coordinates (X_M, Y_M) are known in the reference system joined with the part, it can be determining the "virtual pole" defined as: "the intersection point between the normal to the known profile of the part and the C_1 centrode" [8].

Assuming that the parametric equations of the workpiece's profile, Σ, are:

$$\Sigma: \begin{vmatrix} X = X(u); \\ Y = Y(u); \end{vmatrix} \tag{1}$$

u being a parameter which the Σ profile, the normal to this profile at the current point M, is given by the equation:

$$\overrightarrow{N_{XY}} = \left(X_M + \lambda \cdot \dot{Y}_u\right) \cdot \vec{\imath} + \left(Y_M - \lambda \cdot \dot{X}_u\right) \cdot \vec{\jmath}. \tag{2}$$

In equation (2), λ is a scalar size that represents the modulus of the vector $\overrightarrow{N_{XY}}$.

IOP Conf. Series: Materials Science and Engineering 1235 (2022) 012020 doi:10.1088/1757-899X/1235/1/012020

As stated: "the virtual pole is the intersection between the normal $\overrightarrow{N_{XY}}$ and the C_1 centrode" [7,8]. The intersection of the vector $\overrightarrow{N_{XY}}$ with the circle of radius R_{rp} allows the elimination of the parameter λ and thus, determining the position of the virtual pole P_v in the XY system:

$$C:\left|\begin{array}{l} X=-R_{rp}\cdot\cos\varphi_1; \\ Y=R_{rp}\cdot\sin\varphi_1. \end{array}\right. \quad (3)$$

φ_1 being the rotation angle of the part.

From (2) and (3) results:

$$\begin{cases} X_M+\lambda\cdot\dot{Y}_u=-R_{rp}\cdot\cos\varphi_1; \\ Y_M-\lambda\cdot\dot{X}_u=R_{rp}\cdot\sin\varphi_1; \end{cases} \quad (4)$$

In essence, the relation (4) represents a relationship between the independent parameters u and φ_1, meaning that it is the enwrapping condition.

According to the principle of the "virtual pole" method, the point M_{XY} on the part profile will come into contact with the point $M_{\xi\eta}$ on the tool profile, when the enwrapping condition is accomplished, so when the part will rotate with the angle φ_1, given by the relation (4). At the same time, by definition, the tangency point between the tool profile and that of the part will be also on the contact line.

This observation makes it possible to determine the coordinates of the M_{xy} point in the global reference system, meaning the determination of the tangency point on the contact line.

Next, it must be considered that, for the two profiles to be in contact, the tool must move with the part. The two movements are correlated, and the connection between them is obtained from the rolling condition: $\varphi_2=i\cdot\varphi_1$.

Thus, the current point coordinates, in the tool reference system, can be determined if its movement is considered.

In this way, if the point M_{XY} pass across the entire profile Σ the coordinates of the points belonging to the tool's profile can be determined. These points will be in successive contact with the current point, so the profile of gear cutter can be determined.

3. Correction of the cutter tool

To demonstrate the practical application of the method, it is considered that the generating profile of a gear shape cutter tool was obtained by scanning in the form of a point cloud.

The coordinates of the points that form this cloud are given in matrix form:

$$S:\begin{pmatrix} \xi_1 & \xi_2 & \dots & \xi_n \\ \eta_1 & \eta_2 & \dots & \eta_n \end{pmatrix}^T. \quad (5)$$

A first step consists in checking the profile of the part obtained by generating with the cutting profile S. The current point $M_{\xi\eta}$ is considered, belonging to the point cloud and characterized by coordinates: $M_{\xi\eta}(\xi_i,\eta_i)$.

In order to be able to apply the "virtual pole" method, it is necessary the discretization of the S profile, meaning that is to be approximated by line segments constructed between the points that form the point cloud.

The parametric equation of such a segment is:

$$S_i:\left|\begin{array}{l} \xi=\xi_i+l_i\cdot\sin\alpha_i; \\ \eta=\eta_i+l_i\cdot\cos\alpha_i, \end{array}\right. \quad (6)$$

where $l_i=\sqrt{(\xi_{i-1}-\xi_i)^2+(\eta_{i-1}-\eta_i)^2}$ and $\alpha_i=\arctan\left(\frac{\xi_i-\xi_{i-1}}{\eta_i-\eta_{i-1}}\right)$.

In this case, at the current point $M_{\xi\eta}$, the normal can be defined at the segment S_i, in the form:

$$\bar{n}_i=(\xi_i+\dot{\eta}_i)\cdot\bar{i}+(\eta_i-\dot{\xi}_i)\cdot\bar{j}. \quad (7)$$

Intersecting the normal with the C_2 centrode, the coordinates of the virtual pole are obtained in the system $\xi\eta$:

$$\begin{cases} R_{rs} \cdot \cos\varphi_2 = \xi_i - \lambda_i \cdot \sin\alpha_i ; \\ R_{rs} \cdot \sin\varphi_2 = \eta_i + \lambda_i \cdot \cos\alpha_i . \end{cases} \tag{8}$$

φ_2 being the rotation angle of the tool.

From (8), results the value of the angle φ_2 for which the current point $M_{\xi\eta}$ will process:

$$\varphi_2 = \alpha_i + arccos\left(\frac{\xi_i \cdot \cos\alpha_i + \eta_i \cdot \sin\alpha_i}{R_{rs}}\right). \tag{9}$$

The coordinates of the point $M_{\xi\eta}$ in the global reference frame x_0y_0 are:

$$x_0 = \omega_3(\varphi_2) \cdot \xi \tag{10}$$

and in the xy system they will be given by:

$$x = x_0 + A. \tag{11}$$

Given the absolute movement of the part, the current point in the XY system will reach in:

$$\begin{vmatrix} X_i = x_i \cdot \cos\varphi_1 + y_i \cdot \sin\varphi_1 ; \\ Y_i = -x_i \cdot \sin\varphi_1 + y_i \cdot \cos\varphi_1 . \end{vmatrix} \tag{12}$$

Subsequently, the position of the point generated on the profile Σ, $M_i(X_i, Y_i)$, can be modified by following this algorithm in reverse order.

4. Application. Correction of a line segment type profile

It is assumed that a line segment profile is known by measurement, shown in Figure 2. It is admitted that the point M_{i-1} is the last point in the tolerance field and M_i the first point outside this field.

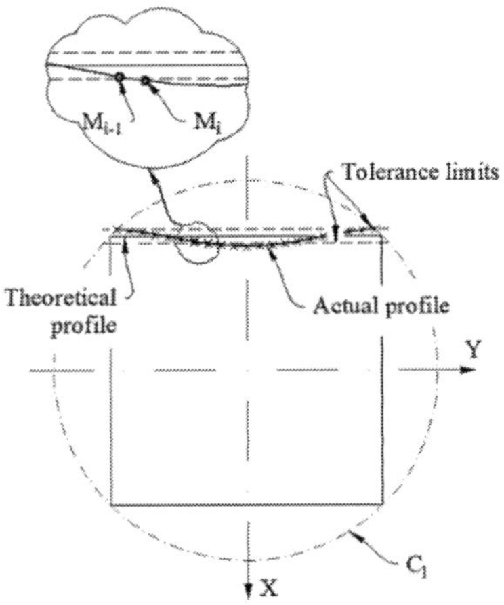

Figure 2. Discreetly known profile.

For the profile to be returned within tolerance limits, the position of each point outside these limits must be changed.

It is considered the current point M_i, of coordinates X_i = -26.324 mm and Y_i = -5.207 mm. To correct the profile, the coordinates must be changed as follows: $X_{i1} = X_i + \Delta X_i$ = -27.536 and $Y_{i1} = Y_i + \Delta Y_i$ = -4.990, R_{rp} = 40 mm, R_{rs} = 60 mm. Thus, the point M_i is replaced by the point M_{i1}.

Note: The correction values ΔX_i = -0.212 mm and ΔY_i = 0.217 mm were chosen so that the profile does not show sudden changes in curvature.

It is also admitted that the previous point, M_{i-1}, has a correct position, so it does not require changes. Therefore, normal to the segment $\Sigma_i = \overline{M_{i-1}M_{i1}}$ of equations:

$$\Sigma_i \begin{vmatrix} X = X_{i1} + l_{i1} \cdot sin\alpha_{i1}\,; \\ Y = Y_{i1} + l_{i1} \cdot cos\,\alpha_{i1,} \end{vmatrix} \tag{13}$$

becomes:

$$\bar{N}_\Sigma = \left(X_{i1} + \lambda \cdot \dot{Y}_{l_{i1}}\right) \cdot \bar{\imath} + \left(Y_{i1} - \lambda \cdot \dot{X}_{l_{i1}}\right) \cdot \bar{\jmath}. \tag{14}$$

The enwrapping condition (4) leads to:

$$\varphi_1 = -\alpha_{i1} + arccos\left(\frac{X_{i1} \cdot cos\alpha_{i1} + Y_{i1} \cdot sin\alpha_{i1}}{R_{rp}}\right). \tag{15}$$

When this point reaches the contact line, its coordinates in the fixed reference system xy will be:

$$\begin{vmatrix} x_{i1} = X_{i1} \cdot cos\,\varphi_i - Y_{i1} \cdot sin\varphi_i\,; \\ y_{i1} = X_{i1} \cdot sin\,\varphi_i + Y_{i1} \cdot cos\,\varphi_i\,, \end{vmatrix} \tag{16}$$

and in the system x_0y_0:

$$M_{x_0y_0} \begin{vmatrix} x_{0i1} = x_{i1} + A_{12}\,; \\ y_{0i1} = y_{i1}. \end{vmatrix} \tag{17}$$

Therefore, taking into account the absolute movement of the tool, its profile must be corrected so that it passes through the point:

$$T_i \begin{vmatrix} \xi_{i1} = x_{0i1} \cdot cos\,\varphi_2 - y_{0i1} \cdot sin\varphi_2\,; \\ \eta_{i1} = x_{0i1} \cdot sin\,\varphi_2 + y_{0i1} \cdot cos\,\varphi_2\,. \end{vmatrix} \tag{18}$$

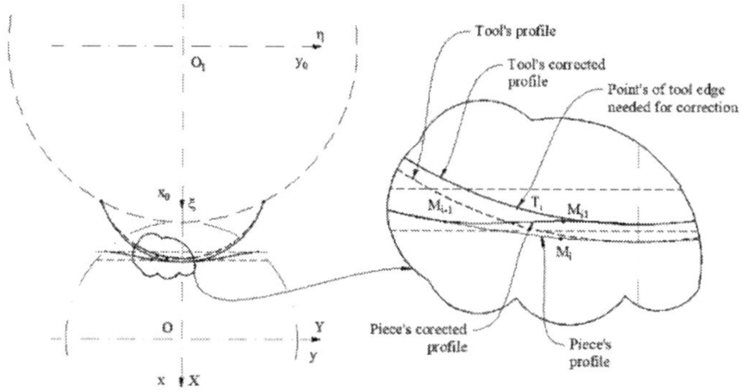

Figure 3. Profile correction.

The coordinate values of the points in the four reference systems are given in table 1.

Table 1. Contact point coordinates in the reference systems used.

Reference system	The point	The coordinates	
		X	Y
XY	M_i	-26.324	-5.207
	M_{1i}	-27.536	-4.990
xy	M_{xy}	-27.978	-0.620
x_0y_0	M_{x0y0}	72.022	-0.620
$\xi\eta$	T_i	71.563	-8.147

5. Conclusions

The paper shows the application of the "virtual pole" method in the case of profiling the gear shape cutter tool which generate by external rolling.

The method is applied to calculate the position of a corrected point on the tool profile intended to generate a line segment profile.

The numerical application developed demonstrates the ability of the method to simply solve the problems related to the corrective profiling for tools which generate by "rolling method".

The method can also be applied to other types of tools such as the rack tool or the rotary cutter tool.

Acknowledgments
This work is supported by the project ANTREPRENORDOC, in the framework of Human Resources Development Operational Programme 2014-2020, financed from the European Social Fund under the contract number 36355/23.05.2019 HRD OP /380/6/13 - SMIS Code: 123847.

ORCID iDs
G A Morosanu https://orcid.org/0000-0001-5427-8292
V G Teodor https://orcid.org/0000-0002-3764-4243
R S Craciun https://orcid.org/0000-0001-5741-3752

6. References

[1] Bordeanu M, Frumuşanu G and Oancea N 2018 *MATEC Web of Conferences* **178** 01001 IManE&E 2018 https://doi.org/10.1051/matecconf/201817801001
[2] Slătineanu L, Dodun O, Beşliu Băncesu I, Coman I, Ghionea A and Mihordea T 2018 *MATEC Web of Conferences* **178** 01006 IManE&E 2018 https://doi.org/10.1051/matecconf/201817801006
[3] Litvin F L 1994 *Gear Geometry and Applied Theory* (New York: Prentice Hall)
[4] Radzevich S P 2018 *Theory of Gearing: Kinematics, Geometry, and Synthesis* (Boca Raton, CRC Press) https://doi.org/10.1201/9780429505195
[5] Oancea N 2004 *Surfaces generation through winding* (in Romanian) vol I (Galaţi: University Foundation Press)
[6] Teodor V G, Baroiu N, Susac F and Oancea N 2016 *IOP Publishing: Materials Science and Engineering* **161** 012028 https://doi.org/10.1088/1757-899X/161/1/012028
[7] Costin G A, Teodor V G and Oancea N 2020 *SCDS-UDJG 2020* Galati
[8] Costin G A and Teodor V G 2019 *SCDS-UDJG 2019* Galati
[9] Costin G A and Teodor V G 2020 *7th Int. Conf. of Materials and Manufacturing Engineering* MATEC web of conferences vol **318** https://doi.org/10.1051/matecconf/202031801001
[10] Baroiu N, Berbinschi S, Teodor V G, Susac F and Oancea N 2017 *IOP Conf. Series: Materials Science and Engineering* **227** 012013 https://doi.org/10.1088/1757-899X/227/1/012013

IOP Conf. Series: Materials Science and Engineering 1235 (2022) 012021 doi:10.1088/1757-899X/1235/1/012021

Influence of the processing conditions of 6061 aluminium alloy thick plates on the distortion of the machined finished parts

M Stoicănescu[1], M Petre[2], G Popa[2] and N C Drăghici[2]

[1] Transilvania University of Brasov, Romania
[2] ALRO SA, 1st Milcov, Slatina, Romania

stoican.m@unitbv.ro

Abstract. The distortion in machining aluminium alloy parts is one of the serious challenges for the aluminium industry, and the residual stresses resulting in the processing steps are the main cause. Two 6061 aluminium alloy samples of parts, representing the same part with a special configuration, in two different stages of machining, were analysed to establish the influence of the processing conditions that can determine the changes in characteristics in the final machining process. Metallographic analysis, porosity determinations by the liquid penetrant inspection, tensile tests and determination of residual stresses by the Hole-Drilling method were performed. All the results obtained were correlated with the processing conditions of aluminium alloy plates. They indicated the role of the processing conditions on the residual stresses and distortion in the machined finished aluminium alloys parts.

1. Introduction

The origin of residual stresses is found in heat treatment, metalworking, welding, machining or a combination of these processes. The distortion in machining aluminium alloy parts is one of the challenging problems in the aluminium industry, and the residual stresses resulting in multi machining steps are considered the main cause [1].

In the present work, the origin of the residual stresses for the subsequently machined parts is most probably determined by the type of rolling direction: longitudinal vs transverse.

According to the literature [2], the stress level is in compression near the surface of the aluminium alloy plate compared to the interior of the plate. Also, residual stresses are different in the rolling or longitudinal direction vs the transverse direction. In addition, the absolute values for the residual stresses in the longitudinal direction are greater than those in the transverse direction. Therefore, to obtain plates with lower residual stresses, hot rolling in the transverse direction is recommended.

It is observed that the highest absolute values for compression are not reached on the plate surface, but mainly towards the center of the plate. This is due to the crystallographic orientation of the material grains, the thermal gradient along with the thickness and the usual planes of the precipitate [3,4].

In general, the typical residual stresses after quenching describe an M-shaped curve. The main negative aspect of the high residual stresses is the variation of the dimensional size and shape for aluminium alloy plates, and implicitly on the subsequently machined parts.

In this paper, two 6061 aluminium alloy samples of parts, representing the same part with a special configuration, in two different stages of machining, were analysed to establish the influence of the processing conditions on the level of the residual stresses.

The 6061 alloy is one of the most common for general purpose use. This alloy has good corrosion resistance, weldability, workability, machinability and is very commonly extruded. Part of the industrial equipment used for the research presented in this paper was recently commissioned in the ALRO plant in Slatina, Romania, respectively: *Independent equipment for the research of aluminium alloy quenching process* and *Independent equipment for the research of aluminium alloy plates ageing process*. ALRO produces a wide range of thick plates used in high technology industrial sectors. A significant part is the hot-rolled thick plates of 6061 aluminium alloy. These products are mainly used for mechanical processing by milling in parts with various complexities. Samples of parts machined by a customer, in two different phases (figures 1 – 2), were received at ALRO, to be analysed and established the processing conditions that lead to changes in characteristics in the final machining processes.

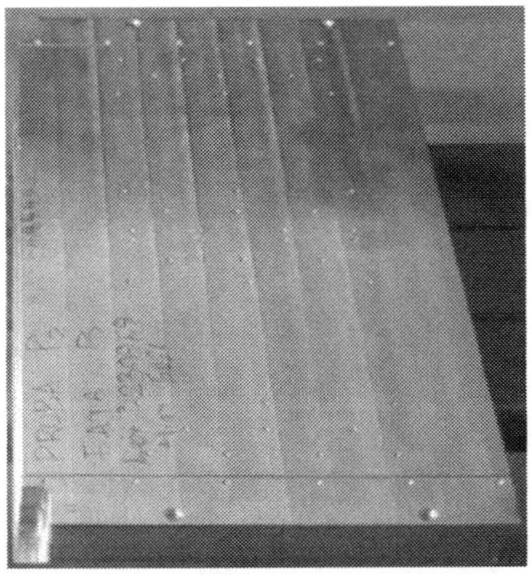

Figure 1. Sample P1 – initial milling on all sides and threaded holes.

Figure 2. Sample P2 – advanced processing phase in which deep milling was performed on one of the sides.

The two pieces with dimensions 97 x 300 x 500 mm represent the same finished product in two intermediate phases by milling processing: sample P1 (figure 1) – initial milling on all sides and threaded holes and sample P2 (figure 2) an advanced processing phase in which deep milling was performed on one of the sides, resulting in pins with 5 mm diameter.

2. Identification of the product delivered

From the data received from the customer together with the package with the two pieces, it resulted that these samples come from a plate with dimensions 100 x 2000 x 4000 mm, 6061 aluminium alloy, temper state T651. The plate was obtained from a cast slab with the chemical composition according to table 1.

Table 1. Chemical composition of 6061 aluminium alloy slab (mass fraction, %).

Si	Fe	Cu	Mn	Mg	Cr	Zn	Ti	Other Elements		Al
								Each	Total	
0.7	0.43	0.3	0.064	1.05	0.15	0.14	0.023	≤0.05	≤0.15	97.14

From the analysis of the processing data, several important aspects are identified for the present analysis, namely:

- The plate delivered had a final width of 2000 mm, which means that it was obtained by transverse rolling (rolling direction perpendicular to the direction of slab casting);

IOP Conf. Series: Materials Science and Engineering 1235 (2022) 012021 doi:10.1088/1757-899X/1235/1/012021

- In the electronic application, the value of 1.9% for the permanent elongation achieved after the stretching with the 5000tF Stretcher was registered. The value is accepted within the specification limits (1.5 - 3%);
- In the *Visual and Dimensional Control Bulletin* issued for each plate, the following values obtained were registered:
  - Longitudinal flatness deviation – 2.96 mm, compared to max. 6.94 mm;
  - Transverse flatness deviation – 1.85 mm, compared to max. 3.47 mm.

3. Results and discussion

3.1. Flatness analysis

Determination of the deviations from the flatness was performed using a metrological verified leveller and a standard gauge tool. 50 x 50 mm grids were drawn on the main sides of the 2 samples and the deviations from the flatness were measured in the two directions, longitudinal and transversal.

Sample P1, representing the part in the initial phase of milling processing did not show flatness deviations on the main sides.

Figures 3 and 4 show the values obtained at the deviation from flatness for sample P2. It can be noted that sample P2 showed large deviations from flatness (max. 1.15 mm – longitudinal and 0.45 mm – transversal).

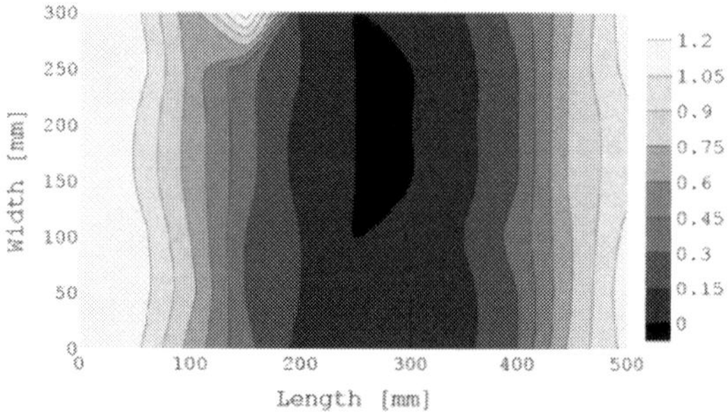

Figure 3. Sample P2 – flatness deviations on the longitudinal direction.

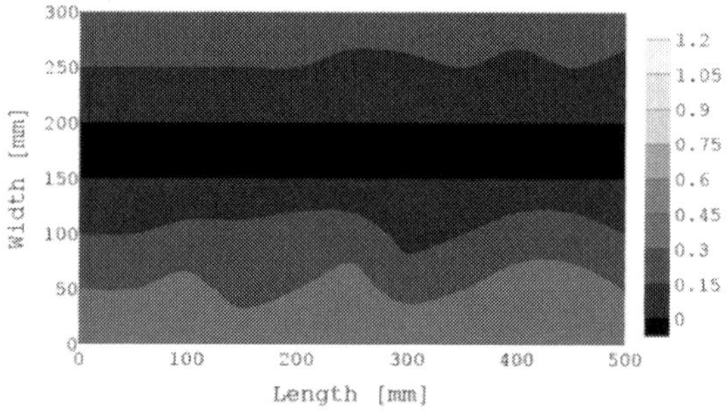

Figure 4. Sample P2 – flatness deviations on the transverse direction.

It can be supposed that, compared to sample P1, the asymmetric deep milling (on one of the sides) determined a change of flatness, with maximum differences for longitudinal flatness on the ends of the plate length, and for transverse flatness on the ends of the plate width, presumably, can be due to residual stresses in the material.

3.2. Residual stresses

The residual stresses were measured using the hole-drilling method, which is one of the most used techniques [5]. This method is standardized [6]). The residual stresses measured on the x (longitudinal direction) and y (transverse) directions and the dependence on the penetration depth are graphically represented in figures 5 – 7. It is observed that for sample P1, the recorded values for the residual stresses are generally negative (compressive stresses), and for sample P2 (in the advanced processing phase), the stresses are positive (tensile stress).

IOP Conf. Series: Materials Science and Engineering 1235 (2022) 012021 doi:10.1088/1757-899X/1235/1/012021

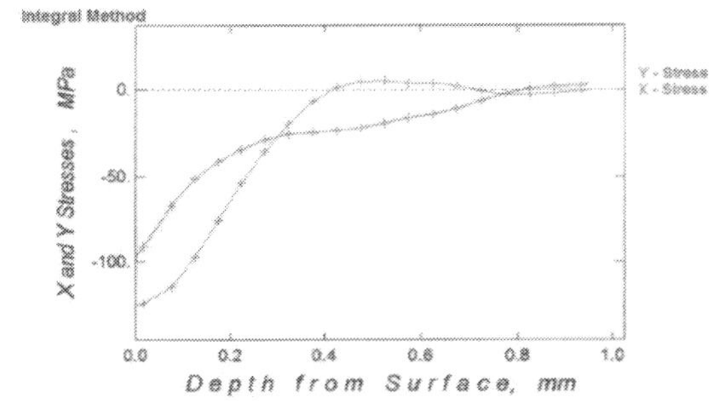

Figure 5. Integral method graph for P1 sample, bottom side.

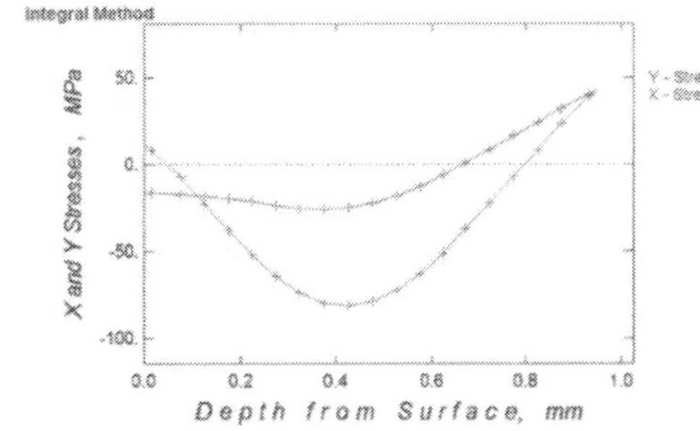

Figure 6. Integral method graph for P1 sample, top side.

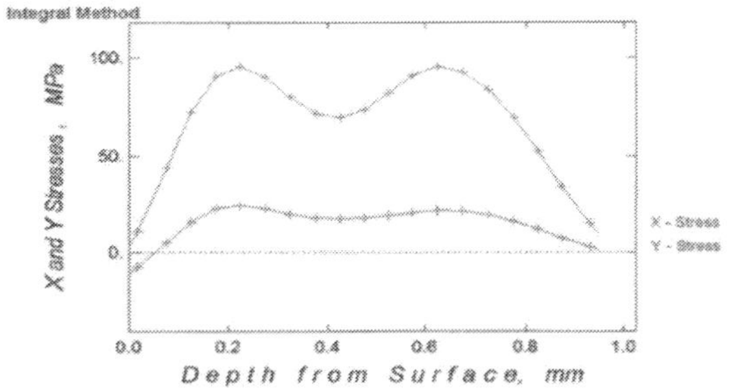

Figure 7. Integral method graph for P2 sample.

3.3. Tensile Tests

On the P1 sample, specific specimens were drawn for the tensile tests, as well as in 3 areas on the sample thickness (figure 8). The values obtained in the tensile tests are shown in figure 9. The ultimate tensile strength (Rm), the yield tensile strength ($Rp_{0.2}$) and the elongation (A_{50mm}) were determined on longitudinal (L) and transversal (T) direction.

Both in the L direction (longitudinal) and in the T direction (transversal) very different results were obtained for mechanical properties in the samples taken from the surface (L1, L3, T1, T3) – top & bottom, compared to the samples taken from the plate axis (L2 and T2) – middle, the differences being very large, approx. 85 MPa for Rm in the L direction and 45 MPa for Rm in the T direction.

Figure 8. Drawing of traction samples on sample *P2*.

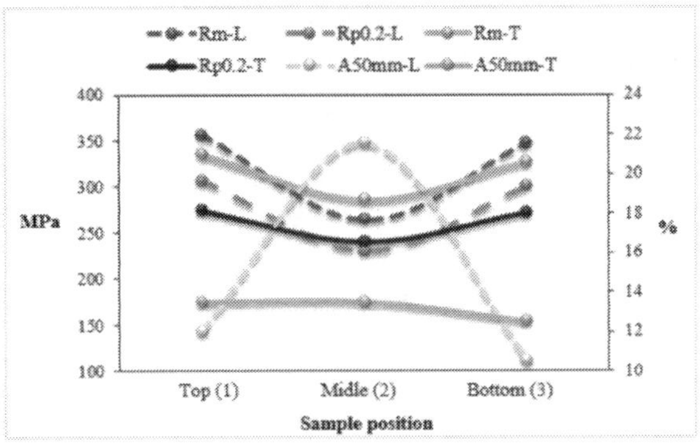

Figure 9. The evolution of the mechanical properties along with the thickness and the rolling direction.

IOP Conf. Series: Materials Science and Engineering 1235 (2022) 012021 doi:10.1088/1757-899X/1235/1/012021

Figure 10. Image with P1 samples during analysis with liquid penetrant

The registered differences denote an accentuated anisotropy of the material along with the thickness.

3.4. Liquid penetrant inspection

The porosity by the penetrant liquid inspection for the P1 sample in the two directions, L (longitudinal) and T (transverse), was determined. Figure 10 shows that the analysed samples didn't show structure defects of the porous type.

3.5. Microstructural analysis

After performing the tensile tests on specimens taken from sample P2, the ends of the specimens were kept for processing in samples for metallographic analysis, namely: the grain size, phase distribution and presence of oxide films. Determination of the average grain size was performed on samples L1, L2, T1 and T2 taken from the tensile specimens, on the longitudinal section (L-ST), samples L1 at a/4 in thickness and L2 at a/2 in thickness, and respectively on cross-section (LT-ST), samples T1 at a/4 in thickness and T2 at a/2 in thickness. Figures 11 – 14 represent the shape of the grains. The result of the determinations of the average grain size and the granulation class in which it falls according to the specific norm, ASTM E112, was class 1.5 for L1 and L2 samples; and class 1.0 for T1 and T2 samples.

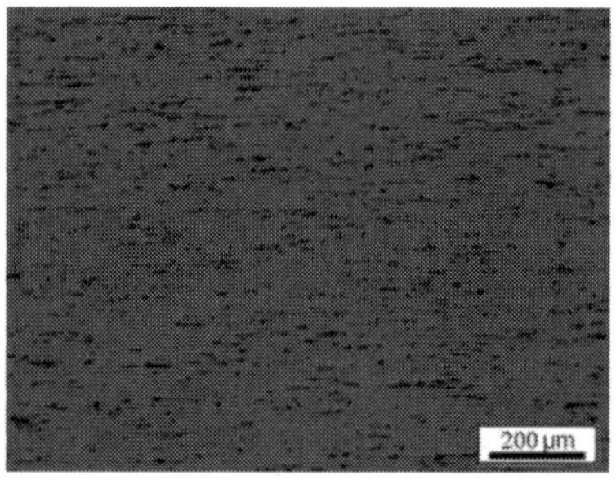

Figure 11. Grain structure for T1 sample.

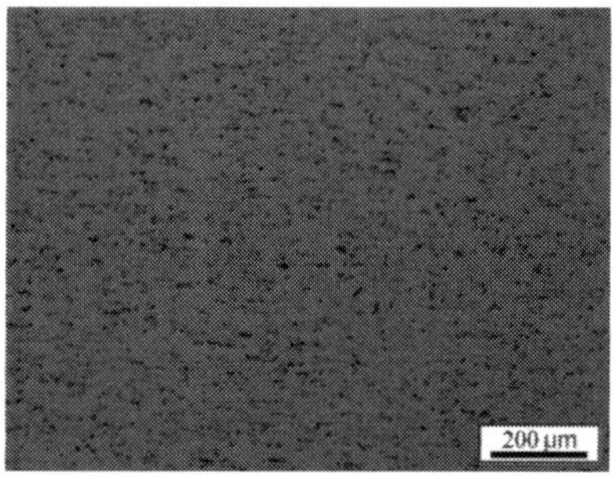

Figure 12. Grain structure for L1 sample.

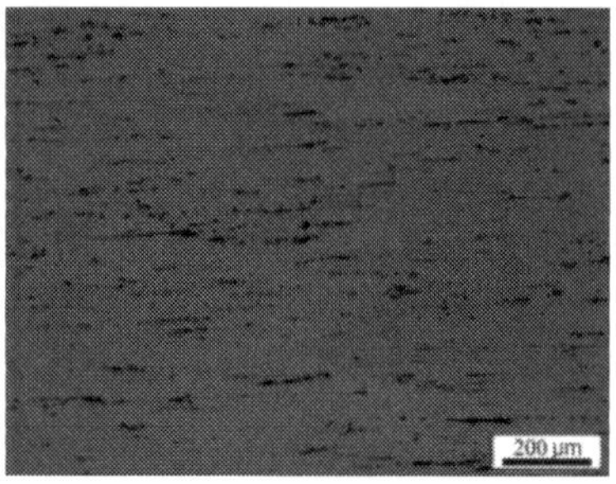

Figure 13. Grain structure for T2 sample.

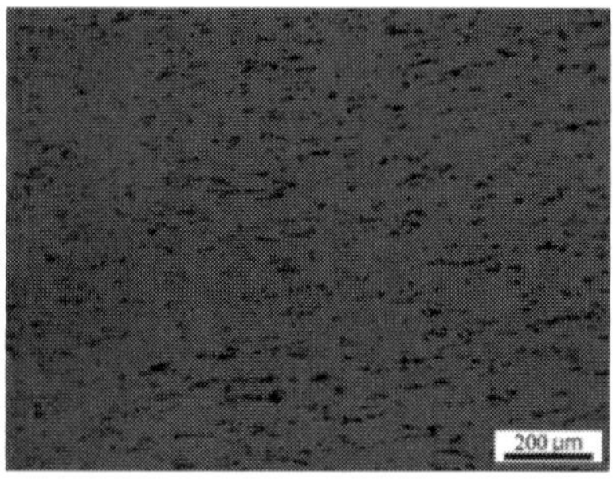

Figure 14. Grain structure for L2 sample.

The average grain sizes in the samples from the bottom of the plate side, compared to the values obtained in the samples from the plate axis (middle) were similar, respectively in the same class, according to ASTM E112. The microstructural analysis of the test specimens showed a uniform distribution of phases in the analysed section. No inclusions were identified in the analysed samples.

4. Conclusions

It was received from a customer two samples of parts with distortions, representing the same finished product, in two intermediate stages of milling to be analysed for establishing the causes of distortions. Compared to the sample P1 (the piece only cut from the plate produced by ALRO and milled superficially on all the 6 sides), in which the measurements of the deviations from the flatness were zero the deep asymmetric milling of the sample P2 (on one of the sides) determined a change of the flatness with maximum differences of 1.15 mm in the corners of the side. Determination of the residual stresses highlighted that the sample P2, an advanced processing phase of sample P1, showed deviations from flatness, as well as changes in the level of residual stresses. Liquid penetrant inspection for porosity determination on specific samples taken from sample P1 highlighted that the analysed samples did not show defects of porous type. The metallographic analysis showed normal differences between the average grain sizes in the samples from the plate side, compared to the values obtained in the samples from the plate axis, a uniform distribution of phases in the analysed section, without identifying inclusions. Analyses and determinations performed on the samples received did not highlight the causes of the distortions. The conditions for obtaining by hot rolling of the 6061-alloy plate, namely, large final width by transverse rolling, with smaller reductions per pass than longitudinal rolling, as well as the advanced complexity of the part processed by milling, proved to be favourable conditions for obtaining a product which, after the final mechanical processing, can lead to dimensional non-conformities, the possible cause being a certain degree of residual stresses of the material before processing.

Acknowledgments

Part of the cost of the industrial equipment used to obtain the results presented in this work was funded by European Union through Competitiveness Operational Programme, Priority Axis 1 Research, Technological Development and Innovation, within the project "*Investments in the R&D Department of ALRO aiming at improving the research infrastructure for the aluminum alloy heat treated plates with high qualification industrial applications*", based on the Funding Contract no. 42/05.09.2016.

ORCID iDs

M Stoicanescu https://orcid.org/0000-0003-3138-1830
M Petre https://orcid.org/0000-0002-1414-588X
G Popa https://orcid.org/0000-0002-4968-4170
N C Drăghici https://orcid.org/0000-0001-7987-1412

5. References

[1] Li J and Wang S 2017 *Int J Adv Manuf Technol* **89** 997–1012 https://doi.org/10.1007/s00170-016-9066-6
[2] Rashed H M M A 2018 *Fundamentals of Aluminium Metallurgy, Recent Advances* Control of Distortion in Aluminium Heat Treatment (Woodhead Publishing Series in Metals and Surface Engineering) chapter 13 pp 495–524 https://doi.org/10.1016/B978-0-08-102063-0.00013-8
[3] Prime M B and Hill M R 2002 *Scripta Materialia* **46** (1) 77–82 https://doi.org/10.1016/S1359-6462(01)01201-5
[4] Chakrabarti D J, Weiland H, Cheney B A, Staley J T 1996 *Through thickness property variations in 7050 plate,* Aluminium Alloys—ICAA5 Trans Tech Publications p 1085–90 https://doi.org/10.4028/www.scientific.net/MSF.217-222.1085
[5] Schajer G J 2013 *Practical Residual Stress Measurement Methods* John Wiley & Sons, Ltd. https://doi.org/10.1002/9781118402832
[6] ASTM E837 – 13a, Standard Test Method for Determining Residual Stresses by the Hole-Drilling Strain-Gage Method, ASTM International West Conshohocken PA, 2013 www.astm.org

IOP Conf. Series: Materials Science and Engineering 1235 (2022) 012022 doi:10.1088/1757-899X/1235/1/012022

Establishing the machining process for complex aluminum alloy part made from plate raw material

Ş A Moldovan, M Banica and V Năsui

Technical University of Cluj-Napoca, North University Center of Baia Mare, Romania.

moldovan.stefanadrian@gmail.com

Abstract. This paper presents a technical solution to establish the machining process for complex structural products made from aluminum alloy plates used in the aerospace industry. Those products have a very tight tolerance for geometrical deviations and positioning characteristics. To achieve a good product and ensure the machining process, repeatability is necessary to solve two issues: first, the temperature dilatation impact, and second the raw material behavior. To solve these problems, several tests were performed for products used in the aerospace industry. For deviations identification, all products were measured using a coordinate measuring machine, and the resulting report data was used for statistical analysis. Using this information, appropriate processing strategies were established. Through several tests and recording all dimensions changes during the milling process, good products were obtained, and it was ensured repeatability of the machining process.

1. Introduction

The aerospace industry has one of the most critical quality perspectives for the products made in this domain. The product quality needs to be maintained by creating a robust machining process to ensure production repeatability. All the aircraft components represent a part of an assembly; those are connected between them and define the resistant structure. Figure 1 represents the geometrical model of the aircraft fuselage [1].

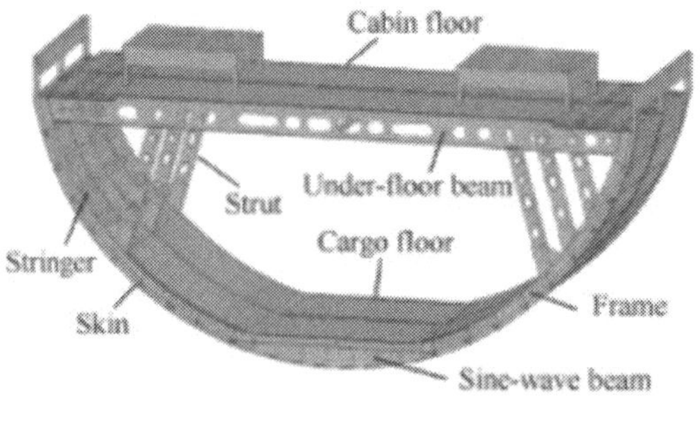

Figure 1. Geometrical model of the aircraft fuselage.

This article presents a solution for the machining strategies to avoid a scrap of the machined parts due to the geometrical deviations. The research is performed on the floor beams parts shown in figure 2, which are made from aluminum alloy plates and part of the structural aircraft. The geometrical deviations were identified through several machined tests. Analyzing the measurements dates was established that the machining strategy needs to be changed. All dates were collected using the C.M.M. (coordinate measuring machine), where have been measured all the parts machined. The aim of this research is to control the raw material behavior and combat the temperature dilatation impact.

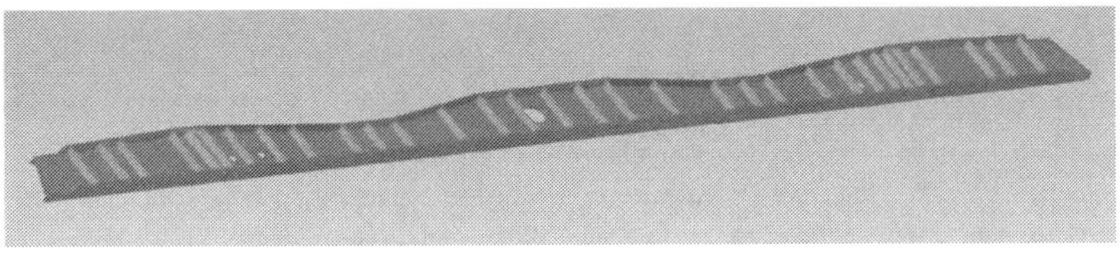

Figure 2. Floor beam part geometry.

2. Defining the machining process influence

All parts are machined on a specific clamping device, created to be suitable to constrain all the customer requirements on a 5 axis C.N.C. (computer numerical control) machine Bavius PBZ HD 600 shown in figure 3. Bavius PBZ HD is a profile machining center that offers high material removal rates for cutting aluminum profiles and solid material and stable and precise machining [3].

Figure 3. Bavius PBZ HD 600.

The research was developed using raw material laminated plates made from aluminum alloy 7050, T7451 temper, with 5160 mm length shown in figure 4. The produced parts have thin walls, and the section has a high resistance to shocks and breakages used to have the right balance for strength on the fuselage. The geometrical tolerances are very tight ± 0.254 mm from the part datum, which is the middle rib.

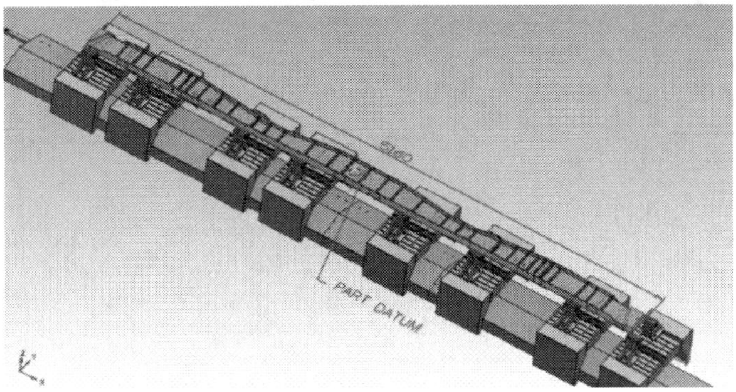

Figure 4. Machined part.

3. Machining process results

After the first four pieces, machined in the same condition, the geometric deviations were measured in C.M.M. In table 1 is shown the statistical analysis of the ribs' position to the middle rib.

Table 1. First machining process results.

Rib position on X-axis [mm]	P1	P2	P3	P4
2289.05	-0.453	-0.448	-0.453	-0.453
2142.87	-0.375	-0.358	-0.363	-0.363
2047.62	-0.374	-0.336	-0.377	-0.374
1791.91	-0.362	-0.351	-0.365	-0.356
1493.65	-0.318	-0.349	-0.267	-0.310
1388.75	-0.278	-0.278	-0.270	-0.272
1241.68	-0.259	-0.309	-0.263	-0.245
1025.4	-0.184	-0.149	-0.157	-0.195
904.75	-0.111	-0.147	-0.147	-0.154
769.87	-0.110	-0.113	-0.110	-0.118
514.22	-0.078	-0.155	-0.075	-0.197
307.85	-0.051	-0.056	-0.150	-0.048
168.15	-0.053	-0.061	-0.063	-0.056
-165.1	-0.035	-0.087	-0.144	-0.037
-304.8	-0.131	-0.098	-0.173	-0.192
-511.17	-0.153	-0.110	-0.105	-0.113
-766.83	-0.134	-0.173	-0.113	-0.158
-901.7	-0.170	-0.173	-0.170	-0.185
-1022.35	-0.181	-0.178	-0.178	-0.183
-1238.63	-0.229	-0.229	-0.227	-0.146
-1385.7	-0.256	-0.264	-0.259	-0.256
-1533.53	-0.300	-0.295	-0.150	-0.302
-1788.86	-0.228	-0.311	-0.238	-0.306
-2044.57	-0.303	-0.292	-0.222	-0.355
-2139.82	-0.386	-0.318	-0.224	-0.394

In figure 5 is a visible vectorial trend generated by the measured deviations by C.M.M. To identify the influence factor of the process, the Ishikawa methodology was used. These were identified as main factors: temperature influence and raw material behavior due to the machining strategy.

For an accurate corrective action, it was determined to eliminate the factors one by one. The first factor chosen to be eliminated the temperature influence was. We used the temperature compensation function incorporated in the Bavius PBZ HD machine [4]. This function compensates for the difference between normal temperature (20 °C) and measured temperature by using the temperature sensors applied on the machine frame [2].

Using the temperature compensation was a visible improvement, but the values are near to the limits. Going forward to the second factor defined by Ishikawa methodology, the machining strategy needs to be changed to eliminate the material behavior. The previous strategy applied was a simple one starting from one side of the part and finishing in the opposite direction. The strategy is changed and applied to the intermediate sequences, one pocket from the left end - one pocket from the right end, till the middle of the part. The strategy is shown in figure 7.

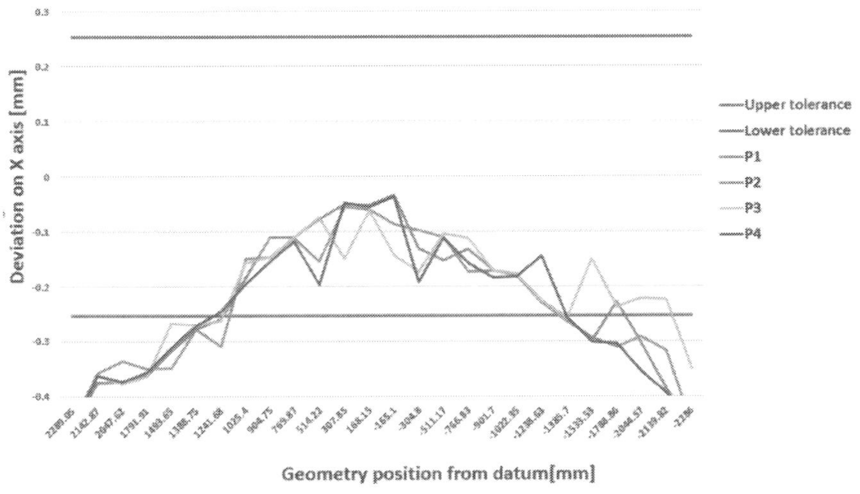

Figure 5. First machining approach.

After this implementation was machined, the second batch of four pieces, the measurement results are visible in figure 6.

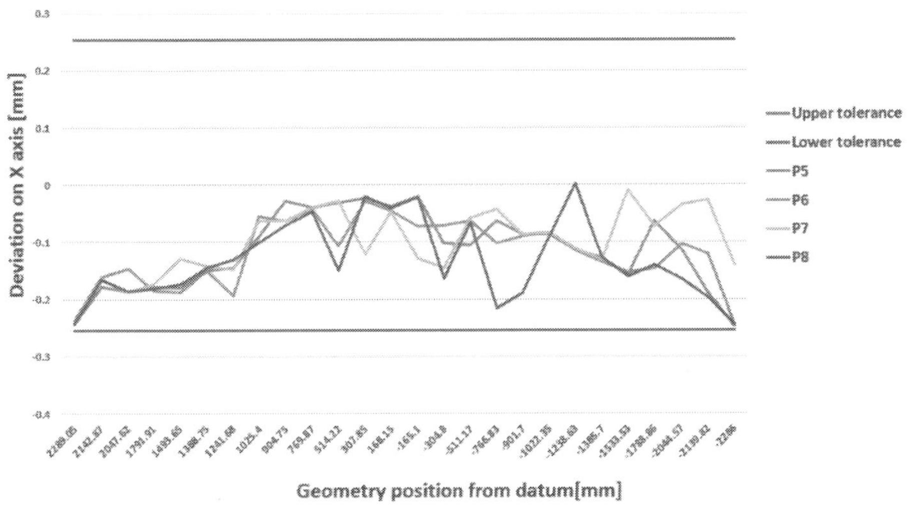

Figure 6. Second machining approach with temperature compensation.

Figure 7. The new machining strategy.

After the third batch of four pieces machined with the new C.A.M. (computer-aided manufacturing) strategy, the deviations trend become closer to the nominal value. In figure 8, it is visible that the raw material behavior is eliminated.

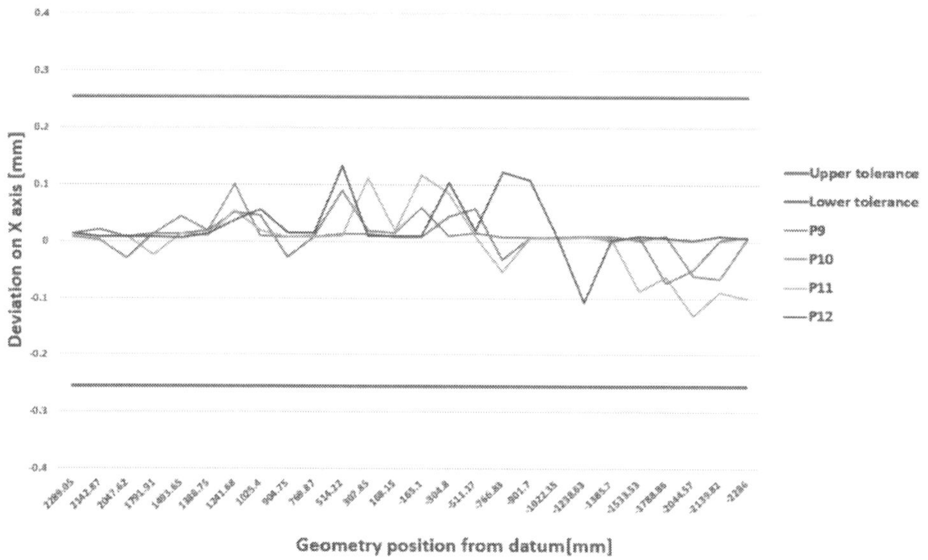

Figure 8. Third machining approach with temperature compensation and updated C.A.M. strategy.

4. Conclusion

This research defined a new methodology to compensate for the temperature influence and combat the raw material behavior to ensure repeatability of the machining process.
The future development is to apply this solution to the different alloys and temper by obtaining the specific characteristics of the raw material used in the machining process.

Acknowledgments

This research could not be completed without the machining department from S.C. Universal Alloy Corporation Europe S.R.L.

ORCID IDs

Ş A Moldovan https://orcid.org/0000-0003-0768-1366
M Banica https://orcid.org/0000-0003-3232-7043
V Nasui https://orcid.org/0000-0001-5371-1576

5. References

[1] Yiru R, Jinwu X, Jianqiang Z and Zhangping L 2016 Crashworthiness analysis of aircraft fuselage with sine-wave beam structure *Chinese J. Aeronaut.* **29** (2) 403-10 https://doi.org/10.1016/j.cja.2016.02.002

[2] Original Instruction Manual 2021 *Training Guide Bavius machine centre PBZ HD* (Baia Mare: UACE) Chapter 6 pp 120

[3] Definition of C.N.C. machine capability PBZ HD https://www.bavius-technologie.com/en/products/profile-machining-centers-pbz/pbz-hd/ Accessed 17.03.2021

[4] Function Manual 2016 *SINUMERIK 840D sl / 828D Extended Functions* (Baia Mare: UACE) chapter 4.2 pp 224-227

IOP Conf. Series: Materials Science and Engineering 1235 (2022) 012023 doi:10.1088/1757-899X/1235/1/012023

Research of the dependence of geometric parameters of holes on electroerosive grinding modes with a changing polarity of electrodes

R Strelchuk[1], O Shelkovyi [1], Y Gutsalenko[1], C Iancu[2], V Subbotina[1], S Knyazev[1] and O Volkov[1]

[1] National Technical University "Kharkiv Polytechnic Institute" Kirpichova Str. 2, Kharkiv, Ukraine, 61002
[2] "Constantin Brancusi" University of Targu Jiu, Republic Blvd 1, Targu Jiu, Romania, 210152

r.m.strelchuk@gmail.com

Abstract. In this paper are presented the results of electrical discharge diamond grinding with with difference polarities for processing various materials and calculates geometric parameters of erosion holes. The purpose of the study is to determine the volume of holes during electrical discharge diamond grinding using the mathematical method of planning experiments. The experimental plans is a combination of statistical theory and mathematical model, which is very useful in the modelling and analysing the technical problems. As a result of experimental studies, a mathematical model of the volume of holes from electrical modes of the process of electrical discharge diamond grinding was built. The mathematical formula of geometric parameters of holes obtained can be used to determine the volume of holes for different electrical quantities of discharge diamond grinding. It is found that the most significant factors that affect the geometric parameters of holes are the voltage of the igniting pulse and the pulse duration. The dependence of electrical parameters on the hybrid grinding process is investigated. There is a certain volume of holes, at which the discharge acts most efficiently, and the process performance is the most optimal. During the operation of the pulse generator, it is preferable to maintain a certain value of geometric parameters of holes, which results in an unhindered release of metal, since the effect of electrical erosion increases.

1. Introduction

The process of electroerosive grinding is improved due to the generation of electric discharges in the cutting zone, which affect the processed material and the working surface of the diamond wheel on the conductive bundle [1]. This helps to increase processing performance and maintain a high cutting capacity of the tool [2]. The process of electrical discharge diamond grinding with with difference polarities under various electrical conditions has not been studied, and the cutting zone of hybrid processing has not been studied. To reveal the physical essence of the process of electroerosive diamond grinding, in which the overall impact consists of individual phenomena occurring in micro volumes in a fraction of a second (diamond grinding, electroerosive processing), it is necessary to investigate the influence of electrical modes on the geometric parameters of erosion holes. In the case of diamond grinding, it is necessary to study the micro-cutting with a separate grain [3]. Transferring the results obtained in these studies to the real process of mass micro-cutting and scratching, one can get an overall picture of the phenomena that occur during diamond grinding. Similar methods apply in the study of electroerosive processes [4].

In the process of electroerosive grinding, electric discharges occur in the cutting zone during the breakdown of the interelectrode medium between the chips or metal surface and, when the chips slide along the bundle, discharges initiated by the explosive electroerosion of bridging elements, as well as discharges associated with the rupture of the electrical circuit in places where the chips are separated from the processed material at the end of micro-cutting [5]. All these electroerosive processes are associated with electrical discharges, which can have different conditions of initiation and formation and differ in their nature, parameters, and intensity of action on the electrodes [6].

When investigating the dependence of electrical modes on the hybrid grinding process, it was found that for specific conditions, there is an optimal value of interelectrode gap, at which the discharge action is most efficient and the removal of material from the electrodes is the greatest. As the gap increases or decreases, the volume of holes decreases. Therefore, when working with the interelectrode gap, it is necessary to take into account the effect of the discharge on the bundle – the necessary removal shall be provided, and to reduce the effect of electroerosive processes on the treated surface, where the requirements for the quality of processing can be a limiting factor [7].

By studying the phenomena accompanying single electrical discharges, it is possible to describe the integral process of electroerosion treatment, provided that the additivity of the process is preserved. Electroerosive shaping is the result of the successive appearance of a large number of holes on the treated surface resulting from a single electric discharge.

2. Research Methodology

To determine the influence of electrical modes on geometric parameters of holes, we carried out a physical simulation of the process on a special installation (Figure 1) consisting of a mechanical device for conducting discharges (position 1) and an attachment (position 2) that allows selecting a single discharge when massive pulses arrive at the interelectrode gap.

The graphite electrode, a spherical tool with a radius of 0.5 mm, and the test sample are installed in the appropriate holders, electrically isolated from the device parts. The sample is vertically moved by a coarse feed mechanism, which is a screw pair.

The precise feed mechanism, fixed in the upper bracket, ensures the movement of the graphite electrode through a special nut with a fine thread and the installation of the necessary interelectrode gap. The value of the latter is determined by an hour-type indicator with a measurement unit of 1 μm. The gap is counted from the short-circuit position of electrodes.

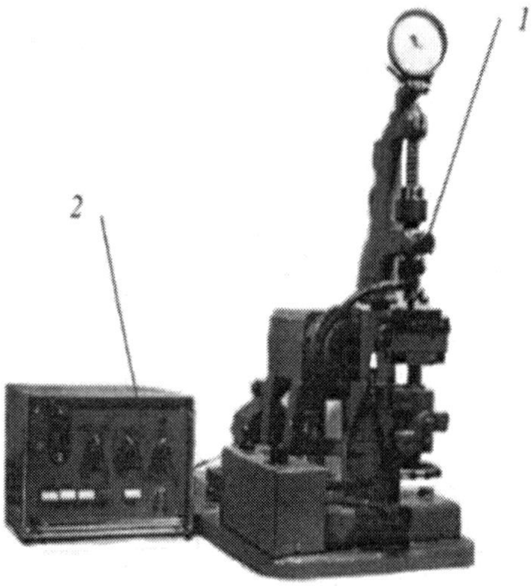

Figure 1. General view of a special installation for conducting single discharges:

To determine the short-circuit position, a device is applied to the pulse generator, the functional diagram of which is shown in Figure 2.

At the breakdown of the interelectrode gap, the pulse of the discharge current causes a voltage drop on the non-inductive shunt of the ammeter of a special pulse generator. This voltage is applied to the device input. The first oscillation of the amplified signal triggers a single-shot multivibrator, which forms a rectangular pulse.

The latter enters the differentiating circuit, where it is converted into two pointed pulses of opposite polarity. The positive pulse is "cut off" by the diode, and the remaining pulse of negative polarity goes to the separation stage and controls the operation of the blocking trigger. In one stable state, the latter does not interfere with the normal operation of the master generator of the pulse voltage source.

The complex of experimental studies included the study of the dependence of geometric parameters of erosion holes on electrical characteristics of the process. The high-resolution profilography method was used to determine the diameter *d*, mm, and depth *h*, mm, of the erosion hole (Figure 3a). The profilograms were recorded on the Hommelwerke T8000 device in two mutually perpendicular directions.

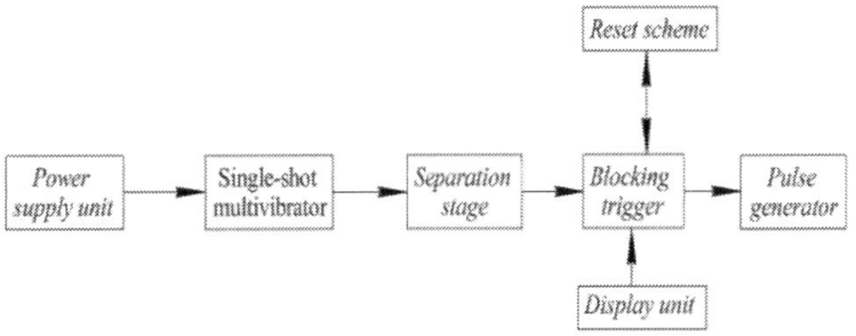

Figure 2. Functional diagram of the device to the pulse generator for the allocation of single discharges.

When a pulse is received, the trigger is transferred to another stable state and the output of the master generator is shunted. In this case, the pulse voltage does not enter the interelectrode gap and discharges stop. The discharge path is monitored visually by the LED indicator. The reset scheme is used to bring the blocking trigger to its original state. The attachment is connected to the pulse generator using a shielded cable. The high-resolution profilography method was used to determine the diameter *d*, mm, and depth *h*, mm, of the erosion hole. The profilograms were recorded on the Hommelwerke T8000 device in two mutually perpendicular directions.

The surface profilogram was used to measure the diameter d and depth h of the erosion hole (Figure 3b) The complex of experimental studies included the study of the dependence of geometric parameters of erosion holes on electrical characteristics of the process.

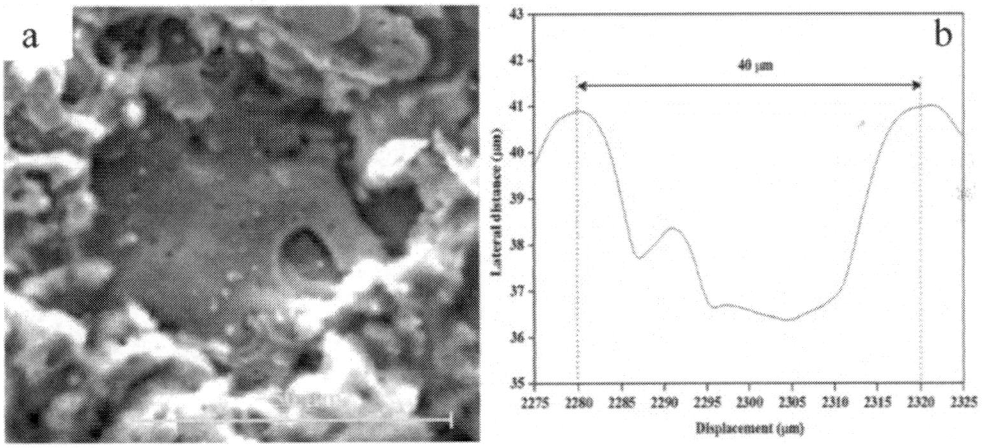

Figure 3. SEM erosion hole (a), hole surface profilogram (b)

The studies were carried out on special samples having the shape of a rectangular parallelepiped with dimensions of 14×14×5 mm, made of a hard alloy of the *VK6* brand (Tungsten carbide *WC* 94%, Cobalt Co 6%). The voltage – *U*, V; the maximum electric discharge current– *I*, A; the pulse duration – *t*, sec., were selected as independent variables (factors) during the experiment. Throughout the experiment, each parameter received 3 levels of coding values, the actual value of each parameter at the coding levels as shown in Table 1. The optimization parameter was the volume of the hole *V*, mm^3, calculated by the formula (volume of the sphere):

$$V = \frac{\pi}{6} d^2 h \tag{1}$$

The functional dependence of the hole volume on the studied factors was represented as a regression equation. To set up the experiment, a complete factor plan of type 2^3 was implemented in three randomized series, which made it possible to independently assess the influence of each factor [8]. Experimental matrix of 8 experiments is shown in Table 2.

The estimates of regression coefficients were determined by the formula:

$$b_i = \frac{\sum_{i=1}^{N} x_{iu} \bar{y}}{N} \tag{2}$$

where x_{iu} – the value of x_i factor in the uth experiment; $\bar{y}$ – the average value of the optimization parameter in the same experiment; N – the number of experiments.

Table 1. Value of input parameters at the coding levels

Parameter	U, V	I, A	t, s
Value at the coding level	X_1	X_2	X_3
1	45	22	7
–1	5	2	1
0	25	12	4

The significance of coefficients was estimated by comparing the absolute value of the coefficient with the value of the confidence interval. The analysis showed that coefficients b_0, b_1, b_2, b_3 can be considered significant with a 95% confidence probability [9]. The regression coefficients for paired interactions and higher-order interactions are statistically insignificant. Thus, the regression equation is as follows:

$$V = 62.84 - 2.14U - 4.03I - 8.16t \tag{3}$$

Table 2. Experimental matrix

experiment No.	U, V	I, A	t, s	V, mm^3
1	5	2	1	16.9
2	45	2	1	35.5
3	5	22	1	84.3
4	45	22	1	140.2
5	5	2	7	89.1
6	45	2	7	141.6
7	5	22	7	428.6
8	45	22	7	765.8

The adequacy of the model, that is, its ability to predict the actual response values with the required accuracy, was tested by the Fisher criterion:

$$F = \frac{S_{ad}^2}{S_{\{y\}}^2} \tag{4}$$

where $S_{\{y\}}^2$ – the variance of the experiment reproducibility; S_{ad}^2 – the residual variance or the variance of adequacy. The calculated value of the Fisher 5.89 criterion was less than the table value at the 5% significance level, so the resulting model is adequate [10].

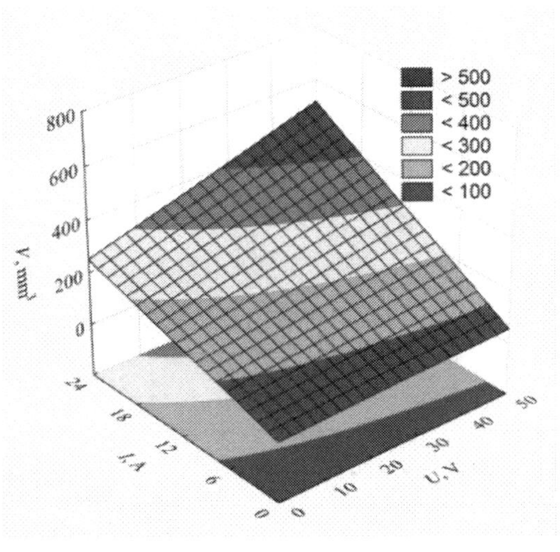

Figure 4. Response surface: V – (U, I). Calculation conditions: t = 4 s

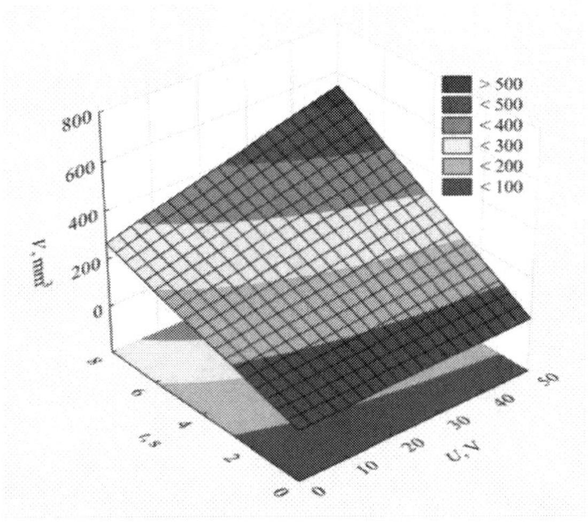

Figure 5. Response surface: V – (U, t). Calculation conditions: I = 12 A

3. Results

Figures 4-5 show a graphical interpretation of the resulting mathematical model after experimental studies. The analysis of the obtained dependences shows that the volume of the hole increases (Figure 4) with the increase in the voltage and current amplitude. This is due to the expansion of the discharge channel, since the diameter of the hole in the first approximation is equal to the diameter of the through-conduction channel at the end of the discharge. With the increase in the voltage and the amplitude of the discharge current, the average power introduced into the erosion gap increases, and the density of the heat flow increases, which, in turn, leads to the increase in the depth of the hole and the volume of the bundle removed by a single discharge.

An increase in the pulse duration (Figure 5), all other things being equal, leads to an expansion of the discharge channel and, as a result, to an increase in the diameter of the hole. This causes a slight decrease in the heat flux density q. At the same time, the thermal energy per unit surface, defined as qt, increases. This leads to an increase in the thickness of the molten layer of the processed material and, accordingly, to an increase in both the depth and volume of the hole.

4. Conclusions

The paper studies the process of of electrical discharge diamond grinding with difference polarities in the cutting zone for processing various materials and calculates geometric parameters of the erosion hole. As a result of the study, geometric parameters of the hole during electroerosive grinding were determined using the mathematical method of planning experiments. The experimental plans is a combination of statistical theory and mathematical model, which is very useful in the modelling and analysing the technical problems. As a result of experimental studies, a mathematical model of the volume of holes from electrical modes of the process of electrical discharge diamond grinding was built. The mathematical formula of geometric parameters of holes obtained can be used to determine the volume of holes for different electrical quantities of discharge diamond grinding. It is established that the most significant factors that affect the geometric parameters of holes are the voltage of the igniting pulse and the pulse duration.

The dependence of electrical parameters on the hybrid grinding process is investigated. There is a certain volume of holes, at which the discharge acts most efficiently, and the process performance is the most optimal. During the operation of the pulse generator, it is preferable to maintain a certain value of geometric parameters of holes, which results in an unhindered release of metal, since the effect of electrical erosion increases.

ORCID IDs

R Strelchuk https://orcid.org/0000-0002-7221-031X
O Shelkovyi https://orcid.org/0000-0002-7414-4854
Yu Gutsalenko https://orcid.org/0000-0003-4701-6504
C Iancu https://orcid.org/0000-0003-1178-0578
V Subbotina https://orcid.org/0000-0002-4497-4419
S Knyazev https://orcid.org/0000-0001-6422-3658
O Volkov https://orcid.org/0000-0001-8797-0322

5. References

[1] Sharma R, Gupta A, Vates U K and Singh G K 2019 Electrical discharge diamond grinding (EDDG): A review Advances in Engineering Design *Lecture Notes in Mechanical Engineering* (Pleiades Publishing) 523–33 https://doi.org/10.1007/978-981-13-6469-3_49

[2] Choudhary S, Doon R and Jha S K 2019 Prediction of the material removal rate and surface roughness in electrical discharge diamond grinding using best-suited artificial neural network training technique *Adv. Intell. Syst. Comput.* **697** (Springer Verlag) 487–95 https://doi.org/10.1007/978-981-13-1822-1_45

[3] Zhu G, Bai J and Huang Y 2014 Effect of micro-EDM on diameter consistency of micro-hole arrays *Key Eng. Mater.* **609–10** 1489–93 https://doi.org/10.4028/www.scientific.net/KEM.609-610.1489

[4] Sivasankar S, Jeyapaul R and Bhanu Prasad V V. 2012 Performance study of various tool materials for electrical discharge machining of hot pressed ZrB_2 *Multidiscip. Model. Mater. Struct.* **8** 505–23 https://doi.org/10.1108/15736101211281614

[5] Zhai W J and Wang C 2004 Study on characteristics and mechanism of silicon nitride in electrochemical grinding *Mocaxue Xuebao/Tribology* **24** 360–3

[6] Mookam N, Sunasuan P, Madsa T, Muangnoy P and Chuvaree S 2021 Effects of Graphite and Boron Carbide Powders Mixed into Dielectric Fluid on Electrical Discharge Machining of SKD 11 Tool Steel *Arab. J. Sci. Eng.* **3** https://doi.org/10.1007/s13369-020-05156-4

[7] Strelchuk R, Trokhymchuk S, Sofronova M and Osipova T 2020 Revealing patterns in the wear of profile diamond wheels *Eastern-European J. Enterp. Technol.* **3** 30–7 https://doi.org/10.15587/1729-4061.2020.203685

[8] Ya Ropyak L, Vytvytskyi V S, Velychkovych A S, Pryhorovska T O and Shovkoplias M V 2021 Study on grinding mode effect on external conical thread quality *IOP Conf. Ser.: Mater. Sci. Eng.* **1018** 012014 https://doi.org/10.1088/1757-899X/1018/1/012014

[9] Yadav R S and Yadava V 2018 Hybrid design-based modelling and multi-objective optimization of hybrid machining of hybrid metal matrix composites *Proc. Inst. Mech Eng C: J. Mech. Eng. Sci.* **233** 2275-301 https://doi.org/10.1177/0954406218786324

[10] Nath Yadav R, Yadava V and Singh G 2012 International Journal of Manufacturing Technology and Management Composite' presented at 4th International and 25th All India Manufacturing Technology *Int. J. Adv. Manuf. Technol.* **28** 14–6

Experimental results on the quality of the surface on turning with inclined tangential cutting line

M V Puriciuc, R Cioară and C Pisarciuc

Department of Engineering and Industrial Management, Transilvania University of Brasov, 29 Eroilor av., Brasov, Romania

cioarar@unitbv.ro

Abstract. The authors devised an innovative processing method, called turning with inclined tangential cutting line. To validate the concept a suitable tool was made, and experimental research was performed in order to evaluate the results from the perspective of the quality of the generated surfaces. The research program involved the processing of various materials with operating modes characterized by the same depth of cut, but different speeds and feed rates. The roughness of the obtained surfaces was measured using a high-performance electronic roughness meter. For comparison, processing under similar conditions was performed using a normal tool. This paper presents a number of experimental results and relevant conclusions supported by them. Also, some directions of further research are highlighted. The authors considered both logical and necessary outcomes after critical analysis of results and in terms of extending the parameters that characterize the designed process.

1. Introduction

Turning is a commonly used processing process. It is therefore justified the special interest for the identification of the optimal conditions that would lead to the achievement higher performances.

One of the parameters that is often of great importance is the quality of the surfaces generated by turning. The quality of the surface is expressed by its roughness. At turning, the roughness depends on the material of the machined part, the material and geometry of the active part of the tool with which the machining is done, the parameters of the working process (speed, feed, depth of cut), the nature and quality of the coolant. Machining technique is also important.

It is common to turn cylindrical surfaces by the conventional processing method. In this case, the active part of the tool, most often removable, the top face is slightly inclined to the cutting plane or reference plane, horizontal plane that contain the axis of the workpiece and the tool nose and is limited by the main cutting-edge and the end-cutting edge. Adequate importance is given to the values of the side-cutting edge angle, the end-cutting edge angle, the shape angle, and back rake angle.

The importance of these aspects is frequently identified in the research presented in the scientific literature. Analysed performances are the quality of the obtained surface, dimensional accuracy, the vibrations generated by the process, the intensity of tool wear, etc.

Experimental research most often focuses on the processing of a particular material such as alloy steel EN 47 [1], AISI H11 steel [2,3], X38CrMoV5-1 (AISI H11) steel [4,5], AISI H13 die tool steel [6], AISI 1020 steel [7], 42CrMo4 (AISI 4140) alloy steel [8], 205Cr115 (AISI D3) tool steel [9] and aluminium alloy 6061 [10,11]. In mentioned references, the goal is to establish the optimal combination of working parameters that will lead to the achievement of a specified objective. In this sense, various methods are used to model and optimize the machining process. To reduce complexity and to obtain faster results, the Taguchi method of designing experiments and the analysis of variance is often used [12]. Puh *et al.* [13] study through multi-objective investigations the optimal combination of parameters cutting speed, feed rate and depth of cut that assure the minimum roughness of machined surface in the condition of maximum material-removal rate. The research is using the Grey-based Taguchi method, and the significance of parameters of the cutting process has been evaluated by the analysis of variance. Radovanović *et al.* [14] study the optimal turning parameters at finishing of medical 316LVM stainless steel, the aim of the research being the minimum surface roughness. The study takes into consideration four parameters, insert radius, depth of cut, feed and cutting speed, and for the design of the experiments, the Taguchi method with L_{27} orthogonal matrix is used. Because disparate results are hard to be interpreted, several other research papers are identified referring to the optimization of cutting parameters for surface roughness on turning [15-18]. It is also useful to predict the roughness of the turned surface based on the values of cutting parameters and vibration signals measured during machining [19,20]. The values of the cutting forces measured in real-time also serve to adapt the working parameters in order to obtain the desired quality of the generated surface [21]. Artificial neural network and multiple regression method are frequently used for process modelling and to predict the roughness of the processed surface [22-25].

2. Experimental set-up

The authors of this paper have identified a new turning technique, named turning with the tangential cutting line. In order to validate this concept, an experimental research programme was initiated to validate the concept and provide information about the quality of surfaces generated in this way.

The first step was the design of a turning tool with a tangential cutting line, as presented in figure 1. This new tool was obtained starting from right-hand lathe tool, ISO 6 (DIN 4980) whit active part made of carbide P30. The front of the tool was cut off and welded to another in a way that the long edge of the carbide plate is, on the one hand, contained in the front plane of the tool, perpendicular to the axis of its body, and, on the other hand, is inclined with respect to the support plane of the tool body at an angle of 70° approximately.

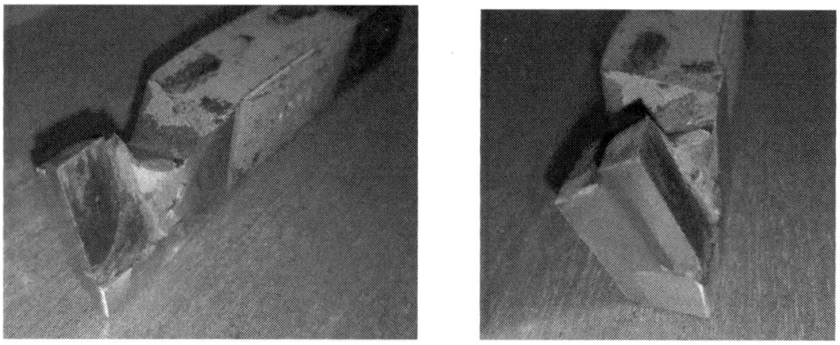

Figure 1. Turning tool, with inclined tangential cutting line.

The relative position between the turning tool with tangential cutting line and the workpiece is presented in figure 2. In order to be able to compare the results obtained when turning with the tangential cutting knife described with those obtained using a usual turning scheme, machining was performed with the same parameters using a right-side tool ISO 2 (DIN 4972), with active part from the same P30 carbide.

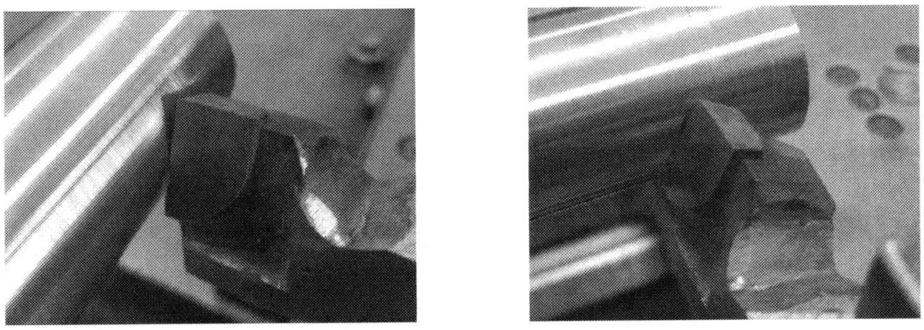

Figure 2. The relative position between the turning tool with tangential cutting line and the workpiece.

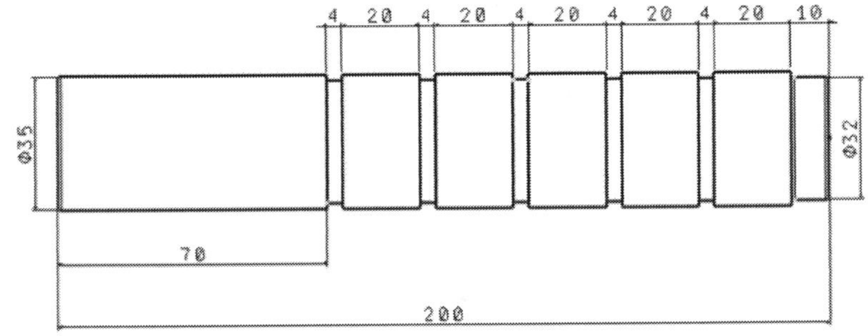

Figure 3. Initial geometric characteristics of specimens.

The experimental research was performed on specimens made of carbon steel (OL50 / E295) and 6061 aluminium alloy (duralumin). The main aim was to identify the influence of cutting speed on the roughness of the surface obtained by turning, but also the influence of the feed values. The research program considered five values for speed, n ∈ {160; 250; 400; 630; 1000}, in rpm, and three values for feed-rate, f ∈ {0.12; 0.16; 0.20}, in mm/rot. At all tests, the depth of cut was the same, t = 0.5 mm.

Only one specimen was used for each material, successive processing causing a gradual reduction in their diameter. The initial geometric characteristics of specimens are presented in figure 3.

The tests were performed without cooling on an SN 320 lathe. For each knife and for each material, three passes were made, one for each value of the feed. At each pass, the speed of the specimens took all five values, one for each sector of it. An image during machining with an ISO 2 tool is shown in figure 4 and an image during machining with a knife with an inclined tangential edge is shown in figure 5. A professional electronic roughness meter, TESA-rugosurf 10-G, was used to measure the roughness values (figure 6). For each machined surface, roughness measurement was made along three equiangular disposed generatrix (0°, 120° and 240°) and the average value was calculated.

Figure 4. Turning with tool ISO 2, steel OL50 specimen, feed f = 0.20 mm, depth of cut t = 0.5 mm, speed n = 1000 rpm.

Figure 5. Turning with tangential tool, steel OL50 specimen, feed f = 0.20 mm, depth of cut t = 0.5 mm, speed n = 1000 rpm.

Figure 6. Example of roughness measurement with the TESA-rugosurf 10-G roughness meter.

3. Results and discussion

The test results, only for the roughness Ra, are summarized in tables 1 - 4. Table 5 graphically reflects the experimental results. All graphs show comparatively the values of the obtained roughness, in identical technological conditions (except for the cutting speed, which has deviations due to variations in the diameter of the specimen), on turning with conventional ISO 2 tool and with tangential cutting line. Each time the trend of roughness evolution depending on the speed is highlighted, in order to identify the optimal technological combination of parameters.

Table 1. Roughness values at turning steel with conventional ISO 2 tool, depth of cut t = 0.5 mm

f	Ø specimen	n (rpm)				
(mm/rot)	(mm)	160	250	400	630	1000
0.12	34	4.630	5.672	4.426	1.962	1.986
0.16	33	8.501	8.061	4.454	2.352	2.336
0.20	32	10.700	10.792	6.028	4.655	4.704

Table 2. Roughness values at turning steel with tool with tangential cutting line, depth of cut t = 0.5 mm

f	Ø specimen	n (rpm)				
(mm/rot)	(mm)	160	250	400	630	1000
0.12	31	4.103	0.733	0.438	0.960	1.854
0.16	30	5.573	0.861	0.979	1.100	1.955
0.20	29	7.180	0.904	0.508	0.913	1.081

Table 3. Roughness values at turning aluminium with conventional ISO 2 tool, depth of cut t = 0.5 mm.

f	Ø specimen	n (rpm)				
(mm/rot)	(mm)	160	250	400	630	1000
0.12	28	7.031	6.161	6.891	5.454	5.948
0.16	27	7.736	7.477	7.751	6.427	4.713
0.20	26	7.752	8.760	8.312	6.884	6.173

Table 4. Roughness values at turning aluminium with tool with tangential cutting line, depth of cut t = 0.5 mm

f	Ø specimen	n (in rpm)				
(mm/rot)	(mm)	160	250	400	630	1000
0.12	31	1.566	2.071	1.938	1.371	1.595
0.16	30	2.511	1.845	1.575	1.957	2.395
0.20	29	3.300	3.230	3.376	2.349	1.670

It should be noted that for all combinations of test parameters - materials, feeds and speeds (hence cutting speeds) - the surfaces obtained by turning with a tangentially cutting edge have lower roughness, most often spectacularly lower than on surfaces obtained by turning with a conventional ISO 2 tool. At machining the steel specimen with a tangential tool, the roughness is minimal for speeds in the range 250 - 630 rpm (cutting speeds in the range 22.5 - 63 m/min) for all three feed values. The best values were obtained for the speed of 400 rpm (approx. 37 m/min), for all three feed values. For the mentioned range of cutting speed, the feed values do not determine significantly different values of the roughness of the generated surfaces.

IOP Conf. Series: Materials Science and Engineering 1235 (2022) 012024 doi:10.1088/1757-899X/1235/1/012024

Table 5. Variation of roughness in relation to speed and tool type (ISO 2 tool and tangential cutting line tool) for three feed values, for specimens of OL50 steel and duralumin

Feed	Specimen of OL 50 steel	Specimen of aluminium alloy 6061
0.12 mm/rev	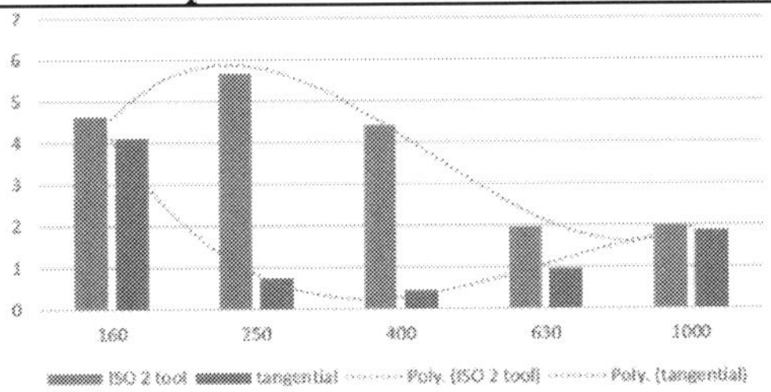	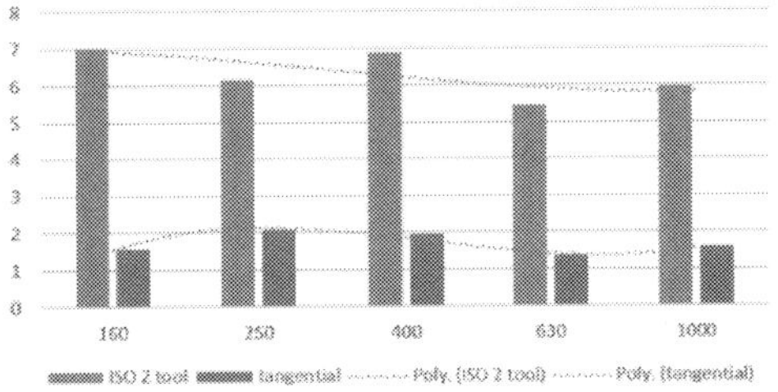
0.16 mm/rev	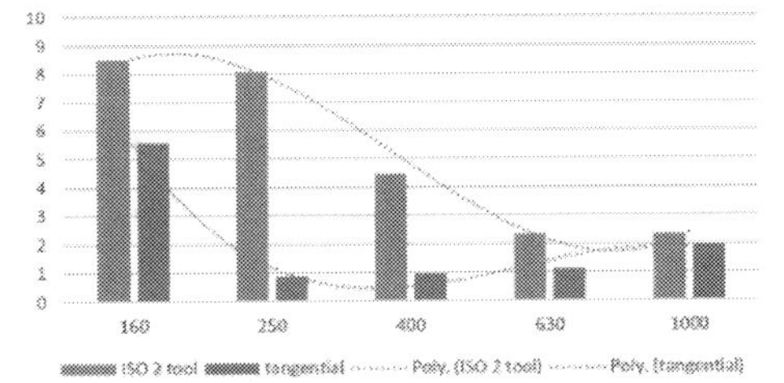	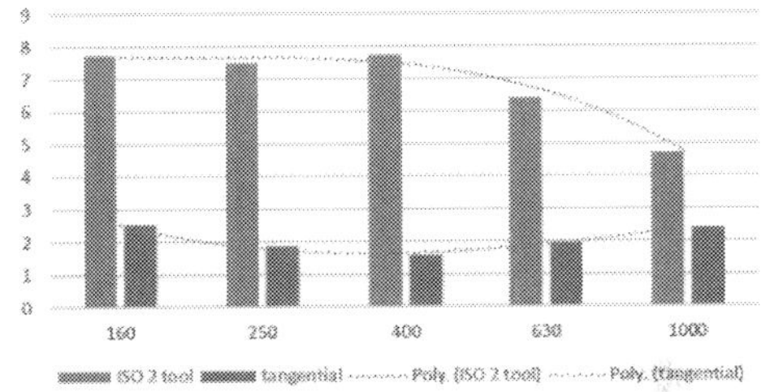
0.20 mm/rev	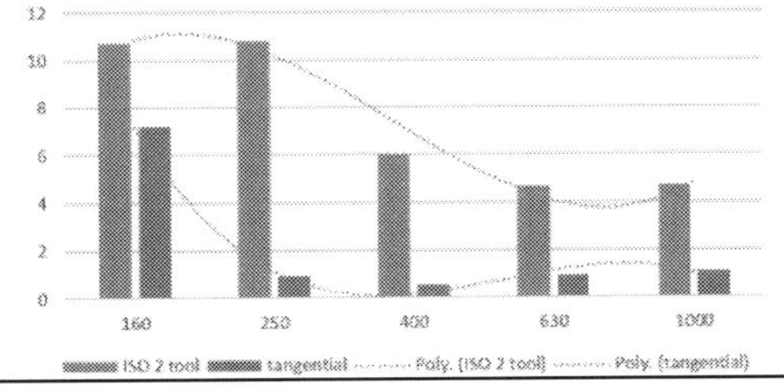	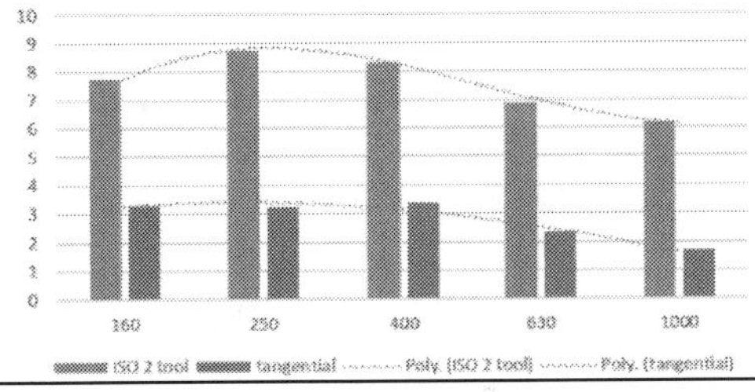

It is also noted that, for the same range of cutting speed, the roughness of the surfaces obtained by turning with tangential cutting line is significantly lower than that of the surfaces obtained by turning with a traditional tool, type ISO 2.

When turning the specimen from aluminium alloy 6061, for all speeds and for all feeds the value of roughness was, when turning with a tangential cutting line, consistently lower, 2 - 3 times or more, than when turning with a classic ISO 2 tool. When the tangential tool is used for turning aluminium 6061, there is only a small variation of roughness depending on the speed. As the value of the feed increases, so does the value of roughness, at least for low and medium speed values. On turning with a tangential tool, the roughness is lower for the steel specimen than it is for aluminium alloy specimen for all feeds and speeds, except for the minimum speed selected for the tests, n = 160 rpm.

4. Conclusions

Turning is a frequently used machining process, which justifies the interest in identifying optimal technological conditions that lead to superior performance in terms of quality of surface and dimensional accuracy. The paper presents a different approach from the traditional one which, as highlighted in the article, led to positive results. For this reason, a new processing method has been designed, called turning with a tool with an inclined tangential cutting line. For testing purposes, a lathe tool suitable for the method was made and used for processing with different parameters (two materials, feed rate and machining speed). The parameter taken into account to compare the new process with the conventional turning was the surface roughness. For comparison, processing was conducted with the same parameters but using a traditional lathe ISO 2 tool. According to the findings presented, for any combination of parameters the results obtained were very good, the roughness when machining with a tool with inclined tangential cutting line being better or much better than in the case of the samples machined with an ISO 2 tool. The research can be expanded, taking into account a wider set of working parameters. It is also necessary to study the influence of the cutting edge angle of inclination over the quality of the generated surfaces.

ORCID iDs
M V Puriciuc https://orcid.org/0000-0001-6246-7608
R Cioară https://orcid.org/0000-0002-9756-2776
C Pisarciuc https://orcid.org/0000-0003-0595-1668

References

[1] Sharma L, Sharma J P and Sharma N 2016 *Int. J. Adv. Res. Mech. Eng. & Technol.* **2** (4) 14 https://doi.org/10.13140/RG.2.2.23103.56484
[2] Benlahmidi S, Aouici H, Boutaghane F, Khellaf A, Fnides B and Yallese M A 2016 *Int. J. Adv. Manuf. Technol.* **89** 803-20 https://doi.org/10.1007/s00170-016-9121-3
[3] Dureja V S, Gupta V K, Sharma V S and Dogra M 2009 *J. Manuf. Science Eng.* **223** 441
[4] Fnides B, Aouici H, Yallese M A 2008 *J. Mechanika* **2** 73
[5] Aouici H, Yallese M A, Belbah A, Ameur M F and Elbah M 2013 *Sadhana* **38** 429
[6] Kumar P and Chauhan S R 2015 *Arab J. Sci. Eng.* **40** 1471-85 https://doi.org/10.1007/s13369-015-1606-0
[7] Sonowal D, Sarma D, Barua P B and Nath T 2017 *IOP Conf. Ser.: Mater. Sci. Eng.* **225** 012186 https://doi.org/10.1088/1757-899X/225/1/012186
[8] Marinković V and Madić M 2011 *Sci. Res. Essays* **6** (16) 3474
[9] Stanimir A, Zamfirache M and Eftenie N C 2011 *Fiability & Durability* **1** (7) 104
[10] Abdallah A, Rajamony B and Embark A 2014 *IOSR J. Eng.* **04** (10) 01
[11] Ranganath M S 2014 *Int. J. Adv. Res. Innov.* **1** 62
[12] Nalbant M, Gökkaya H and Sur G 2007 *Mater. Design* **28** 1379-85 https://doi.org/10.1016/j.matdes.2006.01.008
[13] Puh F, Jurkovic Z, Perinic M, Brezocnik M and Bulian, S 2016 *Technical Gazette* **23** 377-82 https://doi.org/10.17559/TV-20150526131717
[14] Radovanović M, Slatineanu L, Janković P, Petković D and Madić M 2015 *Appl. Mech. Mater.* **809-810** 153-8 https://doi.org/10.4028/www.scientific.net/AMM.809-810.153
[15] Kumar N and Uppal N 2013 *Mechanica Confab* **2** (4) 45
[16] Govindan P and Vipindas M P 2014 *Int. J. Mech. Eng. & Rob. Res.* **3** 89
[17] Rewatkar S A, Hatekar Y K and Shahu S R 2018 *Int. J. Eng. Develop. & Res.* **6** 540
[18] Ilham M, Suparjon and Sanjiwani Y 2020 *Int. J. Mech. Eng. Tech. & Applic.* - MECHTA **1** 43 https://doi.org/10.21776/MECHTA.2020.001.02.2
[19] Upadhyay V, Jain P K and Mehtain N K 2013 *Measurement* **46** 154-60 https://doi.org/10.1016/j.measurement.2012.06.002
[20] Hessainia Z, Belbah A, Yallese M A, Mabrouki T and Rigal F R 2013 *Measurement* **46** 1671-81 https://doi.org/10.1016/j.measurement.2012.12.016
[21] García-Plaza E, Núñez P J, Mata F and Sanz A 2012 *Adv. Mat. Res.* **498** 157-62 https://doi.org/10.4028/www.scientific.net/AMR.498.157
[22] Asiltürk I and Çunkas M 2011 *Expert Systems with Applic.* **38** (5) 5826-32 https://doi.org/10.1016/j.eswa.2010.11.041
[23] García-Plaza E, Núñez P J, Salgado D R, Cambero I, Herrera Olivenza J M and García Sanz-Calcedo J 2013 *Procedia Eng.* **63** 599-607 https://doi.org/10.1016/j.proeng.2013.08.245
[24] Karayel D 2009 *J. Mat. Proces. Tech.* **209** 3125-37 https://doi.org/10.1016/j.jmatprotec.2008.07.023
[25] Solarte-Pardo B, Hidalgo D and Yeh S-S 2019 *Appl. Sci.* **9** 479 https://doi.org/10.3390/app9030479

IOP Conf. Series: Materials Science and Engineering 1235 (2022) 012025 doi:10.1088/1757-899X/1235/1/012025

Influence of the cutting parameters on the flank surface roughness at thread whirling of low carbon steel worm shafts

G Creţu and F Negoescu

Department of Machine Manufacturing Technology, "Gheorghe Asachi" Technical University of Iaşi, Blvd D. Mangeron 59A, 700050 Iaşi, Romania

cretu_gh@yahoo.com, gcretu@tcm.tuiasi.ro

Abstract. Thread whirling is a machining process applied mainly for its high material removal rate, ensuring the thread cutting in one single machining pass. The process presents other advantages as continuous and accurate machining, by tilting the whirling ring to the helix angle, machining long threads of deep thread parts as worm shafts, feed screws, etc. This paper presents experimental research results on the whirling process, respectively the influence of the cutting parameters specific to the whirling process on the roughness of the flank surface, in the case of worm shafts made of low carbon steel. The empirical mathematical models were obtained and discussed.

1. Introduction

Worm manufacturing is of current interest for several industries, from worm shafts from industrial machinery to medical worm screw implants [1]. Depending on the production type, the manufacturing technology is different, from specialized whirling machines [2] used for batch, flow, or mass production to adapted subassemblies to Swiss-type turning lathe [1] for small dimensions of worm screws production or classical machine tools as lathe [3-5].

There are several studies on improving the manufacturing process for thread whirling or emphasizing the influences and variations regarding the process parameters. Thus, Son and Lee [2] applied finite element analysis in Ansys for highlighting the thermal distribution and the thermal stress in the whirling thread device, demonstrating the tool displacement that may occur due to the thermal stresses. An extended study was covered by He *et al.* [6] to estimate the specific cutting energy in the whirling process. As a result, they give data for selecting the optimal machining parameters for low values of specific cutting energy, good material removal rate, and good machining precision and accuracy. The model was developed for whirling milling, and it was validated by a set of experiments carried out on a whirling milling machine for the case study of a ball screw shaft [7]. Their experimental setup measured the cutting forces and the cutting power, determining the force coefficients and respectively the energy specific to the cutting of worm thread by whirling milling.

Machining worm shafts and long threads by whirling can be achieved on lathes, using threading devices mounted instead of the turning tool holder, or on special thread whirling machine tools [8]. Given the cost of such special whirling machine tools for job-shop production, the worm shafts are often machined by using lathe-fitted whirling heads.

Machining the worm shafts by thread whirling devices is one of the most productive methods of machining such parts. Among the vortex machining methods, the one with internal tangency was preferred due to its advantages in the cutting process [9].

This paper presents some experimental results on the study of the surface quality of worm flanks. The influence exerted on the arithmetical mean deviation R_a of the assessed flank profile by the input variable for the considered whirling process.

2. Experimental setup

Experimental research was conducted in laboratory conditions at the "Gheorghe Asachi" Technical University of Iasi, on a universal lathe type SNA 560 x 1000, by using a special in-house made device for thread whirling, presented in Figure 1.

This device is rigidly clamped on the lathe carriage slide through a fixing plate. The whirling head has four slots for cutters, and their cutting tips are circle aligned with eccentricity e_x from the lathe spindle, respectively, regarding the workpiece axis. The rotation of the whirling head is ensured by a belt transmission from a dc motor. The rotation speed n_s of the whirling head is given by the motor rotation speed and by the belt pulleys transmission ratio.

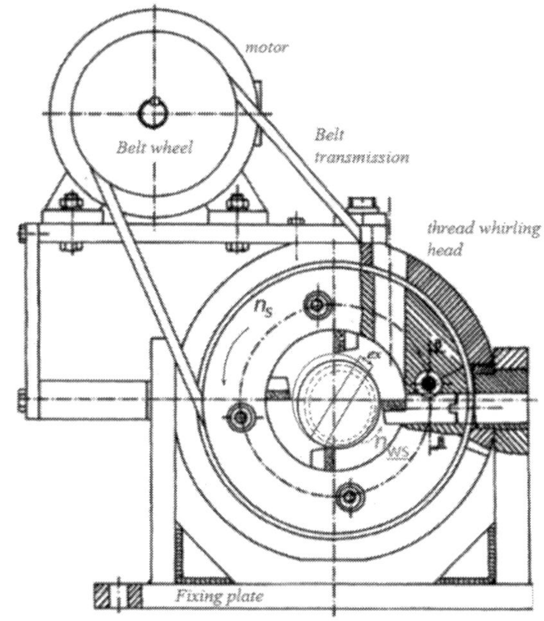

Figure 1. Thread whirling device adapted for a universal lathe (modified after [3]).

The primitive diameter of the belt groove of the driven pulley from the whirling device is D_{pd} = 216 mm. Several belt pulleys were used on the motor spindle, with the following primitive diameters: D_{p1} = 188 mm, D_{p2} = 162 mm, D_{p3} = 133 mm, D_p4 = 92 mm, D_{p5} = 57 mm, to obtain different rotational speeds of the whirling head. Correspondingly, the rotation speed of the main axis of the device can be of n_{s1} = 1241 rpm, n_{s2} = 1070 rpm, n_{s3} = 878 rpm, n_{s4} = 614 rpm and n_{s5} = 376 rpm.

The speed corresponding to the cutting tips when machining the reference diameter can be given by:

$$v_s = \frac{p \cdot (D_{vs} + 2.5 \cdot m_x) \cdot n_s}{1000} \text{ [m/min]}. \quad (1)$$

The rotational speed range of the lathe spindle was modified, from the range of 16 ÷ 1600 rpm was reduced to the range 2.4 ÷242 rpm by introducing an intermediate shaft between the lathe drive motor and its gearbox.

The rotational speeds used in the experiments are 2.4 rpm, 3 rpm, 3.76 rpm, 4.74 rpm, 6 rpm, 7.52 rpm, and 9.48 rpm. The total cutting depth is given by the thread depth of the machined worm, which is calculated with the relation (2).

$$h = 2.25 m_x \text{ [mm]}. \quad (2)$$

Tests were performed for three different moduli: m_{x1} = 2 mm, m_{x2} = 3 mm and m_{x3} = 4 mm. For these modulus values the total cutting depths were t_{p1} =4.5 mm, t_{p2} = 6.75 mm and t_{p3} = 9 mm.

Another parameter that can be considered, derived from independent and constructive parameters, is the feed per tooth f_z, given by the relation (3):

$$f_z = \frac{p \cdot d_{a1} \cdot n_m}{n_s \cdot z_s \cdot \cos g} \text{ [mm/tooth]}, \quad (3)$$

where: z_s - the number of roughing cutting tools.

For a given worm (d_{al}, g), a constant number of roughing cutting tools z_s and a certain rotational speed of the whirling head, the feed per tooth f_z is directly proportional to the speed of the part n_{ws}. It is sometimes preferred to characterize the cutting process by this parameter, as it shows the loading of each tooth during the machining process.

The experimental tests were performed on the non-alloy quality structural steel E335 - EN 10025-2:2019, with a hardness of 224 HB.

Some operating characteristics of the thread whirling device used are as follows: - the motor used is of the ASI 90 L - 24 - 4 type, which develops a nominal power of 1.5 kW, has a rotational speed of 1425 rpm, and a nominal load efficiency of 76%; - the inner diameter of the main shaft is 116 mm; - the minimum diameter of the cutting tools tips aligning circle is 20 mm; - the maximum diameter of the cutting tools tips aligning circle is 80 mm; - the maximum angle of inclination of the whirling head in the support: 18°.

The constructive parameters of the whirling head and the cutters are established according to the characteristics of the machined worm. The shape of the active part of the cutting tools is identical to the shape of the worm hollow in the normal section on the average helix of the hollow.

A special sharpening device was designed and built to ensure the precision of sharpening all the cutting tools of the thread whirling device. It achieves the following angles for the cutting part of the tool: - relief angle of the side cutting edges: 5°14', - end relief angle of the front cutting edge: 15°, - rake angle for all three cutting edges: 0°.

The roughness R_a of the relief and rake surfaces of the cutting tools, resulting from sharpening, is between 0.17 µm and 0.25 µm.

The positioning of all the side cutting edges has to be on the same cone which can be described by rotating a side edge around the axis of rotation of the cutting tools holder - the whirling ring is an essential condition for the correct operation of the worm thread whirling device. To ensure this condition, the axis of the grooves in which the cutting tools are mounted in the whirling ring may have a 0.015 mm limited deviation to the normal plane at the rotation axis of the whirling ring. Precise positioning in the radial direction was performed with the optical device mounted on the motor support plate of the thread whirling device.

The machining tests were realized both in the feed direction and against the feed to study the influence of the feed type. To compensate for the clearance from the lathe gearbox, when threading in the feed direction, a mechanism was used for ensuring additional friction at the level of the lathe spindle shaft, higher than the moment created by the cutting forces.

When machining against the feed, the chips are formed in more difficult conditions. Their thickness increases from zero to a maximum value, with a much lower speed than when machining in the normal feed direction.

3. Results and discussions

Experiments have shown that the thermal energy resulting from the cutting process is absorbed mainly by the chips detached during the machining, as also mentioned in the specialty literature [2,3,6,7,9,10], 80% of the thermal energy being transferred through the chip, 10% to the cutting tools and 10% to the workpiece. Due to the rotation of the whirling ring and the fact that the tools are cutting less than a quarter of one complete rotation of the whirling ring, the temperature of the cutting tools remains low.

The thermal energy transmitted to the machined workpiece is also low because when machining with normal cutting parameters (speeds less than 200 m/min and feed per tooth between 0.1 and 0.2 mm/tooth), the speed of heat propagation in the workpiece material is lower than the longitudinal movement speed of the cutting tools. The workpiece temperature in the machining area increases significantly only at low feed rates per tooth and at high cutting speeds.

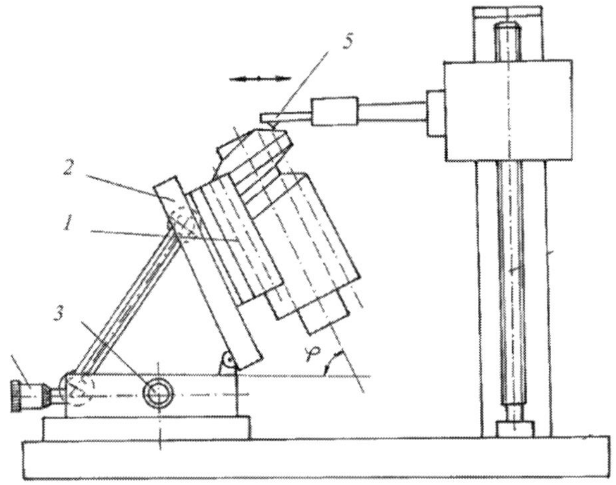

Figure 2. Schematic of the worm shaft positioning for flank surface roughness measurements on the radial direction (modified after [3])

The tests performed with high cutting speeds and relatively reduced feed showed that the workpiece heats up considerably, registering a higher heat-affected zone, and the fast increase of the temperature from the cutting area has direct negative effects on the durability of the tool. The experimental condition and the results of the roughness measurements are given in table 1.

The roughness measurement was performed with a "Surtronic" type device produced by Taylor - Hobson. For positioning the roughness tester and the worm shaft to perform the measurement, there were used the clamping fixtures and the table of an existing profilograph type Profilometer 252.

To measure the roughness in the radial direction, the worm shaft is positioned on the prismatic support *1* of the table movable plate *2* of the device (figure 2) and fixed with screws using the T channels from the movable plate *2*.

Table 1. Experimental conditions and results for thread whirling of E335 steel.

No	D_{vs} mm	m_x mm	q	n_s rpm	v_s m/min	n_m rpm	f_z mm/tooth	g [°]	Ra [μm]
1	36	2	12	1241	160	3.76	0.07	4.76	0.35
2	36	2	12	1241	160	6	0.11	4.76	0.63
3	36	2	12	1241	160	9.48	0.17	4.76	0.38
4	36	2	12	1070	138	4.74	0.08	4.76	0.65
5	36	2	12	1070	138	7.52	0.13	4.76	0.25
6	36	2	12	1070	138	9.48	0.17	4.76	0.25
7	36	2	12	878	113	3.76	0.09	4.76	0.37
8	36	2	12	878	113	6	0.15	4.76	1.2
9	48	3	10	1070	186	3	0.09	5.71	0.35
10	48	3	10	1070	186	4.74	0.14	5.71	0.21
11	48	3	10	1070	186	6	0.17	5.71	0.38
12	48	3	10	878	153	3.76	0.1	5.71	0.15
13	48	3	10	878	153	6	0.21	5.71	0.27
14	48	3	10	878	153	7.52	0.24	5.71	0.3
15	48	3	10	614	107	3	0.15	5.71	0.35
16	48	3	10	614	107	4.74	0.24	5.71	0.7

Since the roughness tester allows the roughness measurement on the curved surfaces with a relatively large radius of curvature, there were performed measurements of roughness on the direction tangential to the profile. To measure the roughness in the tangential direction, the movable plate of the table is inclined at an angle $\varphi = p/2 - \gamma$, and the prismatic on which the measuring worm is fixed is inclined at an angle equal to the normal pressure angle $\alpha_n = 20°$. Using the micrometre screws 3 and 7 of the table, the worm is positioned so that the stylus tip moves in an approximately tangent direction to the reference diameter of the worm.

As one can notice, the surface roughness of the worm shaft flanks is comparable to that obtained by grinding. A program [11] for determining the mathematical dependence relation for multidimensional strings was used for obtaining the mathematical relations for the dependence of the surface roughness depending on the modulus m_x, the cutting speed v_s and the feed per tooth f_z.

The dependence of the roughness on the modulus, the tool speed, and the feed per tooth when machining in the feed direction in a single pass is expressed with the relation (4).

$$R_a = -0.4323 + \frac{0.9276}{m_x} + \frac{75.6863}{v_s} - \frac{0.0079}{f_z} \text{ [mm]} \quad (4)$$

The variation of the roughness depending on the cutting speed of the tool and the feed per tooth can be observed in figures 3 and 4.

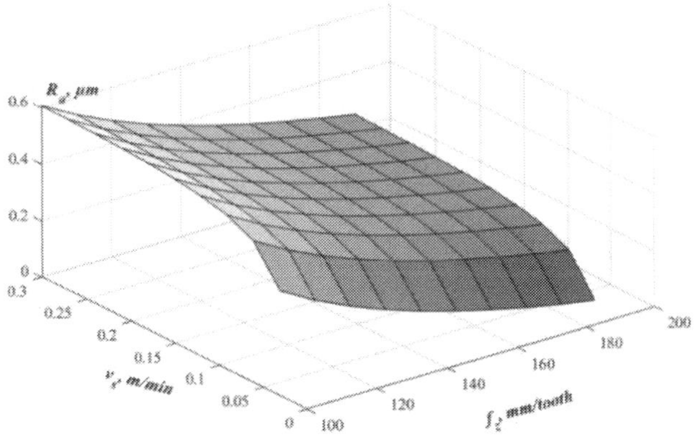

Figure 3. Roughness variation, depending on tool speed, v_s, and feed per tooth, f_z for modulus m_x = 3 mm when machining in the feed direction.

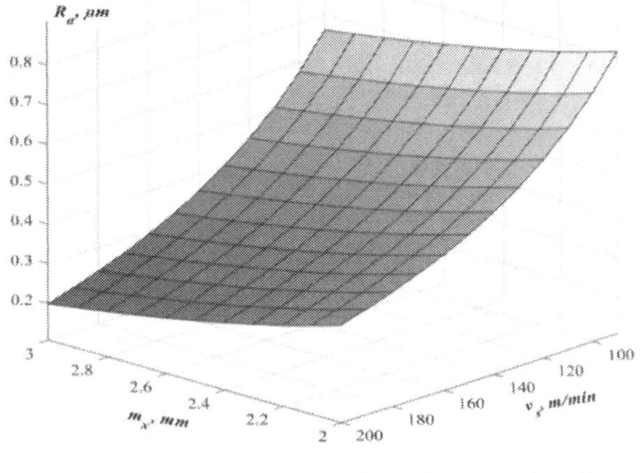

Figure 4. Roughness variation depending on the axial modulus and tool speed, for feed per tooth of f_z = 0.14 mm/tooth when machining in the feed direction.

4. Conclusions

The worm shafts machining technology using whirling threading devices is one of the most productive manufacturing processes. The cutting parameters setup is established according to the objective pursued, taking into account the limitations imposed by the specific conditions of the whirling process.

Each constructive solution for thread whirling device allows the machining of certain dimensions of worm shafts; for this reason, to widen the size range of machined worm shafts, it is necessary to build sets of devices on groups of part sizes.

The accuracy of the worm shafts obtained by this machining method depends on both the accuracy of the used device and the precision of the lathe on which it is mounted.

The good surface quality of the worm flanks can be achieved by the alignment of the cutters, their geometric precision, and adequate machining parameters.

ORCID iDs

G Cretu https://orcid.org/0000-0002-2036-3592
F Negoescu https://orcid.org/0000-0003-4187-0275

References

[1] Soshi M, Rigolone F, Sheffield J and Yamazaki K 2018 Development of a directly-driven thread whirling unit with advanced tool materials for mass-production of implantable medical parts *CIRP Annals* **67** (1) 117-20 https://doi.org/10.1016/j.cirp.2018.03.016

[2] Son J H and Lee Y M, 2008 Thermal Analysis of Whirling Unit Attached to Thread Whirling Machine *MSF* **575-578** 93-7 https://doi.org/10.4028/www.scientific.net/msf.575-578.93

[3] Cretu G 1997 *Contributions regarding the worm shaft machining technology using the thread whirling process* (in Romanian) Ph D Thesis, Gheorghe Asachi Technical University of Iasi, Romania

[4] Cretu G 2013 Contributions on design and accomplishing of the whirling thread cutting devices that could be adapted for ordinary lathes *Adv Mater Res* **837** 67–70 https://doi.org/10.4028/www.scientific.net/amr.837.67

[5] Cretu G and Coteaţă M 2013 Contributions on the influence of the cutting parameters and constructive geometrical parameters of the tool on the precision of the worm gears manufactured using whirling thread cutting *Appl Mech Mater* **371** 54-8 https://doi.org/10.4028/www.scientific.net/AMM.371.54

[6] He Y, Wang L, Wang Y, Li Y, Wang S, Wang Y, Liu C and Hao C 2019 An analytical model for predicting specific cutting energy in whirling milling process *J Clean Prod* **240** 118181 https://doi.org/10.1016/j.jclepro.2019.118181

[7] Wang L, He Y, Li Y, Wang Y, Liu C, Liu X and Wang Y 2019 Modeling and analysis of specific cutting energy of whirling milling process based on cutting parameters *Proc CIRP* **80** 56-61 https://doi.org/10.1016/j.procir.2019.01.028

[8] Cretu G 2017 Contributions to systemic analysis for worm screw production using thread whirling devices *IOP Conf Ser Mater Sci Eng* **227** 012032

[9] Cretu G 2015 Aspects of power consumption in alloy steels processing using thread whirling *Int. J. of Modern Manufacturing Technologies* ISSN 2067–3604 VII(2)

[10] Cretu G, Rusu B and Nedelcu D 2003 Comparative study regarding the processing of metric and trapezoidal threads produced by OLC 45 by turning, milling and whirling thread cutting *Meridian Ingineresc* **2** 85-7 ISSN 1683-853X

[11] Cretu G 1992 *Fundamentals of the experimental research Handbook for the laboratory activities - in Romanian* (Iasi: Institutul Politehnic)

IOP Conf. Series: Materials Science and Engineering 1235 (2022) 012026 doi:10.1088/1757-899X/1235/1/012026

Application of zirconium in dentistry for creating dental crowns

D Pavlova[1]**, R Filchev**[2] **and T Dovramadjiev**[2]

[1] Dentaprime Dental Clinic, St. St. Constantine and Helena, 27th Str. Nr. 1, 9006, Villa Area, Varna, Bulgaria
[2] Technical University of Varna, Departments of Communication Engineering and Technologies, and Industrial Design

tihomir.dovramadjiev@gmail.com

Abstract. Modern dental healthcare is unthinkable without the use of biocompatible materials. They must have the necessary mechanical properties to ensure the strength of dental structures, as well as be as acceptable as possible by the human body. In recent years, there has been an increase in the application of zirconium dioxide (ZrO_2) material to metal ceramics in the manufacture of dental crowns, bridges and others. This has its natural continuation of the evolution of dental materials, which significantly improve human health not only in the oral area but also in complex, where the goal is to minimize the external contact of biological tissue with certain metal components. The present study shows the application of zirconium in the development of crowns on dental implants. Technological working hours are optimized. The process is followed in detail, highlighting important points that accompany the practical part of making the dental crown through the use of modern technologies in design. To objectively present the information in the study, a set of photographic material is presented, visualizing in detail important moments in the production of the zirconium crown.

1. Introduction

Good dentistry is a priority in the overall care of human health. The human oral area is often in contact with the external environment. It is subjected to invasive influences by bacteria that penetrate by ingesting certain substances through food, smoking, etc. Inflammation, infection and other side effects on biological tissues, teeth and gums sometimes occur. In certain cases, this may be due to other factors related to improper dentistry, trauma during dental procedures, poor metabolism in humans, etc. In the context of these important points, the conclusion stands out that a holistic approach, personal culture and social activities related to the protection of people's dental health are needed. The creation and development of professional networks and organizations engaged in human health research is a priority. Such a network is CIII-HU-1506-01-2021 Ergonomics and Human Factors Regional Educational CEEPUS Network 9 European partners: HU (General Coordinator), AT, BG, HR, PL, RO, RS, SI, SK [1]. It is necessary to study and create interdisciplinary links between ergonomics / human factors and design processes, using a specialized methodology [2]. Also, the ergonomics application, compatible with eHealth, provides good opportunities for improving healthcare [3]. Considering these examples, the highlighted opinion is that the inclusion of the advantages of these scientific fields in a single concept significantly improves health care and in particular dental treatment. This starts with studying the properties of the materials used in dental practice [4,5] and localizing the risk factors associated with the applicable materials [6,7]. In recent years, the beneficial biocompatible use of zirconium has been increasingly proven [8,9]. It is successfully used for the production of dental crowns in combination with metal implants (for example, titanium alloys or zirconium-based, etc.) [10-14]. The application of modern technologies in the design, which include CAD / CAE / CAM systems [15-19], is necessary for the production of zirconium dental models. The primary process of digitization takes place by obtaining three-dimensional data through 3D scanning and conventional design in a computer environment. This helps to refine the exact final geometry of 3D models first in a virtual environment and subsequently materialize the samples through 3D printers and CNC machines [20-26]. The refinement of the many stages and the interaction of the different units, scientific fields and interdisciplinarity in a single optimized system is concluded in the combination of the so-called Dentistry 4.0 similar to Industry 4 [27].

2. Materials and Methods

In the present study, the process of making a real zirconium crown with the help of modern technologies in design is traced in detail.

2.1. Materials

Materials including Ti for the dental metal implant (model of the Swiss company Tri [28]) and zirconium dioxide (ZrO_2) for the crown made in the laboratory of Dentaprime dental clinic - Varna [29] were used in the specific case development. The crown pattern is made of a disk zirconium template.

2.2. Methods

When creating final models of zirconium crowns, a set of specific activities can be applied, but in general, it is desirable to adhere to a single concept, covering stages in sequence. Figure 1 shows a methodology for making the zirconium crown:

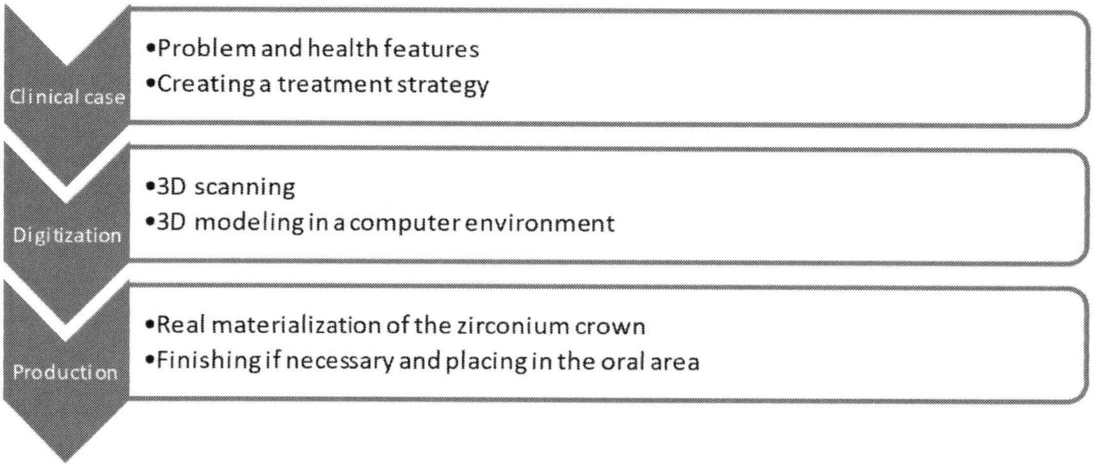

Figure 1. Methodology for making the zirconium crown

3. Technical Realization

Following the methodology of Figures 2 - 6, images are presented containing key moments from the making of a zirconium crown in a real clinical case. Respectively Figure 2 shows the Clinical case and set the task for restoration of the 36th tooth, Figure 3 and Figure 4: CAD process, Figure 5: CAM process and Figure 6: Laboratory process (incl. CNC and 3D machine work), and a completed model as well.

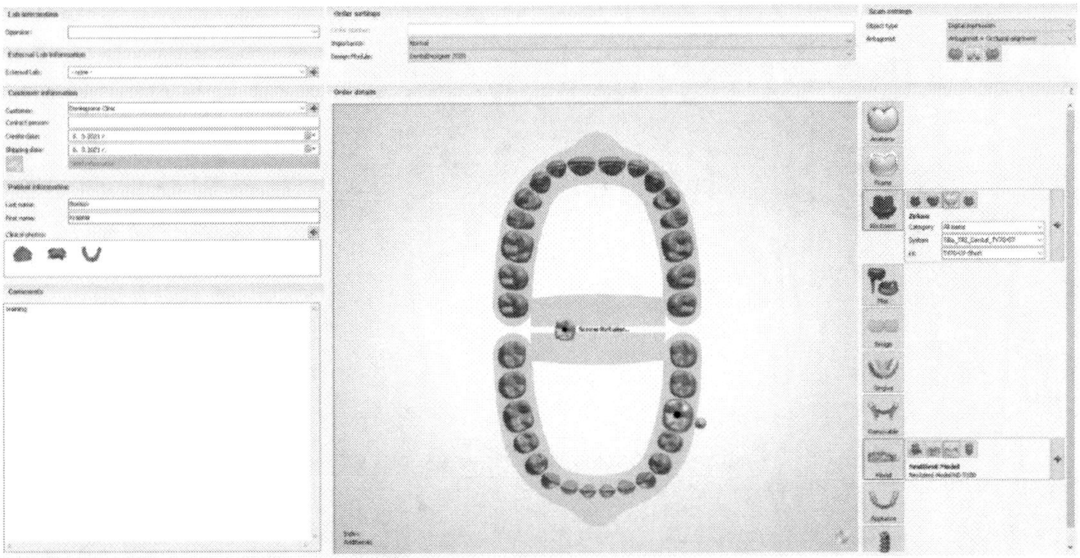

Figure 2. Clinical case and the task of restoring the 36th tooth with a zirconium crown on an implant (Material Ti; short base; Tri-TV70-07 (Dental Software 3Shape [30])

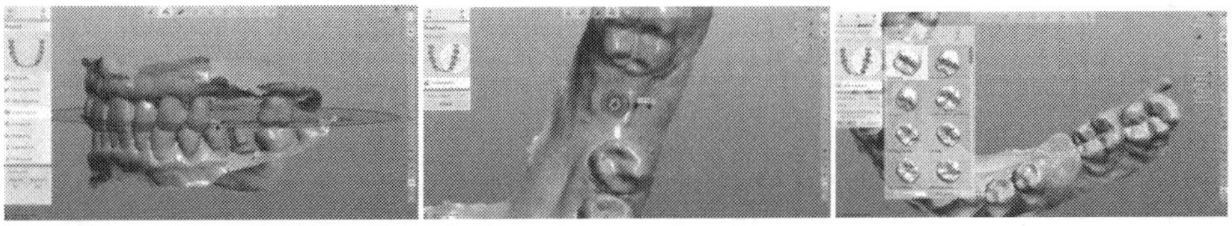

Figure 3. CAD process. From left to right: a digital imprint of the upper and lower jaw (bite + scan body model); directions; selection of the anatomical shape of a tooth from 3Shape software library

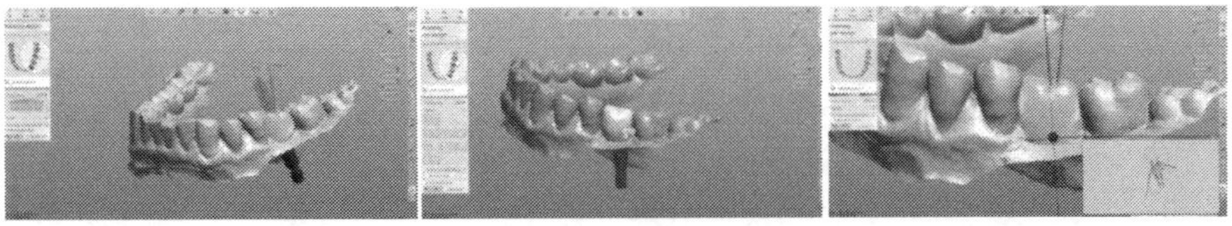

Figure 4. From left to right: defining the position; anatomical design; transection.

Figure 5. CAM process (Millbox software [31]). From left to right: blank (on a zirconium disk); 3D geometry of the crown with cavities; refinement of the work area.

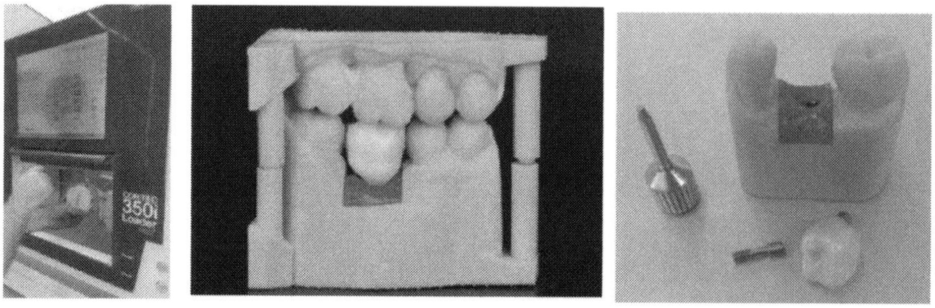

Figure 6. Laboratory process. From left to right: making the crown using a CNC machine; 3D working models in the process of operation (specification: the jaw is 3D printed); completed process + ready model of zirconium sintered and glazed crown.

The obtained final result has a good quality of execution, as the developed model of zirconium crown is successfully placed on the client.

4. Conclusion

This report presents in detail the practical use of modern design technologies applicable in dentistry. A detailed overview of the current technical means as of the date of writing the report has been made.

The report summarizes the advantage of CAD / CAM systems, as well as specialized software over conventional production of dental crowns. Good practices are related to:

- The applied technical means used in the research are with a high degree of efficiency, as their application in real production is proven.
- The shared experience is useful for a wide range of specialists working in the field of modern dentistry.

The authors of the study have tried to extract some of the most important moments accompanying the process of constructing a zirconium crown in a real clinical case. Scientific information is also reported in a balanced way (through images and text). The present study presents the important moments in the development of zirconium dental models. The information is useful for a wide range of users, students, trainees and professionals, as well as all Stakeholders.

Acknowledgements

This paper (result) is (partially) supported by the National Scientific Program "Information and Communication Technologies for a Single Digital Market in Science, Education and Security (ICTinSES)" (grant agreement DO1 205/23.11.18), financed by the Ministry of Education and Science."

ORCID iDs

D Pavlova https://orcid.org/0000-0002-0188-1888
R Filchev https://orcid.org/0000-0001-8795-7048
T Dovramadjiev https://orcid.org/0000-0003-0734-1345

References

[1] Szabó G, Balogh Z, Dovramadjiev T, Draghici A, Gajšek B, Jurčević Lulić T, Reiner M, Mrugalska B and Zunjic A 2021 *Acta Tech. Napocensis, Ser. Appl. Math* **64** (1-S1) 201-212
[2] Sevilla G and Sáenz L M 2016 *Advances in Ergonomics in Design. Advances in Intelligent Systems and Computing* vol **485** (Cham, Springer) 505-13 https://doi.org/10.1007/978-3-319-41983-1_46
[3] Okkonen J 2020 *Finnish Journal of EHealth and EWelfare* **12** (4) 316–26 https://doi.org/10.23996/fjhw.94451
[4] O'Brien W 2002 Dental Materials and Their Selection - 3rd Ed. Quintessence Publishing Co
[5] Powers J and Wataha J 2013 Dental materials: Properties and manipulations. Tenth edition, Elsevier . http://lccn.loc.gov/2015039912
[6] Shahi S, Özcan M, Dizaj S, Sharifi S, Husain N, Eftekhari A and Ahmadian E 2019 *Toxicol.*

Mech. Methods **29** (5) 368-77 https://doi.org/10.1080/15376516.2019.1566424
[7] Tillberg A 2008 https://www.diva-portal.org/smash/get/diva2:142221/FULLTEXT01.pdf
[8] Bollen C and Al-Masri M 2020 *Implant dentistry today* https://eacim-ceramic-implantology.com/wp-content/uploads/2020/03/IDT-March-Bollen.pdf
[9] Lorusso F, Noumbissi S, Francesco I, Rapone B, Khater A and Scarano A 2020 *Materials* **13** (23) 5534 http://dx.doi.org/10.3390/ma13235534
[10] Scarano A, Stoppaccioli M and Casolino T 2019 *BMC Oral Health* **19** 286 https://doi.org/10.1186/s12903-019-0988-x
[11] Madfa A, Al-Sanabani F, Al-Qudami N, Al-Sanabani J and Amran A 2014 *The Open Biomaterials Journal* **5** 1-9 http://dx.doi.org/10.2174/1876502501405010001
[12] Agustín-Panadero R, Román-Rodríguez J, Ferreiroa A, Solá-Ruíz M and Fons-Font A 2014 *J. Clin. Exp. Dent.* **6** (1) 66-73 https://dx.doi.org/10.4317%2Fjced.51304
[13] Alfawaz Y 2016 *J. Contemp. Dent. Pract* **17** (5) 418-22 https://dx.doi.org/10.5005/JP-JOURNALS-10024-1865
[14] Hansen T, Schriwer C, Øilo M and Gjengeda H 2018 *J. Dent.* **72** 14–20, https://doi.org/10.1016/j.jdent.2018.01.013
[15] Miyazaki T and Hotta Y 2011 *Aust. Dent. J.* **56** 97–106 https://doi.org/10.1111/j.1834-7819.2010.01300.x
[16] Panagiotis K, Konstantinos K and Markopoulos A P 2020 *Machines* **8** 13 https://doi.org/10.3390/machines8010013
[17] Susic I, Travar M and Susic M 2017 *IOP Conf. Series: Mater. Sci. Eng.* **200** 012020, https://doi.org/10.1088/1757-899X/200/1/012020
[18] Mrugalska B, Dovramadjiev T, Pavlova D, Filchev R, Stoeva M, Bozhikova V and Dimova R 2021 *Procedia Manuf.* **54** 296-301 http://dx.doi.org/10.1016/j.promfg.2021.09.002.
[19] Dovramadjiev T, Pavlova D and Radeva J 2021 Information and Communication Technology Application in Healthcare with Computer-Aided Design of Immediate Partial Dentures. *Advances in Human Factors and Ergonomics in Healthcare and Medical Devices. AHFE 2021. Lecture Notes in Networks and Systems* vol **263** (Cham, Springer) https://doi.org/10.1007/978-3-030-80744-3_26
[20] Yeğin E and Atala M H 2020 *Int. J. Implant Dent.* **6** 69 https://doi.org/10.1186/s40729-020-00267-x
[21] Balamurugan S 2020 Conference: 31st National Convention of Production Engineers.At: The Institution of Engineers (Karnataka state, Bangaluru, India)
[22] Abdulla A M, Ali H K and Jamel R S 2020 *Al–Rafidain Dent. J.* **20** (1) 95-113, http://dx.doi.org/10.33899/rden.2020.164542
[23] Theodoros K, Linah M A, Gallucci G O, Weber H and Wismeijer D 2009 *Int. J. Oral Maxillofac. Implants* **24** 110-117
[24] Alekseev K V, Blynskaya E V, Tishkov S V, Alekseev V K and Ivanov A 2020 *Russ. J. of Biotherapy* **19** 13-21 https://doi.org/10.17650/1726-9784-2019-19-1-13-21
[25] Prasad S, Kader N, Sujatha G, Raj T and Patil S 2018 *J. 3DPrint. Med.* **2** (3) 89–91 https://doi.org/10.2217/3dp-2018-0012
[26] Shin S, Lim J, Kang Y, Kim J, Shim J, Kim J 2020 *Materials* **13** 5433 https://doi.org/10.3390/ma13235433
[27] Dobrzanski A L and Dobrzanski B L 2020 *Processes* **8** 525 https://doi.org/10.3390/pr8050525
[28] Tri Dental Implants, Switzerland, https://tri-implants.swiss/
[29] Dentaprime Dental Clinic, Varna, Bulgaria, https://www.dentaprime.com/
[30] Dental Software 3Shape CAD/CAM software. https://www.3shape.com/en/software-overview
[31] Millbox software – Dental CAM system. https://www.cimsystem.com/dental/products/millbox-overview/

Absorbing materials based on magnetically controlled Fe-Al microparticles

M D Sosnin and I A Shorstkii

Advanced technologies and new materials laboratory, Kuban State Technological University, 2 Moskovskaya st., Krasnodar, Russian Federation

i-shorstky@mail.ru

Abstract. Rotating permanent magnetic field (RPMF) method was used to obtain an absorbing material based on hexagonally packaged Fe-Al particles. A new kind of material - absorber in the microwave range was obtained. As the main characteristic, the reflection, absorption and attenuation spectra of the electromagnetic radiation of Fe-Al composites were obtained and analysed. Reflection characteristics of Fe-Al composite materials using the RPMF method were improved.

1. Introduction

Materials for absorbing electromagnetic waves are being investigated as coatings against electromagnetic interference in self-covering technologies, and their properties of electromagnetic waves reflection and absorption have been reported in the literature [1–9]. These composite materials mainly consist of metal oxides, alloy–epoxy composites, and magnetic powder materials in different size ranges which have generated particular interest. Major characteristics such as complex permeability ($\mu'-j\mu''$) and permittivity ($\varepsilon'-j\varepsilon''$) which determine the reflection and attenuation characteristics, magnetic materials such as Fe and Al powders are among the suitable candidates for use in the absorption of electromagnetic waves in the GHz band, since these materials have high magnetization values at high frequencies.

Several factors such as morphology, especially particle packing, are crucial for determining the properties of electromagnetic absorption. Currently, various methods are used to produce materials-absorbers of various morphologies: nanowires, nanoparticles [10], flake, nanodisks [11] and three-dimensional tubes, spheres, flower structures [12,13]. As it was reported in our previous paper tetrahedral packing structures shows good absorption characteristics in a wide frequency range [14,15].

Many investigations have been undertaken to improve reflection characteristics, specifically the application of other metals on the core surface. Coating processes commonly apply Al on ferrite surfaces due to the excellent corrosion resistance of Al and its alloys [16]. These features make Fe–Al compounds attractive materials for structural and coating applications at elevated temperatures and in hostile environments [17].

Mechanical alloying (MA) is a powder processing technique used to synthesize a variety of non-equilibrium and equilibrium materials starting from blended elemental and compound powders at room temperature in a ball-mill [18]. Unfortunately using the MA method is not possible to regulate connection forces of core and shell. The rotating permanent magnetic field (RPMF) is a new kind of technology, which allowed it to provide a coating process of non-magnetic materials of metal surfaces [19].

In the current study, a novel RPMF method was used for a Fe(Al) composite materials synthesis and its absorption characteristics evaluation.

2. Materials

Iron spherical particles with an average size of 60 ± 10 μm were used as a main fabricating composite material compound. As a second compound of fabricating composite material aluminium nanoparticles (60 ±5 nm) were used. Paraffin was used as a matrix of composite materials. All materials were purchased from a local powder metallurgy company (Krasnodar, Russia).

3. Methods

Preparation of Porous Fe/Al Composite.

Fe(Al) microparticles were prepared by the rotating permanent magnetic field (RPMF) method described in [14,19]. Briefly, a Fe(Al) sample synthesis is shown graphically in Figure 1. The process consists of two steps: magnetically alloying of Fe particles by Al using rotating magnetic field with N-N orientation and composite material formation of a tetrahedral-type particles packaging using rotating magnetic field with N-S orientation. Finally, the product was cut into samples measuring 23x10 mm with a thickness of 5 mm. During the routing process, the condition of the direction of the magnetic field along the wrong side of the samples was observed.

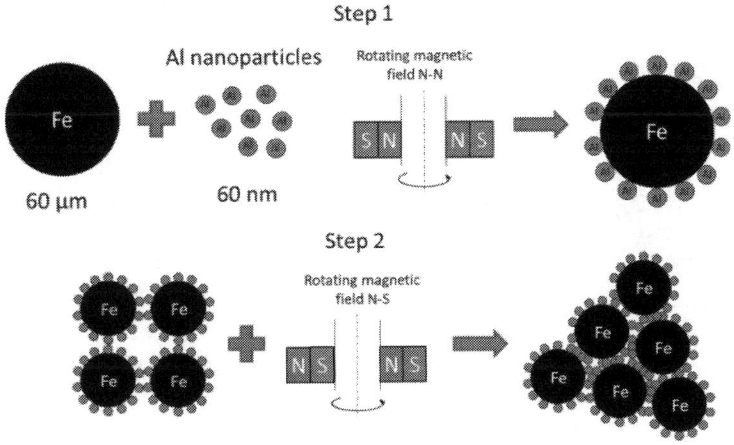

Figure 1. The coating process of Fe particles by Al nanoparticles using rotating magnetic field with magnets direction of N-N (step 1) and synthesis of composite material using rotating magnetic field with magnets direction of N-S (step 2). Adopted from [19]

Characterizations

The phase analysis was verified using powder X-ray diffraction (XRD) (Shimadzu XRD–7000 S diffractometer) using Cu Ka radiation (λ=1541 Å) with a scanning voltage of 40 kV, a scanning current of 30 mA.

Scanning electron microscopy JSM 7500 (Jeol, Japan) (operating at an accelerating voltage of 10.0 kV and equipped with energy-dispersive X-ray spectroscopy) was used to observe morphological features and sizes.

The S parameters, including S11, S12, S21 and S22, will be checked by the Keysight ENA E5080B vector network analyzer using the transmission line method. Rectangular samples (23x10 mm) and waveguide V1-23 x 100 mm with a self-made flange were used for measurements. Internal Keysight ENA software was used for the ε′, ε″, μ′, and μ″ values calculation. Finally, the RL value with different thicknesses can be calculated by using the method, described in the literature [20,21].

IOP Conf. Series: Materials Science and Engineering 1235 (2022) 012027 doi:10.1088/1757-899X/1235/1/012027

Modelling Fe-Al structure transition from cubic packaging to tetrahedral-type was carried out using the program Vizimag V 3.18.

4. Results and discussion

Particles under the influence of a rotating permanent magnetic field tend to form chains along the lines of force (Figure 2).

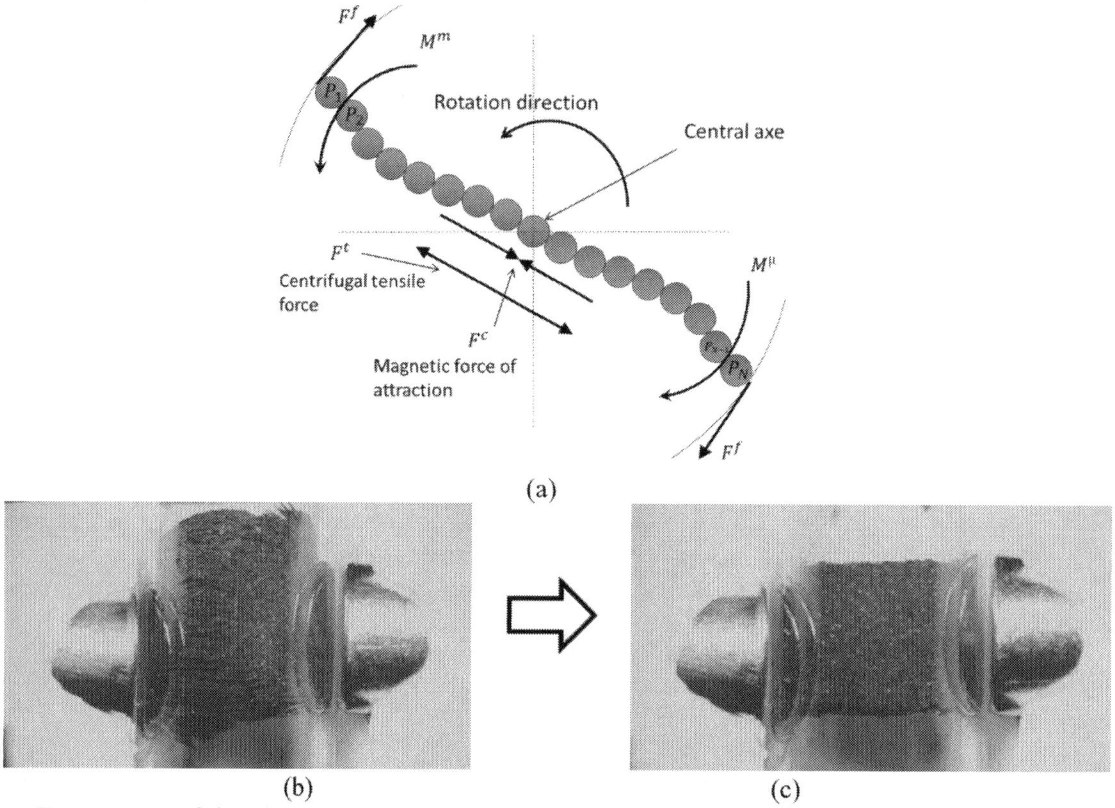

Figure 2. Particles chain under influence of rotating magnetic field (a), particle alloy before rotating (b) and after rotation (c).

In the case of a rotational magnetic field, the formed particles tend to change their orientation relative to the field (Figure 2). The magnetic moment M^m and the opposite moment of viscosity M^μ arise on a chain of particles when a rotational magnetic field is applied to it, as shown in Figure 2 and are expressed by the equations:

$$M^m = \frac{\mu_0 - \mu_r}{4\pi}\frac{3|\vec{m}|^2 N^2}{2(2a)^3}\sin(2\alpha)$$

$$M^\mu = \frac{4}{3}N\pi a^3\,\frac{2N^2}{\ln\left(\frac{N}{2}\right)}\eta\omega$$

In this equation, α – defines the angle between the rotational field and the magnetization vector of the particle, N – the number of particles in the chain, μ_r-the relative permeability of the medium, n-the viscosity of the medium, and ω-the angular velocity of the chain.

The tensile force F^t acting on the chain arises as a result of the centrifugal force of rotating particles (which is very small compared to the magnetic force) and does not contribute to the disintegration of the chain. The force of attraction F^c arises as a result of the dipole interaction between the particles and tends to hold the particles together. The friction force F^f, between the vessel wall and the particle, occurs due to the coefficient of friction of the particle against the wall of the working container and is one of the forces that destroy the chain of particles. The angle α is insignificant in the absence of friction $f_l=0$ with the vessel wall and the particles quickly change their orientation under the influence of a rotational magnetic field. However, when the value $f_l>0$, the angle of deformation of the chain also changes. According to [22], this coefficient has a destructive character for a chain of particles at fl>1, which means that the angle α reaches its critical value and the chain is destroyed under the influence of displacement. However, it is worth noting that this angle depends on the magnetization parameter μ of the particles themselves. So, in the work [23], the factor of tortuosity of granules was neglected.

When modelling a monolayer array of particles in a rotational magnetic field, the Vizimag software constructed the structure of particles in an external magnetic field with the simulation parameters shown in figure 3. The chain of particles is carried along by the external magnetic field even with a slight rotation. When modelling an array of particles, n=78 spheres, the effect of the most unused zones of the array of particles is interesting (dark area Figure 3). When the rotation is applied, the dark area of the pattern decreases, which is explained by the involvement of a larger number of particles in the movement, with a constant change in the set of particles in the chain, thereby confirming the theory of dense packing presented in [15].

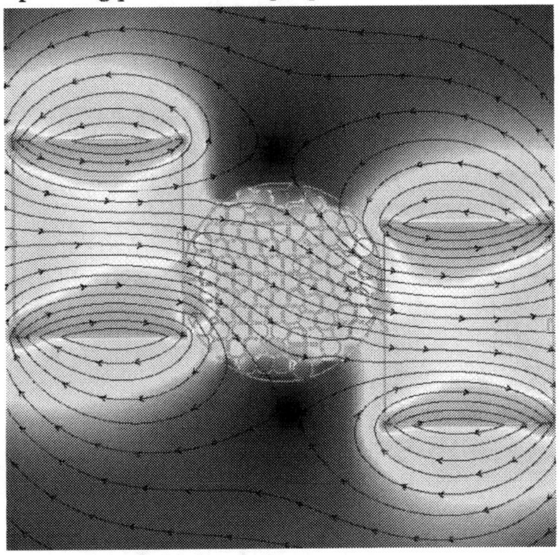

Figure 3. Simulation of magnetic field flux of two permanent magnets with spherical particles interaction

Morphology

The morphological characteristics of the Fe(Al) sample are shown in Figure 4. It can be seen that the Fe(Al) sample is a porous Fe(Al) structure with a core-shell with hexagonal packing of particles. The EDX results demonstrate the presence of the Al nanoscale layer in the Fe(Al) sample in a mass ratio of 11:89, which corresponds well to the thin layer of shell particles. The formed array of Fe(Al) particles, in accordance with the action of a rotating magnetic field [15], obtained a porous structure with pores of approximately 15 microns in diameter. All pores were filled with Al nanoparticles.

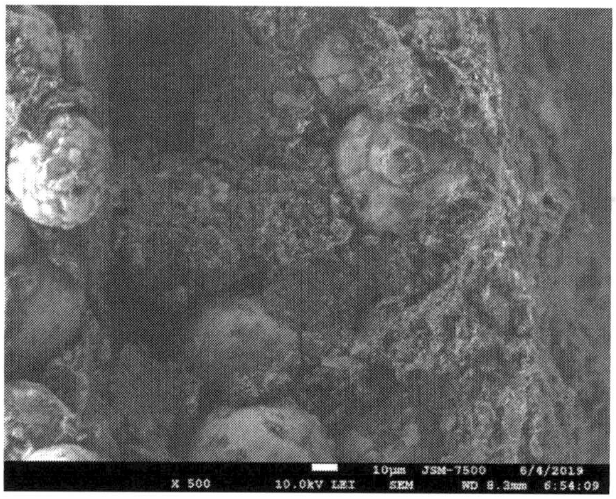

Figure 4. SEM image of Fe(Al) particles array in a paraffin matrix.

Absorption characteristics

Figure 5 shows the calculated value of the reflection losses of a composite material sample made according to the proposed technology. In general, a reflection loss value of less than -10 dB can attenuate about 90% of the incident electromagnetic wave, and thus $RL_{min} < -10$ dB is considered an ideal material for absorbing microwaves. It can be observed that RL increases with increasing frequency range for a composite material sample. The resulting sample has excellent microwave absorption properties both in the minimum RL value and in the frequency width. The optimal RL_{min} of -27 dB is achieved at a frequency of 12.0 GHz.

In comparison with literature data of absorption characteristics for Fe based composite materials, our RL values in the frequency range from 8 to 12 GHz demonstrates stronger absorption characteristics [24]. The attenuation coefficient indicates that Al shell, as a result of cladding a spherical particle of Fe according to the proposed method, has Foucault currents. Thus, the hypothesis proposed in our previous paper [19] is approved.

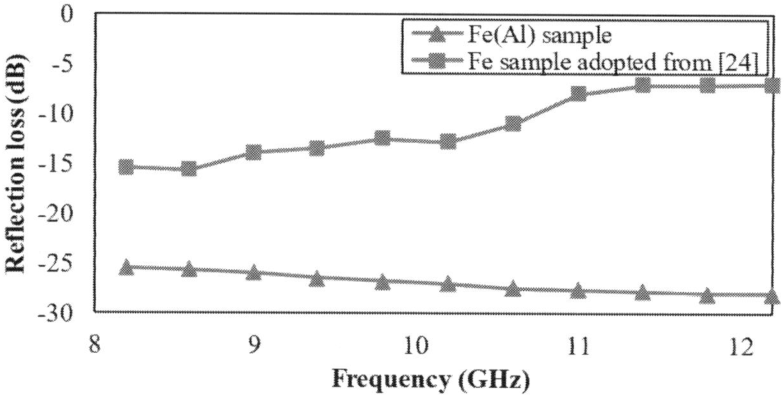

Figure 5. Reflection loss curves of the Fe(Al) composite material and Fe samples with nano-plates [24].

5. Conclusions

In this study, we have developed new Fe(Al) structures with a porous core-shell with hexagonal packing of particles using a rotational magnetic field. The mechanism of Fe particles coating by aluminium nanoparticles using mechanical friction was described. Suggested hypothesis about the effect of the surface charge density in the Al nanoparticles layer on the magnetization of the Fe core was tested based on electromagnetic radiation reflection spectra. The calculated reflection losses imply that Fe(Al) composite materials with a porous core and shell have attractive electromagnetic radiation absorption properties. Experimental results based on the reflection coefficient values confirm the proposed hypothesis for a composite material made of Fe(Al) particles.

The most effective frequency bandwidth was observed in the range of 8-12 GHz for the composite material thickness of 5 mm. An optimal frequency value for porous Fe(Al) composite material reached at -27 dB, which is higher than -15.2 dB for composite materials based on Fe particles from the literature. Meanwhile, the Fe(Al) structure with hexagonal particle packing promotes the scattering of electromagnetic waves and provides improved microwave absorption properties.

Funding
"The innovation project was carried out with the financial support of the Kuban Science Foundation in the framework of the Commercilizable scientific and innovation projects competition № НИП-20.1-63/20".

ORCID iDs
M D Sosnin https://orcid.org/0000-0001-6275-6274
I A Shorstkii https://orcid.org/0000-0001-5804-7950

6. References

[1] Peng Y, Nie J, Zhang W, Bao C, Ma J and Cao Y 2015 Preparation of soft magnetic composites for Fe particles coated with $(NiZn)Fe_2O_4$ via microwave treatment *Journal of Magnetism and Magnetic Materials* **395** 245–50 https://doi.org/10.1016/j.jmmm.2015.07.071
[2] Jian X, Wu B, Wei Y, Dou S X, Wang X, He W and Mahmood N 2016 Facile synthesis of Fe_3O_4/GCs composites and their enhanced microwave absorption properties *ACS applied materials & interfaces 8* (9) 6101-9 https://doi.org/10.1021/acsami.6b00388
[3] Niu M, Pham-Huy C and He H 2016 Core-shell nanoparticles coated with molecularly imprinted polymers: a review *Microchimica Acta* **183** 2677-95 https://doi.org/10.1007/s00604-016-1930-4
[4] Luo F, Luo Z, Hu W, Li G, Li Y, Liu X and Wang J 2019 Ultra-low inter-particle eddy current loss of Fe_3Si/Al_2O_3 soft magnetic composites evolved from $FeSiAl/Fe_3O_4$ core-shell particles *Journal of Magnetism and Magnetic Materials* **484** 218-24 https://doi.org/10.1016/j.jmmm.2019.04.027
[5] Zhang C, Chen H, Ma M and Yang Z 2015 Facile synthesis of magnetically recoverable Fe_3O_4/Al_2O_3/molecularly imprinted TiO_2 nanocomposites and its molecular recognitive photocatalytic degradation of target contaminant *Journal of Molecular Catalysis A: Chemical* **402** 10-6 https://doi.org/10.1016/j.molcata.2015.03.008
[6] Duan W, Li X, Wang Y, Qiang R, Tian C, Wang N, Han X and Du Y 2018 Surface functionalization of carbonyl iron with aluminum phosphate coating toward enhanced anti-oxidative ability and microwave absorption properties *Applied Surface Science* **427** part A 594-602 https://doi.org/10.1016/j.apsusc.2017.08.034
[7] Komarov F F, Milchanin O V, Parfimovich I D, Grinchenko M V, Parhomenko I N, Tkachev A G and Bychanok D S 2017 Absorption and Reflectance Spectra of Microwave Radiation by an Epoxy Resin Composite with Multi-Walled Carbon Nanotubes *Journal of Applied Spectroscopy* **84** (4) 596-602 https://doi.org/10.1007/s10812-017-0516-1
[8] Serebryannikov S V, Cherkasov A P, Serebryannikov S S, Dolgov A V, Eremtsova L L and Konshin P I 2020 Radio-Absorbing Composites for Microwave Devices *Bulletin of the Russian Academy of Sciences: Physics* **84** (5) 609-12 https://doi.org/10.3103/S1062873820050275

[9] Komarov F F, Zukowski P, Kryvasheyeu R M, Munoz E, Koltunowicz T N, Rodionova V N and Togambaeva A K 2015 Effects of surfactant and fabrication procedure on the electrical conductivity and electromagnetic shielding of single-walled carbon nanotube films *Phys. Status Solidi A* **212** 425-32 https://doi.org/10.1002/pssa.201431493
[10] Green M, Tian L, Xiang P, Murowchick J, Tan X and Chen X 2018 FeP nanoparticles: a new material for microwave absorption *Materials Chemistry Frontiers* **2** 1119-25 https://doi.org/10.1039/C8QM00003D
[11] Li X, Wang L, You W, Xing L, Yu X, Li Y and Che R 2019 Morphology-controlled synthesis and excellent microwave absorption performance of $ZnCo_2O_4$ nanostructures via a self-assembly process of flake units *Nanoscale* **11** (6) 2694-702 https://doi.org/10.1039/C8NR08601J
[12] Liu L, He N, Wu T, Hu P and Tong G 2019 Co/C/Fe/C hierarchical flowers with strawberry-like surface as surface plasmon for enhanced permittivity, permeability, and microwave absorption properties *Chemical Engineering Journal* **355** 103-8 https://doi.org/10.1016/j.cej.2018.08.131
[13] Li C, Ge Y, Jiang X, Sui J, Zhang Z and Yu L 2019 Lightweight three-dimensional Fe_3O_4/carbon micro-flowers with tunable microwave absorption properties *Journal of Alloys and Compounds* **798** 414-23 https://doi.org/10.1016/j.jallcom.2019.05.250
[14] Shorstkii I A and Yakovlev N 2020 Method of Absorbing Material Formation Based on Magnetically Controlled Particles of Fe_3O_4 *Inorg. Mater. Appl. Res.* **11** 1236–43 https://doi.org/10.1134/S2075113320050317
[15] Shorstkii I and Yakovlev N 2019 Synthesis of magnetically controlled Fe_3O_4 composites and their enhanced microwave absorption properties *Mater. Res. Express* **6** 046104 https://doi.org/10.1088/2053-1591/aafc58
[16] Zhou Z, Xie F and Hu J 2008 A novel powder aluminizing technology assisted by direct current field at low temperatures *Surface and Coatings Technology* **203** (1-2) 23-7 https://doi.org/10.1016/j.surfcoat.2008.07.021
[17] Deevi S C, Sikka V K and Liu C T 1997 Processing, properties, and applications of nickel and iron aluminides *Progress in Materials Science* **42** (1-4) 177-92 https://doi.org/10.1016/S0079-6425(97)00014-5
[18] Farahbakhsh I, Zakeri A, Manikandan P and Hokamoto K 2011 Evaluation of nanostructured coating layers formed on Ni balls during mechanical alloying of Cu powder *Applied Surface Science* **257** (7) 2830-7 https://doi.org/10.1016/j.apsusc.2010.10.071
[19] Shorstkii I and Sosnin M 2021 Microwave Absorption Properties of Fe_3O_4 Particles Coated with Al via Rotating Magnetic Field Method *Coatings* **11** (6) 621 https://doi.org/10.3390/coatings11060621
[20] Nicolson A M and Ross G F 1970 Measurement of the intrinsic properties of materials by time-domain techniques *IEEE Transactions on instrumentation and measurement* 19 (4) 377-82 https://doi.org/10.1109/TIM.1970.4313932
[21] Weir W B 1974 Automatic measurement of complex dielectric constant and permeability at microwave frequencies *Proceedings of the IEEE* **62** (1) 33-6 https://doi.org/10.1109/PROC.1974.9382
[22] Bezzubtseva M M 2018 The probability of crushing particles of material in magnetic liquefied layer of electromagnetic mechanoactivation *European Scientific Conference* – 2018 – pp 22-4
[23] Sandulyak A V, Sandulyak A A and Ershova V A 2009 On the model of channel-by-channel magnetization of a granular medium (with a radial permeability profile of a quasi-continuous channel) *Tech. Phys.* **54** (5) 743-5 https://doi.org/10.1134/S1063784209050235
[24] Yang P, Yu M, Fu J and Wang L 2017 Synthesis and microwave absorption properties of hierarchical Fe micro-sphere assembly by nano-plates *Journal of Alloys and Compounds* **721** 449-55 https://doi.org/10.1016/j.jallcom.2017.05.296

IOP Conf. Series: Materials Science and Engineering 1235 (2022) 012028 doi:10.1088/1757-899X/1235/1/012028

Study of filtering capacity of some ceramic composites on bentonite base

S Butuc (Anghel), S Matei, M Stoicanescu and A Crisan

Transilvania University of Braşov, Material Science Departament, 29 Eroilor Blvd, 500036, Braşov, România

stoican.m@unitbv.ro, sorin.butuc@unitbv.ro

Abstract. The paper presents the results of a study of the filtering capacity of some filters from ceramic composite materials made on laboratory installation. The filtration process involved applying a pressure gradient to the filter by means of a centrifugal pump from the filter installation. The ceramic composite materials used to make the filters have a bentonite-based matrix. To achieve the porous structure, the matrix was reinforced with SiC and Al_2O_3 ceramic powders and to control the pore size, additional Al metal particles were added. The stages of obtaining and characterizing the filters, respectively of testing their filtering capacity of the suspended particles from the rainwater are presented distinctly. In conclusion, the dependencies between a series of structural and geometric characteristics of the filtering capacity were established.

1. Introduction

Natural raw materials, such as clay, feldspars, quartz sand, etc., gave rise to the manufacture of the first ceramic products. Along with these raw materials, there appeared synthetic materials, such as alumina, and the need for products featuring special uses and properties led at times to the full use of synthetic raw materials. Ceramic products are relatively inexpensive and widespread. For their properties, ceramic products have been used in many fields. Clay-based ceramic materials have been used by purifying and storing water at points of use [1,2]. Ceramic filters are used for water purification [3].

Mustapha *et al.* [4] consider the use of porous ceramic for wastewater treatment an effective technology. Agbo *et al.* [5] consider water filtration to be the passing of water through a permeable ceramic material. Various types of clay, sawdust and colloidal silver can be used to manufacture ceramic filters [6]. Akosile *et al.* [6] investigated ceramic filters based on clay materials and their properties in terms of improving the quality of drinking water. Ceramic filters feature low weight, accessibility, low manufacturing cost [7]. The use of ceramic filters for water treatment causes no changes in water temperature, taste, etc., as opposed to the chemical or thermal disinfection procedures [7]. There have been manufactured ceramic filters made of clay, diatomite and sawdust [8], a mixture of 80% clayey soil and 20% rice bran [9,10], intended to retain water suspensions and purify it. Chaukura *et al.* [11] propose the manufacture of layered "biscuit" ceramic filters using biomass and clay. This type of filter is effective in eliminating the total hardness of water, of dissolved solid suspensions, and it regulates turbidity [11]. The performance yielded by this type of filter is comparable to the sand filtration step in a conventional water treatment plant. Water turbidity [10,11] can be the main indicator in characterizing the filter life. Nazarova *et al.* [12] obtained a macroporous ceramic filter which can be used to remove abrasive impurities from liquids.

This paper is a study on obtaining some materials based on bentonites reinforced with ceramic and metallic particles, prepared in filter-type test samples in order to determine their filtration capacity.

2. Method used and materials

The matrix of the composite material consists of powdered bentonite (53% montmorillonite (Na,Ca) 0.33 $(Al,Mg)_2(Si_4O_{10})(OH)_2{\cdot}nH2O$; 1.3% Fe_2O_3; 45.7% SiO_2) - providing good absorption properties [13,14]. SiC and Al_2O_3 ceramic powders [15] were used as reinforcement materials in order to ensure the porosity of the composites [15,16]. The higher the proportion of these particles, the thinner the layer of bentonite covering them, thus increasing the volume of the pores in the composite [17]. Al metallic powders were added to control the pore size [15-18]. The grain size featured by SiC ceramic powders is 60-80 μm, by Al_2O_3 is of approximately 50 μm, and by Al metallic particles is below 65 μm [15].

24 g bentonite mixed with 36 g SiC, respectively 24 g bentonite mixed with 36 g Al_2O_3 were weighed analytically. There was obtained 40% bentonite and 60% SiC, respectively Al_2O_3. Thus, there were obtained two groups of materials, denoted in the paper by P and S. The materials were very well mixed by homogenisation for 15 minutes and following the homogenisation, 7.2% distilled H_2O for 60 g of the base mixture, i.e. 12% distilled H_2O for 100 g mixture, was added to each group of materials. 3%, respectively 9%, Al metallic particles were added to each group of materials, continuing their mixing until complete homogenisation. The samples based on bentonite and SiC were denoted by P, and the 3%, respectively 9%, Al metallic particles percentage was indicated. Similarly, the samples based on bentonite and Al_2O_3 were denoted by P, and the 3%, respectively 9%, Al metallic particles percentage was indicated. The materials obtained were compacted in a mould by applying a 20 kN force. The materials were pressed using the WDW-150S-type universal testing machine. There were obtained cylindrical samples, 2 for each type of material, with a diameter of 16 mm and a height between 14 and 16 mm.

Following the compaction of the cylindrical samples, to remove humidity, the batch of samples was dried in the open air at a temperature of T = 20-23 °C, for 48 h. Afterwards, the samples were sintered [19] at a temperature of T = 1100 °C in a Nabertherm type oven according to the following treatment cycle: heating from T = 40 °C to T = 1100 °C for 71 min, holding at T = 1100 °C for 30 min, followed by cooling from T = 1100 °C to T = 40 °C for 90 min.

3. Experimental results and discussions

3.1 Compressive strength

The sintered samples were compression tested on the WDW-150S universal testing machine. It develops test forces between 1 and 150 kN, for the study of the behaviour of samples under compression, traction, bending stresses, at temperatures between -120 and 1200 °C. The test results are presented in the form

Table 1. Results of sintered samples under compression stress.

Type sample	F_{bc} [kN]	F_{sc} [kN]	F_{pc} [kN]	R_{bc} [MPa]	R_{sc} [MPa]	R_{pc} [MPa]	E_c [GPa]
P1-3AL	14.82	7.63	7.99	77	39	41	1.85
P2-3AL	17.44	2.48	10.35	89	13	53	2.97
P1-9AL	16.97	2.62	10.37	87	13	53	3.38
P2-9AL	14.54	3.50	9.50	75	18	49	2.92
S1-3AL	12.02	-	12.02	61	-	61	3.03
S2-3AL	13.01	-	13.01	67	-	67	3.42
S1-9AL	17.97	-	17.76	91	-	89	4.61
S2-9AL	21.89	3.91	21.43	109	19	107	5.57

of diagrams (e.g. Figure 1) containing all the values of the main quantities specific to the type of test. The results of the compression tests conducted in this study are presented in Table 1. In table 1 Fbc is the maximum strength withstood by the tested sample; Fsc is the compression breaking strength; Fpc is the initial compressive strength; Rbc is the strength of the material to the maximum force; Rsc is the strength of the material before breaking; Rpc is the initial strength of the material to compression and Ec isYoung's modulus under compression. Fsc, Fpc, Rsc respectively Rpc are intermediate values displayed by the machine. The Fbc value differences are due to the dimensional variations of the samples.

An example of a diagram recorded following the compression test is shown in Figure 1.

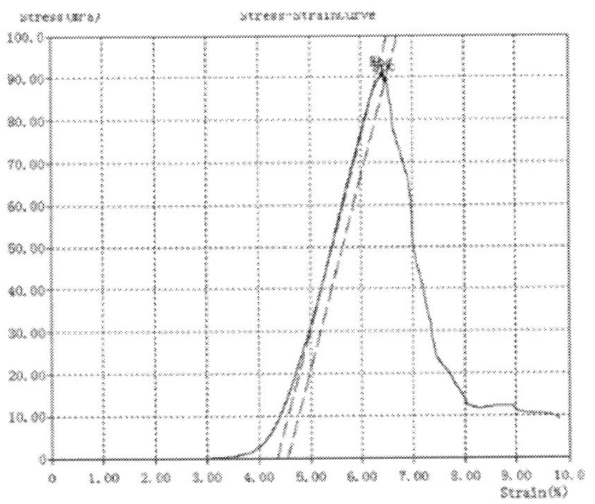

Figure 1. The compression diagram was obtained for the S2-9Al-type sample.

3.2 Porosity

To meet the objective proposed in this paper, there was determined the porosity of the samples. The porosity was determined by measuring the volume of the pores. The method considers an initial sample in a graduated cylinder [13] with an initial volume of V0 = 30 ml of distilled water. Then, the V1 value was read and recorded, where V1 is the volume of the water obtained immediately after the immersion of the sample in the cylinder. The sample was kept in the water in the cylinder for 30 minutes. At the end of this period, the V2 value of the water in the graduated cylinder was read and recorded, where V2 is the volume of the water after holding the sample in the graduated cylinder for 30 minutes. The sample was then removed from the graduated cylinder. The porosity was determined using the expression (1).

$$P = \frac{V1-V2}{V1-V0} \cdot 100 \qquad (1)$$

P was used to denote the porosity as a measurement of the volume of the open pores in the material. Two determinations were conducted for each batch of composites, as presented in Table 2, and the average of these values is shown in Figure 2 and Figure 3. Similar values have been found in previous research [17]. The number of determinations was limited to two considering the appropriate constancy of the values.

Table 2. Porosity and compressive strength values.

Sample type	V_0 [ml]	V_1 [ml]	V_2 [ml]	P [%]	Compressive strength [MPa]
P1-3AL	30.00	33.20	32.60	18.75	77
P2-3AL	30.00	33.10	32.50	19.35	75
P1-9AL	30.00	33.20	32.40	25.00	87
P2-9AL	30.00	33.30	32.50	24.24	85
S1-3AL	30.00	33.20	32.70	15.62	61
S2-3AL	30.00	33.50	32.90	17.14	63
S1-9AL	30.00	33.50	32.80	20.00	109
S2-9AL	30.00	33.60	32.90	19.44	91

From the study of the results of the samples analysed in terms of porosity and compressive strength, there were graphically represented the average values for two samples from the same batch of composites. The average values are shown in Figure 2 and Figure 3.

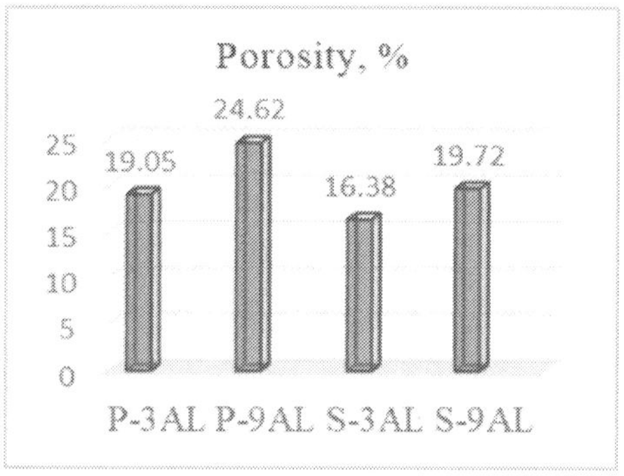

Figure 2. Graphical representation of the average porosity for the analysed ceramic composite samples

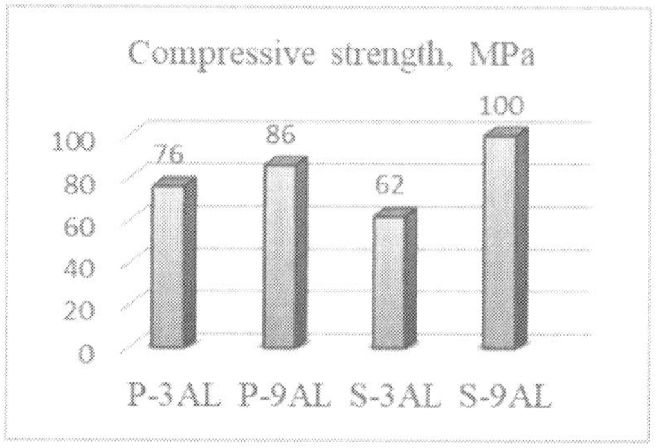

Figure 3. Graphical representation of the average compressive strength for the analysed ceramic composite samples

The porosity and mechanical strength variations of composites according to the content of Al particles are given by their oxidation during sintering in air, a process accompanied by an increase in volume and the generation of internal compression stresses. Previous studies on this type of composite have confirmed this fact [13].

3.3 Study on the filtration capacity

A laboratory installation was used to study the filtration behaviour of the samples. This allows the filtration process to be performed by applying a pressure gradient in the sample area using a centrifugal pump. The pressure difference between the two faces of the sample determines the passage of the fluid through the sample and the retention thereby of the suspended particles. Having regard to the results of the porosity and compressive strength determinations, the samples with 9% Al particle content were selected to study the filtration capacity.

To remove moisture, the samples were dried at a temperature of T = 100 °C in the Nabertherm oven. Afterwards, the samples were weighed analytically using the Radwag-type analytical balances with 10-5 mg precision. The free flow of the installation was measured before the filtration through samples. The free flow was determined by measuring the time in which a predetermined volume of 600 ml of rainwater was discharged without the composite sample being inserted into the installation. The general flow formula was applied (expression 2).

$$Q=\frac{\Delta V}{\Delta t} \tag{2}$$

A volume V1= 600 ml of rainwater was discharged in a time of t = 14.56 seconds using the centrifugal pump fitted in the laboratory installation. The free flow of the installation was determined by applying the formula (2) and there was obtained the Q1 = 2.47 l/min value, and Ql was used to denote the free flow of the installation.

For the analysed samples, the filtration procedure involved measuring the water discharged at regular intervals, thus determining the flow values in four working stages.

Upon the insertion of the samples into the filtration installation, the flow values following the filtration for 5 minutes, 10 minutes, 15 minutes and 20 minutes were calculated using the formula (2) and they are given in Table 3.

Table 3. Variations in filtration flow rates of composite samples.

Sample type	Q [l/min]			
	t=5 [min]	t=10 [min]	t=15 [min]	t=20 [min]
S-9Al	0.358	0.156	0.0533	0.0284
P-9Al	0.436	0.164	0.0622	0.0408

The graphical representation of the variations of the flows recorded at regular time intervals indicated continuous decreases of the flows. The flow rate of a ceramic filter decreases over time due to clogging with suspended particles [21]. The various values of the recorded flows are influenced by the porosity of the material. Following filtration, the samples were dried to remove excess moisture. The samples were dried in a Nabertherm oven at T = 100 °C. The weights of the samples recorded before and after filtration are shown in Table 4.

Table 4. Weights of the analysed samples after filtration.

Sample Type	Weight after sample drying in the oven before filtration (g)	Weight after sample drying in oven after filtration (g)	Weight increase (%)
S-9Al	7.27921	7.37810	1.35
P-9Al	7.97662	8.14665	2.13

The variation of the flows recorded at regular intervals depending on the nature of the ceramic filter based on silicon carbide and alumina were represented graphically in Figure 4.

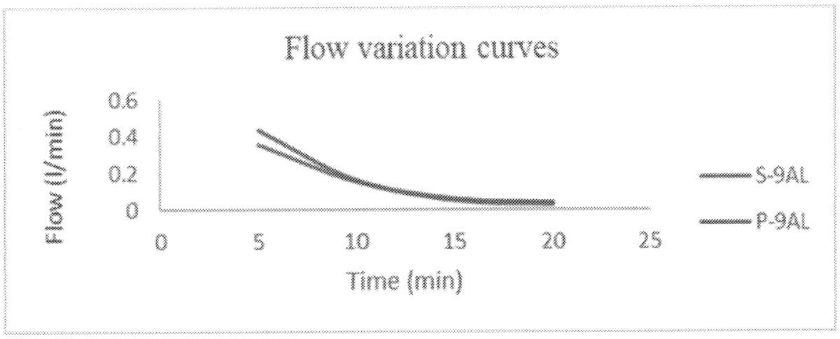

Figure 4. Variation of recorded flows.

There can be noted that, by retaining the filtering particles, the physical characteristics of the filter are modified to a great extent, mainly in terms of porosity. The weight increase of the samples used as filters confirms this process. The correlation between these indicators can be used in a later stage of dimensional optimization of the filters for certain conditions of use.

4. Conclusions

Filter samples were made using materials based on bentonite reinforced with SiC, Al_2O_3 ceramic particles and Al metallic particles. A review of the compression tests conducted shows that, for the same materials, the strength of the material increased with the increase of the percentage of Al metallic particles. The compressive strengths are high enough to preserve the geometry and dimensions of the finished products during processing and use. They are comparable to those of materials presented in similar studies [13,17,19,20]. The average porosity of the composites obtained ranged between 15% and 25%. Higher porosity values were featured by the samples made of bentonite reinforced with silicon carbide ceramic particles. During the filtration process, there was noted a relatively rapid decrease in the porosity of the composites due to the retention in the filter pores of the particles suspended in the filtered liquid. Future research will consider the detailed analysis of this process, to optimise the filtration characteristics of composites in association with the dimensional optimization of filters, for certain conditions of use, and their metallographic analysis to establish their morphology.

ORCID iDs

S Butuc (cas Anghel) https://orcid.org/0000-0002-9680-8420
S Matei https://orcid.org/0000-0002-8040-9409
M Stoicanescu https://orcid.org/0000-0003-3138-1830
A Crisan https://orcid.org/0000-0002-3921-8096

References

[1] Ajibade F, Akosile S, Otawatuyi O, Ajibade T, Lasisi K, Adewumi J, Babatola J and Oguntuase A 2019 *Res. in Engin.* **2** 100011 https://doi.org/10.1016/J.RINENG.2019.100011
[2] Hunter P 2009 *Envirom. Sci. Tech.* **43** 8991-7 https://doi.org/10.1021/es9028217
[3] Yang X, Huang G, Zhang P, Yao Y, Li Y and Zhou S 2021 *Sci. of the Total Envir.* **786** 147424 https://doi.org/10.1016/j.scitotenv.2021.147424
[4] Mustapha S, Oladejo T J, Muhammed N M, Saka A A, Oluwabunmi A A, Abdulkabir M and Joel O O 2021 *Sci. Afr.* **11** e00705 https://doi.org/10.1016/j.sciaf.2021.e00705
[5] Agbo S, Ekpunobi E, Onu C and Akponie K, 2017 *Int. J. Multidiscip.Sci.Eng.* **6** 18-23
[6] Akosile S I, Ajibade F O, Lasisi K H, Ajibade T F, Adewumi J R, Babatola J O and Oguntuase A M 2020 *Sci. Afr.* **7** e00218 http://dx.doi.org/10.1016/j.sciaf.2019.e00218
[7] Martens O and Nelson A 2011 *Int. J. Sci. Adv. Tech.* **1** 225-233
[8] Ekpunobi U, Agbo S and Ajiwe V 2019 *Journ. of Envir. Chem. Engin.* **7** 102791 https://doi.org/10.1016/j.jece.2018.11.036
[9] Mahmudul M, Tanveer S and Nakajima J 2019 *Envir. Techn. & Innov.* **14** 100319 http://dx.doi.org/10.1016/j.eti.2019.100319
[10] Salvinelli C, Elmore A, Reidmeyer M, Drake K and Ahmad K 2016 *Water Res.* **104** 28-33 https://doi.org/10.1016/j.watres.2016.07.076
[11] Chaukura N, Chiworeso R, Gwenzi W, Motsa M M, Munzeiwa W, Moyo W, Chikurunhe I and Nkambule T 2020 *Appl. Clay Sci.* **185** 105409 https://doi.org/10.1016/j.clay.2019.105409
[12] Nazarova E, Alimova D, Mikhaylov V, Krivoshapkina E F and Krivoshapkin P 2019 *Cer. Int.* **45** (7A) 8767-73 https://doi.org/10.1016/j.ceramint.2019.01.201
[13] Matei S and Crisan 2020 A *Comp. termopl. Comp. termorez.* EUT Brasov
[14] Lee H, Rukmanikrishnan B and Lee J 2019 *Int. Jour. of Biolog. Macr.* **141** 538–44 https://doi.org/10.1016/j.ijbiomac.2019.09.021
[15] Hubadillah S, Harum Z, Othman M, Ismail A and Gani P 2016 *Jour.of Asian Cer. Soc.* **4** 164-77 https://doi.org/10.1016/j.jascer.2016.02.002
[16] San O and Ozgur C 2007 *Journ. of Membr. Sci.* **305** (1-2) 169-75 https://doi.org/10.1016/j.memsci.2007.08.001
[17] Matei S, Stoicanescu M, Varga B, Tiron E and Crisan A 2021 *Andv.in Mech.Eng* **13** https://doi.org/10.1177%2F16878140211011888
[18] Xu C 2005 *Cer. Int.* **31** (4) 537–42 https://doi.org/10.1016/j.ceramint.2004.06.019
[19] Choi H-J, Kim J-U, Kim H-S, Kim S H and Lee M-H 2015 *Cer. Int.* **41** (8) 10030-7 https://doi.org/10.1016/j.ceramint.2015.04.090
[20] Youmoue M, Fongang R T, Sofack J C, Kamseu E, Melo U C, Tonle I K, Leonelli C and Rossignol S 2017 *Cer. Int.* **43** (5) 4496-507 https://doi.org/10.1016/j.ceramint.2016.12.101
[21] Hurst C 2019 Options for Providing Microbiologically Safe Drinking Water *The Structure and Function of Aquatic Microbial Communities* (Cham: Springer) chapter 8 pp 185-260 https://doi.org/10.1007/978-3-030-16775-2_8

IOP Conf. Series: Materials Science and Engineering 1235 (2022) 012029 doi:10.1088/1757-899X/1235/1/012029

Composite materials with polychromatic effects used in electrical installations

Mircea Florin Drăghici and Maria Stoicanescu
Transylvania University of Brasov Romania; Faculty of Materials Science and Engineering

mircea.draghici@unitbv.ro; stoican.m@unitbv.ro

Abstract. The article presents the importance of using composite materials with polychromatic and heat-sensitive effects that can be used in electrical components and have an important role in transmitting visual information on various parameters of electrical components or materials in which they are used. The main condition to use any composite material to be electrical isolator is the low conductivity. The polymers are the near with these requirements and it can be used in such domain. This paper proposes to use in some applications (with a method, which will be studied in detail in future research) a basic polychromatic and heat-sensitive polymer as an electrical insulator, which has good mechanical and durability properties and at the same time provides information on the state of the material in real-time (thermal properties, electrical properties, etc.). The connection between the material features and the domain in which this material is being used is very important. The research is based on both theoretical knowledge and practice (how to obtain the material and its properties).

1. Introduction

This paper tries to bring into discussion an original topic, that of the preparation of new composite materials for specific applications in the field of electrical installations.

Materials used in electrical engineering are generally classified into several types of materials: conductive materials, semiconductor materials, electrical insulating materials, magnetic materials [1].

Any technical system must meet a mandatory requirement, namely, to operate in direction to perform all the functions for which it was created, keeping the quality parameters within given limits, and the frequency of repairs to be such as not to reduce the economic efficiency of operating the system respectively [2]. Incorporating inorganic fillers in the electrical insulating polymers ensures both lower cost and adequate properties such as thermal, mechanical, and electrical properties required by the power industry. [3].

The composite materials proposed to be analysed are materials that have polychromatic properties and thermal acuity, the article proposes to use these properties of these materials in transmitting visual information about temperature directly, without using any type of device for interpreting information. To use composite materials in the field of electrical installations, in the structure of some electrical components (electrical appliance housings, cable insulators / electrical conductors, etc.) these materials must meet certain specifications such as high voltage puncture resistance, of the order of kV; resistance to high temperatures, resistance to open flame for a certain period (with delay or increased delay in flame propagation); low conductivity; mechanical resistance; anticorrosive resistance to the action of the main chemical reagents used in constructions (cement, lime, plaster, etc.); high elasticity; the high lifespan of the material, but also the polychromatic property and thermal sensitivity, comparable to the lifespan of the electrical appliance for which the material is used.

The transformer stations along with those of distribution, sub-distribution and the final electricity users have a primordial role for the distribution of the electric energy within the energy system. [4].

Cables/ conductors are one of the main and important assets of power distribution and transmission systems.

At the moment, the entire energetic system supply have many years working hours, all theirs electric equipment are outdated, during function they had small updating procedures and, sure they need to be replaced in future [5].

In order to achieve continuity in the power supply we must take into consideration the safety of the electric circuit.

The direct information about the technical condition of electrical cables can be obtained by observing and analysing continuously of several specific functional parameters.

In any cable under-voltage and carrying electricity to the receiver, a constant temperature change is noticed, because of the electromagnetic energy becomes thermal energy.

Due to the Joule-Lenz effect, there is a spontaneous flow of heat that migrates from the inside of the cable to its surface. The evolutions of the temperature in the electric cables can be motorized for a long period, continuously, or just occasionally as required.

At the moment, continuous monitored in the electric system supply are only the transformers, transport cables connections and the protection circuit switch breakers [1].

Due to excessive overheating of the cables supply can occur negative changes in them parameters like shortening functioning time and voltage losses along the cable.

In order to improve the safety and the technical qualities of the power circuit this article offers the solution that the power supply of electrical receivers be made with electrical cables that have an insulating sheath made of composite materials with polychromatic properties, sensitive to the change of the cable temperature, their coating changing its colour depending on the temperature of the inner conductors.

2. Supervision solutions of thermal stresses for electrical low voltage cables.

Temperature variations in the cable can be detected by carefully monitoring them, thus avoiding problems of any kind that may occur in the system [6]

The thermal stress for the low voltage cables causes a negative influence in every aspect, and by default, on the maintenance of the electrical supply cables.

Cables hyperthermia puts in danger the normal operation, shortening its lifespan (operations), and on the other hand, hypothermia causes an oversized installing, unjustified economically. Hot zones of the low voltage supply cables are not always visible, and therefore, when the low voltage supply cables must perform a major task, it is possible that this operation cannot run due to the thermal overload of the low voltage supply cables.

The energy dissipated as heat in the unit of time for a homogeneous conductor is from the relation:

$$P = Ri^2 \tag{1}$$

If the length of the conductor is known, the cross-section s is constant and the current density has a constant value.

P- power dissipated, R- resistance, i- the intensity of the current.

The transfer of heat into metals is made by conduction, convection and radiation.

When in the electrical installations appear the thermal requests due to the overloads to which they are accidentally subjected for a longer time or when short circuits occur, the modifications being due to the effect of the Joule-Lenz transformation, the electrical cables degrade quickly [7].

The temperature of the electrical equipment is calculated with the relation (2) as follows:

$$\theta = \vartheta + \theta a \qquad (2)$$

Where:

θ- temperature, ~~θa~~ ambiance temperature, ϑ raised temperature (caloric effect)

In table 1 are presented the temperatures at the nominal values θad and defect θkad of the aluminum and copper cables at a certain time tk the stability in operation is given by the values of θk inferior to θad.

Table 1. Electric conductor materials Al, Cu - Restricts of temperature

No.	Low voltage cable – used material	Constructive type	θad [°C]	θkad [°C]
1	Copper	Simple or painted electric supply conductor	60	200
		Overhead electric supply conductor	80	170
2	Aluminium	Simple or painted electric supply conductor	60	180
		Overhead electric supply conductor	80	130

If we consider that the material is entirely homogem, the ambient temperature is constant, the temperature in the cable has the same value along the entire length at a certain time, in case of overload of this circuit we can study the transient heating and failure regime [7].

Thermal stability is defined as the resistance of cables subjected for a short period of time to stress due to short circuits without visibly damaging them [7,8].

Are possible few methods to supervise the temperature which are based on different phenomena and physical effects showing us with some devices: the thermographic camera; thermometers; thermoresistors; thermocouples; etc.

High precision information for controlling the temperature of the cables both on the surface, under the insulation or inside them can be obtained by continuously supervising the temperature [9].

2.1. Method using Finite Element Model

FEM makes possible the analysis of temperature estimation, using mathematical models.

Mathematical modeling is done using systems of differential equations. The Finite Element Model uses various methods in calculating variations to approximate a solution by minimizing an associated error function [10].

If the cables are placed on the surface, the current carrying capacity in the cable increases. This result is based on the specialty study done by a group of researchers on the XLPE cable mounted on the ground and/ and buried at different depths [11].

2.2. Method using thermocouples.

This method of monitoring the low voltage cables/ conductors is using an architecture designed specifically. The measurement of the multiple temperatures is made by several thermocouples mounted in different points along the low voltage supply cables/conductors. The information regarding temperatures measured from the thermocouples corresponding to different points is converted and processed in a terminal computer system.

IOP Conf. Series: Materials Science and Engineering 1235 (2022) 012029 doi:10.1088/1757-899X/1235/1/012029

The thermocouples made are from different materials Nickel-Chrome and Nickel-Aluminium which can measure temperatures within the range 0 to +1260 °C; the measured signals are then processed with digital converters with cold junction compensation and sent via SPI communication to a microcontroller,

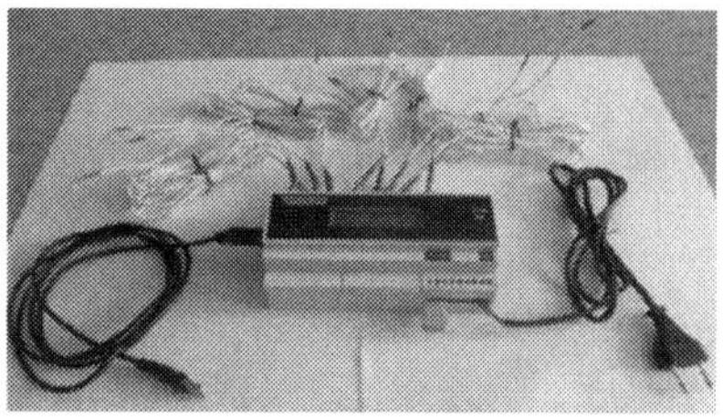

Figure 1. Sample of the temperature logger

The microcontroller (temperature logger) will display on the LCD screen the temperature of the selected thermocouple. Simultaniesly, the converted data containing the acquired temperatures are sent to a computer via USB cable where they can be seen, processed like alarms, or saved.

The most important advantage of this method is that: by using a monitoring temperature data logger with a PC interface connection can be stored, processed data and the received information can be used to schedule predictive maintenance for low voltage cables/ conductors [12].

2.3. Method using infrared thermographic monitoring.

This method of monitoring, using infrared thermographic monitoring, represents a modern temperature monitoring method for the low voltage supply cables/ conductors. Usually, this method is used in maintenance programmed times by the installation user and has the great advantage that in a shorter time we can obtain information by the measured values of the temperature.

The great minus of this solution is that: the temperature is influenced by the solar radiation, shade, humidity, etc. these lead to the appearance of incorrect values of the temperatures read on the analyzed surface of the cable [13,14].

Using an infrared camera, we can see in the figure below (Fig. 2) the temperature to which the low voltage electrical cables are subjected, which supply electrical receivers from a factory production unit (the picture was captured by the author with an IR Camera device).

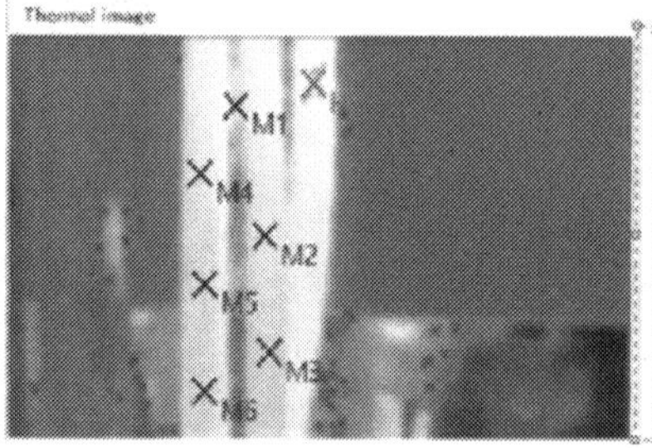

Thermal image markers:

No	Temp. [°C]	Emiss.	Refl.Temp. [°C]
M1	28.6	0.95	20.0
M2	23.8	0.95	20.0
M3	27.7	0.95	20.0
M4	28.7	0.95	20.0
M5	27.7	0.95	20.0
M6	27.3	0.95	20.0
M7	15.9	0.95	20.0
CS1	10.3	0.95	20.0
HS1	31.4	0.95	20.0

Figure 2. Image of infrared camera

The Infrared Cameras provides diagnosis impossible to get with differed means. Due to this camera, it become possible to highlight the hidden issues existing before the damage occur, more over it improves safety.

The different equipment analyzed using the infrared thermographic camera highlight instantaneous temperature values. This method is easy to apply, very efficiency and is not destructive [15].

2.4. Method using composites materials in the isolations shield of the cables/ conductors with polychromatic and heat-sensitive effects.

Similar to the infrared thermographic monitoring of temperature for the supply low voltage cables/ conductors this new method, which will be studied in detail in future research, have the advantage that the information regarding temperature surface cables/ conductors is instantly obtained, preparation of the supervised surface is not necessary.

The most important advantage is for see an anomaly of the cables/ conductor's temperature is not needed any device, the cable/ conductor is changing the colour of the isolation surface with the temperature, instantly when the maximum accepted temperature was raised.

IOP Conf. Series: Materials Science and Engineering 1235 (2022) 012029 doi:10.1088/1757-899X/1235/1/012029

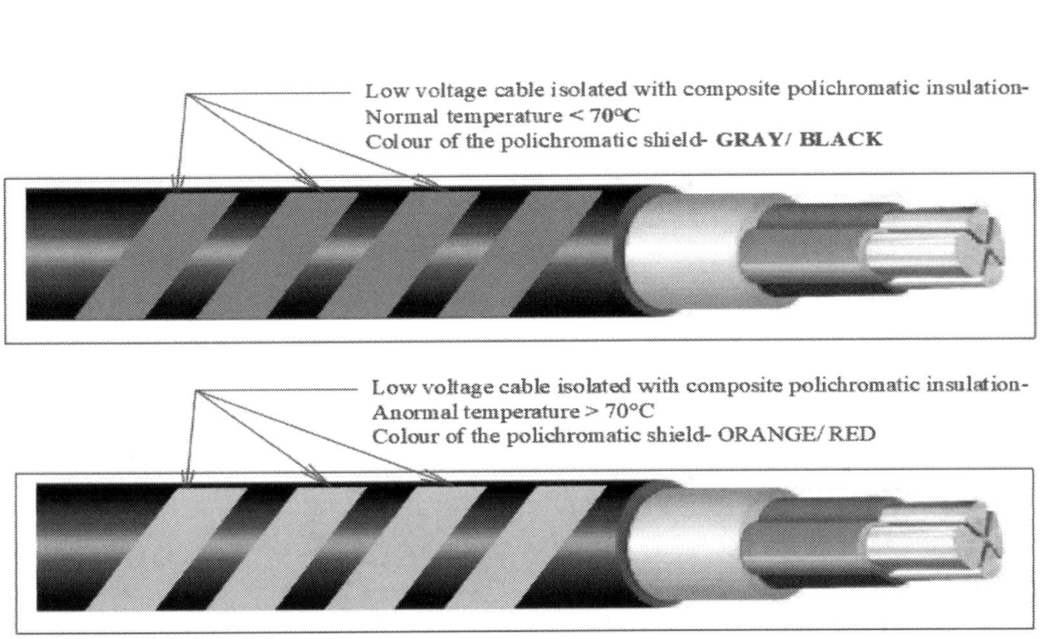

Figure 3. Electrical supply low voltage cables isolated with a polychromatic composite shield. (it represents the theoretical method proposed by the article)

Polychrome composite materials have been invented recently, one of which is the polychrome red phosphorus flame-retardant nylon composition [16].

Some materials used to insulate cables/conductors like glass, paper and Teflon, which have high resistivity, are proper electrical insulators. Other materials with lower thermal resistivity may be used in order to prevent significant current leakage at commonly used voltages and are therefore used as insulation for electrical wiring and cable. Examples include rubber-type polymers and most plastics that can be heat-resistant or thermoplastic.

3. Conclusion

The proposal of this article is the following: Introduction of a composite material with polychromatic effect in the sheath of low voltage electrical cables/conductors, on their surface, having high electrical insulation properties, high flexibility, material that could be applied over the classic insulating sheath of these cables (classic PVC sheath, rubber, etc.), this solution, which will be studied in detail in future research, would bring substantial benefit to the safe operation of the use of these cables.

For three-phase low-voltage cables with several conductors (400 V), the ideal place for installing cable insulation made of material with polychromatic effects and sensitivity to temperature changes is on the outer surface of the common insulation of conductors, this composite material covering conductors in turn electrical, mechanical strength, high flexibility and high thermal conductivity.

Most polymers have an inherently very low electrical conductivity, so polymers are commonly used as electrical insulation materials. The use in some applications of a basic polychromatic and heat-sensitive polymer as an electrical insulator, which has good mechanical and durability properties and at the same time provides information on the state of the material in real-time (thermal properties, electrical properties, etc.).

The structure-property relationship of the composite materials used in energy engineering represents the central point of interest having the major influence in their design and realization, having followed with predilection the electrical, thermal and mechanical properties.

ORCID iDs
M F Draghici https://orcid.org/0000-0001-5979-9441
M Stoicanescu https://orcid.org/0000-0003-3138-1830

References

[1] Ivan E, Husu A G and Olariu M I 2009 *Materials used in Electrical Engineering* (in Romanian, Târgovişte: Bibliotheca)
[2] Băjenescu Titu I 2006 *Technical System Fiability* (in Romanian, Bucureşti: Matrixrom)
[3] Han J and Garrett R 2008 *Overview of polymer nanocomposites as dielectrics and electrical insulation materials for large high voltage rotating machines* (NSTI-Nanotech)
[4] Matthews F L and Rawlings R D 1999 Overview. *In Composite Materials: Engineering and Science, 2nd ed.* (CRC Press, Woodhead Publishing Limited: Cambridge, UK)
[5] Adăscăliţei A A 2015 *Theory of electrical circuits* (In Romanian, Iaşi: Ed Performantica)
[6] Yan M and Wu J 2013 Direct measurement method for operation temperature of power cable conductor based on electromagnetic coupling *Gaodianya Jishu/High Voltage Engineering* **39** (11) 2664-9 https://doi.org/10.3969/j.issn.1003-6520.2013.11.013
[7] IEC 60287-2-1:2015 *Electric cables - Calculation of the current rating* - Part 2-1: Thermal resistance - Calculation of the thermal resistance
[8] Enescu D 2020 *Thermal Assessment of Power Cables and Impacts on Cable Current Rating,* Energies **13** *20*
[9] Bascom E C R 2014 *Considerations for advanced temperature monitoring of underground power cables*, IEEE PES T&D Conference and Exposition Chicago, IL, USA
[10] Gray M A 2008 Multiphysics Technologies *Computing in Science & Engineering* **10** (2) 82-4 https://doi.org/10.1109/MCSE.2008.59
[11] Luo J, Liu Y and Luo Y 2005 Temperature calculation of power cable conductor in real-time *Proceedings Electrical Insulation Conf and Electrical Manufacturing Expo*, Indianapolis, IN, USA pp. 26-9 https://doi.org/10.1109/EEIC.2005.1566251
[12] Chen K, Yue Y and Tang Y 2021 Research on Temperature Monitoring Method of Cable on10 kV Railway Power Transmission Lines Based on Distributed Temperature Sensor *Energies* **14** 3705 https://doi.org/10.3390/en14123705
[13] Hohmann M1999 *Infrared thermometry for electrical maintenance* **98** 9 Electrical Construction and Maintenance https://www.ecmweb.com/content/article/20887643/infrared-thermometry-for-electrical-maintenance
[14] Stoynova A V and Bonev B B 2017 Infrared survey in electrical preventive maintenance *XXVI Int Sci Conf Electronics ET* pp 1-4 https://doi.org/10.1109/ET.2017.8124392
[15] Andritoi D, Luca C, Corciova C and Ciorap R 2019 The use of thermography as a prediction element in the maintenance of medical equipment *6th Int Conf on Advancements of Medicine and Health Care through Technology, Meditech 2018* ed. S Vlad (NM Roman, Springer, New York) **71** 73-8 https://doi.org/10.1007/978-981-13-6207-1_12
[16] Zhou S, Chen B, Huang Z and Zheng J 2018 A kind of polychrome red phosphorus flame-retardant nylon composition and preparation method thereof *CN105504797B* https://patents.google.com/patent/CN105504797B/en

IOP Conf. Series: Materials Science and Engineering 1235 (2022) 012030 doi:10.1088/1757-899X/1235/1/012030

Cementite, still an advanced material: a brief overview

E Bauer Grosse

University of Lorraine, Institut Jean Lamour - Campus Artem - 2 allée André Guinier-BP 50840 F-54011 Nancy Cedex, France

elisabeth.bauer-grosse@univ-lorraine.fr

Abstract. This paper proposes to give recent advances regarding cementite, a long-known iron carbide. Illustrations for which it remains an advanced material are given in various fields such as metallurgy, catalysis, energy, environment and medicine.

1. Introduction

Cementite (mineral name: cohenite) considered here is an iron carbide (IC) often called θ or θ-Fe_3C known since 1885 [1]. Its unit cell, first determined in 1940 [2] is orthorhombic, containing 12 atoms of iron and 4 atoms of carbon (see figure 1). In the Pnma space group, its lattice parameters, subject to small variations are around a=0.508 nm, b=0.675 nm and c=0.452 nm. Cementite was qualified in 2010 as 'ancient advanced material' [3] because it ranks amongst the oldest materials to serve mankind, bringing over time a lot of innovations in various fields. Early 2020, an important critical review [4] was devoted to cementite as a phase. Many properties are up to date and it can be seen how cementite studies span the disciplines. The present paper, inspired by [3-4 and references therein], is focused on man-made cementite seen here as means to improve various performances or as a solution to solve various problems specific to these last years in different fields shown in figure 1 such as metallurgy and catalysis for well-established industrial applications and such as energy, environment and medicine for more recent applications.

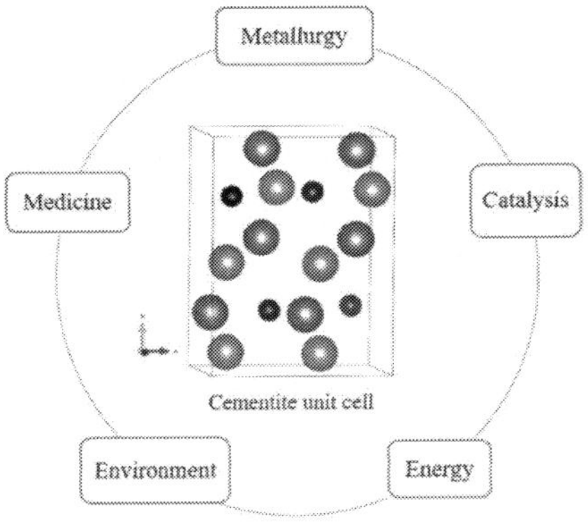

Figure 1. Cementite unit cell surrounded by a selection of application fields for which this material plays a role mentioned in this paper.

The cell is modelled through the Visualization for Electronic and Structural Analysis (VESTA) software [5]. Large grey balls show iron atoms and small black show carbon atoms.

2. Role of cementite in well-established industrial applications

Its role is extremely important in some well-established applications. Two examples are described here. The first relates to the field of metallurgy and the second to the field of catalysis.

2.1. *In metallurgy*
Man-made cementite has its roots in metallurgy. Its name, given in 1885 by Osmond and Werth [1] comes from its action cementing the grains in hypereutectoid steel. Since then, cementite has never ceased to play a role in the design of steels and cast irons, helping to produce novel grades with enhanced properties. It can be done through the choice of its fraction, distribution and morphology in the material microstructure. Research and development are fruitful in this field. Table 1 gives an idea of results dating from the first semester of this year. They concern different properties improved thanks to the spheroidization of the existing cementite through adapted processes in various iron-based matrices of as-received industrial steels with different carbon contents.

Table 1 Enhanced properties in various steels mostly due to the role of cementite (among 2021 refs).

As-received steels	Process	Cementite (θ) state after	Enhanced properties	Ref
Low-carbon	Extrusion + annealing	S-θ in pearlite colonies	strength-ductility-wear resistance	[6]
Medium carbon	Heat-treatment	Submicron-sized granular θ	Corrosion resistance	[7]
high carbon wire	HVLC	Nano-sized S-θ in martensite	Strength- ductility- balance	[8]
High carbon HSLA	Heat-treatment	θ nano-sized precipitates	Fire-resistance/ long-term service	[9]
high carbon PM	Heat-treatment	Micron-sized S-C in bainite	Impact toughness	[10]
Ultrahigh carbon PM	Warm deformation	Submicron-sized S-θ in ferrite	Superplastic forming	[11]

HSLA: High strength low alloy; PM: powder metallurgy; HVLC: high-voltage low-current; S: spherical

2.2. *In catalysis*
There is currently a huge need to develop high-performance catalysts to activate a number of chemical reactions for industrial applications. Historically, man-made cementite was found among products of iron-based catalysts used in the Fischer-Tropsch synthesis (FTS) first developed in 1925 to produce synthetic liquid fuels and valuable chemicals. With similar catalytic activity to noble metals, cementite was successfully tested in 1956 in FTS [12] and then in other applications such as the carbon nanotubes (CNTs) synthesis [13] in 2008. Nowadays, cementite is part of the materials used in the design of numerous iron-based catalysts. A lot of information can be found on the subject in [4,14-15] where cementite is often chosen as an illustration among other ICs. The cementite-based catalysts developed in recent years are complex composite or hybrid materials obtained with synthesis methods often fit for each application. Three examples are given here. In 2010, a mesoporous Fe_3C sponge was synthesized and its usefulness was exemplified by its catalytic activity and high stability in ammonia decomposition [16]. In 2016, graphene encapsulated Fe_3C in CNTs catalyst (Fe_3C@G-CNT) has proven its effectiveness in selective hydrogenation reaction [17]. In 2017, an encapsulated catalyst with SAPO (silicoaluminophosphate molecular sieves)-34 (layers number) zeolite shell has been synthesized over Fe_3C nanoparticles via surface modification of Si-OH groups. Called Fe/C@Si-SAPO, this FT cementite-based catalyst is highly selective for the production of light olefins [18]. Many other cementite-based catalysts have been developed over the past decade to activate reactions such as oxygen reduction reaction (ORR), hydrogen evolution reaction (HER), oxygen evolution reaction (OER) [14], occurring in particular in new energy conversion or storage devices (§ 3.1.2 and 3.1.3).

3. Role of cementite in more recent applications
Man-made cementite is involved in more recent applications by helping to provide solutions to societal challenges. Illustrations are given here in the case of clean energy, environment and medicine.

3.1. *In clean energy*
There is also a huge need to develop high-performance catalysts and electro-catalysts in the field of energy. Three examples are given in the following subsubsections.

3.1.1. In FTS. FTS has received renewed interest in recent years due to the necessity to decrease global dependency on fossil fuels. According to a report published in 2021 [19], the importance of this process as a source of sustainable and renewable energy requires further contribution towards designing iron-based FTS catalysts with highly active and selective IC as the predominant phase of which cementite may be part.

3.1.2. In energy conversion. Fuel cells, metal-air batteries and water splitting for hydrogen production have been regarded as possible solutions for safe and convenient power generation applications, requiring to development of new electrocatalysts [14]. Among other materials, various complex nanocomposites of cementite and carbonaceous materials are explored. Two examples are given for their high performances. The first one is an electro-catalyst composed of nanocrystalline Fe_3C encapsulated in graphitic shells, supported with vertically aligned graphene nanoribbons, designated as Fe_3C-GNR [20]. The second one is an electro-catalyst composed of Fe_3C nanoparticles encapsulated at the tip of nitrogen-enriched carbon nanotubes, which were aligned on one-dimensional nitrogen-doped carbon nanofibers, called Fe_3C@NCNTs [21]. Thematic will be a focus in future research.

3.1.3. In energy storage. With the consumption of fossil energy, the demand for electrical energy is increasing, requiring the development of energy storage devices such as dye-sensitized solar cells (DSSCs), rechargeable lithium-ion batteries (LIBs) and lithium-sulfur batteries (Li-SBs). In DSSCs, cementite is involved in achieving design for counter electrodes (CEs) more advantageous than costly Pt-based CEs. Fe_3C/C composite [22], graphene-wrapped iron carbide nanoparticles called Fe_3C@G [23] and Fe_{1-x} S nanoparticles anchored on Fe_3C-encapsulated and N-doped carbon nanotube hybrids called Fe1-xS/Fe_3C-NCNT [24] are new Fe_3C-containing nanoarchitectures for highly efficient CEs in DSSCs. In LIBs and Li-SBs, graphite-based materials are commonly used as anodes but there are not suitable for high-performance batteries. As the introduction of cementite into the carbon matrix improves the capacity for lithium storage [25], new cementite-based anodes have been designed. The most recent results (2020-21) relate to LIB anodes made up of uniformly inserted Fe_3C nanoparticles in biomass-derived porous carbon [26] and to Li-SB cathodes made up of well-dispersed ultrafine Fe_3C nanocrystals embedded in ordered micro-mesoporous carbon spheres [27].

3.2. In environment

Cementite also acts in environmental remediation. The case of waste treatment to remove organic pollutants in water and the case of mitigation of electromagnetic (EM) pollution are described here.

3.2.1. In waste treatment. The opportunity to use cementite, in this case, was first reported in 2015 [28] where graphite carbon encapsulated cementite nanoparticles exhibited excellent adsorption properties to remove methylene blue from wastewater. Since then, among other heterogeneous Fe-based materials used for peroxymonosulfate (PMS) or peroxydisulfate (PDS) activation [29], cementite-based materials are developed as catalysts for organics removal, presenting advantages such as high efficiency, nontoxicity and low cost. Table 2 gives examples of cementite-containing catalysts developed these five last years to activate PMS or PDS for different organic pollutants removal.

Table 2. Examples of Fe_3C-containing catalysts to activate PMS/PDS for organic pollutants removal.

Catalyst name	Supplementary information about carbon	Pollutant	Ref	Year
Fe_3C@NCNT	NCNT for nitrogen-doped carbon nanotube	Phenol	[30]	2017
Fe_3C@NCNT-900	900 for the synthesis temperature in Celsius degrees	Sulphamethoxazole	[31]	2019
Fe_3C@NCNTs/GNS	GNS for graphene nanosheet	Bisphenol A	[32]	2019
Fe_3C@CFs	CFs for carbon fibres	Acid Red 1	[33]	2019
Fe_3C/NC	NC for nitrogen-doped carbon sheet	Ibuprofen	[34]	2019
Fe_3C@NSCNT	NSCNT for nitrogen-sulfur-codoped carbon nanotube	Bisphenol A	[35]	2021

Fe_3C: cementite @ or /: encapsulated or embedded

3.2.2. In mitigation of EM pollution. The opportunity to use cementite, in this case, was first reported in 2009 [36] with magnetic Fe_3C/dielectric carbon nanocomposites exhibiting a very interesting EM wave absorption ability in the frequency range of 0.9–9.0 GHz compared to the traditional iron nanowires used as a microwave absorber. Since then, one of the aims is to extend the frequency range having an excellent EM wave absorption ability. Table 3 indicates the results obtained these last years on new hybrid cementite-carbon based magnetic-dielectric materials with the frequency range selected for the excellence of EM wave absorption ability.

Table 3. Examples of cementite-containing microwave absorber materials

Design of containing cementite microwave absorber materials	Frequency range	Ref	Year
Carbon-coated Fe_3C nanocapsules	1 to 18 GHz	[37]	2014
Fe_3C/Helical carbon nanotubes hybrid	2.9 to 9.9 GHz	[38]	2016
Solid or macroporous Fe_3C/N-doped carbon nanofibers CNFs	1 to 18 GHz	[39]	2018
Porous carbon matrix with inlaid Fe_3C/Fe_3O_4 micro-particles	5.07 to 18.00 GHz	[40]	2019
Fe_3C/C or Fe_3C/Fe/C with flower-like structures formed by two-dimension flakes	3.68 to 18.00 GHz	[41]	2019
Fe/Fe_3C-C core-shell nanoparticles	Around 6.2 GHz	[42]	2019
Fe_3C@carbon composites	13.6 to 18.0 GHz	[43]	2020

3.3. In medicine
ICs nanoparticles were considered in 2017 as an innovative nanoplatforms for biomedical applications [44]. Among these, cementite is particularly involved in magnetic hyperthermia (MH). Cementite nanoparticles with magnetic and colloidal properties for this application were fabricated in 2019 [45] and their suitability was demonstrated in 2020 [46]. Iron and cementite nanoparticles encapsulated in a carbon matrix are also candidates for MH application [47]. Cementite is also involved in biosensing applications. For example, as a component of electrochemical biosensor, ultrafine Fe_3C nanoparticles embedded in N-doped graphitic carbon sheets (Fe_3C@NGCSs) can simultaneously determine ascorbic acid, dopamine, uric acid and xanthine [48]. Another example is given by the discovery of the excellent and simultaneous triple-enzyme mimetic property (peroxidase-like, oxidase-like, and catalase-like activity) of Fe_3C decorated carbon nanofibers (Fe_3C/C NFs), leading to promising applications especially in biosensing and biomedicine [49].

4. Conclusion
This brief overview shows that cementite can play a beneficial role such as an enhancement of the properties in metallurgy and improved activity in industrial catalysis. It can show great potential in the design of new catalysts, electro-catalysts and electrodes in devices and processes for clean energy. It may furthermore act as a catalyst for organic removal or as a microwave absorber in environmental remediation. It also shows that it could be used in medicine in several biomedical applications. Cementite thus remains a promising material for future applications.

ORCID iDs
E Bauer Grosse https://orcid.org/0000-0002-9944-1185

5. References
[1] Osmond F and Werth J 1885 Théorie cellulaire des propriétés de l'acier *Ann. mines* **8** 5-84
[2] Lipson H and Petch N J 1940 The crystal structure of cementite *J. Iron Steel Inst.* **2** 95-106
[3] Giordano C, Kraupner A, Wimbush S C and Antonietti M 2010 *Small* **6** 1859-62 https://doi.org/10.1002/smll.201000437
[4] Bhadeshia K H D H 2020 *Int. Mater. Rev.* **65** (1) 1-27 https://doi.org/10.1080/09506608.2018.1560984
[5] Momma K and Izumi F 2011 *J. Appl. Cryst.* **44** 1272–6 https://doi.org/10.1107/S0021889811038970
[6] Demirtas M and Sekban D M 2021 *Metall. Res. Technol.* **118** (2) 217 https://doi.org/10.1051/metal/2021021
[7] Liu H, Wei J, Dong J H, Chen Y Q, Wu Y M, Zhou Y T, Babu S D, and Ke W 2021 *J. Mater. Sci. Technol.* **61** 234-46 https://doi.org/10.1016/j.jmst.2020.05.031
[8] Mallick M, Mitra S K, Basak D, Roy N K and Maity J 2021 *Philos. Mag.* **101** (5) 555-75 https://doi.org/10.1080/14786435.2020.1849843
[9] Wei L J, Ji X M, Yu Y S, Misra R D K, Liu P C and Wang XM 2021 *Mater. Lett.* **291** 129448 https://doi.org/10.1016/j.matlet.2021.129448
[10] Güral A and Altuntas O 2021 *Mater. Chem. Phys.* **259** 124203 https://doi.org/10.1016/j.matchemphys.2020.124203

[11] Nikiel P, Szczepani S and Korpala G 2021 *Metals* **11** (2) 328 https://doi.org/10.3390/met11020328
[12] Shulz J F, Hall W K, Dubs T A and Anderson R B 1956 *J. Am. Chem. Soc.* **78** (2) 282-5 https://doi.org/10.1021/ja01583a009
[13] Yoshida H, Takeda S, Uchiyama T, Kohno H and Homma Y 2008 *Nano Lett.* **8** (7) 2082-6 https://doi.org/10.1021/nl080452q
[14] Ye Z, Zhang P, Lei X, Wang X, Zhao N and Yang H 2018 *Chem. Eur. J.* **24** 8922-40 https://doi.org/10.1002/chem.201706028
[15] Wang X, Zhu K, Ju Y, Li Y, Li W, Xu J and Hou Y 2019 *J. Magn. Magn. Mater.* **489** 165432 https://doi.org/10.1016/j.jmmm.2019.165432
[16] Kraupner A, Antonietti M, Palkovits R, Schlicht K and Giordano C 2010 *J. Mater. Chem* **20** 6019–22 https://doi.org/10.1039/C0JM00774A
[17] Shi J, Wang Y, Du W and Hou Z 2016 *Carbon* **99** 330-7 https://doi.org/10.1016/j.carbon.2015.12.049
[18] Qiu T, Wang L, Lv S, Sun B, Zhang Y, Liu Z, Yang W and Li J 2017 *Fuel* **203** 811-6 https://doi.org/10.1016/j.fuel.2017.05.043
[19] Otun K O, Yao Y, Liu X and Hilderandt D 2021 *Fuel* **296** 120689 https://doi.org/10.1016/j.fuel.2021.120689
[20] Fan X, Peng Z, Ye R, Zhou H and Guo X 2015 *ACS Nano* **9** (7) 7407-18 https://doi.org/10.1021/acsnano.5b02420.
[21] Zhao Y, Zhang J, Guo X, Fan H, Wu W, Liu H and Wang G 2017 *J. Mater. Chem. A* **5** 19672-9 https://doi.org/10.1039/C7TA05936A
[22] Kim J, Kang J S, Jeong J, Son Y J, Lee M J, Kang J, Lim A, Park H S and Sung Y E 2018 *J. Power Sources* **396** 213-9 https://doi.org/10.1016/j.jpowsour.2018.06.015
[23] Xu H, Zhang C, Yao J, Pang S, Zhou X and Cui G 2018 *J. Photochem. Photobiol. A-Chem.* **355** 48-54 https://doi.org/10.1016/j.jphotochem.2017.11.030
[24] Chen M, Wang G C, Shao L L, Yuan Z Y, Qian X, Jing Q S, Huang Z Y, Xu D L and Yang S X 2018 *ACS Appl. Mater. Interfaces* **10** (37) 31208-24 https://doi.org/10.1021/acsami.8b08489
[25] Chen S, Wu J, Zhou R, Zuo L, Li P, Song Y and Wang Li 2015 *Electrochim. Acta* **180** 78-85 https://doi.org/10.1016/j.electacta.2015.08.100
[26] Zhao C, Fan R, Murugesan B, Xu Y, Ma J, Yao J, Krucinska I, Cai Y and Gao J 2021 *J. Alloys Compd.* **881** 160661 https://doi.org/10.1016/j.jallcom.2021.160661
[27] Jin Z *et al.* 2020 *Chem Eng. J.* **388** 124315 https://doi.org/10.1016/j.cej.2020.124315
[28] Wang X, Zhang P, Wang W, Lei X, Zou B and Yang H 2015 *RSC Adv.* **5** 27857-61 https://doi.org/10.1039/C5RA00624D
[29] Li J, Yang L, Lai B, Liu C, He Y, Yao G and Li N 2021 *Chem. Eng. J.* **414** 1286474 https://doi.org/10.1016/j.cej.2021.128674
[30] Wang C, Kang J, Liang P, Zhang H, Sun H, Tade M O and Wang S 2017 *Environ. Sci.: Nano* **4** 170-9 https://doi.org/10.1039/C6EN00397D
[31] Shang Y, Chen C, Zhang P, Yue Q, Li Y, Gao B and Xu X 2019 *Chem Eng J* **375** 122004 https://doi.org/10.1016/j.cej.2019.122004
[32] Ma W, Wang N, Du Y, Tong T, Zhang L, Lin K Y A and Han X 2019 *Chem Eng J* **356** 1022-31 https://doi.org/10.1016/j.cej.2018.09.093
[33] Gao S, Zhou H, Xia Y, Liu X, Yao Y, Wang W and Chen H 2019 *Catal. Sci. Technol.* **9** 4365-73 https://doi.org/10.1039/C9CY00756C
[34] Zhang G, Ding Y, Nie W and Tang H 2019 *J. Environ. Sci.* **78** 1-12 https://doi.org/10.1016/j.jes.2018.10.002
[35] Fu H, Luo H, Lin Q, Zhong Q, Huang Z, Wang Y and Wu L 2021 *Chemical Engineering Journal* **409** 128201 https://doi.org/10.1016/j.cej.2020.128201
[36] Terada M, Itoh M, Liu J R and Machida K 2009 *J. Magn. Magn. Mater.* **321** (9) 1209-13 https://doi.org/10.1016/j.jmmm.2008.11.001
[37] Tang Y, Shao Y, Yao K F and Zhong Y X 2014 *Nanotechnology* **25** 035704 https://doi.org/10.1088/0957-4484/25/3/035704
[38] Tang H *et al.* 2016 Composites Part B: Engineering **107** 51-8

https://doi.org/10.1016/j.compositesb.2016.09.003
[39] Liu H, Li Y, Yuan M, Sun G, Liao Q and Zhang Y *Sci. Rep.* **8** 16832 https://doi.org/10.1038/s41598-018-35078-z
[40] Lou Z, Yuan C, Zhang Y, Li Y, Cai J, Yang L, Wang W, Han H and Zou J 2019 *J. Alloys Compounds* **775** 800-9 https://doi.org/10.1016/j.jallcom.2018.10.213
[41] Peng S, Wang S, Hao G, Zhu C, Zhang Y, Lv X, Hu Y Jiang W 2019 *J. Magn. Magn. Mater.* **487** 165306 https://doi.org/10.1016/j.jmmm.2019.165306
[42] Kuang D, Hou L, Wang S, Luo H, Deng L, He J and Song M 2019 *Appl. Surf. Sci.* **493** 1083-9 https://doi.org/10.1016/j.apsusc.2019.07.073
[43] Gu W, Zheng J, Liang X, Cui X, Chen J, Zhang Z and Ji G 2020 *Inorg. Chem. Front.* **7** 1667-75 https://doi.org/10.1039/D0QI00099J
[44] Yu J, Chen F, Gao W, Ju Y, Chu X, Che S, Sheng F and Hou Y 2017 *Nanoscale Horiz.* **2** 81-8 https://doi.org/10.1039/C6NH00173D
[45] Gangwar A, Varghese S S, Meena S S, Prajapat C L, Gupta N and Prasad N K 2019 *J. Magn. Magn. Mater.* **481** 251-6 https://doi.org/10.1016/j.jmmm.2019.03.028
[46] Varghese S S, Meena S S, Viswanadh, M K, Gangwar A, Neogi K, Muthu M S and Prasad N K 2020 *J. Mater. Sci.-Mater. Electron.* **31** 10772-82 https://doi.org/10.1007/s10854-020-03628-9
[47] Avakyan L, Manukyan A, Bogdan A, Gyulasaryan H, Coutinho J, Paramonova E, Sukharina G, Srabionyan V, Sharoyan E and Bugaev L 2020 *J Nanopart. Res.* **22** 30 https://doi.org/10.1007/s11051-019-4698-8
[48] Chen Y, Zhang X F, Wang A J, Zhang Q L, Huang H and Feng J J 2019 *Microchim. Acta* **186** 660 https://doi.org/10.1007/s00604-019-3769-y
[49] Chen S, Wang Y, Zhong M, Yu D, Wang C and Lu X 2019 *ACS Biomater. Sci. Eng.* **5** (3) 1238-46 https://doi.org/10.1021/acsbiomaterials.8b01552

IOP Conf. Series: Materials Science and Engineering 1235 (2022) 012031 doi:10.1088/1757-899X/1235/1/012031

Construction of surface layers with special properties in the hardening of steels

O Volkov[1], S Knyazev[1], V Subbotina[1], C Iancu[2], Yu Gutsalenko[1], O Shelkovyi[1] and R. Strelchuk[1]

[1] National Technical University "Kharkiv Polytechnic Institute", Kirpichova Str. 2, Kharkiv, Ukraine, 61002
[2] "Constantin Brancusi" University of Targu Jiu, Republic Boulevard 1, Targu Jiu, Romania, 210152

r.m.strelchuk@gmail.com

Abstract. Designing surface layers with special properties at strengthening of steels gives the chance to receive effective tools of a special purpose. The technological process of hardening by boron saturation of the jewellery tool - crimpings for fixing jewellery stones from carbon steel (0.7% C) is offered. Boron saturation of steel crimpings for fixing jewellery stones was performed before traditional hardening and low tempering. The microhardness of the surface layer of boron saturated steel has been more than doubled: at a depth of up to 50 μm – up to more than 16 GPa against 6 GPa for traditional hardening technology. Between the upper hardened layer and the unaltered inner metal, a transitional structure zone with tempering martensite is observed. This zone has a thickness of about 100 μm and an average microhardness in the range from 10 GPa to about 14 GPa.

1. Introduction

The condition of the tool significantly affects the quality of the product that can be obtained with it. Over time, due to use, tool wears out and loses its geometry, which is reflected by the deterioration of indicators of the product. To increase the material's resistance to wear, surface hardening methods, including diffusion borating, are effective [1].

It is known that boron layers form relatively thick diffusion layers and have a high level of microhardness [2–4]. Arborization techniques include fighting with the furnace and the off-peak method of heating the processed products.

In small-scale production, the easiest way to implement cutting is the method of powder arborization in a container using furnace heating. It is also suitable for fighting pastes when using both furnace heating and off-peak processing [1,5,6]. Out-of-peak processing can include methods of heating with exothermic mixtures, laser and induction heating [1,7]. Exothermic heating is provided by simpler equipment, but its flow is difficult to regulate and control. Laser heating requires certain equipment[7,8]. Induction heating also requires equipment, but an order of magnitude simpler and cheaper.

The problem of borating in various ways is considered quite widely. However, there are still several issues that need to be studied. In several potential technical objects of borating, a separate group can be composed of such, a loss of wear resistance of which leads to a change in the formative geometry of the working contact surface with a loss of required performance capabilities. In this case, the plastic deformation of the counter-body in the working operation of the product can be both a side undesirable effect [9] and the main task of its use, as in our case presented below.

Furnace boron saturation remains the least expensive among possible alternatives from the point of view of equipment. With small-scale production or strengthening, it has a faster payback period compared not only with more technically complex laser heating but also with other approaches to wear-resistant modification of metal surfaces using sophisticated equipment, such as ion implantation [10].

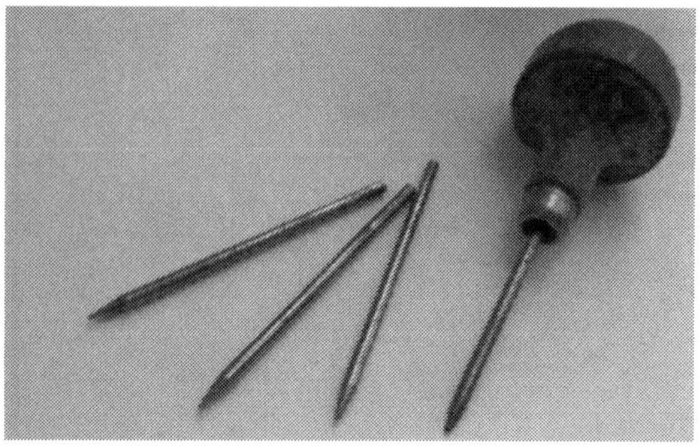

Figure 1. Jewellery punches

There is a wide range of products with small forms for local strengthening. such small-sized products include jewellery tool, which is a punch for fixing jewellery stones (figure 1).

The punch is a tool that quickly fails. It is used to install jewellery stones in products. Its conical working part has a hemispherical hole. During operation, the border of hemispherical hole loses its sharpness. Quality of work with a blunted root screw decrease. Accordingly, there is a need for such a root screw for a new one. In this regard, the use of borating to further strengthen this tool is investigated in this work, with metallographic studies of the obtained surface layers. This is necessary for a clear understanding of their morphology, which will allow you to control the properties of the resulting layers. Therefore, this study is aimed precisely at forming surface layers with special properties when strengthening steel.

2. Experiment

In the course of the study, we studied the roots which are made of CT 70 steel, the chemical composition is shown in table 1.

Table 1. Chemical composition of CT 70 steel, %

C	Mn	Si	S	P	Cr	Ni	Cu
0.66-0.73	0.17-0.33	0.17-0.33	≤ 0.028	≤ 0.03	≤ 0.2	≤ 0.25	≤ 0.25

Punches are supplied for use after quenching at a temperature of 820 °C (water) and low tempering at 180°C.

Furnace borating of punch samples was carried out in a special container.

A pre-coating by spreading consist of boron carbide powder with particle size of 60 μm and a CMC binder in the form of glue. was previously applied to the surface of the root turns. Prepared samples were placed in a boron saturation container and filled with graphite powder.

The container with samples was placed in an oven preheated to 920 °C for 3.5 hours. After cooling in calm air, the samples were cleaned of scale, and their quenching was carried out at a temperature of 820 °C (water), and a low release of 180 °C.

Before performing metallographic analysis, microblocks of samples were made at an angle of 45° to the central axis. Conventional and colour etching was used. Conventional etching was performed in a 4% alcohol solution of HNO_3 according to the standard method. Colour etching was performed when heated to 45 °C in a reagent: 1% $C_6H_2OH(NO_2)_3$ + 33% NaOH + 66% H_2O. In this case, the boride

FeB should be coloured brown, and Fe_2B – yellow. The metallographic analysis is performed on a Zeiss AXIO Vert.A1 metallographic microscope at 200x magnification.

Microhardness investigation was done using a PMT-3 microhardness meter according to the standard scheme at a static load of 100 g for indentation of the diamond tip in the form of a four-sided pyramid with an angle at the top of 136°. The loading time of the indenter was 7 seconds for each measurement.

3. Results

Visualization of changes in the microstructure by metallographic analysis (figure 2) and data on microhardness from the cross-section of samples allow us to determine the depth of the layer with a reinforced structure after furnace borating.

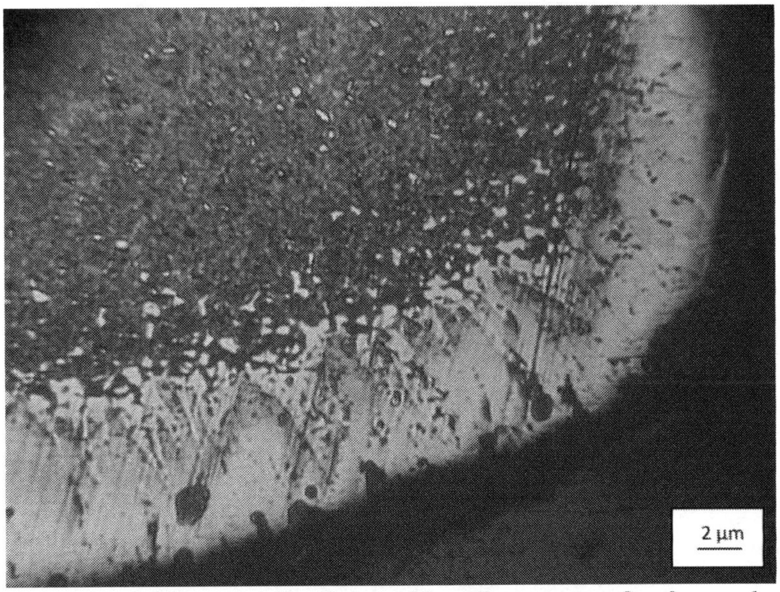

Figure 2. Microstructure of the punch for fixing of jewellery stones after furnace boron saturation, quenching and tempering

Analysis of the microhardness over the cross section of the samples showed its increase to the surface. On the basis of measurements and metallographic studies after color etching, it is possible to divide the modified layer into two zones along the cross section of the sample (figure 3).

The surface zone 50 µm thick is visually a white layer, microhardness 16,800 MPa. In this zone, the boron atoms activated under the influence of heating from the mixture formed a solid solution of boron in iron and iron diboride Fe_2B.

Below the white layer there is a transition zone with a thickness of about 100 µm and a microhardness in the range of 10,170–12,672 MPa (see figure 4), in which scattered particles of a solid boron solution are observed. Their concentration in the bulk decreases as they approach the core with an unchanged structure, into which the influence of borating has not spread, with an initial microhardness of about 6,000 MPa.

Thus, the surface structure of the sample mainly consists of a layer of Fe_2B borides and a smaller number of fragments of a solid boron solution in iron. It partially penetrates in the base metal zone, the structure of which is martensite of tempering.

Analysis of all obtained results allows us to state that after furnace boron saturation, significant changes are observed in the structure of samples made of CT 70 steel for the punch for fixing jewellery stones.

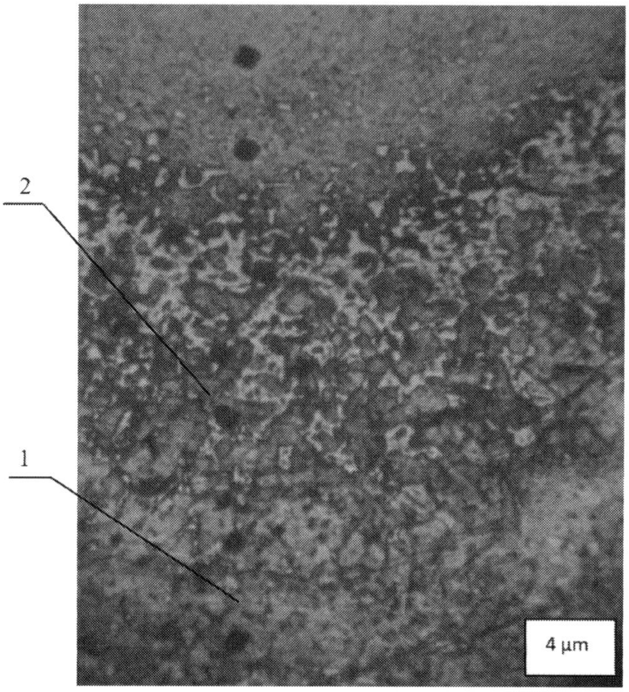

Figure 3. Microstructure of the punch for fixing of jewellery stones after furnace boron saturation, quenching and tempering (after colour etching): 1 – Borides Fe_2B; 2 – Solid boron solution in iron

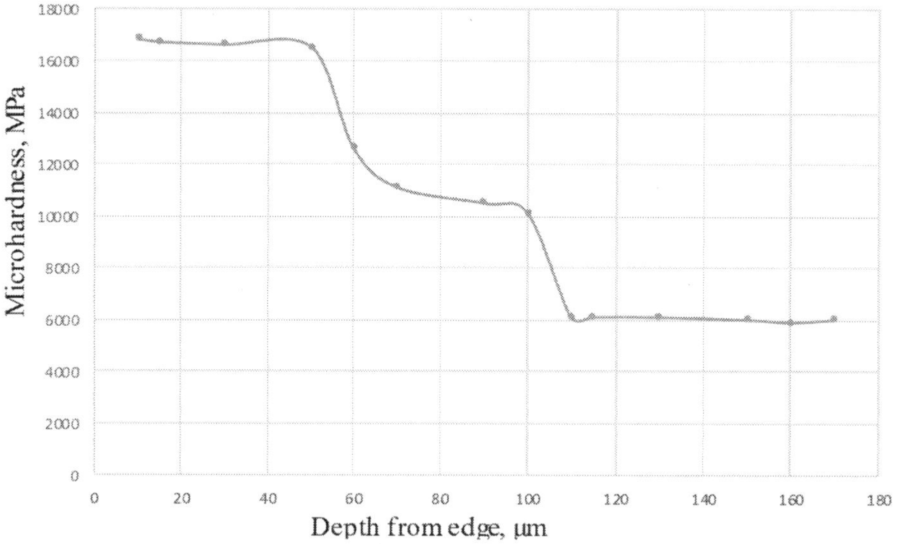

Figure 4. Microhardness on the depth of the sample after furnace borating

The technological approach considered in this study made it possible to strengthen the surface layer of the punch for fastening jewellery stones.

Thus, the microhardness of the surface layer increased from 6 GPa to 16.8 GPa, which is almost 2.8 times higher than the initial microhardness level. Obtaining such a high level of hardening of the surface of the working part of the considered tool can contribute to a significant increase in its resource.

IOP Conf. Series: Materials Science and Engineering 1235 (2022) 012031 doi:10.1088/1757-899X/1235/1/012031

4. Conclusion

Thus, a technological approach is proposed for strengthening a jewellery tool – a punch for fixing jewellery stones. Structural changes in the working parts of punch samples during furnace boron saturation are described. Effective strengthening of the surface layers of the working part of the punch is confirmed. With the use of the proposed development, a significant increase in the resource of this tool in jewellery practice should be expected.

ORCID iDs

O Volkov https://orcid.org/0000-0001-8797-0322
S Knyazev https://orcid.org/0000-0001-6422-3658
V Subbotina https://orcid.org/0000-0002-4497-4419
C Iancu https://orcid.org/0000-0003-1178-0578
Yu Gutsalenko https://orcid.org/0000-0003-4701-6504
O Shelkovyi https://orcid.org/0000-0002-7414-4854
R Strelchuk https://orcid.org/0000-0002-7221-031X

Reference

[1] Kulka M 2019 *Current Trends in Boriding* (Cham: Springer International Publishing) https://doi.org/10.1007/978-3-030-06782-3
[2] Bektes M, Calik A, Ucar N and Keddam M 2010 Pack-boriding of Fe-Mn binary alloys: Characterization and kinetics of the boride layers *Mater. Charact.* **61** 233–9 https://doi.org/10.1016/j.matchar.2009.12.005
[3] Shunqi M, Guriev A M, Ivanov S G, Guriev M A and Lygdenov B D 2019 Complex saturation of corrosion-resistant steels AISI430 and AISI303 with boron and carbon *IOP Conference Series: Materials Science and Engineering* **684** 012009 https://doi.org/10.1088/1757-899X/684/1/012009
[4] Keddam M, Kulka M, Makuch N, Pertek A and Małdziński L 2014 A kinetic model for estimating the boron activation energies in the FeB and Fe 2 B layers during the gas-boriding of Armco iron: Effect of boride incubation times *Appl. Surf. Sci.* **298** 155–63 https://doi.org/10.1016/j.apsusc.2014.01.151
[5] Dybkov V L 2017 Boriding of Chromium Steels *Powder Metall. Met. Ceram.* **55** 650–5 https://doi.org/10.1007/s11106-017-9852-1
[6] Chernega S, Poliakov I, Grinenko E and Krasovskiy M 2016 *The project "TEMPUS-MMATENG*
[7] Kusuhara T, Morimoto J, Abe N and Tsukamoto M 2010 Laser Boronizing of Stainless Steel with Direct Diode Laser *J. High Temp. Soc.* **36** 295–9 https://doi.org/10.2351/1.5061188
[8] Mikołajczak D, Kulka M and Makuch N 2016 Laser borided composite layer produced on austenitic 316L steel *Arch. Mech. Technol. Mater.* **36** 35–9 https://doi.org/10.1515/amtm-2016-0007
[9] Kowalik M, Rucki M, Paszta P and Gołębski R 2016 Plastic defomations of measured object surface in contact with undeformable surface of measuring tool *Meas. Sci. Rev.* **16** 254–9 https://doi.org/10.1515/msr-2016-0031
[10] Narojczyk J, Morozow D, Narojczyk J W and Rucki M 2018 Ion implantation of the tool's rake face for machining of the Ti-6Al-4V alloy *J. Manuf. Process.* **34** 274–80 https://doi.org/10.1016/j.jmapro.2018.06.017

Analytical crash analysis of heavy truck cabin structure according to regulation ECE-R29

A Ioniţă, M D Iozsa, C Stan and A M Manea

Automotive Engineering Dept., Faculty of Transports, Politehnica University of Bucharest, Splaiul Independentei nr. 313, sector 6, Bucuresti, Romania, RO-060042

alexandra.ionita930111@gmail.com

Abstract. The safety of the occupants is the prime consideration for all commercial vehicles' structure design. The truck cabin must be designed in such a way that the deformation patterns of the cabin structure during an impact can assure sufficient survival space for occupants. The objective of this study is to develop a CAD method to see the behaviour of the cab structure under the impact loads and the dynamic responses of the cab structure during two complex tests in line with ECE-R29regulation. In this paper, a finite element model of a lightweight cabin is built by using its Computer-Aided Design (CAD) model (which was developed by using CATIA software) and subjected to tests by using Ansys software. The test was made in order to evaluate the energy management and the cab strength under the loads. Finally, a solution is suggested to improve the cab's response for optimization design in order to validate an ECE R29 test.

1. Introduction

The safety of the occupants of trucks is a prime consideration for cabin structure design. That's why the truck cabin must be designed in such a way that the deformation patterns of the cabin structure during an impact can assure sufficient survival space for occupants [1].

In Europe, Volvo Trucks Accident Research Team (ART) has investigated accidents involving heavy goods vehicles (HGV) since 1969. Based on this systematic analysis of statistical data covering all HGV brands, they classified HGV accidents into three categories [2]: **Type A** Accidents causing fatalities or severe injuries to heavy goods vehicle occupants, **Type B** Accidents causing fatalities or severe injuries to car occupants, **Type C** Accidents causing fatalities or severe injuries to vulnerable road users.

Because of numerous type A accidents new and improved tests had to be created and implemented. Therefore, for ECE Regulation 29 cab safety tests were designed to assure enough survival space for truck cabin occupants.

The regulation contains three structural tests front impact (Test A), A-pillar impact (Test B), and roof strength (Test C) and is performed on a heavy truck depending on the weight and class.

Among different types of heavy good truck collisions, frontal and rollover impact are the severest because of these accidents, a significant number of injuries and deaths occurred to the truck occupants.

The ECE Regulation 29 was first developed in the early 1970s and was last updated with the 02 series of amendments in October 2002 and again in 2009 the cab Strength Informal Group of the Working Party on Passive Safety (GRSP) proposed a 03 series of amendments and in 2011 it has been approved. It is proposed that the existing roof strength test be modified to include an initial impact test to an upper edge of the cabin side at 20° pendulum [3].

Roof strength in rollover accident on ECE R29-03 is shown in Figure 1. The test depicts the whole process of 180° rollover accidents including two steps, the first step C1 is dynamic pre-loading of vehicles of category N_2 and N_3 with a gross mass exceeding 7500 kg to test the strength of the roof and the upper of A pillar when 90° rollover occurs. The impactor is rigid, and the mass shall not be less than 1500 kg and the energy shall be 17.6 kJ (see P1 on figure 1). In the second step C2 a static load is applied by the loading device to the roof of the cab, corresponding to the maximum mass authorised for the front axle of the vehicle, subject to a maximum of 98 kN (see P2 on figure 1) [3].

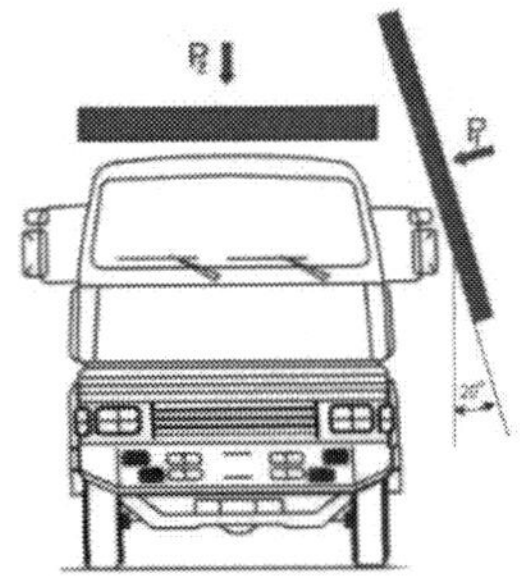

Figure 1. Roof strength test (Test C) ECE Regulation No. 29.

2. Methodology

2.1 CAD Model

To carry out the study, the geometric modelling of a heavy truck day cabin was performed, starting from a MAN truck structure. From a structural point of view, the cabin is an integrated construction consisting of the following components: the floor, front wall, back wall, sidewalls and roof. Due to the low computing power for modelling, but also to the fact that the impact studied was on the upper part of the cabin, and the lower part according to existing specialized studies [4-6] did not suffer deformations being caught by its chassis frame. A heavy truck cabin structure without a floor was used, thus reducing the computation time (see figure 2). The initial model was made in the Catia V5 program and later imported in the Ansys specialty program for mechanical simulations where the impactor according to the standard was added, at 20°, with a mass of 1562 kg.

Figure 2. The geometric model of the heavy truck cab structure.

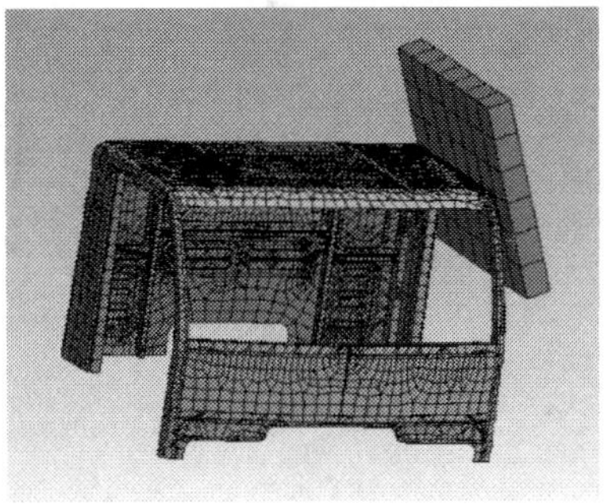

Figure 3. FE model of the heavy truck cab structure.

2.2 Finite element methods

The finite element analysis (FEA) is based on the idea of building a complicated structure with simple blocks or dividing a complicated structure into manageable and smaller pieces.

FEA provides a way of virtually testing a product design using explicit analysis. The requirement for a crash simulation is regulated through specific accident scenarios and safety targets. The objective is to allow energy absorption to take place throughout the vehicle. Using the finite element method, the automotive manufacturers aim to move simulation to the early design stage and thus reduces costs drastically. Vehicle crash analysis is probably the well-known simulation in automotive design. The emphasis is now on vehicle response to validate design strategies and permit change before available physical test [1].

As the discretization is done with small elements, the calculation error decreases considerably and it will be better to observe if there is plastic deformation of the structure in the impact area. However, a discretization with small elements can lead to very long calculation times of the explicit dynamic analysis, counted in days. For the studied structure, elements with a size of 20 mm were used in the impact area of the heavy truck structure, for better accuracy of the results, while on the rest of the structure the elements used had a size of 80 mm. A total of 8739 items with 7840 knots and a calculation time of 23 h resulted in the Ansys program (Figure 3).

2.3. Material modelling

Depending on the requirements that appear in each sheet or beam, some manufacturers adopt the option of using several different materials [1]. The analyses performed are dynamic, and the material of the structure was established from the Ansys library. The cabin was designed to use the same material for all profiles of the strength structure: "Structural Steel Non-linear" from Ansys library with the following properties: Young's modulus $E = 2 \cdot 10^7\ N/m^2$; Tensile Yield Strength: $\sigma_c = 370$ MPa. The impactor is rigid, with a mass of 1562 kg.

3. Initial conditions

For the roof strength test, two tests are performed: test C1 which simulates the lateral overturning of the cab, followed by the test C2 which simulates the total overturning of the cab. In the case of test C1, an impactor was modelled, a rigid body in the shape of a plate with a mass of 1562 kg, and the contact surface of the impactor makes an angle of 20^0 with the vertical longitudinal plane according to the indications in the regulation. The minimum impact energy required by the ECE R29 regulation is 17.6 kJ, so the impact velocity of the rigid body will be [1]:

$$E_i = \frac{m \cdot v^2}{2} = 17{,}6\ kJ \quad (1)$$

$$v = \sqrt{\frac{2 \cdot E_i}{m}} = 4{,}784\ m/s \quad (2)$$

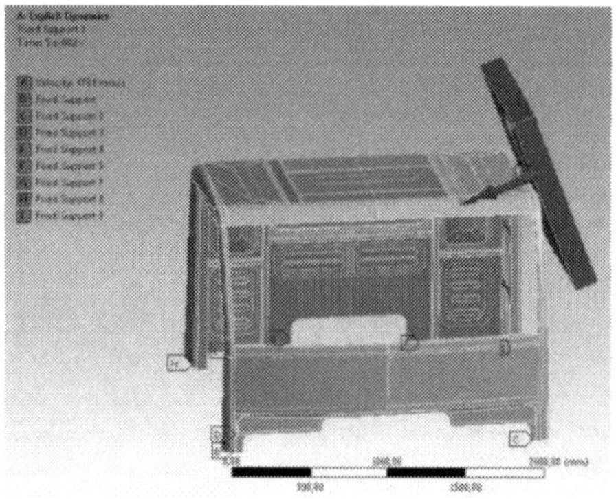

Figure 4. Initial conditions of UNECE R29 impact.

Figure 5. Optimization of the rear panel.

The simulation analysis of the cab structure under pendulum side impact was carried out with Ansys software, as shown in figure 4 and the anchor points with the floor were considered fixed. After the first tests on the initial structure, some changes were made on the rear panel to improve it by increasing the thickness of the inner panel and decreasing the thickness of the outer panels. The weight of the improved cabin structure was slightly changed. In section 4 were presented the test results of both cabs, in parallel.

IOP Conf. Series: Materials Science and Engineering 1235 (2022) 012032 doi:10.1088/1757-899X/1235/1/012032

4. Results and discussion

The survival space of the occupants is vital, so that during the impact, the deformed structure of the heavy truck must not touch the occupant. To determine this space, the distance at which the driver's head is located and the structure before the impact is established and will be verified whether this distance is maintained also after the impact has occurred.

The impact resulted in a total deformation of 209 mm in the case of the initial structure (Figure no. 6) and deformation of 126 mm in the case of the improved structure is shown in figure no 7. Following the initial analysis, an improvement of the structure was made [6] (Figure no 5). The study shows that the maximum displacement of the cab corner that occurred at the time of impact is 209 mm which means it doesn't get into the survival space of the occupants which is at about 295 mm [4].

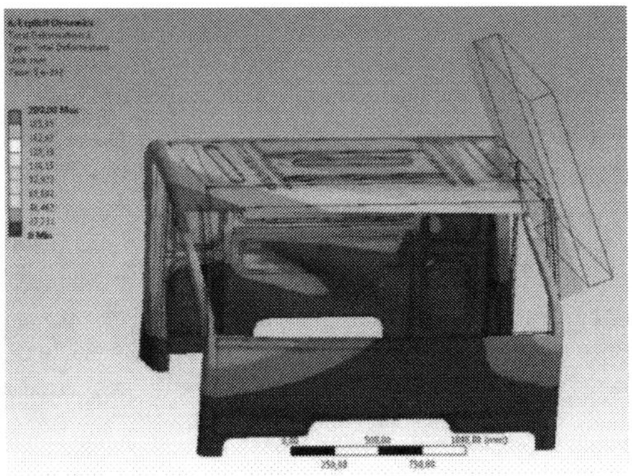

Figure 6. Total initial deformation of the cabin structure.

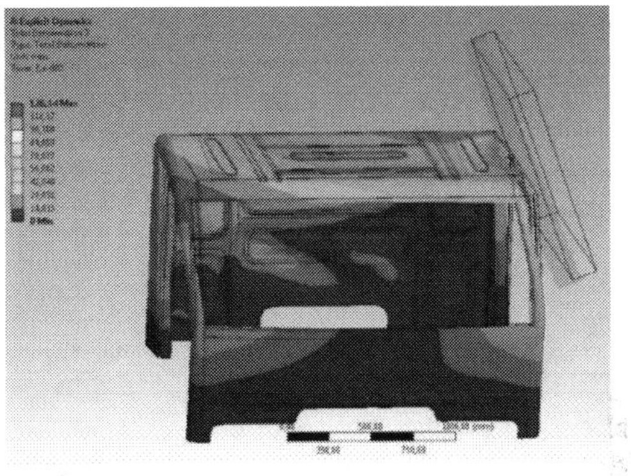

Figure 7. Total improved deformation of the cabin structure.

Analysing the energy balance of the initial structure test in the figure no. 8, the kinetic energy remains constant for few cycles and after the impact, it suddenly goes to approximately zero velocity before starting to increase again, a sign that the impactor at that point began to move backwards. As it is figured in the graph due to the deformation this kinetic energy was absorbed by the frame and was converted into internal energy. Figure 8 shows the graph of the behaviour of energies with a number of cycles.

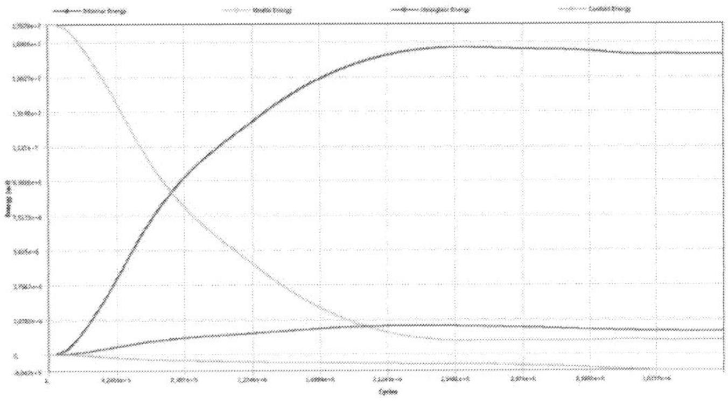

Figure 8. Energies balance of the initial structure.

In figures 9 and 10 we can analyse the equivalent stress that occurs in the cabin structure during the impact process and we can see that in the case of the initial cabin structure the tensile yield strength limit of 370 MPa has been exceeded, being 479 MPa, while in the case of improved cabin structure the limit of 367 MPa wasn't reached. In the case of the initial cab will happens plastic deformities in the red and orange areas (Figure 9).

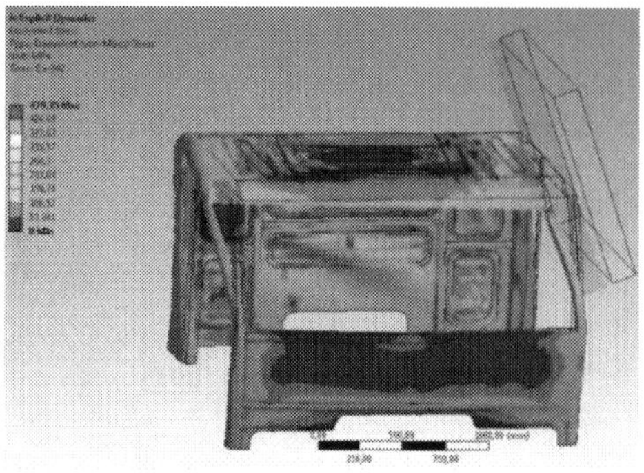

Figure 9. Equivalent stress in the initial cabin structure.

Figure 10. Equivalent stress in the improved cabin structure.

Equivalent stress analysis can be used as a scalar indicator to determine material failure. This result based on the equivalent tensile stress is the theory state that the failure at a particular location occurs when the energy of distortion reaches the same energy for failure in tension. In this case, yielding occurs when the equivalent stress reaches the yield strength of the material in simple tension. After the force of impact, the result of the stress was shown in figure 9 for the initial cabin and in figure 10 for the improved cabin.

In figures 11 and 12 are shown the maximum points. In both cases, elastic deformations and plastic deformations are present. It can also be seen that at the beginning of the impact the equivalent stress in the structure exceeds the yield strength of the material and after the structure begins to deform the equivalent stress decreases near the value of the yield strength, being around 367 MPa.

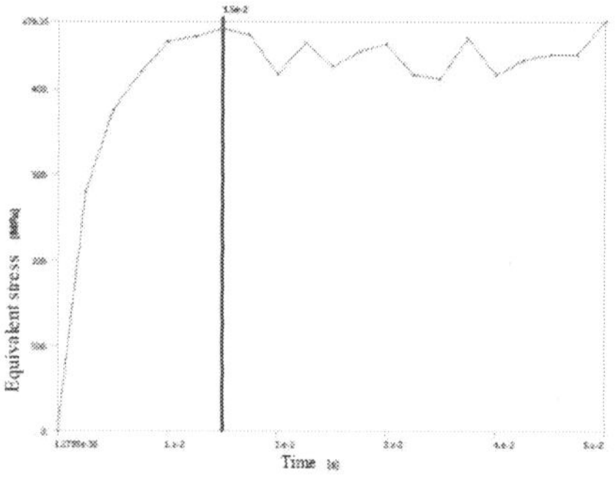

Figure 11. Graph of the equivalent stress and time in the initial cabin structure.

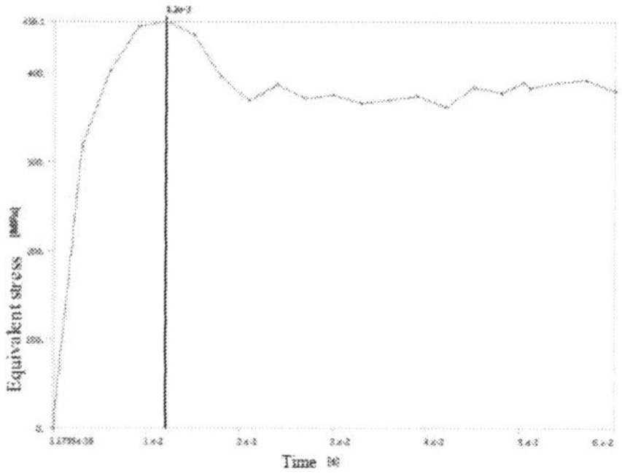

Figure 12. Graph of the equivalent stress and time in the improved cabin structure.

In figure 13 is shown the equivalent plastic strain of the initial cabin structure in the upper corner of the rear wall. The value of the equivalent plastic strain in the case of the improved cabin is 0.247 mm (Figure 14) with 0.019 mm less than in the case of the initial cab where it was 0.266 mm.

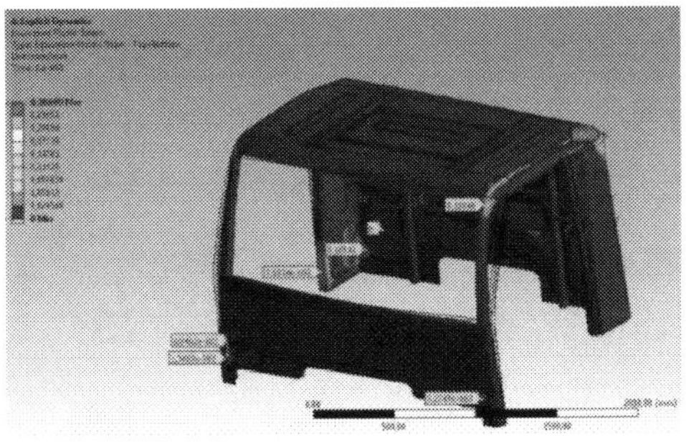

Figure 13. Equivalent plastic strain in the initial cabin structure.

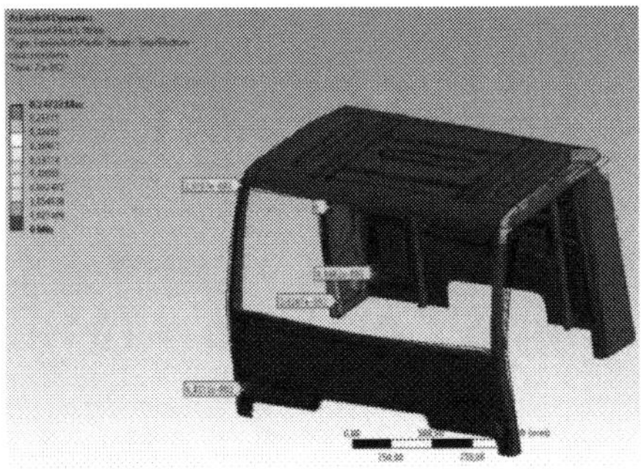

Figure 14. Equivalent plastic strain in the improved cabin structure.

In figure 11 and figure 12 is shown the stress curve of material, there is a point called 'yield point'. After the stresses beyond this yield point, the material undergoes non-elastic deformation. There is a permanent defomation after this yield point and is called plastic deformation in materials science. In both cases, the value of equivalent plastic strain appeared in the upper corner of the rear wall.

5. Conclusions

The main objective of the study is to estimate the proportion of damage to the truck driver caused by the deformation of the structure that occurs at the time of impact. The initial structure is an already existing truck structure, which has been measured and introduced into the test software. Following initial tests, the cabin structure suffered major deformations close to the passenger survival space, but after it was strengthened by changes of thickness on the rear wall, it had a better response to requests.

Taking into account the results we can say that the heavy truck structure passes and complies with the conditions imposed by the regulation in terms of survival space.

ORCID iDs
A Ionita https://orcid.org/0000-0002-8211-6407
M D Iozsa https://orcid.org/0000-0002-6547-8633
C Stan https://orcid.org/0000-0002-2753-8877
A M Manea https://orcid.org/0000-0002-3976-8848

6. References

[1] Iozsa D 2009 *Studies on the real-time analysis of the behavior of bodies on impact, in order to improve the passive security offered to passengers*, Project Report ID-140, IDEI Program
[2] Kockum S, Örtlund R, Ekfjorden A and Wells P 2017 Volvo Trucks Safety Report 2017 Volvo Truck Accident Research Team
[3] *** Regulation No 29 of the Economic Commission for Europe of the United Nations (UN/ECE) —Uniform provisions concerning the approval of vehicles with regard to the protection of the occupants of the cab of a commercial vehicle
[4] Parnell T K, White C V and Day S E Finite Element Simulation of 180° Rollover for Heavy Truck Vehicles https://www.researchgate.net/publication/237104640_Finite_Element *Simulation_of_180_Rollover_for_Heavy_Truck_Vehicles*
[5] Dobrovolny C S and Schulz N 2016 Development of a simplified finite element approach for investigation of heavy truck occupant protection in frontal impacts and rollover scenarios *14th International LS-DYNA - Session Automotive* 2016 Detroit
[6] Morello L, Rossini L R, Pia G, Tonoli A 2011 *The Automotive Body Volume I Components Design*, Springer, Netherlands, XXII, 668 https://doi.org/10.1007/978-94-007-0513-5

Modelling and simulation automatic braking systems for vehicles

Gabriel Badea, Marius Toma, Dan Alexandru Micu, Gheorghe Frățilă and Ștefan Saragea

Automotive Engineering Department, Politehnica University of Bucharest, Bucharest RO-060042, Romania

gabriel.badea.ro@gmail.com

Abstract. An automatic braking system is an important future step in attending the autonomous vehicle. This system increases the active safety of a vehicle, significantly reducing road accidents, and lowering the effects of accidents. The paper will present the steps that will develop the autonomous braking systems in conformity with the Society of Automotive Engineers (SAE) from the USA. As well the paper will present the modelling and simulating of the automatic braking systems of a vehicle from the middle class using LMS Amesim software.

1. Introduction

One of the biggest concerns in the automotive industry is the safety of vehicles. Vehicle manufactures have the desire to produce a safe vehicle that is capable of reducing the gravity of an impact and in the best case to avoid the collision. Starting with 2022 in Europe all the new cars sold need to be equipped with the automatic braking system. [1] The number of car fatalities accidents is still very high (more than 20 000/year) and the target set by the European Union it is not yet reached [1].

One of the ADAS functions that will reduce the number of accidents is the automatic braking system, AEB. The vehicle will be equipped with control systems that will assist the driver in braking moments. This would significantly decrease the amount of damage produced on the car from a technical perspective and monetary as well [2]. The ADAS system can be equipped with a neural network as a controller. They are universal functions that are learning based on examples, so they are capable of adapting in real life [3].

The scope of the paper is to study the dynamics of the vehicle in the braking moment from the ADAS system.

2. Amesim modelling

For simulation was used the Amesim software, version 2019 V2. For this modelling, I used pre-configured components from different library's, like 'Powertrain', 'Signal,Control'.

In figure 1 one can see the Sketch window of the Amesim software, where you can add, change components for the model. One of the main components of the model is VDABS01, which is a simplified model of the ABS.

In figure 2 is presented the variation of the braking force that is used by the components used in Amesim, VDABS01.

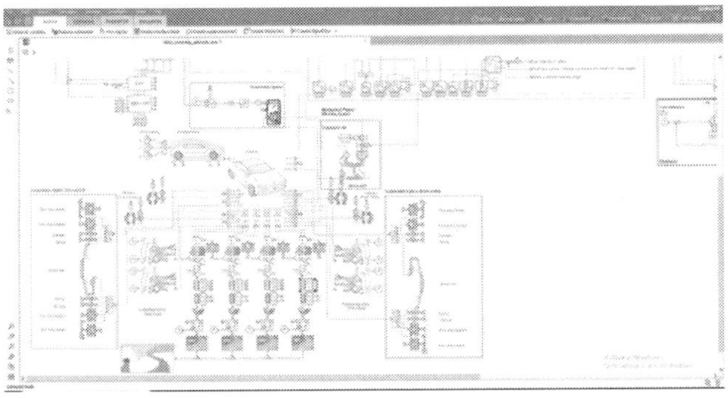

Figure 1. Sketch Model [4]

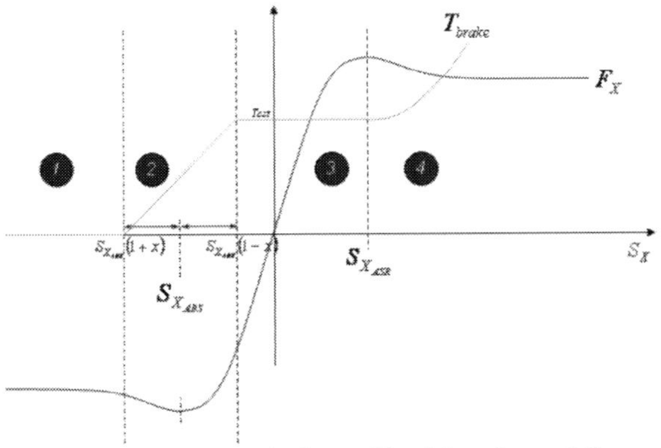

Figure 2. Variation of braking force [4]

Table 1. The principal characteristic of the vehicle

Parameters	Value	units
Wheelbase	2912	mm
Front track width	1627	mm
Backtrack width	1618	mm
Length	4943	mm
Width	1874	mm
The free radius of the tire	338	mm
Weight	1725	kg

Below is described each phase of the variation:

1) ABS total regulation: The estimated tire longitudinal slip is lower than $S_{X\ ABS}(1+x)$, the regulation released the input brake torque down to zero.

2) ABS partial regulation: The estimated tire longitudinal slip is in the regulation margin interval $[S_{X\ ABS}(1+x), S_{X\ ABS}(1-x)]$, the regulation released the input brake torque using a simple linear function concerning S_X.

3) No regulation: The estimated tire longitudinal slip is in the interval $[S_{X\ ABS}(1-x), S_{X\ ASR}]$, the input brake torque is applied.

4) ASR total regulation: The estimated tire longitudinal slip is greater than $S_{X\ ASR}$, the input brake torque is applied with additional braking torque using a quadratic function concerning S_X.

The following symbols were used:

- S_X - tire longitudinal slip,
- S_{XABS} - ABS longitudinal slip regulation value,
- x - ABS regulation margin,
- S_{XASR} - ASR longitudinal slip regulation value,
- F_X - tire longitudinal force function of tire longitudinal slip,
- T_{brake} - brake torque with ABS/ASR regulation (with constant input brake torque, *Test*).

The vehicle used as a model is Audi A6, generation C7. The main parameters of the model are presented in table 1.

3. Presenting results obtain with simulation

For this simulation, we will do multiple tests depending on vehicle speed and the adhesion coefficient.

The main parameters set for the AEBS system are:

a) Target decelerating when the automatic braking is activated: 5.5 m/s^2

b) Target decelerating when the automatic braking is deactivated: 2 m/s^2.

c) Time after the brake is released: 0.5 s.

d) Time after the brake starts to be activated after the target decelerating is achieved: 0.12 s.

3.1. First Test

The target speed of the vehicle is 90 km/h and the adhesion coefficient is set to equivalent to the dry asphalt. In figure 3 we can observe two braking decisions, they are following the AEBS parameters. These decisions are observed as well in the figure 5.

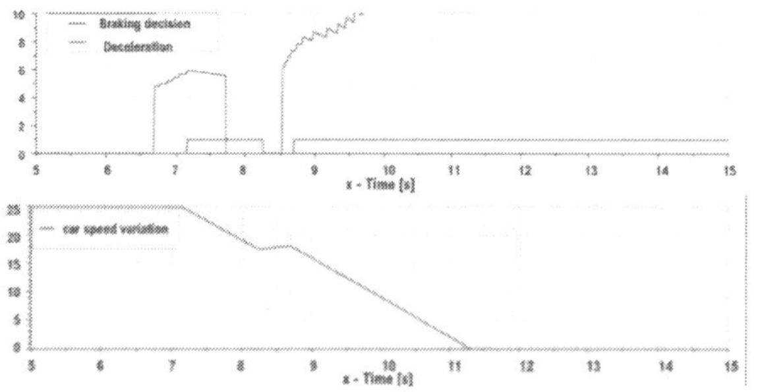

Figure 3. The decision of braking in relation to deceleration and variation of speed, *test 1 (90km/h, dry asphalt)*.

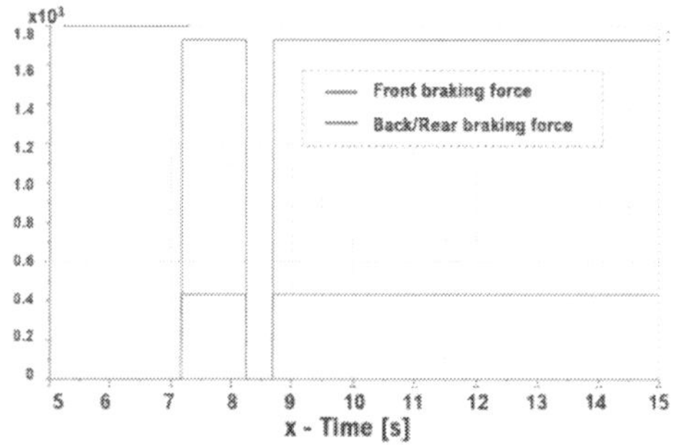

Figure 4. Variation of braking force, *test 1 (90km/h, dry asphalt)*.

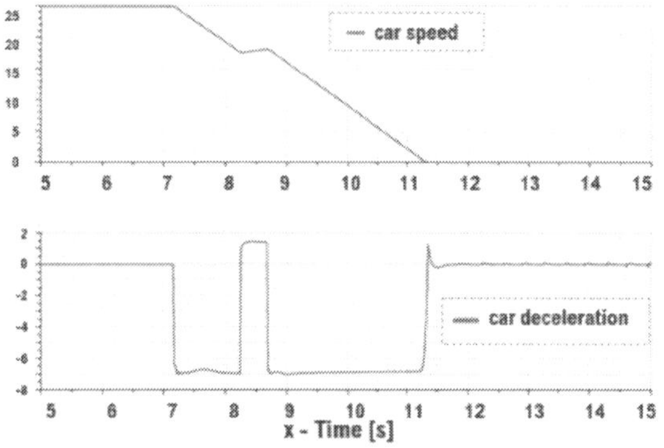

Figure 5. Speed variation in relationship with deceleration, *test 1 (90km/h, dry asphalt)*.

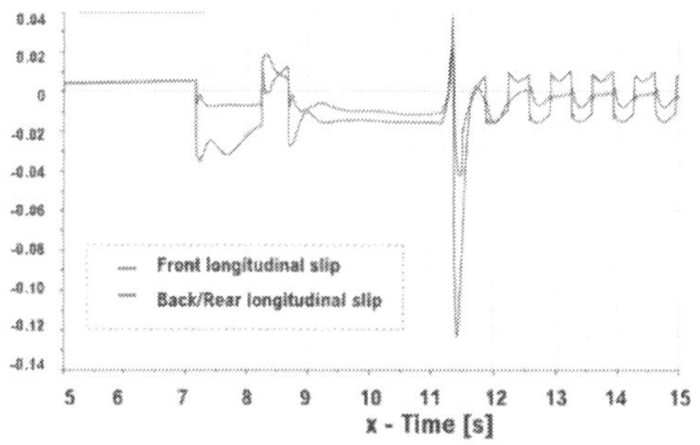

Figure 6. Variation of longitudinal slip, *test 1 (90km/h, dry asphalt)*.

In the braking force graph, figure 4, we can observe the missing intervention of the ABS system, this happens because we do not have a great longitudinal slip, as seen in figure 6. In this test, the vehicle was not able to stop in time and not impact the obstacle.

3.2. Second Test

The target speed of the vehicle is 90 km/h and the adhesion coefficient is set to equivalent to the wet asphalt. In figure 7 and figure 9 we can observe only one break decision.

In the braking force graph, figure 8, we can observe the intervention of the ABS system for front and back, this happens because we have a great longitudinal slip, as seen in figure 10. In this test, the vehicle was not able to stop in time and not impact the obstacle.

3.3. Third Test

The target speed of the vehicle is 60 km/h and the adhesion coefficient is set to equivalent to the dry asphalt. In figure 11 we can observe two braking decisions, they are following the AEBS parameters. These decisions are observed as well in figure 13.

In the braking force graph, figure 12, we can observe the missing intervention of the ABS system, this happens because we do not have a great longitudinal slip, as seen in figure 14. In this test, the vehicle was able to stop in time and not impact the obstacle.

Based on real-life situations, right now, a vehicle from this category is able to break in time in order to avoid an accident at this speed. We can conclude that the simulation is not far from the actual capacities of the vehicle.

IOP Conf. Series: Materials Science and Engineering 1235 (2022) 012033 doi:10.1088/1757-899X/1235/1/012033

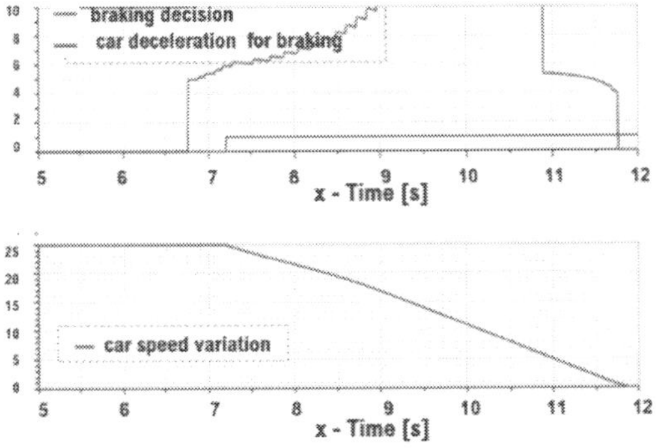

Figure 7. The decision of braking in relation to deceleration and the variation of speed, *test 2 (90km/h, wet asphalt).*

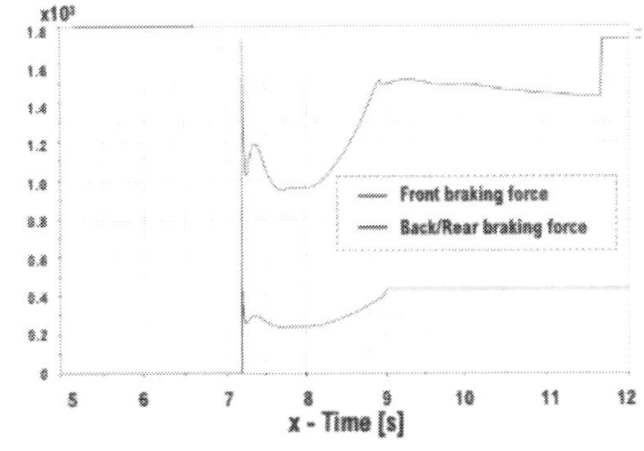

Figure 8. Variation of braking force, *test 2 (90km/h, wet asphalt).*

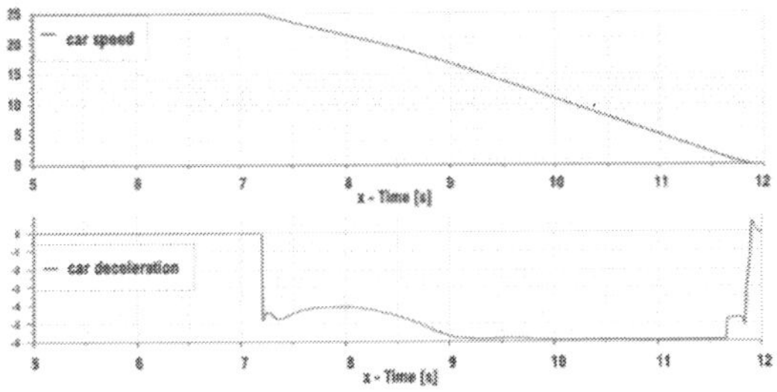

Figure 9. Speed variation in relationship with deceleration, *test 2 (90km/h, wet asphalt).*

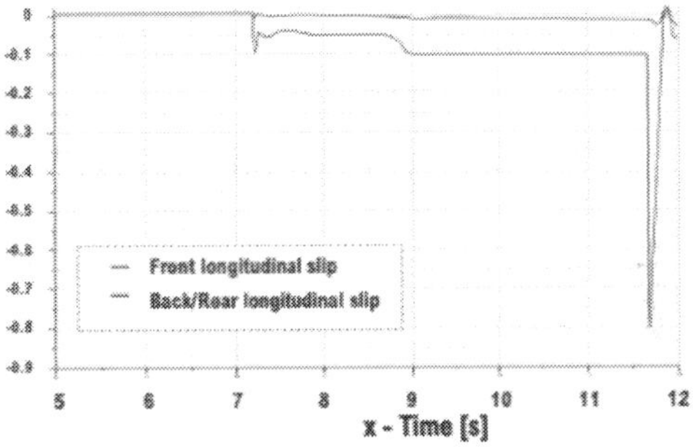

Figure 10. Variation of longitudinal slip, *test 2 (90km/h, wet asphalt).*

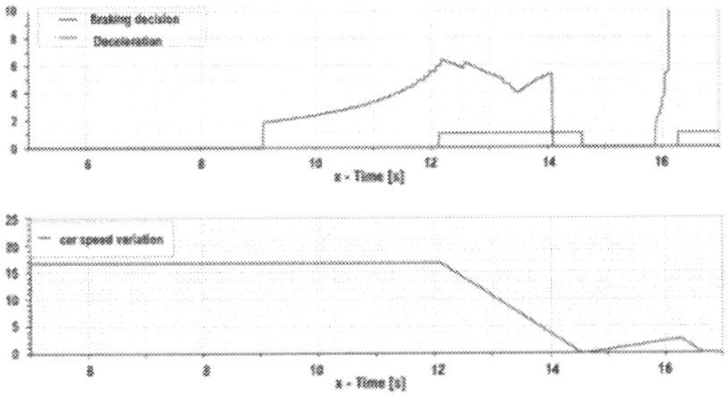
Figure 11. The decision of braking in relation to deceleration and the variation of speed, *test 3 (60km/h, dry asphalt).*

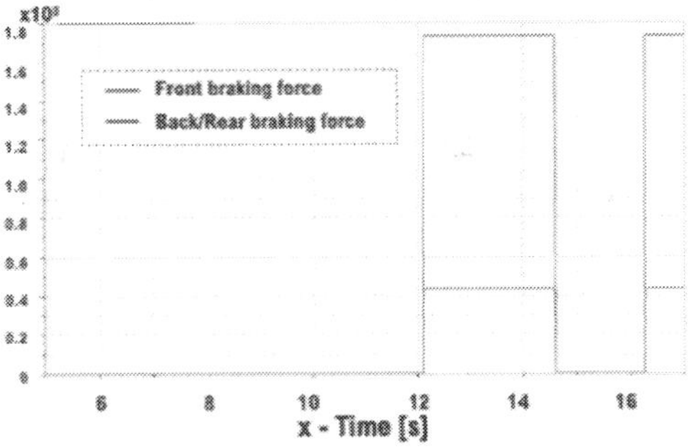

Figure 12. Variation of braking force, *test 3 (60km/h, dry asphalt).*

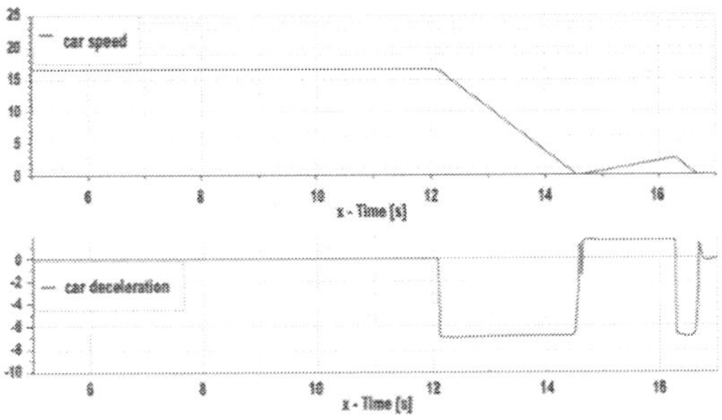

Figure 13. Speed variation in relationship with deceleration, *test 3 (60km/h, dry asphalt).*

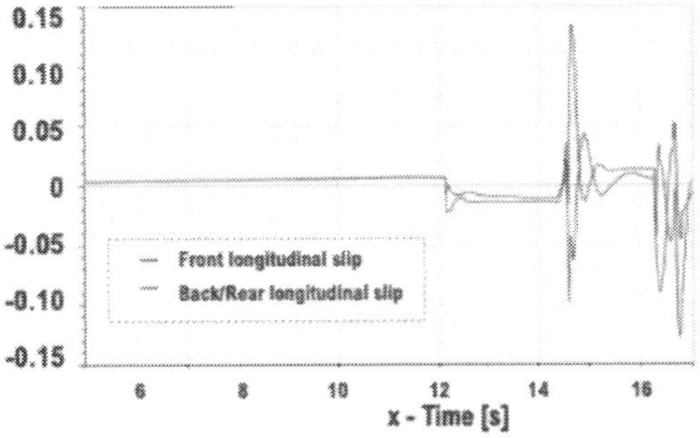

Figure 14. Variation of longitudinal slip, *test 3 (60km/h, dry asphalt).*

IOP Conf. Series: Materials Science and Engineering 1235 (2022) 012033 doi:10.1088/1757-899X/1235/1/012033

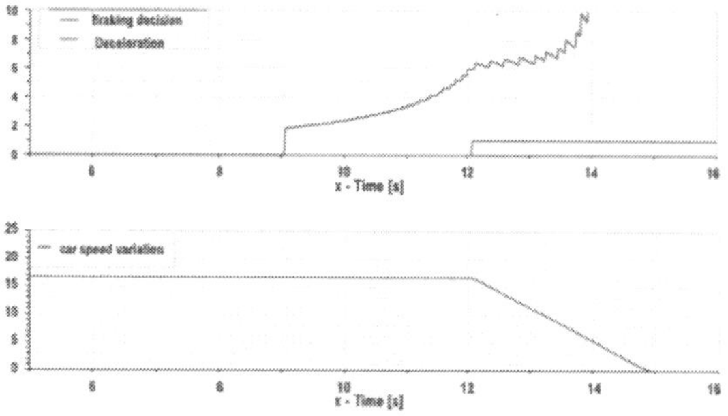

Figure 15. The decision of braking in relation to deceleration and the variation of speed, *test 4 (60km/h, wet asphalt)*.

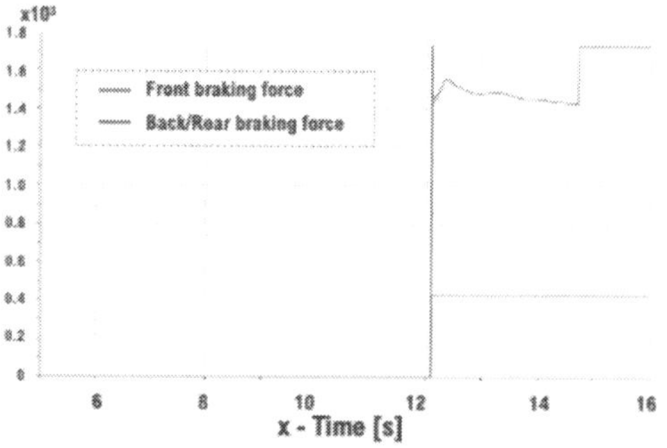

Figure 16. Variation of braking force, *test 4 (60km/h, wet asphalt)*.

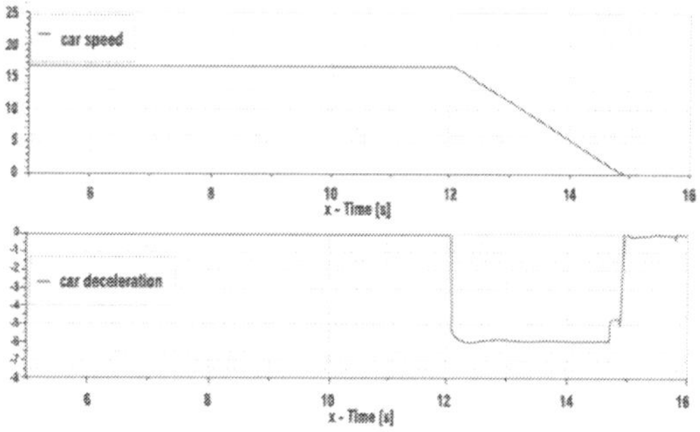

Figure 17. Speed variation in relationship with deceleration, *test 4 (60km/h, wet asphalt)*.

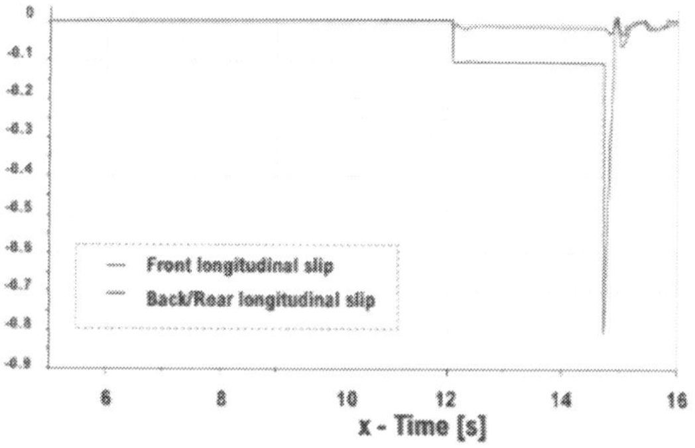

Figure 18. Variation of longitudinal slip, *test 4 (60km/h, wet asphalt)*.

3.4. Fourth Test

The target speed of the vehicle is 60 km/h and the adhesion coefficient is set to equivalent to the wet asphalt. In figure 15 and figure 17 we can observe only one break decision.

In the braking force graph, figure 16, we can observe the intervention of the ABS system only for the front, this happens because we have a great longitudinal slip, as seen in figure 18. In this test, the vehicle was able to stop in time and not impact the obstacle.

If in dry conditions at this speed, the vehicle was able to avoid the accident for the wet conditions is not possible. In real life, the vehicle is able to do that. We can conclude that the simulation can be further improved for wet conditions to be closer to a real-life model.

4. Conclusion

With the help of simulation, we can analyze the performance of the AEB system and try different configurations of the model until the optimal model is created.

In the simulated speeds, we can see that the system can stop without colliding with the obstacle. Based on the speeds simulated, we can assume that the AEB system is advantageous in towns and low speeds zone when it can avoid an impact.

We can observe that the system's performance is in relationship with the adherence coefficient of the road. Future research can be done with the vehicle running on snow.

Acknowledgements

This work has been funded by the European Social Fund from the Sectoral Operational Program Human Capital 2014-2020, through the Financial Agreement with the title 'Scholarships for entrepreneurial education among doctoral students and postdoctoral researchers (Be Antreprenor!)', Contract no. 51680/09.07.2019 – SMIS code: 124539

ORCID IDs
G Badea https://orcid.org/0000-0002-2395-0098
M Toma https://orcid.org/0000-0002-8588-1964
D A Micu https://orcid.org/0000-0003-1433-5648
G Fratila https://orcid.org/0000-0002-6740-5403
S Saragea https://orcid.org/0000-0002-9578-0841

References
[1] https://ec.europa.eu/transport/road_safety/specialist/statistics_en Accessed on 14.12.2020
[2] Deepak C, Guruprasath P, Kumar S H and Anthony V L 2017 Intelligent Braking System in Automobile Applications *Int Conf on Latest Innovations in Applied Science, Engineering and Technology* 2017 pp 103-8 Corpus ID: 212517133
[3] Bamigboye O O and Obaje S E 2016 Intelligent automatic car braking control system using neural network classifier *Int J of Engineering Inventions* **5** (6) 51-6 Corpus ID: 212587386
[4] LMS. AMESim Help

An aerodynamic theoretical study on exterior accessories of intervention vehicles

L Ilea, D Iozsa and G Fratila

University Politehnica of Bucharest, Autovehicule Rutiere Department, 313 Splaiul Independentei St., Romania

laurentiu.ilea@yahoo.com

Abstract. In the last decades, an extended set of restrictions and requirements for the automotive industry were imposed due to the increased concern for the environment and the population's health. Their purpose is to gradually reduce the level of emissions. Personal and commercial vehicles represent an important percent of total industry emissions footprint [1], however, intervention vehicles also bring a significant contribution in this area. The current paper proposes a theoretical study from an aerodynamic point of view of generic exterior equipment such as warning lights, sound devices, etc. for different body types such as sedan, hatchback, SUV, trucks. While the base vehicle complies with standard regulations regarding the emissions, the paper will highlight the significant impact brought by these elements when integrated on the vehicle, outside the manufacturer's development process. The paper will also show the relationship between equipment's dimension and position and the aerodynamic coefficient, though in fuel consumption and overall emissions performance.

1. Introduction

The emissions requirements for the automotive industry initially implemented through guidelines and reinforced after by financial penalisation, reshaped the CO_2 emissions strategy for each manufacturer. While a few decades ago, during development, the main focus was on the driving experience and endurance, nowadays, to reduce global emissions levels, great efforts are done to optimise the main contributors to fuel consumption: engine management, weight, rolling resistance and aerodynamics.

The purpose is to develop and implement a technology that enables the respect of CAFÉ (Corporate Average Fuel Economy) target which is assessed and updated frequently by the authorities. If the sold vehicles fail to comply, the penalisation that must be paid is 95 euros per extra gram of CO_2, per vehicle [2]. For the big manufacturers, this can easily transform into hundreds of millions of euros, therefore the stake to invest in the research and development branch, to be able to mass-produce 'green' vehicles is essential.

The current paper has the purpose of highlighting, from an aerodynamic perspective, the influence of exterior accessories for intervention vehicles on fuel consumption and therefore on their CO_2 emissions.

2. Intervention vehicles

Intervention vehicles are part of the multi-stage transformed vehicles category. These vehicles are purposed to allow transport and intervention in specific cases as emergencies, public maintenance works, etc. For this reason, in coherence with legislation, each vehicle in this category must have mandatory and specific warning lights, along with acoustic signals, plus essential mechanical and electrical accessories.

Most manufacturers have dedicated developing branches for this segment. The OEM (Original Equipment Manufacturer) develops a 'base vehicle' that can accommodate a great diversity of accessories and transformations, like warning signals, exterior air conditioning units, volume extenders for extended interior space, pick-up, just to name just a few, as exemplified in figure 1. After purchase, these 'base vehicles', are generally transferred to a second party company that manages the transformation process according to specific client requirements [3].

The main difficulty is that the accessories can have a great variety of shapes, dimensions, mounting positions, the reason for which the fuel consumption and emissions of the possible final transformed vehicles may vary between what can be very different values, due to very different influences on aerodynamic coefficient.

Figure 1. Examples of intervention vehicles.
(a – warning lights, b – pickup, c – volume extenders, d – exterior air conditioning unit)

3. Aerodynamic of exterior accessories for intervention vehicles

The aerodynamic performance of vehicles can be determined and homologated by different methods (wind tunnel, decelerating time on track) and is assessed with 'Cd', the aerodynamic coefficient on X longitudinal direction. All passenger vehicles commercialised in European Union must comply with WLTP (Worldwide Harmonised Light Vehicle Test Procedure) and RDE (Real Driving Emissions) standards which means that the Cd must be determined in a wind tunnel, in the presence of an independent witness authority, to assure the transparency regarding the best and worst level of aerodynamic performance declared by the manufacturer for a given model.

For more accuracy, some manufacturers evaluate the aerodynamic performance by multiplying the aerodynamic coefficient (Cd) with the projected frontal area of the vehicle (A). For each 0.020 m^2 of CdA, under WLTP measuring conditions, the vehicles are assessed to consume 0.1 litres of combustible less and to produce with 1 gram of CO_2 per kilometre less. Considering that at speeds above 100 km/h the aerodynamic force is the main resistance to overcome and the context of the emission, it is thus explained why recently the vehicles tend to have a more aerodynamic body and also a big range of technical elements that help reduce the aerodynamic force.

CFD (computational fluid dynamic) studies showed that the influence of a generic longitudinal ramp of warning lights has various influences, depending on vehicle body type, warning lights position, shape, size, just to name a few. The first step of the study is to assess the performance and analyse the airflow for the base vehicle, in its standard configuration and without any external accessories. To have a better image, it was chosen to evaluate a generic sedan (code 'L'), hatchback (code 'B') and SUV (code 'H') body type, as indicated in figure 2. For an image on the possible performance dispersion, the warning lights will be mounted in advanced position (code 'i'), median position (code 'ii') and retreat position (code 'iii'), as presented in figure 3. It is mentioned that due to the big diversity of warning lights arbitrary equipment was chosen with shape and dimension described in figure 4. Base data obtained with a CFD software, based on the Lattice-Boltzmann method, presented in table 1.

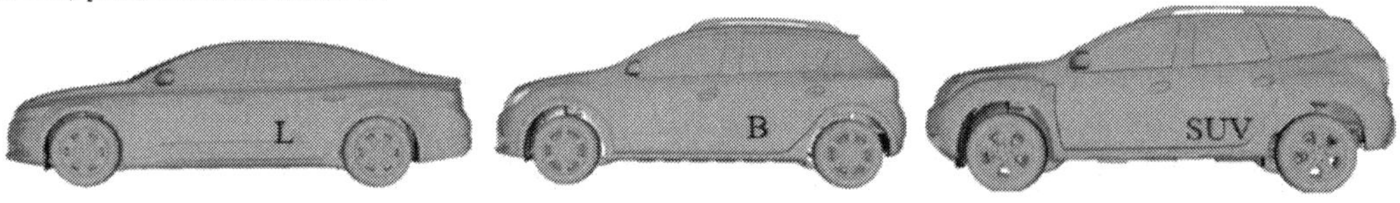

Figure 2. 3D model of vehicles arbitrary chosen for 'warning lights' study.

Figure 3. 'Warning lights' mounting positions.

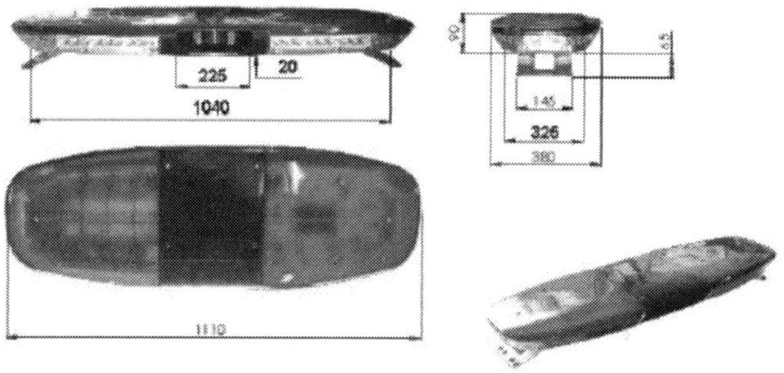

Figure 4. 'Warning light' shape and dimension, used for the current study.

Table 1. Base data used for 'warning lights' influence on aerodynamic performance.

Studied case	**L (A= 2.240 m²)**	**L - i**	**L - ii**	**L - iii**
CdA [m²]	0.659	0.905	0.753	0.796
Δ CdA [m²]	-	+ 0.246	+ 0.094	+0.137
Δ CdA [%]	-	+ 37	+14	+20
Studied case	**B (A= 2.390 m²)**	**B - i**	**B - ii**	**B - iii**
CdA [m²]	0.836	0.996	0.910	0.905
Δ CdA [m²]	-	+0.160	+0.074	+0.069
Δ CdA [%]	-	+19	+8.8	+8.2
Studied case	**SUV (A= 2.445 m²)**	**SUV - i**	**SUV - ii**	**SUV - iii**
CdA [m²]	1.000	1.127	1.059	1.059
Δ CdA [m²]	-	+0.127	+0.059	+0.059
Δ CdA [%]	-	+12	+5.8	+5.8

The values in table 1 are obtained with a CFD software applying the Lattice Boltzman mathematical model. The testing environment respects the real dimensions and proprieties of S2A wind tunnel, from 2 Montigny-le-Bretonneux, France [1]. The environment has a fine mesh near the vehicle (cell with the dimension of 2 mm) and becomes scarcer as recordings are made at a bigger distance from the vehicle. It is mentioned that fine recordings are made also behind the vehicle to have a good representation of wake. The measurements are made with an airspeed of 165 km/h, correlated with current homologation legislation for aerodynamic performance.

It is concluded that for a 'sedan' vehicle, the worst encountered situation is with warning lights in an advanced position when the aerodynamic resistance increases by 37%, as compared with aerodynamic resistance of the same vehicle, but without warning lights. The same advanced position of the warning lights ramp is also the worst case for a generic 'hatchback' and 'SUV' vehicle, but the CdA is increased with plus 19% and 12%. The different impacts can be explained by the fact that the aerodynamic performance of the 'base vehicle is the best for the 'sedan' and worst for the 'SUV', with the 'hatchback' between the two. So, the exterior accessory has the biggest impact on the best CdA vehicle and less impact on the worst CdA vehicle. The influence of the warning lights ramp on airflow is shown in figure 5, where for each position (advanced, median, retreat) the vorticity magnitude is displayed on 'sedan' vehicle (a), 'hatchback' vehicle (b) and 'SUV' vehicle (c).

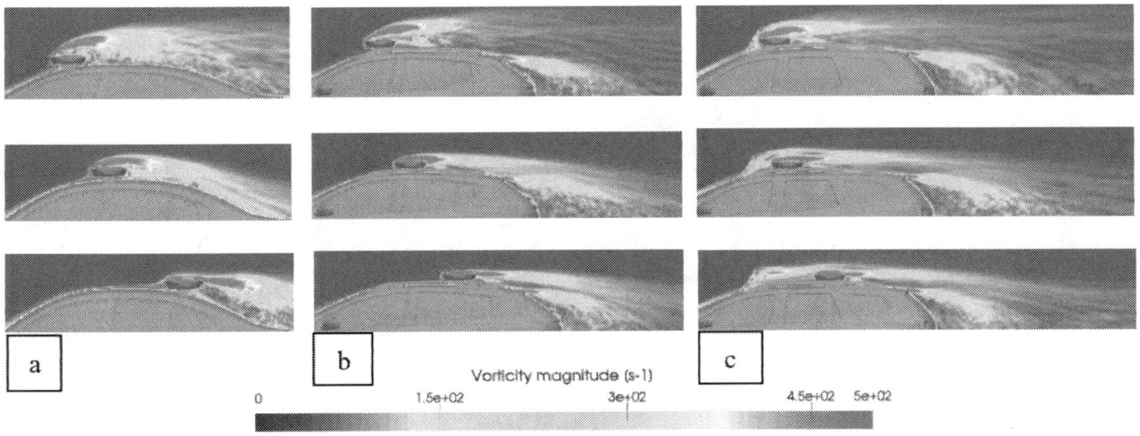

Figure 5. Vorticity magnitude for 'sedan', 'hatchback' and 'SUV" body vehicles, with warning lights ramp.

For cylindrical warning lights, with reduced volume, the impact is smaller, but the same tendency is kept.

The 'pick-up' transformation can bring a degradation of 0.050 m^2 to 0.080 m^2, the value obtained using the same CFD method on a single cab 'SUV' body, without warning lights. This can be translated into degradation of 2.5 grams to 4 grams of CO_2. If the warning lights ramp is added, the above increase of 12% will significantly change due to very different airflow and rear wake shape and size. In this case, the addition of warning lights may bring a degradation between 0.100 m^2 and 0.200 m^2, depending on the equipment parameters, but also the height between the cab roof and ramp.

All above values were obtained in a virtual wind tunnel, using a fluid dynamics solver based on Lattice Boltzmann [4][5] mathematical model, while the studied vehicles are 3D models arbitrary chosen. The CdA results presented may vary depending on studied vehicle design, warning lights geometry and position, the parameters used during simulation setup and run.

4. Conclusions

Using the CFD tools, the current paper highlights the not negligible negative impact of external accessories on aerodynamic performance and thus, on vehicle emissions. By choosing an arbitrary vehicle and warning lights accessory it is proven that the CdA value is increased. In the studied cases, the main conclusion was that a given accessory has the biggest impact on the most aerodynamic body type and the least impact on the worst aerodynamic body type. When introducing an exterior accessory, the biggest influence will be noticed on the wake that is equilibrated, but more sensible to perturbators factors, like warning lights placed on the roof. In all studied body type cases, the advanced mounting position is less favourable because high-speed airflow from the windshield is projected directly into the warning lights ramps, inducing a big wake.

All the values and associated conclusions can be applied for the specifically used geometry. Any other designs may induce different CdA values, but the same general behaviour is expected. Another source of errors in the software parameters and mathematical model used. While the technology made great advancements and is used by the automotive industry, it must be considered that each approximation induces a small error which leads to a difference between real-world and simulated tests. The CFD method remains a very important tool but must be carefully used, and results analyzed and finally validated in real-world scenarios.

The current work will be continued by simulating the impact of extended volume shapes, sound warning accessories and other objects that are frequently mounted on intervention vehicles and induce a big increase in aerodynamic performance and the emissions they finally produce during the exploitation.

ORCID iDs
L Ilea https://orcid.org/0000-0001-6329-004X
D Iozsa https://orcid.org/0000-0002-6547-8633
G Fratila https://orcid.org/0000-0002-6740-5403

References
[1] Holman C, Harrison R and Querol X 2015 Review of the efficacy of low emission zones to improve urban air quality in European cities *Atmospheric Environment* **111** 161-9 https://doi.org/10.1016/j.atmosenv.2015.04.009
[2] Continental Automotive *Passenger Cars / Light and Medium Duty Vehicles* 2019 (Worldwide Emission Standards and Related Regulations, Emission Booklet)
[3] Society of Motor manufacturers and traders limited 2011 *Multi-Stage Build Vehicle Approval Seminars* (SMMT Driving the motor industry)
[4] Ricot D 2016 *Turbulent flows using Lattice Boltzmann Method application on automotive configurations* MUSAF 16, ONERA, Toulouse
[5] Qian Y H, D'Humanieres D, Lallemand P 1992 Lattice BGK Models for Navier-Stokes Equation *EPL* **17** 479 https://doi.org/10.1209/0295-5075/17/6/001

IOP Conf. Series: Materials Science and Engineering 1235 (2022) 012035 doi:10.1088/1757-899X/1235/1/012035

Software strategy for internal combustion engine and electric motor control on a hybrid electric vehicle equipped with belt starter generator and automated manual transmission

Ştefan Saragea, Marius Toma, Dan Alexandru Micu, Gheorghe Frăţilă and Gabriel Badea

Automotive Engineering Department, Politehnica University of Bucharest, Bucharest RO-060042, Romania

stefansaragea@gmail.com

Abstract. In order to fulfil the latest EU emission standards, the average amount of CO_2 emissions for each vehicle sold has to be reduced to avoid fines for manufacturers. One of the simplest and cheapest ways of hybridization is the belt starter generator (BSG), which makes it suitable for an entry-level range of vehicles. This paper presents a software strategy developed for internal combustion engine (ICE) and belt starter generator (BSG) control on a vehicle equipped with automated manual transmission and also the drivability, comfort and emission performances improvements expected by using this strategy.

1. Introduction

In order to fulfil the latest EU emission standards, the average amount of CO_2 emissions for each vehicle sold has to be reduced to avoid fines for manufacturers. The latest European Union regulations (EC regulation 443/2009) sets a target of 95 g CO_2/km for the average emissions of the new car fleet for each car manufacturer starting with the year 2021.

In figure 1 is presented the average CO_2 emissions for vehicles sold in European Union for each manufacturer. As can be seen from the average of CO_2 emissions, all manufacturers are above the 95 g CO_2/km, and further improvements in ICE efficiency and increasing the level of hybridization is needed to achieve the target.

One of the simplest and cheapest ways of hybridization is the belt starter generator (BSG), which makes it suitable for an entry-level range of vehicles.

The software strategy presented in this paper is developed for both hybrid vehicles equipped with an internal combustion engine (ICE) and belt starter generator (BSG) with manual or automated-manual gearboxes, and also for conventional vehicles equipped only with ICE.

Man	R	Specific emissions target	Average CO2 emissions
BMW GROUP pool	1,003,404	140.8	125.6
DAIMLER AG pool	985,163	139.7	136.7
FCA-TESLA pool	984,048	129	115.7
FORD-WERKE GMBH pool	978,961	133.7	130.8
HONDA MOTOR EUROPE LTD pool	116,980	130.8	125
HYUNDAI pool	552,340	126.9	124.2
KIA pool	497,564	128.7	122.8
MG-SAIC pool	14,102	140.1	124
MITSUBISHI MOTORS pool	139,438	133.2	115.6
PSA-OPEL pool	2,507,547	125.7	113.9
RENAULT pool	1,653,093	126.3	118.1
SUZUKI pool	254,016	123.1	120.6
TATA MOTORS JAGUAR LAND ROVER pool	214,955	178	157.6
TOYOTA-MAZDA pool	1,035,435	129	108
VW GROUP PC pool	3,722,198	133.2	123.9
NISSAN INTERNATIONAL SA	391,479	130.5	114.3
VOLVO CAR CORPORATION	326,205	148.3	131.9

Figure 1. Average CO_2 emission of each manufacturer for vehicles sold in Europe in 2019 [1]

2. Strategy presentation

The strategy was constructed using LMS Imagine.Lab AMESim program [2].

The software strategy was developed to be used for both conventional vehicles (ICE only) and hybrid vehicles equipped with an internal combustion engine (ICE) and belt starter generator (BSG) with manual or automated-manual gearboxes.

The strategy was divided into 4 parts:

2.1. Strategy activation

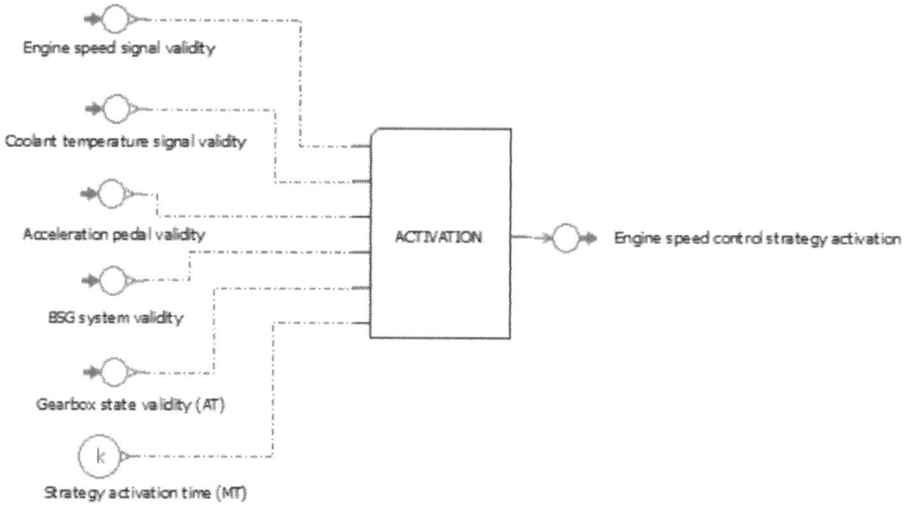

Figure 2. Strategy activation conditions.

In figure 2 are presented the conditions for the activation of the internal combustion engine (ICE) speed control during gear shifts. To have a smooth operation of ICE, the strategy will be activated only if all the following signals are valid:

- Engine speed signal – the engine speed is the main parameter of the strategy
- Coolant temperature – the coolant temperature is needed because the parameters of the strategy can be calibrated in the function of coolant temperature value. The coolant temperature has an impact on engine torque loses (ICE friction torque) which is an important factor in strategy functionality (conventional vehicles, ICE only)
- Acceleration pedal position – the acceleration pedal position is needed to calculate the engine speed setpoint during a gear shift. For low acceleration pedal positions, a smoother engine speed evolution can be used and for high acceleration pedal positions, faster engine speed decreasing must be achieved.
- BSG system – for vehicles equipped with BSG, the BSG is the main system that controls the ICE engine speed evolution during a gear shift.
- Gearbox state – for vehicles equipped with Automatic Manual Transmission, the shifting time is necessary for engine speed setpoint during gear shift calculation.

For a vehicle equipped with a manual gearbox (MT), the duration of the strategy activation can be set to avoid an engine speed hook if the driver stays with the clutch pedal pressed (sailing). The duration of the strategy activation can be set in the function of acceleration pedal position before strategy activation and selected gear.

The validity of the input signals is generated by the ICE control unit using self-diagnosis for each of the presented sensors. If one of the input signals is not valid (for example sensor malfunction, not plausible signal), the strategy will not be activated. In the case of vehicles equipped with Automated Manual Transmission, the gearshift will work on a backup strategy.

2.2. Engine speed setpoint during gear shift

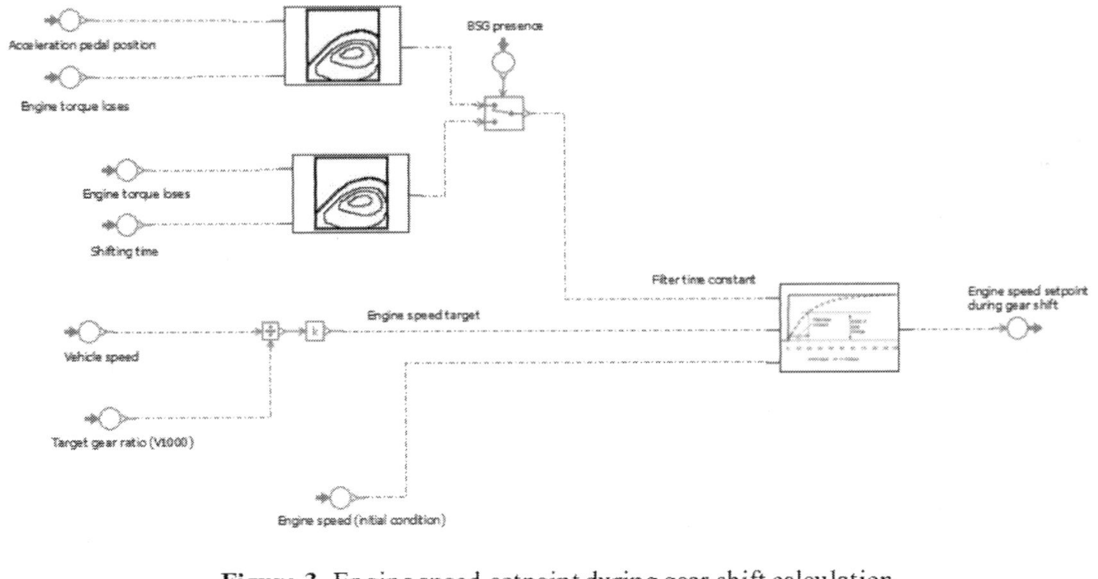

Figure 3. Engine speed setpoint during gear shift calculation.

In figure 3 is presented the engine speed setpoint calculation during a gear shift. The initial condition is the engine speed at which the strategy is activated. The target engine speed is calculated from the vehicle speed and the target gear ratio (V1000 - vehicle speed at 1000 min^{-1}). The engine speed setpoint transition is made using a calibrated filter (first-order filter). The filter time constant is calculated in function of vehicle type:

IOP Conf. Series: Materials Science and Engineering 1235 (2022) 012035 doi:10.1088/1757-899X/1235/1/012035

- for mild hybrid vehicles (ICE + BSG), the filter can be calibrated in the function of engine torque loses and acceleration pedal position. For low acceleration pedal position, for smoother gear shifts, the filter can be increased to achieve smoother gear shifts. For the high acceleration pedal position, the filter can be reduced to achieve faster gear shifts to increase vehicle performance.
- for conventional vehicles (only ICE), the filter can be calibrated in the function of engine torque loses and shifting time.

2.3. PI regulator

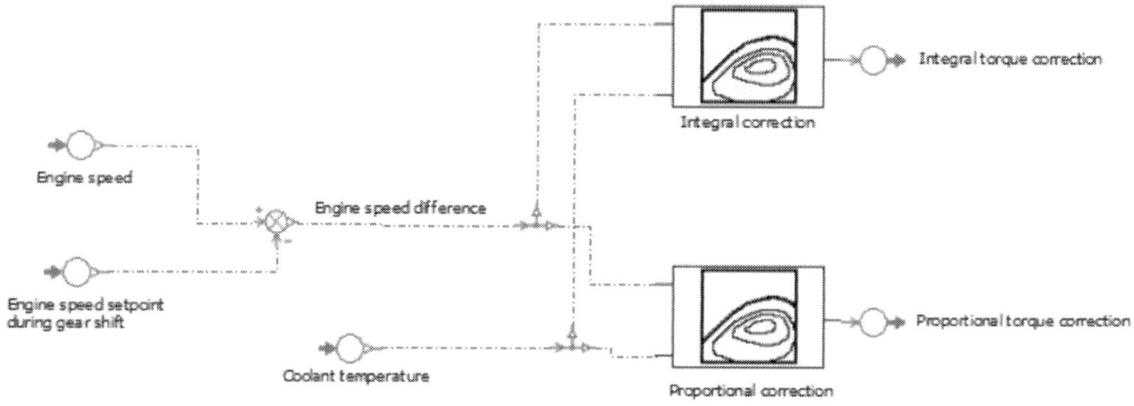

Figure 4. PI regulator.

In figure 4 is presented the PI regulator used to control the engine speed during a gear shift.

When the strategy is activated, the engine speed is compared to the engine speed setpoint (presented in 2.2). In function of the engine speed difference and the coolant temperature, the proportional and integral correction factors are calibrated to achieve smooth engine speed evolution during a gear shift. The proportional and integral correction factors will be calibrated as follows: in case of low coolant temperature, the factors will be lowered due to combustion low quality and high engine friction. Regarding the engine speed difference, the factors will be increased in coherence with the engine speed difference.

2.4. Torque correction

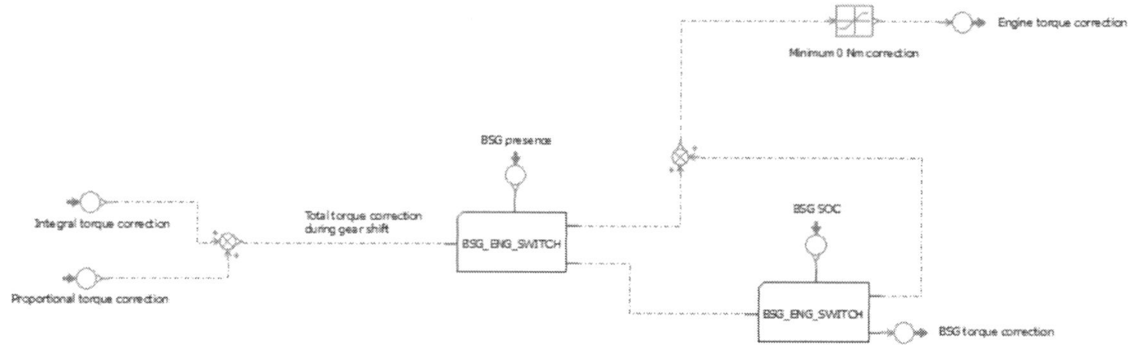

Figure 5. Torque correction.

In figure 5 is presented the torque correction calculation. The proportional and integral corrections are combined and in the function of the vehicle, the type is sent to ICE or BSG.

- For Mild hybrid vehicles equipped with BSG, the torque calculated during gear shift is sent to the BSG. In the case of positive torque, the BSG acts as an electrical motor and increases internal combustion engine speed. In the case of negative torque, the BSG acts as a generator and decreases internal combustion engine speed to follow the engine speed setpoint. In case of the low state of charge (SOC) for the battery, the BSG is deactivated from the strategy and the engine speed is controlled by the ICE only.
- For ICE vehicles (conventional vehicles), the torque correction is limited only to positive torque. The negative torque is not considered because in the case of low engine speeds if high negative torque is requested, there is a risk of engine stall. The disadvantage of using only the positive torque is that in the case of ICE with high inertia and low torque consumption (auxiliary torque loses, friction torque loses and pumping torque loses), the engine speed decrease is low (500-600 min^{-1}) [3] and to achieve smooth gear shift, the shifting time must be increased. For manual transmission applications or in the case of AMT at high engine load (shifting time is reduced to a minimum), a software strategy improvement can be considered, increase the load on the alternator or activate the air conditioning compressor during gear shift to increase the engine brake.

3. Expected results

The main improvement expected using this strategy is smoother vehicle longitudinal acceleration during a gear shift.

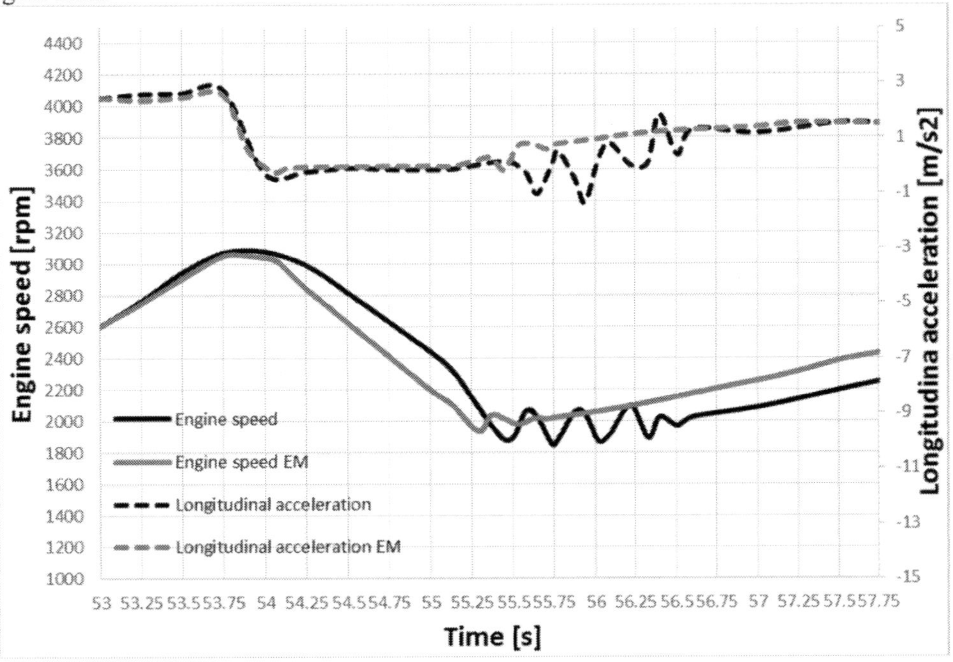

Figure 6. Engine speed and longitudinal acceleration variation during gear change from 2nd to 3rd gear. [3]

In figure 6 are presented the improvements expected to be obtained using the software strategy presented above during gearshifts for mild hybrid vehicles equipped with a belt starter generator (BSG). Using the BSG to control the ICE speed during gearshifts, the maximum car body longitudinal oscillation was 0.9 m/s^2 (Longitudinal acceleration), compared to a maximum longitudinal acceleration of 2.5 m/s^2 for the vehicle without ISG and engine speed control strategy during gear shift from 2nd to 3rd gear.

For conventional vehicles (only ICE), as presented in a previous paper [4], by controlling the engine speed during the gear shift, the longitudinal acceleration oscillation was reduced by 50%, from 3 m/s^2 to about 1.5 m/s^2.

4. Conclusion

This paper is presented a software strategy used to control the internal combustion engine speed during gear shift to synchronize the ICE speed to the input shaft of the gearbox. The strategy was developed to be used for mild hybrid vehicles equipped with Belt Starter Generators (BSG), for conventional vehicles (ICE only) and also for vehicles equipped with manual transmissions and automated manual transmissions (AMT).

By using the mild-hybrid system with BSG, the vehicle CO_2 emissions can be reduced by up to 20% [5,6]. A reduction of 20% in CO_2 emissions can lead to the emission target (95 g CO_2/km in 2021) fulfilment for most manufacturers, compared to 2019 average CO_2 emissions (presented in Figure 1).

ORCID iDs

S Saragea https://orcid.org/0000-0002-9578-0841
M Toma https://orcid.org/0000-0002-8588-1964
D A Micu https://orcid.org/0000-0003-1433-5648
G Fratila https://orcid.org/0000-0002-6740-5403
G Badea https://orcid.org/0000-0002-2395-0098

References

[1] https://www.eea.europa.eu/data-and-maps/indicators/average-co2-emissions-from-motor-vehicles
[2] LMS. AMESim Help. AMEHelp, 2012
[3] Saragea Ș, Badea G, Frățilă G 2021 Drivability improvements obtained using a mild hybrid electrical system on a vehicle equipped with a manual gearbox. *IOP Conf. Ser. Mater. Sci. Eng.* **1037** 012052 https://doi.org/10.1088/1757-899X/1037/1/012052
[4] Saragea Ș, Oprean IM, Badea G, Frățilă G 2018 Modeling the Torque Oscillations During Gear Shifts on Automated Manual Transmission *Proc. of the 4th Int. Congress of Automotive and Transport Engineering* pp 634-641 https://doi.org/10.1007/978-3-319-94409-8_74
[5] Saragea Ș, Oprean I M, Bățăuș M, Frățilă G 2019 Driveability improvements obtained using a mild-hybrid electrical system on a vehicle equipped with an automated manual transmission, *Science and management of automotive and transportation engineering 4th edition* (Craiova)
[6] Saragea Ș, Badea G, Frățilă G 2020 Fuel consumption and emission improvements obtained using a mild hybrid electrical system on a vehicle equipped with a manual gearbox. *IOP Conf. Ser.: Mater. Sci. Eng.* **1037** 012053 https://doi.org/10.1088/1757-899X/1037/1/012053

IOP Conf. Series: Materials Science and Engineering 1235 (2022) 012036 doi:10.1088/1757-899X/1235/1/012036

Theoretical and experimental research of an automotive gasoline engine fuelled with butanol

C Sandu[1], C Pana[1], N Negurescu[1], A Cernat[1], D Fuiorescu[1], R Georgescu[1] and C Nuțu[2]

[1] Department of Thermotechnics, Engines, Thermal and Frigorific Equipment, University Politehnica of Bucharest, Splaiul Independenței 313, Bucharest 060042, Romania
[2] Department of Automotive, University Politehnica of Bucharest, Splaiul Independenței 313, Bucharest 060042, Romania

sandu_cristian2007@yahoo.com

Abstract. Among alcohols used for spark ignition engine fuelling, butanol can be a viable alternative fuel due to its very good combustion proprieties. Butanol has a high miscibility with the gasoline and stable blends can be assured. The higher oxygen content and the higher combustion speed of butanol comparative with gasoline are proprieties which may improve the combustion process and may lead to the reduction of greenhouse gases level such as CO_2 and pollutants like CO, HC, and NO_X. A theoretical and experimental study was conducted on a spark-ignition engine with 1.5L displacement fuelled with n-butanol-gasoline blends at different volume percentages and different excess air coefficients at the 55% engine load of and 2500 min-1 engine speed. Starting from a predefined model in AMESim was developed and calibrated a physical-mathematical model for the process's simulation from the engine cylinder. The objective of the paper is to get a good representation of the real engine in numeric to have a better prediction of the engine performance at the different operating conditions. The theoretical and experimental investigations show important advantages of n-butanol in blend with gasoline: the increase of the thermal efficiency due to the stable engine operation at lean mixtures use.

1. Introduction

Nowadays, regarding fossil fuel reserves, a globally large fossil fuel consumption is not sustainable for more than a few decades. For the transportation sector, special attention was given to the study of the use of alcohol-gasoline blends at spark ignition engines, alcohols like methanol, butanol, ethanol, etc being used. According to studies done by Gautam and Martin [1,2] and Yacoub *et al.* [3], the use of blended conventional fuels such as gasoline or diesel with alcohols may improve the thermal efficiencies of the internal combustion engines and reduce the fuel consumption. A study conducted by Wu *et al.* [4] concluded that the use of butanol, produced from corn, to fuel a spark-ignition engine may lead in the transportation sector to a decrease of gasoline consumption by 39-56% and greenhouse emissions by up to 48% on a life-cycle basis.

Butanol is less corrosive than ethanol making it ideal for fuelling systems that use classic fuel [6]. Two studies conducted by Liu *et al.* [7,8] concluded that the use of n-butanol has led to lower soot

IOP Conf. Series: Materials Science and Engineering 1235 (2022) 012036 doi:10.1088/1757-899X/1235/1/012036

concentrations when compared to biodiesel. Also, Liu *et al.* [9] have studied multiple fuels and their effect on pollutant emissions and have concluded that oxygenated blends have a low impact on engine emissions. Another study by Yuqiang Li *et al.* [10] analysed the effects of fuelling a spark ignition engine with isopropanol-n-butanol-ethanol and gasoline. The blending fuel used was 30% IBE (isopropanol-n-butanol-ethanol) and 70% gasoline. Results show an improvement of the thermal efficiency by 4.3% and lower pollutant emissions, 4% for carbon monoxide, 15-20% reduction of unburnt hydrocarbons and in some cases a reduction of nitrogen oxides by up to 18.6%. When analysing the combustion process at fuelling with butanol blends, J. Serras-Pereira *et al.* [11] concluded that butanol is more difficult to combust in cold engine conditions. Butanol as a standalone fuel was used in a homogenous charge compression ignition (HCCI) engine. Studies were limited to using n-butanol and engine performance, measured by IMEP, was limited by combustion stability at lean conditions [12]. Low NO_x and soot emissions levels were achieved without utilizing exhaust gas recirculation (EGR) at low to mid-engine loads, [13]. In a study by Alasfour *et al.* [14] a 7% power loss was observed when fuelling a spark ignition engine with a blend of 30% butanol and 70% gasoline. Alasfour *et al.* [15] also studied the effect of the 30% butanol-gasoline blend on emissions, noticing a decrease in pollutants level for CO and HC for λ between 0.9 and 1.05 and also a decrease by 9% for NO_x. Increasing the engine spark timing will lead to higher NO_x emissions was observed by Alasfour in his study *NO_x emission from a spark-ignition engine using 30% iso-butanol – gasoline* [16].

The objective of this paper is to determine the effects of fuelling a spark ignition engine with 10% n-butanol and 90% gasoline blend on energetic performances, emissions and cyclic variability over 250 engine cycles. The engine was firstly fuelled with pure gasoline to set up the reference at the engine load χ=55%, engine speed of n=2500 min^{-1} and different air-fuel ratios (λ). After setting up the reference on the bench using gasoline and blended mix, a numerical simulation of the engine is done using AMESim and the results are compared. The numeric model is pre-defined in the AMESim libraries and will be used and adapted to simulate the spark ignition engine on the bench [5].

2. Bench setup description and modelling

2.1. Engine bench description

In this section the engine test bench is described and starting from this description, the numeric modelling of the engine and its components is adapted. Engine (1) is a Cielo Nubira spark ignition engine (A15MF) coupled to an AVL dynamic brake testbed (4). The dyno braking moment (torque) is controlled with a control cabinet (3). During testing, the dyno will heat up so it's cooled with the cooling system (2). The engine is also equipped with a radiator cooling system (9). The amount of air in the engine is controlled with a throttle body (5) and the airflow is measured with an airflow transducer (13). The fuel flow is measured with a flow transducer installed on the fuel system circuit. The cylinder pressure is measured with a piezo transducer, the signal is amplified and sent to an acquisition system, a PC in this case (11). The supercharging pressure is measured with a manometer. The exhaust gasses go through the catalytic converter (8) and into the gas analyser (10). Additionally, the excess air coefficient is also measured by the gas analyser.

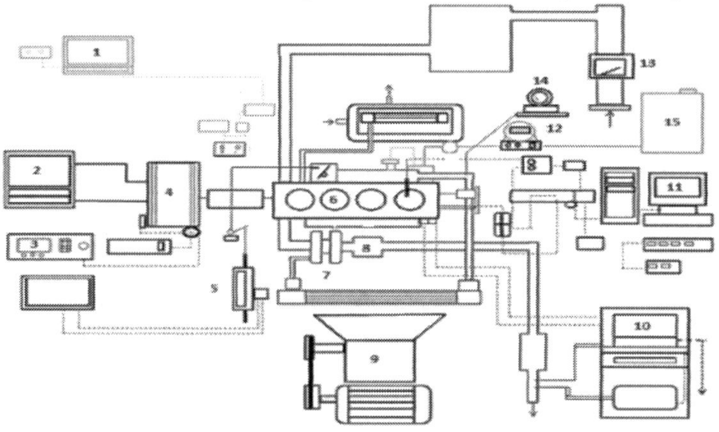

Figure 1. Engine Bench Setup.

The principal components of the test bench are 1 – ECU, 2 – dyno cooling system, 3 – dyno control cabinet, 4 – Eddy current dyno, 5 – throttle body actuator, 6 – A15MF engine, 7

– turbocharger, 8 – catalytic converter, 9 – radiator cooling system, 10 – gas analyser, 11 – PC of the data acquisition system, 12 – fuel mass flow flowmeter, 13 – airflow meter, 14 – boost pressure transducer, 15 – fuel tank.

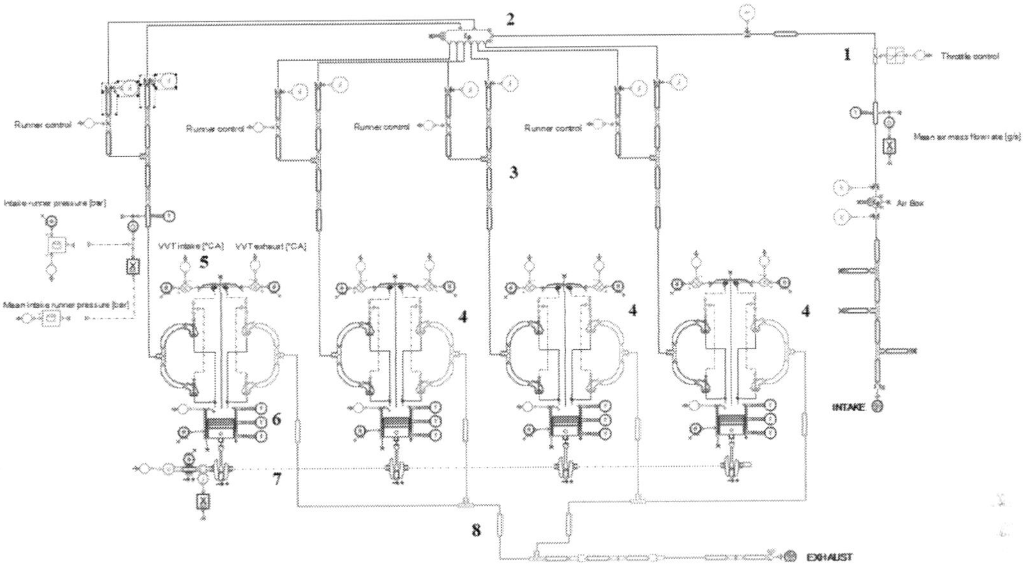

Figure 2. Overview of the numeric engine model [5].

In figure 2 the following were numbered 1 - throttle body, 2 - intake manifold, 3 - air intake ducts, 4 - bent ducts before the cylinder, 5 - intake and exhaust valves, 6 - piston group, 7 - crankshaft, 8 -exhaust ducts.

The model uses several lookup tables to define certain parameters such as engine speed, spark advance, and throttle opening.

2.2. Engine combustion process model

The sub-model used to simulate the engine is defined as **ENGCCSI10** in the AMESim libraries [5]. The model represents a combustion chamber with variable volume, dynamic pressure and heat exchange with the cylinder walls. It also takes into consideration the mixing of three gases: air, fuel and burnt gases. The fuel ($C_xH_yO_z$) is defined by three parameters: x, y, z and the heating value (PCI). The model is presented in figure 3.

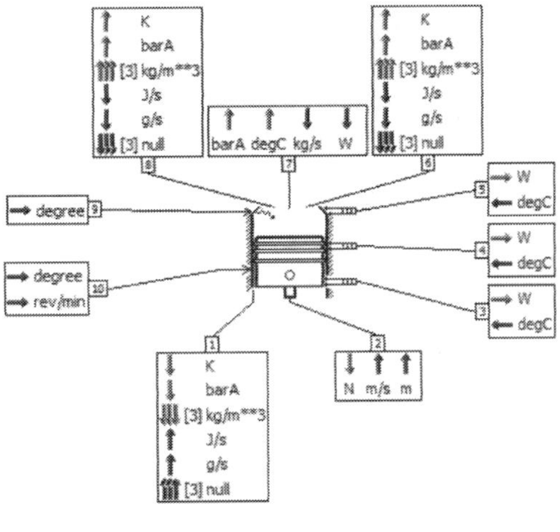

Figure 3. Representation of the engine model [5].

The heat release model is based on the Wiebe-Myamoto expression. The ignition of the air-fuel mixture is defined at 360 CAD. The stoichiometric air-fuel mixture is defined in this model as PCO. The equation to determine the heat released during combustion is presented below [5]:

$$Q_{comb} = mfb \cdot PCI$$ thee ratio of (1)

where mfb is theratioof fuel if the fuel/air ration is less than one, $\lambda \leq 1$ and the (mass of air/ PCO) if the fuel/air ration is greater than one $\lambda > 1$; PCI is the heating value [J/kg]

The dynamic pressure is determined as follows [5]:

$$\frac{dP}{dt} = \rho \cdot \left(\frac{dr}{dt}T + r\frac{dT}{dt}\right) + \frac{d\rho}{dt}rT \quad (2)$$

To solve the equation above the following need to be determined:

- T and temperature evolution dT/dt;
- perfect gas constant r and its evolution dr/dt;
- density ρ and density evolution dρ/dt.

3. Simulation and results

To simulate the engine working at 2500 min^{-1} and engine load of 50%, the input parameter k as one of the main input signals of the simulation is equal to 2.65. Using k=2.65 the brake mean effective pressure of the engine is around 10 bar.

In the figure below (figure 4) the law of combustion is presented in two cases: 1) resulted from the measurements and calculations from the engine on the bench and 2) resulted from the numeric simulation of the model. The law of combustion at fuelling with pure gasoline is presented with the blue line (a). It is noticed that almost the same amount of heat is released in both cases, around 1100 J. In the real engine, mounted on the test bench, the combustion develops rapidly at the beginning of the combustion process versus the simulated process. One possible source of error is that the engine on the bench is also turbocharged and thus the engine filling is different vs a naturally aspirated engine. The ignition delay of the mixture was not measured on the bench but the simulated model assumes a default value of 10 CAD. The combustion duration in the model is defined statically as 70 CAD.

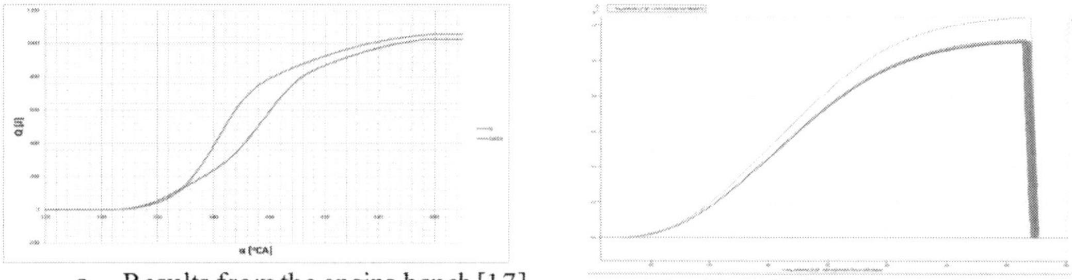

a. Results from the engine bench [17].

b. Results from the engine simulation.

Figure 4. Comparison of the law of combustion between the engine on the bench and the simulated model. The duration of combustion is 70 CAD.

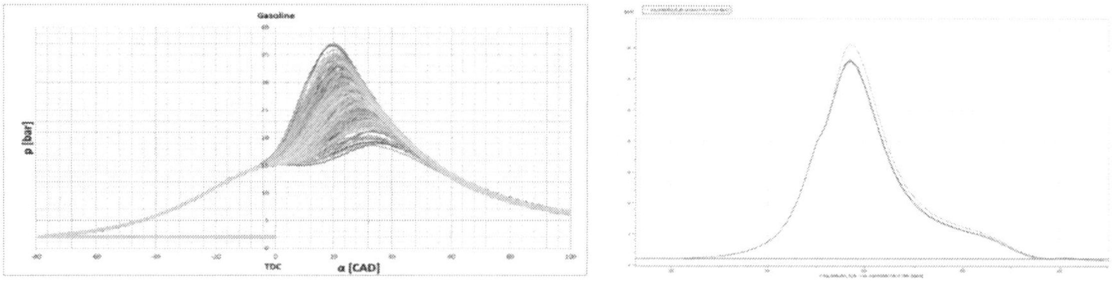

Figure 5. Comparison of pressure diagrams between a) engine on the bench and b) simulated model. The combustion duration is 80 CAD.

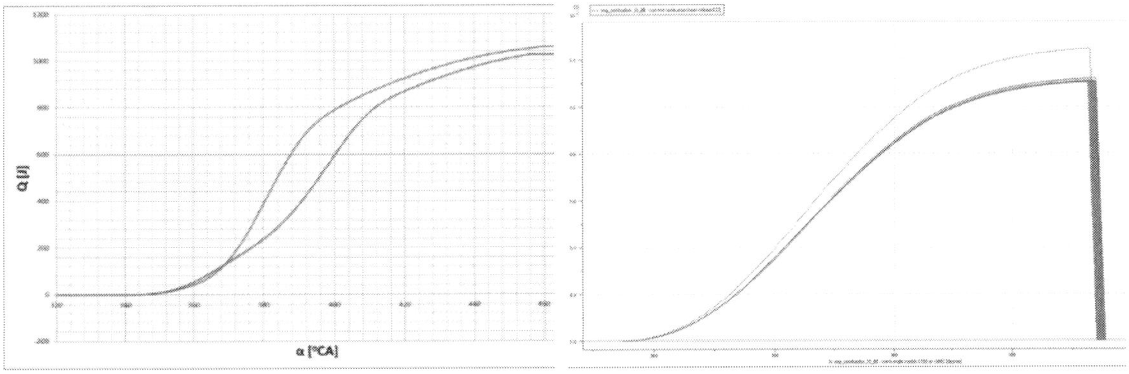

Figure 6. Comparison of the law of combustion between the engine on the bench and the simulated model. The duration of combustion is 80 CAD.

4. Conclusions

Results between the engine bench and the numeric model have been compared in this paper. At first glance, the simulated results are not too different from the real engine, nor are they identical. Many possible sources of error can influence the output of the numeric model.

The comparison of pressure diagrams is presented in figure 5. The maximum pressure varies from cycle to cycle, between 35 and 40 bar for the simulated model and between 25-30 for the engine on the bench. The dispersion in values is more evident in the engine on the bench (a) the combustion process is more unstable vs the simulated model. One reason why the combustion process on the bench is more unstable could be the excess air coefficient λ which is always fluctuating. The numeric model only computes with stoichiometric air-fuel mixtures.

Increasing the burning duration from 70 CAD in the numeric model to 80 CAD (figure 6) will lead to similar pressure diagrams to the engine on the bench. The difference in total heat released between the two cases is 6% and the average duration 5%. It is noticed that almost the same amount of heat is released in both cases, around 1100 J. These differences can be considered acceptable but they can be reduced with further model optimization. The mixture ignition delay was estimated at 10 CAD in the simulation. On the bench, the angles at which maximum pressure is achieved varies between 380 CAD and 390 CAD. The numeric model achieves maximum pressure at 385 CAD.

A major difference between the engine on the bench and the numeric model is that the bench is turbocharged while the model is naturally aspirated. This is because modelling the turbocharger would add even more complexity to the model. The engine filling is different which could explain why the engine burns the mixture faster vs the numeric model.

Acknowledgements

The authors would like to thank AVL List GmbH Graz, Austria, for providing the possibility to use the Simulation Software AVL BOOST.

This work has been funded by:

-the European Social Fund from the Sectoral Operational Programme Human Capital 2014-2020, through the Financial Agreement with the title "Scholarships for entrepreneurial education among doctoral students and postdoctoral researchers (Be Entrepreneur!)", Contract no. 51680/09.07.2019 - SMIS code: 124539

-a grant of the Romanian Ministry of Research and Innovation, CCCDI – UEFISCDI, project number PN-III-P1-1.2-PCCDI-2017-0404 / 31PCCDI/2018, within PNCDI III

ORCID iDs

C Sandu https://orcid.org/0000-0002-9813-7261
C Pana https://orcid.org/0000-0002-4629-6551
N Negurescu https://orcid.org/0000-0003-3107-6836
A Cernat https://orcid.org/0000-0003-0551-0758
D Fuiorescu https://orcid.org/0000-0001-6365-9847
R Georgescu https://orcid.org/0000-0003-2213-4256
C Nuțu https://orcid.org/0000-0001-9085-6053

5. References

[1] Gautam M, Martin II D W 2000 *Proceedings of the IMechE Part A: J. Power and Energy* **214** (5) 497-511 https://doi.org/10.1243/0957650001538047
[2] Gautam M, Martin II D W and Carder D 2000 *Proceedings of the IMechE Part A: J. Power and Energy* **214** (2) 165-82 https://doi.org/10.1243/0957650001538263
[3] Yacoub Y, Bata R and Gautam M 1998 *Proceedings of the IMechE Part A: J. Power and Energy* **212** (5) 363-79 https://doi.org/10.1243/0957650981536934
[4] Wu M, Wang M, Liu Jand Huo H 2008 *Biotechnol Progress* **24** 1204-14 https://doi.org/10.1002/btpr.71
[5] AMESim Application Libraries 2013 version 13.0.0 / Rev13 - 35759-31626
[6] Doğan O. 2011 *Fuel* **90** 2467-72 https://doi.org/10.1016/j.fuel.2011.02.033
[7]] Liu H, Huo M, Liu Yu, Wang X, Wang Hu, Yao M, and Lee C F 2014 *Fuel* **133** 317-25 https://doi.org/10.1016/j.fuel.2014.05.038
[8] Liu H, Bi X, Huo M, Lee CF, Yao M. 2012 *Energ. Fuel* **26** (3) 1900-11 https://doi.org/10.1021/ef201720d
[9] Liu H, Li S, Zheng Z, Xu J, Yao M. 2013 *Appl. Energy.* **112** 246-56 https://doi.org/10.1016/j.apenergy. 2013.06.023
[10] Li Y, Meng L, Nithyanandan K, Lee T H, Lin Y, Lee C-FF and Liao S 2016 *Fuel* **184** 864-72 https://doi.org/10.1016/j.fuel. 2016.07.063
[11] Serras-Pereira J, Aleiferis P G and Richardson D 2013 *Combust Sci Technol* **185** (3) 484-513 https://doi.org/10.1080/00102202.2012.728650
[12] He B-Q, Yuan J, Liu M-B and Zhao H 2014 *Fuel* **115** 758–64 https://doi.org/10.1016/j.fuel.2013.07.089
[13] Zheng M, Han X, Asad U and Wang J. 2015 *Energy Conversion and Management* **98** 215–24 https://doi.org/10.1016/j.enconman.2015.03.098
[14] Alasfour F N 1997 *Appl. Therm. Eng.* **17** (6) 537-49 https://doi.org/10.1016/S1359-4311(96)00069-5
[15] Alasfour F N 1998 *Appl. Therm. Eng.* **18** (5) 245-56 https://doi.org/10.1016/S1359-4311(97)00081-1
[16] Alasfour F N 1998 *Appl. Therm. Eng.* **18** (8) 609-18 https://doi.org/10.1016/S1359-4311(97)00082-3
[17] Sandu C, Pană C, Negurescu N, Cernat A, Nuțu C and Georgescu R 2020 *E3S Web Conf., 180 (2020) 01010* https://doi.org/10.1051/e3sconf/202018001010

IOP Conf. Series: Materials Science and Engineering 1235 (2022) 012037 doi:10.1088/1757-899X/1235/1/012037

Concept development of an independent electronic system for vibroacoustic diagnostics of vehicles

N A Nitu, V Caunii and C Bujoreanu
Gheorghe Asachi Technical University of Iasi, Department of Mechanical Engineering, Mechatronics and Robotics, 59A Blvd. Mangeron, 70005 - Iasi, Romania

carmen.bujoreanu@gmail.com

ABSTRACT: This paper proposes the concept of electronic noise and vibration diagnostic system for vehicles that aims to avoid wrong or unreasonable replacement of their components in the repair process. Our concept for a vibroacoustic diagnostic system consists of a set of sensors that receive noise and vibration, judiciously arranged and connected to a DAQ (Data Acquisition System). This allows the ECU (Electronic Control Unit) to accurately determine, by three-dimensional detection, the area in which vibroacoustic anomalies are manifesting (due to current or incipient defects). The data storage memory works according to the first in - first out principle, retaining only significant information to save resources. Establishing a mathematical model is one of the focuses for this concept for predicting the failure of various components of the vehicle based on vibroacoustic and then developing an artificial intelligence program capable of identifying and classifying new behaviours are the steps which we follow in the development of this tool. Also, the benefits of early detection of a subassembly failure are found in increasing traffic safety.

1. Introduction

The century in which we live has among its basic defining characteristics the time factor, and hence we notice that the trend of systematization and efficiency of methods and technologies related to the operation of vehicles and their maintenance and repair is constantly evolving [1].

For proper repairing, in the automotive industry are used two main types of diagnostic procedures, one is based on dedicated facilities with special tools and the other one is onboard diagnostic [2]. According to the vehicle complexity, it could be applied a direct and simple diagnostic procedure (like dimension measurement of a component, a hydraulic or pneumatical pressure, or an electrical resistance) or a complex procedure that involve an electronic network of vehicle control units [3].

At present, design processes of complex strategies of diagnostic are based on the fact that in the profile industry, there is a strong lack of highly qualified and trained personnel, and one solution is a partial replacement of their skills with appropriate software & hardware. As a consequence, these diagnostic technologies shall not require highly educated people for operating them and to have a high rate of success. An example is the evaluation of wear status of the mechanical component from the gearbox: classically, it is necessary to remove it from the vehicle and to dismantle it for measurement, but in the actual stage o technology is only need to connect a dedicated electronic diagnostic computer to the gearbox control unit and to interrogate the measured values which show values in millimetres or percentage for the mechanical components. For that, the gearbox control unit is measuring the electrical resistance of position sensors and transform them into millimetres by interpolation [4]. Similar is happening with the pneumatical system from a commercial vehicle, wherein the classical method it is necessary to use a box of manometers for measuring the value of pressure, and now this is done by the electronic control unit of the system, which is interrogating electrical pressure sensors allocated for every pneumatical circuit. Both examples of actual diagnostic methods have in common the principle of measuring an electrical value, which is software transformed by the electronic control unit into mechanical or pneumatical value for different components of the vehicles.

Reviewing the literature, we found that an independent electronic system capable of vibroacoustic to diagnostic the state of a defect or a potential defect of the entire vehicle would be of interest. Segmentation on the elements subject to the defect state (which can affect other subsystems of the vehicle) requires high accuracy of the diagnostic, avoiding mistakes and eliminating unnecessary extra time. The purpose of our study in this paper is to design a methodology for the development of such an independent electronic system for vehicle diagnostic.

2. Methodology

This concept of electronic vehicle vibroacoustic diagnostic system aims to avoid erroneous or unreasonable replacement of elements, for example in the suspension, exhaust system, steering, gearbox, drivetrain and other systems of the vehicle in the repair process. But, at the same time, by predicting the failure of a subassembly we could support the increase of traffic safety degree.

The implementation of the empirical human method of vibroacoustic diagnostic in an independent electronic system presume in principle the following steps:

a) defining the matrix of sensors for vibroacoustic analysis;
b) creation of a vibroacoustic database characterizing the vibroacoustic behaviour of the relevant components of the vehicle;
c) establishing a mathematical model for predicting the failure of the various components of the vehicle by vibroacoustic;
d) development of an artificial intelligence software capable of identifying and classifying new behaviours (not initially recorded in the vibroacoustic database), to ensure the continuous development of the vibroacoustic diagnostic method constantly adapted to the evolution of technology (e.g. transition from vehicles powered by internal combustion engines to hybrid or fully electric).

Creating the database requires clearly defining two basic components:

a) a passive component involving the experimental determination of the normal state of operation of the vehicle as a whole, by acquiring and recording noise and vibration. Then the spectral analysis is realized to identify the behaviour of the various components (engine, transmission, suspension, body, etc.) on their natural frequencies and, respectively, the characteristic sound level;
b) an active component involving the application, experimentally controlled, of disturbing excitations to determine the frequency - amplitude manifestation of the vibroacoustic behaviour of the studied system and the creation of a reference base. Consequently, the vibroacoustic diagnostic system can differentiate the state of good functioning from the defective one of the different components of the diagnosed vehicle.

The scientifically creating process of the database can involve the following four main aspects [5]:

a) classification of noise and vibration sources;
b) detection of rare vibroacoustic events;
c) detection of vibroacoustic events in real audio life;
d) large-scale detection of poorly monitored sound events for vehicles.

An important topic referring to the database is the development of the vibroacoustic recording method that takes into account the way of the noise and vibration transmission, namely [6]:

a) airborne transmission;
b) structure-borne transmission through component;
c) mix transmission.

IOP Conf. Series: Materials Science and Engineering 1235 (2022) 012037 doi:10.1088/1757-899X/1235/1/012037

The way of vibroacoustic transmissibility involves specific physical phenomena: absorption, reflection, refraction, diffraction, distortion, attenuation and amplification of the signal. From here starts the selection of the sensors used as well as their initial calibration procedure.

An extremely important detail is the state of motion of the studied vehicle from a vibroacoustic point of view:

a) stationary vehicle, with the engine running at idle, medium or high revolutions;
b) moving vehicle: at a constant speed, accelerating or decelerating respectively.

In this context, acoustic signals could be an advantage in terms of vibrations, as their non-intrusive behaviour allows the ability to capture overlapping signals from several components and the ability to detect faults that affect pure airborne transmission. But it must take into account that the acoustic signals are more affected by the ambient noise which may mask the signal of interest.

For the real traffic conditions of motor vehicles, a robust monitoring implementation scheme could involve:

a) the acquisition of the signal;
b) several post-processing operations such as the correct selection of time histories, mediation and digital filtering to capture the signal of interest;
c) implementation of signal processing techniques taking into account the characteristics of the signal and the type of vehicle / mechanical component at which the signal is measured (means rotating or alternative machines with simple or complex mechanisms);
d) the extraction of several characteristics to be analysed to assess the health condition of the internal combustion engine or to detect some incipient defects and to determine the causes of their occurrence [6].

As a mathematical model, we can use the Fourier series for the analysis of periodic vibrational signals [6].

The Green function together with its experimentally determined gradient can be applied to measurements of sound transmission through the airborne by making the connection between the sound pressure level, the particle speed and the volume velocity [7].

3. Device design concept

The design of the system starts from the analysis of sound, noise and vibration parameters in existing or prototype vehicles. The sensor array for recording vibroacoustic behaviour may consist of accelerometers, microphones and optical encoders. Moreover, the recent development of low-cost MEMS microphones can help the acoustic monitoring technology to be implemented in onboard diagnostic systems (figure 1 and figure 2).

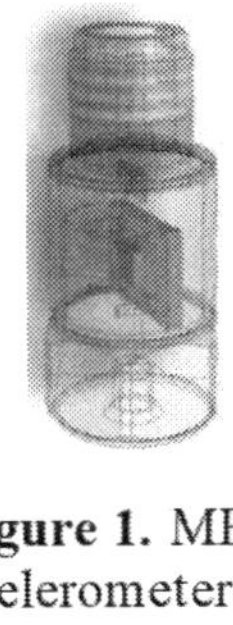

Figure 1. MEMS accelerometer with printed circuit board design [8].

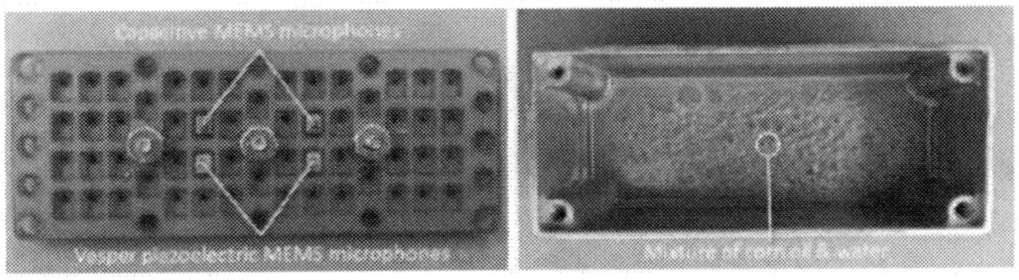

Figure 2. MEMS microphone resistant to environmental factors (dust, water, etc.) [9].

It is desirable that through the mixed-use of the recording elements to be able to validate the vibroacoustic response scheme of the studied element. The optimal selecting method of sensors used to capture noise and vibration is affected by the following:

- type of signal to be acquired, namely noise or vibration;

- target component to be monitored;
- environmental conditions in which they are to be located (on the chassis exposed to water/dust/shocks / variable temperature in extreme limits, etc.);
- possible influence of disturbing factors, such as loud ambient noise or electromagnetic interference;
- frequency range to be monitored (in principle 20 Hz - 20 kHz which is the useful one for the human ear);
- type of target failures;
- measurement accuracy;
- number of axes on which it measures;
- the way of fixing and weather protection on the studied system;
- dimensions and mass of the sensor.

Table 1. Characteristics to be followed in choosing sensors for vibroacoustic monitoring [8].

Measurement	Sensor	Key Information	Target Faults
Vibration	Piezo accelerometer	Low noise, frequencies up to 30 kHz	Bearing condition, gear meshing, pump cavitation, misalignment, imbalance, load condition
Vibration	MEMS accelerometer	Low cost/power/size, frequencies up to 20 kHz+	Bearing condition, gear meshing, pump cavitation, misalignment, imbalance, load condition
Sound pressure	Microphone	Low cost/power/size, frequencies up to 20 kHz	Bearing condition, gear meshing, pump cavitation, misalignment, imbalance, load condition
Sound pressure	Ultrasonic microphone	Low cost/power/size, frequencies up to 100 kHz	Pressure leaks, bearing condition, gear meshing, pump cavitation, misalignment, imbalance

Table 2. Mechanical specification of sensors for vibroacoustic monitoring [8].

Sensor	Size	Number of Axes	Vibration Mechanical Package	Industry Standard Interface	Integrated Features	Mechanical Attach	Environmental Tolerance
Piezo Accelerometer	Medium	1 to 3	Yes	Yes	No	Yes	Excellent
MEMS Accelerometer	Small/Medium	1 to 3	Yes	Yes	Yes	Yes	Excellent
MEMS Microphone	Small	1	No	No	No	Noncontact	Good
MEMS Ultrasonic	Small	1	No	No	No	Noncontact	Good

The decision to choose the signal analysis techniques for the vibroacoustic diagnostic method must meet some basic requirements such as:

a) nature of the signals to be measured: whether they are periodic or random / whether they are stationary or non-stationary;
b) which techniques should be used about the nature of the signals and which is the most appropriate field of analysis;
c) what information can be obtained through those signals using the appropriate technique / how efficient the chosen technique is.

IOP Conf. Series: Materials Science and Engineering 1235 (2022) 012037 doi:10.1088/1757-899X/1235/1/012037

The following step should be the creation of a data library designed for predicting the failure or even the vehicle immobilization with enough time for informing the user and to be taken the proper measures. The electronic unit responsible for this will be integrated into the electronic network of the vehicle and will provide information to users (driver, transport coordinator and vehicle manufacturer).

For an efficient way of informing, the system will be able to generate error codes to be displayed on board the vehicle, but which can be transmitted remotely to other users by telemetry / FMS (Fleet Management System). As a construction, the system, shown in figure 3 and figure 4, respectively, consists of a set of sensors (a) that receive noise and vibration (microphones, vibration sensors, etc.) judiciously placed and connected to a data acquisition system type DAQ [10] that allows the computer ECU (Electronic Control Unit) (b) to accurately determine the area where these vibroacoustic events occur (due to current or incipient defects) by three-dimensional detection. Well-known solutions such as LabVIEW [11] can be used as information processing software.

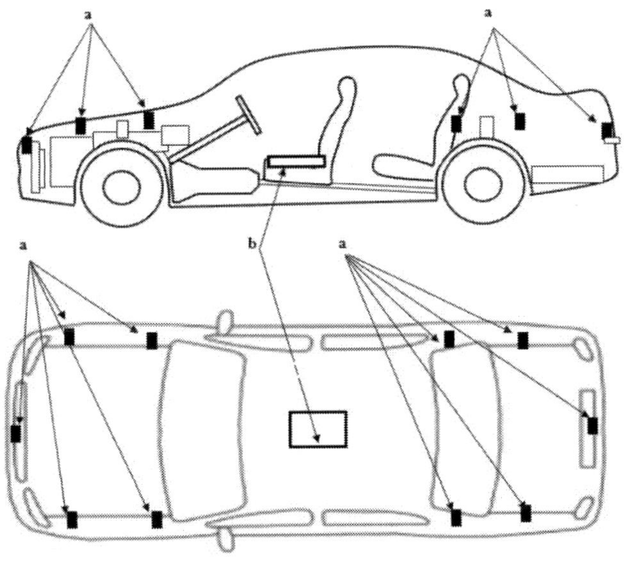

Figure 3. Location of the sound, vibration and noise sensors (a) and computer (b) on a car – side and upper view.

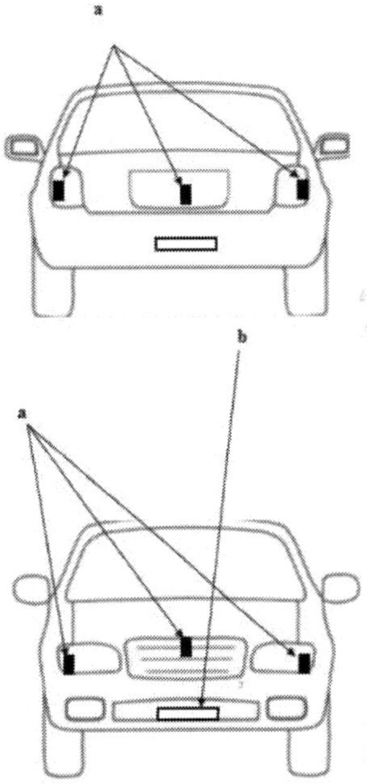

Figure 4. Location of the sound, vibration and noise sensors (a) and computer (b) on a car, rear and front view.

With the acquired information, the ECU analyses the parameters of the vibroacoustic events: frequency, intensity, the moment when they appear, and according, classifies them. Then, the events are identified and ECU sets an error code with the corresponding freeze frame (conditions in which the anomaly/defect appears). The data should be available and viewed by the user in an online/offline system (computer, tablet or dedicated tester) using a diagnostic interface. If a noise event is new, it is stored and recorded in the ECU with related parameters, including the area in which it occurred for analysis. The memory in which the data is stored will work according to the first in - first out principle (the first data entered is the first data out), retaining only the significant data for saving resources. The ECU parameters will be programmable, with the option to be configured according to the needs of the user and to be adapted to any vehicle, having a procedure for learning of normal operation. This procedure is also useful in the development phase of the vehicle by the manufacturer, who also establishes the error codes according to the defects that have occurred.

Another important feature of the system is that it can detect noises and vibrations from a level imperceptible to the human ear, being able to diagnose the malfunction of an element, subassembly or assembly from the early stage of failure, warning the user in time to prevent immobilization on long

IOP Conf. Series: Materials Science and Engineering 1235 (2022) 012037 doi:10.1088/1757-899X/1235/1/012037

durations of the vehicle for repairs or chain failures. The alert level can be programmed by the manufacturer according to his needs. The simplicity of this system is the operation based on the principle of superposition [12] which can be reduced from complex to simple by analogy with the musical area, where we can imagine a complex orchestra in which starting from this principle, it searches the instrument which is not working properly.

4. Conclusions

The electronic system for vibroacoustic diagnostic of the vehicles is conceptual described in our paper and can bring important advantages: increase the diagnose speed as the faulty component or subassembly is precisely detected; reduce the downtime of the vehicle for repair because it will accurately diagnose the defective component; detect the low level of noise caused by the failure of an component, assembly or subassembly because it detects noise or vibration from an early stage, at a low intensity imperceptible to the human ear; warn the driver by an indicator lamp or displaying a warning message on a screen about a malfunction; send messages to other computers and receive messages from them, with the choice to be integrated into any vehicle; ability to work in different configurations: stand-alone mode, multiple configuration, master-slave configuration, with or without communication with other modules; programming possibility according to the number of axles, engine, gearbox, steering, braking system, traction type and propulsion system (internal combustion engine, electric, hybrid or hydrogen) and depending on the type and arrangement of the sensors; increase of the safety, efficiency and economy of transport / operation of the vehicle.

The device could be practically adapted to any mechanism, because it can learn noises and vibrations during normal operation, differentiating them from those that occur in the event of a defect.

ORCID iDs

N A Nitu https://orcid.org/0000-0002-4842-5441
V Caunii https://orcid.org/0000-0001-9316-5645
C Bujoreanu https://orcid.org/0000-0002-5680-2788

References

[1] Petruni A, Giagloglou E, Douglas E, Geng J, Leva M C and Demichela M 2019 *Safety Science* **119** 229-339 https://doi.org/10.1016/j.ssci.2017.05.007
[2] Stratulat M and Andreescu C 1998 *Automotive Diagnostics* (*in Romanian*) Bucuresti: Science and Technic Society SA
[3] Selig M, Shi Z, Ball A and Schmidt K 2012 *Journal of Physics: Conference Serie* **364** 012013, 25th Int. Congress on Condition Monitoring and Diagnostic Engineering https://doi.org/10.1088/1742-6596/364/1/012013
[4] Newnes Sensor Technology Handbook 2005 (Chapter 15 - *Position and Motion Sensors)* ed Wilson J S pp 321-409 https://doi.org/10.1016/B978-075067729-5/50055-0
[5] Mesaros A, Heittola T, Diment A, Elizalde B, Shah A, Vincent E, Raj B and Virtanen T 2017 *D case 2017 Challenge Setup: Tasks, Datasets and Baseline System - DCASE* - Workshop on Detection and Classification of Acoustic Scenes and Events, Munich, Germany
[6] Delvecchio S, Bonfiglio P and Pompoli F 2018 *Mechanical Systems and Signal Processing* **99** 661–83 https://doi.org/10.1016/j.ymssp.2017.06.033
[7] Wolff O, Tijs E and de Bree H 2009 A PU probe array-based panel noise contribution analysis whilst driving *SAE Technical Paper* 2009-01-2123 https://doi.org/10.4271/2009-01-2123
[8] Murphy C 2020 *Choosing the Most Suitable Predictive Maintenance Sensor* Analog Devices, www.analog.com
[9] www.vespermems.com
[10] www.ni.com/en-us/shop/data-acquisition
[11] https://www.ni.com/en-us/shop/labview
[12] Gilbert P U P A 2022 Superposition, beats and harmony *Physics in the Arts (Third Edition)* (Academic Press) chapter 15 pp 257-70 https://doi.org/10.1016/B978-0-12-824347-3.00015-7

IOP Conf. Series: Materials Science and Engineering 1235 (2022) 012038 doi:10.1088/1757-899X/1235/1/012038

Study regarding the reduction of energy, vibrations and noises generated by the car fan

A M Vranau, V Caunii and C Bujoreanu

Gheorghe Asachi Technical University of Iasi, Department of Mechanical Engineering, Mechatronics and Robotics, 59A Blvd. Mangeron, 70005 - Iasi, Romania

cbujorea@tuiasi.ro

Abstract. The ventilation and air conditioning system is a complex and energy-consuming one. So far there is a lot of research in this area, but most of it is related to the study of soundproofing materials behind the dashboard and the modification of the air ducts. Our paper presents a more efficient constructive solution, which will reduce energy, noise and vibration. The purpose is to study air distribution systems with multiple fans. The number of small fans will be directly proportional to the seats in the car (depending on the type of car). The difference between the system proposed by us and the classic one is that the fans will start depending on the seats occupied by the passengers. Favourable results of studies can lead to high passenger comfort. Also, vibration reduction solutions for ventilation and air conditioning systems are all the more necessary in the case of silent electric cars.

1. Introduction

The ventilation and air conditioning system (HVAC) is a large consumer of energy, fuel and generator of pollutants. Along with the electric power steering system, the robotic gearbox and the auxiliary systems in the car's equipment create great difficulties for car manufacturers due to the high energy they consume. This lost energy generates a decrease in engine power.

However, the HVAC system is responsible for achieving thermal comfort for car occupants. Studies show that inside a car the feeling of thermal comfort is reached when the airspeed has values of about 1.25 m/s, the humidity is between the threshold of 30% and 70%, and the temperature is between 22-25 °C [1].

In the past, the system took up a lot of space and did not offer many functions, but now, manufacturers have managed to develop compact systems to control temperature, airflow, recirculation functions and various options for eliminating unpleasant odours. The air conditioning system has become a standard option for most manufacturers because it no longer involves high costs.

The first automatic control module of the HVAC system was introduced in 1964. In modern cars, the control is held by this module, which, through the sensors available in each car, it receives values that lead to the creation of thermal comfort inside (exterior temperature sensors, air quality sensor, passenger compartment temperature sensor). The control of indoor temperatures is closely related to the parts of the human body that are exposed (there will be different temperatures in the limbs, torso or head). Another task that the HVAC system has to fulfil is one related to passenger safety, namely, defrosting the windshield.

The air conditioning system that is the subject of this paper is a simple one, without additional comfort functions.

2. Presentation of the HVAC system

We reiterate the idea, as we said in the introduction, that the system is simple. This system equips the Dacia Logan car. The construction of the ventilation and air conditioning system is shown in the figure below [1].

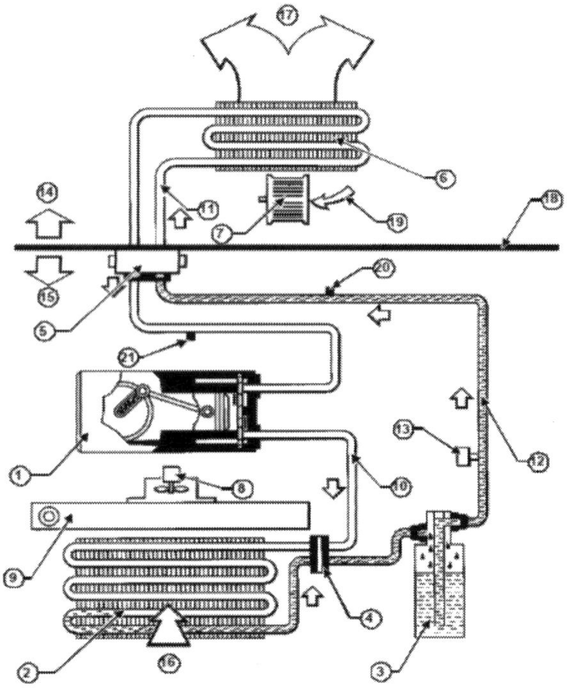

Figure 1. Description of the air conditioning system. [1]

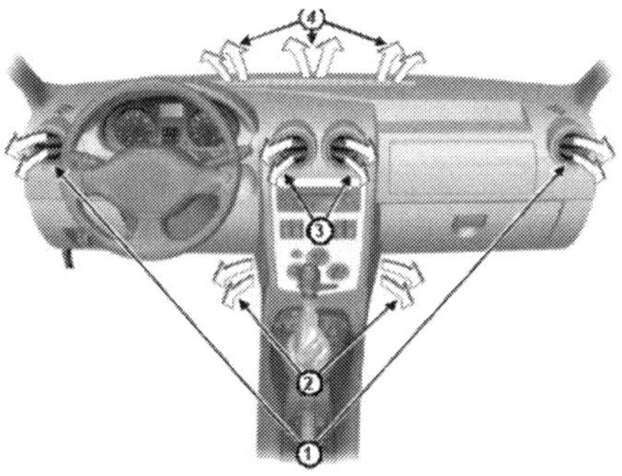

Figure 2. Ventilation grilles in the passenger compartment. [1]

The components of the HVAC system shown in figure 1 are 1 - compressor; 2 - condenser; 3 - dehydrating bottle; 4 - Pressure sensor; 5 - Thermostatic expansion valve; 6 - Vaporizer; 7 - Interior fan motor; 8 - Cooling fan motor; 9 - Engine radiator; 10 – High-pressure liquids; 11 – Low-pressure steamers; 12 – High-pressure steamers; 13 - Pressure sensor; 14 - Interior; 15 - Engine compartment; 16 - Outdoor air; 17 - To the air mixing box; 18 - Table; 19 - Outdoor or recycled air; 20 - High pressure valve; 21 - Low pressure valve.

Inside, the car in the work is equipped as follows (figure 2): 1 - side aerators, 2 - front leg heating outlet, 3 - central aerators, 4 - windshield demisting.

3. Noise and vibrations generated by the AC installation

Unfortunately, although the role of the air conditioning system is to increase human thermal comfort, it also has a negative role that contributes to the increase of technical discomfort caused by sound.

One of the noise and vibration generating components is the cooling fan. These sounds lead to a state of dissatisfaction on the part of the passengers and a decrease in the degree of concentration [2]. Especially for larger cars, the noise generated by the car fan often covers the noise of the engine, tire, running gear, transmission.

This is a big quality problem that leads to the creation of an unsatisfactory image of the brand/model [3]. In addition to these grievances, there is a problem of legal regulation of the sound level, which in case of malfunction exceeds the allowed level.

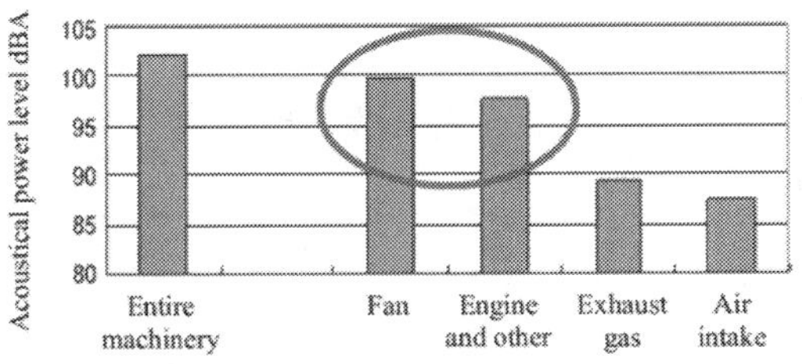

Figure 3. Contribution of the fan to the sound level. [3]

Figure 4. Transfer fan from PC components. [6]

There are many studies in this area because noise and vibration are a problem that affects both manufacturers and car users. A 1980 work by Gray presents the reconstruction of a cooling fan based on techniques from the design of silent marine propellers [3].

The reduction of noise and energy consumed by the components of HVAC systems is even more important for electric cars, where the noise is very easily noticed in the absence of the heat engine [4].

They are usually called vibrations, unwanted movements that produce mechanical noises or relatively high stresses. According to the "Explanatory Dictionary of the Romanian Language" (DEX - 1998), vibration is a "periodic movement of a body or particles of an environment, performed around a balanced position"[5]. Each body with mass and elasticity can vibrate. We must keep in mind that vibration is one of the first signs of a problem in a system or machine.

The vibration level of a system can help us identify its operating status. Common causes of vibrations are tolerance of workmanship and contact between moving parts. Given the familiarity with the air conditioning system and the knowledge of vibration sources, the goal is to model a simple air conditioning system, easy to adapt to almost any car and capable of creating thermal comfort for passengers with minimal resources.

4. Constructive solution

The solution for optimizing the performance of the HVAC system is based on the fact that in most cases, only the driver is in the car, and the idea is to obtain the feeling of thermal comfort only for the area occupied by it and not for the entire volume of the passenger compartment.

This approach to improving the thermal comfort inside the passenger compartment proposes a system that detects which seats are occupied on the seats and starts adjusting the temperature and flow of the air conditioning according to the occupant's desire and to obtain a feeling of thermal comfort as high as possible. The implementation of the system was done as follows:

The main intake fan of the HVAC distribution system has been replaced with a 120 mm fan, with a power consumption of approximately 4 W super quiet in the PC components industry. The initial fan of the car has a power of about 500 W according to the manufacturer's specifications. The chosen fan will be called a transfer fan and will be adjusted with a PWM (Pulse Width Modulation) system. The sound produced by it is about 35 dB [6], compared to the initial fan which produces a noise of about 90 dB.

In each vent was installed a cooling fan with a diameter of 30 mm ordered PWM. The advantage of these fans is the easy connection with the rest of the components, low noise and reliability.

A 200 N pressure sensor was mounted in each seat to detect if the seat is occupied (corresponds to a weight of approximately 20 Kg). Four temperature sensors were installed in the ceiling corresponding to each occupied place. The assembly was performed according to the figure below. The sensors are connected to the NI 6009 acquisition board [7].

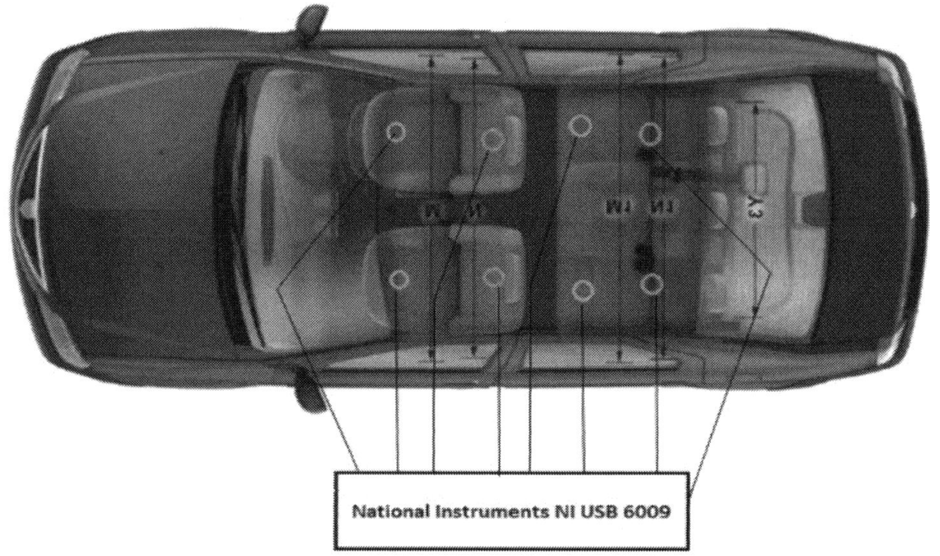

Figure 5. Temperature and weight sensors connected to the data acquisition board.

The images below show the initial configuration of the car (figure 6), air conditioning using air dampers, and the proposed solution, mounting the fans in the air vents (figure 7).

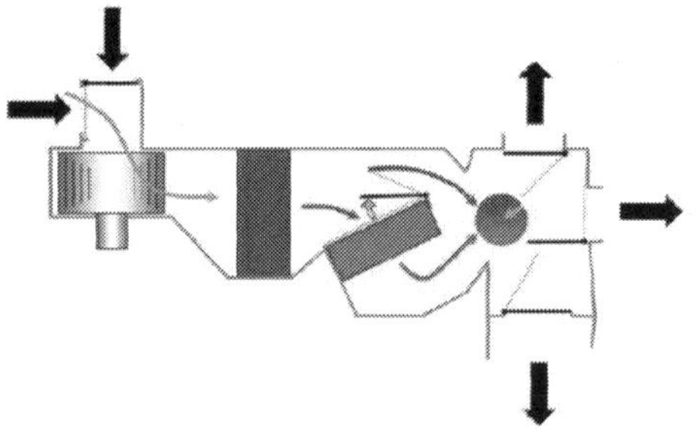

Figure 6. The original configuration of the HVAC system.

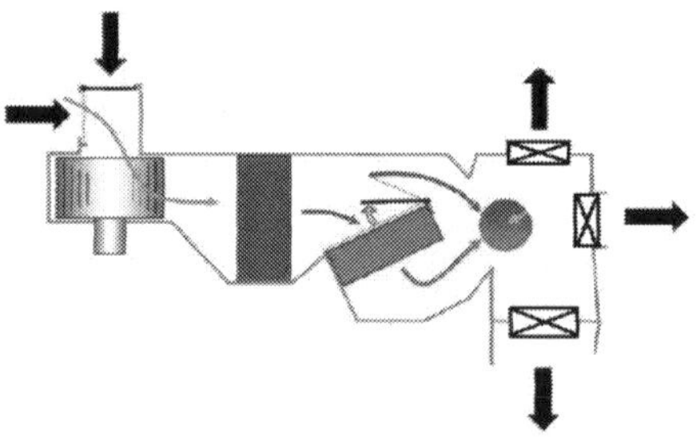

Figure 7. HVAC system with fans.

For testing, verifying and validating the experimental results of HVAC system operation in the two situations (driver's seat air conditioning and the entire passenger compartment), a test manikin equipped on each body segment with KTY 81 type temperature sensors was produced by NXP Semiconductors will be used. The NI USB 6009 acquisition board produced by National Instruments will be used for this purpose to measure and acquire the data of 8 temperature sensors. The sensors were mounted on a mannequin that simulates a man behind the wheel of a car and must measure the temperature on different body segments (head, torso, upper and lower limbs).

As it can be seen, on each important segment of the body, a temperature sensor was mounted in a hemisphere to isolate it from the influence of radiant surfaces. Temperature sensors of the KTY81 series have a positive temperature coefficient of resistance and are usually used in measurement and control systems.

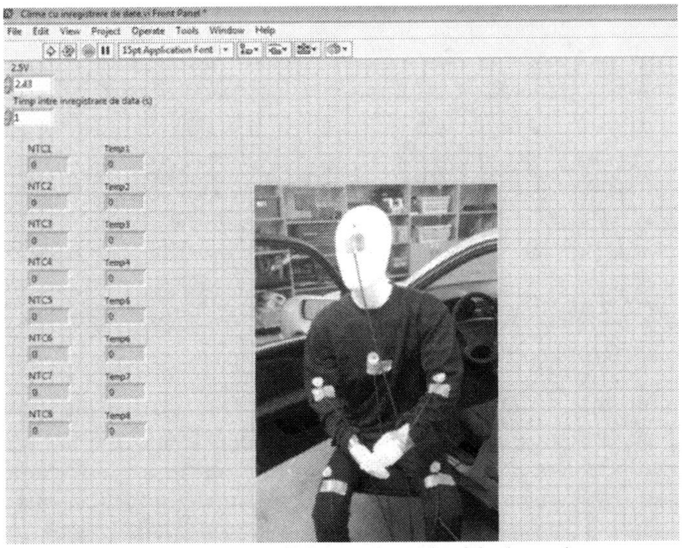

Figure 8. Test manikin equipped with thermistors.

The sensors are encapsulated in the plastic box and their characteristics are presented in the table below:

Table 1. KTY81 sensor errors. [8]

Ambient temperature [°C]	*Ambient temperature [°F]*	*Temperature coefficient [%K]*	*Temperature error [K]*
0	32	0.85	±1.91
10	50	0.83	±1.67
20	68	0.8	±1.41
25	77	0.79	±1.27
30	86	0.78	±1.39
40	104	0.75	±1.64
50	122	0.73	±1.91
60	140	0.71	±2.19

It is observed that in the range of temperatures of interest 0-60 °C the sensors can have a maximum ± 3 Kelvin error. A Fluke TIS 10 thermal imaging camera will be used to verify the results.

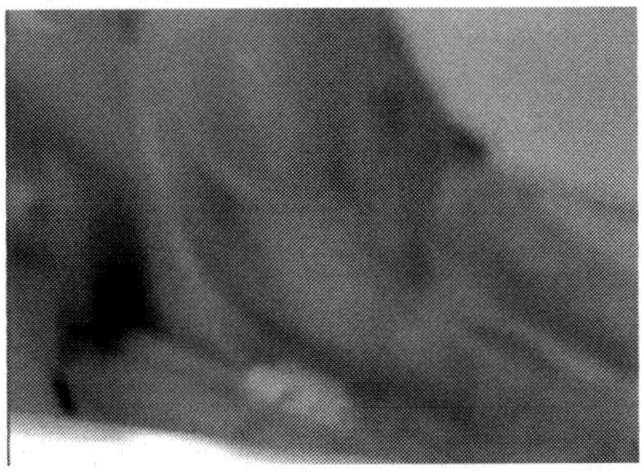

Figure 9. Cooling the manikin in the initial configuration.

The improvement of the performances from the point of view of the energy consumption is obvious because in 80% of the cases according to the official statistics the car has a single occupant.

Thus, in the initial configuration, the consumption due to the operation of the fan is over 500 W, to which is added the load on the motor due to the operation of the air conditioning compressor, and of the alternator that supplies electricity to the HVAC system. The electromagnetic clutch consumes 50 W. The power of the compressor is 2050 W according to the manufacturer's table [9]. In total, we get 2600 W which is 3.49 HP (Dacia Logan used in tests has a power of 75 HP) which is 4% of engine power. Reducing to 1/4 this power we obtain a fuel saving of 3% of the total fuel consumption which is quite significant.

The fan of the HVAC air conditioning system is a rather noisy system and it affects the quality of the interior comfort through the background noise it produces.

By replacing it with smaller silent fans used in the calculation technique to cool the computers and the boards in their component, much lower background noise will be produced. Much lower background noise will increase the quality of interior comfort in terms of noise, improving the quality of sound auditions and telephone calls. This will allow you to avoid expensive audio systems that extract background noise, filter it and mix it with the sound of the 180-degree offset audio system to cancel it. To optimize and reduce the noise produced by the air conditioning system, piping sections with different diameters can be used, which will be passive noise attenuators.

5. Conclusions

Noise reduction systems are not available for small and medium-class cars and the proposed solution can be easily implemented. Simplification of the climate assembly by the ease of mounting/adapting the fans to the initial system (3 pins - 2 for power supply and 1 for speed change) and electronic system control. The energy consumed by the fans is much lower than the initial configuration. By reducing the energy consumed, fuel consumption is also reduced. With all these decreases, the degree of comfort increases because the level of noise and vibrations is attenuated (the air dampers disappear).

ORCID iDs
A M Vranau https://orcid.org/0000-0002-8201-4945
V Caunii https://orcid.org/0000-0001-9316-5645
C Bujoreanu https://orcid.org/0000-0002-5680-2788

6. References

[1] Dacia Logan Repair Manual 2004. Air conditioning system (*in Romanian*)
[2] Razak N F D, Sani M S M, Azmi W H and B Zhang 2017 Noise and vibration analysis for automotive radiator cooling fan *4th Int. Conf. of Mechanical Engineering Research IOP Conf. Series: Materials Science and Engineering* **257** 012083 https://doi.org/10.1088/1757-899X/257/1/012083
[3] Allam S and Abom M 2015 Noise Reduction for Automotive Radiator Cooling Fans *Int. Conf. on Fan Noise, Technology and Numerical Methods, Institution of Mechanical Engineers* **2015**
[4] Bennouna S, Naji S, Cheriaux O, Moreau S, Ouedraogo B and Ville J M 2015 Aeroacoustic prediction methods of automotive HVAC noise *SAE Technical Paper* 2015-01-2249 https://doi.org/10.4271/2015-01-2249
[5] Explanatory Dictionary of the Romanian Language https://dexonline.ro/
[6] https://gembird.com/item.aspx?id=3335
[7] User Guide NI 6008/6009 Multifunction DAQ https://www.ni.com/pdf/manuals/371303n.pdf
[8] KTY81Datasheet https://www.alldatasheet.com/
[9] Service Manual Valeo TM21 Compressor 2017
[10] http://www.valeocompressors.com/media/brochures/TM21_service-manual.pdf

IOP Conf. Series: Materials Science and Engineering 1235 (2022) 012039 doi:10.1088/1757-899X/1235/1/012039

Research regarding the effects of a diesel engine's control unit retrofitting, focussing on its performance and pollution

C C Suciu[1], I Ionel[1], D Ostoia[1] and S V Igreț[2]

[1] Politehnica University Timisoara, Faculty of Mechanical Engineering, Piata Victoriei Nr 2, 300006, Timisoara, Romania
[2] Aurel Vlaicu University, Faculty of Engineering, B-dul Revoluției, nr. 77, 310130, Arad, Romania

ioana.ionel@upt.ro

Abstract. Due to the stricter and stricter European restrictions on diesel engines, fewer manufacturers produce vehicles fuelled by diesel and since not anyone affords new vehicles and the out-of-use diesel vehicle waste would be huge. New ways of improving the pollution levels on currently registered vehicles must be found. The article presents a novel approach focusing on this scope, starting with the basic actual engine, and developing by retrofitting up to mechanical and software upgrades, aiming to raise the output of mechanical characteristics and simultaneous reduction of exhaust pollution. In short, the paper is focusing on research regarding the effects of a diesel engine's control unit retrofitting.

1. Scope

The European perspectives are imposing stricter and stricter diesel engine efficiency standards to reduce carbon dioxide, NO_x and other particulate emissions, especially greenhouse gases, according to with latest European emission standards [1].

The direction in which the manufacturers are going at is clear; the question is which consideration may occur concerning the lower emission class vehicles already on the road? Experiments trying to lower the emissions were accomplished; a few examples include retrofitting a cooled exhaust gas recirculation system [2], which is changing the parameters of the compression-ignited engine's control unit [3], by modifying the fuel injection system [4], or by using biofuel [5,6]. This research aims to provide a modality of improving both the power output and emissions characteristics of a series produced euro 3 diesel engine.

2. Proposal and results

Before presenting the improvements, understanding what it focuses on through the research is essential.

In table 1 are presented the engine characteristics [7].

The performance measurements according to the manufacturer are related to a top speed of 180 km/h, with acceleration from 0 to 100 km/h for 12.4 seconds [8].

Table 2 presents the engine's components.

Table 1. Engine characteristics.

Number of cylinders	4
Cylinder's layout	Straight/ inline
Cylinder bore [mm]	79.5
Cylinder stroke [mm]	95.5
Compression ratio	19.5
Number of valves	8
Valve's position	OHC
Engine code	ALH
Fuel type	Diesel
Engine's position	Front, Transversal
Engine's displacement [cm^3]	1,896
Intake	Turbocharged + intercooler
Max power output [kW]	66 @ 3,750 rpm
Max torque output [N·m]	210 @ 2,100 rpm
Transmission layout	Front-wheel drive
Gearbox	Manual
Number of gears	5
Fuel tank volume [l]	55

Table 2. Engine components.

Fuel pump	Bosch VP37
Fuel pump's plunger [mm]	10
Injector manufacturer	Bosch
Injector's nozzle size [microns]	0.184
Turbocharger	Garret GT1749V
ECU	Bosch EDC15

Before actually extracting the data, a description of the electronic control unit is presented first. The engine control unit (ECU) controls several actuators in internal combustion engines, to ensure optimum engine performance. This is achieved by collecting values from a multitude of sensors in the engine compartment, interpreting the data as multidimensional maps, and adjusting the actuators accordingly. The components in the engine compartment, before being ordered by the ECU, are dynamically controlled by mechanical and pneumatic systems. For obtaining the maps, FGTech Galletto 4 and the related program, as described in figure 1, were used.

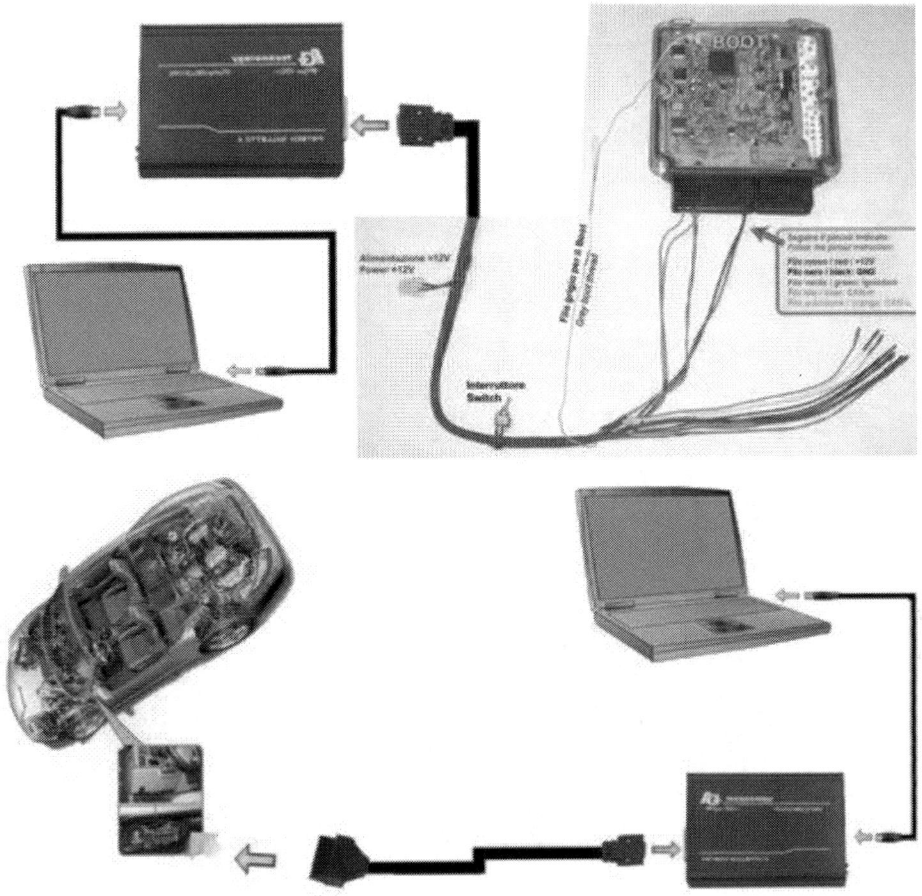

Figure 1. FGTech Galletto with possible uses.

The onboard diagnostics (OBD) cable is inserted in the OBD socket in the passenger compartment of the vehicle, as shown at the bottom of figure 1. The software that is delivered with the device is accessed and thus the ignition is turned on. In the software, it chooses the vehicle type, for which it intends to read or write the file with the maps, following of course by saving or writing of the generated data on the engine control unit.

2.1. Solution proposed for the engine upgrading retrofit
After obtaining the data from the ECU, engine upgrades are performed.

2.1.1. Flywheel, pressure plate and clutch
From the factory, the car is equipped with a double-mass flywheel. The reason for choosing to use it for a retrofit is linked to the idea of reducing the vibrations transmitted by a single-mass flywheel. The problem encountered is connected to its limit when it comes to high values of engine torque, which led to accelerated wear. It was replaced with a single mass flywheel, characterised by a reduced mass, its weight being about 6 kg. The benefits of such a flywheel are: (i) increased acceleration, more efficient engine brake, and (ii) a lower fuel consumption at constant speeds.

Together with the conversion of the flywheel, it proposes also that the clutch must be replaced, as well. It was opted for an organic disc with Kevlar inserts and a hardened pressure plate, for better control in urban traffic. The advance of the flywheel comes in accelerations, where less fuel is used. This has a high impact on urban traffic and when launching the vehicle at the traffic lights.

2.1.2. Injection system
For higher performance, components of the injection system have been improved, such as:
i) Enlarged injector nozzles: 0.270 micrometres with 5 holes, instead of 0.180 micrometres with holes.
ii) Enlarged injection pump plunger to 12 mm, from 10 mm.

These two proposals allow that the fuel might be injected at a higher pressure and the maximum quantity of fuel injected/stroke was, in consequence, also increased [9,10].

2.1.3. Turbocharger and air intake system
One also opted for a larger variable geometry turbocharger, Garrett GT1749VB (presented in figure 2), with a maximum output of 200 kPa, in acceleration bursts. It notes that the exhaust gallery is delivered with the turbine housing, being cast together.

Figure 2. Garrett GT1749VB mounted on the engine.

Starting from the turbocharger's compressor to the intake manifold, the intercooler's standard rubber tubes were replaced with an aluminium pipe route with a diameter of 70 mm. The side-mounted factory intercooler was replaced with a front-mounted one, and the intake manifold was replaced with a larger, more efficient one presented in figure 3.

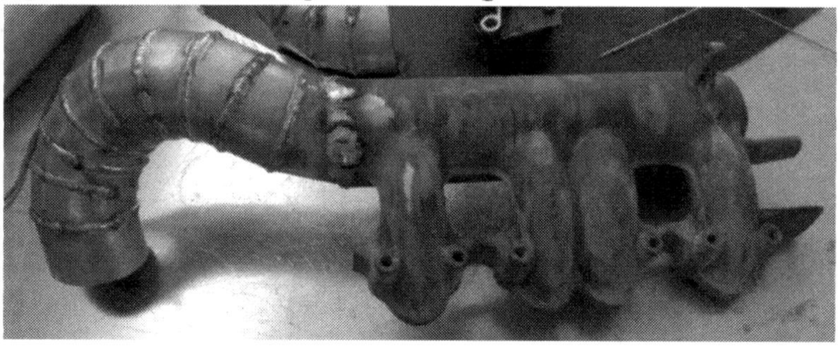

Figure 3. Intake manifold proposed for the retrofit.

For the comparison tests, temperature values were measured using the VCDS diagnose interface. Measurement results presented a very large difference in temperature values. Compared to the initial intercooler, at an ambient temperature of 28 °C, at a constant speed, with the upgraded system, an increase by 2-4 °C, for an ambient temperature of 29-30 °C, during an acceleration regime, is recorded, while with the factory setup it reached temperatures over 50 °C.

2.2. Measurements and interpretation

To modify the maps as efficiently as possible, the dynamometer stand MAHA LPS 3000 presented in figure 4, was used [11].

Figure 4. MAHA LPS 3000.

On the dynamometer, measurements were performed, from where data was obtained regarding the maximum power and torque developed. Measurements consisted of power runs, from where the graph from figure 5 was obtained as well as in constant speed running mode, 2000, 2500, 3000, 3500 revolutions/minute in 4th gear.

In figure 5, generated by MAHA LPS 3000 dynamometer, are presented the measurements results, after software optimizations and mechanical upgrades.

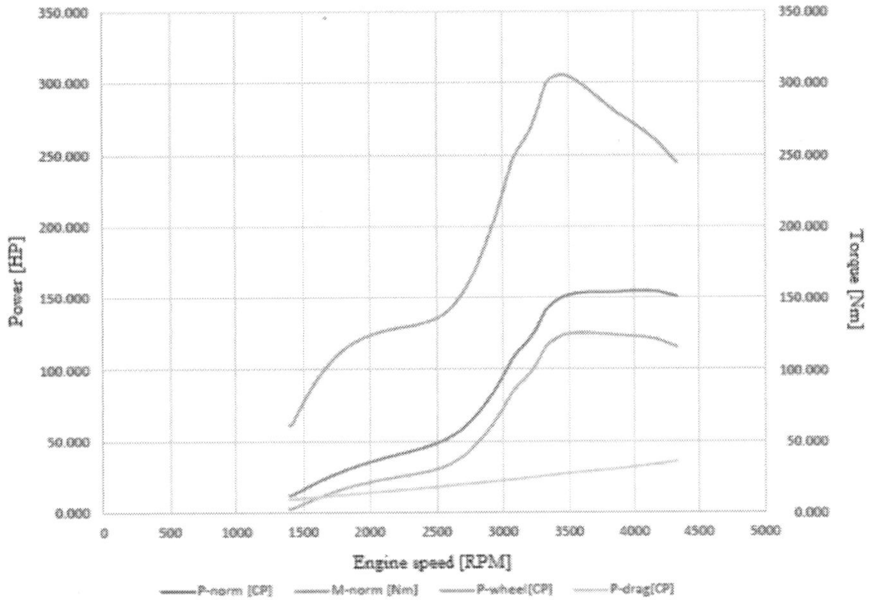

Figure 5. Measurement results, after software and hardware upgrades.

The axis and coloured lines from figure 5 are to be interpreted as it follows: engine speed [rpm] is represented on the X-axis, measured power in horsepower (hp) unit of measurement, torque and drag is represented on the Y axis, engine's power output is represented by the blue curve, engine torque is represented by the orange curve, engine's power at the wheels is represented by the grey curve, the force of drag [hp] is represented by the yellow curve. The values obtained are converted into an international unit of measurement from the common units in table 3.

Table 3. Engine characteristics.

Characteristics	Value
Power – engine [kW]	113.9
Power – wheel [kW]	90.1
Power – drag [kW]	23.7
Engine speed [rpm]	4050
Vehicle speed [km/h]	144.7
Torque – engine [N·m]	305.6
Engine speed [rpm]	3,455
Vehicle speed [km/h]	123.3

Concerning the gas measurements, it indicates that they were performed at different engine speeds: 2,000, 2,500, 3,000, 3,500 rpm, covering for 60 seconds, each set of research.

Table 4 presents the engine and vehicles characteristics for each of the measurements.

Table 4. Measurement input data.

Characteristics, in specific units	2,000 rpm	2,500 rpm	3,000 rpm	3,500 rpm
Injection quantity [mg/stroke]	14	15	18.6	19.6
Acceleration pedal position [%]	39	40.8	47.8	49.4
Vehicle speed [km/h]	72	86	104	123
Fuel consumption [l/h]	4	5.4	8.1	10.2
Coolant temperature [°C]	91.8	92.7	90.9	99.9
Fuel temperature [°C]	57.6	58.5	57.6	67.7
Intake air temperature [°C]	49	54	57.6	72
Boost – specified [kPa]	117.3	122.4	136.6	146.8
Boost – actual [kPa]	117.3	122.4	137.7	146.8

The gas analyser results are presented in Table 5, in association with the results indicated in Table 4.

Table 5. Gaz analyser output data.

Characteristics, in specific units	2,000 rpm	2,500 rpm	3,000 rpm	3,500 rpm
CO [% vol.]	0.02	0.022	0.012	0.01
CO_2 [% vol.]	4.3	5.3	6	8.9
HC [ppm vol.]	6	6	7	8
O_2 [% vol.]	14.5	13.3	11.7	9.2
Air excess ratio	3.353	2.752	2.206	1.724
NO_x [ppm vol.]	163	156	226	379
CO_{corr} [%vol]	0.07	0.06	0.03	0.02

Reporting to the input data extracted in table 4, the engine load from 2,000 rpm to 3,500 rpm of constant engine speed is not high, because the difference in throttle usage is 10.4% and the maximum value reached is under 50% of its total possible usage (100%). Due to the lighter flywheel, the inertia is lower and so the necessary quantity of fuel necessary in higher speeds is also higher, which can be observed by the increase in injected fuel quantity/stroke. This can be also tracked by looking at the boost pressure, rising significantly at high engine speed. The initial mechanical stress on the engine's internal components is also reduced due to the lightened flywheel, therefore the overall lifespan predicted by the manufacturer does not suffer negative effects. The output recorded by the gas analyzer proves that at lower engine speeds of 2,000 rpm, the excess air ratio is 94.4% higher than at 3,500 rpm. NO_x is at its lowest value at an engine speed of 2,500 rpm and since urban traffic, does not daily exceed that value, proves that the updates on the engine are beneficial to the case. The way the power is delivered thru the engine's speed does not impact its reliability and in daily drive situations, there are no real necessities of exploiting the engine at full potential. The limit in terms of forces to be applied on the engine's mechanical components are known, therefore only the ones necessary for an upgrade have been replaced [12-15]. The results are important for older vehicles, as, by the proposed retrofitting, it is possible to benefit from important reductions in terms of pollution exhaust levels, as well as simultaneous fuel reduction.

3. Conclusion

One demonstrated that an engine that develops 66 kW, according to initial factory specifications, has reached over 110 kW, by benefitting from changes concerning both the engine control unit and mechanical ones. All results are essential and reached, without exceeding the values imposed by the law in terms of emissions limits, for a vehicle having the right to drive on public roads. Overall, the way the car behaves is notable improved, in both acceleration and braking terms. The average fuel consumption also decreased, meaning that the pollution is reduced. By reducing the levels of emissions and changing the pollution class of a vehicle one can also reduce the waste and number of vehicles needed to be replaced.

Acknowledgements

The research is part of the PhD program of the first author at the Politehnica University of Timisoara.

ORCID iDs

C C Suciu https://orcid.org/0000-0002-8754-534X
I Ionel https://orcid.org/0000-0002-2999-1210
D Ostoia https://orcid.org/0000-0002-6054-7062
S V Igreț https://orcid.org/0000-0001-5649-0617

IOP Conf. Series: Materials Science and Engineering 1235 (2022) 012039 doi:10.1088/1757-899X/1235/1/012039

References

[1] https://ec.europa.eu/clima/policies/eu-climate-action_en Accessed 04.07.2021

[2] Bikas G and Zervas E 2007 Nonregulated Pollutants Emitted from Euro 3 Diesel Vehicles as a Function of Their Mileage *Energy & Fuels* **21** (5) 2731-6 https://doi.org/10.1021/ef070036d

[3] Tzamkiozis T, Ntziachristos L and Samaras Z 2010 Diesel passenger car PM emissions: From Euro 1 to Euro 4 with particle filter *Atmospheric Environment* **44** (7) 909-16 https://doi.org/10.1016/j.atmosenv.2009.12.003

[4] Landhäußer F *et al.* 2014 *Electronic Diesel Control (EDC)* In: Reif K. (eds) Diesel Engine Management. Bosch Professional Automotive Information. (Wiesbaden, Springer Vieweg) pp 220-71 https://doi.org/10.1007/978-3-658-03981-3_20

[5] Makareviciene V, Sendzikiene E, Pukalskas S, Rimkus A and Vegneris R 2013 Performance and emission characteristics of biogas used in diesel engine operation *Energy Conversion and Management* **75** 224-33 https://doi.org/10.1016/j.enconman.2013.06.012

[6] Lee H, Ha T and Choi H 2016 Experimental verification of optimized NO_X reduction strategies in a decrepit Euro-3 diesel engine retrofitted with a cooled EGR system *J Mech Sci Techno* **30** 2873–80 https://doi.org/10.1007/s12206-016-0546-2

[7] Georgi B, Hunkert S, Liang J and Willmann M 1997 Realizing Future Trends in Diesel Engine Development *SAE Technical Paper* 972686 https://doi.org/10.4271/972686

[8] https://www.ultimatespecs.com/car-specs/Volkswagen/3118/Volkswagen-Golf-4-TDI-90.html Accessed 16.07.2021

[9] Hasegawa T, Maehara H, Tanishige R, Iwasaki T, Kobayashi T and Matsumoto Y 2000 Study on the Fuel Spray and Combustion of the Variable Orifice Nozzle (VON) for Direct Injection Diesel Engines *SAE Transactions* **109** 985-94 http://www.jstor.org/stable/44634282

[10] Babu P S S, Lakshminarayanan P A, Setty K L S, Selvakumar N A, Karthik R, Srinivasan G, Nain A, Pendalwar S and Mahesh P 2010 Euro-3 Compliant Diesel Engine Using a Cost-Effective Fuel Injection Pump without Electronics *SAE Technical Paper* 2010-01-1504 https://doi.org/10.4271/2010-01-1504

[11] https://www.maha.co.uk/roller-set-for-function-and-performance-dynamometer-lps-3000-r-100-2.htm Accessed 15.07.2021

[12] Chen Y and Borken-Kleefeld J 2016 NOx Emissions from Diesel Passenger Cars Worsen with Age *Environmental Science & Technology* **50** (7) 3327-32 https://doi.org/10.1021/acs.est.5b04704

[13] Schipper L and Fulton L 2013 Dazzled by diesel? The impact on carbon dioxide emissions of the shift to diesel in Europe through 2009 *Energy Policy* **54** 3-10 https://doi.org/10.1016/j.enpol.2012.11.013

[14] Rimkus A, Žaglinskis J, Rapalis P and Skačkauskas P 2015 Research on the combustion, energy and emission parameters of diesel fuel and a biomass-to-liquid (BTL) fuel blend in a compression-ignition engine *Energy Conversion and Management* **106** 1109-17 https://doi.org/10.1016/j.enconman.2015.10.047

[15] Rimkus A, Melaika M and Matijošius J 2017 Efficient and Ecological Indicators of CI Engine Fuelled with Different Diesel and LPG Mixtures *Procedia Engineering* **187** 504-12 https://doi.org/10.1016/j.proeng.2017.04.407

IOP Conf. Series: Materials Science and Engineering 1235 (2022) 012040 doi:10.1088/1757-899X/1235/1/012040

Nondestructive evaluation of nanostructured components using Eddy current microsensor

N Iftimie[1], R Steigmann[1], D Faktorova[2] and A Savin[1]

[1] National Institute of Research and Development for Technical Physics, NDT Department, 47 Mangeron Boulevard, 700050, Iasi, Romania
[2] Faculty of Electrical Engineering and Information Technology, University of Žilina, Slovak Republic

steigmann@phys-iasi.ro, dagmar.faktorova@fel.uniza.sk

Abstract. Nanostructured materials (NMs) are used in the manufacture of microsensors and some MEMS devices, include nanoscale wires, carbon nanotubes, biomaterials, etc. Nondestructive evaluation of structural integrity of NMs can be carried out by optical (SEM, EBDS) and electromagnetic methods (eddy current). Due to small dimensions of the objects to be tested, the eddy current sensors used in detection of possible discontinuities requires miniaturization so that the answer in amplitude and phase shall allow their localization. In the case of thin tubes made of 304 hard stainless steel, with outer diameters 332μm and wall thickness 38 μm used in medical devices, intended for special catheters, their small dimensions and possible discontinuities require encircling sensor working at high frequencies of the order of MHz. The paper presents the investigation of small diameter tubes using an Eddy current microsensor and the results obtained at the detection of artificial flaws in order to increase the probability of detection (POD), for a high reliability coefficient.

1. Introduction

Nanostructured materials are nowadays used in various applications, starting from small diameter tubes for medical applications to thin-wall tubes for nuclear and power engineering and not only.

In medical applications, AISI 304 SS is used due to especial mechanical and chemical properties, being suitable for easy cleaning and sterilization due to smooth surface quality [1].

The tubes used in medical applications are redrawn to obtain the desired dimensions and temper to obtain seamless tubing. The small diameter tubes used as guide wires and stents, needles, painless injection needles, catheters generate a market of medical engineering with a global revenue over USD 700 billion as expected in 2024 [2], especially in the actual situation where the requirements of medical supplies increase for vaccination and medical testing. Even some of these are for single use, their need to be carefully tested to reach the high standards for human body interaction. The most usual flaws in these kinds of very thin-walled long metal tubes are cracks, pits, delamination, inclusions, scratches, appeared during their fabrication.

The methods for non-destructive testing of thin tubes are various, starting from ultrasonic testing using immersing method [3], ultrasound methods by synthetic aperture focusing technique [4], or using Lamb waves [5]. Also, eddy currents, and X-ray, mechanical vibration are usually methods for thin tubes testing. The use of eddy currents testing to determine the flaws in these tubes has gained interest in the last years due to no direct contact possibility, high reliability and environmentally friendly [6].

The minimal requirements for testing of small diameter tubes are high sensitivity to detect flaws at the limit of detection (at least 10% from the wall thickness), possibility to locate the flaw with high accuracy, classification of flaws upon their nature, etc. Due to redrawn – cold working, the austenitic steel became slightly magnetic so that the electromagnetic testing became suitable. The eddy current method using a microsensor is imposed to achieve the requirements presented above. The paper presents the results obtained at the testing of small diameter tubes with thin walls using a special designed encircling microsensor.

2. Materials and methods

The studied samples are represented by two small diameter tubes with thin walls having outer diameter 332 μm and inner diameter 256 μm, length 1490 mm (the measures are specified by the producer in the factsheet), made from hard stainless steel AISI 304, intended to be used in special catheters (Figure 1). The features, given by the producers are: tensile strength 540 – 750 MPa, density 8.00 g/cm^3, modulus of elasticity 193 GPa, electrical resistivity 0.72 x 10-6 Ω.m, thermal conductivity 16.2 W/m.K. The chemical composition of AISI 304, with Fe in balance, is presented in Table 1, [7].

Table 1. Chemical composition (%)

	C	Si	Mn	P max	S	N	Cr	Ni
%wt.	<0.07	<1.0	<2.0	0.045	<0.015	<0.11	17÷19.5	8÷10.5

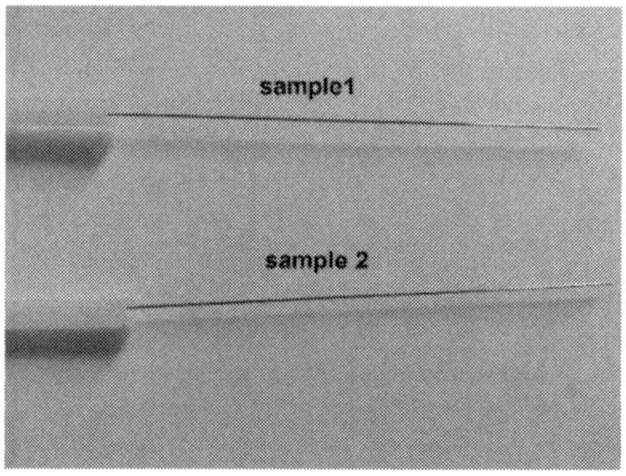

Figure 1. Studied samples.

The small dimensions of the wall thickness require frequencies in the range of MHz, knowing that the penetration depth δ depends on the working frequency f, $\delta = \frac{1}{\sqrt{\pi \mu \sigma f}}$ where σ is the electrical conductivity and μ is the magnetic permeability of the sample's material. The standard depth of penetration is the depth at which the eddy current density is 36.8% (1/e) of the material surface value. The detection of discontinuities by eddy currents are based on the changing of conductivity and the redistribution of eddy current flows. Maxwell's equations are for known region

$$\begin{aligned} \nabla \times \bar{E}_0(\bar{r}) &= -j\omega\mu_0 \bar{H}_0(\bar{r}) \\ \nabla \times \bar{H}_0(\bar{r}) &= (\sigma_2 + j\omega\varepsilon_0)\bar{E}_0(\bar{r}) \end{aligned} \quad (1)$$

For flawed region

$$\begin{aligned} \nabla \times \bar{E}_f(\bar{r}) &= -j\omega\mu_0 \bar{H}_f(\bar{r}) \\ \nabla \times \bar{H}_f(\bar{r}) &= (\sigma_f + j\omega\varepsilon_0)\bar{E}_f(\bar{r}) \end{aligned} \quad (2)$$

The perturbation of the electromagnetic field is given by $\bar{E}_0 - \bar{E}_f$, $\bar{H}_0 - \bar{H}_f$. Deriving a vector wave equation for $\bar{E}_0 - \bar{E}_f$ [8]

$$\begin{aligned}\nabla\times\nabla\times(\bar{E}_0-\bar{E}_f) &= -j\omega\mu_0\nabla\times(\bar{H}_0-\bar{H}_f)= \\ &=(\omega^2\mu_0\varepsilon_0 - j\omega\mu_0\sigma_2)(\bar{E}_0-\bar{E}_f) - j\omega\mu_0(\sigma_2-\sigma_f)\bar{E}_f\end{aligned} \quad (3)$$

Considering the last term in (3) as a source, the formal solution for the perturbed field $\bar{E}_0 - \bar{E}_f$ became

$$\bar{E}_0(\bar{r}) - \bar{E}_f(\bar{r}) = j\omega\mu_0 \iiint_{V_{flaw}} \vec{G}_{22}\left(\bar{r},\bar{r}'\right)\bar{E}_f(\bar{r}')(\sigma_2(\bar{r}')-\sigma_f(\bar{r}'))d\bar{r}' \quad (4)$$

where $\vec{G}_{22}(\bar{r},\bar{r}')$ is the dyadic Green's function for the known region and the volume of integration is the flawed region for which $\sigma_2 - \sigma_f \neq 0$ [9,10].

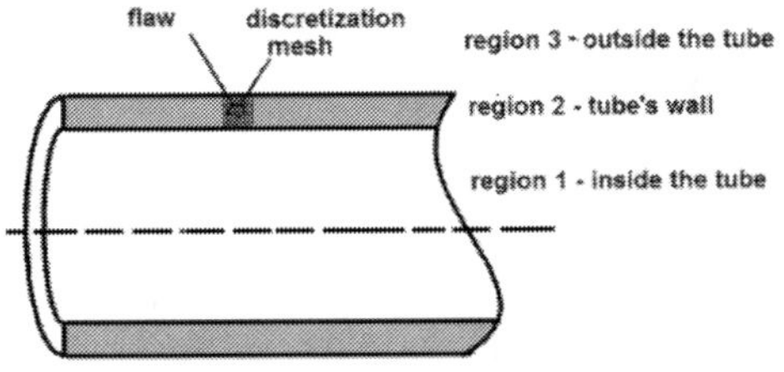

Figure 2. Region of interest.

In order to develop the model, considering the region of interest consists of three parts: the interior of the tube, the tube wall and the region exterior to the tube, (Figure 2). In eddy current method, electromotive force (emf) that is induced in a probe coil is measured.

According to scattering theorems [9], the field created by transducer in tube's wall, in flawed region, could be write as

$$\bar{E}_2(\bar{r}) = \bar{E}_d(\bar{r}) - j\omega\mu_0\sigma_2(\bar{r}) \int_{V_{anomalous}} \vec{G}(\bar{r},\bar{r}')\chi(\bar{r}')\bar{E}_d(\bar{r})d\bar{r}' \quad (5)$$

where $\bar{E}_d(\bar{r})$ is the electric field scattered by flaw and $\chi(\bar{r}')$ is contrast function, having value 1 if r' is in center of one discretizing cell and 0 if r' is outside the anomalous region. Equation (5) is an 2nd order Freedholm equation. Finally, the discrete variant of 2nd order Freedholm equation given by (5) is solved, and the field $\bar{E}_d(\bar{r})$ scattered by the flaw in centers of the cells in which the flaw was discretized can be determined, and according to differential Ohm law

$$\bar{J}_d = \sigma_2\bar{E}_d \quad (6)$$

the current density of the discontinuities equivalent sum is calculated.

The electric field scattered by flaw inside the tube [10]

$$E_{3d}(\bar{r}) = j\omega\mu_0 \iiint_{V_{anomalous}} \vec{G}_{21}(\bar{r},\bar{r}')\bar{J}_d(r')dr' \quad (7)$$

with $\vec{G}_{21}(\bar{r},\bar{r}')$ is dyadic Green calculated in [11]. Then the electromotive force induced in reception coil is calculated

$$e = \oint_C \bar{E}_{1d}(\bar{r})d\bar{r} \quad (8)$$

where C is the contour of reception coil. The approach mode of the analytical model, based on Green's dyadic functions, as well as the used discretization based on moments method, allow the solving of direct problem for encircling eddy current transducer.

3. Experimental set-up

The dimension of controlled objects as well as those of the flaws imposes the usage of high control frequency, about ~5 MHz. In the case of the tubes, the minimum perceptible flaw, at 98% probability of detection (POD), for 98% reliability coefficient shall be slot with 25x5x10 μm [12]. The experimental set-up is presented in Figure 3.

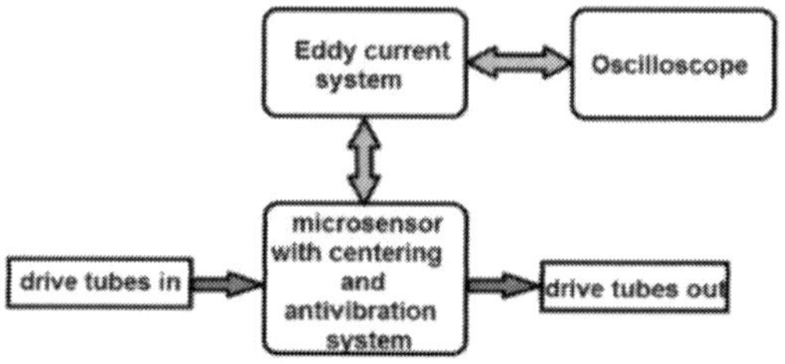

Figure 3. Experimental set-up.

The Eddy current microsensor is differential send receiver type, having a system for guiding the tube inside it in order to avoid vibration and to line-up with the axis. The measurements were carried up at 4 MHz frequency, the signal fed to the emission coil having 10Vpp injection, gain 65, phase 169°, the signal being cut-off with a low pass filter at 20 Hz. After optimal parameters determination, the phase was adjusted so that the noise due to tube vibration shall be projected on the second channel of the oscilloscope. The processed signals are visualized on the first channel, the signals being recorded and plotted with a code developed in Matlab.

4. Results and discussions

In the practice of nondestructive testing, when the transducer encircles the product, due to eccentricity possibility, vibration can occur during the tube displacement and the separation of the signal in 2 channels is required.

The projection of the vibration noise on the one of the channels is presented in Figure 4. It is observed that the other channel is slightly influenced by the noise [13].

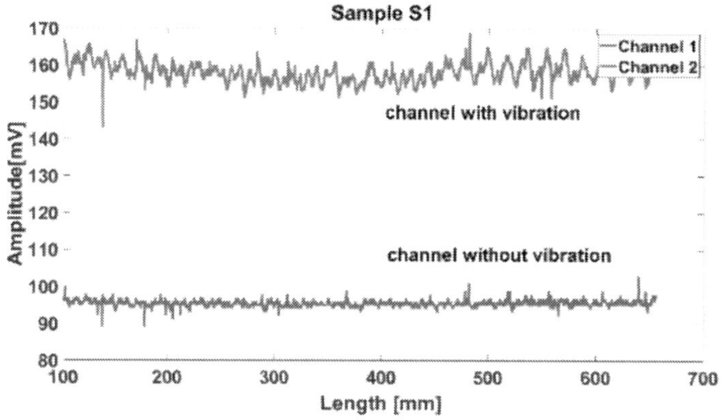

Figure 4. Noise due to vibration of the tube in the interior of transducer.

In Figure 5 is presented the results of examination for the region 0 – 656 mm of the sample S1. The flaw is clearly emphasized in the region between 530 - 545 mm from the beginning of tube. Figure 6 presents the result of examination for the region 656 - 1490 mm. No flaw was detected.

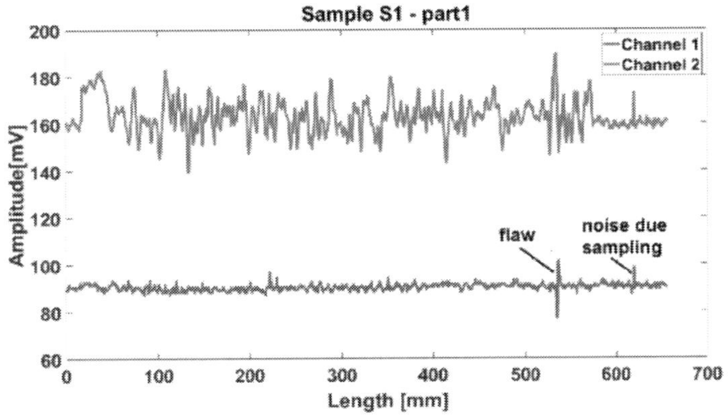

Figure 5. First region of sample S1.

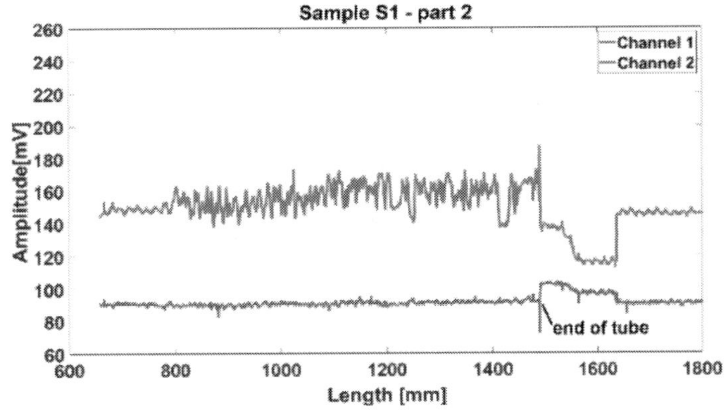

Figure 6. Second region of sample S1.

For signal to noise ratio (SNR) evaluation, the amplitude of the signal due to the discontinuity was measured and the maximum amplitude of the noise on the channel on which the measurements were made, was estimated. 10 displacements were made, with the tube rotated with randomly angle.

The results of the control are presented in Table 2.

Table 2. The results of sample S1 control.

No.crt.	Signal amplitude [mVpp]	Noise amplitude [mVpp]	SNR
1	107	4	26.75
2	96	6	16
3	969	5	19.2
4	104	4	26
5	85	6	14.17
6	93	4	23.25
7	100	6	16.67
8	111	7	15.86
9	103	5	20.6
10	100	3	33.3

The statistical data analysis for the results from Table 2 are: Average of signal's amplitude [mVpp]: 99.5; Standard deviation (SD) of signal's amplitude [mVpp]: 7.44; Average amplitude of noise amplitude [mVpp]: 5; SD of noise's amplitude [mVpp] 1.247. Also SNR of 21.18 is acceptable for SD of SNRs 6.09.

For the second tube, the parameters of the control were kept. Because the signal due to flaw on sample S2 has smaller amplitude, the representation scale was increased so that the signal should be clearly visible. Figure 7 presents the examination result for the region 0 - 700 mm. This region does not contain flaws. In Figure 8 is presented the results for the region 700 - 1490 mm. The flaw was emphasized in the region between 1005 - 1020 mm.

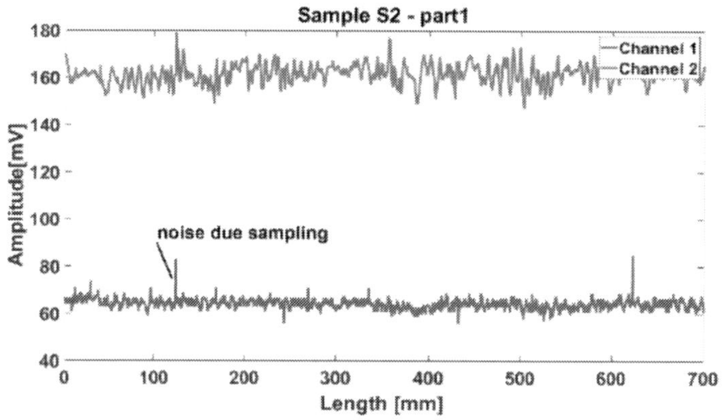

Figure 7. First region of sample S2.

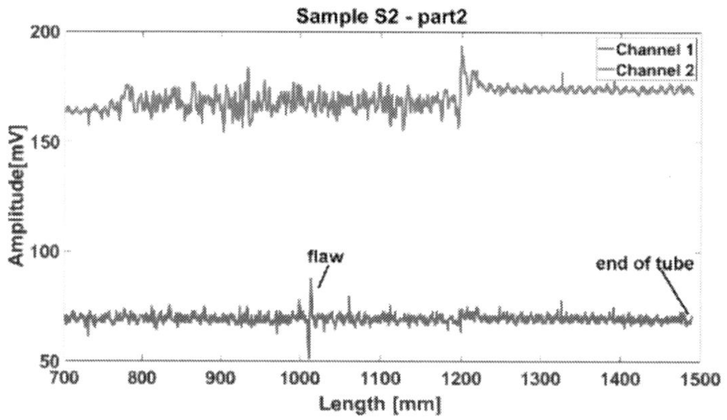

Figure 8. Second region of sample S2

10 displacements were made, with the tube rotated with randomly angle (Table 3).

Table 3. The results of sample S2 control.

No.crt	Signal amplitude [mV_{pp}]	Noise amplitude [mV_{pp}]	SNR
1	67.7	4	16.925
2	62.1	4	15.525
3	67.9	5	13.58
4	60.7	6	10.116
5	63.8	4	15.95
6	60.7	4	15.175
7	60	5	12
8	55.2	6	9.2
9	54.8	4	13.7
10	69.1	6	11.516

The statistic data analysis for the results from Table 3 are: Average of signal's amplitude [mV_{pp}]: 62.2; SD of signal's amplitude [mV_{pp}]: 4.75; Average amplitude of noise amplitude [mV_{pp}]: 4.8; SD of noise's amplitude [mV_{pp}] 0.87; SNR is 13.37 and SD of SNR is 2.6.

5. Conclusions

In order to evaluate the quality of small diameter tubes with thin walls, eddy current method using microsensor is adequate to emphasize the discontinuities, especially when they are along the generatrix. For evaluation of the external surface of the tubes, a send-receive encircling sensors in differential connections has been developed and used with reliable results. According to ASME code Sect V, art 8, the flaws which present a SNR higher than 3 can be emphasized using electromagnetic method of eddy current. The obtained results at the testing of S1 and S2 samples, shows that the flaws are clearly emphasized and SNR are 21.18 ± 6.09 for sample S1 and respective 13.37 ± 2.6 for sample S2.

The method is limited only by the vibration due to small dimensions of tubes, compared with their length and require a careful analysis of tube vibration at passing through microsensor.

Acknowledgements
This work was supported by the Ministry of Research, Innovation and Digitization, Romania under CCCDI-UEFISCDI, project number PNCDI III-P1-1.2-PCCDI-207-03239/60PCCDI 2018—MedicalMetMat and Nucleus Program MAGNA2022, Project PN 19 28 01 02.

ORCID iDs
N Iftimie https://orcid.org/0000-0001-9614-0101
R Steigmann https://orcid.org/0000-0003-1618-7650
D Faktorova https://orcid.org/0000-0001-6417-1042
A Savin https://orcid.org/0000-0001-9863-3110

References

[1] Nutt MJ ed. 2003. *Stainless steels for medical and surgical applications* **1438** West Couskokocken PA: ASTM International https://doi.org/10.1520/stp1438-eb
[2] Ugalmugle S and Swain R Automated Suturing Devices Market https://www.gminsights.com/industry-analysis/automated-suturing-devices-market
[3] Xiao X, Gao B and Tian GY 2019 *IEEE Int. Instrumentation and Measurement Technology Conf.* (I2MTC) pp. 1-5 https://doi.org/10.1109/I2MTC.2019.8826968
[4] Thavasimuthu M, Rajkumar KV, Jayakumar T, Kalyanasundaram P and Raj B 1999 *Review of Progress in Quantitative Nondestructive Evaluation* 1987-93 (Boston Springer) https://dr.lib.iastate.edu/handle/20.500.12876/61692
[5] Yang S, Wang M and Yang L 2019 *Applied Sciences* ***9*** (17) 3576 https://doi.org/10.3390/app9173576
[6] Gao P, Wang C, Li Y, Wang L, Cong Z and Zhi Y 2017 *Nondestructive Testing and Evaluation* **32** (2) 133-51 https://doi.org/10.1080/10589759.2016.1149583
[7] https://www.azom.com/
[8] Bowler N 2019 *Eddy-current nondestructive evaluation* (New York Springer) https://doi.org/10.1007/978-1-4939-9629-2.
[9] Chew WC 1990 Waves and Fields in Inhomogeneous Media (Renhold, NY Van Nostrand)
[10] Grimberg R, Udpa L, Savin A, Steigmann R and Udpa S 2005 *Research in Nondestructive Evaluation* **16** (2) 79-100 https://doi.org/10.1080/09349840591004932
[11] Grimberg R, Udpa L, Savin A, Steigmann R, Palihovici V and Udpa SS 2006 *NDT & E International* **39** (4) 264-71 https://doi.org/10.1016/j. ndteint.2005.08.004
[12] Silk MG, Stonenham AM and Temple JA 1987 *The reliability of nondestructive inspection* Adam Hilger Eds.
[13] Albert CL 1996 *Fundaments of industrial control Practical Guides for measurement and control Instrument society of America*, Ed DA Coggan (Portland: ISA, Book News)

Computer-aided methods for furniture decoration with traditional motifs of textile heritage

A Lungu, L Gurău, S Georgescu and C Coşereanu

Faculty of Furniture Design and Wood Engineering, "Transilvania" University of Brasov, B-dul Eroilor, nr. 29, 500036, Brasov, Romania

cboieriu@unitbv.ro

Abstract. The present paper introduces computer-aided methods for the transposition of ornaments inspired from the textile heritage into the surface furniture decoration. Using digital technology, the furniture industry may bring into modern life the traditional motifs, contributing thus to the preservation of the cultural heritage from the region Țara Bârsei, located in Transylvania. Two motifs inspired by the traditional costumes from South-Eastern Transylvania were digitized using CorelDraw software. They were imported in AutoCAD and the information was transferred both to CNC router and laser equipment. The CNC routing of ornaments was simulated with two types of tools and two processing methods (engraving and carving) using CAD-CAM-CAE software and the models were afterward processed on wood panels by CNC router and laser equipment and then visually analysed, concluding which method and tool is suitable for each type of ornament, so to transpose better the original motif on furniture wood surface.

1. Introduction

A specific concern of each generation is to perpetuate the cultural heritage to the future, bringing it in the modern life through digital technologies. One way to save traditional ornaments is to introduce them in various industries such as textiles, embroidery on clothing and footwear, interior design, furniture design, digital printing or graphics and design of modern products. There are several methods applied for the decoration of the furniture surfaces, such as carving, painting, marquetry or wood pyrography or engraving, but among these there are two methods that offer the advantage of advanced technology and they are laser jet engraving of wood and CNC routing of wood.

Aesthetic evaluation of laser engraving work was carried out on radial and tangential veneer cut from beech (*Fagus sylvatica*), pine (*Pinus sylvestris*), chestnut (*Aesculus hippocastanum*), mahogany (*Swietenia macrophylla*) and *Sapelli* wood, and the best aesthetic perception was achieved on light coloured wood species, because of the contrast between the pattern and background [1]. Different image conversion methods and colour image reproduction in burned wood surface were used to analyze the colour shades obtained through pyrography with various intensities of laser beam [2]. Apart from studies of aesthetics and colour reproduction, investigation on cutting width, processed surface quality, heat affected zone and wood surface burning were also conducted by researchers [3], and a mathematical relation between wood surface roughness and laser parameters have been determined [4]. Laser beam focal length, cutting speed and power are the important parameters to be considered not only for laser engraving, but also for laser cutting of wood and wooden based materials, such as plywood [5].

Various species of wood were subjected to tests regarding CNC routing process and determination of optimum parameters to be used for an adequate quality of the processed areas. The quality of the routed surfaces in spruce wood (*Picea abies*), poplar wood (*Populus alba*), plywood and medium-density fibreboard (MDF) was visually analysed after scanning them with high resolution, and VCarvePro 9.519 software developed by Vectric was used to design the tool path, select the tools and parameters and simulate the routing process, transferring the information to CNC Router [6]. Other researchers have tested the CNC milling process on European black pine (*Pinus nigra* Arnold) [7], alder wood (*Alnus Glutinosa*) [8], oak wood (*Quercus Robur*) [9], evaluating the processed surface roughness for various parameters, such as depth of cutting, feeding rate, milling pattern and milling direction. A study conducted on MDF [10] concluded that to ensure an adequate reaction to the progression of tool wear during any MDF milling process, a workpiece rather than a tool should be measured. Edge-glued panels (EGP) have been also investigated for CNC router process, using five parameters: cutting direction, cutting depth, cutting width, feed rate and spindle rotation feed [11].

2. Material, methods and equipment

2.1. Traditional motifs

Historical and ethnographic region located in the SE part of Transylvania, namely Ţara Bârsei is a multiethnical one, and for this reason it is known to have a rich textile heritage. In Braşov, the main town of this region, there is a men community (named "Juni"), organized in an old part of the town named Şchei. These men preserves and still wears the popular costumes during the traditional events that take place anually. The main piece of their traditional costumes is a shirt made of white canvas with decorative stitches, decorated mainly with phytomorphic und cosmomorphic motifs, including the urn of flowers, a stylised version of the universal "tree of life" motif. A part of this shirt decorated with urn of flowers is presented in Figure 1(a). A peculiarity of the Saxon woman costume from Ţara Bârsei is the black patch tied under the chin with velvet ribbons, embroidered with gold thread and decorated with floral motifs, as shown in Figure 1(b). The sewing techniques used for the two motifs are totally different, the first one is sewing in crosses on one side (a) and the second one is full embroidery (b). These two motifs were collected from pictures taken from objects belonging to private collections, where they are very well conserved.

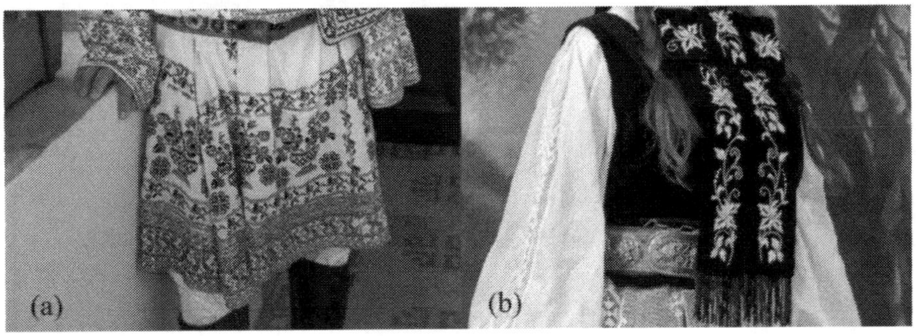

Figure 1. Traditional motifs from textile objects: (a) – Model 1, from Şcheii Braşovului; (b) – Model 2, from Saxon woman costume.

2.2. Approach methods and equipment

The scientific research presented in the paper includes the vector graphic design of the motifs using professional graphics software CorelDraw, followed by export of the file to AutoCAD software and import it afterwards in VCarvePro 9.519 software, developed by Vectric, for the simulation of CNC routing on wood surface, selecting the tools and processing methods.

IOP Conf. Series: Materials Science and Engineering 1235 (2022) 012041 doi:10.1088/1757-899X/1235/1/012041

The simulation process indicated the appropriate tool and processing method to be applied, in order to obtain the closest variant of the ornament to the original motif, as shape and accuracy. The selected combination of tool and method was afterwards transferred to CNC Router and the ornaments were processed on wood surface. For a comparison from the aesthetic point of view, the same ornaments have been processed on laser beam equipment.

A CNC Router, model ISEL GFV, Eiterfeld, Germany was used for milling the ornaments and OmniBEAM 150 Laser Machining Tool (LMT), manufactured by COHERENT, INC., Santa Clara, California, United States with nitrogen assist gas was used for laser engraving the ornament on wood surface.

Maple wood (*Acer pseudoplatanus*) panels of 300 x 200 x 18 mm were used as ornamentation base. Before processing the ornaments, the panels were sanded with 80 grit size, so to obtain a good flatness of the surface, in the range ± 0.15 mm.

The selected traditional motifs were designed using CorelDraw software, and the results are shown in Figure 2 (a1 and b1). The CDR vector file formats were converted to DXF files in AutoCAD and the results are shown in Figure 2 (a2 and b2). These files have the suitable format for importing them in VCarvePro 9.519 software for the simulation process.

The simulation software allows two type of processing methods on wood surfaces, namely "Engrave", with a constant cutting depth of 3 mm for the contour, which can be applied both to closed or opened contours of the drawing, and "V-Carve", with cutting depth between 1 mm and 3 mm for the surface and 3 mm for the contour, which can be applied only for closed contours of the drawings. In case of thin contour lines, the two methods can be combined.

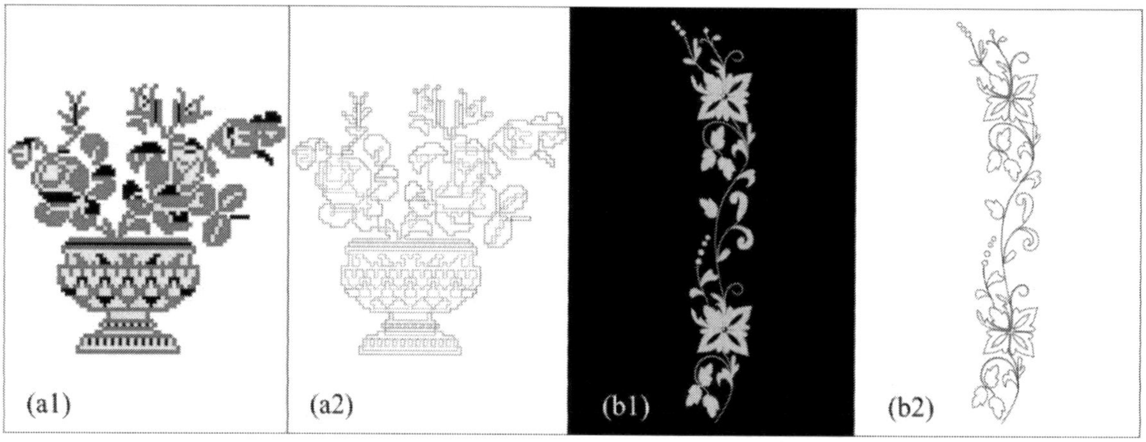

Figure 2. The drawings of the traditional motifs: (a1, a2) – Model 1, (b1, b2) – Model 2.

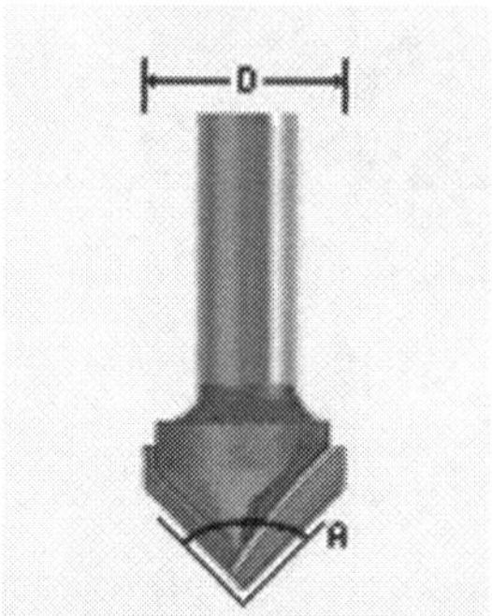

Figure 3. Tool used for router milling of wood.

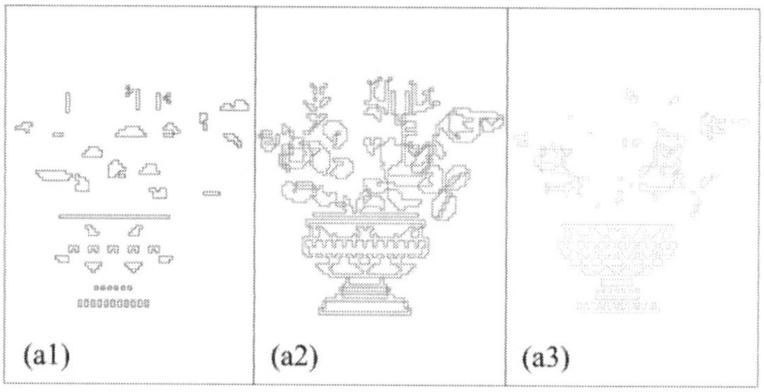

Figure 4. Decomposition of Model 1 into colours: (a1) – black, (a2) – red, (a3) – white.

V groove router bits tools (Figure 3) angled 90° and 120° respectively, have been used for the simulation and CNC routing. Cutting depth, feed speed and spindle rotation speed have been set up, as presented in Table 1. These tools are recommended for engraving and plane bas-relief, which is the type of ornament suitable to transfer the textile motifs on wood surface. These types of ornaments require low cutting depths and high feed speeds during processing them on wood surfaces, and these characteristics are fulfilled by the selected tools.

Table 1. Milling parameters set up.

Included angle A (degrees)	Diameter D (mm)	Spindle speed (r.p.m.)	Pass depth (mm)	Feed rate (m/min)	Plunge rate (m/min)	Clearance pass (mm)
90	32	15000	3	6	6	0.5
120	32	15000	3	6	6	0.5

The DXF format files resulted from AutoCAD have been also imported in LaserLink™ software, which creates files that runs the Coherent OmniBEAM in the LMC format, containing the tool paths and the process parameters that run an efficient laser machining job. Because the original motif of Model 1 has three colours in its composition, it was decomposed in three files, one for each colour, as seen in Figure 4, and for each file different parameters were set up for laser beam, in order to obtain different intensities of the colour resulted from burned wood. These parameters are presented in Table 2.

Table 2. Laser engraving parameters set up.

Motif	Power (% of Laser Power)	Power (W)	Feed, (m/min)
Model 1 – black	20	30	5.4
Model 1 – red	15	22.5	5.4
Model 1 – white	7	10.5	5.4
Model 2	15	22.5	5.4

3. Results and discussion

The simulation process has shown that V-Carve method was not a properly one, because the motifs have not been processed completely, because of the presence of open contours in the drawing. With combination of V-Carve and Engrave methods, the ornaments looked very crowded.

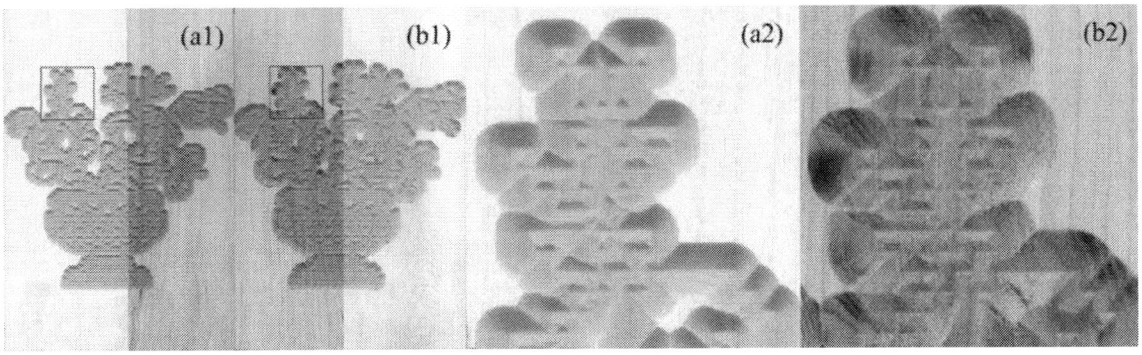

Figure 5. Router milling of Model 1: (a1) – Engrave 90°; (a2) – Detail for Engrave 90°; (b1) – Engrave 120°; (b2) – Detail for Engrave 120°.

Applying Engrave method for the real processing on CNC Router, had as result the images in Figures 5 and 6, a1 and b1, where noticeable differences were observed for using V-tools with different angles of 90° and 120° (as seen in details a2 and b2 from Figures 5 and 6). The trace of the 120° angled tool on the ornament looked wider and more crowded than that of the 90° angled tool.

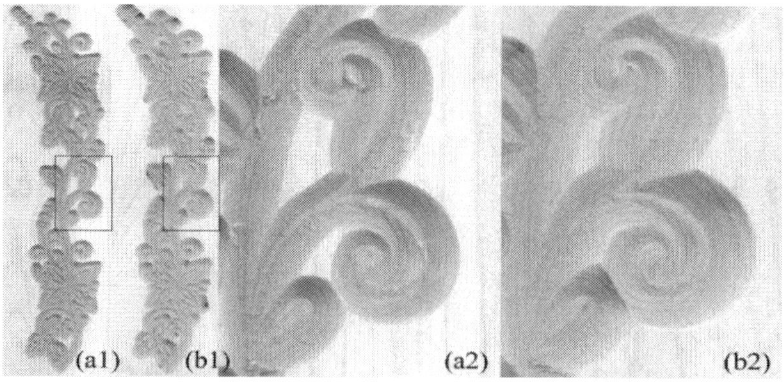

Figure 6. Router milling of Model 2: (a1) – Engrave 90°; (a2) – Detail for Engrave 90°; (b1) – Engrave 120°; (b2) – Detail for Engrave 120°.

BeamHMI software was used for the laser engraving of the two motifs. The resulted ornaments are presented in Figures 7 and 8. The three layers of the Model 1 were processed one by one without moving the wood piece and the total recorded processing time was the sum of the three ones (black + red + white = 13:47 + 25:02 + 18:33 min:s). The processing times for each model and method are shown in Table 3.

Figure 7. Laser engraving for Model 1.

Figure 8. Laser engraving for Model 2.

Table 3. Processing times.

Model	Method	Tool	Processing time, min:s
Model 1	CNC Router engraving	V-bit 90°	14:52
	CNC Router engraving	V-bit 120°	14:52
	Laser engraving	Laser	57:22
Model 2	CNC Router engraving	V-bit 90°	6:30
	CNC Router engraving	V-bit 120°	6:30
	Laser engraving	Laser	27:50

Processing times are connected to productivity first, and tool wear in case of CNC routing. Even if the processing times are longer for Laser engraving than in case of CNC engraving, the laser ornaments win through their aesthetic appearance, so the selected method is a compromise between aesthetic and costs.

4. Conclusions

This study presents the methods used to transfer the traditional motifs collected from the textile heritage to the decoration of the furniture wood surfaces, based on advanced software tools, and to perpetuate the traditional motifs and their symbol values to next generations.

The simulation software of CNC Router offers the possibility to select the appropriate tools and methods and to eliminate the unsatisfactory variants. The experimental work provides the complete image of aesthetic, surface quality and technological costs. The final decision is a compromise between them.

Further experimental work will investigate the quality of the processed surfaces in terms of roughness measurements for various species of wood and engineered wood products.

Acknowledgments

The authors thank Dan Petrea for his work with milling processing and Dr. Alexandru Stănescu for providing the photos with the two models.

ORCID iDs

A Lungu http://orcid.org/0000-0001-8266-5230
L Gurău http://orcid.org/0000-0001-9998-6671
S Georgescu http://orcid.org/0000-0002-3660-0265
C Coşereanu http://orcid.org/0000-0002-5805-8315

References

[1] Chernykh M, Kargashina E and Stollmann V 2018 *Acta Facultatis Xylologiae* **60** (1) 121-7 https://doi.org/10.17423/afx.2018.60.1.13
[2] Petru A 2015 Image Conversion for laser pyrography *Pro Ligno* **11** (4) 646-53
[3] Liu Q, Yang C, Xue B, Miao Q and Liu J 2020 *BioResources* **15** (3) 5366-78 https://doi.org/10.15376/biores.15.3.5366-5378
[4] Gurău L and Petru A 2018 *BioResources* **13** (4) 8168-83 https://doi.org/10.15376/BIORES.13.4.8168-8183
[5] Gabdrakhmanov A T, Bobrishev A A and Shafigullin L N 2019 *IOP Conf. Series: Materials Science and Engineering* 481 012045 https://doi.org/10.1088/1757-899X/481/1/012045
[6] Paiu I, Androne A and Coşereanu C 2020 *Pro Ligno* **16** (3) 60-7
[7] Hazır E and Koc H 2016 Optimization of wood surface machining parameters in CNC routers: Response surface methodology (RSM) approach *International Journal of Scientific Research Engineering & Technology* **5** (10) 494-501 Corpus ID: 189759102
[8] Aghakhani M and Khazaeian A 2014 *Asian Journal of Agriculture and Food Sciences* **2** (6) 548-53 https://ajouronline.com/index.php/AJAFS/article/view/2073
[9] Kazlauskas D, Jankauskas V, Bendikienė R, Keturakis G and Mačėnaitė L 2017 *Mechanicka* **23** (3) 469-72 https://doi.org/10.5755/j01.mech.23.3.18482
[10] Laszewicz K, Górski J, Wilkowski J and Czarniak P 2013 *Wood Research* **58** (3) 451-64
[11] Sütçü A 2013 *BioResources* **8** (1) 795-805 https://doi.org/10.15376/biores.8.1.795-805

IOP Conf. Series: Materials Science and Engineering 1235 (2022) 012042 doi:10.1088/1757-899X/1235/1/012042

CAM optimization in moulds retrofitting

Adrian But[1] **and Lucian Gal**[2]

[1] Politehnica University of Timişoara, Mechanical Faculty, Department of Materials and Manufacturing Engineering, B-dul Mihai Viteazu nr.1, Timişoara 300222, Romania

[2] "Aurel Vlaicu" University of Arad, B-dul Revolutiei nr 1.

adi.but@gmail.com

Abstract. Because the moulds are using in mass production, in time, defects appear, and their wear leads a big number of rejects parts. The cost to manufacturing a new mould is high, so, it is often used to retrofitting used moulds. The paper presents the moulds manufacturing steps correction, with applications and a computer aided manufacturing solution on CNC machine tools. Locating the defective area, choosing the most efficient milling strategies, protecting the parts area without defects in the mould during processing, paying attention to possible collisions during work or in positioning the cutting tools, the CNC programmer is fully engaged and have a big responsibility. The paper come with innovative CAM solutions to optimize the mould manufacturing process.

1. Introduction

Molds are tools used in mass production. Their processing is difficult due to their geometric complexity particularities, the high technological manufacturing processes what must be use, the material from which they are made.

The fact that they are required in thousands (when the resistors are inserted manually - 3000-4000 cycles) or tens of thousands of cycles (15000-20000 cycles - when the resistors are introduced by robots), the fact that errors can occur due to improper use by the operators (wrong introduction of the resistances in the mold), due to the non-observance of the casting parameters, the molds are destroyed, they wear out.

Replacing it with a new mold is expensive and therefore we try to repair it (of course, at a lower price), to bring it into operating parameters, or to make some changes.

Example of a mold to be reconditioned:

The areas marked in red are the defects reported by the customer on the castings (Figure 1a)

After the defects have been identified (Figure1.b), the mold is cover with welding in the defective areas, especially in the nests.

For laser welding, an electrode wire of Ø 0.5 with a hardness between 52 and 54 HRC is used. The molds have a hardness of 45 HRC but because they are not de-tensioned this type of electrode wire is used.

Welding (Figure 2a) is done in such a way that takes into account the correct welding, especially in the area of the edges but also to save electrode wire.

Figure 1.a)

Figure 1.b)

The risks that may arise when welding can be the following:

- improper use of the electrode wire which leads to damage to the welded area and / or detachment of the weld from the welded area
- not cleaning the plan before welding can lead to welding detachment
- the use of electrode wire with a hardness less than 45 HRC for welding can lead to detachment of the weld from the welded area
- the welding seams must be uniform and fill their middle area to avoid air gaps as in Figure 2b.

The mold is cleaned and positioned in the CNC machine tool according to the CAD drawing (Figure 3.b)

Figure 2.a) Image with correct welding

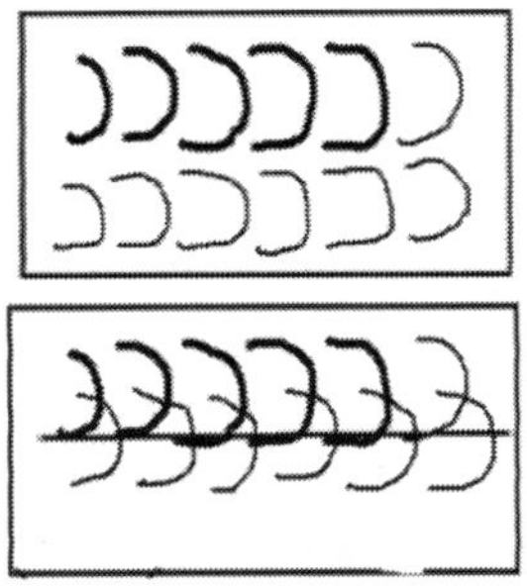

Figure 2.b) positions of welding seams

The origin X, Y, Z is taken over with the help of the RENISHAW electronic probe, and after the origin of the part has been established according to the CAD drawing (Figure 3.a), the NC program is conceived using CAM programs.

IOP Conf. Series: Materials Science and Engineering 1235 (2022) 012042 doi:10.1088/1757-899X/1235/1/012042

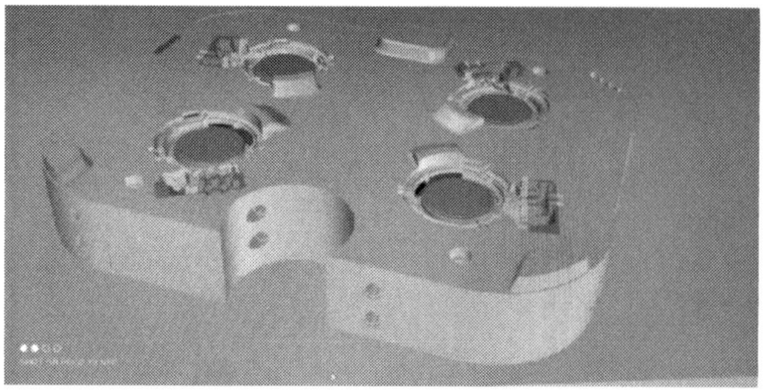

Figure 3.a) CAD mold presentation **Figure 3.b)** The mold fixing

2. CNC Technology

The programs are optimized according to the processed area both in terms of surface quality, processing time and saving cutting tools.

An Ø6 diameter with R0.5 end mill cutting tool was chosen as an example to change the cutting parameters and see how they influence the manufacturing time for faceing.

2.1. Test 1

Cutting parameters:

- rotational speed S = 4000 rpm, feed rate v = 500 mm / min, working engagement 40% D_c, type of processing: from outside to inside, optimization of the number of tool repositioning, number of Z - passes – 10, Z increment value of 0.1 mm.

The total time of face mill operation t = 14.5 minutes.

The same parameters were used but this time with a percentage of 80% of the tool diameter to see how the machining time is influenced and the following machining time resulted: t = 14.4 minutes. A very small time difference.

2.2. Test 2

For the following test, same cutting parameters are used with a working engagement 80% of the tool diameter, but this time was analysed the influence of manufacturing machining direction (Figure 4).

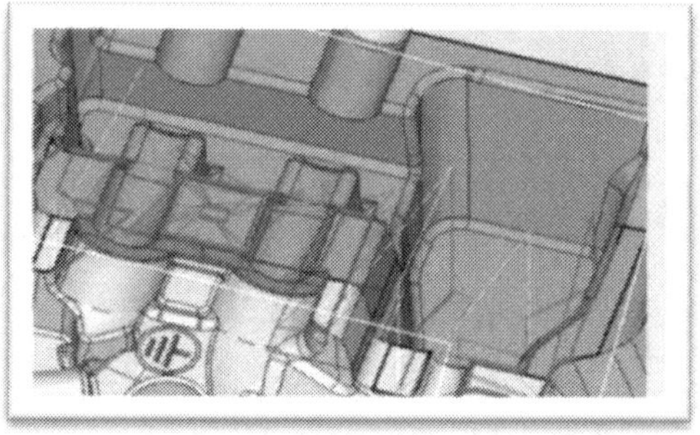

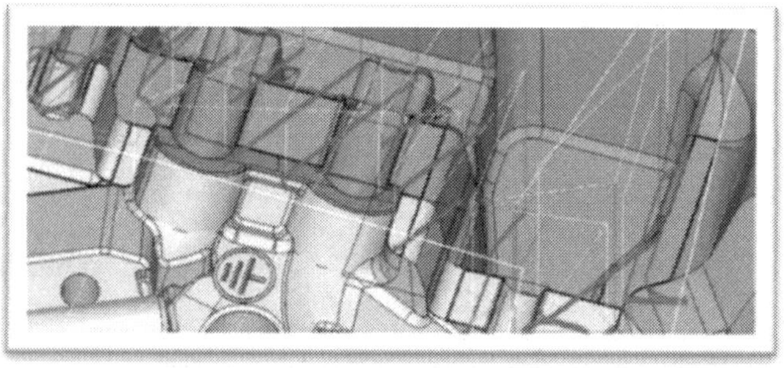

Figure 4.a) Parallel unidirectional **Figure 4.b)** Parallel bidirectional

In unidirectional manufacturing solution the total time is t = 37 minutes, a very large increase in processing time and surface quality remaining the same.

In bidirectional manufacturing solution, the total time is t = 19.5 minutes, a significant reduction in time but the quality of the surfaces remaining the same. (Figure. 5)

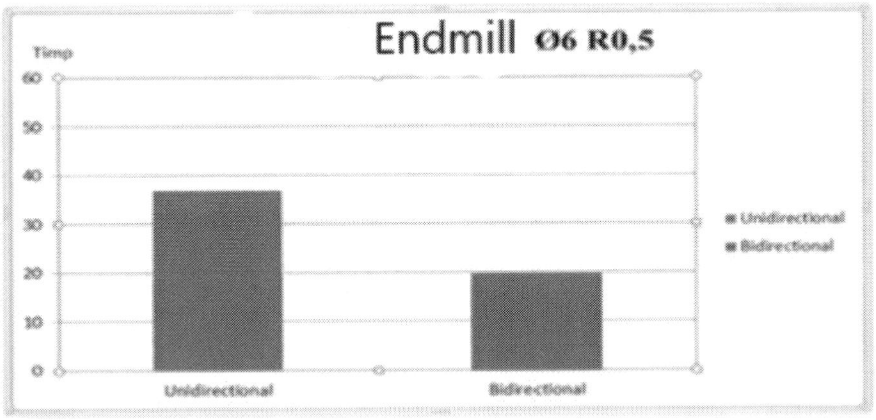

Figure 5. Histogram results for unidirectional t=37 s and bidirectional t=19,5 s

2.3. Test 3

For the following example, was change the diameter of the end mill cutting tool: Ø8 mm diameter with R0.5 using the following cutting parameters: rotational speed S = 3500 rpm, feed rate F = 700 mm / min, working engagement 80%, type of processing from outside to inside, optimization of the number of tool repositioning, number of Z - passes of 10, Z increment value of 0.1 mm.

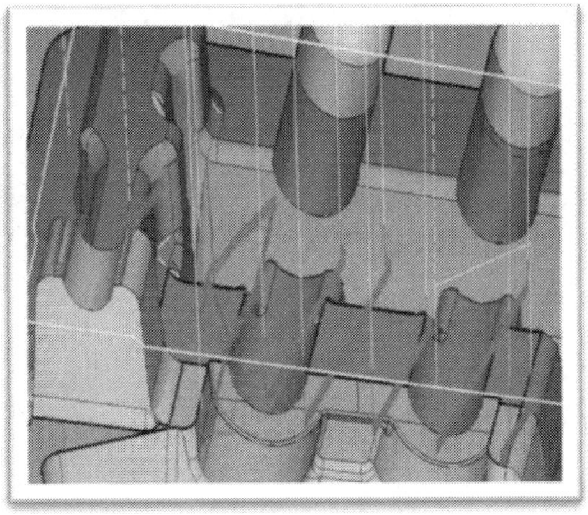

Figure 6.a) Parallel unidirectional

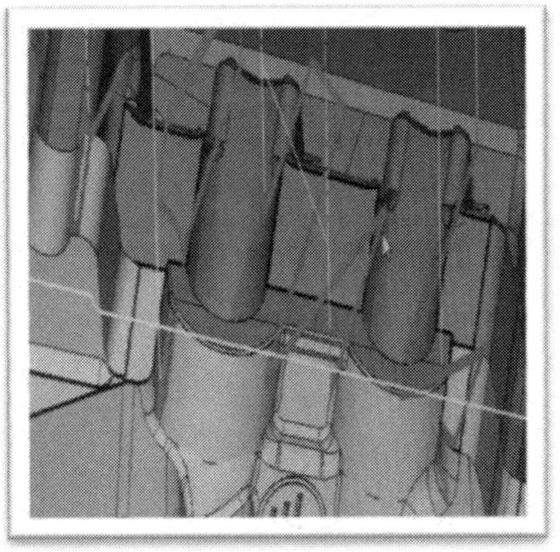

Figure 6.b) Parallel bidirectional

The machining time decreased considerably t = 10.3 minutes (bidirectional) and the surface quality remained the same compared to machining with Ø6 R0.5 milling cutter, machining direction from outside to inside and tool utilization percentage of 80%.

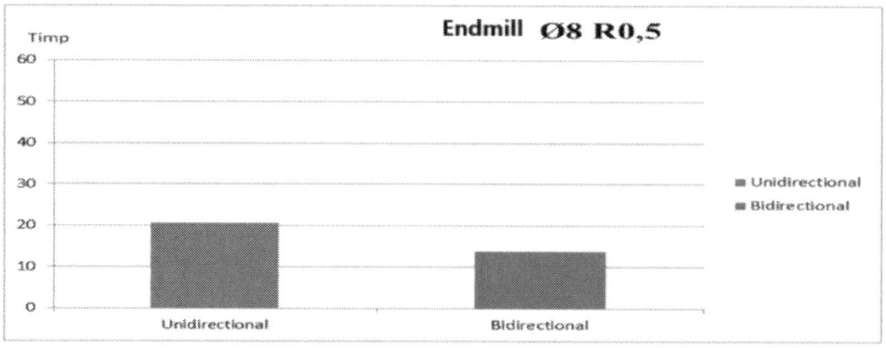

Figure7. Histogram results for unidirectional t=21 s and bidirectional t =13,5 s

2.4. Test 4 - Influence of machining direction for curved surfaces

Different processing direction angles were chosen for the curved surface. This processing method aims at the result on the quality of the processed surface. A spherical milling cutter with a diameter of Ø3 mm was used (Fig.8.) Cutting parameters: rotational speed S = 7500 rpm, feed rate F = 400 mm / min, increment value of 0.1 mm, 45° angle.

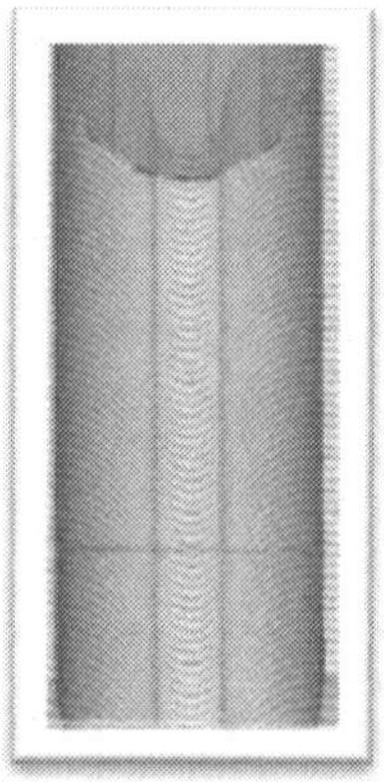

Figure 8. The simulation trajectory and the results at 45°

The surface quality is wrong and requires grinding because the resistance can be caught wrong when closing the mold leading to its cracking.

The quality of the surfaces processed for gripping the resistance and of the water tube must be very well processed so that when closing the mold, they are not caught incorrectly.

In the following example the same parameters are used but the processing angle of the surface is changed to 10 °.

Was observed that the time has increased by the order of seconds, but the processed surface has improved considerably and no further processing is required.

Figure 9. The simulation trajectory and the results at 10°

3. Conclusions

The processing times can be improved by adjusting the following parameters but also depending on the geometry of the processed surface:
- the speed of advance that influences the quality of the processed surface but also the processing time
-the diameter of the cutting tool that influences the processing times
- percentage of use of the diameter of the cutting tool that influences the times necessary for the processing of the respective area
- the processing direction that influences the processing times
-increment value that influences the processing area, surface quality and processing times
- number of Z passes that influence the processing time of a surface but also its quality

ORCID iDs

L Gal https://orcid.org/0000-0002-8548-3165
A But https://orcid.org/0000-0002-4032-1381

4. References

[1] Hazarika M and Dixit U S 2015 *Setup planing for machining* (Springer International Publishing) https://doi.org/10.1007/978-3-319-13320-1
[2] But A 2009 *Advanced machine-tools and manufacturing systems* (Timisoara: Politehnica)
[3] *Metal Cutting Technology* Technical guide
[4] *** SPINWORKS -Pokolm cutting tools catalog - 2016
[5] Kumar C and Deb S 2012 Generation of optimal sequence of machining operations in setup planning by genetic algorithms *Journal of Advance Manufacturing Systems* **11** (1) 67-80 https://doi.org/10.1142/S0219686712500059
[6] Sormaz D N, Arumugam J and Ganduri C 2007 Integration of rule-based process selection with virtual machining for distributed manufacturing planning *Process planning and scheduling for distributed manufacturing* (London Springer) chapter 3 pp 61-90 https://doi.org/10.1007/978-1-84628-752-7_3
[7] Nikolaus K 2014 *Manufacturing: Self-Organizing Factories* Siemens-Digital Factories
[8] Bahemia H and Squire B 2010 A contingent perspective of open innovation in new product development projects *Int. J. Innovative Management* **14** (4) 603-27 https://doi.org/10.1142/S1363919610002799

FEM-based investigation on machining forces and temperatures induced during AISI-4140 hard turning

Anastasios Tzotzis[1], Nikolaos Efkolidis[1], Dimitrios Tzetzis[2] and Panagiotis Kyratsis[1]

[1] University of Western Macedonia, Dept. of Product and Systems Design Engineering, Kila Kozani, GR50100, Greece
[2] International Hellenic University, School of Science and Technology, Thermi, GR57001, Greece

pkyratsis@uowm.gr

Abstract. Turning is probably one of the most common machining processes used to manufacture parts for the industrial sector. In order to reduce development times and costs during turning, it is necessary to employ methodologies and techniques that enable the precise prediction of the important parameters that relate to the process. Such a methodology is the implementation of Finite Element Method (FEM). In this work, the 3D modelling of AISI-4140 hard turning with ceramic tooling is being presented by utilizing the DEFORM-3D™ Finite Element Analysis (FEA) software system. Nine tests were conducted according to typical cutting conditions as recommended by tool manufacturers; cutting speed of 80 m/min, 115 m/min and 150 m/min and feed of 0.08 mm/rev, 0.11 mm/rev and 0.14 mm/rev, at a constant depth of cut equal to 0.30 mm. The developed friction forces, heat transfer and contact interface stresses between the tool and the workpiece were approximated with the aid of well-established models to simulate the machining mechanisms. Additionally, a validation of the yielded results for the machining components and the temperature distribution was performed by comparing the numerical results with experimental data. Most of the simulations generated results in agreement with the experimental ones. The agreement percentage exceeds 90% in most tests for the investigated components (machining forces and temperature), confirming both the validity and the accuracy of the model.

1. Introduction

Current advances in Finite Element Analysis (FEA) software systems allows for the efficient investigation of machining processes with minimal experimental work and in some cases without the need for physical experimental setups at all.

Oezkaya *et al.* [1] developed a 3D Finite Element Method (FEM) simulation model for the optimization of the flow drilling process on aluminum cast alloy AlSi10Mg. According to the authors, the optimization can be achieved by modifying the tool geometry and analyzing the impact on process conditions such as temperatures, forces and torques. Xiang *et al.* [2] analyzed machining induced wear mechanisms and developed a Finite Element (FE) model focused on tool wear in diamond drilling of SiCp/Al composites. This was achieved by incorporating Usui wear model combined with graphitization induced chemical wear model into the 3D FE model. Davoudinejad *et al.* [3] presented in their work a 3D FE model of the micro end-milling process applied to different cutting conditions. The model considers the effect of tool run-out and the tool edge radius for prediction purposes of the chip formation, temperature distribution and cutting forces. Tzotzis *et al.* [4-6] studied the effects of the drill diameter, cutting speed and feed on the generated thrust forces and chip morphology during drilling of Al7075-T6 alloy by combining physical testing and numerical investigations in three dimensions. Moreover, the authors [7] investigated AISI-4140 hard turning and the development of cutting forces with the aid of a Computer Aided Design (CAD) based approach and the implementation of 3D FEM. The yielded results were verified with well-established statistical tools and were used to build mathematical models for prediction purposes. By using similar strategies, Kyratsis *et al.* [8] employed the features of a commercially-available CAD system and 3D FEM in order to model the turning process of AISI-D3 and examine the effects of typical cutting conditions on the produced machining forces.

In this work, DEFORM-3D™ was utilized to develop a model in three dimensions that investigates the machining forces and the temperature distribution that occur when hard turning AISI-4140. The obtained results were processed and used to build a 3D FE model. To validate the model, the simulated values and the equivalent experimental ones were compared. The comparison results indicate that the FE model of AISI-4140 hard turning can yield reliable results.

2. Materials and Methods

2.1. *Setup of the machining process*

A CAD-based approach was followed to develop a more simplistic model of the operation and to obtain the critical parameters such as the cutting angles and the positioning of the tool. The tool-holder and the cutting tool were assembled so that the cutting angles can be derived. Moreover, the tool model was built via a developed application [9] that is used to automatically generate tool models of high quality. The used components have ISO designation numbers PCBNR2525M12 and CNGA120408 respectively. The CNGA-ceramic tools are rhombic-shaped inserts with 80° included angle and are used for machining in general hardened steels. Additionally, a cylindrical bar with a diameter of 72 mm was the design scheme for the workpiece which was positioned to the assembly accordingly. The materials that represent the workpiece is AISI-4140 steel, whereas the equivalent material for the tool is ceramic. Since the cutting angles are the ones related to the tool-holder specifications and the tool's positioning, it is derived that the lead angle is $kr = 75°$ and both the rake (γ) and inclination (λ) angles are negative and equal to −6°.

Figure 1a illustrates the positioning of the tool on the workpiece and the definition of the machining force components, which are the tangential (F_t), the radial (F_r) and the feed (F_a). In addition, Figure 1b depicts the feed direction, as well as the formed lead angle of the tool.

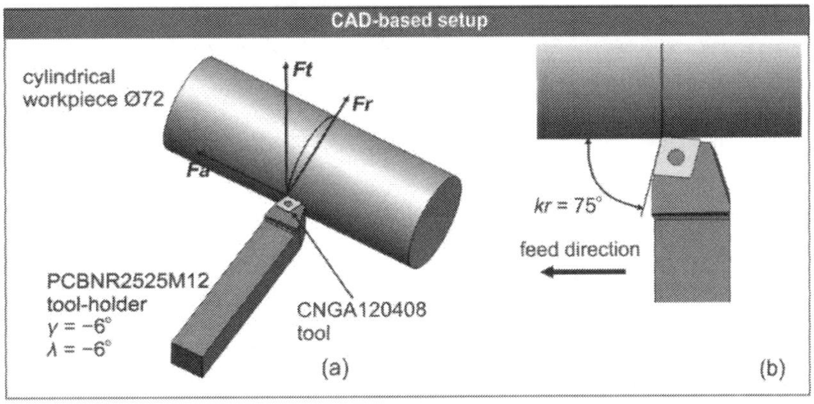

Figure 1. The setup of the tool-workpiece interface; the machining force vectors (a) and the feed direction (b).

2.2. Layout of the simulation process

DEFORM-3D™ ver. 12, FEA system was utilized to perform nine numerical simulations of the hard turning process in three dimensions. To maintain simulations at reasonable times, the FE model was simplified and the simulations were allowed to run for a short amount of time after the occurrence of the steady state. Specifically, the workpiece model was designed to be a part of the full cylindrical bar resembling an arc, with a diameter of 72 mm and an angle of 45° (Figure 2a). The defined workpiece geometry along with the model of the tool constitute the analysis domain of the process. Furthermore, the workpiece model was set to be plastic and designed with a pre-cut surface according to the applied depth of cut (Figure 2b). It was meshed with about 90,000 to 140,000 elements according to the value of feed. Moreover, a refined mesh was applied on the surface of interest between the tool and the workpiece, by using a 7:1 size ratio (Figure 2b). Contrarily, the tool was set to be fully rigid with the best possible mesh size (approximately 50,000 elements). Figure 2c illustrates the CAD model of the cutting tool and points out the 4:1 mesh size ratio that was applied to the cutting edge of the tool. This refinement can lead to reduced simulation times and yield more accurate results since this area is the one that is always in contact with the cut surface of the work during machining.

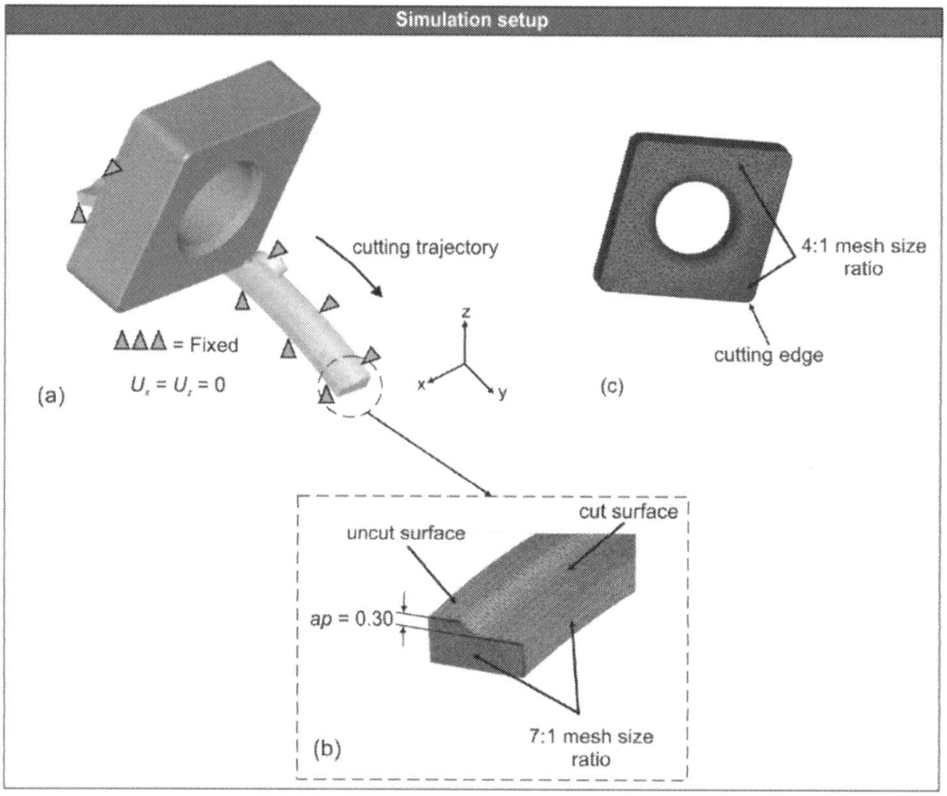

Figure 2. The simulation process; the cutting trajectory (a), the meshed workpiece section (b) and the meshed tool (c).

For the modelling of the material's flow stress behaviour, the generalized Johnson-Cook material model was utilized in the present numerical study. This model is widely applied when dealing with conditions that occur during machining. Equation 1 describes the mechanisms of the aforementioned model [9]. The constants that were used in the model are available in the literature. Specifically, Table 1 includes the thermo-mechanical properties of the AISI-4140 steel, whereas Table 2 presents the used model constants that define the function of the model for this case.

$$\sigma = \left(A + B\varepsilon^n\right)\left(1 + C\ln\frac{\dot{\varepsilon}}{\dot{\varepsilon}_0}\right)\left[1 - \left(\frac{T - T_0}{T_m - T_0}\right)^m\right] \tag{1}$$

Table 1. Thermo-mechanical properties of AISI-4140 [7].

AISI-4140 properties				
Young's modulus [GPa]	Density [kg/m^3]	Poisson's ratio	Thermal expansion [μm/m°C]	Thermal conductivity [W/mK]
212 @ 20 °C 192 @300 °C 164 @600 °C	7850	0.30	11.9 @ 20 °C 13.6 @300 °C 14.9 @600 °C	41.7 @ 20 °C 41.4 @300 °C 34.1 @600 °C

Table 2. Model constants of AISI-4140 [7].

Constitutive model						
A [MPa]	B [MPa]	C	n	m	T_0 [°C]	T_m [°C]
106	1167	0.0352	0.1424	0.763	20	1547

The first part of Equation 1 represents the strain hardening properties of the material, the second part represents the strain sensitivity and lastly part three represents the thermal softening properties of the material. Where A is the initial yield stress, B is the strain hardening modulus, C is the strain rate dependence coefficient, ε is the plastic strain, n denotes the strain hardening exponent, m denotes the thermal softening coefficient, $\dot{\varepsilon}$ represents the plastic strain rate, $\dot{\varepsilon}_0$ represents the reference plastic strain rate, T stands for the reference temperature, T_0 stands for the bulk temperature of workpiece material and finally T_m stands for the melting temperature of workpiece material. In this investigation, a reference strain rate equal to 1/s was used. The modelling of the material fracture due to the developed stresses was achieved with the use of the normalized Cockcroft and Latham criterion [10]. In the normalized version of the criterion, the maximum principal stress is normalized by the effective stress. The approximation of the development of the frictional forces due to the contact pressure at the tool-workpiece interface can be carried out with the aid of several advanced models [11]. However, in order to maintain the simplicity of the FE model and at the same time to yield results of acceptable accuracy, the Coulomb's law was employed. This model enables the simple approximation of the generated frictional forces at the sliding zone, which is of great importance when investigating machining. The friction coefficient for this study was set to $\mu = 0.577$.

Finally, the movement controls were set for the tool-workpiece assembly. The workpiece was fixed in both X and Z axis. Contrarily, the tool was set to travel along the path that is dictated by the cutting trajectory (Figure 2a). Additionally, the boundary conditions for the heat exchange with the environment were also defined by attributing the heat transfer condition to all surfaces of the workpiece. Regarding the coefficient values for heat transfer, the default values were applied to the model for convection, as well as for conduction.

3. Results and discussion

Nine simulation runs were performed by combining three levels of cutting speed (80 m/min, 115 m/min and 150 m/min) and three levels of feed (0.08 mm/rev, 0.11 mm/rev and 0.14 mm/rev) at constant depth of cut equal to 0.30 mm. The produced simulated results of the machining components were smoothened by applying the first order exponential smoothing and then compared to equivalent experimental results that are available in the bibliography [12]. The nine combinations of the cutting conditions is shown in Table 3. Figure 3 visualizes the comparison between the results of the nine numerical tests and the equivalent experimental ones. Specifically, Figure 3a, 3b, 3c and 3d illustrate the comparison of the results for each one of the force components respectively (radial, tangential and feed). According to Figure 3, all nine tests yielded results that are in correlation with the experimental values. Even though some discrepancies are present, the level of agreement is close to 80% in most tests and in some cases exceeds 90%. In the present study, these discrepancies are mostly present in the tangential component graph (Figure 3b). In contrast, the case of the resultant force demonstrates an increased level of accordance due to the fact that the radial force component also displays a good level of agreement between the numerical and the experimental results, which at the same time is the main contributor to the calculated resultant force induced during hard turning.

Table 3. Standard order of the simulation runs.

Run	V_c [m/min]	f [mm/rev]	ap [mm]
1	80	0.08	0.30
2	80	0.11	0.30
3	80	0.14	0.30
4	115	0.08	0.30
5	115	0.11	0.30
6	115	0.14	0.30
7	150	0.08	0.30
8	150	0.11	0.30
9	150	0.14	0.30

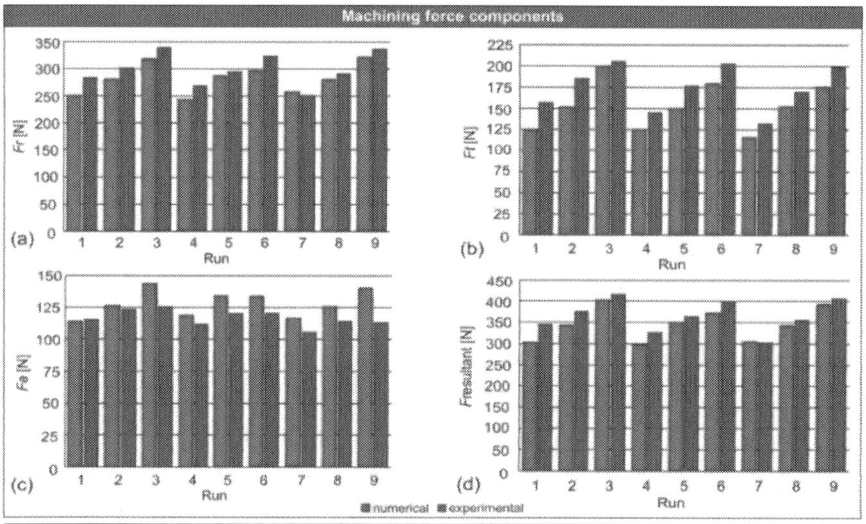

Figure 3. Comparison graph of the numerical and the experimental results; the radial component (a), the tangential component (b), the feed component (c) and the resultant of the three components (d).

By validating the numerical results, the temperature distribution on the workpiece cut surface was investigated. As expected, the maximum temperature appears at the contact interface between the tool's cutting edge and the cut surface of the workpiece. The distribution pattern of the temperature, which includes 8 zones, is shown in Figure 4 and is noticed in all nine tests. Moreover, higher values of feed seem to have a rather small effect on the temperature values, acting increasingly. Figure 4 depicts the temperature results for the test with cutting speed equal to 150 m/min and feed equal to 0.14 mm/rev at ambient temperature of 20 °C, where the maximum temperature was found to be 781 °C at the contact interface. Furthermore, the chip curling at step 1370/2540 was found to have temperatures between 495 °C and 590 °C.

IOP Conf. Series: Materials Science and Engineering 1235 (2022) 012043 doi:10.1088/1757-899X/1235/1/012043

4. Conclusions

The present investigation introduces a 3D FE model for turning. Nine turning simulation runs were carried out with the implementation of DEFORM-3D™. The selected workpiece material is AISI-4140 steel and the used tool a non-coated ceramic, namely CNGA120408. Moreover, the selected cutting conditions derive from the combination of three levels of cutting speed V_c = 80 m/min, 115 m/min and 150 m/min, as well as three levels of feed f = 0.08 mm/rev, 0.11 mm/rev and 0.14 mm/rev respectively.

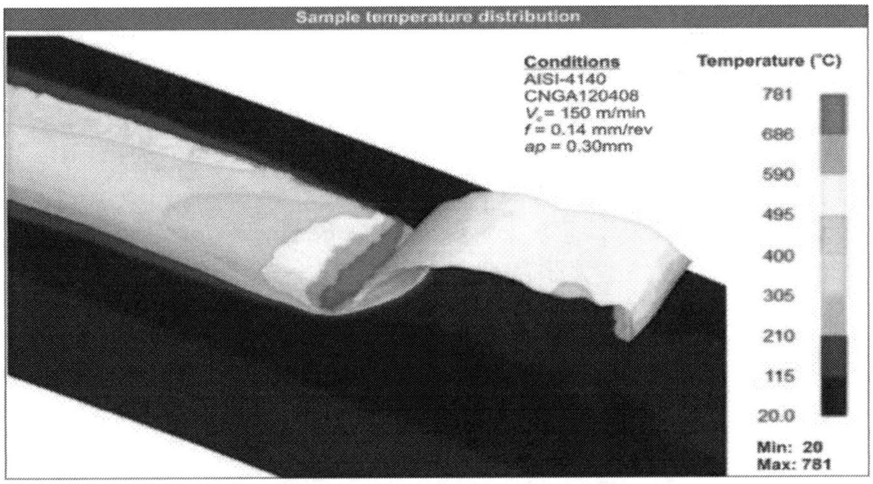

Figure 4. Sample temperature distribution during machining of AISI-4140.

The validity of the generated numerical results for each of the machining force components were validated via experimental work. The comparison analysis revealed a satisfying level of agreement between the simulated and the experimental values. Finally, next conclusions are deduced:

- Feed strongly affects the generated machining forces, especially the resultant. Shifting feed from 0.08 mm/rev to 0.11 mm/rev increases the resultant machining force by approximately 14.5% on average. Similarly, an increase in feed from 0.11 mm/rev to 0.14 mm/rev boosts the resultant machining force by about 12.6%.
- The radial force F_r seems to be the component that affects the most the resultant machining force.
- The agreement between the numerical and the experimental results is close to 90%.
- The average maximum temperature on the tool-workpiece interface for the nine simulations exceeds 800 °C.

ORCID iDs
A Tzotzis https://orcid.org/0000-0002-6942-9636
N Efkolidis https://orcid.org/0000-0001-9225-0311
D Tzetzis https://orcid.org/0000-0001-5006-5759
P Kyratsis https://orcid.org/0000-0001-6526-5622

References

[1] Oezkaya E, Hannich S and Biermann D 2019 *Int. J. Mater. Form.* **12** 477-90 https://doi.org/10.1007/s12289-018-1429-0
[2] Xiang J, Pang S, Xie L, Gao F, Hu X, Yi J and Hu F 2018 *Materials (Basel)* **11** 252 https://doi.org/10.3390/ma11020252
[3] Davoudinejad A, Tosello G, Parenti P and Annoni M 2017 *Micromachines* **8** (6) 187 https://doi.org/10.3390/mi8060187
[4] Tzotzis A, Markopoulos A P, Karkalos N E, Tzetzis D and Kyratsis P 2021 *IOP Conf. Ser. Mater. Sci. Eng.* **1037** 012009 https://doi.org/10.1088/1757-899X/1037/1/012009

[5] Tzotzis A, García-Hernández C, Huertas-Talón J L and Kyratsis P 2020 *Mech. Ind.* **21** 4 https://doi.org/10.1051/meca/2020046
[6] Tzotzis A, Markopoulos A, Karkalos N and Kyratsis P 2020 *MATEC Web Conf.* **318** 01038 https://doi.org/10.1051/matecconf/202031801038
[7] Tzotzis A, García-Hernández C, Huertas-Talón J L and Kyratsis P 2020 *Strojniški Vestn. - J. Mech. Eng.* **66** (7-8) 467-78 https://doi.org/10.5545/sv-jme.2020.6784
[8] Kyratsis P, Tzotzis A, Markopoulos A and Tapoglou N 2021 *Machines* **9** (1) 4 https://doi.org/10.3390/machines9010004
[9] Tzotzis A, Garcia-Hernandez C, Huertas-Talón J L and Kyratsis P 2020 *J. Manuf. Mater. Process.* **4** (4) 104 https://doi.org/10.3390/jmmp4040104
[10] Cockcroft M G and Latham D J 1968 *J. Institute Met.* **96** 33-9
[11] Melkote S N, Grzesik W, Outeiro J, Rech J, Schulze V, Attia H, Arrazola P J, M'Saoubi R and Saldana C 2017 *CIRP Ann.* **66** (2) 731–54 https://doi.org/10.1016/j.cirp.2017.05.002
[12] Tzotzis A, García-Hernández C, Huertas-Talón J L and Kyratsis P 2020 *Micromachines* **11** (9) 798 https://doi.org/10.3390/mi11090798

IOP Conf. Series: Materials Science and Engineering 1235 (2022) 012044 doi:10.1088/1757-899X/1235/1/012044

Field robots development in the aspect of achieving the goals of sustainable agriculture

S Sobocki[1], J Wojciechowski[1], S Legutko[2], J Mac[1], M Zawada[1] and S Szymczyk[1]

[1] Łukasiewicz Research Network - Industrial Institute of Agricultural Engineering, 31 Starołęcka street, 60-963 Poznań, Poland

[2] Poznań Uniwersity of Technology, Faculty of Engineering Management, 3 Piotrowo street, 60-965 Poznań, Poland

sebastian.sobocki@pimr.lukasiewicz.gov.pl

ABSTRACT. The article deals with the issues related to the use of field robots in agricultural production in the aspect of the use of a sustainable farming system. At the beginning, the very essence of sustainable agriculture is presented and the problems that remain to be solved are identified, thus determining the market demand for autonomous field robots. The next part of the article presents the main division of agricultural robots and an overview of the most interesting solutions from the last few years that have been implemented by private companies and research centres from around the world. The review of available solutions makes it possible to identify the main challenges related to the implementation of robotic projects into working in difficult field conditions that are faced by the engineers who design them. The projects of autonomous robots used for work in agriculture, implemented by the Łukasiewicz Research Network - Industrial Institute of Agricultural Engineering in cooperation with consortium members, are also presented. At the end of the article, there is a summary and conclusions, which in a general sense define the need for further research on field robots, so as to achieve their fully autonomous work in the near future.

1. Introduction

Forecasts of the estimated number of people in the world indicate that there will be around 10 billion of us by 2050. Many people wonder how to meet this growing demand for food, when there are already areas of hunger in the world where about 690 million people are malnourished [1]. The solution may be bigger industrialization of agriculture through the growth of using fertilizers and plant protection products, but how will this affect food quality and directly our health? The growing awareness of the necessity of eating healthy food in societies forces us to find production methods that will make it possible to increase its efficiency while achieving the best possible quality. The article deals with the application of robots, which are able to meet many challenges of modern agriculture, in agricultural production.

2. Sustainable agriculture - problems identification

The idea of sustainable agriculture (figure 1) has been known for several decades, over which it has significantly evolved. Żmija [2] and Roszkowska-Mądra [3] described in detail the essence and causes of this idea as well as the stages of development and implementation of the sustainable agriculture strategy in the EU over the last dozen or so years. According to them, sustainable agriculture can be defined as a form of effective and, at the same time, safe production that allows the cultivation of high-quality food, while maintaining the principles that do not cause environmental degradation. The concept of sustainable development assumes both striving to improve the living conditions of the rural population and running a business in these areas, while not compromising the specific rural resources, such as the natural environment, landscape and cultural heritage. Zegar [4] determines that this approach seeks to reconcile the laws of nature and economics by finding the optimal value between them, in line with the statement that the environment is harmed not by agriculture itself, but only by certain agricultural technologies.

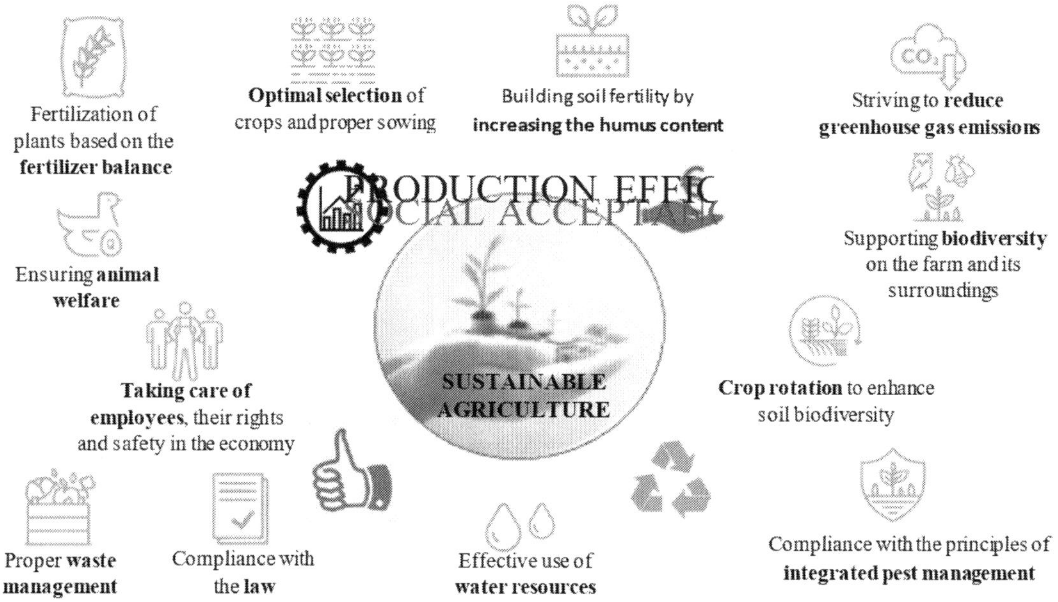

Figure 1. The main goals of sustainable agriculture

Despite the identification of the need to introduce the principles of sustainable agricultural production many years ago, as a result of the existence of certain limitations, meeting the needs of the dynamically growing number of the world's population, which, according to UN data [5], is to increase to about 10 billion by 2050 (figure 2), it is still mainly carried out through intensive agricultural production which is the use of highly industrial agriculture. Population growth will increase global food demand. Research by Ray *et al.* [6,7] showed that in order to satisfy the food needs of such a large population, plant production should be doubled by then.

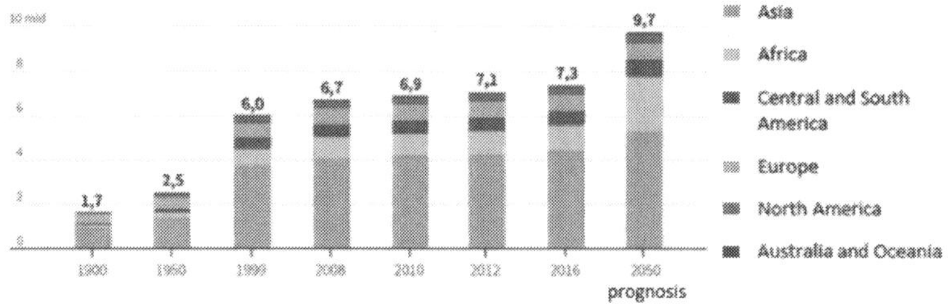

Figure 2. The growth of the world population and on individual continents [5]

With limited cultivation areas on Earth (figure 3.), ensuring that enough food is available for a rapidly growing population becomes a huge challenge. It should also be noted that the growing demand for biofuels and competition for the use of limited arable land between crops dedicated to the subsequent production of biofuels and crops dedicated to food production make it even more difficult to ensure food security.

IOP Conf. Series: Materials Science and Engineering 1235 (2022) 012044 doi:10.1088/1757-899X/1235/1/012044

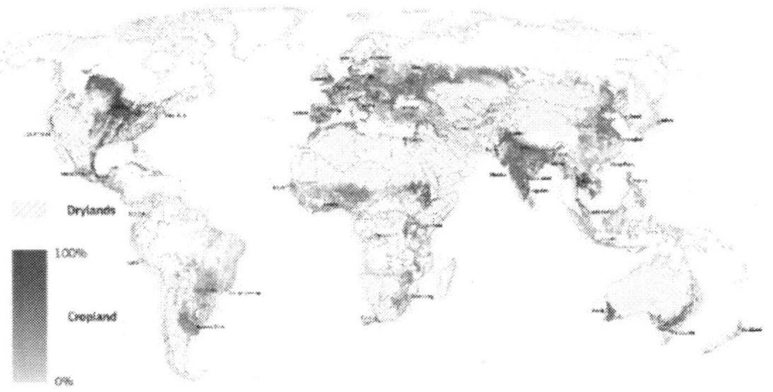

Figure 3. Areas on land with favorable cultivation conditions [8]

Most of the countries with agricultural lands (USA, Russia, European countries) has a low or even negative rate of natural increase. Poland also belongs to this group. According to recent data, unemployment in Poland is a thing of the past. For several years, the Polish economy has been experiencing a significant lack of labour, including agriculture, which is lacking around 550,000 seasonal workers [9]. An important factor is also the fact that there is a demand for seasonal workers for a period of only 45-70 days a year, which discourages potential candidates from taking up this type of work. Poland's demographic forecasts are exactly the opposite of the direction of the world demography. The negative birth rate will result in the reduction of employees, especially for agriculture.

Industrial agriculture means the use of a large amount of fertilizers and plant protection products. Currently, 20 times more artificial fertilizers are used than before the Second World War, but it should be noted that this consumption continues to grow (figure 4, figure 5).

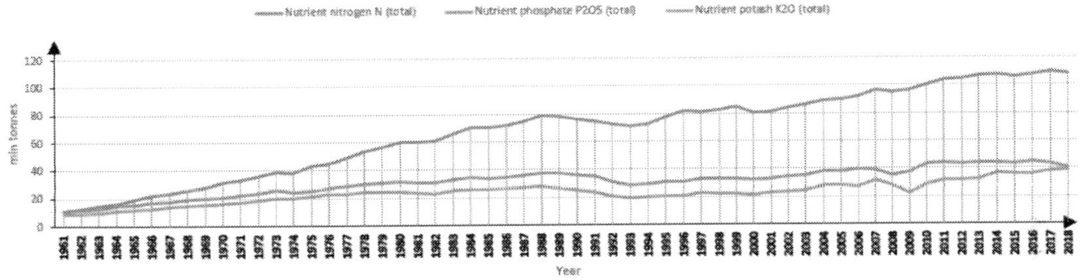

Figure 4. World fertilizer consumption in recent years [10]

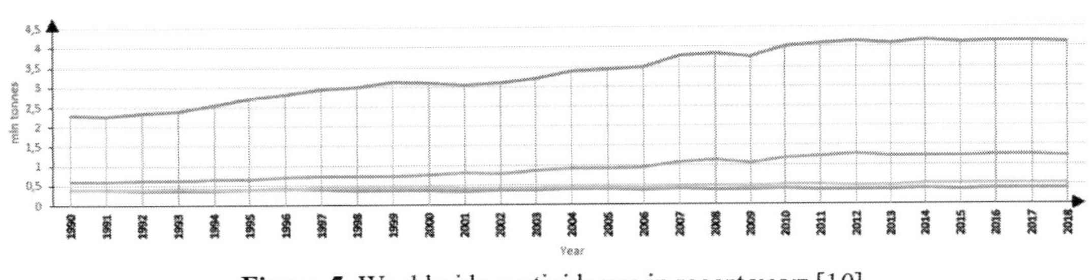

Figure 5. Worldwide pesticide use in recent years [10]

Poisoning the soil with mineral fertilizers, as a result of inept and excessive use of them, may lead to deterioration of the quality of crops, cause the disappearance of soil microflora activity, in particular, adversely affect the nitrification processes and atmospheric nitrogen fixation processes. Excessive use of potassium in plants growing on over-fertilized soils leads to dangerous animal diseases (e.g. pasture tetany). Liming soils can cause chlorosis of leaves, disrupting carbohydrate and protein metabolism.

Over-fertilization with nitrogen fertilizers is also dangerous. Too high a nitrate content in plants is harmful to humans and animals.

Resistance is the hereditary ability of some individuals (biotypes) of a given species of weed to survive and reproduce, despite the use of an effective dose of herbicide that effectively eliminates the rest of the population. In successive generations, as a result of the use of the same herbicides for several years (with the same active substance or with the same mechanism of action), the population of resistant biotypes increases to a level where a given species previously considered sensitive is no longer eliminated. It turns out that often the low effectiveness of the herbicide or its complete lack, results from the mistakes made: performing the treatment at too low temperatures, using too low doses of herbicides, treating weeds in advanced stages of development, or using a poorly selected or faulty technique.

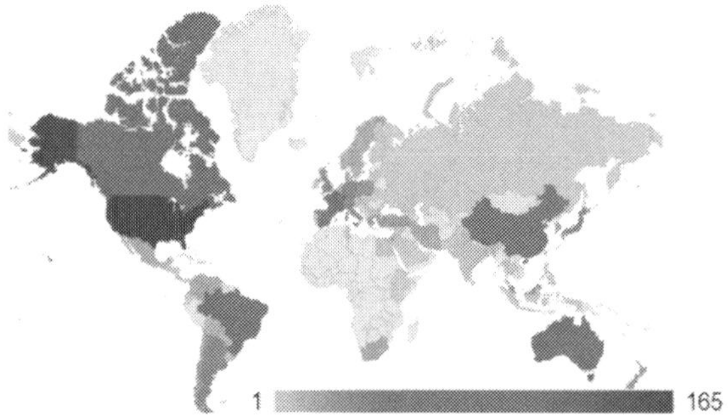

Figure 6. The number of unique herbicide resistance cases globally [11]

The accumulation of toxic substances in plants causes contamination of all links in the food chain. The movement of chemicals from soil to water causes eutrophication of surface and groundwater. Soil poisoning with heavy metals (nickel, mercury, cadmium, arsenic, lead), and then their accumulation in plant tissues causes irreversible changes in plant organisms, reduces the growth of plant mass, and reduces yield. In humans, an excess of heavy metals can cause atherosclerosis and cancer.

3. Field robots as a solution to the problems of modern agriculture

The presented problems create the need to automate plant production using autonomous agricultural robots. They will make it possible to maintain the current production efficiency while reducing the demand for human labour - working 24 hours a day. The use of precise navigation and control systems for working tools inside the robots, using GPS data and data from vision systems identifying the position of crops in the field and distinguishing them from weeds, will enable a precise, optimal, depending on the needs of a given plant and its development stage, use of fertilizers and protection measures. By using external weather data, they will enable timely identification of the needs of the entire crop or even a specific plant when the problem (e.g. disease) affects only a small group of the population.

Based on the analysis of the global robot market, their main producers can be distinguished, which are: AGCO Corporation, AgEagle Aerial Systems Inc., Blue River Technology, John Deere & Company, IBM, Lely, Naïo Technologies, Precision Hawk and Trimble Inc. Based on their product portfolio, field robots can be divided into:

- small field robots performing single agrotechnical procedures,
- autonomous multifunctional platforms with replaceable working modules,
- robotic machines aggregated with an agricultural tractor,
- autonomous tractors or conventional tractor navigation systems.

The group of small robots (figure 7.) is in most cases represented by drones, the main task of which is to properly scan the field [12], so as to enable the development of maps, for example, the fertility of the field, the shortage of fertilizers, the occurrence of diseases, pests, etc. Drones are less often used as sprayers [12,13], which are capable of carrying out the appropriate protective treatments without causing the negative effects of crop and soil compaction (as is the case with conventional sprayers), but their disadvantage is their low efficiency due to the possibility of taking up only a small amount of crop

IOP Conf. Series: Materials Science and Engineering 1235 (2022) 012044 doi:10.1088/1757-899X/1235/1/012044

protection agents. The group of small robots also includes small vehicles dedicated mainly to weeding between rows of crops [14] or dosing plant protection products and fertilizers [15,16].

Figure 7. Small field robots: AgEagle Drone [12], VoloDrone [13], Smart Weeding Robot [15], Robot Oz [14], Rowbot [16]

The group of autonomous multifunctional platforms (figure 8.) is currently the most strongly developed and most widely represented area in field robotics. These are devices for performing several agrotechnical operations through the possibility of replacing appropriate working modules, which most often include: seeders, active weeder for inter-rows and rows, selective sprayers. The group of these robots is a representation of a conventional tractor and agricultural machine attached to the three-point linkage, but these robots have a smaller and more compact structure, which has a positive effect on their weight and reducing formation of subsoil layer.

Figure 8. Autonomous driving platforms: Robotti [16], Dino [13], Etarob [18]

In most cases, robotic machines aggregated with an agricultural tractor (figure 9.) are active hoes and selective sprayers used to implement precise application of plant protection products and liquid fertilizers. Their operation requires a conventional agricultural tractor, driven by an operator or navigation systems.

Figure 9. Robotic machines aggregated with a tractor: Blue River Technology [19], Garford [20], Robovator [21]

The last group is mainly the navigation systems of conventional tractors, because the autonomous tractors themselves are only prototypes not put into mass production (figure 10.). Navigation systems are displays and controls integrated into the steering system of a conventional tractor that help to accurately monitor and map terrain information in real time [25]. These systems require the tractor operator to control the correctness of work, so that in the event of an undesirable operation he can take over the control of the vehicle.

Figure 10. Concepts of autonomous tractors: Case IH [22], John Deere [23], Kubota [24]

4. Realizations of Łukasiewicz Research Network - Industrial Institute of Agricultural Engineering

Łukasiewicz Research Network - Industrial Institute of Agricultural Engineering has also applied projects related to the implementation of autonomous robots for work in agriculture (figure 11). The first project - Autonomous Technology of Biomass Transport [26,27] concerned the development of the construction of a tracked vehicle, which is used to transport biomass in difficult marshy terrain. The vehicle cooperates with the operator of another mowing and harvesting vehicle, who calls it in to tranship the collected biomass. The vehicle decides by itself and selects the most advantageous route, it connects to the mowing vehicle and replaces the empty container with the full biomass.

The second project - the Agrorob field robot is a typical autonomous mobile platform for sowing and cultivating wide-row crops [28,29]. The project consisted in developing a platform structure for the following treatments: sowing, mechanical weeding and spraying.

Figure 11. Autonomous field robots designed by Łukasiewicz Research Network - Industrial Institute of Agricultural Engineering: ABT platform [26, 27], AgroRob [28, 29]

The knowledge gained during the implementation of previous autonomous robot projects will contribute to the development of the new robot: "Inteligentny robot spełniający wymogi rolnictwa precyzyjnego", i.e. a newly acquired project co-financed by the European Union from the European Regional Development Fund under the Smart Growth Operational Program.

5. Problems and limitations - challenges related to the implementation of field robots on the market

The limitations in implementing the principles of sustainable agricultural production described by Antoszek [30] relate to social barriers (low level of wealth of the inhabitants), economic (limited investment opportunities), intellectual (low level of education and environmental awareness) and ethical barriers (giving consent and tolerating inappropriate behaviour). These limitations in the case of field robots were extended by Doruchowski [31], identifying additional ones related to technological barriers.

Autonomous field robots use various types of sensors: passive cameras, active laser scanners, radars, sonars, inertial sensors, encoders and spectral sensors for the processes of perception, location and motion planning in an open and unpredictable environment in which they work. The unpredictability of the working environment (changing weather conditions, contamination of sensors, strange / unexpected behaviour of people in the vicinity of the robot, etc.) makes the process of processing this information in real time very difficult and can sometimes fail. The processing of information from the aforementioned sensors is based on probabilistic methods, and often on machine learning, in particular the use of neural networks. Some unreliability of the current methods of controlling such vehicles affects the safety of their use, which may lead to an accident or even a catastrophe, thus significantly limiting the use of such vehicles on a large scale in today's agriculture.

6. Conclusions

As a result of the analysis of the solutions available on the market for field robots, the most important conclusion is that, there is a huge need for further work and research aimed at the proper operation of control systems and information processing from specific sensors in order to achieve the reliability and safety of such robots.

The use of autonomous field robots for work in agriculture and the use of sensors and decision algorithms in their construction will additionally allow in the near future to:

- identify the wear of working tools, so that they can be replaced with new ones at the right time [32], it will reduce financial losses to a minimum by eliminating unnecessary downtime;
- estimate of the achieved values of crops, which will enable the maximization of profits and the possibility of planning further projects;
- control the physical condition of each plant (using information about the current and expected weather situation), allow for the elimination of the disease at an early stage of its occurrence, when it affects, for example, only a small part of the crop population, thus significantly reducing the use of fungicides;
- reduce the dosage of herbicides or, use intelligent hoeing, mechanical weeding of fields, completely eliminating the need for their use;
- eliminate the need for people to perform hard work in unfavourable environmental conditions (rain, heat, dust), which will reduce the incidence of accidents in agriculture and occupational diseases.

Acknowledgments

This study was supported by project co-financed by the European Union from the European Regional Development Fund under the Smart Growth Operational Program. Project implemented as part of the National Center for Research and Development, competition: 6/1.1.1/2019/SG OP Fast Track "Inteligentny robot spełniający wymogi rolnictwa precyzyjnego" POIR.01.01.01-00-1230/19-00.

ORCID iDs

S Sobocki https://orcid.org/0000-0003-4730-9331
J Wojciechowski https://orcid.org/0000-0003-0561-3941
S Legutko https://orcid.org/0000-0001-8973-5035
J Mac https://orcid.org/0000-0002-0888-9268
M Zawada https://orcid.org/0000-0002-2552-5984
S Szymczyk https://orcid.org/0000-0002-0081-3111

7. References

[1] https://www.unicef.org/reports/state-of-food-security-and-nutrition-2020 accessed 19.04.2021

[2] Żmija D 2014 Sustainable Development of Agriculture and Rural Areas in Poland *Studia Ekonomiczne* **166** 149-158 – *in Polish*

[3] Roszkowska-Mądra B 2009 The Concepts of Agricultural and Rural Development in the European Union *Gospodarka narodowa. The Polish Journal of Economics* **235** (10) 83-102 https://doi.org/10.33119/GN/101247

[4] Zegar JS 2010 The category of the rural development optimum - the present challenges *Roczniki Nauk Rolniczych* **G 97** (3) 301-12 - *in Polish*
[5] https://www.cia.gov/library/publications/the-world-factbook accessed 19.04.2021
[6] Ray D K, Ramankutty N, Mueller N D, West P C and Foley J A 2012 Recent patterns of crop yield growth and stagnation *Nature Communications* **3** 1293 https://doi.org/10.1038/ncomms2296
[7] Ray D K, Mueller N D, West P C and Foley J A 2013 Yield trends are insufficient to double global crop production by 2050 *PLOS ONE* **8** (6) e66428 https://doi.org/10.1371/journal.pone.0066428
[8] https://wad.jrc.ec.europa.eu/globalagriculture accessed 19.04.2021
[9] https://www.pb.pl/rosnie-zapotrzebowanie-na-pracownikow-sezonowych-w-rolnictwie-856430 accessed 19.04.2021
[10] http://www.fao.org/faostat/en/#data accessed 19.04.2021
[11] http://www.weedscience.org accessed 19.04.2021
[12] https://ageagle.com/ accessed 19.04.2021
[13] https://www.volocopter.com/ accessed 19.04.2021
[14] https://www.ecorobotix.com/ accessed 19.04.2021
[15] https://www.naio-technologies.com/ accessed 19.04.2021
[16] https://www.rowbot.com/ accessed 19.04.2021
[17] https://www.agrointelli.com/ accessed 19.04.2021
[18] https://maskor.fh-aachen.de/en/projects/etarob/ accessed 19.04.2021
[19] https://bluerivertechnology.com/ accessed 19.04.2021
[20] https://garford.com/ accessed 19.04.2021
[21] https://www.robovator.com/ accessed 19.04.2021
[22] https://www.caseih.com/northamerica/en-us/Pages/campaigns/autonomous-concept-vehicle.aspx accessed 19.04.2021
[23] https://www.deere.co.uk/en/agriculture/future-of-farming/ accessed 19.04.2021
[24] https://www.kubota.com/news/2020/20200115_1.html accessed 19.04.2021
[25] https://agriculture.trimble.com/ accessed 19.04.2021
[26] Zembrowski K, Sobocki S, Pawłowski T, Rakowicz A and Siczyński Ł 2016 Autonomous platform technology for biomass transport *Int. Sci. J. of Mechanization in Agriculture* **62** (2) 3-7 https://stumejournals.com/journals/am/2016/2/3
[27] Zembrowski K, Sobocki S, Rakowicz A and Siczyński Ł 2016 Strength analysis of the supporting structure of the abt vehicle during container reloading *J. of Research and Applications in Agricultural Engineering* **61** (2) 134-9
[28] Szczepaniak J, Grzechowiak R, Wojciechowski J, Maleszka M, Radkowski S, Mączak J and Jasiński M 2018 Field tests of an autonomous field robot for wide-row cultivation *Int. Sci. J. of Mechanization in Agriculture* **64** (3) 72-5 https://stumejournals.com/journals/am/2018/3
[29] Zawada M, Ciechanowski M, Szulc T, Szychta M, Smela A and Kamprowski R 2019 Vision systems in contemporary agriculture *Technika rolnicza ogrodnicza leśna* **64** 13-6 – *in Polish*
[30] Antoszek J 2002 The sustainable development conception and the possibilities of its implementation in rural areas in the province of Lublin (on the basis of survey) Lublin: *Annales UMCS* **57** (11) 197-209
[31] Doruchowski G 2005 Elements of precision agriculture in plant protection *Agricultural Engineering* **9** (6) 131-9 – *in Polish https://ir.ptir.org/artykuly/pl/66/IR(66)_1291_pl.pdf*
[32] Jasiulewicz-Kaczmarek M, Legutko S and Kluk P 2020 Maintenance 4.0 technologies – new opportunities for sustainability driven maintenance *Management and production engineering review* **11** (2) 74-87 https://doi.org/10.24425/mper.2020.133730

Research on the automation of internal logistics activities in an assembly line

I G Pascu[1], G C Neacsu[1], E L Nitu[1], R Beloiu[2] and A Gavriluta[1]

[1] University of Pitesti, Faculty of Mechanics and Technology, Manufacturing and Industrial Management Department, Targul din Vale Street No.1, Romania.
[2] University of Pitesti, Faculty of Electronics, Communications and Computers, Electronics, Computers and Electrical Engineering Department, Targul din Vale Street No.1, Romania.

iuliana.pascu@upit.ro

Abstract. The need for automation and digitization has led to a rapid evolution of internal logistics processes and systems. This paper presents the automation mode logistics activities in an assembly line, by integrating a robotic arm. For the integration of the robotic arm, several steps are followed, consisting in designing the assembly line and the robotic arm using the Process Simulate software, simulating the operation of the line for different scenarios for integrating the robotic arm and analysing them. The study is completed with the realization in the laboratory of the assembly line with automated logistics activities. At the end of the project, it is desired to use the automated assembly line in the university environment, for the formation of students' competencies in the field of use of digital techniques and industrial logistics.

1. Introduction

On the strength of evolving technologies and the trend towards automation, manufacturing companies need to introduce new production concepts in order to increase their performance and remain competitive [1]. Most enterprises are undergoing a process automation with the development of the industrial revolution. The focus of these transformations is mainly on production, so terms such as "Factory of the Future" or "Smart Factory" are used similarly to this concept of automation [2]. The smart factory is one of the key constructions of Industry 4.0 [3].

The basic idea of the 4.0 industrial revolution is to create machines and robots that will be equipped with processors, motion sensors, which will be able to communicate with each other, but also with other production equipment [4]. The Industry 4.0 concept was first presented at the "Hannover Fair" in 2011 [5]. As a consequence of Industry 4.0 and new technologies, the concept of logistics 4.0 uses in the logistics process technologies such as: Internet of Things, sensor network, cloud computing, big data, robots, augmented reality [6].

Robots are increasingly used in production systems to improve the automation of manufacturing processes and are used for activities such as: material handling, assembly, distribution [7,8]. This will further improve productivity, reduce manufacturing cycle time, but also costs [7].

2. Research objective and working methodology

The objective of the study is to integrate a robotic arm into an existing assembly line in the Lean Learning Factory laboratory of the University of Pitesti. The Lean Learning Factory laboratory comprises an assembly line (figure 1a) on which the assembly process of an experimental product - steering wheel (figure 1b) is carried out through manual activities, allowing the obtaining of a diversified range of products.

The existing assembly line in the laboratory is made of modular and flexible equipment, which can allow different configurations of workstations, as well as the application of different methods of supply with components and different forms of interoperable transfer. In the current configuration, the line consists of four workstations, power systems and modular, flexible transfer, which is carried by the assembly manual activities investigational product [9].

To carry out this study, the following steps will be followed:

- virtualizing existing assembly lines: development of the assembly line virtual environment;
- identifying the right operation for the integration of the robotic arm;
- experimentation integration of the robotic arm in the existing assembly line.

a)

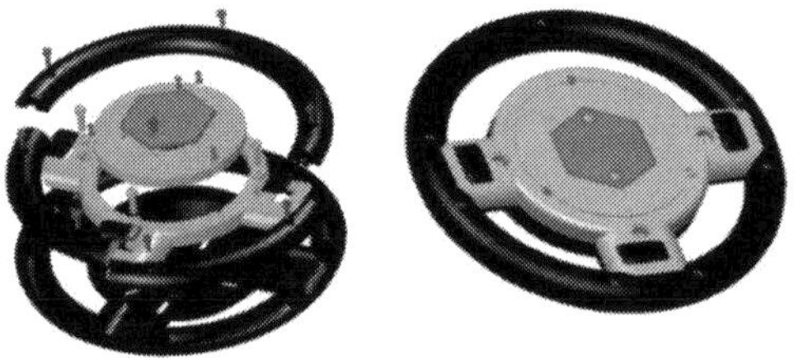

b)

Figure 1. Modular-flexible assembly line (a); experimental product - steering wheel (b).

The IRB 120 robotic arm (figure 2), which is to be integrated in the assembly line, is one of the new generations of industrial robots made by ABB. The payload of this robotic arm is 3 kg, ideal for manufacturing industries that use flexible automation based on robots.

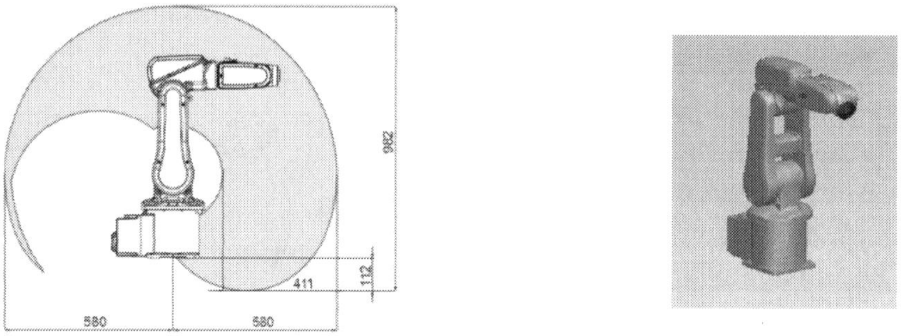

Figure 2. Robotic arm IRB 120.

3. Virtualization of the existing assembly line

Virtualization assembly line will use the Siemens software packages existing lab LLF. Process Simulate is part of the Tecnomatix software package produced by Siemens. It is a software that allows the design, analysis, simulation and optimization of manufacturing processes, having as advantages:

- Increasing the quality of the process by emulating realistic processes throughout their life cycle;
- Optimization of cycle times by simulation;
- Minimize production risk by similar different manufacturing scenarios;
- Validation of the commissioning of the production in a virtual environment;

- Ensuring safe ergonomic processes.

To design virtual assembly line was used PlantSimulation Tecnomatix software design and simulation of manufacturing processes both 2D and 3D, developed by Siemens. For the design stage of the assembly line, the following activities were performed:

- Research on the technical and functional data of the existing assembly line;
- 3D design workstations forming the assembly line in the software and importing PlantSimulation their Process Simulate software;
- 3D design components assembled in software CATIA V5 and import them into Process Simulate software.

For 3D design workstations forming assembly line was used PlantSimulation software. The workstation and the conveyor were built from cylindrical bars and cubes from the PlantSimulation software library. An example of a virtualized workstation with conveyor is shown in Figure 3.

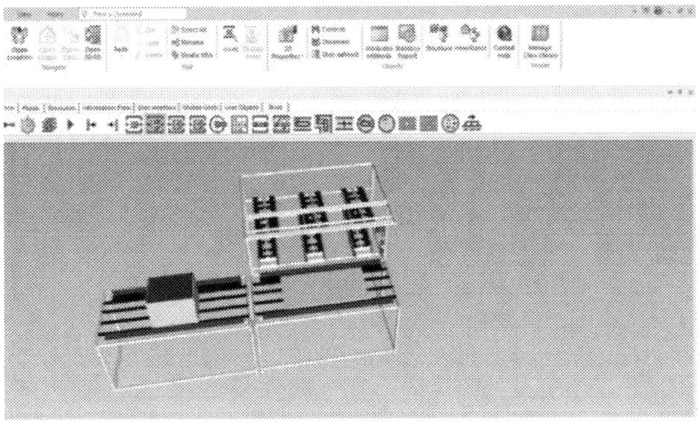

Figure 3. Workstation with conveyor.

After being made in the Plant Simulation software, they were saved in *.jt* format so that they could be imported into the Process Simulate software. Before importing the workstation with conveyor into the Process Simulate software, a virtual workspace was created and the type of components were defined. Their import was done by the command to insert from the toolbar of the program. After bringing the workstation with conveyor, they were duplicated to achieve the configuration of the existing assembly line in the laboratory.

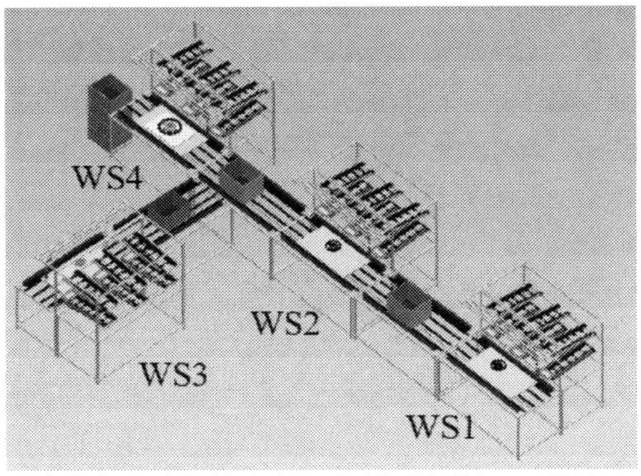

Figure 4. Assembly line with components.

The 3D design of the assembly components: lower steering body, upper steering body, storage box and buttons were made using CATIA V5 software, after which they were converted to *.jt* format to be imported into the Process Simulate software. In figure 4 is shown the assembly line resulting in Process Simulator, after the component import.

4. Identify proper operation for integration of the robotic arm

In order to identify the operation in which the integration of the robotic arm is efficient, an analysis was performed from the point of view of the productivity of the production line and of the activities carried out in the workstations. The parameters used in the analysis were: duration of products' processing in the workstation, duration of handling of components in the workstation, duration of interoperable transfer of products between workstations (Table 1), information on the location of jobs and the size of the interoperable stock.

Table 1. The values of the parameters used in the analysis.

Workstations	Duration of the product in PL (VA activities)	Duration handling components PL (NVA activities)	Duration of interoperable transfer of products between PL	Total duration of the operation
WS 1	14.9	8.68	3.39	26.97
WS 2	33.69	19.23	3.39	56.31
WS 3	24.97	22.6	3.39	50.96
WS 4	14.9	19.22	3.39	37.51

It can be seen that in workstations 2 and 3 there are activities with no added value (NVA) with the longest duration, as well as activities with added value with high value. For these reasons, the durations of the operations carried out in these workstations have the longest durations. It follows that, to equalize the duration of the operation and, thus, increase the productivity of the assembly line can be considered three scenarios for the integration of the robotic arm:

integration of the robotic arm in the workstation 2;
integration of the robotic arm in the workstation 3;
integration of the robotic arm within workstations 1 and 2, by alternately servicing them.

4.1. Analysis of the integration of the robotic arm at the workstation 2

For the integration of the robotic arm in the workstation 2, a technical and functional analysis was performed to determine the positioning of the robotic arm in the workstation. Following this analysis, it was found that the robotic arm cannot enter the workstation shelf to take over the components, cannot be positioned on the same side with the operator for safety reasons and must be positioned in the work area of the workstation. due to the constraints of the rotational radius of the robotic arm. Thus, it was decided to eliminate the component shelf and integrate the robotic arm in its place to be in the work area. The necessary components were placed in storage boxes within the radius of rotation of the robotic arm. Figure 5 shows the robotic arm integrated in workstation 2 of the assembly line.

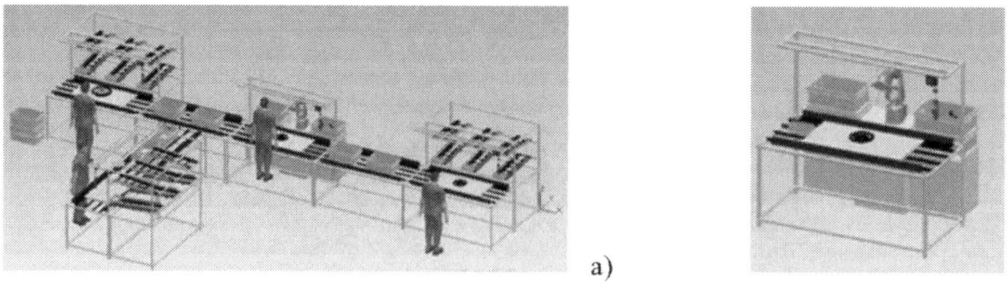

a) b)

Figure 5. Assembly line with robotic arm integrated in workstation 2.

Following the integration of the robotic arm in workstation 2, a simulation of the operation was performed using the Process Simulate software, which resulted in an operation duration of 49.76 seconds, shorter than in the initial situation, in which lasted 56.31 seconds.

4.2. Analysis of the integration of the robotic arm at the workstation 3

For the integration of the robotic arm in the workstation 3, the same technical and functional analysis was performed to determine the positioning of the robotic arm in the workstation. The same conclusions were obtained as in item 2. Thus, instead of the shelf, it was decided to place the robotic arm, and the components necessary for this workstation were positioned in storage boxes in the radius of rotation of the robotic arm. In Figure 6 is shown the robotic arm integrated in the work station 3 of the assembly line.

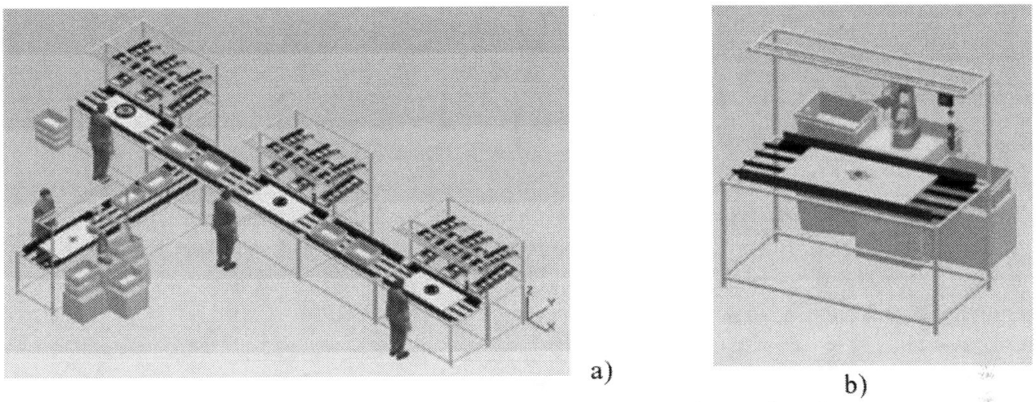

Figure 6. Assembly line with robotic arm integrated in workstation 3.

Following the integration of the robotic arm in workstation 3, a simulation of the operation was performed using the Process Simulate software, which resulted in an operation duration of 32.9 seconds, shorter than in the initial situation in which the duration was 50.96 seconds.

4.3. Analysis of the integration of the robotic arm in workstations 1 and 2

For the integration of the robotic arm in workstations 1 and 2, the same type of technical and functional analysis was perfomed to determine the positioning of the robotic arm in the workstation. Following this analysis, it was found that the robotic arm cannot serviced two posts simultaneously. Thus, it was decided to eliminate and replace workstations 1 and 2 with a rotary table for their alternative service. The necessary components were placed in storage boxes near the operator, but also in the radius of rotation of the robotic arm. In Figure 7 shows the robotic arm in the working positions 1 and 2 of the assembly line.

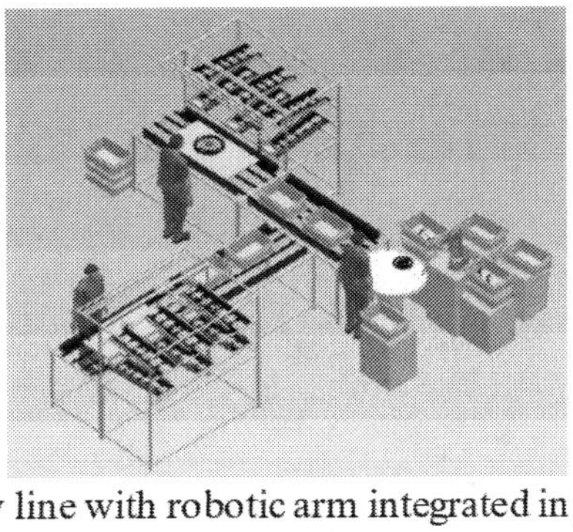

Figure 7. Assembly line with robotic arm integrated in positions 1 and 2.

After the integration of the robotic arm and the rotating mass in workstations 1 and 2, a simulation of the process carried out at the new workstation was performed, using the Process Simulate software, which resulted in an operation duration of 72.26 seconds, less than the cumulative duration of the two operations from the initial situation, of 83.28 seconds.

Following the analysis of the 3 scenarios, the following were found:

- the areas of the assembly line for the first two scenarios changes (27 m^2), while if the last scenario surface of the assembly line is less (20 m^2);
- the durations of the operations in which the robotic arm was integrated decreased, in the case of scenario 2 (integration of the robotic arm at the workstation 2) registering the smallest decrease, of 18.06 s;
- in case of scenario 3, the two operations initially performed by 2 operators, will be performed by a single operator;
- in the case of each of the 3 scenarios, it is necessary to ensure the safety of the operators so that there is no risk of accidents.

Based on the previous conclusions, it is estimated that the implementation of the robotic arm in the third scenario is more efficient because the assembly line will occupy a smaller area, will be served by only 3 operators, and the cumulative duration of the first two operations was decreased by 11.02 s.

5. Experimenting the integration of the robotic arm in the existing assembly line

In order to experience the integration of the robotic arm in the existing assembly line in the Lean Learning Factory laboratory at the University of Pitesti, the operator's security problem was solved by purchasing two optical barriers, which will be integrated in the robot and rotary table automation system. For using the robot arm, a gripper with pneumatic suction cups has been designed and carried out, Figure 8, which makes handling easy, fast and safe product components. Also, the components necessary for the construction of the rotary table were purchased, all of which will be functionally interconnected. Once completed these assemblies will experience the workings of the assembly line to validate the solution set about theoretical and numerical simulation.

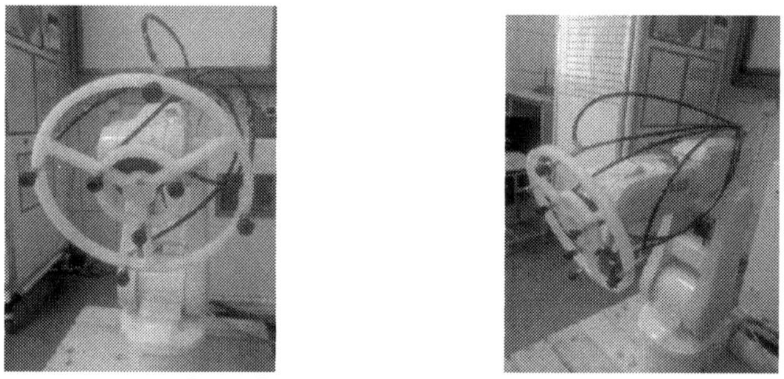

Figure 8. The gripper mounted on the robotic arm with suckers.

6. Conclusions and perspectives

This paper briefly presented the steps taken to identify the best way to integrate a robotic arm into an assembly line with mostly manual activities. Three scenarios were developed and methods of theoretical analysis, 3D design and numerical simulation were used using the ProcessSimulate software.

The chosen solution consists in the use of a rotary table, which will allow the use of the robotic arm for two workstations, which will be served by a single operator. It also turned out that this scenario will reduce the area occupied by the assembly line, and the cumulative duration of the first two operations will decrease.

The study will continue by analysing the possibility of integration of other innovative logistics systems (automatic transfer, automatic feeding, etc.), by optimizing the organization of the new automated workstation, interconnecting all automated systems within the line and by effectively implementing these solutions in the Lean laboratory Learning Factory from the University of Pitesti. Thus, it will be possible to use it for presentation for teaching purposes, but also for experimenting with the concepts specific to Industry 4.0.

Acknowledgments
This work was carried out within the *Research project on the automation of logistics activities in an assembly line by using a robotic arm*, no. CIPCS-2020-05, funded by the University of Pitesti.

ORCID iDs
I G Pascu https://orcid.org/0000-0001-8992-8232
G C Neacşu https://orcid.org/0000-0002-4131-0991
E L Niţu https://orcid.org/0000-0003-4571-3488
R Beloiu https://orcid.org/0000-0002-5815-4008
A Gavriluţă https://orcid.org/0000-0001-9121-8475

7. References

[1] Weckenbord C, Kieckhafer K, Muller C, Grunewald M and Spengler T 2020 Balancing of assembly lines with collaborative robots *Business Research* **13** 93-132 https://doi.org/10.1007/s40685-019-0101-y
[2] Kayikci Y 2018 Sustainability impact of digitization in logistics *15th Global Conference on Sustainable Manufacturing Procedia Manufacturing* **21** 782-9 https://doi.org/10.1016/j.promfg.2018.02.184
[3] Osterrieder P, Budde L and Friedli T 2020 The smart factory as a key construct of industry 4.0 A systematic literature review *International Journal of Production Economics* **221** 107476 https://doi.org/10.1016/j.ijpe.2019.08.011
[4] Kopacek P 2018 Development Trends in Cost Oriented Production Automation *IFAC Papers Online* **51** (30) 39-43 https://doi.org/10.1016/j.ifacol.2018.11.242
[5] Rojko A 2017 Industry 4.0 Concept: Background and Overview *International Journal of Interactive Mobile Technologies* **11** (5) 77-90 https://doi.org/10.3991/ijim.v11i5.7072
[6] Radivojevic G and Milosavljevic L 2019 The concept of logistics 4.0 *4th Logistics International Conference* pp 283-292
[7] Daniyan I, Mpofu K, Ramatsetse B and Adeodu A 2020 Design and simulation of robotic arm for manufacturing operation in the railcar industry *30th International Conference on Flexible Automation and Intelligent Manufacturing Procedia Manufacturing* **51** 67-72 https://doi.org/10.1016/j.promfg.2020.10.011
[8] Ali M, Simic M and Imad F 2017 Calibration method for articulated industrial robots *International /conference on Knowledge Based and Inteligent Information and Engineering Systems KES2017 Procedia Computer Science* **112** 1601-10 https://doi.org/10.1016/j.procs.2017.08.246
[9] Gavriluţă CA, Niţu E, Gavriluţă AC, Rizea A, Anghel D and Belu N 2018 Designing a layout for an assembly line used in the automotive industry *Proceedings of the 6th RMEE Management Conference* 646-654 ISSN 2247-8639 Todesco Publishing House Cluj-Napoca

IOP Conf. Series: Materials Science and Engineering 1235 (2022) 012046 doi:10.1088/1757-899X/1235/1/012046

Frequency optimization of nozzle guide vane in the low-pressure turbine system

R Robak[1,2] M Szczepanik[3] and S Rulik[1]

[1] Silesian University of Technology, Faculty of Energy and Environmental Engineering, Konarskiego 18, 44-100 Gliwice, Poland
[2] AvioPolska Sp. z o.o., Michała Grażyńskiego 141, 43-346 Bielsko-Biała, Poland
[3] Silesian University of Technology, Faculty of Mechanical Engineering, Konarskiego 18A, 44-100 Gliwice, Polska

rafal.robak@avioaero.it

Abstract. The aim of the article is to present the approach for modal optimisation of nozzle guide vane in the low-pressure turbine system of aircraft engine. Described analysis assumption related to the finite element model definition, meta-model estimation and optimisation process itself. Goal function has been determined basing on Campbell diagram. Typical approach to solve modal analysis for mentioned structure is full three-dimensional model that is time consuming. For optimisation purposes employed two dimensional one in order to make the process more robust (element types with support of Fourier series). This approach is reducing degree of freedom quantity and is still capable to consider non-axisymmetric structure behaviour. The article provides relationship between geometrical features of nozzle and frequency change.

1. Introduction

Optimization process is crucial for successful design and provide balance between technical and program requirements like durability, weight, performance, maintainability, or component cost. Modal analysis [1], [2] is one of the key elements in design phase of the nozzle guide vane in low pressure turbine. The aim of the analysis is to verify natural frequency and modal shapes of the system against operating range and excitation (forcing function). Such verification is fundamental to prevent high cycle fatigue failure mode or excessive wear at contact interfaces. Finite element method analyses [3] are typically used in the new product introduction phase before product validation through certification phase. In following article presented approach of modal analysis for two-dimensional finite element model to reduce analysis lead time. Optimization process uses meta-model (mathematical transfer function) determined on DOE (design of experiment) method to pass from geometrical parameters to natural frequency value and cross section area.

2. Parametric model and FEM preparation

Model includes three geometrical parameters related to the shell thicknesses on turbine casing as presented in the figure 1.

Optimization process needs to be robust therefore approach uses simplified model of nozzle guide vane with adjacent structure (turbine casing) that contributes in the system mode response. Nozzle guide vane has been modelled by means of the plane stress with thickness elements and turbine casing with axisymmetric type of elements. Thicknesses have been determined basing on outer and inner band radiuses and perimeters for all nozzle segments while airfoils in addition were scaled basing on empty to full volume ratio.

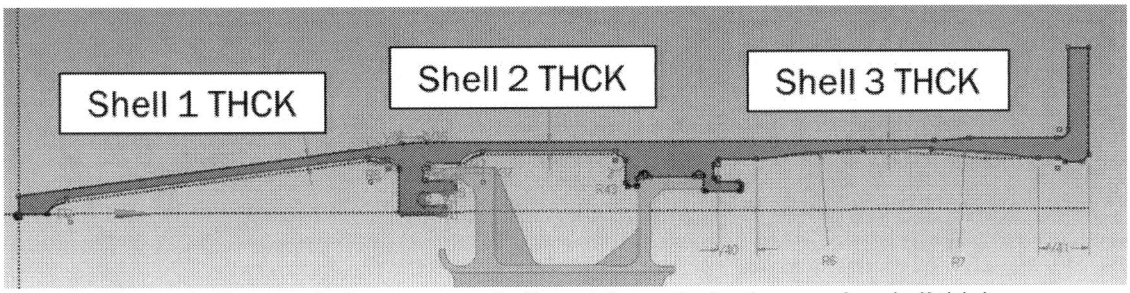

Figure 1. Geometrical parameters dedicated for optimization - casing shell thicknesses

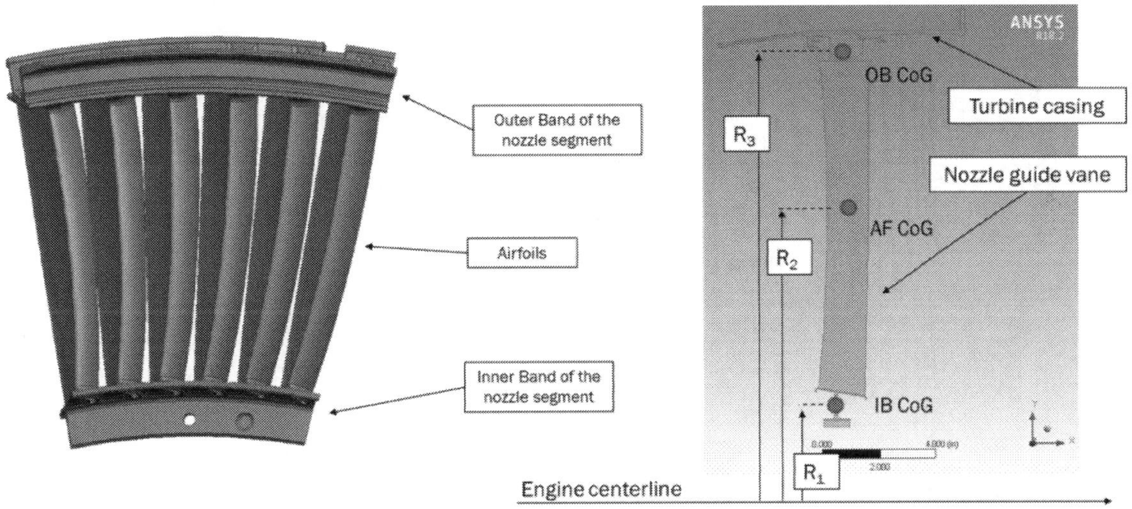

Figure 2. Approach for two-dimensional analysis

Temperature of the casing and nozzle guide vane are set to ambient temperature. Cut faces (forward and aft flange of the turbine casing) have been constrained in all available degree of freedom for the selected type of quad element; two translations set to zero (radial and axial).

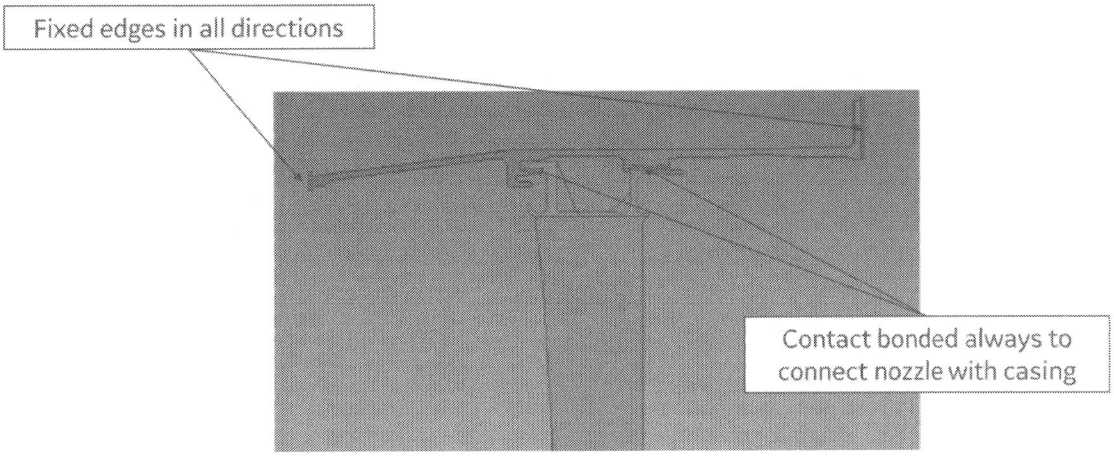

Figure 3. FEM model constrains and contact between nozzle and casing

3. Goal function definition

The aim of the optimization process is to avoid crossing of natural frequency of the analysed system in the operating range with the forcing function. Desired scenario is to make a mode placement above frequency of maximum rotational speed with one per revolution excitation. Figure 4 shows notional Campbell diagram that illustrates the natural frequency of the system with negative slope caused by the increased temperature that reduces the stiffness and so the frequency. Engine order is the function of the rotational speed (one per revolution excitation). Crossing of two lines indicates resonance condition or condition with risk of high wear on mating surfaces (contact interfaces between nozzle hook and turbine casing rails).

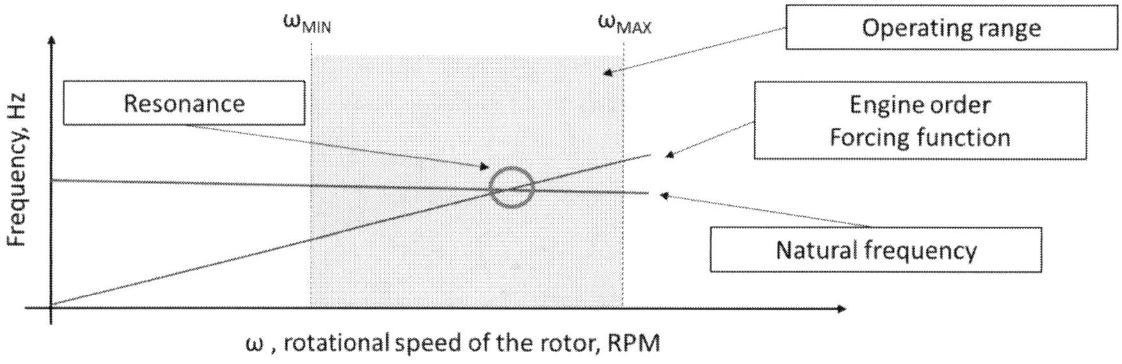

Figure 4. Notional Campbell diagram

Figure 5. First system mode of the nozzle-casing system

The goal function in the multi-objective optimization algorithm was set to maximize the first system mode frequency (see figure 5) and minimize the model area (derivative of the system weight). For what concern single objective algorithms to maximize the first system mode frequency.

Second and following system modes of the structure are not in the scope of interest since will be outside operating range and will not interact with excitation from shaft imbalance.

Table 1. Geometrical parameters with ranges

Parameter	Lower bound, in	Upper bound, in
Shell A	0.050	0.150
Shell B	0.050	0.150
Shell C	0.050	0.125

4. Optimization process and results

Process is basing on the commercially available solution implemented in the Ansys environment called Design Explorer [8]. For defined geometrical parameters selected lower and upper bound in order to ensure feasible turbine casing geometry. Ranges have been reported in the table 1.

To estimate transfer function between geometrical parameters and frequency or area following design points have been evaluated as presented in the table 2.

Table 2. Design of Experiment - Design matrix

Design point	Shell A, in	Shell B, in	Shell C, in
1	0.100	0.100	0.0875
2	0.050	0.100	0.0875
3	0.150	0.100	0.0875
4	0.100	0.050	0.0875
5	0.100	0.150	0.0875
6	0.100	0.100	0.050
7	0.100	0.100	0.125
8	0.059348	0.059348	0.057011
9	0.14065	0.059348	0.057011
10	0.059348	0.14065	0.057011
11	0.14065	0.14065	0.057011
12	0.059348	0.059348	0.11799
13	0.14065	0.059348	0.11799
14	0.059348	0.14065	0.11799
15	0.14065	0.14065	0.11799

Graphical representation of the design matrix reported in the figure 5 by blue dots in two planes AB and AC. Orange square represents optimization domain.

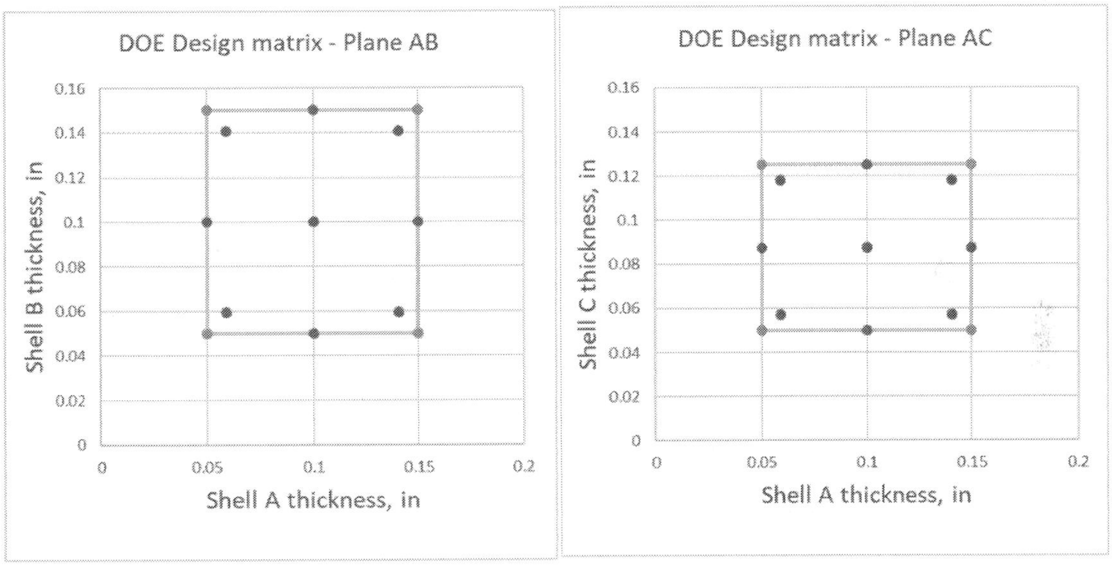

Figure 6. Design of experiment Design matrix – graphical representation

First result in the process is correlation matrix that describes influence strength between design variables (turbine shell thicknesses) and resulting parameters (first system mode of the nozzle-casing and model surface area).

Results reveal strong positive correlation between shell thickness B and natural frequency of the first system mode. In addition, shell thickness A has strong positive correlation with model surface area. Shell thickness C is not significantly contributing in the selected responses.

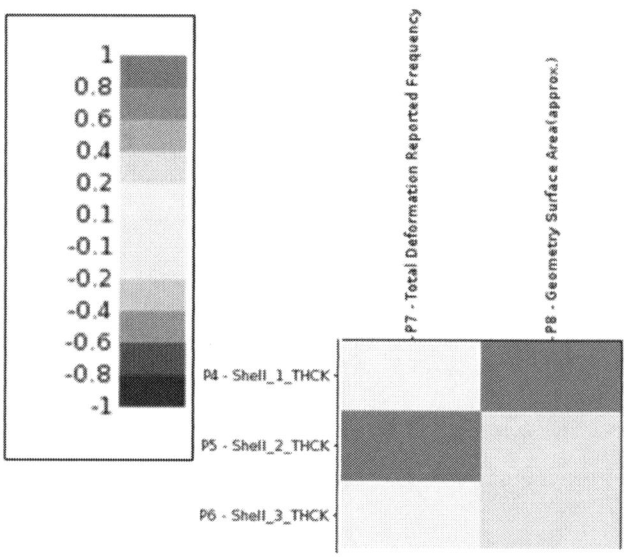

Figure 7. Correlation matrix between geometrical parameters and responses

5. Optimization summary

Results have been summarized in the table 3 that describes different optimization algorithms, configuration and shows the best candidate to match defined goal. First algorithm called screening is randomly selecting from domain (dimensional thicknesses on shells A, B and C) 1000 geometrical configurations and is down-selecting candidate with the highest frequency and the lowest model area.

This method is a brute force approach since is investigating entire domain without specific rule. Multi-Objective Genetic Algorithm is more sophisticated approach and based on iterative process that starts from initial population of chromosomes. Population is subjected to following routines like selection and genetic operators (crossover and mutation). Solution is converging in iterations reaching termination criteria (no further results improvement). The last algorithm is the nonlinear programming by quadratic lagrangian that is gradient method to determine the optimal solution. In this particular case only one goal has been defined and as presented in the table obtained the highest frequency shift with respect to baseline configuration (254 Hz).

Table 3. Results of the optimization process

Algorithm	Screening	Multi-Objective Genetic Algorithm [4-6]	Nonlinear Programming by Quadratic Lagrangian [7]
Goal	Maximize frequency Minimize model surface area	Maximize frequency Minimize model surface area	Maximize frequency
Configuration	1000 samples	100 samples, maximum 20 iterations. 354 evaluations	59 evaluations
Candidates	Shell A= 0.05635 Shell B=0.1485 Shell C=0.0547	Shell A = 0.0533 Shell B = 0.1474 Shell C = 0.0547	Shell A = 0.150 Shell B = 0.150 Shell C = 0.110
Responses	265 Hz 18.794 in^2	264.4 Hz 18.784 in^2	277.2 Hz 19.143 in^2

Multi-objective genetic algorithm shows similar results in terms of response (natural frequency of the first system mode) to the screening method however is using 65% less evaluation cases.

Implemented gradient method (nonlinear programming by quadratic lagrangian) is finding a maximum response however neglecting totally surface area that is 2% higher than genetic algorithm case.

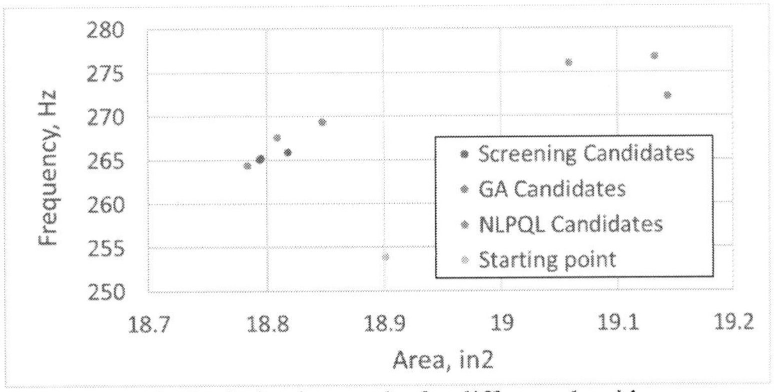

Figure 8. Optimization results for different algorithms

6. Summary and conclusions

For defined shell thickness domain, the maximum frequency shift obtained is equal to 9.0% by setting all geometrical parameters in upper limit (maximum stiffness condition). This solution from weight standpoint is not desired since is greater comparing to baseline. Remaining candidates proposed by genetic algorithm and screening method shows 4.3% - 6.3% frequency shift and at the same time weight reduction with respect to the baseline configuration. Results have been reported in figure 8 considering three candidates for each algorithm.

For what concern methods where two goals were defined outcome depends on the goal importance. Final selection shall be made by design engineer from pareto front.

Analysis clearly indicates casing shell thickness in between nozzle hooks that influence the first system mode response among three defined design variables. The most effective method in considered process is the Multi-Objective Genetic Algorithm that run 354 evaluation cases and show similar result as screening method that is more time consuming. Optimization is working on the meta-model (mathematical model, transfer function that for given input parameters geometrical shell thicknesses is providing output response so frequency and model area) therefore is very robust and provides guidance for design engineer (correlation matrix reported in figure 7). Transfer function has been obtained from the design of experiment approach. In the presented study accounted only geometrical parameters associated with the low-pressure turbine casing. In order to reach higher impact on the system mode next step is to define parameters related to the nozzle itself, in particular outer band hooks leaning and thicknesses, forward to aft hook axial distance and gusset thicknesses.

ORCID iDs
R Robak https://orcid.org/0000-0003-3849-2533
M Szczepanik https://orcid.org/0000-0001-7983-9463
S Rulik https://orcid.org/0000-0001-9274-3722

References

[1] Geradin M and Rixen D J 2015 Mechanical Vibrations (Wiley)
[2] Case J and Chilver A H 1986 Strength of materials and structures (Baltimore: Edward Arnold)
[3] Zienkiewicz O C 2013 The Finite Element Method: Its Basis and Fundamentals (Elsevier) https://doi.org/10.1016/C2009-0-24909-9
[4] Exler O, Schittkowski K and Lehmann T 2012 A comparative study of numerical algorithms for nonlinear and nonconvex mixed-integer optimization *Mathematical Programming Computation* **1** 383-412 http://klaus-schittkowski.de/minlp_comp_study.pdf

IOP Conf. Series: Materials Science and Engineering 1235 (2022) 012046 doi:10.1088/1757-899X/1235/1/012046

[5] Exler O, Lehmann T and Schittkowski K 2012 MISQP: A Fortran Subroutine of a Trust Region SQP Algorithm for Mixed-Integer Nonlinear Programming - User's Guide Report Department of Computer Science, University of Bayreuth https://www.schittkowski.de/downloads/numerical/MISQP.pdf
[6] Exler O amd Schittkowski K 2007 A trust region SQP algorithm for mixed-integer nonlinear programming *Optimization Letters* **1** 269-80 https://doi.org/10.1007/s11590-006-0026-1
[7] Gill P E, Murray W, Saunders M A and Wright M H 1984 Sequential Quadratic Programming Methods for Nonlinear Programming. In: Haug E.J. (eds) *Computer Aided Analysis and Optimization of Mechanical System Dynamics. NATO ASI Series (Series F: Computer and Systems Sciences)* **9** (Berlin: Springer) https://doi.org/10.1007/978-3-642-52465-3_23
[8] https://www.ansys.com/

Eigenfrequency simulation for a roller screw used at feed system for ultrasonically aided micro-electrical discharge machining

Bogdan-Ionut Cristea, Liviu Daniel Ghiculescu and Cristian Deopale

Department of Manufacturing Engineering, "Politehnica" University of Bucharest, Bucharest, Splaiul Independentei 313, Romania

cristea.ibogdan@gmail.com

Abstract. The current paper focuses on the problem of resonance between the own frequency (eigenfrequency) of the roller screw and the loads attached to it, and the frequency used by the command of the feed system to actuate the stepping motor in the process of ultrasonically aided micro-electrical discharge machining. If the roller screw assembly enters resonance, meaning equality of own frequency and the actuating frequency, it starts to vibrate with uncontrollable amplitude, introducing instability of the machining process. The frequency at which the command system operates is limited to a minimum of 50 Hz (for technological machining) and a maximum of 30 000 Hz (for rapid feed), limited by the maximum torque of the motor. The main frequencies in the operating domain are simulated in Comsol, to see if the roller screw resonates with the command system. The vibrations modes were also simulated, focusing on the torsional vibrations, induced by the command of the feed system, the case of this study. The results pointed out that the roller screw does not enter resonance.

1. Introduction

Natural frequencies or eigenfrequencies are certain discrete frequencies at which a part or system is prone to vibrate. If a system is actuated by an exterior periodical force, whose frequency is equal or close to one of the natural frequencies of the system or parts of it, the produced vibration has relatively big amplitudes, even for relatively small amplitudes of the perturbing force. This represents the resonance state of the system [1].

This could be the case of the feed system for ultrasonically aided micro-electrical discharge machining – microEDM+US, which requires good control of the process in terms of positioning precision and the time required to achieve the commanded position. Considering that the microEDM+US process takes place within a very narrow space, the frontal gap, less than 5 μm, between the surface of the tool, which vibrates with ultrasonic frequency with an amplitude of around 1-2 μm, and machined surface, the problem of the resonance of the feed system becomes critical [2].

Therefore, the paper studies if the perturbing force induced by the stepping motor that actuates the feed system could produce resonance by its command frequency overlapped on the own frequency of some parts of the system, especially the roller screw, the most important part of the system, concerning the effective linear increment. The study approaches the natural frequency of the roller screw, using a dedicated module for eigenfrequency calculation, considering chiefly the torsional vibration that could be induced by the stepping motor. Another study [3] addresses the modeling and vibration mode analysis of a ball screw drive using a dynamic model to examine the resonant modes of the system that could be excited, degrading the system's accuracy. In lead screw drives, numerous elements contribute to the flexibility of the system, and this flexibility contributes to vibratory behaviour in any mechanical system [4]. Torsion vibrations were studied by [5] using finite element method (FEM). It was found that the table position, relative to the lead-screw, changes all the natural frequencies of the lead-screw, similar to the case approached in the present paper.

1.1. Roller screw design

This paper tests if the roller screw assembly enters resonance with the frequency used by the command system of a motor within a feed system designed for microEDM+US, presented in figure 1. Due to the complexity of the nut, for the economy of high calculation resources, the simulation only contains the roller screw and the three rollers – named roller screw assembly.

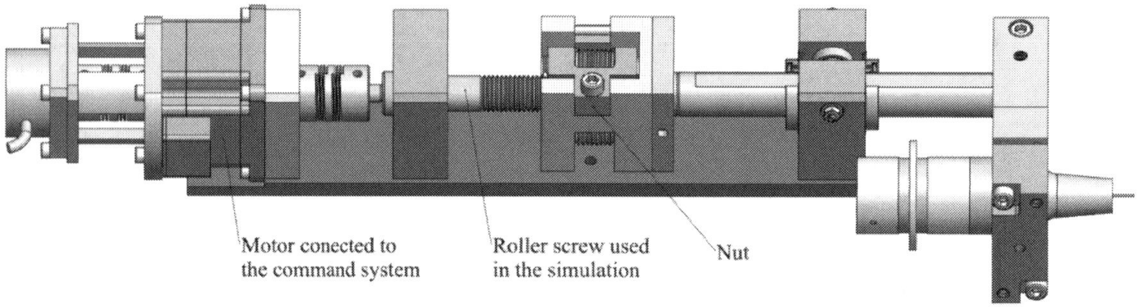

Figure 1. Feed system for ultrasonic aided micro electrical discharge machining.

1.2. Motor Frequency

In order to determine the frequencies at which the roller screw is subjected to resonance, the motor used in the assembly needs to be characterized. This is a high accuracy stepper motor [6], figure 2 shows the speed and torque characteristics. To ensure a maximum acceleration of the motor, the torque must be maximum. Maximum acceleration is used to minimize the response time of the feed system in case of critical situations.

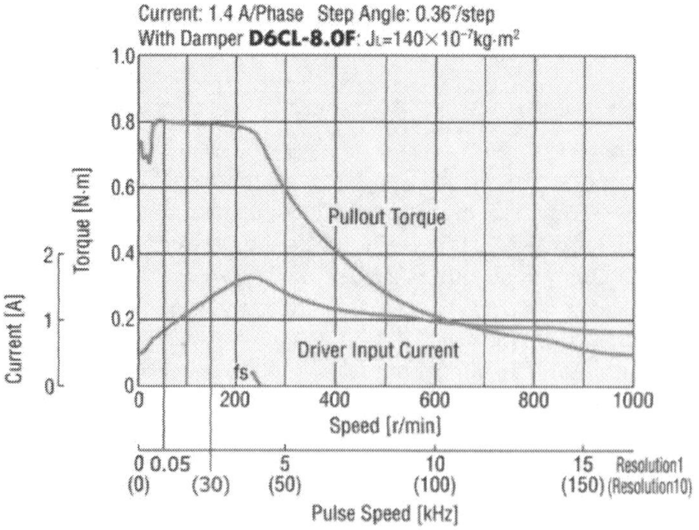

Figure 2. Speed – torque characteristics of the stepper motor CRK564PMBP [6].

The frequency needed for the motor to operate at maximum torque is between 0.05 – 30 kHz, as seen in figure 2, highlighted in blue. The higher the frequency, the faster the rotor of the motor operates.

The maximum speed is the one that the system can never exceed in any given situation. At microEDM+US, it is used to avoid short-circuit and idle machining time, when the electrode is too far from the part to initiate electrical discharges used for machining.

The minimum speed is used only to determine the origin of the axis and the origin of the part reference as well as to provide the displacement for technological feed. Both speeds are calculated using formulas 1 and 2:

$$v_{max} = i * f_{max} \text{ [mm/s]} \tag{1}$$

$$v_{min} = i * f_{min} \text{ [mm/s]} \tag{2}$$

where: i – linear increment of the feed system [mm]; f_{max} – maximum frequency of the pulses [Hz]; f_{min} – minimum frequency of the pulses [Hz];

At every pulse, the feed system executes a movement of 0.2 µm, so $i = 2 * 10^{-4}$ mm. The maximum frequency at which the torque is maximum is 30 000 Hz, and the minimum is 50 Hz, so f_{max}= 30 000 Hz and f_{min}= 50 Hz.

Replacing the values in formulas 1 and 2, the maximum speed, v_{max} = 360 mm/min, and the minimum speed, v_{min} = 0.6 mm/min, of the motor are obtained.

1.3. S speed profile

The motor, due to its programmed software, accelerates and decelerates using an "S" speed profile, as shown in figure 3, to avoid shocks caused by sudden acceleration and deceleration.

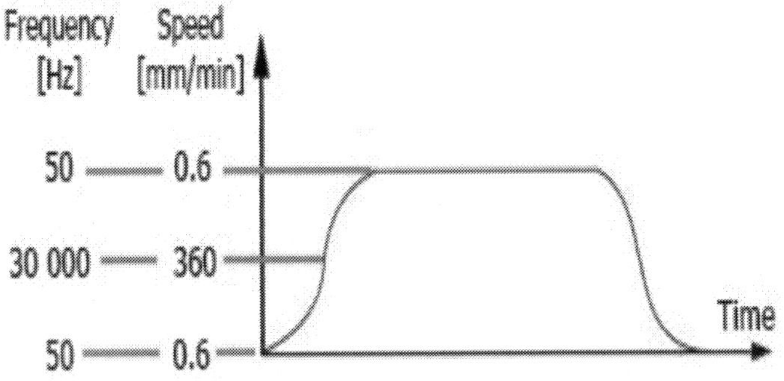

Figure 3. Speed profile (S), used by the motor.

While machining, the command system mainly uses the frequency of 50 Hz. In case of a short-circuit or out of range for sparking, the frequency of 30 000 Hz is used to retract the feed. The same frequency is used for rapid advance movement towards the workpiece. Therefore, the 30 000 Hz and 50 Hz frequencies are the most used, and they will be simulated to test for resonance.

2. Simulation

In order to determine the resonance between the roller screw assembly and the frequencies of the command system, the roller screw (figure 4) is exported from Solidworks 2016 and imported in Comsol Multiphysics 5.5, solid mechanics module, with the eigenfrequency study.

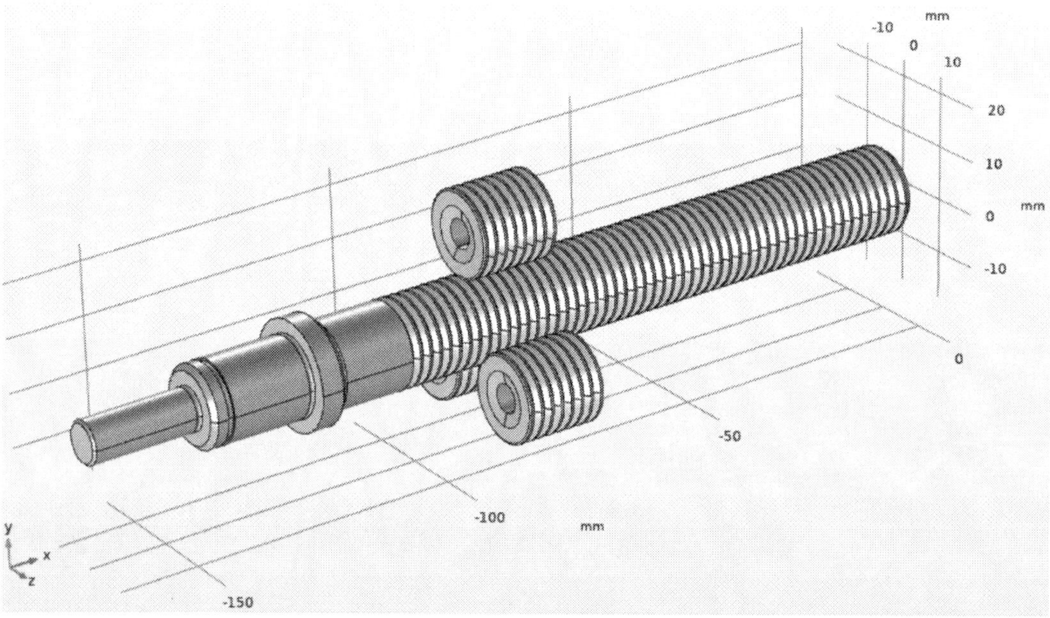

Figure 4. Roller screw assembly imported in Comsol.

Afterward, the material of the roller screw is added in Comsol (figure 5), 50CrMo4 – recommended by SKF [7], with the characteristics presented in table 1.

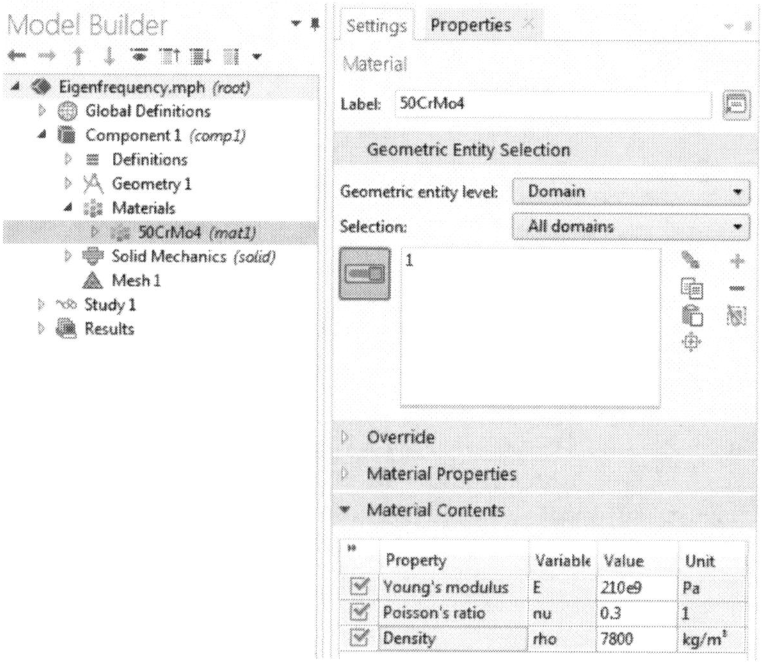

Figure 5. Adding 50CrMo4 as the material of the roller screw.

Table 1. Material characteristics for 50CrMo4

Name	Density [Kg/m^3]	Young's modulus [Pa]	Poisson' ratio [-]
50CrMo4	7 800	210 * 10^9	0.3

The meshing is set to fine, with a maximum element size of 13.2 mm, and a minimum size of 1.65 mm. Meshing consists of splitting the geometry of the roller screw in a finite number of elements.

Complete mesh	
Mesh vertices:	20186
Element type:	All elements
Tetrahedra:	84553
Triangles:	24554
Edge elements:	9863
Vertex elements:	814
Domain element statistics	
Number of elements:	84553
Minimum element quality:	0.01013
Average element quality:	0.5954
Element volume ratio:	1.174E-6
Mesh volume:	34680.0 mm³

Figure 6. Element statistics.

The reason is that the solution is firstly found in small simple elements, and then assembled on the entire geometry.

The higher the meshing quality, the more precise the simulation becomes. Figure 6 shows the number of elements and the minimum quality of each.

In figure 7, a visual representation of the mesh quality is displayed, where number 1, color-coded green is an element with ideal quality, and 0, color-coded red is an element with the worst quality. Quality is measured in terms of distortion, based on the deviation from a simple shape, which penalizes elements with large or small angles, compared to the angles in an ideal element.

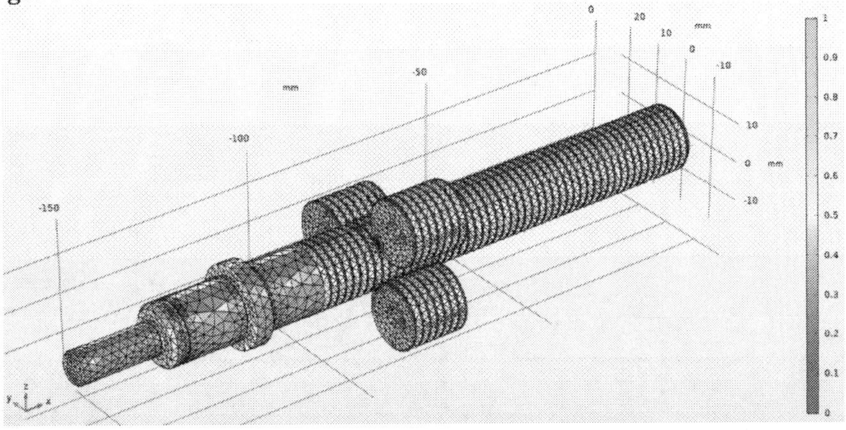

Figure 7. Mesh quality of the roller screw.

After the meshing and the material allocation, the configuration of the eigenfrequency study begins, as shown in figure 8.

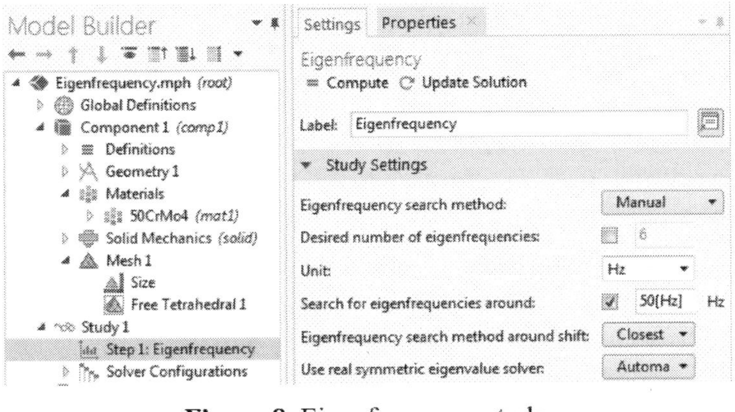

Figure 8. Eigenfrequency study.

Knowing that the minimum frequency used by the command system is 50 Hz, that values is typed in "search for eigenfrequencies around". Afterward, the solution is processed, by clicking on "compute". The result is presented in figure 9, where a frequency of 50 Hz has the closest value of 0.015 Hz, therefore the roller screw does not enter resonance with the command system's frequency.

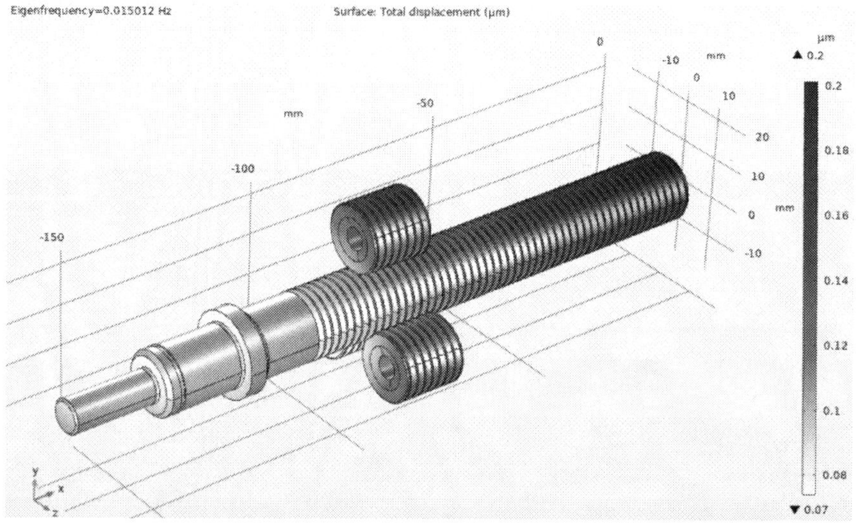

Figure 9. Eigenfrequency study of the roller screw on a frequency of 50 Hz.

However, the rollers on the roller screw are not stationary, so for the study to be relevant, the eigenfrequency needs to be tested when the rollers are in several different positions (figure 10).

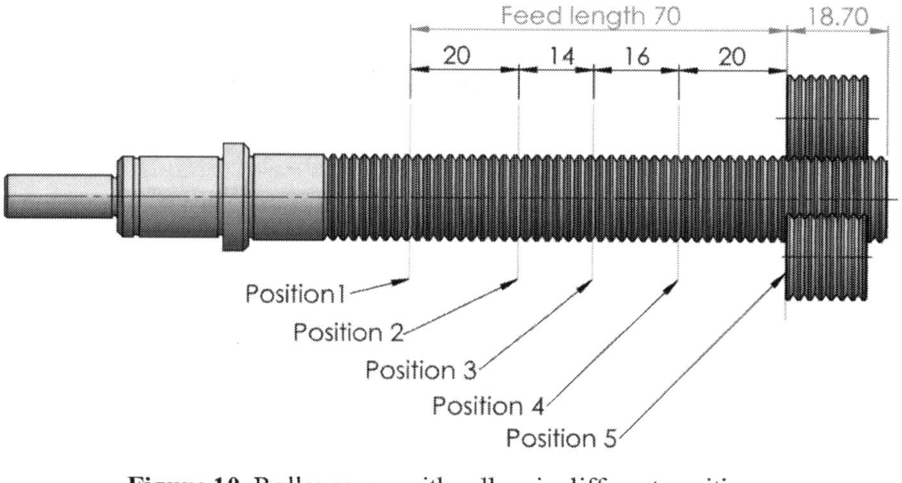

Figure 10. Roller screw with rollers in different positions.

3. Results and discussion

For every position, the frequency is tested for resonance. The set of values (eigenfrequency) have been added in table 2, for all positions, in order to check if the roller screw assembly enters resonance. The testing domain frequency is 50 Hz, respectively 30 000 Hz, the most common frequencies used during machining.

So, one can see from table 2, the values marked as closest to perturbing frequencies due to actuation of stepping motors has the minimum difference of more than 49,9 Hz in case of 50 Hz, for the technological feed, and more than 20 Hz in case of 30 000 Hz, for the rapid feed, critical situations in the micro EDM process. The position of the rollers changes the natural frequencies of the roller screw, as seen in table 2. But, taking into account that these positions are transitory rather than static, it can be stated that the designed feed system does not enter resonance during machining. And if it enters it is only for a brief period, and should not affect the feed system's accuracy.

Table 2. Roller screw's eigenfrequency for different testing frequencies and roller positions

Testing frequency [Hz]	Eigenfrequency [Hz]				
	Position I	Position II	Position III	Position IV	Position V
50	0.0031i	0.0038i	0.0012i	0.0012i	0.0037i
	0.0042i	0.0049i	0.0022i	0.0028i	0.0109i
	0.0051i	0.012i	0.0047i	0.0107i	0.0009
	0.0161i	0.0025	0.0111i	0.0018	0.0025
	0.0042	0.0072	0.0022	0.0052	0.0054
	0.015	**0.012**	**0.011**	**0.009**	**0.011**
Difference [Hz]	*49.985*	*49.988*	*49.989*	*49.991*	*49.989*
30 000	**30 405**	23 763	27 689	**31 057**	26 404
	30 685	**30 218**	29 658	31 846	26 462
	32 693	30 885	**29 682**	33 020	**30 027**
	32 851	33 068	31 609	33 163	31 127
	32 910	33 488	33 935	33 713	32 904
	33 456	33 711	34 179	33 932	34 185
Difference [Hz]	*405*	*218*	*318*	*1 057*	*27*

After dynamic modelling tests using the player command from Comsol, the frequencies in table 2, indicated only longitudinal vibration modes, but the stepping motor is expected to induce torsional vibration modes. So, from this point of view, there is no danger to enter the resonance during machining.

4. Conclusions

Eigenfrequency simulation of the feed system is of utmost importance for microEDM+US process, considering that the material removal process takes place in a very narrow working gap of less than 5 μm. Therefore, induced vibration (due to resonance) negatively influences machining accuracy. Based on simulation with finite element method, the roller screw assembly does not enter in resonance with the command system, as the testing frequency and the eigenfrequency of the assembly are not equal, or even close in any position of the rollers. Further research will be done on the physical feed system for the confirmation of computerized results.

Acknowledgments
This work was supported by a grant from the Ministry of Research, Innovation and Digitization, CNCS/CCCDI – UEFISCDI, project number PN-III-P2-2.1-PED-2019-0367, within PNCDI III.

ORCID iDs
B I Cristea https://orcid.org/0000-0002-4503-4373
L D Ghiculescu https://orcid.org/0000-0002-9923-2180
C Deopale https://orcid.org/0000-0002-7533-1897

References

[1] Rades M 2008 *Vibrații mecanice* (in Romanian) ed Printech (Bucharest) chapter 1 pp 1–6
[2] Jahan M P 2015 *Electrical discharge machining (EDM)* ed Nova Publisher (New York) chapter 5 pp 171–4
[3] Vicente D A, Hecker R L, Villegas F J and Flores G M 2012 *Int. J. Adv. Manuf. Technol.* **58** 257-65 https://doi.org/10.1007/s00170-011-3375-6
[4] Orang V A 2009 Friction-induced Vibration in Lead Screw Systems *University of Waterloo* chapter 8 pp 185
[5] Yong Z, Fangyu P and Jihong C 2007 *Int. Conf. Adv. Mecht. Auto.* 2194-9 https://doi.org/10.1109/ICMA.2007.4303892
[6] Oriental Motor General Catalogue Stepping motor and Drive Package DC Input C-121
[7] SKF Group 2014 *A selection guide* PUB MT/P1 14489 EN

Finite element analysis for testing safety-belt anchorages

A M Manea, M D Iozsa, C Stan and A Ioniță

Automotive Engineering Dept., Faculty of Transports, POLITEHNICA University of Bucharest, Splaiul Independentei nr. 313, sector 6, Bucuresti, ROMANIA, RO-060042

mannea_annamaria@yahoo.com

Abstract. The passive safety of the vehicle is ensured by designing optimized restraint systems to protect the occupants during the accident. The safety-belts have the role of restraining the occupants of the car on the seat during an impact or in the event of a rollover. Their retention prevents or minimizes the impact between the occupants and the interior components of the passenger compartment. Two simplified models of seat structures are proposed to determine the strength of the safety-belts anchorages. The test of the anchorage for safety-belts of the two seat models is performed in accordance with Regulation no. 14 UNECE. For the two seat models, a three-dimensional resistance structure was made. The 3D models of the seats were inserted in the FEA software. The analysis of the models is done by the finite element method with the Static Structural module of the FEA software. A tensile force is applied during the test on the safety-belts anchorages. The traction force shall be applied in the direction corresponding to the seating position, at an angle of 10°±5° above the horizontal, in a plane parallel to the median longitudinal plane of the vehicle. During the tests the safety-belt anchorages shall withstand the entire period of application of the traction force and the upper safety-belt anchorage shall not be displaced during the test forward of a transverse plane passing through the R point of the seat.

1. Introduction

The seat belt is an assembly of straps with a safety buckle. The role of the seat belt is to restrain the occupants of the car during an impact or in the event of a rollover. Thus, by retaining them, the impact between the occupants and the interior components of the passenger compartment is prevented or reduced as much as possible [1,2].

The permissible field of the seat belt anchorages is defined by the American specifications FMVSS 210 and the European Directives (CEE-UN) [3].

The new seat belts are designed to increase safety by equipping them with a pretensioning mechanism and a force transducer. They help to improve occupant protection by maintaining body movement in the passenger's chest area during impact. In the first part of the impact the pretensioning mechanism instantly restrains the occupant, applying a restraining force in the chest area and stops the tendency of the occupant to move forward. When the holding force reaches a maximum value, the mechanism can control the movement of the chest, with the help of a force transducer, further applying a constant pressure on the chest. These restraint systems of the occupant slow down the movement of the body and increase its stopping distance to the limit of contact with the elements of the interior components of the passenger compartment [3].

2. Geometric modelling of the seats

The work, by modelling and simulation with finite elements it is desired to check the resistance of the safety-belt anchorages points of two different models of seats.

During the tests the safety-belt anchorages shall withstand the entire period of application of the traction force and the upper safety-belt anchorage shall not be displaced during the test forward of a transverse plane passing through the R point of the seat. The R point is the seat reference point, defined by the vehicle manufacturer for each seat and established in relation to the three-dimensional reference system [3]. The resistance structures of the seats are independent. The tests are performed individually for each seat.

The test of the anchorage points for safety-belts of the two seat models is performed in accordance with Regulation no. 14 UNECE [3]. By regulation no. 14 UNECE uniform provisions are given concerning the approval of vehicles with regard to safety-belt anchorages.

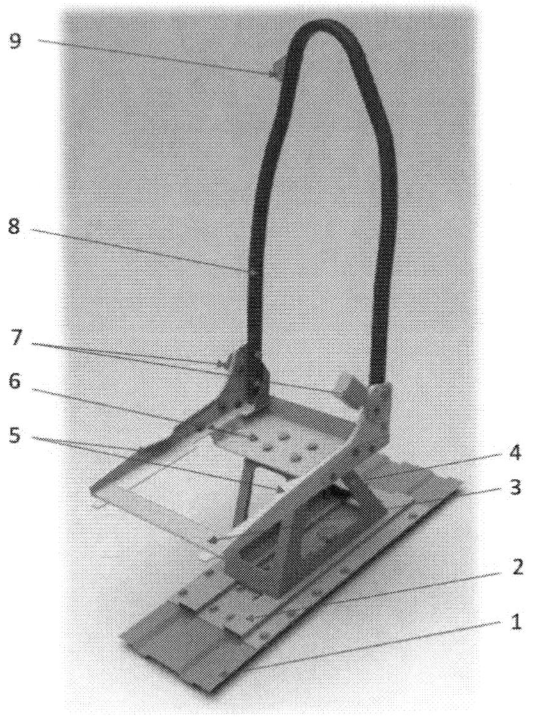

Figure 1. Model 1 car seat.
1 - Automotive floor; 2 - Seat support on the floor of the vehicle; 3 - Transverse front plate; 4 - Seat support; 5 - Side seat plates; 6 - Rear seat plate; 7 - Lower anchor points; 8 - Backrest frame; 9 - Upper anchor point plate.

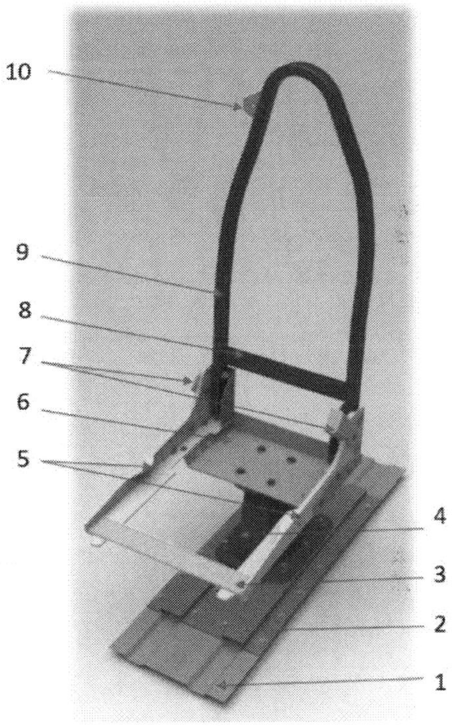

Figure 2. Model 2 car seat.
1 - Automotive floor; 2 - Seat support on the floor of the vehicle; 3 - Transverse front plate; 4 - Seat support; 5 - Side seat plates; 6 - Rear seat plate; 7 - Lower anchor points; 8 - Backrest stiffening plate; 9 - Backrest frame; 10 - Upper anchor point plate.

If the anchor points resist, the models are validated in accordance with the requirements imposed by the Regulation no.14.

Figures 1 and 2 show the three-dimensional resistance structures of the seat models. The resistance structures of the seats are made simplified due to the lack of hardware equipment. Simplified structures have also been developed to facilitate the calculation process and reduce the simulation times.

The geometries of the 3D models of the seats were used in the FEA software where, with the help of the Static Structural module, the analysis of their dynamic behaviour was performed.

2.1. Material modelling and finite element network

The modelling of the material from which the two seat structures are made is necessary to perform the strength calculations. The material was chosen from the FEA software library.

The thickness of the material of the main structural elements of the two seats is different depending on the functional characteristics of the elements.

Two types of materials are chosen for modelling the seat model 1 material: *"Structural Steel Non-linear 3"* (Density: ρ=15000 kg/cm^{-3}; Young's Modulus: E=2×10^5 MPa; Poisson's Ratio: μ=0.3; Tensile Yield Strength: σc=700 MPa) and *"Structural Steel Non-linear 2"* (Density: ρ=15000 kg/cm^{-3}; Young's Modulus: E=2×10^5 MPa; Poisson's Ratio: μ=0.3; Tensile Yield Strength: σc=400 MPa).

Material modelling for seat model 2: *"Structural Steel Non-linear"* (Density: ρ=10600 kg/cm^{-3}; Young's Modulus: E=2×10^5 MPa; Poisson's Ratio: μ=0.3; Tensile Yield Strength: σc=500 MPa).

The finite element network generated for the seat model 1 consists of 180772 nodes and 81181 elements and for the chair model 2 consists of 184722 nodes and 83845 elements.

2.2. Initial conditions

The initial conditions imposed on the two seat models for testing the safety-belt anchorages are:
- the floor of the vehicle is considered a fixed support;
- the Standard Earth Gravity is applied on both models;
- the forces corresponding to each seat model are applied to the three seat belt anchorages;
- the traction force is applied in a direction corresponding to the seating position of the passenger at an angle of 10° to the horizontal axis;
- the backrests of the seats were considered at an angle of 15° to the vertical plane.

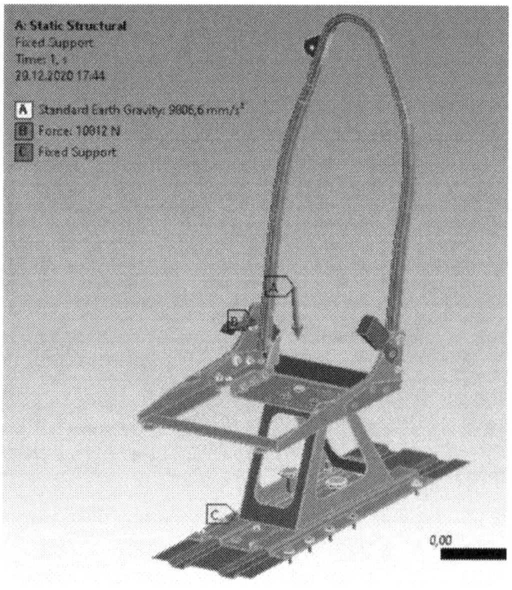

Figure 3. The initial conditions of the seat model 1.

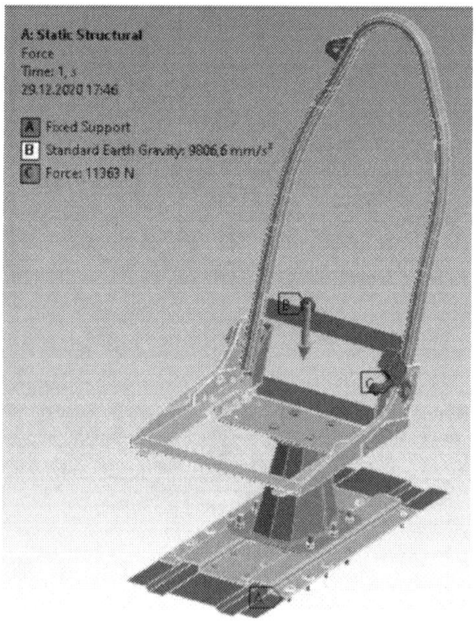

Figure 4. The initial conditions of the seat model 2.

According to [3], two test forces are applied simultaneously to the anchor points during the test. The forces applied to traction devices are for vehicles of categories M3 (passenger vehicles, which have more than eight seats, without the driver's seat, and a maximum permissible technical mass of less than 5000 kg) and N3 (goods vehicles with a maximum permissible mass exceeding 12000 kg) [5]. For these vehicles the value of each traction force will be 450 ± 20 daN. In addition, these traction forces will be supplemented with an additional force equal to the full mass of the seat, multiplied by 6.6 times.

Table 1 shows the main parameters used in the tests on the anchorage points of the seat belts located in the structure of the two types of seats.

Table 1. The main parameters of the seat models

Model Type	Mass (kg)	Weight (daN)	F_s (daN)	F_t (daN)	F_{total} (daN)
Seat 1	28	27.4	181	2*450 ±20	1081±40
Seat 2	36.2	35.4	560	2*450 ±20	1134±40

Where: $F_s = 6.6xG$ [daN] - additional force; F_t [daN] - traction force; $F_{total} = F_s + F_t$ [daN] - total force.

Anchor points must withstand the specified load for at least 0.2 s.

3. Results and discussion

The results of the tests at the anchorage points of safety-belt will be evaluated by the values corresponding to the Equivalent plastic strain of the material and by the values corresponding to the total deformation.

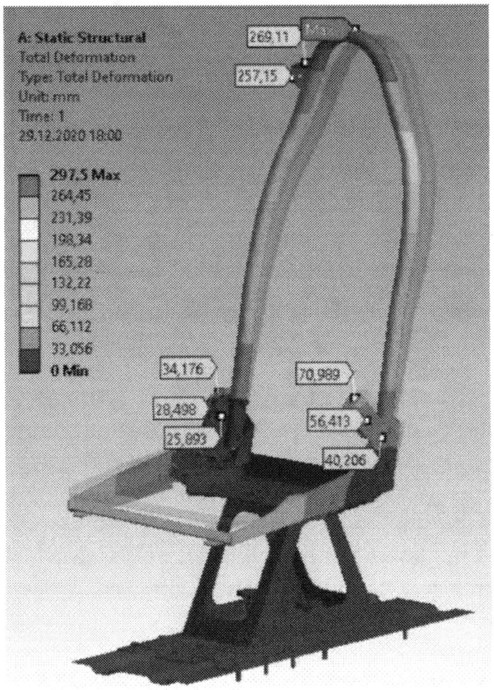

Figure 5. Total deformation of the seat model 1.

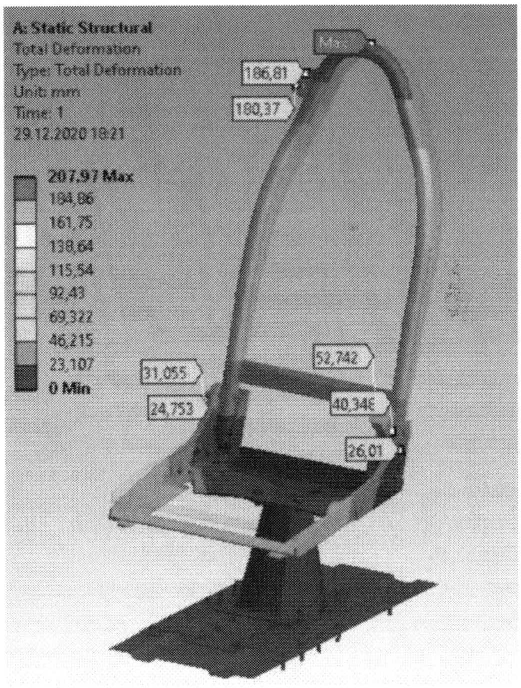

Figure 6. Total deformation of the seat model 2.

Figures 5 and 6 show the total deformation of the seat belt anchorages of the seat model 1 and 2, respectively. It can be seen from the colours grid that the value of the maximum deformation of the seat model 1 is = 297.5 mm (figure 1). This value is recorded on the upper frame of the backrest. Also,

IOP Conf. Series: Materials Science and Engineering 1235 (2022) 012048 doi:10.1088/1757-899X/1235/1/012048

from the figure can be seen, according to the samples taken from the model, the values of the deformations appeared in the anchorage points of the seat belts.

In figure 6 it can be seen from the colours grid that the value of the total deformation is = 207.97 mm. According to the samples taken from the model, the values of the deformations appeared in the anchorage points of the seat belt can be read.

Figures 7 and 8 show the total deformation resulting from the test carried out on the anchorage points of the safety-belt of model 1 and 2, respectively in the transverse plane. It can be seen that the upper anchorage of the safety-belt did not exceed during the test the transverse plane inclined by 10° to before passing through the R point of the seat.

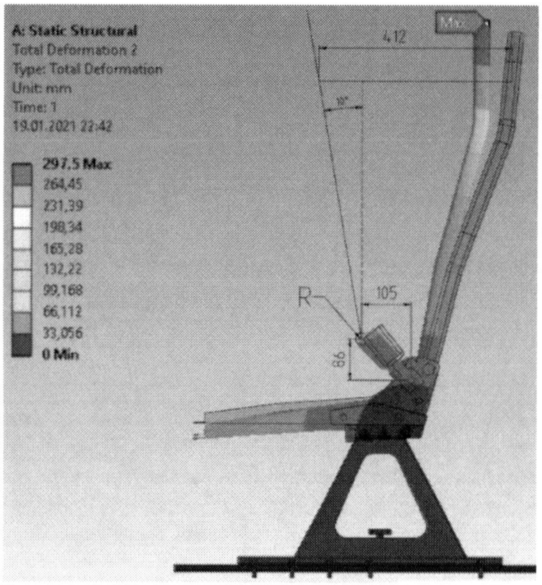

Figure 7. The deformation the effective upper safety belt anchorage points of the safety belt in relation to the transverse plane on the model 1.

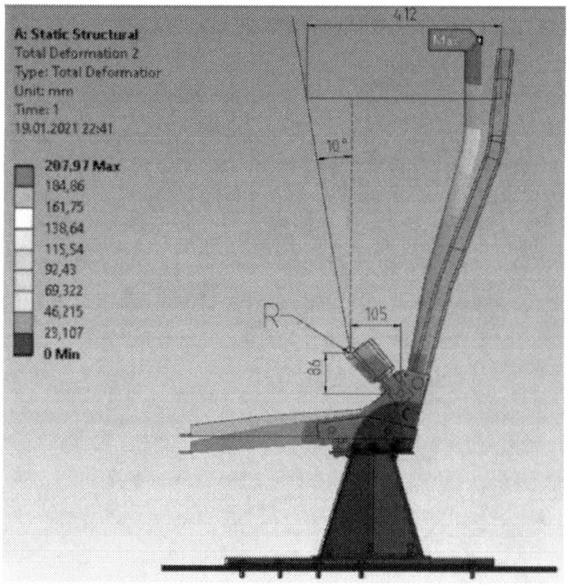

Figure 8. The deformation the effective upper safety belt anchorage points of the safety belt in relation to the transverse plane on the model 2.

Figures 9 and 10 show the Equivalent plastic strain of the seat belt anchorages of the seat model 1 and 2, respectively. From the colours grid it can be seen in the case of model 1, that the value of the Equivalent plastic strain is about ε = 0.105 mm. This value is recorded on a seat side plate assembly screw. In the case of model 2, it can be seen, in figure 10, the Equivalent plastic strain of the seat belt anchorage. From the colours grid that the value of the Equivalent plastic strain is approximately ε = 0.083 mm. This value is recorded on the side plate of the seat.

In both cases, according to the samples taken from the model, the values of the Equivalent plastic strain appeared in the anchorage points of the seat belt are zero, no specific plastic deformations appear.

4. Conclusions

The objectives of this article have been achieved. Following the tests on the anchorage points of the seat belts of the two seat models, all the anchorage points withstood the application of traction forces on them. During the tests the applied force was supported by the anchor points of the two seat models throughout.

The upper anchorage point of the safety belt must not exceed during the test a transverse plane inclined 10° forwards and passing through the R point of the seat.

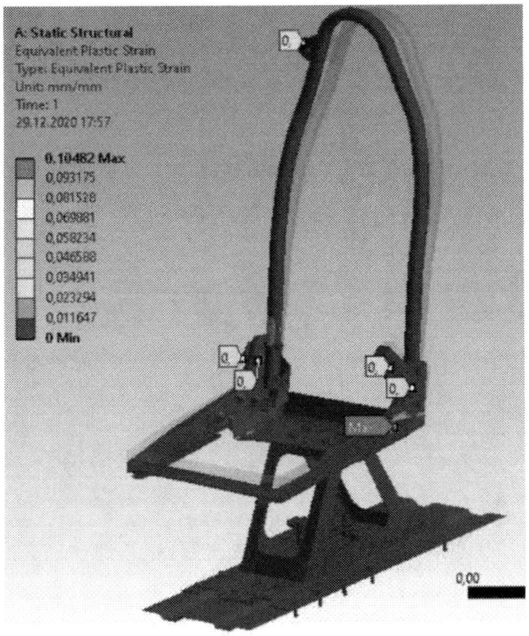

Figure 9. Equivalent plastic strain of the seat belt anchorages of the seat model 1.

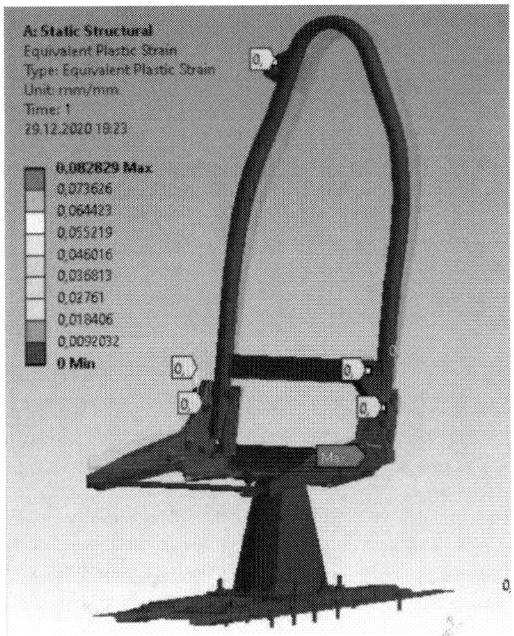

Figure 10. Equivalent plastic strain of the seat belt anchorages of the seat model 2.

Thus, following the seat belt anchorage tests, the upper seat belt anchorages of the two seat models did not exceed during the test the transverse plane inclined 10° forwards and passing through the R point of the seat.

By modelling and simulation with the finite element method, the safety-belt anchorages of the two seat models have fulfilled the uniform provisions according to Regulation no.14.

The topic addressed in the article can be extended and used in complex modelling on passive passenger safety.

It is considered the validation of the results obtained and presented in the paper by subsequent experimental determinations.

ORCID IDs
A M Manea https://orcid.org/0000-0002-3976-8848
M D Iozsa https://orcid.org/0000-0002-6547-8633
C Stan https://orcid.org/0000-0002-2753-8877
A Ioniță https://orcid.org/0000-0002-8211-6407

5. References

[1] Huang M 2002 *Vehicle Crash Mechanics 2nd ed.* SAE International, CRC Press Boca Raton FL
[2] *Bosch Automotive Handbook* SAE USA (Automotive (BOSCH) Handbook - PDF Drive)
[3] Regulation No 14 of the Economic Commission for Europe of the United Nations (UNECE) - Uniform provisions concerning the approval of vehicles with regard to safety-belt anchorages, ISOFIX anchorages systems, ISOFIX top tether anchorages and i-Size seating positions
[4] Iozsa D 2009 *Studies on the real-time analysis of the behavior of bodies on impact, in order to improve the passive security offered to passengers* Project ID-140 IDEI Program, Bucharest
[5] https://www.rarom.ro/cs-uploads/categorii_vehicule.pdf

IOP Conf. Series: Materials Science and Engineering 1235 (2022) 012049 doi:10.1088/1757-899X/1235/1/012049

Numerical analysis of the stresses appeared around some lifting ears from a cylinder used in the petrochemical industry

A Iamandei, R G Ripeanu, L Stanciu, I Popa and S Vasilescu

Petroleum- Gas University of Ploiesti, Bucharest Blvd, no.39, Romania

vserban@upg-ploiesti.ro

Abstract. In the petrochemical industry there are many devices with cylindrical shapes such as: columns and horizontal or vertical vessels. All these devices, during the fitting or transporting processes must be raised using some special ears. Using the same methodology with those presented in [1,2,4,5], the paper presents the stresses that appear around some raising ears (made from curved beams) welded on the cylindrical shell. The connection between the ears and cylinder are analysed in two cases: when the ear is welded directly on the cylinder and when the ear is welded on a reinforcement pad welded on the cylinder. Maximum stresses obtained in certain elements of the cylinder (for different thicknesses of the shell) are compared with the allowable limits of the materials.

1. Introduction

During the raising or transporting process any industrial device needs some ears/buttons that can be loaded with external forces to complete the technological process [3]. The analysed device is a cylinder and the corresponding raising ears have the geometry presented in figure 1. In the figure 1a is presented the main view of the cylinder together with the ears (see positions 18 and 19), in figure 1b the geometry of the ear welded directly on the cylinder, and in figure 1c the geometry of the ear welded on a reinforcing pad.

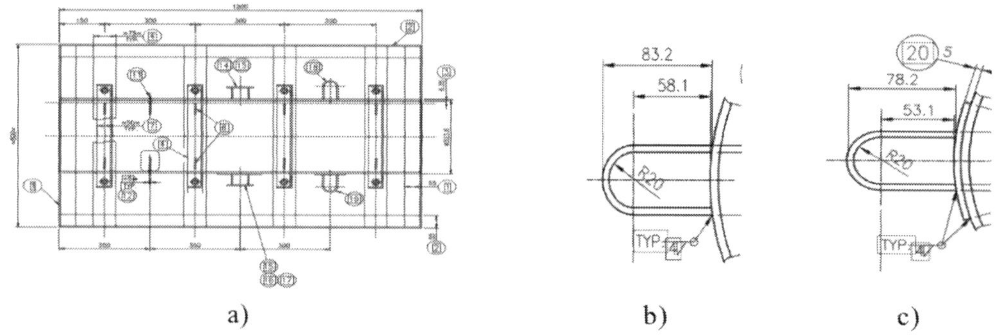

Figure 1. The overall view of the cylindrical shell

IOP Conf. Series: Materials Science and Engineering 1235 (2022) 012049 doi:10.1088/1757-899X/1235/1/012049

The cylinder and the ears are made from S 235 JR [9] and respectively PC 52 [10] materials. The main mechanical characteristics for the above materials are [8-10]:

- S235 JR: the yield limit $R_e = 235$ MPa; the mechanical strength: $R_m = 360$ MPa;
- PC 52: the yield limit $R_e = 350$ MPa; the mechanical strength $R_m = 510$ MPa

The numerical analysis has been carried out using the finite element method (COSMOSM - [7]) and during the raising process three main load cases have been studied:

1- the cylinder has been embedded in the support, in horizontal position and every ear has been loaded with a 1000 N vertical force (figure 2a);
2- the vertical 1000 N loads for every ear when the cylinder is inclined with 45° (figure 2b);
3- the vertical 1000 N for every ear when the cylinder is raised in a vertical position (figure 2c).

In all the load cases presented above both possibilities of welding between the ears and cylinder have been studied: when the ears are welded directly on the cylinder and when between the ear and the cylinder is introduced a reinforced pad (figure 1b).

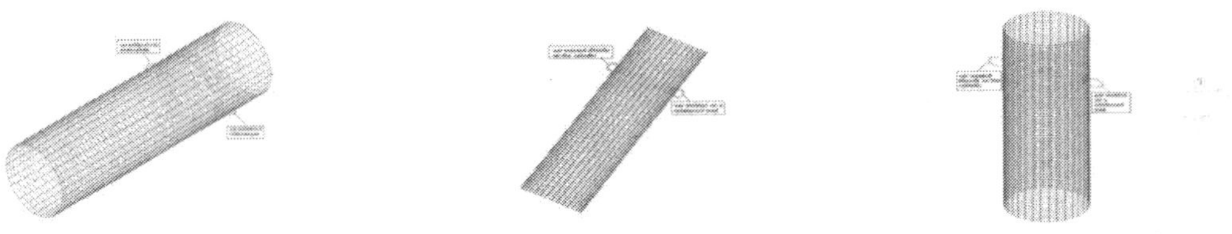

a) load case 1 b) load case 2 c) load case 3

Figure 2. The FEM simulation model

For every load case, for the cylinder has been considered three usual thicknesses for a 12" pipe [8]: 3.97 mm, 4.57 mm and 6.35 mm. The calculation models have 2208 elements and 2238 nodes, including the ears. The models are uniformly meshed, the sizes of elements being chosen in order to cover completely the reinforcing area, including the welding, that has been considered having the same material and thickness as the pad. A separate influence of the welding sizes and shapes will complete the subject for another paper.

2. Numerical results fron FEM simulation

For the first load case when the cylinder has 3.97 mm thickness and it is located in horizontal position, the equivalent stress distributions (Von Misses) are presented in figure 3 (for the ear welded on a reinforcement pad) and in figure 4 (for the ear welded directly on the cylinder).

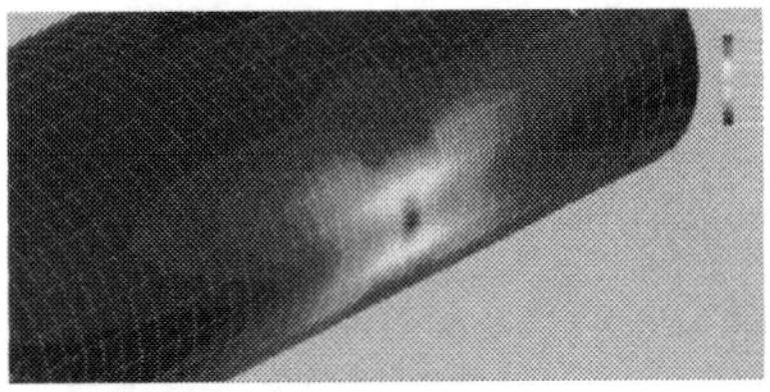

Figure 3. Ear welded on a reinforced pad-3.97 mm thickness

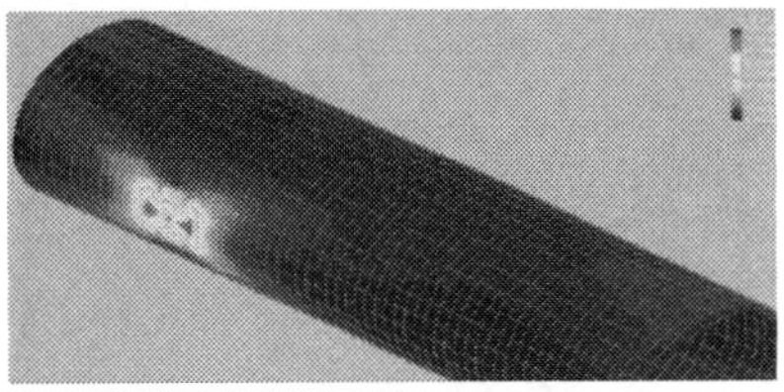

Figure 4. Ear welded directly on the cylinder-3.97 mm thickness

When the cylinder has 4.57 mm thickness, for the same horizontal position the equivalent stress distributions are presented in figures 5 (for the ear welded on a reinforcement pad) and 6 (for the ear welded directly on the cylinder). For 6.35 mm thickness the same stress distributions are presented in figures 7 and 8.

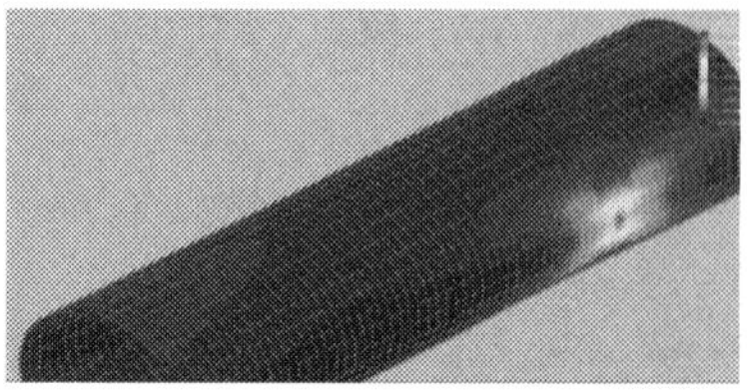

Figure 5. Ear welded on a reinforced pad-4.57 mm thickness

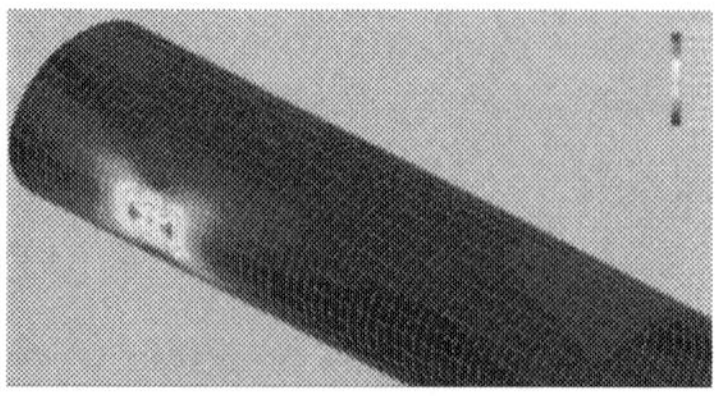

Figure 6. Ear welded directly on the cylinder-4.57 mm thickness

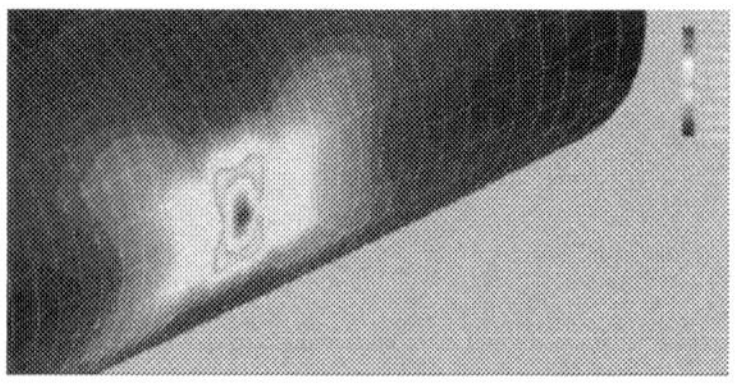

Figure 7. Ear welded on a reinforced pad-6.35 mm thickness

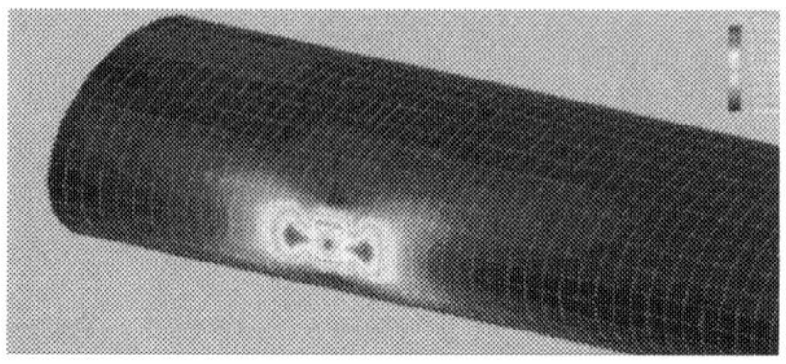

Figure 8. Ear welded directly on the cylinder-6.35 mm thickness

For the second load case when the cylinder has 3.97 mm thickness and is inclined with 45 degrees (in respect with horizontal direction), the equivalent stress distributions (Von Misses) are presented in figure 9 (for the ear welded on a reinforcement pad) and in figure 10 (for the ear welded directly on the cylinder). When the cylinder has 4.57 mm thickness, for the same inclined position the equivalent stress distributions are presented in figures 11 (for the ear welded on a reinforcement pad) and 12 (for the ear welded directly on the cylinder). For 6.35 mm thickness the same stress distributions are presented in figures 13 and 14.

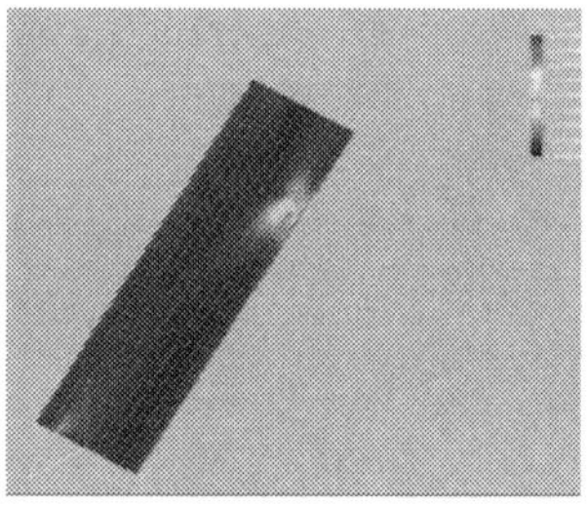

Figure 9. Ear welded on a reinforced pad -3.97 mm thickness

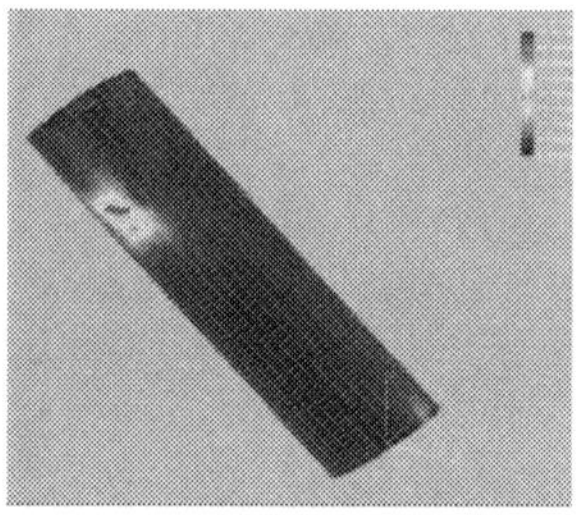

Figure 10. Ear welded directly on the cylinder-3.97 mm thickness

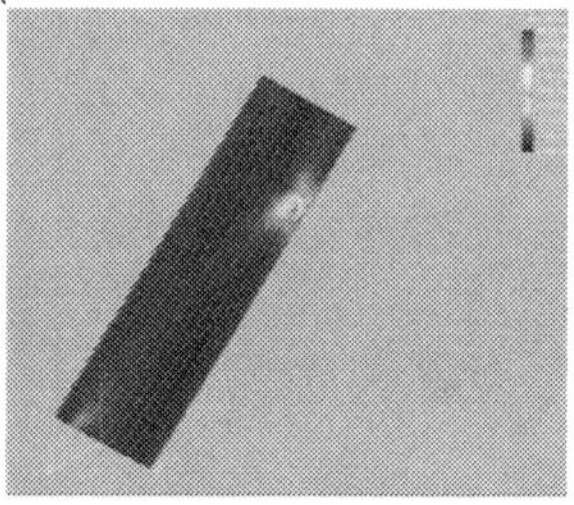

Figure 11. Ear welded on a reinforced pad-4.57 mm thickness

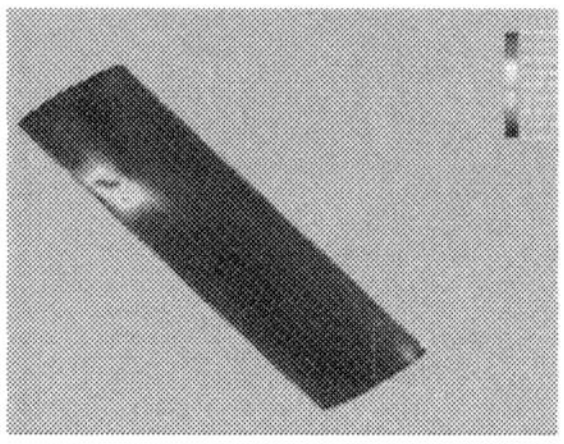

Figure 12. Ear welded directly on the cylinder-4.57 mm thickness

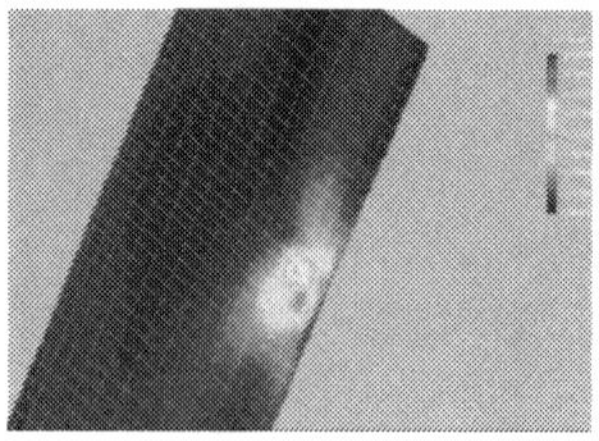

Figure 13. Ear welded on a reinforced pad-6.35 mm thickness

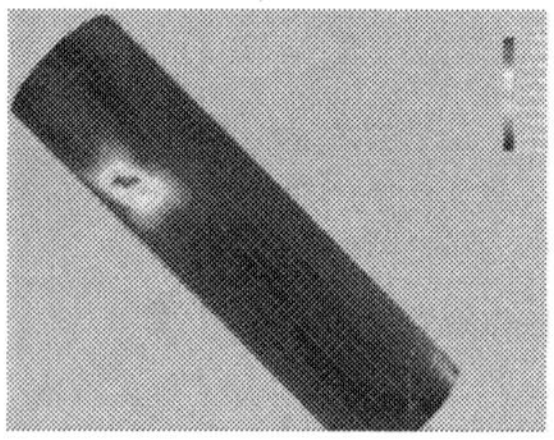

Figure 14. Ear welded directly on the cylinder-6.35 mm thickness

For the third load case when the cylinder has 3.97 mm thickness and it is in vertical position, the equivalent stress distributions (Von Misses) are shown in figure 15 (for the ear welded on a reinforcement pad) and in figure 16 (for the ear welded directly on the cylinder).

When the cylinder has 4.57 mm thickness, for the same vertical position the equivalent stress distributions are presented in figures 17 (for the ear welded on a reinforcement pad) and 18 (for the ear welded directly on the cylinder). For 6.35 mm thickness the same stress distributions are presented in figures 19 and 20.

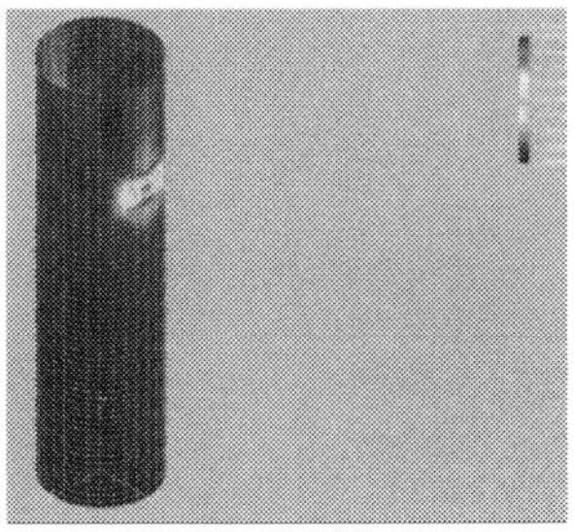

Figure 15. Ear welded on a reinforced pad-3.97 mm thickness

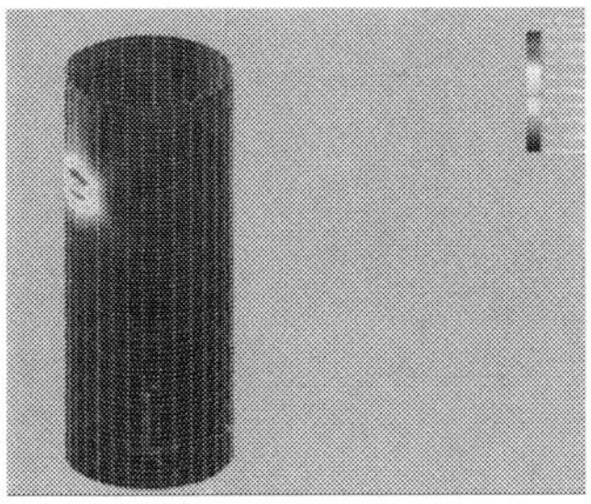

Figure 16. Ear welded directly on the cylinder-3.97 mm thickness

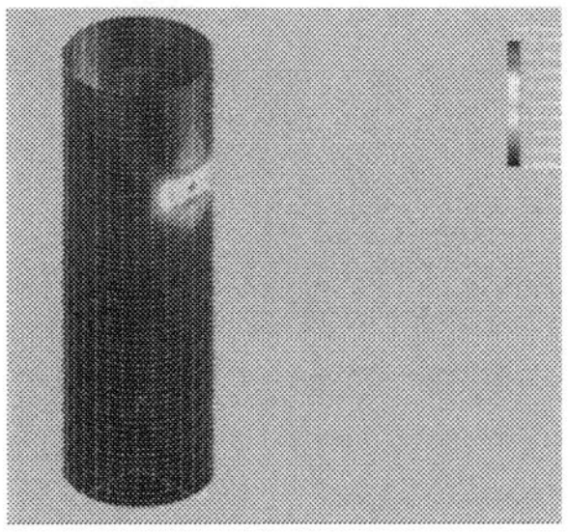

Figure 17. Ear welded on a reinforced pad-4.57 mm thickness

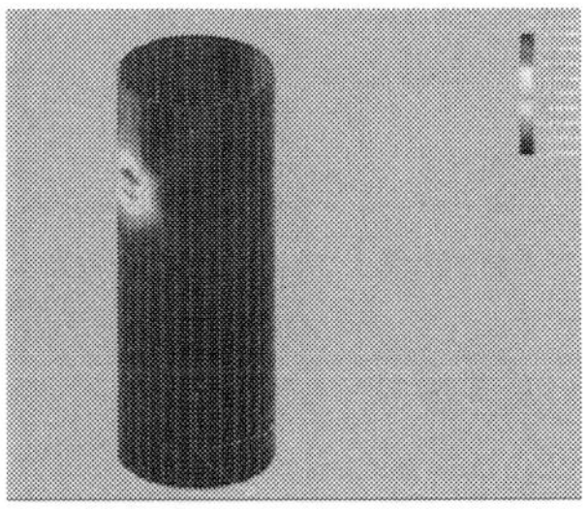

Figure 18. Ear welded directly on the cylinder-4.57 mm thickness

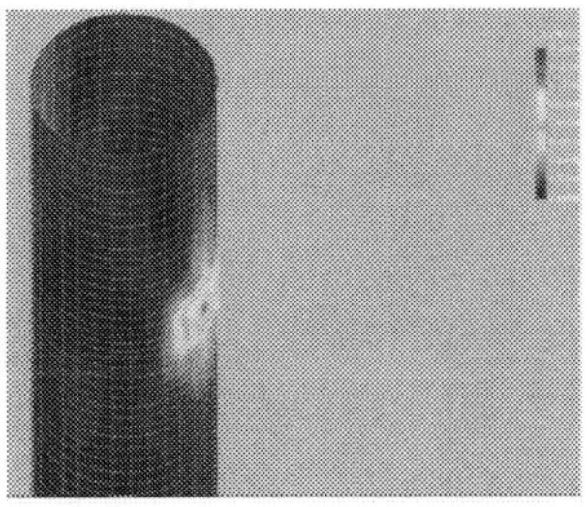

Figure 19. Ear welded on a reinforced pad-6.35 mm thickness

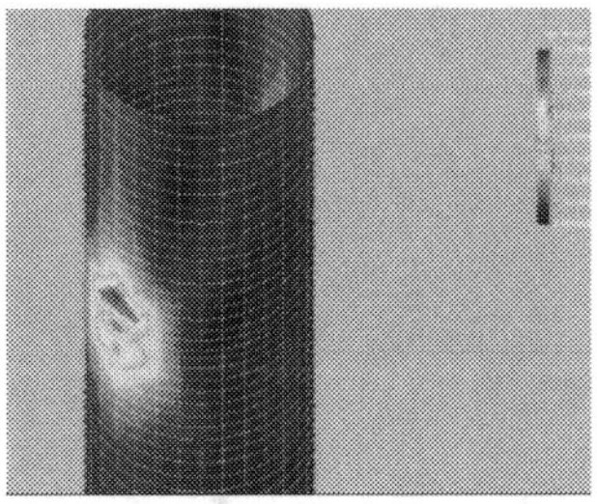

Figure 20. Ear welded directly on the cylinder-6.35 mm thickness

Analysing the stress distributions for different thicknesses of the shell it can be highlighted the following conclusions:

- the maximum equivalent stress values induced by the ears in the cylinder appear for the lowest thicknesses of the shell when the ear is welded directly on the cylinder, for the horizontal position;
- due to the position of the acting point of the raising force, when the cylinder is located in horizontal position, the stress distributions present symmetry in respect with the welding points of the ear; for the other positions of the cylinder the action points of the vertical force are changing their positions and the stress symmetry is disturbed;
- the maximum values of stresses from the ears appear in the welding points and have nearly the same values for all thicknesses of the shell; for this reason, the thickness of the ear can be independently obtained, for the presented case of the value of 12 mm; the maximum stress from the ears is 160 MPa;
- because the maximum value of equivalent stress is 91 MPa (smaller than the allowable limit of the shell material) and is obtained for the cylinder located in horizontal position and for the ear welded directly on the shell, the reinforced pad is not recommended;
- in order to complete the rising process of the cylinder from horizontal to vertical position it is enough to manufacture the ears and to weld them directly on the shell;
- for comparison between numerical and experimental values of stresses [6] the main nodes around the ears (welded directly on the cylinder) are presented in figure 21;

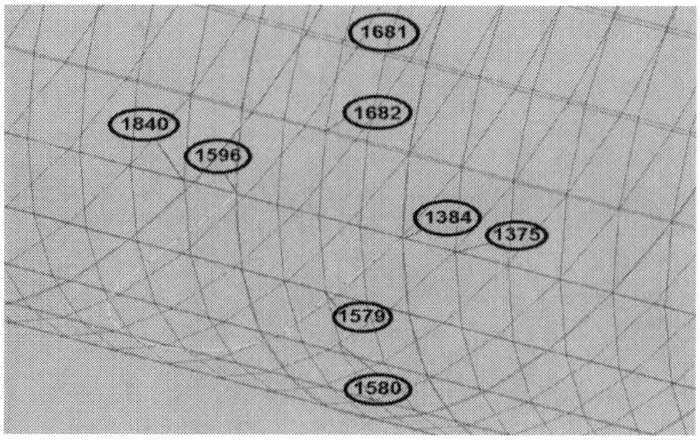

Figure 21. Nodes near the ear

- the main values of axial stresses (that are going to be compared with experimental values) are presented in table 1 for the cylinder with 6.35 mm thickness, both for horizontal (fig. 8) and vertical (fig. 20) positions. For all load cases y axis is vertical and z longitudinal. The corresponding experimental values are resumed from [6] in table 2.

Table 1. Main values of the axial stresses around the ear

Node no.	**Cylinder in horizontal position (ear loaded with a vertical force – 1000 N)**		**Cylinder in vertical position (ear loaded with a vertical force – 1000 N)**	
	σ_y - [MPa]	σ_z - [MPa]	σ_y - [MPa]	σ_z - [MPa]
1384	-	-	-12.60	-11.85
1579	-31.20	-8.64	-	-
1596	-	-	14.80	13.14
1682	-31.20	-8.64	-	-

Table 2. Main experimental values of the axial stresses around the ear

Node no.	**Cylinder in horizontal position (ear loaded with a vertical force – 1000 N)**		**Cylinder in vertical position (ear loaded with a vertical force – 1000 N)**	
	σ_y - [MPa]	σ_z - [MPa]	σ_y - [MPa]	σ_z - [MPa]
1384	-	-	-12.64	-11.02
1579	-30.76	-9.45	-	-
1596	-	-	14.77	13.36
1682	-30.3	-9.30	-	-

3. Conclusions

In the paper are presented the stresses induced by some raising ears in a cylindrical shell located in different raising positions. There are analysed three raising positions: when the cylinder it is in a horizontal position, when is inclined with 45 degrees (in respect with horizontal direction) and when it is in vertical position. In order to resume the main results obtained from the numerical simulations the maximum stress intensity values in the cylinder (Von Mises) from all load cases are presented in table 3 below (for the laboratory cylinder with 6.35 mm thickness).

Table 3. Maximum stress intensity values in the cylinder

Load case	Ear welded on a reinforced pad - cylinder in horizontal position- fig.7	Ear welded directly on the cylinder - cylinder in horizontal position- fig.8	Ear welded on a reinforced pad - cylinder at 45° – fig. 13	Ear welded directly on the cylinder - cylinder at 45° – fig. 14	Ear welded on a reinforced pad - cylinder in vertical position – fig. 19	Ear welded directly on the cylinder - cylinder in vertical position – fig. 20
Max stress intensity - [MPa]	15.48	37.05	11.37	22.52	8.76	15.69

The analysis proves that the maximum levels of stresses induced in cylinder are lower than the allowable limit of the shell and the reinforced pad is not necessarily. The adopted geometry of the ear is very easy to be manufactured and is recommended to be used. The numerical results of the axial stresses can be compared with the experimental values presented in table 2. All the details regarding the experimental work are presented in [6].

ORCID IDs
A. Iamandei https://orcid.org/0000-0002-7512-533X
R. Rapeanu https://orcid.org/0000-0001-8255-8144
L. Stanciu https://orcid.org/0000-0003-4976-3263
I. Popa https://orcid.org/0000-0001-6131-1345
S. Vasilescu https://orcid.org/0000-0002-3761-7376

References

[1] Posea N 1979 *Strength of Materials* (in Romanian: Rezistenţa materialelor) Publishing House Didactică şi Pedagogică Bucharest
[2] Vasilescu S and Talle V 2007 *Strength of Materials* (in Romanian: Rezistenţa materialelor) Publishing House of the Petroleum-Gas University of Ploieşti
[3] Iatan R and Vasilescu I 2002 *Transportation of the Oversized Technological Equipment* (in Romanian: Transportarea utilajelor tehnologice agabaritice) Publishing House Matrix Rom Bucharest
[4] Vasilescu S and Ilinca C 2013 A Strength Calculation of a Nozzle Using Comparative Methods *Proc of the 14th Symp on Experimental Stress Analysis and Materials Testing, Key Engineering Materials* vol 601 https://doi.org/10.4028/www.scientific.net/KEM.601.84
[5] Ilinca C and Vasilescu S 2013 A Special Pipe Support Analysis Proc *of the 14th Symp on Experimental Stress Analysis and Materials Testing, Key Engineering Materials* vol **601** https://doi.org/10.4028/www.scientific.net/KEM.601.80
[6] Iamandei A, Ripeanu R G, Stanciu L, Popa I and Vasilescu S 2021 Experimental Results Regarding Strains and Stresses Appeared Around Some Lifting Ears from a Cylinder Used in Petrochemical Industry submitted *IMANEE 2021*
[7] *** COSMOSM User's Guide
[8] EN 1993-1, Eurocod 3: *Design of steel structures* (in Romanian : Proiectarea structurilor de oţel)
[9] EN 10025-2:2004 *Hot Rolled Products of Structural Steel.*
[10] SR 438-1:2012 *Steel Products for Concrete Reinforcement* (in Romanian: Produse de otel pentru armarea betonului)

IOP Conf. Series: Materials Science and Engineering 1235 (2022) 012050 doi:10.1088/1757-899X/1235/1/012050

Experimental results regarding the strains and stresses appeared around some lifting ears from a cylinder used in the petrochemical industry

A Iamandei, R G Ripeanu, L Stanciu, I Popa and S Vasilescu

Petroleum-Gas University of Ploieşti, Bucharest Blvd, no.39, Romania

vserban@upg-ploiesti.ro

Abstract. During transporting or raising processes any equipment used in petrochemical industry needs some special ears to be easy and safety manipulated. In the paper are presented some experimental results measured around some lifting ears welded on a cylinder. A special experimental device has been manufactured to complete the experiments. The acquisition data have been registered with a modern Quantum system equipped with a Catman Easy software.

1. Introduction

The experimental device used for strain gauges measurements was built to verify the numerical results obtained using the Finite Element Method [1]. A general view of the 12" cylinder (that has exactly the form described in [1] is presented in figure 1a. In figure 1b is presented the Quantum acquisition system equipped with the Catman software [2]. The vertical forces acting the ears during the raising process have been provided by a laboratory crane.

a) b)

Figure 1. The overall view of the laboratory device and acquisition system

To verify the theoretical results presented in [1] some electrical gauges have been located in the neighbourhood of the ears (figure 2) where the position of the transducers is also completed with the number of nodes (from FEA). In figure 2 the general axis system is presented: y represents the vertical direction and z the longitudinal one.

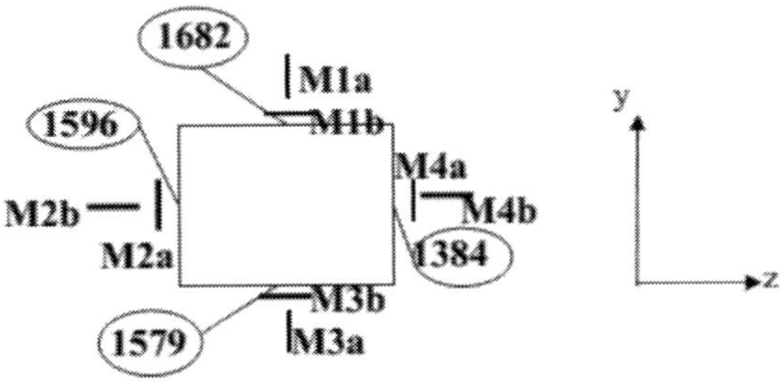

Figure 2 The positions of the transducers

The cylinder (with 6.35 mm thickness) has been located both in horizontal and vertical positions and every ear has been loaded with a vertical 1 kN force. The vertical force has been provided by a laboratory crane. The amplitude of the vertical force from the hook varied from zero to 2 kN (1 kN for every ear) and back to zero in about eleven minutes, like in figure 3 below.

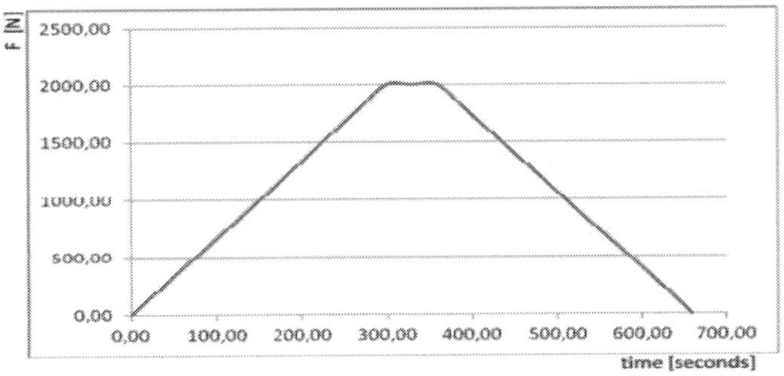

Figure 3 The variation of the hook force during the loading and unloading process

2. Experimental results

The experimental measurements have been registered during the loading and unloading process. With the registered data the following graphs (strains against time) were drawn: in figures 4 and 5 on case the ears are welded directly on the horizontal cylinder (strain gauges M1, M3) and in figures 6 and 7 in case the ears are welded directly on the vertical cylinder (strain gauges M2, M4).

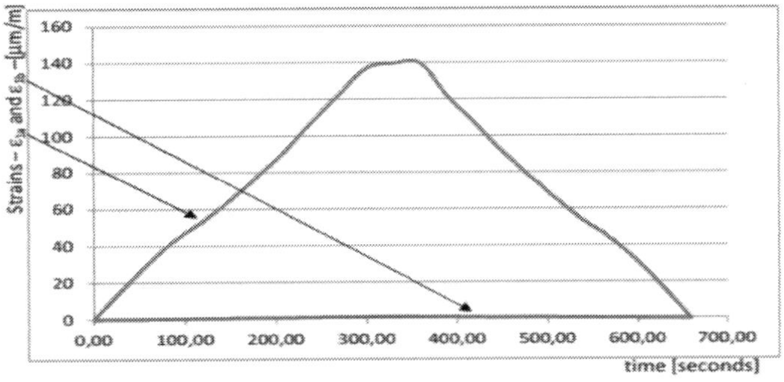

Figure 4. Strains against time (M1a, M1b) – ears welded directly on the horizontal cylinder

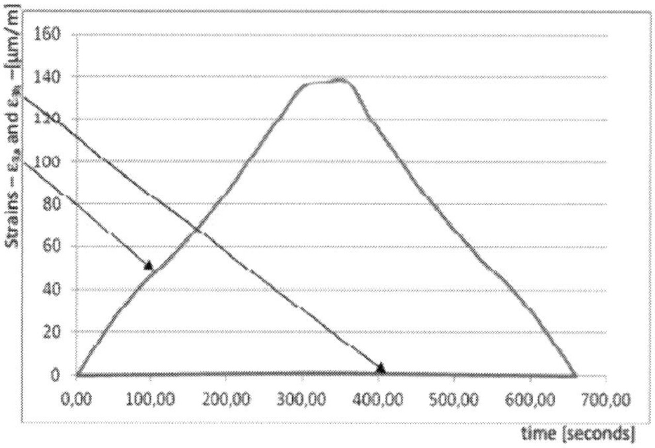

Figure 5. Strains against time (M3a, M3b) – ears welded directly on the horizontal cylinder

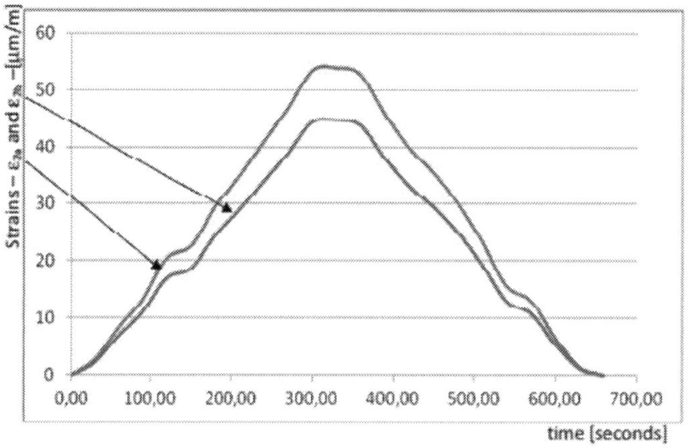

Figure 6. Strains against time (M2a, M2b) – ears welded directly on the vertical cylinder

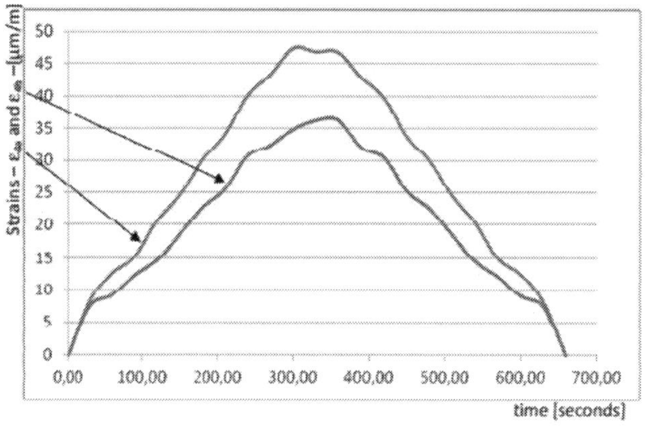

Figure 7. Strains against time (M4a, M4b) – ears welded directly on the vertical cylinder

Analysing the strains against time graphs presented in figures 4-7 the following conclusions can be highlighted:

- in the case of the ears welded directly on the horizontal cylinder it can be noticed a symmetry of the values indicated by the strain gauges M1 and M3; the above symmetry is produced by the application point of the vertical force that is symmetric in relation to the welding points;

- in the case of vertical position of the cylinder the symmetry it is not conserved because the acting point of the force is sliding up during the loading process and does not remain symmetric in relation to the welding points of the ear;
- during the unloading (discharging) process the strain values are nearly the same with those obtained during the loading process; this proves that the entire experiment has been conducted in the elastic range of the material and no remaining deformations appeared

With the values of strains presented above, using the generalised Hooke's law [3-5] the values of corresponding stresses have been calculated. Based on the values of stresses the following time functions have been drawn:

- in figures 8 and 9 the longitudinal and circumferential stresses around the ears welded directly on the horizontal cylinder (M1 and M3 points – figures 8 and 9);
- in figures 10 and 11 the longitudinal and circumferential stresses around the ears welded directly on the vertical cylinder (M2 and M4 points- figures 10 and 11).

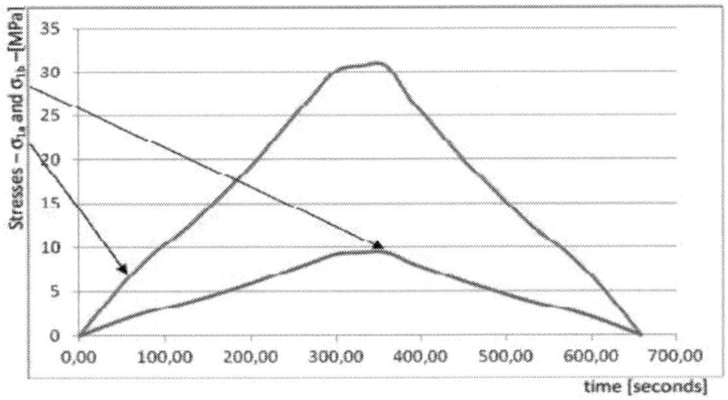

Figure 8. Stresses against time (M1a, M1b) – ears welded directly on the horizontal cylinder

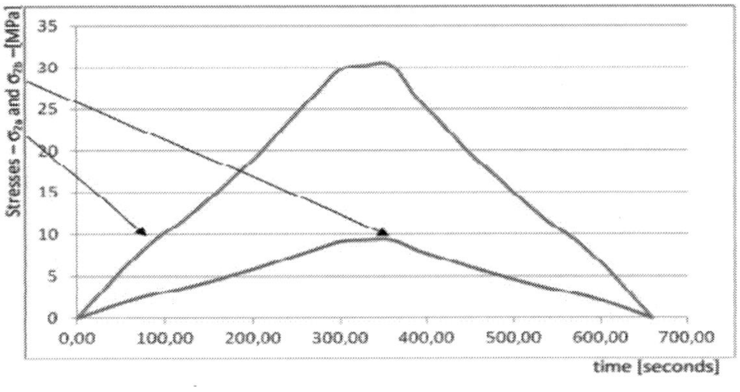

Figure 9. Stresses against time (M3a, M3b) – ears welded directly on the horizontal cylinder

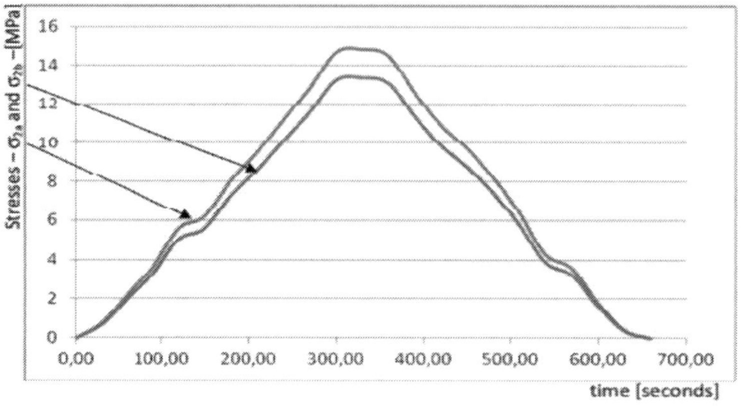

Figure 10. Stresses against time (M2a, M2b) – ears welded directly on the vertical cylinder

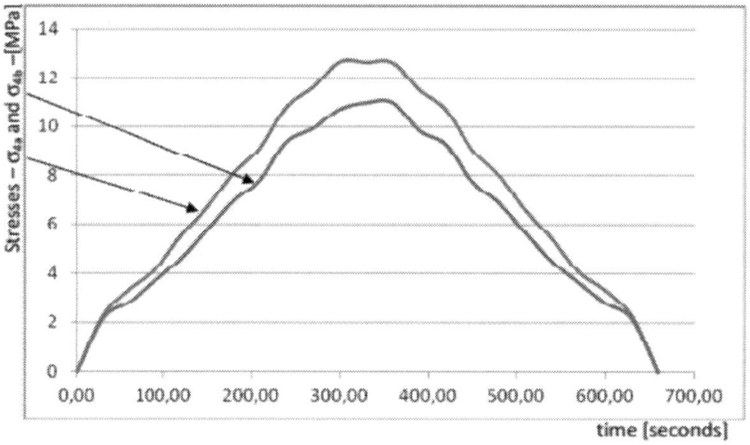

Figure 11. Stresses against time (M4a, M4b) – ears welded directly on the vertical cylinder

Analysing the graphs from figures 8-11 the following conclusions can be highlighted:

- the maximum values of stresses that appear around the ears welded directly on the horizontal cylinder are about 31 MPa (fig. 8 and 9) and present symmetry (in points M1 and M3) due to the symmetric position of the vertical force in relation with the welding points;
- the values of stresses in the case of vertical position of the cylinder foes do not remain symmetrically in relation with the welding points due to sliding of the application point of the vertical force during the raising process; the maximum values of stresses are around 15 MPa;
- during the unloading (discharging) process the values of stresses are nearly the same with those obtained during the loading process; this proves that the entire experiment has been conducted in the elastic range of the material and no remaining stresses appeared.

A synthesis of the experimental data (for the maximum values of stresses) is presented in table 1 below where is realised also a comparison between the experimental and numerical results obtained in [1].

The maximum errors obtained because of strain gauges measurements prove that the entire experiment has been carried out with very good accuracy, the differences between numerical results (presented in [4] and the experimental data being smaller than 8.5%. The maximum relative errors appear for the smaller values of stresses (respectively strains) and are quite normal (expected) because it is known that the strain gauge transducers have higher relative errors in the range with lower loads. The errors that appear in the points M1b, M2b, M3b and M4b can be also explained due to the position of transducers that are not located exactly in the corresponding nodes from FEA [1]. For the other values of stresses, the maximum error is below 2.5%.

Table 1. Main values of the axial stresses near the ear

Theoretical and experimental results	Maximum Stresses							
	M1a	M1b	M3a	M3b	M2a	M2a	M4a	M4b
Experimental results [MPa]	30.76	9.45	30.30	9.30	14.77	13.36	12,64	11.02
Theoretical results [4] [MPa]	31.2	8.64	31.2	8.64	14.80	13.14	12.66	10.75
Relative error [%]	1.4	8.5	2.9	7.1	0.2	1.6	0.1	2.5

3. Conclusions

In the paper are presented the experimental data regarding the stresses induced by some raising ears in a cylindrical shell located in two extreme raising positions (horizontal and vertical). The experimental results are close to the numerical values obtained by using a finite element analysis [1]. This technical solution of making the raising ears from curved beams is easy to be manufactured and is recommended to be used.

ORCID IDs
A. Iamandei https://orcid.org/0000-0002-7512-533X
R. Rapeanu https://orcid.org/0000-0001-8255-8144
L. Stanciu https://orcid.org/0000-0003-4976-3263
I. Popa https://orcid.org/0000-0001-6131-1345
S. Vasilescu https://orcid.org/0000-0002-3761-7376

References

[1] Iamandei A, Ripeanu RG, Stanciu L, Popa I and Vasilescu S 2021 Numerical Analysis of the Stresses Appeared Around Some Lifting Ears from a Cylinder Used in the Petrochemical Industry submitted *IMANEE 2021*
[2] *** CATMAN EASY User's Guide
[3] Iatan R and Vasilescu I 2002 *Transportation of the Oversized Technological Equipment* (in Romanian: Transportarea utilajelor tehnologice agabaritice) Publishing House Matrix Rom Bucharest
[4] Posea N 1979 *Strength of Materials* (in Romanian: Rezistenţa materialelor), Editura Didactică şi Pedagogică Bucharest
[5] Vasilescu S and Talle V 2007 *Strength of Materials* (in Romanian: Rezistenţa materialelor) Publishing House of the Petroleum-Gas University of Ploieşti

IOP Conf. Series: Materials Science and Engineering 1235 (2022) 012051 doi:10.1088/1757-899X/1235/1/012051

Power planetary precessional transmission research regarding acoustical behaviour

Iu Malcoci, N Trifan, O Ciobanu and R Ciobanu

Technical University of Moldova, Faculty of Mechanical Engineering and Transport, Fundamentals of Machinery Design Department, Chişinău, Rep. Moldova.

iulian.malcoci@bpm.utm.md

Abstract. In planetary precessional reducers vibration can appear at bearings, gears, misaligned shafts, unbalanced rotors, couplings. Changes in dynamic process affects not only normal behaviours of mechanical systems but also acoustical behaviour. Regarding this aspects, regular acoustical measurements during the operating speed provide information about any necessary repairs or maintenance. Here vibration research was made by using Bruek&Kjaer Sound Level Meter 2250 Light to record Noise Rating for evaluation of global noise in dB according to ISO TC-43.

1. Introduction

Taking into account the harmful activity of noise, in order to prevent illnesses and accidents at work and to ensure the conditions for achieving acoustic comfort, rules have been developed on limiting the acoustic power of equipment and installations and limiting maximum sound intensity levels to values it is considered that the action of the noises is not harmful or does not disturb [1].

At the beginning, the parameter chosen for the elaboration of the noise limitation norms was chosen the level of loudness in the phones, but alone was not enough to appreciate the harmful effect of the noise. Therefore, a second parameter was chosen, frequency. By representing the variation of the sound pressure level as a function of frequency, the noise curves resulted. The rules were developed according to the workplace resulting in separate rules for industrial noise, noise in homes and cities, noise in transport [7].

1.1. Noise assessment methods

For the evaluation of noise and within acceptable limits, several types of such curves were established: NC (Noise Criteria), RC (Room Criteria), NR (Noise Rating).
NC (Noise Criteria) and RC (Room Criteria) are mainly used in the United States.
NR (Noise Rating) curves are used in Europe, in Romania being called Cz noise curves.
Noise evaluation can also be done globally in dB (A).

NR (Noise Rating) are used to evaluate noise according to ISO and are mainly used in Europe. The Technical Committee No. 43 of the International Organization for Standardization (ISO-TC 43) adopted the system for evaluating the intensity level of a noise through a series of curves developed in the eight-octave bands between 63 and 8000 Hz, considered to represent the levels of intensity producing the same physiological discomfort. These curves, called noise curves Cz (or NR) (according to SR 6156:2020 and partial in SR EN ISO 717-1:2013 standards) are highlighted by a number between 0 and 130 - the number that defines a certain noise curve, corresponds to the sound pressure level, measured in dB, of a 1000 Hz sound.

The Cz curves are given in Figure 1 by sound pressure levels corresponding to 1/1 octave frequency bands.

The steps for evaluating the noise level using the Cz curves are the following:

- the noise level in octave bands is determined;
- draw the working curve, joining the points obtained by measurements;
- the Cz curve for the system for which the measurements were made is the real NC curve that is interested in the highest point of the theoretical curve [1,7].

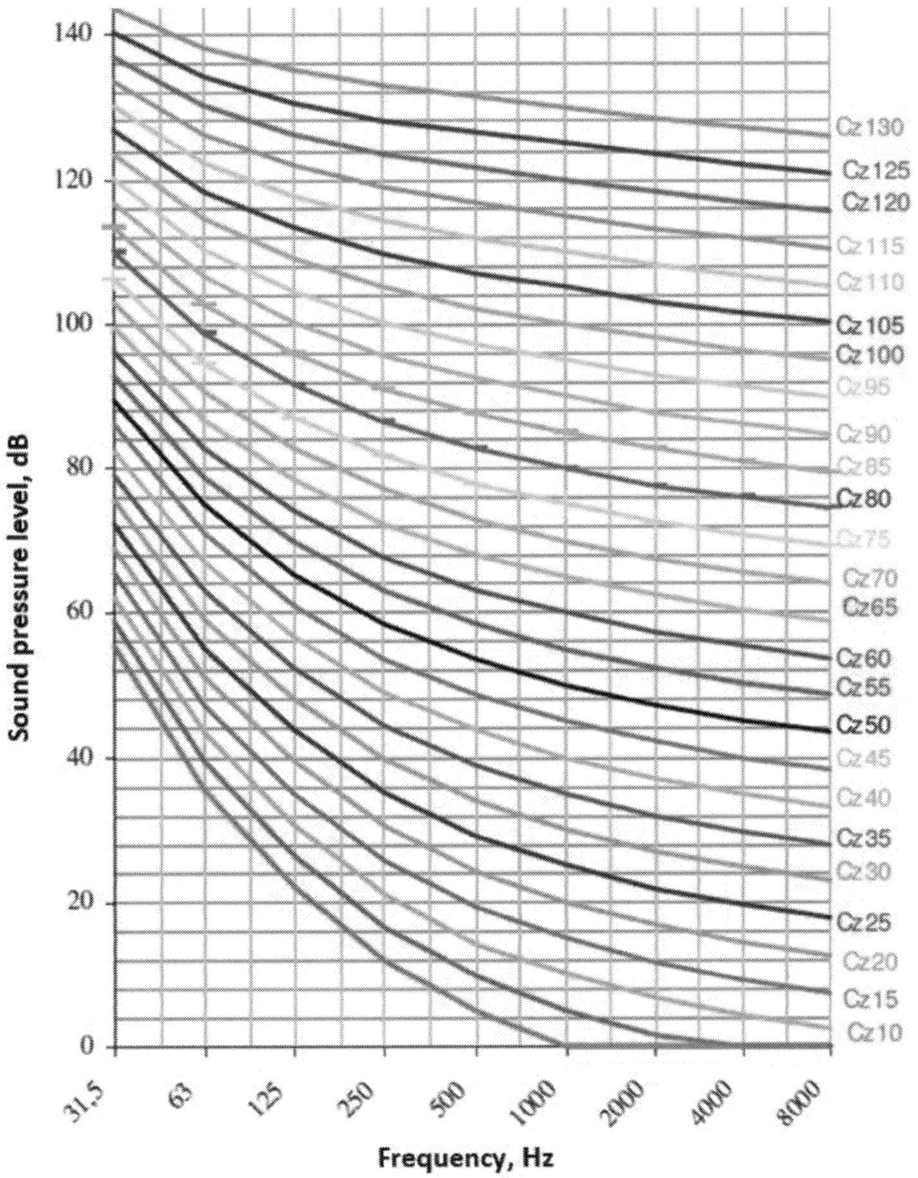

Figure 1. Noise curves Cz.

IOP Conf. Series: Materials Science and Engineering 1235 (2022) 012051 doi:10.1088/1757-899X/1235/1/012051

1.2. Equipment used to measure noise levels.

In a first stage of the analysis of the vibroacoustic behaviour of the precessional planetary gearboxes, quantitative noise level measurements were performed, using a Brüel & Kjær Type 2250 Light noise level meter (Figure 2) [6].

Figure 2. Sound Level Meter Bruel&Kjaer 2250 Light.

Figure 3 shows the typical window for measuring noise curves using the 2250 Light sound level meter, and in Figure 4 the data obtained for the sound pressure level [dB] for the average frequencies [6]:

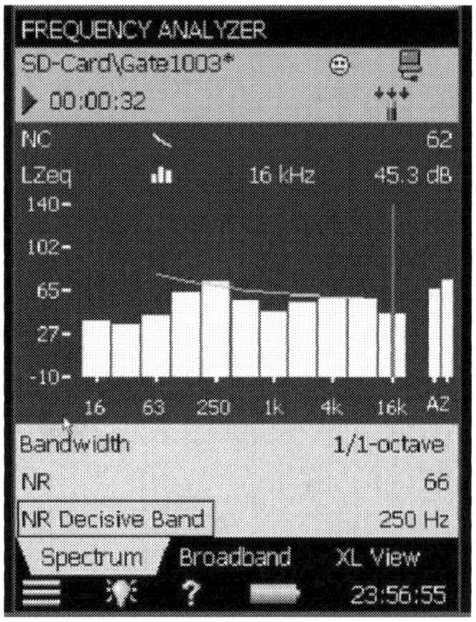

Figure 3. Typical window for measuring noise curves using 2250 Light sound level meter.

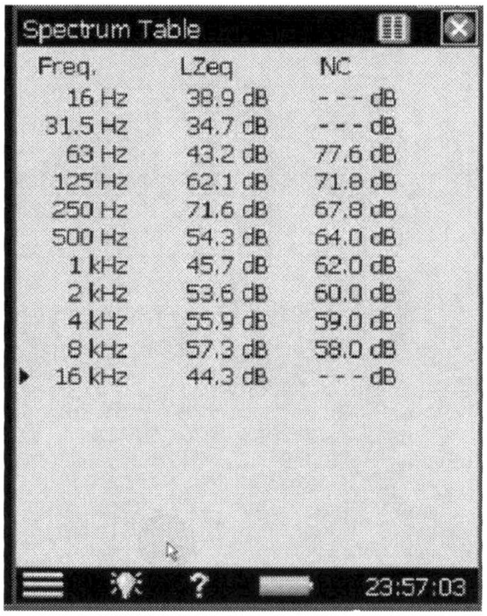

Freq.	LZeq	NC
16 Hz	38.9 dB	- - - dB
31.5 Hz	34.7 dB	- - - dB
63 Hz	43.2 dB	77.6 dB
125 Hz	62.1 dB	71.8 dB
250 Hz	71.6 dB	67.8 dB
500 Hz	54.3 dB	64.0 dB
1 kHz	45.7 dB	62.0 dB
2 kHz	53.6 dB	60.0 dB
4 kHz	55.9 dB	59.0 dB
8 kHz	57.3 dB	58.0 dB
16 kHz	44.3 dB	- - - dB

Figure 4. Data samples for the sound pressure level in dB for average frequencies.

With the data obtained Figure 4 we draw the noise curve Figure 5 the Cz curve for the system for which the measurements were made is the real NC curve that is interested in the highest point of the intersected curves, in this case the Cz 80 curve.

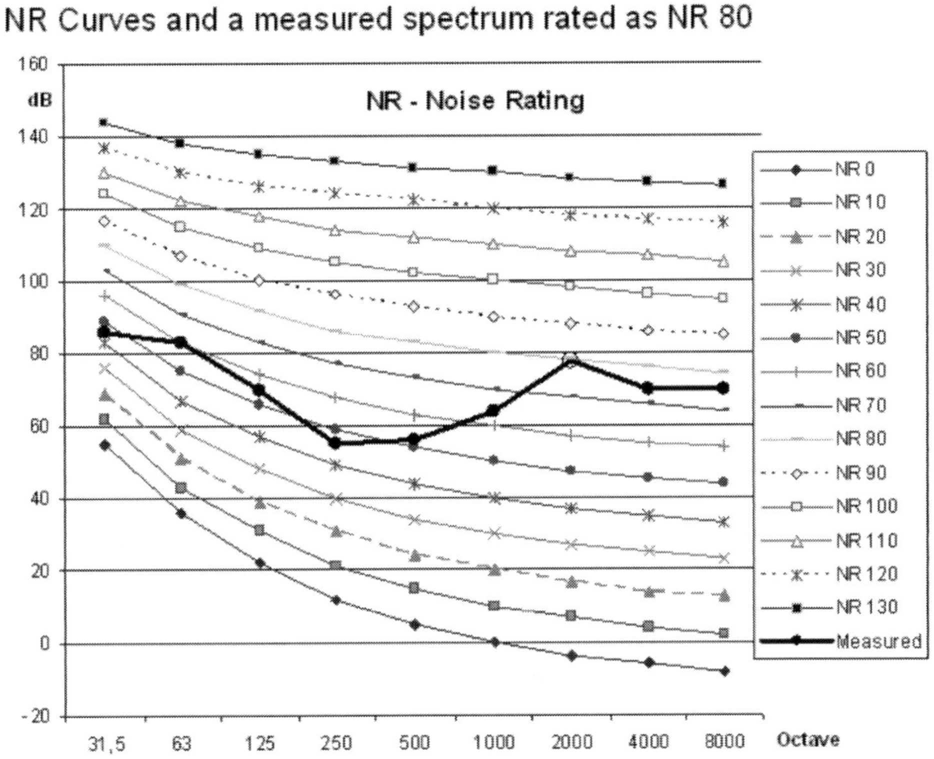

Figure 5. Highest intersected NR Curves - in this sample NR 80.

2. Acoustical research of the precessional planetary power reducer 2K-H Type

The experiments were performed on the test stand (Figure 6) in the laboratory "Fine Mechanics" of the department "Fundamentals of Machine Design" in a closed room with rigid floor [3,4].

Figure 6. Experimental stand for vibro-acoustical research.

The experimental stand consists (Figure 7) from the test table made of aluminium profile GUNT PT500.01 provided with T-shaped channels, in which the electric motor is fixed with P = 0.36 kW at 1800 rpm operating speed, the precessional planetary reducer 2K-H with the ratio of transmission i = -10.5 and maximum torque T2 = 40 Nm, and electromagnetic brake type ПТ-6. The connection between the gearbox shafts and the electric motor was made by means of a compensating coupling with bolts, and the connection between the gearbox shaft and the brake by means of a coupling with elastic bushes [5].

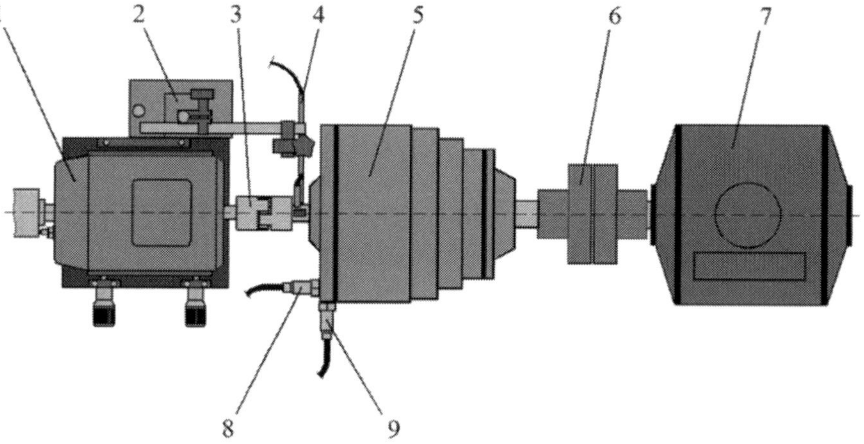

Figure 7. Research stand for the vibrodiagnosis for the 2K-H precession reducer: 1 - electric motor; 2 - magnet holder; 3 - compensating coupling with crabs; 4 - reference sensor; 5 - 2K-H reducer; 6 - coupling with elastic bushes; 7 - brake; 8 - acceleration sensor 1; 9 - acceleration sensor 2.

2.1. Determination of Cz noise curves.

A basic characteristic in order to estimate the acoustics of the reducers is the elaboration of the Cz noise curves. According to the standards and regulations regarding the maximum permissible values, the gearboxes must not exceed the Cz85 curve during operation. In this order, measurements were made with the Brüel & Kjaer 2250 Light sound level meter on the weighting curve (Z), in a closed room in the absence of any background noise (Figure 8) [1,2].

Figure 8. Measurement of Cz noise curves.

IOP Conf. Series: Materials Science and Engineering 1235 (2022) 012051 doi:10.1088/1757-899X/1235/1/012051

The values resulting from the measurements are presented in Figure 10 when the reducer operates (1800 rpm operating speed) without load (Nominal Torque) Tn = 0 Nm and in Figure 9 when the reducer operates under load Tn = 25 Nm. Thus, the noise emitted by the 2K-H precessional planetary gearbox reaches the Cz70 curve without load and the Cz65 curve under load.

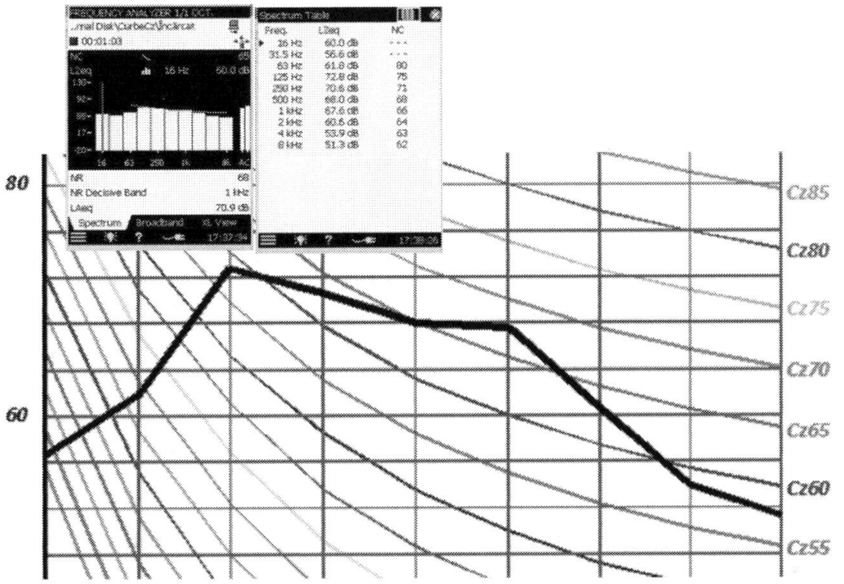

Figure 9. Cz65 curve for the case when the gearbox is operating under load.

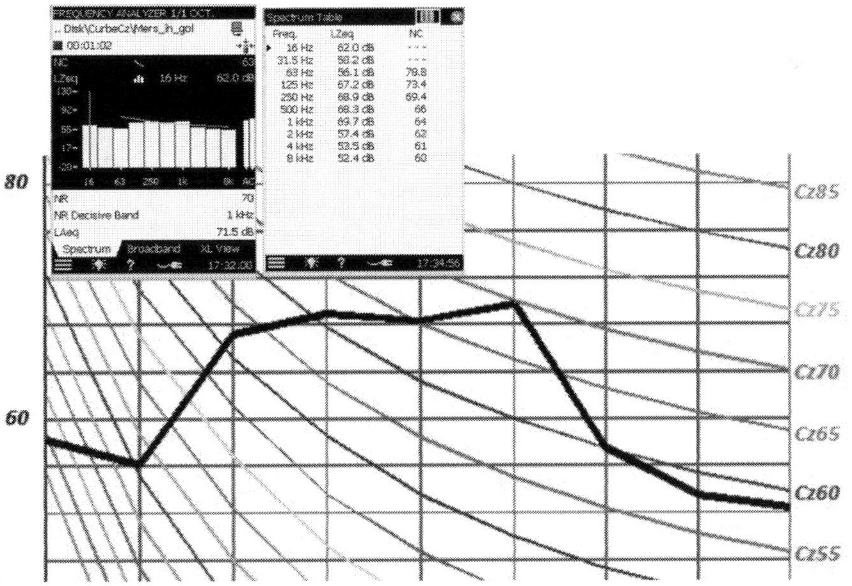

Figure 10. Cz70 curve for the case when the gearbox operates without load.

Conclusions

The results obtained recommend the 2K-H precessional planetary power reducer as a silent reducer with low vibration and noise emission.

Acknowledgments
This paper was elaborated under Moldova State Project 160-PS from 03.01.2020 with title Increasing the competitiveness of precessional transmission by developing and capitalizing „conform" toothed gear and expanding their range of application.

ORCID IDs
I Malcoci https://orcid.org/0000-0002-0263-5232
N Trifan https://orcid.org/0000-0002-5772-1349
O Ciobanu https://orcid.org/0000-0003-0861-3015
R Ciobanu https://orcid.org/0000-0002-4044-7523

References

[1] Malcoci Iu 2015 Vibro acoustic research of planetary precessional transmissions PhD Thesis UTM CZU: 621:534.6(043.2) p 135
[2] Malcoci Iu, Ciobanu O and Ciobanu R 2017 Dinamical systems – Vibro-Acoustical Analysis. Study cases. (Chişinău: „Tehnica-UTM") p 161
[3] Malcoci Iu, Trifan N and Ciobanu R 2020 Precessional gear-box research regarding vibration activity behaviour *IOP Conf. Ser-Mat. Sci.* **749** 012036 https://doi.org/10.1088/1757-899x/749/1/012036
[4] Bostan I, Dulgheru V and Malcoci Iu 2021 Some aspects regarding planetary precessional transmission dynamics *IOP Conf. Ser.: Mater. Sci. Eng.* **1009** 012008 https://doi.org/10.1088/1757-899x/1009/1/012008
[5] Abraham D, Boxhamer J and Mittash P 2011 *Instruction manual GUNT PT500* Gerätebeu GmbH (Barsbütel Germany)
[6] Bruel&Kjaer Sound Level Meter 2250 Light overview: https://www.envi.ro/echipamente-si-produse/sonometre/sonometru-2250/ Accessed 15.05.2021
[7] SR-6156:2020 Acoustics in construction. Protection against noise in civil and socio-cultural constructions. Permissible limits and sound insulation parameters.

IOP Conf. Series: Materials Science and Engineering 1235 (2022) 012052 doi:10.1088/1757-899X/1235/1/012052

Grease lubrication of miniature ball bearings

G Ianuş, D Cojocaru, V Cârlescu, A Tufescu and D N Olaru

Mechanical Engineering and Mechatronic Department, Technical University "Gheorghe Asachi" Iaşi, România

denis.cojocaru@student.tuiasi.ro

Abstract. Estimation of the power loss in miniature ball bearing grease lubricated is a complex problem. Usually the applied loads (radial and axial) have small values and the methodologies recommended by the bearing companies are cannot applied for these conditions. For a ball bearing, some friction processes have differential contribution to the total friction torque and power loss. For very low loads, the lubricant is the most important source for friction torque. In the present paper the authors determined experimentally the friction torque both in a standard 7000C angular contact ball bearing (ACBB) and a modified 7000C ACBB containing only 3 balls without cage, operating with very low axial load and lubricated with lithium soap grease. The experimental values of the friction torque have been correlated with the theoretical Houpert's IVR model developed for hydrodynamic rolling resistances in ball race contacts considering the viscosity of the base oil of grease.

1. Introduction

A lot of mechatronic systems are based on the micro mechanical transmissions. Miniature ball bearings, radial or angular contact are support for rotating spindles. Estimation of the power loss in miniature ball bearing grease lubricated is a complex problem. Usually the applied loads (radial and axial) have small values and the methodologies recommended by the bearing companies are cannot applied for these conditions. So, the SKF complex methodology to evaluate the friction torque in a ball bearing [1] can be used for external loads that exceed the recommended minimum load for every ball bearing.

Some researchers determined experimentally the friction torque in ball and cylindrical bearing using various greases, operating speeds, temperatures [2,3] and compared the experimental values with SKF model. So, Cousseau *et al.* [2] realized a complex experimental study with a 51107 thrust ball bearing loaded with 7000 N with rotational speed between 500 and 2000 rpm. Using six different greases the authors evidenced the role of the different grease composition (thickens and base oil) on friction torque and propose a methodology to determine the friction coefficient in full film condition (μ_{EHD}) based on the SKF friction torque methodology [1] and experimental results.

Kanazawa *et al.* [3] experimentally determined the friction torque in the thrust ball bearing 51107 and in two cylindrical rollers thrust bearings 81107 and 81105 using complex lithium and urea thickens, operating at 200 - 1800 rot/min and loaded with 2000 N.

Considering the experimental values of the friction torques, the authors use the SKF methodology and determine the friction coefficient in the tested bearings. Supplementary, a lot of experiments were realized by authors to determine the film thickness for all tested greases by optical interferometry in a ball-on-glass disc rig. Based on these experiments the authors concluded that for low lubricant parameter λ, the friction coefficient in the greased contacts is lower than the friction coefficient obtained by the base oil of the greases. Increasing of the lubricant parameter (increasing of the speed) the differences can be ignored.

Fischer *et al.* [4] using a ball-on-glass disc tribometer determined by optical interferometry the film thickness for two lithium greases and for corresponding base oils, for rolling speed between 1mm/s to 4m/s. The authors established that for low rolling speed the grease film thickness is maintained at higher level, compared with the film generated by oil as result of thickener, which enter in the ball-race contact. De Laurentis *et al.* [5] determined by ball-on-disc machine the relation between grease composition (thickener and base oil) and friction coefficient in rolling contacts. The authors compared friction coefficient experimentally determined for nine sorts of greases and observe that higher values have been obtained by lithium soap greases.

For a ball bearing, some friction processes have differential contribution to the total friction torqu e and power loss. So, for normal and high load the friction generated by elastic hysteresis in ball races contacts and pivoting friction are dominant. For high-speed and moderate loads, the pivoting and lubricant friction are the parameters that limits the operational speed by increasing of the temperature and decreases the film thickness in ball-races contacts.

For very low loads, the hysteresis losses can be neglected and the lubricant is the most important source for friction torque. A lot of experiments realized by Bălan *et al.* [6], Ianuş *et al.* [7,8], Popescu *et al.* [9] evidenced both for thrust and angular contact ball bearing the essential contribution of the lubricant on the friction torque operating with very low applied loads. As a paradox, for very low loads, the friction torque in a dry ball bearing is lower than in lubrication conditions [7-9].

In the present paper the authors determined experimentally the friction torque both in a standard 7000C angular contact ball bearing (ACBB) and a modified 7000C ACBB containing only 3 balls without cage, operating with very low axial load and lubricated with lithium soap grease. The experimental results were compared with the computed results obtained both with the Houpert's friction torque model [10] and with the SKF model [1].

2. Experimental equipments and procedure

Experimental investigations have been focused to compare the experimental friction torque both for a standard and modified 7000C ACBB with the theoretical models developed by Houpert [10] and also verified for thrust ball bearings lubricated with oil and grease by Bălan *et al.* [6] and Ianuş *et al.* [7,8].

2.1. Experimental equipment

In figure 1 is presented the tested equipment attached to the rotating table of the CETR –UMT 2 Tribometer. In figure 2 are presented the details of the testing components. The standard 7000C ACBB has 8 balls of diameter $d = 4.762$ mm, contact angle α=15° and a phenolic cage guided on inner race. The Modified 7000C ACBB has only 3 balls distributed at 120 degrees and the cage it is removed. A schematic presentation of the modified 7000C ACBB mounted in the test rig is presented in figure 3. The absence of the cage gives the possibility to determine only the influence of the grease friction in the ball-race contacts. As is presented in figure 3, a loading disc mounted on the outer ring generated the axial load $F_a = 3.11$ N corresponding to each normal ball-races contact load $Q = 1.47$ N and $Q = 3.93$ N for standard and modified ACBB, respectively. The maximum Hertzian ball-races pressures in standard 7000C ACBB are 0.38 GPa and 0.5 GPa for outer and inner ball-races contacts, respectively. For modified 7000C ACBB the maximum Hertzian ball-races pressures are 0.53 GPa and 0.69 GPa for outer and inner ball-races contacts, respectively.

2.2. Lubricant characteristics

Both the standard and modified 7000C ACBB have been greased with small quantities of a commercial Shell Gradus S2 V100 2 grease based on lithium soap and mineral oil. The physical typical characteristics are presented in the Table 1.

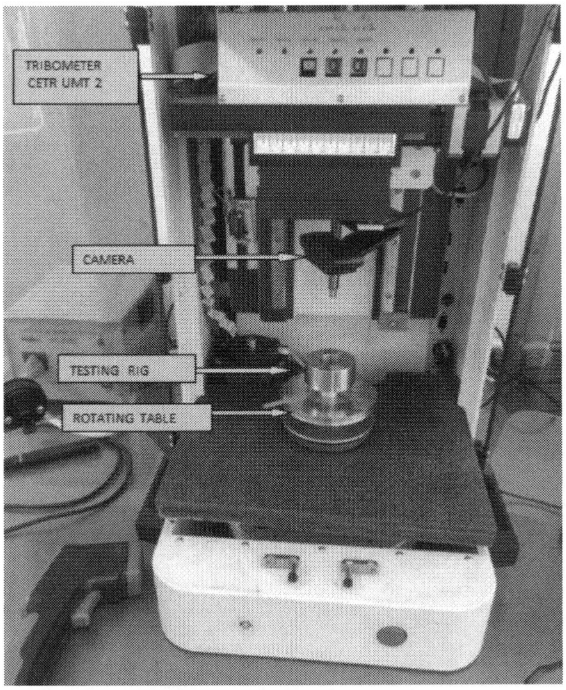

Figure 1. View of the testing equipment

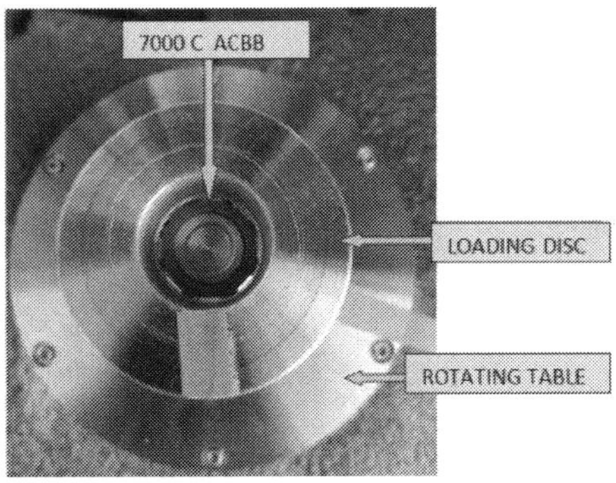

Figure 2. Detail of the testing rig

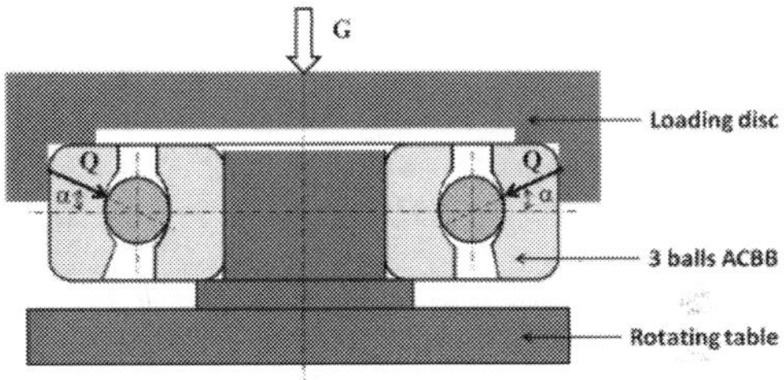

Figure 3. View of the 7000C modified ACBB

Table 1. The most important physical characteristics of grease Shell Gradus S2 V100 2.

Properties	Units	Values
NLGI Consistency		2
Kinematic viscosity of the base oil at 40 °C	mm^2/s	100
Kinematic viscosity of the base oil at 100 °C	mm^2/s	11
Dropping point	°C	180

2.3. Experimental procedure

To determine the friction torque both in standard and modified 7000C ACBB the authors based on spin-down method described in details in [6-9]. In fact, the friction torque is indirectly determined according the time and total angular position obtained in deceleration process of the outer race and disc from an imposed rotational speed until the stopping process. So, all the kinetic energy of the rotating disc and outer race are dissipated in the friction processes between the balls, races and cage for standard ACBB. For modified ACBB the kinetic energy is dissipated only in the six rolling contacts between three balls and races. The procedure follows two steps: (i) in the first step both the rotating table and inner race are rotate at an imposed rotational speed. As result of friction into rolling bearing outer race and loading disc start to rotate until it reaches the speed of the inner race; (ii) In the second step, the rotating table is sudden stopped and the outer race and disc it continues to rotate until the kinetic energy is depleted. All the process of the deceleration is monitored by a video camera and the recording finally processed to establish the total angle and rotating time.

2.4. Evaluation of the friction torque based on the tests

For very low contact loads the most important contribution of the friction in a ball-race rolling contact is generated by the hydrodynamic rolling forces [6-9] and the relation between total friction torque and rotational speed can be approximated by following equation:

(1)

$$Tz(\omega) = K \cdot \omega^n$$

where $Tz(\omega)$ is total friction torque in the ball bearing, K is a constant depending on lubricant viscosity, load and geometry and ω is angular speed.

Based on above mentioned equation, the dynamic equilibrium of the rotating disc and outer race in deceleration process is given by equation:

$$J \cdot \frac{d\omega}{dt} + K \cdot \omega^n = 0 \tag{2}$$

where J is the inertia moment of the rotating mass.

Equation (2) has been solved analytically and the variation of the angular speed $\omega(t)$ and variation of angular position $\varphi(t)$ as function of time are determined by following relations:

$$\omega(t) = \left[\omega_0^{1-n} - \frac{K \cdot (1-n)}{J} \cdot t \right]^{\frac{1}{1-n}} \tag{3}$$

$$\varphi(t) = \frac{J}{K \cdot (2-n)} \cdot \omega_0^{2-n} - \left[\omega_0^{1-n} - \frac{K \cdot (1-n)}{J} \cdot t \right]^{\frac{2-n}{1-n}} \tag{4}$$

where ω_o is angular speed at the starting decelerating process when it is consisted that the time t = 0.

Imposing the experimentally values for total decelerating angle and time φ_{max} and t_{max}, respectively, the nonlinear equations (3) and (4) are solved and parameters K and n have been determined for each angular speed ω_o.

3. Experimental data

In figure 4 are presented the variation of the experimental friction torque for the standard 7000C ACBB, greased with a quantity of approx. 0.17 cm^3 that means a half of the recommended quantity by SKF for precision ACBB [13]. This quantity was adopted to avoid the supplementary friction caused by churning of the grease. Three tests have been realised for various speeds between 100 and 600 rpm at a laboratory temperature of 26 - 28 °C. The experimental values of the friction torque have been compared with the calculated values according to SKF methodology [1] considering the base oil viscosity of 150 mm^2/s at operating temperature. In the SKF methodology have been included only the rolling and sliding components M_{rr} and M_{sl}, respectively.

It can be seen important increasing of the experimental values compared with the values determined by the SKF methodology, the differences are caused by the very low axial load, when the losses caused by elastic hysteresis in the rolling process of the balls are negligible. Also, at very low loads the pivoting frictions depending on the dimensions of the ball-races contact ellipses are negligible.

On the other side, the experimental friction torque values have a good correlation with the IVR model developed by Houpert [10] and verified in other experiments for modified thrust ball bearings lubricated with oil and greases [6-9]. The results confirm the dominant influence of the hydrodynamic rolling resistance on total friction torque.

In figure 5 are presented the experimental values of friction torque for a modified 7000C ACBB with 3 balls and greases with different quantity of grease. The results have been compared with the similar results obtained in dry conditions (without lubricant) resulting the important increases of the friction torque caused by small quantity of grease. Three quantities of grease considered in experiment were 40, 60 and 80 mg. The tests have been realised at the same temperature as the tests with standard ACBB, in a range of rotational speed between 100 and 500 rpm. The results obtained by various quantity of grease in modified 7000C ACBB evidenced increasing of the friction torque with increasing of the grease quantity, especially at low rotational speed as result of churning effects.

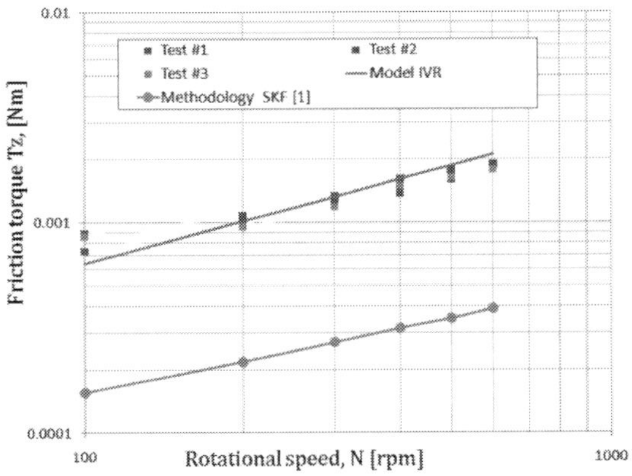

Figure 4. Experimental and theoretical friction torque variations for standard 7000 C ACBB

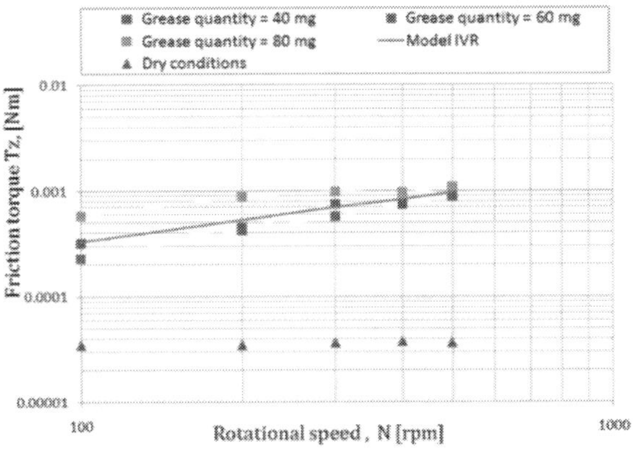

Figure 5. Experimental and theoretical friction torque variations for modified 7000C ACBB

4. Validation of the experimental results

For an ACBB Houpert [10] developed a complex methodology to determine the friction torque considering following components: hydrodynamic rolling resistance, curvature effects, elastic rolling resistance, pivoting effects and additional braking effects caused by cage. In decelerating process must be including the inertial effect also. Details of calculus for all components of the friction torque in a modified 7205B ACBB very low loaded are presented in [9].

For the standard 7000C ACBB the curvature effect, elastic rolling resistance, pivoting friction torques and inertial effects can be neglected compared to the hydrodynamic rolling resistance as result of very low axial load. Also, the friction between the cage and balls and the cage and guiding race have been ignored [10].

Finally, the friction torque generated by Z balls on outer and inner race can be expressed by following equation:

$$Tz = Z \cdot (FR_o + FR_i) \cdot Ri \cdot Ro / dm \tag{5}$$

Ri and *Ro* are the radii of inner and outer races, respectively and are determined by equations:

$$Ri,o = dm \cdot (1 \mp \gamma) / 2 \tag{6}$$

where *dm* is pitch diameter of the bearing and γ is given by equation:

$$\gamma = d \cdot \cos(\alpha) / dm \tag{7}$$

Hydrodynamic rolling resistance developed in ball outer and inner race, FR_o and FR_i respectively are determined by the Houpert's IVR (Iso-Viscous_Regime) equations [10]:

$$FR_{o,i} = 1.213 \cdot E^* \cdot R^2_{y,o,i} \cdot k^{0.358}_{o,i} \cdot U^{0.636}_{o,i} \cdot W^{0.364}_{o,i} \tag{8}$$

where E^* is reduced Young modulus of the ball and races, $R_{y,o,i}$ are the equivalent ball-race radius in rolling direction $U_{o,i}$ and $W_{o,i}$ are the classical dimensionless speed and load parameters determined for outer and inner ball-race contacts [10]. Equation (5) has been solved for imposed axial load of 3.11 N, a base oil viscosity of 150 mm²/s and for various rotational speeds. For the standard 7000C ACBB has been considered Z = 8 and for modified 7000C ACBB has been considered Z = 3. In figure 4 is presented the variation of the IVR friction torque as results of equation (5) and a good correlation with the experimental values has been observed. In figure 5 is presented the variation of the IVR friction torque as results of equation (5) for the modified 7000C ACBB.

5. Conclusions

The friction torque in a greased 7000C angular contact ball bearing in standard and in modified configuration (with 3 balls and no cage) have been realized for very low axial load and using a lithium soap grease. The experimental results confirm the dominant contribution of the hydrodynamic rolling resistance in the friction torque of the miniature ACBB grease lubricated. The authors consider that the friction torque for miniature greased ACBB low loaded and operating at moderate speed can be determined by using the Houpert's model considering especially the hydrodynamic rolling resistance for IVR lubrication regime developed by the base oil of the grease. The authors intend to continue the experiments considering the other grease composition.

Acknowledgments

This paper was supported by Mechanical Engineering Doctoral School of "Gheorghe Asachi" Technical University, Iaşi.

ORCID IDs

G Ianuș: https://orcid.org/0000-0002-2998-0992
D Cojocaru: https://orcid.org/0000-0002-8348-9002
V Cârlescu: https://orcid.org/0000-0001-9242-4967
A Tufescu: https://orcid.org/0000-0003-0208-1986
D N Olaru: https://orcid.org/0000-0001-5989-0798

6. References

[1] https://www.skf.com/binaries/pub12/Images/0901d1968065e9e7-The-SKF-model-for-calculating-the-frictional-moment_tcm_12-299767.pdf Accessed 15.07.2021

[2] Cousseau T, Graça B, Campos A and Seabra J 2011 Friction torque in grease lubricated thrust ball bearings *Tribol. Int.* **44** (5) 523-31 https://doi.org/10.1016/j.triboint.2010.06.013

[3] Kanazawa Y, De Laurentis N and Kadiric A 2020 Studies of friction in grease-lubricated rolling bearings using ball-on-disc and full bearing tests *Tribol. Trans.* **63** (1) 77-89 https://doi.org/10.1080/10402004.2019.1662147

[4] Fischer D, Jacobs G, Stratmann A and Burghardt G 2018 Effect of Base Oil Type in Grease Composition on the Lubricating Film Formation in EHD Contacts *Lubricants* **6** (2) 32 https://doi.org/10.3390/lubricants6020032

[5] De Laurentis N, Kadiric A, Lugt P and Cann P 2016 The influence of bearing grease composition on friction in rolling/sliding concentrated contacts *Tribol. Int.* **94** 624-32 https://doi.org/10.1016/j.triboint.2015.10.012

[6] Bălan M R, Houpert L, Tufescu A and Olaru D 2015 Rolling Friction Torque in Ball- Race Contacts Operating in Mixed Lubrication Conditions *Lubricants* **3** (2) 222-43 https://doi.org/10.3390/lubricants3020222

[7] Ianuş G, Dumitraşcu A C, Cârlescu V and Olaru D N 2016 Friction torque in thrust ball bearings grease lubricated *IOP Conf. Series: Materials Science and Engineering* **147** 012026 https://doi.org/10.1088/1757-899X/147/1/012026

[8] Ianuş G, Cojocaru D, Oprişan M C, Paleu V and Olaru D N 2020 Power loss in grease lubricated ball bearings *IOP Conf. Series: Materials Science and Engineering* **724** 012009 https://doi.org/10.1088/1757-899X/724/1/012009

[9] Popescu A and Olaru D N 2020 Influence of lubricant on the friction in an angular contact ball bearing under low load condition *IOP Conference Series: Materials Science and Engineering* **724** 012040 https://doi.org/10.1088/1757-899X/724/1/012040

[10] Houpert L 1999 Numerical and Analytical Calculations in Ball Bearings, *European Space Agency, ESA-S* 438 283

[11] https://www.skf.com/binaries/pub12/Images/0901d19680495562-Super-precision-bearings-catalogue---13383_2-EN_tcm_12-129877.pdf Accessed 15.07.2021

New approach in solving derivative causality problems in the bond-graph method

M Ibănescu[1] **and R Ibănescu**[2]

[1] Technical University "Gheorghe Asachi" of Iaşi, Department of Structural Mechanics, Bd. Dimitrie Mangeron, nr. 1, Iaşi, Romania
[2] Technical University "Gheorghe Asachi" of Iaşi, Department of Theoretical Mechanics, Bd. Dimitrie Mangeron, nr. 59A, Iaşi, Romania

ribanesc@yahoo.com, ibmih@yahoo.com

Abstract From the bond-graph model of a dynamic system, the state equations can be obtained. When the system contains only energy storing elements in integral causality, the system of state equations is immediately and directly determined. When the model contains at least one energy storing element in derivative causality, the resulted system of differential-algebraic equations arises some difficulties in finding the final form of the system of state equations. The work presents a new method of deducing the state equations in case of bond-graph models with one, or several energy storing elements in derivative causality. This method is based on the kinetic energy of the system and offers the possibility to avoid a difficult mathematical calculus for the transition from a system of differential algebraic equations to the system of state equations.

1. Introduction

The bond-graph method was founded by Henry Painter [1], in 1959. The method has been continuously developed, a lot of scientific articles and books being published till then, and some of the most extended works are [2,3,8-11]. Dedicated software has been conceived for simplifying the application of this method, especially useful in the study of mechanical systems dynamics.

The present work approaches some aspects of bond-graph method applied for modelling plane mechanical systems with one degree of freedom, which contains more inertial elements in derivative causality.

When a system does not contain derivative causality elements, the system of equations which describes its dynamics is a system of differential equations. In case of existing elements in derivative causality, the system of equations is a differential-algebraic one [4-7]. It can be transformed, after some algebraic manipulations, in a differential system of equations. The paper presents a method for finding the system of differential equations based on the calculus of the mechanic system kinetic energy.

2. Method description

A plane mechanical system with one degree of freedom is considered. It has n position parameters $q_j\ j=\overline{1,n}$ of which only q_1 is independent. Between the derivatives of these position parameters, that are velocities, relations of kinematic connections can be written, and they have the shape:

IOP Conf. Series: Materials Science and Engineering 1235 (2022) 012053 doi:10.1088/1757-899X/1235/1/012053

$$\dot{q}_i = k_i \dot{q}_1 \quad i = 1, n \tag{1}$$

where $k_i\, i = 1,..,n$ are constants and $k_1 = 1$.

The differential equation obtained according to the bond-graph method, has the form:

$$\dot{p}_1 = -\sum_{i=2}^{n} k_i \dot{p}_i + \sum_{j=1}^{p} F_j \tag{2}$$

where p_i are the generalized momenta and F_j are the forces applied upon the system elements.

The generalized momenta could be written as

$$p_i = \alpha_i \dot{q}_i \quad i = 1, n \tag{3}$$

By substituting (1) in (3), it is obtained:

$$p_i = \alpha_i k_i \dot{q}_1 = \alpha_1 \frac{\alpha_i}{\alpha_1} k_i \dot{q}_1 = p_1 \frac{\alpha_i}{\alpha_1} k_i \quad i = 1, n \tag{4}$$

Relations (4) are differentiated with respect to time and then, these derivatives are introduced in equation (2). It results:

$$\dot{p}_1 = -\sum_{i=2}^{n} k_i \dot{p}_1 \frac{\alpha_i}{\alpha_1} k_i + \sum_{j=1}^{p} F_j \tag{5}$$

or

$$\dot{p}_1 \left(1 + \sum_{i=2}^{n} \frac{\alpha_i}{\alpha_1} k_i^2 \right)_i = \sum_{j=1}^{p} F_j \tag{6}$$

On the other side, the mechanical system kinetic energy is:

$$E = \sum_{i=1}^{n} \frac{\alpha_i \dot{q}_i^2}{2} \tag{7}$$

When equation (1) is introduced in equation (7), the kinetic energy expression [12-14] becomes:

$$E = \sum_{i=1}^{n} \frac{\alpha_i k_i^2}{2} \dot{q}_1^2 \tag{8}$$

The derivative of the kinetic energy with respect to time is:

$$\dot{E} = \left(\sum_{i=1}^{n} \alpha_i k_i^2 \right) \dot{q}_1 \ddot{q}_1 \tag{9}$$

The generalized momentum p_1 can be written as:

$$p_1 = \alpha_1 \dot{q}_1 \tag{10}$$

and its derivative with respect to time is

$$\dot{p}_1 = \alpha_1 \ddot{q}_1 \tag{11}$$

Taking into account that $k_1 = 1$, equation (9) leads to:

$$\dot{E}=\left(\sum_{i=1}^{n}\alpha_1\frac{\alpha_i k_i^2}{\alpha_1}\right)\dot{q}_1\ddot{q}_1=\alpha_1\ddot{q}_1\left(\sum_{i=1}^{n}\alpha_1\frac{\alpha_i k_i^2}{\alpha_1}\right)\dot{q}_1=\dot{p}_1\left[1+\left(\sum_{i=2}^{n}\frac{\alpha_i k_i^2}{\alpha_1}\right)\right]\dot{q}_1=F_{gen}\dot{q}_1 \quad (12)$$

As the derivative of the kinetic energy equals the power, it results that the coefficient of $\dot{q}_1$ in equation (12) is an equivalent force, named generalised force, which coincides with the expression in the left member of the differential equation (6), resulted from the bond-graph method.

3. Example of method application

It is considered a plane mechanical system, represented in figure 1, consisting in a lever, having the arms a and b in length, while the moment of inertia with respect to the hinge is J. At the ends of the lever, two masses are provided and they are denoted by m_1 and m_2. At the left end of the lever a force $F(t)$ applies and at the right end a spring is attached, its stiffness being k_e, and a damper, whose coefficient of viscous friction is γ. These two elements are connected in parallel. The mechanism is the simplified model of a working beam.

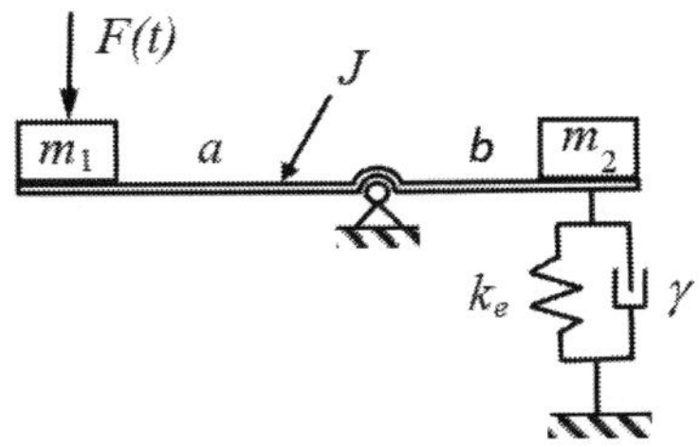

Figure 1. The mechanical system.

The velocity of m_1 is $v_1=\dot{q}_1$, the angular frequency of the lever is $\omega=\dot{q}_2$ and the velocity of mass m_2 is $v_3=\dot{q}_3$, which coincides with the velocity of the spring and damper.
The system has one degree of freedom. As independent parameter is chosen q_1, that is the displacement of mass m_1. The following relations for the kinematic connection can be expressed:

$$\omega=v_1/a \quad and \quad v_2=\omega b=v_1\,b/a \quad (13)$$

By denoting $k_2=1/a$ and $k_3=b/a$, the previous relations become:

$$\dot{q}_2=k_2\dot{q}_1 \quad and \quad \dot{q}_3=k_3\dot{q}_1 \quad (14)$$

The bond-graph model of the system in figure1 is shown in figure 2.
The equations describing the system dynamics deduced from the bond-graph are:

IOP Conf. Series: Materials Science and Engineering 1235 (2022) 012053 doi:10.1088/1757-899X/1235/1/012053

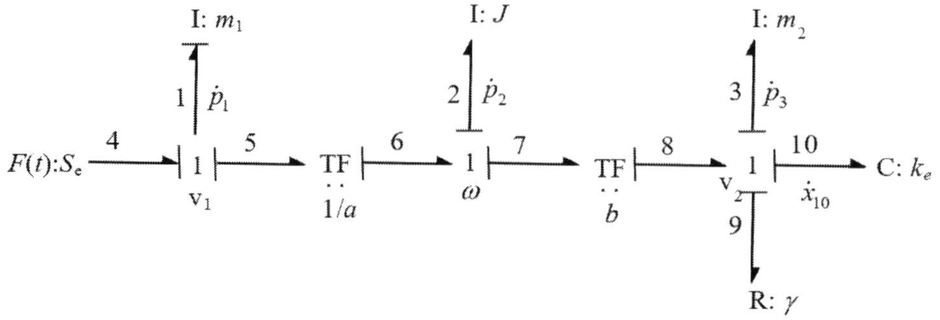

Figure 2. The system bond graph model.

$$\dot p_1 = F(t) - \dot p_2/a - \dot p_3 b/a - k_e x_{10} b/a - \gamma(b/a)^2 (p_1/m) \tag{15}$$

$$p_1 = m_1 \dot q_1 = \alpha_1 \dot q_1 \tag{16}$$

$$p_2 = J\omega = \alpha_2 \omega = \alpha_2 \dot q_2 = \alpha_2 k_2 \dot q_1 = (\alpha_2/\alpha_1) k_2 p_1 \tag{17}$$

$$p_3 = m_2 v_2 = \alpha_3 \dot q_3 = \alpha_3 k_3 \dot q_1 = (\alpha_3/\alpha_1) k_3 p_1 \tag{18}$$

$$\dot x_{10} = (b/a) v_1 = (b/a)(p1/m_1) = (k_3/\alpha_1) p_1 \tag{19}$$

where x_{10} is the displacement of mass m_2, that is, it coincides with the dependent position parameter q_3 and $\alpha_1 = m_1$, $\alpha_2 = J$ and $\alpha_3 = m_2$.

The equations (17) and (18) are differentiated with respect to time:

$$\dot p_2 = (\alpha_2/\alpha_1) k_2 \dot p_1 \tag{20}$$

$$\dot p_3 = (\alpha_3/\alpha_1) k_3 \dot p_1 \tag{21}$$

and they are substituted in equation (15), which is called the equation of generalized momenta.

The following equation, whose variable is p_1, results:

$$\dot p_1 = F(t) - k_2 \dot p_2 - k_3 \dot p_3 - (b/a) k_e x_{10} - (\gamma/m_1)(b/a)^2 p_1 \tag{22}$$

or

$$\dot p_1 \left[1 + (\alpha_2/\alpha_1) k_2^2 + (\alpha_3/\alpha_1) k_3^2 \right] = F(t) - k_3 k_e x_{10} - (\gamma/\alpha_1) k_3^2 p_1 \tag{23}$$

The equations (19) and (23) represent the system of differential equations that models the dynamics of the mechanic system.

The kinetic energy of the mechanical system is:

$$E = m_1 v_1^2/2 + J\omega^2/2 + m_2 v_2^2/2 \tag{24}$$

which could be also written as:

$$E = m_1 \dot q_1^2/2 + J \dot q_2^2/2 + m_2 \dot q_3^2/2 = \alpha_1 \dot q_1^2/2 + \alpha_2 k_2^2 \dot q_1^2/2 + \alpha_3 k_3^2 \dot q_1^2/2 = (\alpha_1 + \alpha_2 k_2^2 + \alpha_3 k_3^2) \dot q_1^2/2 \tag{25}$$

The equation (25) is differentiated with respect to time and it results:

$$\dot{E}=\left(\alpha_1+\alpha_2k_2^2+\alpha_3k_3^2\right)\dot{q}_1\ddot{q}_1=\alpha_1\ddot{q}_1\left(1+(\alpha_2/\alpha_1)k_2^2+(\alpha_3/\alpha_1)k_3^2\right)\dot{q}_1 \tag{26}$$

Because $\alpha_1\ddot{q}_1=\dot{p}_1$, equation (26) becomes:

$$\dot{E}=\dot{p}_1\left(1+(\alpha_2/\alpha_1)k_2^2+(\alpha_3/\alpha_1)k_3^2\right)\dot{q}_1 \tag{27}$$

which is the corresponding power of a generalized force:

$$F_{gen}=\dot{p}_1\left(1+(\alpha_2/\alpha_1)k_2^2+(\alpha_3/\alpha_1)k_3^2\right) \tag{28}$$

It coincides with the left member of equation (23), obtained by using the bond-graph model.

Practically, instead of differentiating the generalized impulses and to replace them in the corresponding differential equation, the kinetic energy is differentiated with respect to time.

This derivative is a power, as a result of multiplying a quantity, called generalized force, by the corresponding velocity of the independent parameter. This generalized force coincides with the left member of the differential equation, expressed in terms of the generalized impulse corresponding to the independent parameter.

4. Conclusions

The mathematical model of systems dynamics provided by the bond-graph method consists in a system of differential – algebraic equations when the model contains storing energy elements in derivative causality. This system can be transformed into a system of differential equations by differentiating the algebraic equations.

The paper presents the use of the method for mechanical system with one degree of freedom and having linear connection equations. These mechanical systems have more position parameters variable in time, of which only one is independent. The chosen independent parameter is that one which corresponds to the inertial element in integral causality.

A differential equation that has as unknown the generalized impulse of the inertial element in integral causality is obtained by using the derivatives of the connection equations. In this equation, the generalized impulse is multiplied by a factor which depends on system characteristics.

The work presents a new method for computing this factor, based on the calculus of the kinetic energy of the system. In this case, the differentiation of the connection equations and their replacement in the differential equation of the generalized impulse are no longer needed.

The mechanical system kinetic energy is expressed in terms of the independent position parameter derivative. After differentiating the kinetic energy with respect to time, the power of a generalized force results and it coincides with the multiplying factor of the generalized impulse in the equation derived by using the bond-graph method.

It is envisaged the extension of the method to more complicated mechanical systems with more degrees of freedom. It is also of interest to see how this method can be applied to non-linear systems. These studies are to be developed in the future.

ORCID IDs

M Ibanescu https://orcid.org/0000-0001-8896-9745
R Ibanescu https://orcid.org/0000-0002-7175-6257

5. References

[1] Paynter H M 1961 *Analysis and Design of Engineering Systems* (Cambridge Massachussetts: M.I.T. Press)

[2] Borutzky W 2010 *Bond Graph Methodology – Development and Analysis of Multidisciplinary Dynamic System Models* (London: Springer-Verlag) http://doi.org/10.1007/978-1-84882-882-7

[3] Ibănescu R and Păstrăvanu O 2001 *Bond-Graph Language in Modeling and Simulation of*

Physical-Technical Systems (Iaşi, Romania: "Gheorghe Asachi" Publishing House)

[4] Ibănescu R and Păstrăvanu O 2002 SOME Aspects regarding the problems generated by the derivative causality in bond-graph models *Bul. Inst. Polit. Iaşi, Section: Mathematics. Theoret. Mech. Phys.* **XLVIII (LII)** 1–2 75–86

[5] Ibănescu R 2011 *Modtech 2011:New face of T.M.C.R. 15th Int. Conf. Modern Technologies, Quality and Innovation,* Vadul lui Voda Republic of Moldova **1-2** 493–6

[6] Ibănescu R 2017 Bond Graphs in System Modeling in *Graph Based Modelling in Engineering*, Mechanisms and Machine Science vol **42** (Springer International Publishing Switzerland) https://doi.org/10.1007/978-3-319-39020-8_1

[7] Ibanescu R and Ibănescu M 2017 Aspects of derivative causality in bond-graph models *MATEC web of Conferences* **112** 06007 https://doi.org/10.1051/matecconf/201711206007

[8] Karnopp D, Margolis D and Rosenberg R 2012 *System Dynamics: Modeling, Simulation and Control of Mechatronic Systems* (New Jersy: John Willey & Sons Inc.) https://doi.org/10.1002/9781118152812

[9] Mughal A M 2016 *Real Time Modeling, Simulation and Control of Dynamical Systems* (Springer International Publishing Switzerland) https://doi.org/10.1007/978-3-319-33906-1

[10] Romero G, Felez J, Morato J and Mera J M 2009 Efficient simulation of mechanism kinematics using bond graphs *Simul Model Pract Theory* **17** (1) 293-308 https://doi.org/10.1016/j.simpat.2008.03.004

[11] Das S 2009 *Mechatronic Modeling and Simulation Using Bond Graphs* (New York: CRC Press) https://doi.org/10.1201/b15831

[12] Gross D, Hauger W, Schroder J, Wall W A and Govindjee S 2011 *Engineering Mechanics 3 Dynamics* (Berlin Heidelberg: Springer-Verlag) https://doi.org/10.1007/978-3-642-53712-7

[13] Palm III W J 2014 *System Dynamics* (New York: McGraw Hill)

[14] Pfeiffer F 2008 *Mechanical System Dynamics* (Berlin Heidelberg: Springer-Verlag) https://doi.org/10.1007/978-3-540-79436-3

IOP Conf. Series: Materials Science and Engineering 1235 (2022) 012054 doi:10.1088/1757-899X/1235/1/012054

How to handle the design preferences with Axiomatic Design

António Gonçalves-Coelho[1,2], João Fradinho[1,2], António Gabriel-Santos[1,2], Miguel Cavique[1,3] and António Mourão[1,2]

[1] R&D Unit for Mechanical and Industrial Engineering - UNIDEMI, NOVA School of Science and Technology, NOVA University Lisbon, Campus de Caparica, 2829-516 Caparica, PORTUGAL
[2] Department of Mechanical and Industrial Engineering - DEMI, NOVA School of Science and Technology, NOVA University Lisbon, Campus de Caparica, 2829-516 Caparica, PORTUGAL
[3] Naval Academy, Base Naval de Lisboa, Alfeite, 2810-001 Almada, PORTUGAL

gonçalves.coelho@fct.unl.pt

Abstract. Picking the best design solution among a small collection of viable alternatives is often a part of engineering design. The traditional method to do it is the weighted multicriteria decision procedure. However, the Axiomatic Design (AD) framework does not employ weighting factors to deal with design preferences. Instead, AD uses the careful placing of each design range. This paper stresses the reasons AD does not include weighting factors—an imaginary flaw that teachers and designers should not attempt to change because that would break the rational soundness of AD teaching and practice. A numerical example supports our rationale.

1. Introduction

The wording "engineering design" refers to a form of creative, multifaceted, and iterative enterprise that promotes the evolution of humanity. There are many pictorial approaches to explain the expression, and one of the best ones is that of figure 1.

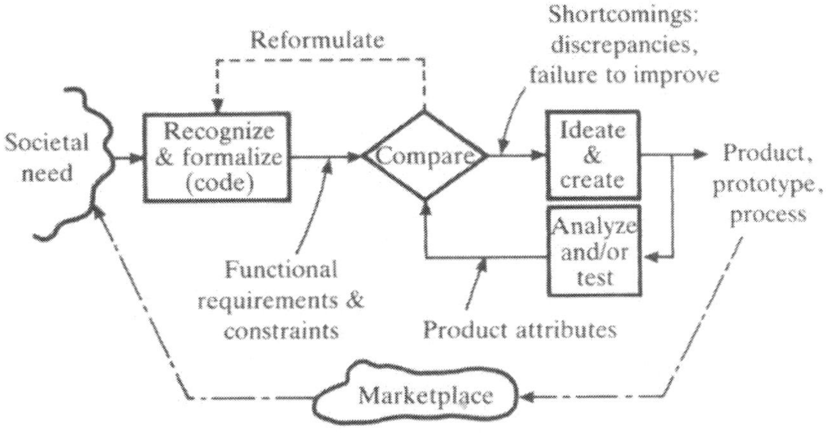

Figure 1. The engineering design after Wilson [1].

The closed-loop paths within figure 1 disclose the iterative nature of the process. Figure 2 depicts engineering design in a much more abridged way.

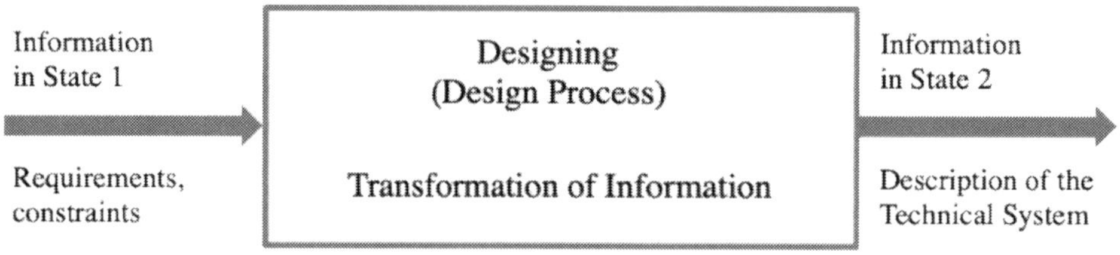

Figure 2. The engineering design after Hubka & Eder [2].

Designers of engineering systems must always consider the different preferences of the multiple stakeholders in their design decisions. For example, a hardware vendor might wish to provide a compact device, and the end-users might prefer easy-to-use systems. In addition, the soon-to-be manufacturer might favor producing an object mainly made of die-stamped parts.

Different methods for decision-making are part of the modern design theory and methodology. This paper aims to show how to deal with the design preferences in Axiomatic Design. Section 2 introduces the basics of the Axiomatic Design (AD). For comparison, section 3 briefly describes the weighted multicriteria decision method (MCDM). Section 4 contains a numerical example, and section 5 is devoted to examining the results attained with both methods. At last, section 6 presents the conclusions.

2. The ABC of Axiomatic Design

Nam P. Suh created the Axiomatic Design (AD) theory in the late 1970s [3] and widely spread it in 1990 through his first textbook on the matter [4]. In the words of Suh, "The ultimate goal of axiomatic design is to establish a scientific basis for design and to improve design activities by providing the designer with a theoretical foundation based on logical and rational thought processes and tools" [5]. In an AD framework, a vector denotes the design object in each of four design domains, as shown in figure 3, and mappings between contiguous design domains depict the design process.

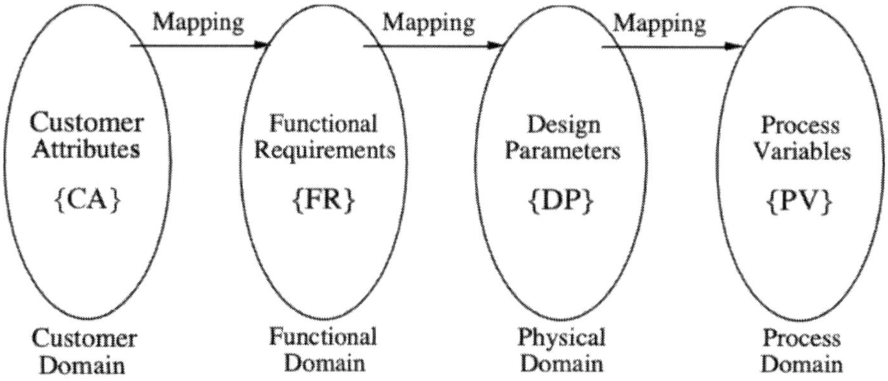

Figure 3. The design world after Suh [5].

The customer attributes vector, {CA}, describes the benefits the customer is expecting. The functional requirements vector, {FR}, arises from the customer attributes and contains the minimum set of requirements that can satisfy the design objectives, stated in precise technical terminology. The functional requirements must be solution-neutral (not biased by solutions imagined by stakeholders). Usually, their phrasing contains action words, such as provide, increase, ensure, etc. [6]. The design parameters vector {DP} specifies the set of elements of the design object meant to satisfy {FR}. It describes the physical features of the design object (aka the embodiment, *e.g.*, parts arrangement, materials, etc.). The process variables vector {PV}, which belongs to the process domain, contains the particularities of the process employed to manufacture the design object. The existing design constraints, (C), do not incorporate figure 1 but represent the economic, social, environmental, political, and technical limits imposed on the design solutions—the input constraints, determined by the stakeholders at the outset of the design process, and the system constraints revealed across the design development. In other words, constraints are limitations to the designers' freedom to search for suitable solutions in the physical and process domains.

Each mapping in figure 3 corresponds to an array of design decisions. For example, the mapping from the functional to the physical domain is denoted by equation (1), where [A] is the design matrix.

$$\{FR\} = [A] \cdot \{DP\},\ A_{ij} = \frac{\partial FR_i}{\partial DR_j}, i = 1..n, j = 1..m \quad (1)$$

In the Axiomatic Design viewpoint, good design decisions conform to the following axioms [7]:

- Axiom 1: Maintain the independence of the functional requirements (FRs).
- Axiom 2: Minimize the information content of the design.

2.1. *The independence axiom*

Concerning axiom 1, the designer must find at least one set of design parameters that fulfill the given functional requirements, which means that the design matrix of equation (1) should be diagonal or at least triangular ($m=n$ in both cases). The diagonal matrix corresponds to an uncoupled design and the triangular to a decoupled design. If $m<n$, then the design matrix is rectangular, and either a coupled design solution results or the FRs cannot be satisfied, as per theorem 1 [7]. The design matrix is also rectangular if $m>n$, but we have a redundant design that is either uncoupled, decoupled, or else coupled, as per theorem 3 [7]. The independence condition is crucial, and more about redundant, independent design solutions can be found in the literature [8].

2.2. *The minimum information axiom*

As for axiom 2, the computation of the information content follows Shannon's definition [9], for whom the probability of success of a one-FR, one-DP design solution is given by

$$I = \log\frac{1}{p} = -\log p \ , \quad (2)$$

where p is the probability of success of the related event. For a one-FR, one-DP design, the probability of success is the ratio of the areas of common range and the system range in figure 4 and is given by

$$p = \frac{\text{Area of the common range}}{\text{Area of the system range}} \quad (3)$$

Figure 4 fully clarifies equation (3). The system probability density function (pdf) of a design solution, p, denotes its probability of success. Vertical straight lines represent the bounds of the design range, and the grey zone is the area within the common range, *i.e.*, the area under the section of the $p(FR)$ curve lying inside the design range. The information content dimensionless unit is the *bit* if base 2 logarithms are used, or the *nat* for natural logarithms.

The information content is given by

$$I = -log \int_{cr} p(FR)dFR \quad (4)$$

and for the whole system range it will be

$$\int_{-\infty}^{+\infty} p(FR)dFR = 1 \quad (5)$$

For a n-FR uncoupled design we have

$$I = -\log\left(\prod_{i=1}^{n} p_i\right) = -\sum_{i=1}^{n} \log p_i \quad (6)$$

The computation of the information content of decoupled and coupled designs is path-dependent, so that "The information contents of coupled and decoupled designs depend on the sequence by which the DPs are changed to satisfy the given set of FRs" as per theorem 7 [7].

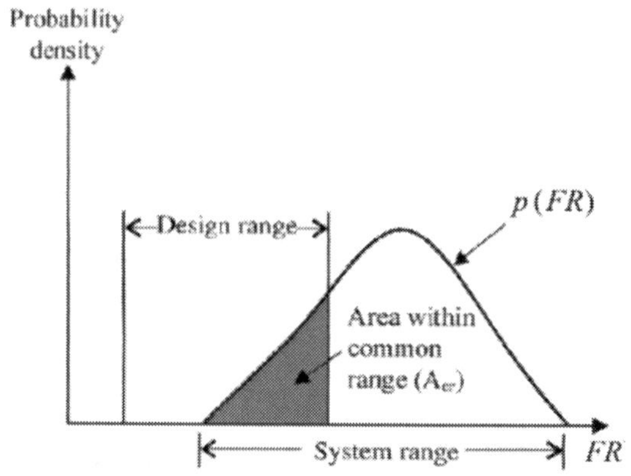

Figure 4. Figuring the information content from the probability density function (adapted from [5]).

Since we know the path for determining the values of the FRs in decoupled design solutions, then employing the fundamental knowledge of conditional probability allows computing their information content. Theorem 17 (Design in the absence of complete information)—which states, "Design can proceed even in the absence of complete information only in the case of a decoupled design if the missing information is related to the off-diagonal elements" [7]—may help in this demanding task. The abovesaid path is undefined in coupled designs so that we cannot compute their information content.

In the AD framework, the minimum information content axiom is the key to select the best of a set of uncoupled and (or) decoupled alternative design solutions. It is worth noting that, by definition, alternative solutions are the ones that share the same FR vector.

2.3. The zigzag decomposition

The design process follows a top-down hierarchical zigzagging pattern between each pair of adjacent design domains, from the most abstract level to a more detailed one, as shown in figure 5. The partly pictured design process in the figure, which top FR is "source of mechanical energy", yields to a two-stroke internal combustion engine.

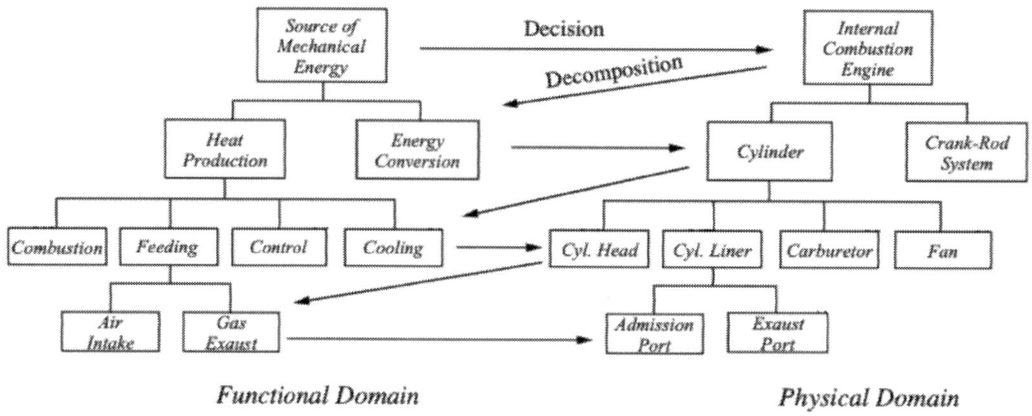

Figure 5. The zigzag decomposition procedure (adapted from [8]).

The design must comply with axiom 1 at each hierarchical level of the decomposition. Zigzag decomposition is unique to AD and should end after attaining the detail required to describe the design object.

3. The weighted multicriteria decision method

Out of the AD context, weighted multicriteria decision method (MCDM) is often the means to choose the best alternative solution.

It is worth noting that the simplest method to deal with such a problem is to sequentially apply the chosen criteria, one at a time, to all the alternative solutions. This procedure allows immediately rejecting the alternatives not complying with at least one of the applicable criteria. However, it has an evident shortcoming: the method does not allow to order the surviving solutions.

The simplest way to determine that ranking is by getting the overall scores for the alternative solutions, C_T, adding their scores associated to each specific criterion, c_i, corrected via weighting factors, w_i, according to the equation:

$$C_T = \sum_{i=1}^{n} w_i c_i, 0 < w_i < 1, \sum_{i=1}^{n} w_i = 1 \quad (7)$$

The criteria to use are indicative values for the functional requirements, FR_i. All the specific criteria are expressed in the same dimensionless scale, for example, 0-10, of type the higher, the better (or inversely, the lower, the better) so that equation (7) yields the overall score for each solution, C_T, denoted in the same dimensionless scale.

Everybody got in touch with some variants of the MCDM method at school since they are used everywhere to grade tests and compare pupils' performance. Employing preferences in the form of weighting factors for the computation of the information content is most probably inspired by this method. But that idea is wrong, and the weighted multicriteria decision method was included here for a better understanding of our claim.

4. A numerical example

The benchmarking of rapid prototyping (RP) machines done almost twenty years ago by Ryder *et al.* [10] suggested this numerical problem. Although its results are outdated, Ryder's work provided the data that inspired this example. The problem here simulated is choosing the best polymer-based additive manufacturing machines to a workshop for mechanical parts with prescribed engineering characteristics. The preselected machines use the following technologies:

- Stereolithography (SLA)
- Fused Deposition Modelling (FDM)
- Laminated Object Modelling (LOM)
- 3-Dimensional Printing (3DP)
- Multi Jet Modelling (MJM)
- Object QUADRA (Obj)
- Droplet on Demand (DoD)

The key factors, or criteria, for selecting the manufacturing technologies were:

i) **Building accuracy.** The exactness of the manufactured parts
ii) **Robustness.** The ability to resist change without modifying its initial configuration
iii) **Building speed.** The speed of the additive manufacturing
iv) **Capital costs.** The cost of machinery, tooling, and workshop infrastructure.
v) **Running costs.** The cost of raw materials, electrical power, labor, etc.

Table 1 lists the capabilities of the machines, with those capabilities assumed as independent in the light of axiom 1. The used values are dimensionless, with the higher, the better.

The scales of items in table 1 have different amplitudes: 0-10 for building accuracy and speed, 0-9 for capital and operating costs, and 0-8 for robustness. All the scales are monotonically increasing, but this characteristic is not mandatory. The different scales' maximum values occur because the original benchmarking included other machines not considered in this work. Besides the different amplitudes, it is worth mentioning that FRs with diverse dimensions are allowed altogether, *e.g.*, speed in ft/s, tensile strength in MPa, production rate in parts per hour, and cost in Euro—because information content

IOP Conf. Series: Materials Science and Engineering 1235 (2022) 012054 doi:10.1088/1757-899X/1235/1/012054

depends on a ratio of areas with same dimensions, as shown by the association of equations (2) and (3) with figure 4. The bounds we considered for the system ranges of the studied machines derive from table 1 and are displayed in table 2.

Table 1. The capability of the rapid prototyping machines (built from [10]).

	Building accuracy	Robustness	Building speed	Capital costs	Running costs
SLA	8	5	6	2	1
FDM	6	8	5	5	4
LOM	6	6	5	5	3
3DP	4	6	9	8	9
MJM	6	2	8	9	7
Obj	8	7	10	8	7
DoD	10	2	2	9	3

Table 2. The bounds of the machines' system ranges.

	Building accuracy		Robustness		Building speed		Capital costs		Running costs	
	min.	max.	min.	max.	min.	max.	min.	max.	min.	max.
SLA	0.0	8.0	0.0	5.0	0.0	6.0	1.5	2.5	0.5	1.5
FDM	0.0	6.0	0.0	8.0	0.0	5.0	4.5	5.5	3.5	4.5
LOM	0.0	6.0	0.0	6.0	0.0	5.0	4.5	5.5	2.5	3.5
3DP	0.0	4.0	0.0	6.0	0.0	9.9	7.5	8.5	8.5	9.5
MJM	0.0	6.0	0.0	2.0	0.0	8.8	8.5	9.5	6.5	7.5
Obj	0.0	8.0	0.0	7.0	0.0	10.0	7.5	8.5	6.5	7.5
DoD	0.0	10.0	0.0	2.0	0.0	2.0	8.5	9.5	2.5	3.5

Table 3 contains the limits of the considered design ranges for the machines under analysis. In terms of duty, four different cases were considered.

Table 3. The design ranges of the studied cases.

		Building accuracy	Robustness	Building speed	Capital costs	Running costs
Case A	min	3.5	3.5	3.5	0.0	0.0
	max	6.5	6.5	6.5	10.0	10.0
Case B	min	7.0	3.5	3.5	0.0	0.0
	max	10.0	6.5	6.5	10.0	10.0
Case C	min	3.5	3.5	7.0	0.0	0.0
	max	6.5	6.5	10.0	10.0	10.0
Case D	min	3.5	3.5	0.0	0.0	0.0
	max	6.5	6.5	10.0	3.0	3.0

Case A depicts general-purpose operation: the design ranges for accuracy, robustness, and building speed are centered close to the midpoint of the system ranges, with an extent (or tolerance) of around ±15% the corresponding scale. The costs are not critical.

Case B differs from case A in that it requires higher accuracy. It reflects the need for a system to produce high-accuracy, one-off parts.

Case C and A are alike, but a higher building speed is required. The associated capital and running costs are not critical. Case C serves the production of small batches.

Lastly, Case D also arises from case A, but low costs are required, and the building speed is not critical. It portrays the regular manufacture of prototypes.

In the real world, the systems' probability distribution functions are often bell-shaped, but uniform functions are used here due to the lack of accurate data, a usual simplification that originates an error in the computed values.

Table 4 displays the information content for the four studied cases, as computed by equation (3) for the hypothesis of uniform distribution. The values used in the computation were the bounds of the system and design ranges from tables 2 and 3.

Table 4. The total information content (bit) of the studied cases.

	Case A	Case B	Case C	Case D
SLA	4.4	6.0	∞	3.2
FDM	4.2	∞	∞	∞
LOM	5.9	∞	∞	∞
3DP	∞	∞	6.4	∞
MJM	∞	∞	∞	∞
Obj	∞	5.7	4.4	∞
DoD	∞	∞	∞	∞

Table 4 is easy to read: The probability of success increases as the information content decreases, and success is fully guaranteed when the information content is zero. Machines with infinite information content are to avoid because they cannot accomplish all the FRs.

The use of the MCDM for the same RP systems is presented next for a matter of comparison. But one has to normalize the data about the robustness, capital costs, and running costs given by table 1 so that the dimensionless scales of all the systems' capabilities are the same. Table 5 includes the results of such conversion for robustness, capital costs and running costs.

Table 5. The capability of the rapid prototyping machines (normalized).

	Building accuracy	Robustness	Building speed	Capital costs	Running costs
SLA	8.0	6.3	6.0	2.2	1.1
FDM	6.0	10.0	5.0	5.6	4.4
LOM	6.0	7.5	5.0	5.6	3.3
3DP	4.0	7.5	9.0	8.9	10.0
MJM	6.0	2.5	8.0	10.0	7.8
Obj	8.0	8.8	10.0	8.9	7.8
DoD	10.0	2.5	2.0	10.0	3.3

Employing equation (7), we used the data of table 5 to evaluate the different RP machines for the equally weighted preferences hypothesis. No more transformations of scale are required since they are all monotonically ascending. Monotonicity of the same sense for all criteria is mandatory in MCDM.

Table 6 displays the attained results, C_T, were c_i and w_i have the same meaning as in equation (7). These results entail the following remark: just by coincidence, the achieved scores (the higher, the better) are not very different from those achieved for Case A via the AD's procedure (*cf.* Table 4, where the lower, the better). The misleading MCDM scores related to the machines discarded through AD are highlighted in table 6 by a light grey background and may look better than the valid ones. They correspond to systems unable to fulfill all the case A requirements, and traditional single-criterion tests would promptly reject the same machines. The above-said coincidence is due to design ranges spanning in the neighborhood of the system range midpoints. Nonetheless, there is no actual similarity between both approaches because they address different issues. The AD procedure deals with relatively well-defined functional contexts, expressed by system and design ranges, while the MCDM tackles vague problems described by indicative values.

Table 6. The computed machine scores for equally weighted preferences.

		Building accuracy	Robustness	Building speed	Capital costs	Running costs	C_T
	w_i	0.2	0.2	0.2	0.2	0.2	
SLA	c_i	8.0	6.3	6.0	2.2	1.1	
	w_ic_i	1.6	1.3	1.2	0.4	0.2	4.7
FDM	c_i	6.0	10.0	5.0	5.6	4.4	
	w_ic_i	1.2	2.0	1.0	1.1	0.9	6.2
LOM	c_i	6.0	7.5	5.0	5.6	3.3	
	w_ic_i	1.2	1.5	1.0	1.1	0.7	5.5
3DP	c_i	4.0	7.5	9.0	8.9	10.0	
	w_ic_i	0.8	1.5	1.8	1.8	2.0	7.9
MJM	c_i	6.0	2.5	8.0	10.0	7.8	
	w_ic_i	1.2	0.5	1.6	2.0	1.6	6.9
Obj	c_i	8.0	8.8	10.0	8.9	7.8	
	w_ic_i	1.6	1.8	2.0	1.8	1.6	8.7
DoD	c_i	10.0	2.0	2.5	10.0	3.3	
	w_ic_i	2.0	0.4	0.5	2.0	0.7	5.6

5. Discussion

The AD's information-based decision-making procedure requires accurate quantitative data about the limits of the design ranges. It also needs data about the boundaries of the system ranges and the corresponding probability density functions. In the lack of reliable data about the pdfs, simplified surrogates may be used, including uniform pdfs. The relative error induced by the substitute pdfs used in this paper is likely similar for the different systems under assessment because they are all of electro-mechanical nature and pursue the same goals. However, the designers can adopt other types of simple pdf, *e.g.*, polygonal, if they find too large the error caused by uniform functions. The fuzzy set theory could help to find proxy distribution functions [11]. One should stress that the AD method's ability to detect unacceptable solutions does not depend on the shape of the probability distribution functions but only on the occurrence of one or more empty intersections of systems and design ranges.

Besides, "All information contents that are relevant to the design task are equally important regardless of their physical origin, and no weighting factor should be applied to them" as per theorem 16 [7]. Therefore, the AD framework excludes weighing factors. Theorem 16 is not surprising since the AD's concept of information content came from Shannon's information theory, and using weighting factors to determine the information contents associated with simultaneous random events is nonsense.

The careful regard on the design ranges' location and span is the right way to reflect preferences in the design decision-making. After all, one cannot sustain that a specific criterion is critical and, at the same time, prescribe for it a wide design range. For example, the design range for the maximum acceleration of a racing car should be small so that the information content of all the alternative solutions is large, which might ease finding the best one. Additionally, it could facilitate rejecting some alternatives if their information content becomes infinite.

The plain MCDM presented here highlights the traditional viewpoint on the link between weighting factors (a concept that is alien to AD) and design preferences. Note that the weighting factors are not independent of each other, as required by axiom 1, making it impossible to change the weighting factors one at a time. However, this method deserves a word of appreciation because it is the *de facto* technique when the decision criteria are qualitative, as it often happens in artificial intelligence and social sciences applications. And yes! It is also helpful in engineering design when design decisions rely on qualitative data, such as expert opinions, in case of which the AD decision technique based on the information content becomes useless.

6. Conclusion

Selecting rapid prototyping machines helped us to remember a procedure alien to the Axiomatic Design framework, which is the use of weighting factors to express design preferences in decision-making processes—as in the case of the weighted multicriteria decision method.

From the practical point of view, the weighted multicriteria decision method might look simpler. But it does not allow a reliable analysis because it is impossible to find out a methodologically correct matching between the weighting factors and the service conditions under study.

The numerical example emphasized that the AD decision method entirely relies on quantitative data about the wanted performance for the design object in the previously defined service conditions.

The chief conclusion is that appropriate position and tolerance of design ranges concerning the corresponding system ranges is the right way to express so-called design preferences since no weighting factors should intervene in an Axiomatic Design framework.

Acknowledgements

The authors acknowledge Fundação para a Ciência e a Tecnologia (FCT, I.P.) for its financial support through the grant UIDB/00667/2020 (UNIDEMI).

ORCID IDs

A Gonçalves-Coelho https://orcid.org/0000-0001-9428-0936
J Fradinho https://orcid.org/0000-0001-5889-771X
A Gabriel-Santos https://orcid.org/0000-0002-1797-2020
M Cavique https://orcid.org/0000-0002-9204-7521
A Mourão https://orcid.org/0000-0002-9470-8477

References

[1] Wilson D R 1980 *An Exploratory Study of Complexity in Axiomatic Design*, PhD Thesis (Cambridge, US-MA: MIT, Department of Mechanical Engineering) p 34
[2] Hubka V and Eder W E 1996 *Design Science: Introduction to the Needs, Scope and Organization of Engineering Design Knowledge*, 2nd Ed. (Berlin: Springer International Publishing) p 4
[3] Suh N P, Bell A C and Gossard D C 1978 On an Axiomatic Approach to Manufacturing and Manufacturing Systems *J. Eng. Ind.* **100** (2) 127–30 https://doi.org/10.1115/1.3439399
[4] Suh N P 1990 *The Principles of Design* (New York: Oxford University Press)
[5] Suh N P 2001 *Axiomatic Design: Advances and Applications* (New York: Oxford University Press) pp 5, 11, 41
[6] Thompson M K 2013 Improving the requirements process in Axiomatic Design Theory *CIRP Ann. Manuf. Technol.* **62** (1) 115–8 https://doi.org/10.1016/j.cirp.2013.03.114
[7] Suh N P 2005 *Complexity: Theory and Applications* (New York: Oxford University Press) pp 23, 44, 45
[8] Gonçalves-Coelho A M, Neştian G, Cavique M and Mourão A 2012 Tackling with Redundant Design Solutions Through Axiomatic Design *Int. J. Precis. Eng. Manuf.* **13** 1837–43 https://doi.org/10.1007/s12541-012-0241-x
[9] Shannon C E 1948 A Mathematical Theory of Communication *Bell Syst. Tech. J.* **27** 379–423, 623–656
[10] Ryder G J, Harrison D K, Green G, Ion W J and Wood B 2002 Benchmarking the rapid design and manufacturing process *3rd National Conference on Rapid Prototyping, Tooling and Manufacturing*, Jun 20–21, London
[11] Kulak O and Kahraman C 2005 Multi-attribute Comparison of Advanced Manufacturing Systems Using Fuzzy vs. Crisp Axiomatic Design Approach *Int. J. Prod. Econ.* **95** (3) 415–24 https://doi.org/10.1016/j.ijpe.2004.02.009

IOP Conf. Series: Materials Science and Engineering 1235 (2022) 012055 doi:10.1088/1757-899X/1235/1/012055

A research of new potential vibration cancellation techniques with the use of piezo stacks

J Harazin and A Wróbel

Wydział Mechaniczny Technologiczny, Politechnika Śląska, Konarskiego 18A, 44-100 Gliwice, Poland

jacek.harazin@polsl.pl

Abstract. Piezoelectric materials offer an ability to exchange energy between electrical and mechanical systems with fair ease, by using the direct and inverse piezoelectric effect. Piezoelectric transceivers, sensors, microphones, actuators or active acoustic noise cancellation devices are some of many applications for piezoelectric materials. Due to the natural ability of the material to convert electricity and mechanical strain, solutions based on piezoelectric materials are often compact and allow applications on micro scales. This article aims at expanding the applicability of piezoelectric materials to dampen vibrations of broader spectrum and bigger amplitudes, in hopes for finding new possible solutions for damping vibrations generated by heavy duty equipment or machines associated with manufacturing processes. The study involves parallel attempts at creating algorithms and mathematical models using classical mathematical algebra and deep learning techniques to actively dampen incoming vibrations with the use of piezoelectric plates. The novelty of this study would involve using piezoelectric plates of different sizes and properties in a single stack, to dampen the frequencies and amplitudes of incoming vibrations.

1. Introduction

Materials exhibiting piezoelectric capabilities feature specific crystalline structures that distribute positively and negatively charged ions in a non-symmetry. Thanks to that specific structure, piezoelectric materials can accumulate charge when mechanical stress is applied and deform when an electric current is run through them. These phenomena are called direct and inverse piezoelectric effects [1,2]. Quartz crystals are mostly recognised by their piezoelectric capabilities, being typically brought up as examples of piezoelectric materials used in modern electronics to provide a constant frequency for precise time measurements. Aside from quartz, the most common piezoelectric materials are piezoceramics and piezoelectric crystals created from different alloy compounds. Those are the two main groups of materials being extensively developed since mid-1900s. Piezoceramics usually contain a mixture of different alloys (notably elements such as lithium, titanium, niobium, barium, zinc, lead) depending on the desired set of material properties [2-4]. Because of the hazards associated with using lead in piezoelectric alloy composites, new methods and materials are being in constant development.

Piezoelectric materials containing polymers are one of the focuses in attempts to decrease the toxicity and widen the range of possible applications [5-7]. Modern industrial applications of piezoelectric materials mostly involve differently sized and shaped piezoelectric plates, foils or sheets and composites or assemblies used in special mechanisms. Devices based on piezoelectric materials have a benefit of being applicable on micro scales, thanks to the piezoelectric phenomena that allow for converting mechanical displacements and electrical currents directly. Due to their rather compact nature, piezoelectric devices are often used in form of transceivers, sensors, microphones, actuators, or active acoustic noise cancellation devices [7-10]. The displacement resulting from applying electric current to piezoelectric plates or foils is very small when compared to the applied voltage. The same can be said about generated voltage when mechanical strain is applied. Those disadvantages are associated with rather poor efficiency of transforming the energy between mechanical and electrical domains. Materials that deal with this problem quite well are piezoelectric crystals which are made from homogenous crystalline structure [11]. They tend to be best at transforming energy but are more expensive when compared to normal piezoelectric plates and are not always an ideal solution. An alternative and cheaper method involves combining simple piezoelectric plates into piezoelectric stacks. A typical piezoelectric stack is made up of multiple piezoelectric slices or plates layered with sheets of insulation. To actuate the stack, electrodes are soldered along both sides of the stack, acting as a positive and negative terminal. Figure 1 represents two variants of a simple piezo stack. Combining multiple layers of piezoelectric

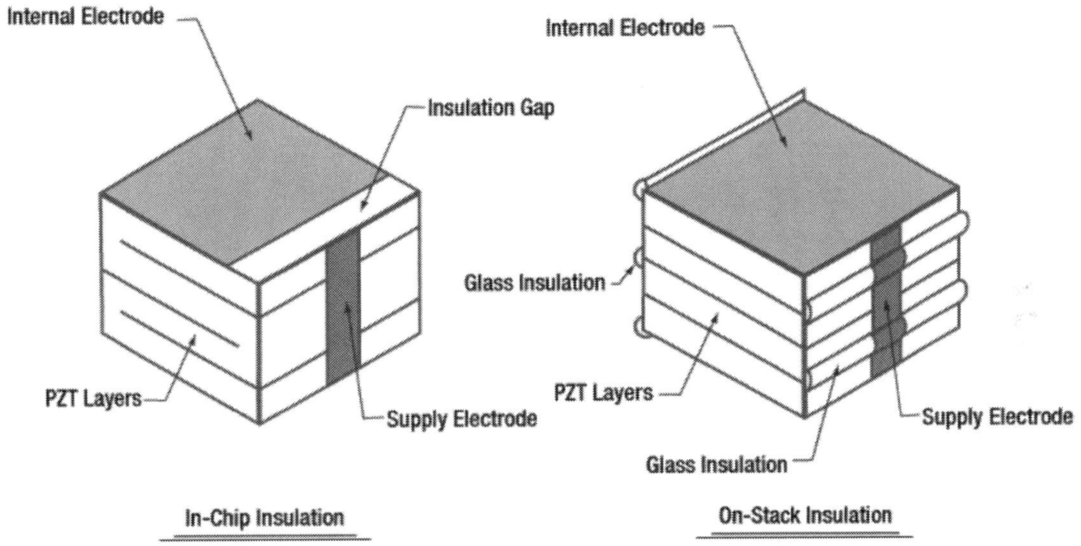

Figure 1. The example of piezoelectric stack and a showcase of its internal composition [12]. The direction of generated displacement depends on the polarisation of plates.

material in such a form helps in increasing the displacement by simply adding the displacements of each layer to the whole. The drawback of such solution is that it takes up more space but can be mitigated by using simple leverage mechanisms in a form called "amplified actuators". The topic of mathematical modelling in solutions using piezoelectric stacks is being continuously researched [13,14] but the focus is being put on piezoelectric stacks that are made from plates of the same thicknesses and material properties.

The novelty of this article revolves around a focus on the possibility of using piezoelectric stacks in active vibration cancellation techniques. A working theory assumes the use of piezoelectric stacks to dampen vibrations with complex spectrum of frequencies by dividing the spectrum between several plates that compose the stack. Each plate would act as a separate "stage" with own set of properties and stimulating currents that deals with a certain part of the spectrum. Several stages would get combined into a stack to achieve both bigger displacement and capability to generate complex vibrations. Such a system could technically provide a bonus in terms of more precise and complex vibration control to the usual functionality of the stack. Mathematical modelling techniques can be used to determine the number of stages and properties of each stage to provide a framework for the operation of the stack. An addition of active modulation by electrical currents with changing frequencies, based on a control circuit utilising mathematical modelling techniques and deep learning methods could provide refined control over the generation of vibrations. Coupled with sensors capable of detecting the frequency of noise generated by machinery and adequate computational power, it may be possible to create an active vibration cancelling device.

IOP Conf. Series: Materials Science and Engineering 1235 (2022) 012055 doi:10.1088/1757-899X/1235/1/012055

2. Methods

The idea of implementing an ability to generate complex vibrations into piezoelectric stack and using it as a form of vibration cancellation device was roughly divided into three main stages. The first stage includes research that must be done to define mathematical models which would allow for determining the sizes and properties of each plate included in designed stack. The second stage involves fine tuning of the designed stack to dynamically adjust to vibrations that have to be cancelled out. For that purpose, we decided to focus on deep learning as a method to gradually fine tune the system response for the incoming vibrations. The third stage includes all the empirical research that must be done to verify and validate all the hypothesis assumed in the project. This involves test stand preparations and gradual testing of each of the steps corresponding to stages of the research. All the stages are being worked on roughly in parallel.

2.1. Mathematical modelling of the desired stack properties

To define a proper mathematical model that would allow the calculation of piezoelectric plate properties and sizes we decided to take the Fourier transform of the vibrations generated by our target device as the entry point for calculations (example of such transform can be seen on figure 2.

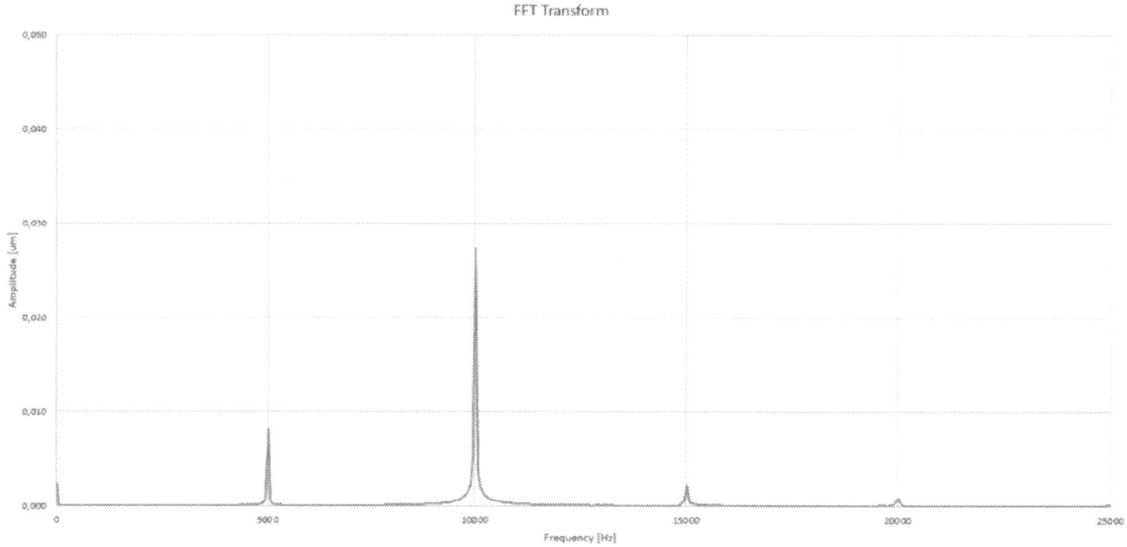

Figure 2. An example of a Fourier transform graph obtained by examining a stack of two piezoelectric plates, each stimulated by a different frequency.

It is possible to determine the transmittance function based on the resonant frequencies taken from the Fourier transform by writing down the resonant and antiresonant frequencies for each peak on the graph. Based on the acquired transmittance, it is then possible to create a cascade model through chain fractioning of the transmittance equation. Chain fractions allow to establish masses and spring loads of a derived cascade model [15,16]. By modifying the resulting model through adding damping factors that correspond to internal damping of the piezoelectric material it is possible to convert it into the Kelvin-Voigt model. After another step of adding the electro motoric forces corresponding to the electrical part of piezoelectric circuit, a model resembling a stack of piezoelectric plates can be formed [17]. A modified Kelvin-Voigt model obtained from the following transformations can be seen on figure 3. In theory, based on the model and a set of boundary conditions in form of piezoelectric material properties (like dielectric constant or Young modulus) or in form of predetermined sizes and shapes of the plates it is possible to calculate the rest of the variables from equations provided by the model. A more thorough analysis of each step will be provided in [16] once the article is published.

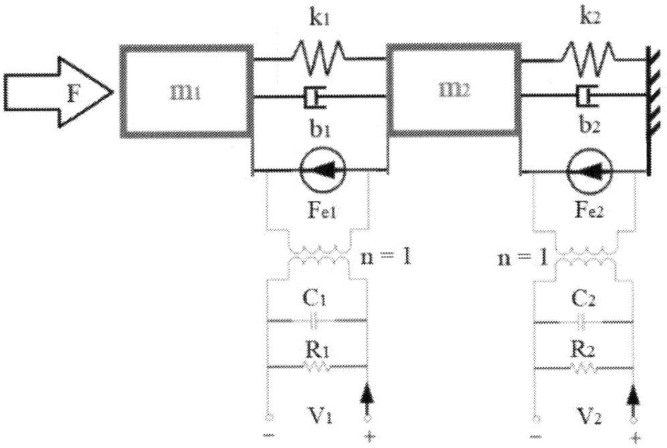

Figure 3. A modified Kelvin-Voigt model representing properties of a stack of two piezoelectric plates with separate electrical circuits [17].

2.2. Fine tuning of frequencies using deep learning algorithms

The process of mathematical modelling can help in determining the initial sizes and properties of piezoelectric plates included in the stack. Assuming real world applications, the target vibration spectrum will be deviating from the average with time which requires some form of real time or near real time adjustment of actuation frequencies to match the spectrum better and prevent situations where the system causes more instability instead of improving it. An approach using the BDLSTM (bidirectional long-short term memory) algorithm was taken [18,19]. So far, the focus has been put on the reduction of hysteresis between the input vibration spectrum and the generated output spectrum. For this purpose, a sample measurement of a working piezoelectric plate was taken in form of a time/displacement graph. The data was cleaned up by creating a time window of considered measurements. The time window was set to 5 seconds. The result was scaled into a min/max function of time and displacement with boundaries set to a range between 1 and -1. Next step included creating sequences for training input. A many-to-one, multi-feature relationship of input and output was considered, where sequences containing time and displacements are used to predict voltage. A time-series of 10 sequences was be chosen. The data was split into training and test datasets, where 80% of the dataset went to the training set and remaining 20% was assigned to the test set. The BDLSTM architecture was prepared based on created datasets. In the input layer of the BDLSTM the LeakyRelu activation function was used, and 300 cells were chosen. Then the hidden layers were added, followed by full dense connected layer. An output layer with tanh activation function was implemented. As a form of evaluation, the model using coefficient of determination R^2 was used where a range between 1 and 0 meant acceptable results.

2.3. Empirical tests

Finally, as a matter of validation and confirmation of the working thesis, a series of tests is being prepared for each step in the research process. The first empirical test involved studying the possibility of combining piezoelectric plates of different shapes and sizes or piezoelectric plates controlled by separate circuits into a single stack [20]. The research also included a study of potential disadvantages of such system and possible complications that must be overcome.

Next planned study will involve testing of the piezoelectric stack output displacement with frequency spectrum close to resonant frequencies for each tested plate in the stack to determine possible benefits and disadvantages in utilizing resonant frequencies to maximize the displacement generated by the stack at each frequency. It is also important to further study the results of using different plates in the stack, in terms of changes in material properties between each layer and its result on the overall hysteresis and displacement of the stack. Next steps will also involve testing of the mathematical algorithm capability to accurately match the best average properties or shapes needed for each stage of the stack to effectively dampen parts of the vibration spectrum and capabilities of the derived deep learning algorithms to dynamically tweak the actuating currents in control circuits to mitigate all the beforementioned possible drawbacks.

3. Results and discussion

Mathematical model focused on calculating the properties and sizes of piezoelectric plates composing designed stack, based on input frequencies is being built based on known methods of synthesis through polynomial chain fractioning and modified Kelvin-Voigt model [15-19]. A publication containing the entire process is still in the review process however and can't be showcased yet. Current concerns involve the accuracy of predicted values resulting from different quality of gathered data and approximations associated with models. The next stage in this part of research will involve tests of displacements generated by real stacks of plates and comparing the results with calculated displacements from the model. Further tests will also include samples gathered from test equipment to get measurements closer to real life applications.

The algorithm using deep learning techniques is showing promising results but was tested with insufficient number of samples and still needs further testing and refining. In terms of the coefficient of determination R^2, a score of 0.99 was achieved when voltage was being predicted based on the displacement. In terms of predicting displacement, based on voltage, the score was at 0.93 and much more error was apparent. It is necessary to introduce a much larger sample size to prove the validity of the model and train it further.

The results obtained from empirical tests conducted so far [20] showed that it is possible to combine different vibration frequencies in a single stack of piezoelectric plates. This assumption was the base for the further development of hypothesis. However, many of the possible sources of interference and factors that may hinder the functionality of this new solution are still not identified and need further testing. Preliminary tests shown that layers of insulation can contribute substantially to the damping factor of each subsequent plate in the stack so it is important to isolate that factor and improve the binding between plates. The effects of utilising resonant frequency for plates in the stack to increase obtained displacements are also still unknown and have to be studied.

4. Conclusions

A concept involving the use of separate circuits and differently sized piezoelectric plates or using plates with different material properties in a stack poses a complex problem of solving multiple fixed and multiple changing variables in the system. The process of synthesizing a solution requires several steps and focus on different aspects of science such as material research, mechanical engineering and artificial intelligence. The proposed concept offers an extension to existing capabilities of piezoelectric stacks and augments their ability of magnifying displacements with an addition of generating complex spectrum of vibrations at the same time. Such solution could be applied in new areas of industry and be an extension to already existing methods of active noise cancellation but applicable in larger scales like industrial equipment or transport. A solution to create such system is still being worked on and requires a lot of time and testing to be considered properly tried and tested though. Theory is still being worked on and preliminary tests have been conducted, but there are still many aspects that remain uncertain and not defined strictly. It is necessary to finalize the process of creating proper mathematical models in next steps of the research and start extensive empirical testing to provide necessary reference and data to validate and verify the assumptions taken in the theoretical approach.

ORCID iDs

J Harazin https://orcid.org/0000-0002-4993-5718
A Wróbel https://orcid.org/0000-0003-4105-2686

5. References

[1] Jiashi Y 2018 *An introduction to the theory of piezoelectricity* (Cham: Springer) https://doi.org/10.1007/978-3-030-03137-4

[2] Soluch W 1980 *Wstęp do piezoelektryki* Wydawnictwo Komunikacji i Łączności

[3] Guo Q, Li F, Xia F, Wang P, Gao X, Hao H, Liu H, Sun H and Zhang S 2020 Piezoelectric ceramics with high piezoelectricity and broad temperature usage range *Journal of Materiomics* **7** (4) 683-92 https://doi.org/10.1016/j.jmat.2020.11.012

[4] Berlincourt D 1981 Piezoelectric ceramics: Characteristics and applications *The Journal of the Acoustical Society of America* **70** 1586-95 https://doi.org/10.1121/1.387224

[5] Huige W *et al.* 2018 An overview of lead-free piezoelectric materials and devices *J. Mater. Chem. C* **6** (46) 12446-67 https://doi.org/10.1039/C8TC04515A

[6] Shrout T R and Zhang S J 2007 Lead-free piezoelectric ceramics: Alternatives for PZT? *J Electroceram* **19** 113–26 https://doi.org/10.1007/s10832-007-9047-0

[7] Maity K and Mandal D 2021 Piezoelectric polymers and composites for multifunctional materials *Advanced Lightweight Multifunctional Materials* ed P Costa, C M Costa, S Lanceros-Mendez (Woodhead Publishing) chapter 7 p 239-82 https://doi.org/10.1016/B978-0-12-818501-8.00001-9

[8] Duan W H, Wang Q and Quek S T 2010 Applications of Piezoelectric Materials in Structural Health Monitoring and Repair: Selected Research Examples *Materials* **3** (12) 5169-94 https://doi.org/10.3390/ma3125169

[9] Li H, Tian C and Deng Z D 2014 Energy harvesting from low frequency applications using piezoelectric materials *Applied Physics Reviews* **1** 041301 https://doi.org/10.1063/1.4900845

[10] Goldfarb M and Celanovic N 1997 Modeling piezoelectric stack actuators for control of micromanipulation *IEEE Control Systems Magazine* **17** (3) 69-79 https://doi.org/10.1109/37.588158

[11] Zhou Q, Lam K H, Zheng H, Qiu W and Shung K K 2014 Piezoelectric single crystal ultrasonic transducers for biomedical applications *Progress in Materials Science* **66** 87-111 https://doi.org/10.1016/j.pmatsci.2014.06.001

[12] Thorlabs on-line brochure https://www.thorlabs.com/newgrouppage9.cfm?objectgroup_id=8040

[13] Adriaens H J M T S, De Koning W L and Banning R 2000 Modeling piezoelectric actuators *IEEE/ASME Transactions on Mechatronics* **5** (4) 331-41 https://doi.org/10.1109/3516.891044

[14] Rodriguez-Fortun J M, Orus J, Buil F and Castellanos J A 2010 General Bond Graph model for piezoelectric actuators and methodology for experimental identification *Mechatronics* **20** (2) 303-14 https://doi.org/10.1016/j.mechatronics.2010.01.004

[15] Białas K, Buchacz A and Dzitkowski T 2009 *Synteza drgających aktywnych układów mechanicznych z tłumieniem w ujęciu grafów biegunowych i liczb strukturalnych* Wydawnictwo Politechniki Śląskiej

[16] Harazin J and Wróbel A 2020 *Synthesis of piezoelectric stack cascade systems by using piezoelements with variable parameters* The 8th Int. Conf. on Modern Manufacturing Technologies in Industrial Engineering. ModTech, June 23rd-27th, 2020, Online edition. Book of abstracts. Red. Constantin Carausu

[17] Sheng-He W and Mi-Ching T 2011 Dynamic modeling of thickness-mode piezoelectric transducer using the block diagram approach *Ultrasonics* **51** (5) 617-24 https://doi.org/10.1016/j.ultras.2011.01.002

[18] Wang G, Yao X, Cui J, Yan Y, Dai J and Zhao W 2020 A novel piezoelectric hysteresis modeling METHOD Combining LSTM and NARX neural networks *Modern Physics Letters B* **34** (28) https://doi.org/10.1142/S0217984920503066

[19] Cui Z, Ke R, Pu Z and Wang Y 2018 Deep bidirectional and unidirectional LSTM recurrent neural network for network-wide traffic speed prediction *Preprint* arXiv:1801.02143v2

[20] Harazin J and Wróbel A 2021 *Empirical analysis of piezoelectric stacks composed of plates with different parameters and excited with different frequencies* The 16th Int. Conf."Mechatronic Systems and Materials", MSM 2021, July 1-2

IOP Conf. Series: Materials Science and Engineering 1235 (2022) 012056 doi:10.1088/1757-899X/1235/1/012056

Theoretical and experimental results of stepped tubes with circular cross-sections

C P Predoiu, S Tabacu, R F Negrea and D Popa

Faculty of Mechanical and Technology, University of Pitesti, Pitesti, Romania

predoiu_paul@yahoo.com

Abstract. The present work focuses on investigating the axial crushing process of circular cross-section tubes with variable diameters using theoretical and experimental methods. In case of an impact, the energy should be consumed by the controlled deformation of the sacrificial structures. The specimens are designed as structures with the same height and have sectors with different diameters and thicknesses. Following the experimental tests, finite element analysis was performed, and the dataset was processed using Matlab. The solution generates a theoretical mean crushing force curve, including the height sectors with different diameters and the mean crushing force curves for experimental and simulation results. The methodology provides the background for a future investigation.

1. Introduction

The current trend in the automotive industry is to minimize cost at both components and assemblies. This trend represents the ongoing research to develop and use new materials, new methods, and technologies for component manufacturing and assembly parts. Intuitive thinking to minimize cost involves using lightweight materials. Still, at the same time, the materials should be strong in case of a collision. Materials are generally characterized usingYoung modulus,shear modulus, and bulk modulus [1,2].

The structure's design has a critical role in case of an impact, because the energy is used by the controlled deformation of the sacrificial structures, defined by tubes with closed profiles. The behavior of thin-walled structures wasfirst studied by Alexander [3] and is representingthe background for the present and future studies [4-8].

In the specialized literature, are presented three types of collapse modes:

- Concertina mode (similar folds)
- Diamond mode (folds with predominant corners)
- Mixed mode (represent a mix of both modes presented above)

Alexander derived the formulas for the prediction of the crushing parameters. Based on the modes presented above the main parameters that affect the axial crushing is the ratio $\frac{D}{t}$ and $\frac{L}{D}$. These parameters were investigated by Abramowicz and Jones [9], Hayduk and Wierzbicki [10]. The Super Folding Element were investigated by Chen and Wierzbicki [11], Wierzbicki and Abramowicz [12] and, structures with triangular cross-section were investigated by Alavi Nia and Parsapour [13], Fan and Lu [14].

Now the diversity is huge and starts from structures with thin walls and with different geometric shapes (circular tube, triangular, hexagonal or rectangular structures, structures with multi-cell, honeycomb), filled structures (aluminium or alloy aluminium, different types of steel, wood, polymer materials, polyurethane foam filled with fiberglass or other materials) till hybrid thin-walled structure.

The mean crushing force (F_m) developed during axial crushing of a structure can be derived by equalizing its work with the energy consumed by the structure during the formation of a fold with a wavelength H[15].

$$F_m \cdot 2H = E_{fold} \tag{1}$$

The maximum value of the force (F_{max}) can be associated with the upper limit from the first fold. Many studies have approached the topic of thin-walled structures with a circular cross-section manufactured from aluminum alloy [15].

The paper analyzes the symmetrical and asymmetrical modes for a set of tubes with circular cross-sections with different diameter sectors made from a drawn pipe, and at the same time will also be analyzed the theoretical and experimental mean crushing force with and without perturbation.

2. Experimental procedure

The behavior of circular tubes under quasi-static axial loading was extensively studied by researchers. A common crushing model for a structure with thin walls is presented.

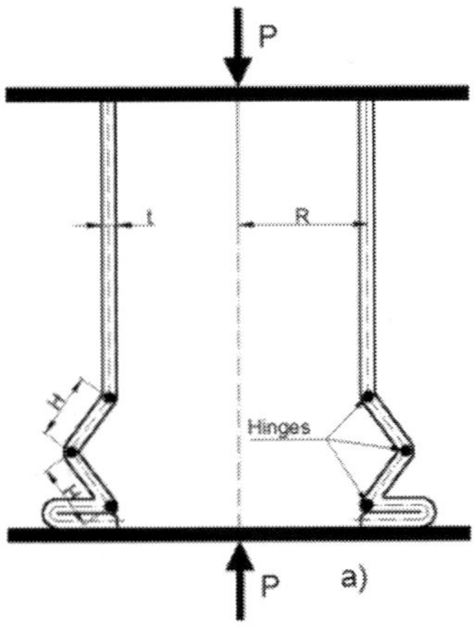

Figure 1. Axi-symmetric collapse mechanism [9].

Considering the factor κ, the effective crush length becomes:

$$F_m \cdot 2H \cdot \kappa = E_b + E_m \tag{2}$$

Where:
E_b – Bending energy
E_m – Membrane energy
Considering that M_0 is fully plastic bending moment / unit length, and L is the length of the circumference, the bending energy (E_b) can be defined as:

$$E_b = 2 \cdot \pi \cdot L \cdot M_0 \tag{3}$$

$$M_0 = \frac{1}{4} \cdot \sigma_0 \cdot t^2 \tag{4}$$

The membrane stretching energy, in the case of circular cross-section structure is:

$$E_{m,tube} = 8 \cdot M_0 \cdot H^2 \cdot \frac{1}{t} \tag{5}$$

The mean crushing force is defined as [15]:

$$F_m = \frac{1}{\kappa} \cdot 22.27 \cdot M_0 \cdot \sqrt{\frac{D}{t}} \tag{6}$$

Based on a study of an analytical solution for the estimate of the mean crushing force of structures with polygonal and star-shaped cross-sections [15], this paper approaches the results of a series of four configurations of circular cross-sections tubes with different sectors, considering the behavior under

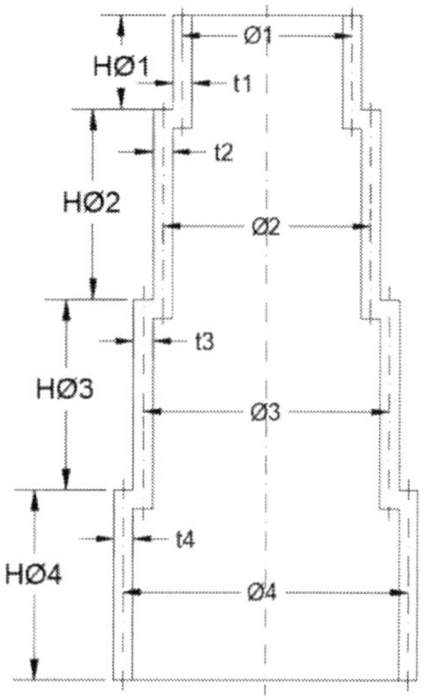

Figure 2. Sample with the description of dimensions.

axial loads. Depending on the geometry, the structures with different cross-sections can be adapted, but in this case, were analyzed only the structures for which we had experimental data in order to be able to compare them.

Table 1 contains the initial dimensions of the samples ($L = 120\ mm$; $\phi_{max} = 35\ mm$).

The material used for the above thin-walled structures with variable diameter sectors is aluminum alloy AA6063. The mechanical properties of the Aluminum alloy used for the experiment are defined by: Young's modulus E = 68.3 GPa, initial yield stress $\sigma_y = 71$ MPa, ultimate stress $\sigma_u = 130.7$ MPa, and Poisson's ratio $\upsilon = 0.3$ [15].

Table 1. The dimensions of the samples.

Samples	H_{ϕ_1} / $t1$ / $\phi_1[mm]$	H_{ϕ_2} / $t2$ / $\phi_2[mm]$	H_{ϕ_3} / $t3$ / $\phi_3[mm]$	H_{ϕ_4} / $t4$ / $\phi_4[mm]$
1st Tube	8 / 1 / 32.6	10 / 1.2 / 32.8	12 / 1.4 / 33	90 / 1.7 / 33.3
2nd Tube	13 / 0.9 / 32.5	13.5 / 1.1 / 32.7	13.5 / 1.2 / 32.8	80 / 1.7 / 33.3
3rd Tube	16 / 1.2 / 32.8	18 / 1.4 / 33	24 / 1.5 / 33.1	62 / 1.7 / 33.3
4th Tube	27 / 0.8 / 32.4	23 / 1.0 / 32.6	30 / 1.2 / 32.8	40 / 1.7 / 33.3

The samples were manufacture from a tube manufactured by extrusion. The specimens were tested under quasi-static conditions using a universal testing machine coupled with a high resolution digital camera for image acquisition.

The data sets generated from the experiment were analyzed with Matlab and LS-Dyna. The following images present the samples before testing.

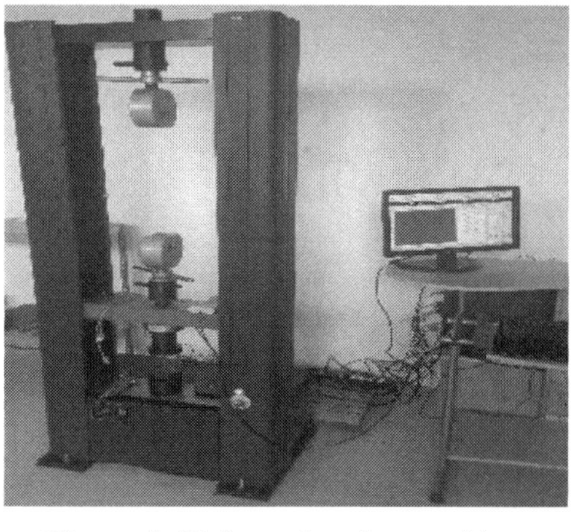

Figure 3. Universal testing machine.

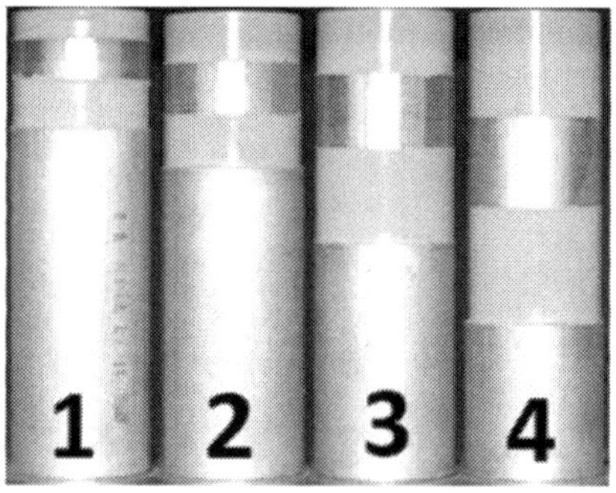

Figure 4. Samples before testing.

3. Results and discussion

After the tests are performed on an universal testing machine, the curves resulting from the experimental analysis are analyzed bellow. Matlab processed the data sets, and the simulations were realized with LS-Dyna. The analysis starts with the samples with different heights of sectors from the bottom to top 90 mm, 12 mm, 10 mm and 8 mm. Their thickness from the bottom is 1.7 mm, 1.4 mm, 1.2 mm and 1.00 mm. The mean diameter of the sectors starts to the bottom is 33.3 mm, 33 mm, 32.8 mm and 32.6 mm. From the experimental results, can be remarked that the tube under axial load, generates a series of peaks. The crushing process was performed on a height of 40 mm from the total height of the sample (120 mm).

For the following graphics in figure 5:

- the black line represents the experimental results of mean crushing force;
- the red line represents the simulation results of mean crushing force without perturbation;
- the red dotted line represents the simulation results of mean crushing force with perturbation;
- the thin red line represents the mean curve for the simulation results without perturbation;
- the thin black line represents the mean curve for the experimental results;
- the horizontal blue line represents the theoretical mean crushing force;
- the vertical blue dotted line represents the sectors of the circular cross-section structure;

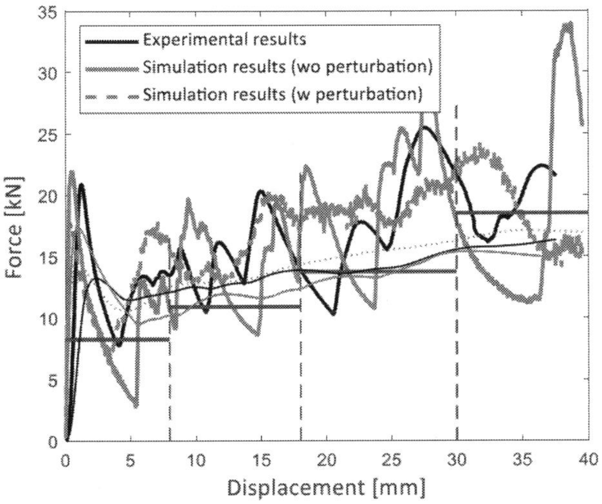

Figure 5. Force – displacement curve for 1st structure.

Was defined a fixed wall (*RIGIDWALL_GEOMETRIC_FLAT_MOTION) and the wall was used to compress the structures axially. To capture the imperfections of the manufacturing process, the coordinates of the node were adjusted (*PERTURBATION_NODE).

Figure 6. Deformation phases of structure – simulation with perturbation.

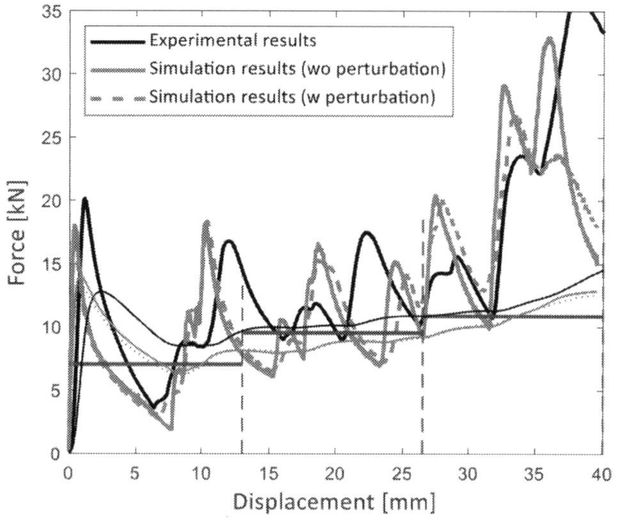

Figure 7. Force – displacement curve for 2nd structure.

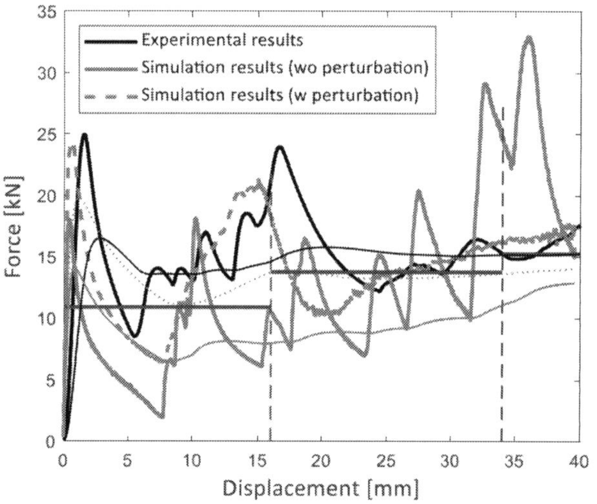

Figure 8. Force – displacement curve for 3rd structure.

The perturbation of the nodal coordinates represent the area of the dimensional precision after samples were manufactured. The keyword *PERTURBATION provides a mean of defining deviation from the designed structure as buckling imperfections. These perturbations can be viewed in LS-PrePost as user-defined fringe plots.

The card used (*PERTURBATION_NODE) includes the type of perturbation (TYPE) which is a spectral field, node-set ID (NSID) which has the value of 101, scale factor (SCL) which has the value of 0.2, CMP (x and y coordinates), ICOORD (the coordinate system used global cartesian), CID (coordinate system ID). Used a spectral field, and the following function (CFTYPE) was defined as Gaussian thus:

$$B(t) = e^{-(at)^2} \tag{7}$$

where:

t – the argument of the function

a – parameter that gives the amplitude of the perturbation

The perturbation was used to cause a deviation comparable with the dimensional precision of the manufacturing process. The results show good accord between the experimental and numerical simulations. A similar value of a peak load was obtained for both studies.

For the simulation case, a comparison of the results for the simulation without perturbation and the simulation with perturbation reveals that the peak force was reached in both cases with a delay, but the loads were achieved for the geometrical model without and with perturbation are identical.

The figure 9 presents the structures from LS-Dyna.

From the Gaussian formula, if the "a" parameter has a small value, then the structure is smooth otherwise, the structure can have a lot of imperfections due to mechanical processing. The next graphic contains only two sectors because the test was performed at 40 mm.

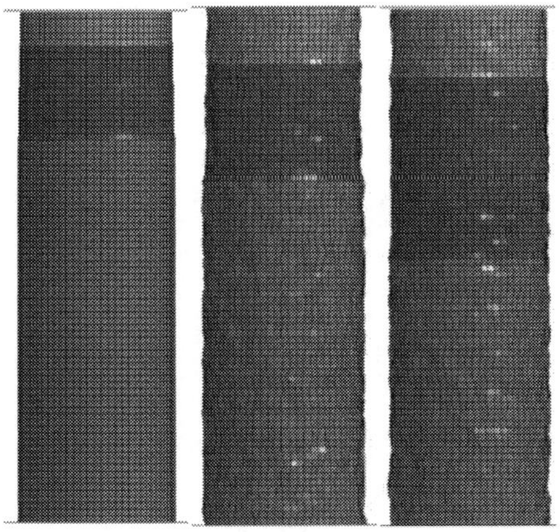

Figure 9. Example of structure:
- First corresponding to the 1st structure without perturbation;
- Second correspond to the 2nd structure with a big perturbation;
- Third correspond to the 3rd structure with a normal perturbation.

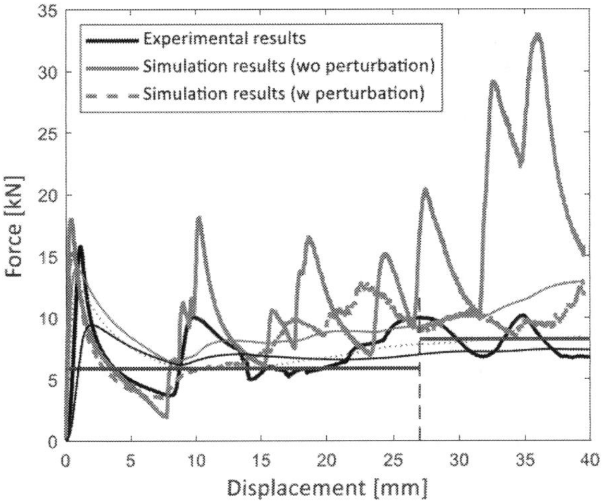

Figure 10. Force – displacement curve for 4th structure.

4. Conclusion

The paper is presenting experimental, numerical, and theoretical results obtained for stepped tubes with a circular section. The benefit of stepped tubes is that they are providing a progressive response under crushing load, a very important requirement for energy absorption devices.

A series of samples were manufactured from a pipe with a circular section. The samples present sectors with different thicknesses, thus the mean crushing force should vary during the collapse of the structure. The tubes were manufactured by turning using a conventional machine. The samples were manufactured from aluminum alloy AA6063. Numerical models were developed to capture the features of the samples. As in most cases, numerical simulation has been used a tool for performance prediction, results were checked with experimental data. To complete the results, the solution of the analytical model was included in the analysis. The analytical model is capable to predict the mean crushing force developed by a structure during axial crushing. The experimental results revealed that the structures are collapsing mostly in a mixed-mode, while the results from the numerical simulation displayed a symmetrical collapse mode. At this point, the influence of the precision of the turning process was included in the simulation as a perturbation applied to the shape of the structure.

The perturbation consisted of a slight deviation of the nodal coordinates. The updated numerical models were subsequently solved and the results included in the analysis. It was found that the precision of the manufacturing process has a major impact on the crushing process as the folding mechanism is influenced by a slight imperfection of the structure (in this case a deviation from nominal thickness). The performances of structures with imperfection are poor compared with the theoretical or ideal structures, thus it becomes important to have a complex investigation of the structures including the performance of the manufacturing process.

ORCID IDs
C P Predoiu https://orcid.org/0000-0003-1625-7380
S. Tabacu https://orcid.org/0000-0002-5736-4724
R F Negrea https://orcid.org/0000-0002-8182-7389
D Popa https://orcid.org/0000-0003-4999-5654

5. References

[1] Evans K E and Alderson A 2000 Auxetic materials: Functional materials and structures from lateral thinking! *Adv. Mater.* **12** (9) 617–28 https://doi.org/10.1002/(SICI)1521-4095(200005)12:9<617::AID-ADMA617>3.0.CO;2-3
[2] Tabacu S, Predoiu P, and Negrea R 2021 A theoretical model for the estimate of plateau force for the crushing process of 3D auxetic anti-tetra chiral structures*Int. J. Mech. Sci.* **199** 106405https://doi.org/10.1016/j.ijmecsci.2021.106405
[3] Alexander J M 1960 An approximate analysis of the collapse of thin cylindrical shells under axial loading *Q. J. Mech. Appl. Math.* **13** (1) 10–15 https://doi.org/10.1093/qjmam/13.1.10
[4] Abramowicz W and Jones N 1997 Transition from initial global bending to progressive buckling of tubes loaded statically and dynamically*Int. J. Impact Eng.* **19** (5-6) 415-37https://doi.org/10.1016/S0734-743X(96)00052-8
[5] Hsu S S and Jones N 2004 Quasi-static and dynamic axial crushing of thin-walled circular stainless steel, mild steel and aluminium alloy tubes *Int. J. Crashworthiness* **9** (2) 195–217 https://doi.org/10.1533/ijcr.2004.0282
[6] Wierzbicki T, Bhat S U,Abramowicz W and Brodkin D 1992 Alexander revisited-A two folding elements model of progressive crushing of tubes *Int. J. Solids Struct.* **29** (24) 3269–88https://doi.org/10.1016/0020-7683(92)90040-Z
[7] Reid S R 1993 Plastic deformation mechanisms in axially compressed metal tubes used as impact energy absorbers *Int. J. Mech. Sci.* **35** (12) 1035–52 https://doi.org/10.1016/0020-7403(93)90054-X
[8] Guillow S R, Lu G, and Grzebieta R H 2001 Quasi-static axial compression of thin-walled circular aluminium tubes *Int. J. Mech. Sci.* **43** (9) 2103–23 https://doi.org/10.1016/S0020-7403(01)00031-5
[9] Abramowicz W, Jones N 1984 Dynamic axial crushing of circular tubes *Int. J. Impact Eng.* **2** (3) 263–81 https://doi.org/10.1016/0734-743X(84)90010-1
[10] Hayduk R J and Wierzbicki T 1984 Extensional collapse modes of structural members *Comput. Struct.* **18** (3) 447–58 https://doi.org/10.1016/0045-7949(84)90065-8
[11] Chen W and Wierzbicki T 2001 Relative merits of single-cell, multi-cell and foam-filled thin-walled structures in energy absorption *Thin-Walled Struct.* **39** (4) 287–306https://doi.org/10.1016/S0263-8231(01)00006-4
[12] Abramowicz W and Wierzbicki T 1983 On the crushing mechanics of thin-walled structures *J. Appl. Mech.* **50** (4a) 727-34 https://doi.org/10.1115/1.3167137
[13] Alavi Nia A and Parsapour M 2014 Comparative analysis of energy absorption capacity of simple and multi-cell thin-walled tubes with triangular, square, hexagonal and octagonal sections*Thin-Walled Struct.* **74** 155–65https://doi.org/10.1016/j.tws.2013.10.005
[14] Fan Z, Lu G, Yu T X and Liu K 2013 Axial crushing of triangular tubes*Int. J. Appl. Mech.* **5** (1) 1350008https://doi.org/10.1142/S1758825113500087
[15] Tabacu S and Ducu C 2019 An analytical solution for the estimate of the mean crushing force of structures with polygonal and star-shaped cross-sections subjected to axial load *Int. J. Mech. Sci.* **161–162** 105010 https://doi.org/10.1016/j.ijmecsci.2019.105010

New applications for shape memory alloys in angular - contact ball bearings preloading systems

Andrei Zamă, Dumitru Olaru and Viorel Paleu

Mechanical Engineering, Mechatronics and Robotics Department, „Gheorghe Asachi" Technical University of Iaşi, Iaşi, 700050, Romania

vpaleu@yahoo.com

Abstract. Shape Memory Alloys (SMA) have been researched extensively in the past forty years. The different available compositions of these alloys can make them suitable for certain applications with specific requirements. The increasing demand for high precision machinery and equipment brought forward the need to search for alloys that that can perform better under stress than conventional materials. This is the case of Shape Memory Alloys. As their name suggests, these alloys can regain their initial shape after being subjected to external stresses, up to a specific degree. Due to the research carried out over the past two decades on these alloys, they became better known and more accessible, being tested in order to be integrated in various applications. One such application consists of using a shape memory alloy formed into disc springs in order to optimize the preloading of angular contact ball bearings. The tests were conducted using a device especially developed for this research, with the purpose of recording the friction torque evolution and the preload induced by disc springs during the operation of the rolling bearings in order to assess the possibility of using elastic elements for active ball bearings preloading systems.

1. Introduction

The progress made in the past few decades in the field of shape memory alloys coupled with the findings in the fields of material science and mechanical engineering, provides the researchers and designers with powerful tools that could overcome the challenges of optimum control for industrial devices. As technology evolves, the requirements for more precise machinery tends to increase as well, making it harder for older machinery to keep up with the technological progress and the environmental regulations. In order to achieve technological advancement, researchers must find new ways to improve the existing elements and to introduce new technology.

Mechanical engineers and industrial designers must collaborate with material science researchers [1] to find new solutions for mechanisms' optimum control. The ball bearings industry benefits from the advances in material science. Many composite materials and new metallic alloys are being tested and used in the ball bearings manufacturing processes and shape memory alloys [2] can be used in various devices and machines that rely on ball bearings [3]. The idea of using SMA as control devices for optimum preload of angular contact ball bearings (ACBB) is new, being advanced by the authors in [4-6].

As a part of these studies, the current paper presents supplementary detail and the results regarding the potential application of Fe-28Mn-6Si-9Cr shape memory alloys [4] in controlling the preload of

IOP Conf. Series: Materials Science and Engineering 1235 (2022) 012057 doi:10.1088/1757-899X/1235/1/012057

ACBB. The experimental results were carried out using a special preloading device, described also in this paper.

2. Experimental section

In order to perform the experiments, we considered using the newly developed Fe-28Mn-6Si-9Cr shape memory alloy to test the effects it induces during the operation of the preload testing device.

Once obtained, the Fe-28Mn-6Si-9Cr shape memory alloy was cold formed into cone shaped disc springs, called modules. [2] A few specimens were tested in order to confirm the behaviour of the alloy. They were subjected to a series of eight compression tests while the modules were placed in different configurations: coupled in a serial configuration, back - to - back and face - to - face.

During the compression cycles, the force was gradually increased and because of the training effect common to shape memory alloys, the shape recovery ratio was determined to be 95%. Temperature plays a key role in the ability of the alloy to revert to its initial shape therefore, this parameter was closely monitored [2].

The second part of the research consisted in testing the possibility to use these modules in order to develop a self-adjustable axial preloading system [4-6]. For this purpose, a tandem face – to - face angular - contact ball bearing, type 7206C was placed inside a preload testing device (figure 1). This type of bearings was chosen because they can accommodate loads in both axial and radial directions, making them suitable for different types of machines that can benefit from an adaptive axial preloading system.

Preloading is required because the bearing's initial internal clearance should be reduced or eliminated during the assembly of the device that they are being used in. In order to ensure a smooth running and high reliability, there are certain cases when these clearances within the bearings need to be further reduced while starting the device and during the actual operation.

Ball bearings in general, are designed with small clearances between their components (the outer ring, the rolling elements and the inner ring – figure 2) in order to ensure a free movement. Sometimes, depending on the application, the internal clearance has to be reduced as much as possible or even cancelled because it can cause the rolling element to skid instead of roll as they should, leading to misalignments between the inner and outer raceways as well as other effects that can cause the bearing to fail, in turn affecting the entire machine that they are being used in. Preloading can be achieved by using two main methods: position preload and constant pressure preload.

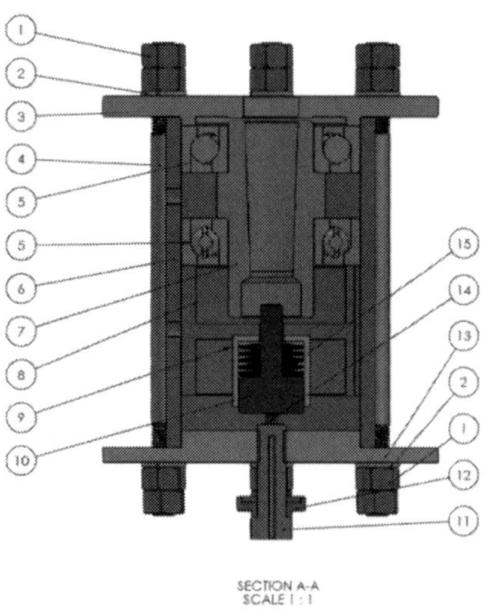

Figure 1. Section view of the preload testing device.

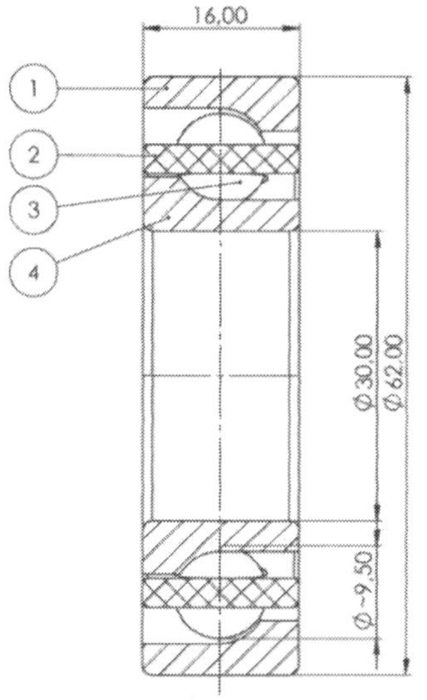

Figure 2. Section view of the 7206C ball bearings.

The preload device mentioned above uses a variation of the constant pressure preload in order to test the shape memory alloy disc springs. A detailed section of the device is given in [4-6]. A section through the device is presented below (figure 3), as well as the preliminary results obtained while testing the SMA disc springs at different initial preload values in regards to the time constraints.

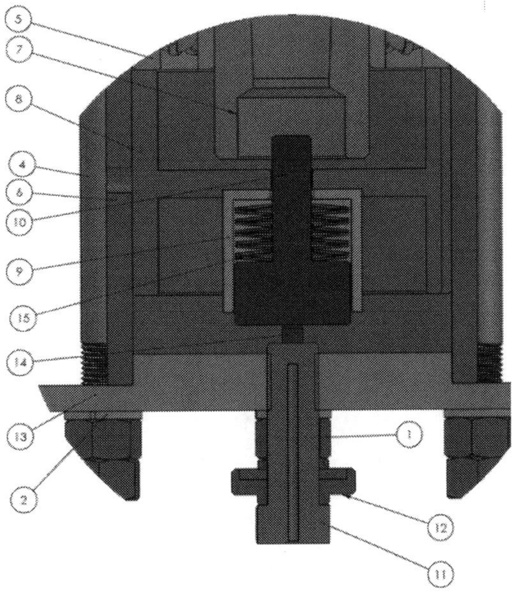

Figure 3. Detailed view of the SMA disc springs and testing principle.

3. Results

The preloading device operates by inducing a uniformly distributed predefined load which is measured through a data acquisition system connected to a load cell (11). The SMA disc springs (15) are centred in and guided with the help of the piston rod (10). The load that SMA made disc assembly exerts on the piston part in turn, transfers the load to the bearing assembly (8) which acts on the outer rings of the 7206C angular contact ball bearings. The first experiments were conducted at low speeds and with grease used as lubricant.

The SMA disc springs were used in different configurations in order to determine the preload that they can maintain while the device was in operation. The tested configurations include variants with paired assemblies of disc springs in back - to - back and front - to - front configurations as well as different numbers of discs.

The results presented below in figures 4 and 5, consist of the load and torque values obtained at low speeds, for the initial tests – 138 rpm and different preload values. For the second test, the speed of the shaft was kept the same as in the first test but the preload was increased by almost 500 N. Also, the testes were performed for an initial equal amount of time to determine if the amount of time the device is in use affects the behaviour of the disc springs. Oher tests were performed at higher speeds and proportional preload as well as tests with low speed and high preload. After gathering data about the behaviour of the springs at low speeds and both low as well as high preload, we gradually increased the shaft speed and the preload in order to observe the effects it has on the bearings.

As proved by the results of tests presented in figure 6, a temperature above 50 °C can activate the shape memory of the disc springs, increasing the preload. In conclusion, this solution can be applied to increase the rotation precision of the shaft of the machine tools, having the main shaft supported by a tandem of the rolling bearings preloaded by SMA discs. The machine may start at an initial preload below the requested optimum one, and after a running-in period the temperature increase over the SMA activation point would establish the requested preload to a higher value, considered the optimum. In such a way, the power loss of the ball bearings supporting the main shaft of precision machine tools could be mitigated, and also the running in period requested to achieve steady-state running conditions.

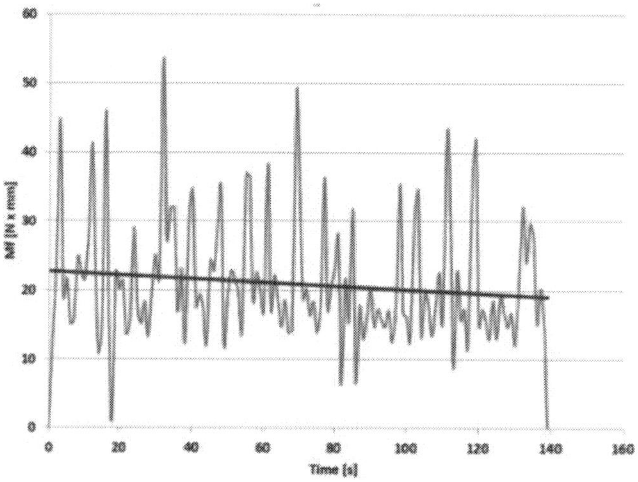

Figure 4. Load and torque versus time – speed: 138 rpm, preload: 144 N.

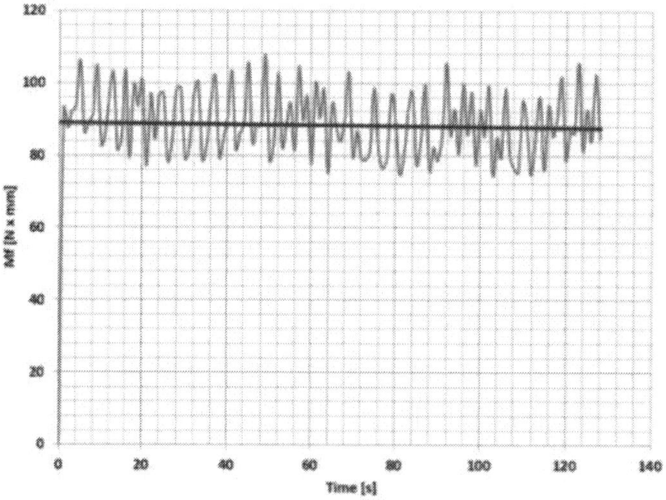

Figure 5. Load and torque versus time – speed: 138 rpm, preload: 640 N.

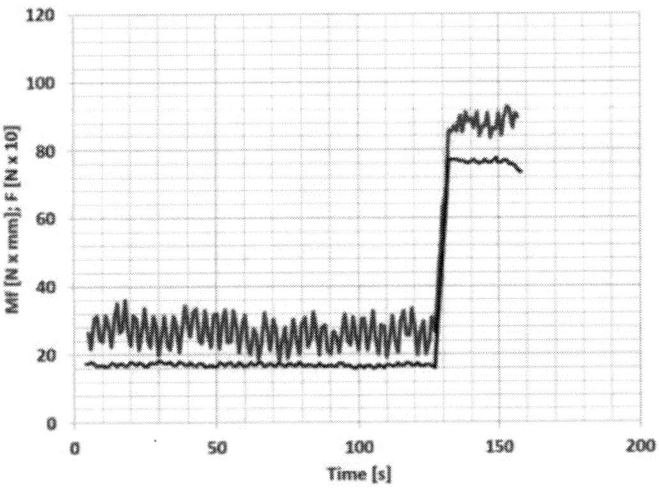

Figure 6. Load and torque versus time –speed: 552 rpm; preload: 170 N.

4. Conclusions

The proposed ball bearings elastic preloading system and the preliminary results show this is a viable concept that can be applied and implemented in practical applications that require an adaptive preloading system. This can be a practical solution for preloading ball bearings of precision machine tools. By adapting the solution presented in this paper and using it in various applications it will be possible to overcome certain actual technical barriers and reduce the operating costs of industrial machinery.

Further and more complex testing is necessary in order to determine the best operating parameters for the Fe-28Mn-6Si-9Cr shape memory alloy and to overcome the technical limitations of operating high speed ball bearings.

ORCID IDs

A Zamă https://orcid.org/0000-0001-5298-7765
D Olaru https://orcid.org/0000-0001-5989-0798
V Paleu https://orcid.org/0000-0003-1763-204X

5. References

[1] Murakami M, Suzuki H and Nakamura Y 1987 Effect of Si on the Shape Memory Effect of Polycrystalline Fe Mn Si Alloys *Transactions of the Iron and Steel Institute of Japan* **27** (3) b.87

[2] Bujoreanu L G, Dia V, Stanciu S, Susan M and Baciu C 2008 Study of Tensile Constrained Recovery Behavior of a Fe-Mn-Si Shape Memory Alloy *Eur. Phys. J. Spec. Top.* **158** 15-20 https://doi.org/10.1140/epjst/e2008-00647-5

[3] Harris T A and Kotzalas M N 2006 *Rolling Bearing Analysis* 5th Ed (Boca Raton: CRC Press) https://doi.org/10.1201/9781482275148

[4] Zamă A, Paleu V, Bujoreanu L G, Paleu C C and Olaru D 2019 Advances in Angular Contact Ball Bearings Testing Machine Design *IJMMT* **SI XI** 3 137-42

[5] Paleu V, Gurău G, Comăneci R I, Sampath V, Gurău C and Bujoreanu L G 2018 A new application of Fe-28Mn-6Si-5Cr (mass%) shape memory alloy, for self-adjustable axial preloading of ball bearings *Smart Mater. Struct.* **27** 075026 https://doi.org/10.1088/1361-665X/aac4c5

[6] Zamă A, Olaru D N and Paleu V 2020 New Concepts in Friction Reduction in Preloaded Assemblies of Angular-Contact Ball Bearings *IOP Conf. Ser.: Mater. Sci. Eng.* **997** 012020 https://doi.org/10.1088/1757-899X/997/1/012020

IOP Conf. Series: Materials Science and Engineering 1235 (2022) 012058 doi:10.1088/1757-899X/1235/1/012058

Adjustable suspension system for lower limb prosthesis

Alina Rodica Spanu[1], Daniel Besnea[1], Edgar Moraru[1], Mazen Elfarra[2] and Oana Chiriac[3]

[1] Department of Mechatronics and Precision Engineering, University Politehnica of Bucharest, Splaiul Independentei 313, Bucharest, Romania.
[2] Hospital Sf. Ioan, Vitan Barzesti 13, Bucharest, Romania
[3] S.C. Activ Ortopedic S.R.L., Petru Maior 61, Bucharest, Romania.

spanu_alina@yahoo.com, alina.spanu@upb.ro

Abstract. The contact between the stump and the lower limb prosthesis has to be provided by its suspension system. The paper presents a solution for an adjustable suspension designed as an inflated cell network. Each cell is supplied with controlled pressure and consequently, a contact force with the skin is developed. By varying the pressure value inside the cell, we may provide variable force, so the compensation for the volume change of the stump could be accomplished. The material the cells are made of should have special properties, the elasticity being the most important. Moreover, the goal of the paper was to approve the opportunity to use the silicone rubber Zhermack for the cell material and a mechatronic system has been realised as experimental set-up.

1. Introduction

The lower limb prosthesis has to enable ambulation as the patient could have a normal life, walking and working. Otherwise, some pains, discomfort, and troubles are experienced, so he will become a traumatized person by his illness. The main role of the suspension in the prosthesis is to firmly attach it to the stump. Consequently, one of the most important requirements of lower limb prosthesis design is the suspension one, whose type and technical features are focused for the wearable fit of the socket. Here we could mention the self-suspending technique, suction and vacuum socket, silicone liners and socket made of polypropylene.

If an improper fit is acting, due to volume variation of the stump during the day, some illness will appear, so a reliable tighten has to be provided. If the contact area is smaller than normal, meaning the existence of clearance with great values, the patient has limited mobility, some unbalanced dynamic forces are acting and the single-leg stance will be the standing position. There are two ways of balancing the body: moving the centre of pressure under the stance foot, named prosthesis with passive stability, or counter-rotation movements around the centre of mass, knowing as active stability. Another embarrassing circumstance will arise, if the socket is very hard adjusted, so tissue breakdown could appear, degenerative tissue ulcer, follicular hyperkeratosis, allergic contact dermatitis, infection, and veracious hyperplasia.

The state-of-the-art describes some solution for avoiding the inconveniences presented before. The adjustable socket [1] is intermittently adjusted before the effects of losing suspension are acting and clinically problems occur. The authors investigated a strategy depending on the adjustment values, while the limb fluid volume is increasing or decreasing, as the sensed distance is varying.

There were fabricated embedded socket with inductive sensors for real-time sensor measurement. A cable connected to an electrical motor has translational motion with increments of 4.5 mm that will be transformed in radial distance 0.4 – 0.5 mm between the socket and the patient tissue. This solution affords a 0.3% volume variation.

From another point of view, the paper [2] pointed out the new design solution for preventing the comfort decreasing based on real-time measurements of limb volume variations. Accordingly, the electromagnets are embedded in the socket and the magnetic material is embedded in the liner, so that the feedback loop was closed by using a pressure sensor and a microcontroller. The force between the socket and the liner is 89 N, while the pressure is 80 kPa.

The pressure distribution over the entire contact area between the socket and the stump could be very carefully monitored by using the load cell with six degrees of freedom [3,4].

The solution described in the paper [5] proposed the air splint prosthetic socket system, which uses the pressure sensor placed over the working area of the socket and the microcontroller for maintaining the pressure inside around 40 kPa as clinical principles require. If the pressure is increasing more, the blood system is interrupted.

Many papers presented the pressure distribution studies by using the sensors inside the socket [4], but they considered only the normal loading. As some practice results concluded, the shear forces, meaning longitudinal and transverse directions, are more detrimental to skin health, but the pressure sensors cannot provide enough information about. Therefore, the load cells are placed at the bottom of the socket or custom transducer was designed.

Another problem is the material used as interface between the stump and the transtibial socket, because of its special physical properties such as silicone, which has a very low friction coefficient and provides a good response in high impact. The International Committee of the Red Cross (ICRC) mentions physical characteristics of the polypropylene socket as guidelines. The liner made of silicone will provide the gripper force between the air splint and the residual limb [5]. At the bottom part of the air splint socket were placed the microcontroller, the electrical motor and the joint for readily access in case of unpleasant or unexpectedly event.

The paper presents a suspension solution for the lower limb prosthesis that is using a number of cells with variable shape made of Zhermack material. The external surfaces of the cell have the main goal of being the contact area between the skin tissue and the socket. The cells are inflated with dry air and the pressure inside is monitored. The material elasticity is an important requirement because of the opportunity of external surface deformation, so the friction forces are provided assuring the contact with the skin.

2. The pneumatic system used for the adjustable suspension

As we have mentioned before, the suspension system is the most important part of the lower limb prosthesis due to its role of assuring a firmly contact with the skin of the patient. The proposed mechatronic solution consists of a well-known number of pneumatic cells placed around the residual limb in a network, so they may be inflated with dry-air pressure by using a pneumatic circuit. The pressure values, controlled with a microcontroller, could produce a deformed shape due to the material elasticity and the contact with the skin is preserved as the friction force occurs. The number of inflated cells and their three-dimensional position around the stump could be controlled too, so that a permanent contact area will be available even though the volume is variable.

The pneumatic systems are widely used for their properties such as high power – to – weight ratio, cleanness, easy maintenance and low cost. The pneumatic components are easier to work with, as well as safer, but we have to take into account the air compressibility and the non-linear flow.

The proposed pneumatic circuit is presented in figure 1. It comprises: 1 – pressure source needed for mathematical modelling; 2 – pneumatic check valve, which is normally closed; 3 – pneumatic compressor; 4 – the pneumatic pressure – reducing regulator with maximum flow coefficient, having the role of maintaining a downstream pressure, by reducing the pressure from the compressor to the pressure value required for the pneumatic cell; 5 – modulated pneumatic orifice having a variable area; 6 – the inflated cell.

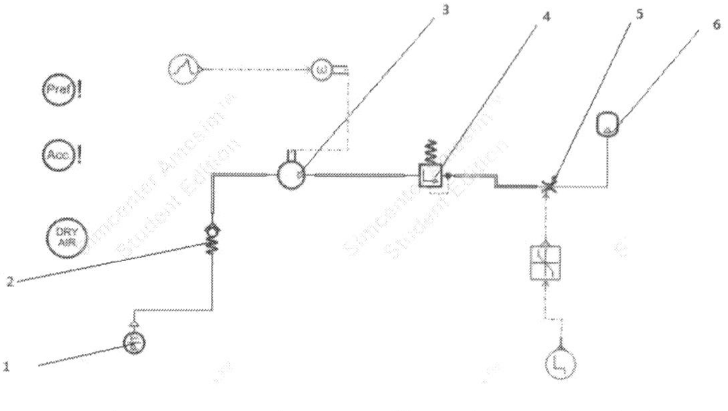

Figure 1. The pneumatic circuit used for the adjustable suspension system.

The pneumatic compressor is actuated by using an electrical motor whose command parameter law is very well known. The area of the modulated pneumatic orifice is controlled too with an imposed law. For the numerical set-up we have used the following values: the absolute pressure controlled with the pneumatic check valve is 1.983 bar, while maximum operating pressure is set 2 bar; the orifice area at maximum opening is 2.5 mm^2; the cell volume is 0.58 m^3; the working temperature is 28 °C.
The simulation results were obtained by using the AmeSim software – student edition and there are presented below. The working time is about 1.2 s for providing the working pressure value inside the cell, as we may conclude by analysing the results shown in figure 2 and figure 3.

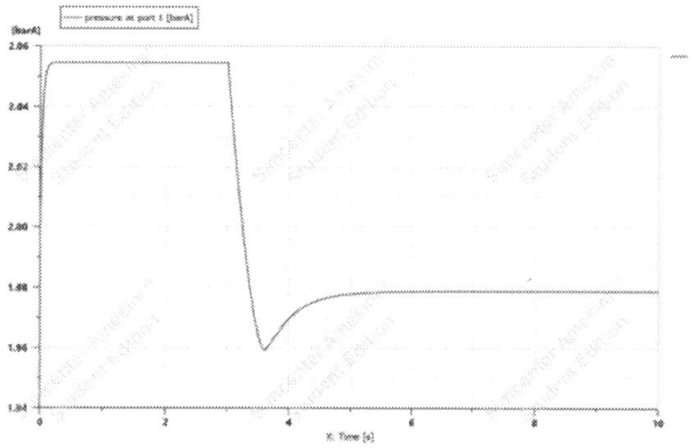

Figure 2. The pressure variation inside the cell

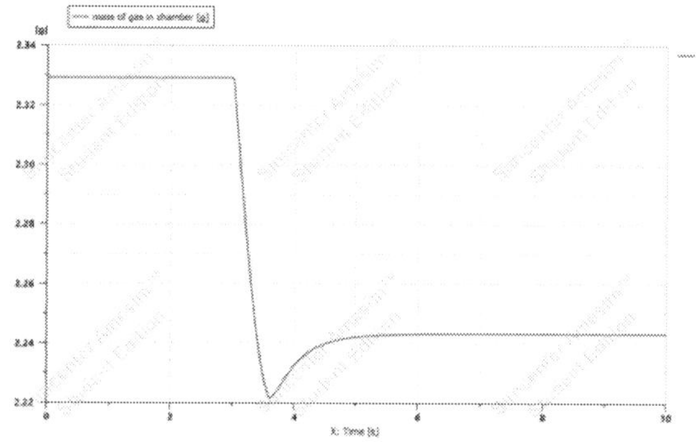

Figure 3. The mass of gas variation inside the cell.

3. The pressure cell manufacturing

The manufacturing of the pressure cell was done by using the rapid prototyping technology and the material was ZA13 RTV Mould Zhermack, affordable for medical devices. This two-component silicon material (base and catalyst) has the property of hardening at room temperature, when the mixing ratio of 1:1 is optimally observed, has a good dimensional stability, it is resistant to high temperatures and aggressive components. It has a great elasticity at working temperature 23 °C and its high-level fluidity is suitable for moulding of a large variety of shapes. The Zhermack material has the tear of 13-14 N/mm, the hardness 13 ShareA and the viscosity 4500 cP, so there are very good mechanical properties suitable for our purpose [7]. In order to obtain the pressure cell, we have designed a mould by using the Catia V17 software, the STL file was obtained and finally it was manufactured on the 3D printer with the FDM procedure as it is shown in figure 4 and figure 5. The printer used was Makerbot Replicator, and as a material polylactic acid was used to make the moulds (this equipment is provided only for use of PLA material). With a maximum execution volume of approximately 7500 cm^3 and a minimum applied layer thickness of 100 microns, this 3D printing equipment is a suitable solution for realizing relative precise components when it comes to thermoplastic extrusion technology. Among the important advantages of this equipment can be highlighted: the easy-to-use software where the extrusion temperature, the execution speed, the degree of filling, the distribution mode of the support structures and other important process parameters are set; the sacrificial layers are easily removed; the possibility of remote control; the quality of the resulting surfaces are relatively acceptable compared to other similar FDM equipment [8]. In our case, we have used the 200 microns layer thickness and 30% infill density, and an extrusion temperature of 200 °C for polylactic acid material in order to obtain necessary components through fused deposition modelling technology.

As the cell core was ready to be used for the pneumatic cell manufacturing, the next stage of the process was the moulding of the pneumatic cell walls. In order to do this, the RTV Zhermack material was poured into the mould with the core [6], as it is presented in the figure 6 and the final result of the cell shape after the solid state is obtained as it is shown in figure 7.

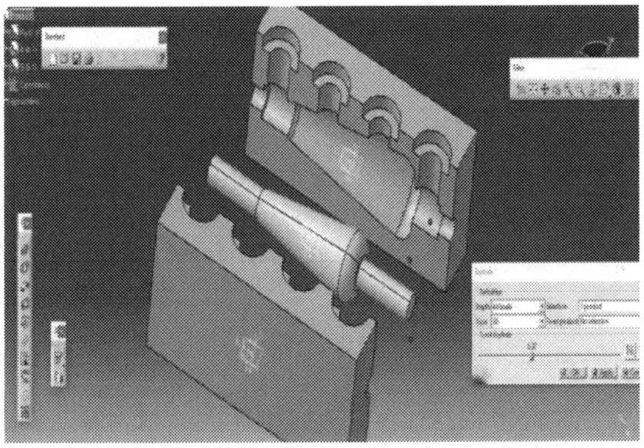

Figure 4. The designed mould for core manufacturing.

Figure 5. The manufactured mould with the core

Figure 6. The material is poured into the mould.

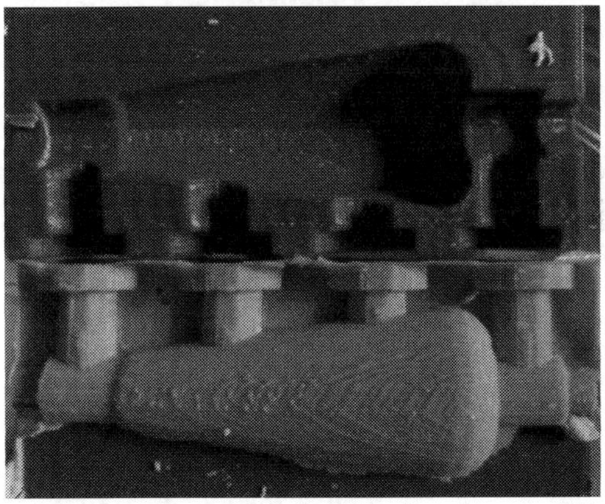

Figure 7. The pneumatic cell after the moulding process.

4. The experimental set – up for verifying the pressure and the contact force

The main goal of this study was to propose a solution for adjustable suspension when the pneumatic cell is inflated with dry air, so the main parameters we have to verify experimentally were the pressure and the contact force with the skin. We have used a pneumatic compressor 1, the pipe 2, pneumatic cell 3, the force sensor 4, the high accuracy micrometric screw 5 (figure 8). The sensor force could be set a very precise movement controlled with the micrometric screw just for establishing the contact with the cell made of silicone rubber Zhermack. For controlling the force developed during the contact, we have used the Arduino Uno microcontroller as it may be seen in figure 8 and figure 9.

IOP Conf. Series: Materials Science and Engineering 1235 (2022) 012058 doi:10.1088/1757-899X/1235/1/012058

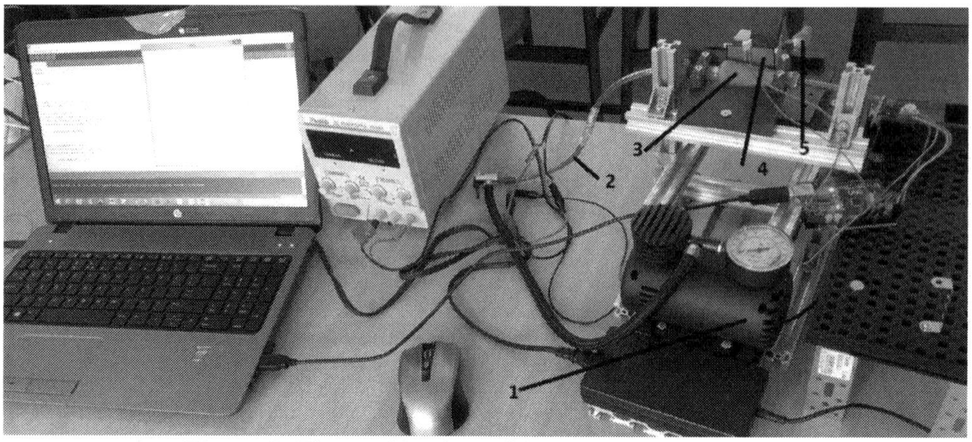

Figure 8. The upside view of the experimental set-up.

A detailed view of the force sensor 1 and the pressure sensor 2 for real time measuring of pressure values developed inside the pneumatic cell, are shown in figure 9. All the experimental process is controlled with the microcontroller Arduino Uno in order to establish the experimentally dependence of force on pressure over time.

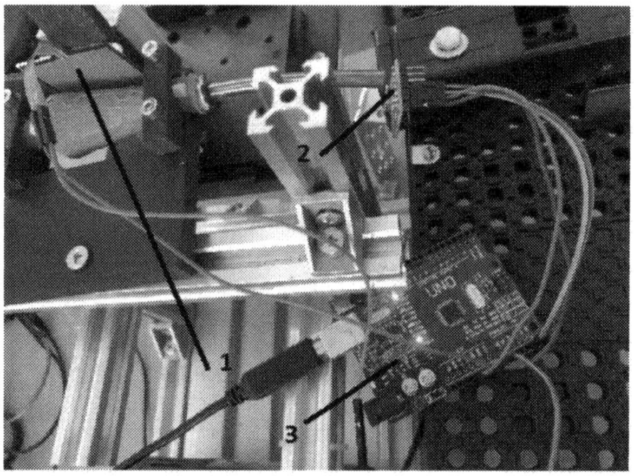

Figure 9. The pressure sensor and the force sensor – detailed view.

Both measured values for the pressure and for the contact force with the skin approved the influence of elasticity of the silicone rubber Zhermack, so that we may consider it appropriate to this purpose. All the experimentally process has been developed at 23 °C. The figure 10 points out the pressure variation over time and in figure 11 we have represented the contact force variation upon time. As we may infer from the latter variation, the contact force is increasing when the pressure is increasing too and it remains constant when the pressure is bigger enough. The shape of the pressure cell should be studied as future work in order to improve the contact force.

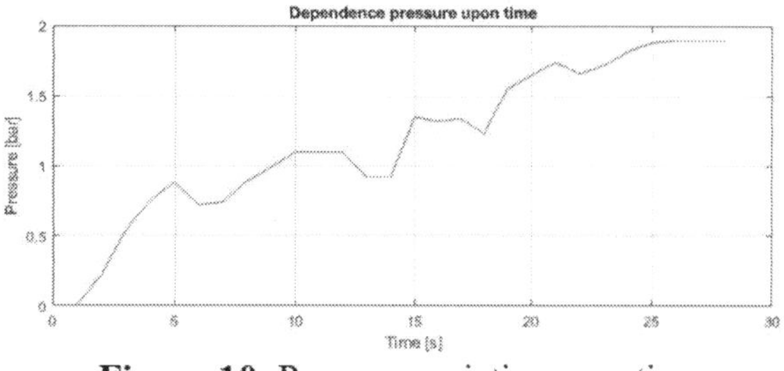

Figure 10. Pressure variation over time.

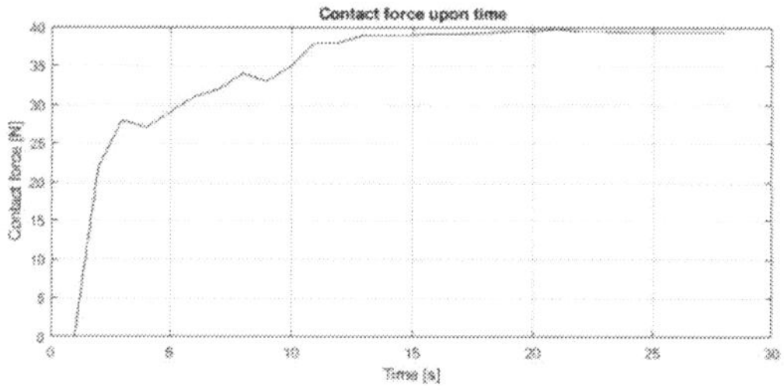

Figure 11. Contact force variation over time.

5. Conclusions

The paper presents a solution for adjustable suspension system based on a network of inflated cells placed around the stump. The property of being adjustable was achieved by inflating the cells with variable pressure as the volume varies during the day. The cells are made of silicone rubber Zhermack that has physical properties suitable for our purpose. The experimental set-up was made as a mechatronic system controlled with Arduino. The measured values pointed out the dependence of contact force upon pressure, so that the material elasticity is good enough.

Acknowledgments

This work was supported by a grant of the Romanian Ministry of Education and Research, CCCDI – UEFISCDI, project number 1PTE/2020, cod PN-III-P2-2.1-PTE-2019-0484, within PNCDI III. The title "Proteză modulară de membru inferior cu sistem de suspensie pneumotronic controlată şi personalizată" („Lower Limb Prosthesis with pneumotronic suspension system controlled and personalized").

ORCID IDs

A R Spanu https://orcid.org/0000-0001-8797-5852
D Besnea https://orcid.org/0000-0002-8513-810X
E Moraru https://orcid.org/0000-0002-5745-8165
M Elfarra https://orcid.org/0000-0002-3682-199X
O Chiriac https://orcid.org/0000-0003-4487-7720

6. References

[1] McLean J, Redd C, Larsen B, Garbini J, Brzostowski J, Hafner B and Sanders J 2019 Socket size adjustments in people with transtibial amputation: Effects on residual limb fluid volume and limb-socket distance *Clinical Biomechanics* **63** 161-71 https://doi.org/10.1016/j.clinbiomech.2019.02.022

[2] Ng J, Johnson J, Miller S, Newton K, Roe T and Woo R 2017 Electromagnetic suspension system for control of limb volume in prosthetics *Procedia CIRP* **65** 180-3 https://doi.org/10.1016/j.procir.2017.04.047

[3] Ali S, Noor A, Arezoo E, Hossein G, Nasrul A and Wan A 2013 Interface pressure in transtibial socket during ascent and descent on stairs and its effect on patient satisfaction *Clinical Biomechanics* **28** (9-10) 994-9 https://doi.org/10.1016/j.clinbiomech.2013.09.004

[4] Lenz A and Bush T 2019 Evaluating shear and normal force with the use of an instrumented transtibial socket: A case study *Medical Engineering and Physics* **71** 102-7 https://doi.org/10.1016/j.medengphy.2019.07.002

[5] Razak A, Osman A, Ali S, Gholizadeh H and Abas W 2015 Comparison study of the prosthetics interface pressure profile of air splint socket and ICRC polypropylene socket for upper limb prosthetics *Biocybernetics and Biomedical Engineering* **35** (2) 100-5 https://doi.org/10.1016/j.bbe.2014.08.003

[6] Berce P, Caizar C, Pacurar R, Radu S, Bratean I and Fodorean I 2014 *Tehnologii de Fabricatie prin Adaugare de Material si Aplicatiile lor* (Manufacturing by Material Adding and their Applications), (Bucuresti, Ed Academiei Romane)

[7] ZHERMAK SpA available at https://www.zhermack.com/, accesed 06.06.2021

[8] MakerBot 3D Printing available at https://www.makerbot.com/, accessed 20.05.2021

IOP Conf. Series: Materials Science and Engineering 1235 (2022) 012059 doi:10.1088/1757-899X/1235/1/012059

Effect of lubricant shear thinning behaviour on major functioning parameters of a pad bearing operating in hydrodynamic lubrication region

C P Salodimitris and P G Nikolakopoulos

Machine Design Laboratory, Department of Mechanical Engineering and Aeronautics, University of Patras, GR-26504 Patras, Greece.

pnikolakop@upatras.gr

Abstract. Type of lubricant used for fluid film bearings is one of the most important factors to consider. Bearing characteristics have to maintain steady throughout a wide range of conditions. However, viscosity, which is one of the primary lubricant properties, highly deviates depending on environmental and operational conditions. Macro-molecular micro-structure of lubricants is strongly correlated to temperature and shear rate, leading to a shear thinning behaviour. This work aims in studying a pad bearing lubricated by a shear thinning oil in hydrodynamic region. Hydrodynamic wedge was modelled for a small fixed pad bearing of 8.5 mm inner radius, 14.5 mm outer radius, 40° angle and 50 μm inlet – outlet film thickness difference. Variety of cases were studied for several runner rotational velocities and minimum film thickness. Lubricant was modelled as a carreau fluid. Load carrying capacity, power losses, friction coefficient, mass flow rate, temperature and viscosity were evaluated. From the aspect of load carrying capacity, this specific pad bearing is optimally operating for rotational velocities of up to 100 krpm. Above this threshold, viscosity highly downgrades due to increase in temperature and shear rate, causing a negative impact in load capacity. Shear thinning behaviour of lubricants is of high importance and should not be considered negligible.

1. Introduction

Thrust pad bearings are included in the general category of fluid film bearings and considered among the most crucial machine elements. A typical thrust bearing consists of several pads. The full understanding and the appropriate design of such a system, is considered as one of the most complex and difficult concepts of engineering. Sufficient load carrying capacity is developed under extreme conditions and provides a stable support under external dynamic forces [1]. These types of bearings take advantage of hydrodynamic wedge and lubricant properties in order to support axial loads and separate the relatively moving surfaces of pad and runner [2]. Physics behind lubrication is extremely challenging and complex preventing many engineers and researchers from achieving a universal solution. Most studies in literature extract conclusions for specific geometries, lubricants or operating conditions in the field of marine, turbomachinery etc. [3,4].

Shifu Yang *et al.* suggested that lubricant properties are strongly influenced by the temperature field of bearing [5]. Temperature rise leads to decrease in viscosity, load carrying capacity and eventually film thickness. Fluctuations for internal temperature of lubricant are attributed to oil inlet temperature, viscous friction and heat flux between oil, pad and runner.

J. Javorova *et al.* noted that present day lubricants include a plethora of viscosity index improvers in the form of high molecular weight polymers [6]. On the other hand, those additives affect other aspects of lubricant behaviour as well. Their macro-molecular micro-structure renders lubricating oils strongly shear thinning with the increase of shear rate. This phenomenon has a great impact on thrust bearing load carrying capacity and friction especially at high rotational velocity. In addition, in order to conclude to Reynolds equation from Navier-Stokes equation, a simplification of Newtonian lubricant is necessary [7]. In other words, Reynold's equation is not applicable in case of a shear thinning lubricated bearing. Therefore, in order to understand and proceed to solution of a pad bearing operating under such conditions, finite element analysis methods have to be implemented.

In the present paper, innovation lies on viscosity models applied to simulate hydrodynamic lubrication. Results regarding load carrying capacity, temperature and viscosity indicate the major mechanisms leading to bearing failure under increased thrust and rotational velocity. Thus, data and methods used lead to results strongly corresponding to reality.

2. CFD Model

2.1. Geometry

In order to study the impact of lubricant shear thinning behaviour on pad bearing functioning parameters, a CFD model has to be established. Flow parameters are calculated under a variety of conditions by using finite element analysis. The software employed for the analysis is ANSYS fluent 2020 student version. Firstly, lubricant geometry and boundary regions are modelled accordingly and considered steady for each simulation. In other words, pad and runner are deemed absolute rigid bodies even though they are not [5]. It is well known that pad deformation strongly affects film geometry and eventually decreases load carrying capacity especially in extreme conditions. Characteristics of the modelled fixed pad geometry are described in table 1. By conducting a brief meshing analysis during other studies in the past it was obvious that model has to be divided in 16 segments in order to achieve better meshing quality as illustrated in figure 1. Concerning boundary regions, surface corresponding to maximum and minimum film thickness are considered the inlet and outlet, respectively. Moreover, top inclined surface and bottom flat surface are identified as the pad and runner, respectively. Finally, the two surfaces corresponding to inner and outer pad radius are considered as lubricant leaks.

Table 1. Parameters of modelled fixed pad geometry

Parameter	Value
Inner radius	8.5 mm
Outer radius	14.5 mm
Inlet – Outlet film thickness difference	50 μm
Pad angle	40°

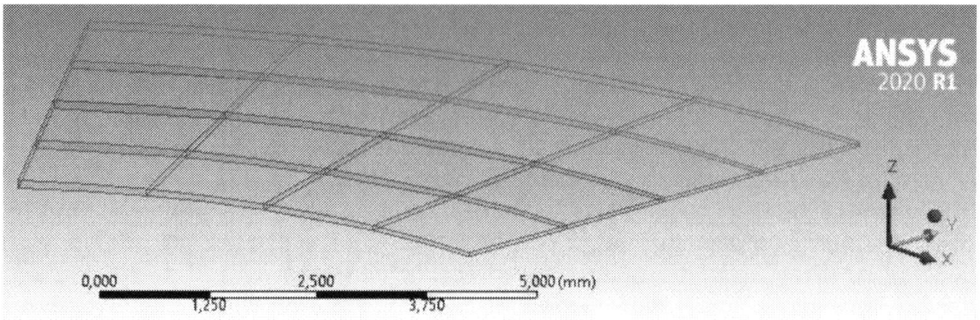

Figure 1. Lubricant geometry modelled in graphics window of Ansys 2020 R1.

2.2. Lubricant

The next modelled characteristic is lubricant. Ansys fluent uses the relatively simple mathematical equation (1) in order to correlate viscosity to temperature and shear rate.

$$\mu(\dot{\gamma}T) = \eta(\dot{\gamma})H(T) \tag{1}$$

Where $(\dot{\gamma})$: shear rate dependent function
$H(T)$: temperature dependent function

There are plenty of lubricant factors to be modelled in order to have an accurate simulation. First of all, the correlation between viscosity and shear rate is assessed. Among the plethora of ways to model such a phenomenon, the most widely used for lubricants is that of Cross or Carreau. In this particular study the carreau model is chosen to describe the function of viscosity to shear rate by using equation (2).

$$\eta = \eta_\infty + (\eta_0 - \eta_\infty)(1 + (\lambda\gamma)^2)^{\frac{n-1}{2}} \tag{2}$$

Where η_∞ : viscosity at high shear rate
η_0 : viscosity at low shear rate
λ : time constant
n : power law index

The next factor to be modelled is the corelation between temperature and viscosity. Although there are still plenty of ways to model this phenomenon, Ansys fluent by default uses a form of Arrhenius law for this kind of correlation, which is described mathematically by equation (3).

$$H(T) = exp\left[a\left(\frac{1}{T} - \frac{1}{T_a}\right)\right] \tag{3}$$

Where T_α : reference temperature
α : activation energy/R

Table 2. Parameters of modelled lubricant

Parameter		Value
Carreau Model	Viscosity at high shear rates, η_∞	0.01 (Pa s)
	Viscosity at low shear rates, η_0	0.02 (Pa s)
	Time constant, λ	3e-06 (s)
	Power law index, n	0.341
Arrhenius Law	Reference Temperature, T_α	363 (K)

	Activation Energy/R, K	3900 (K)
Coefficients	Thermal conductivity, λ	0.135 (W / (m K))
	Specific heat capacity, C_p	1950 (J / (kg K))
	Conduction heat, h	1000 (W / (m^2 K))
Other	Density, ρ	0.88 kg/m^3

Some typical values for a SAE 10W-50 lubricant modelled as a carreau fluid are illustrated in Table 2. Data are derived from the study of S. Azeez and V. Bertola [2]. Furthermore, in order for the data to comply with lubricants from industry, it is considered that they correspond to a temperature of about 90 °C. In addition, coefficients for thermal conductivity, specific heat capacity, heat conduction between lubricant, pad and runner and density are deemed constant and illustrated in table 2. Values are obtained from literature [8] and correspond to a wide variety of lubricants. Finally, temperature at inlet, outlet, leaks, pad and runner is specified to 50 °C. To conclude, figure 2 illustrates lubricant viscosity as a function of temperature and shear rate.

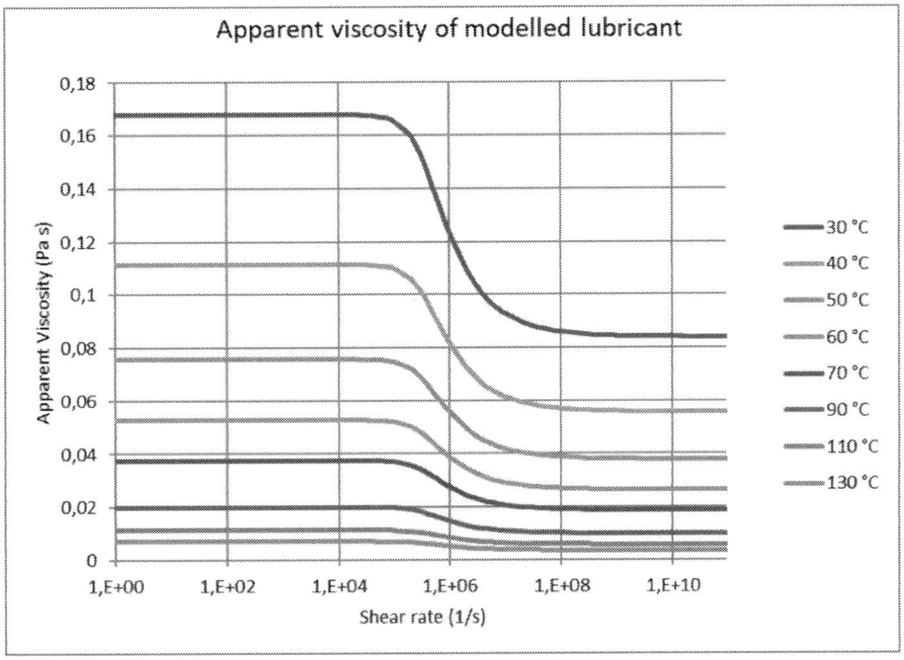

Figure 2. Apparent viscosity of modelled lubricant corresponding to SAE 10W-50 as a function of shear rate for several temperatures.

3. Results

A number of cases is investigated through finite element analysis. Runner rotational velocity of 50, 100, 150 and 200 krpm and minimum film thickness of 20 to 45 μm with 5 μm increment are applied. Bearing is considered to operate at a pressure of 100 kPa and sink into a bath of lubricant at 50 °C as previously stated. It is interesting that bearing develops a threshold with regards to load carrying capacity, which is unique for modelled geometry and lubricant. Results are summarized graphically.

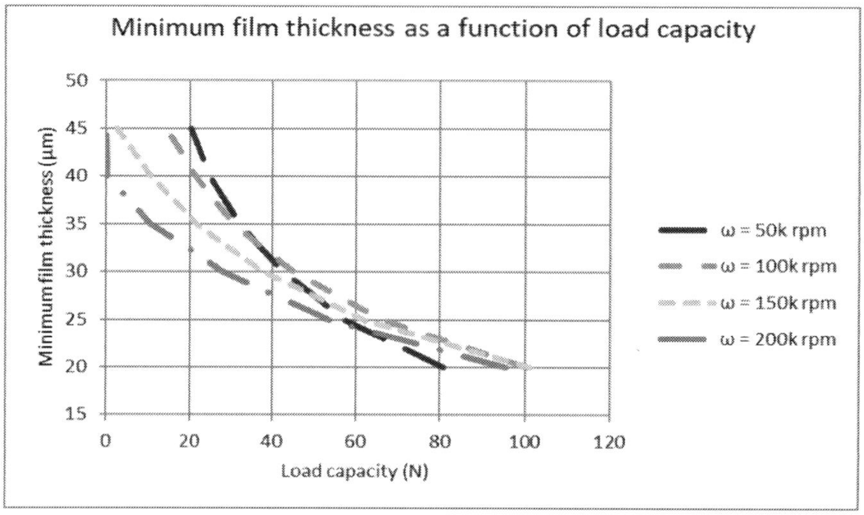

Figure 3. Minimum film thickness as a function of load carrying capacity.

According to figure 3, by comparing film thickness for different velocities it appears that bearing operates optimally up to 100 krpm. Further increase in runner rotational velocity results in film thickness reduction. This phenomenon is caused by the extreme viscosity downgrade due to increased viscous heating.

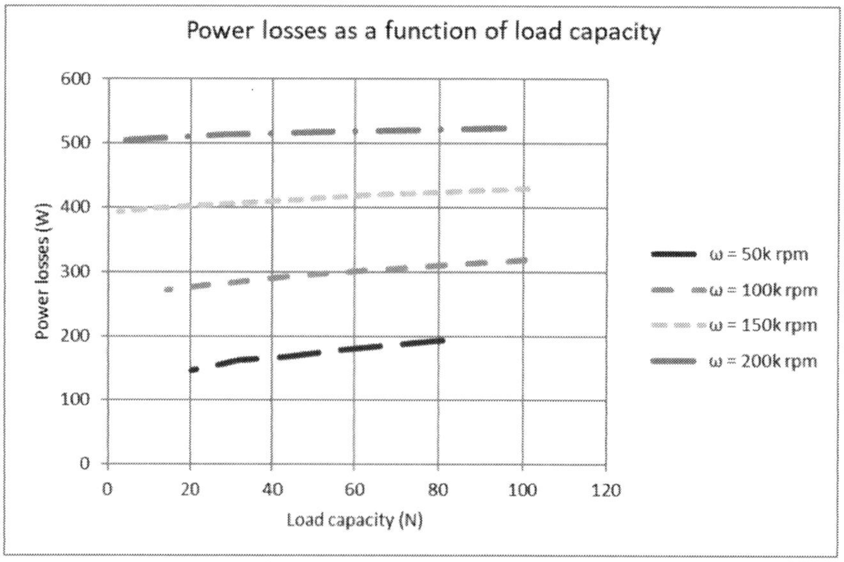

Figure 4. Power losses as a function of load carrying capacity.

In accordance with figure 4, power losses increase alongside thrust and velocity due to increase of friction forces. It appears that for a velocity of 200 krpm, losses slightly exceed 0.5 kW for just a small pad bearing.

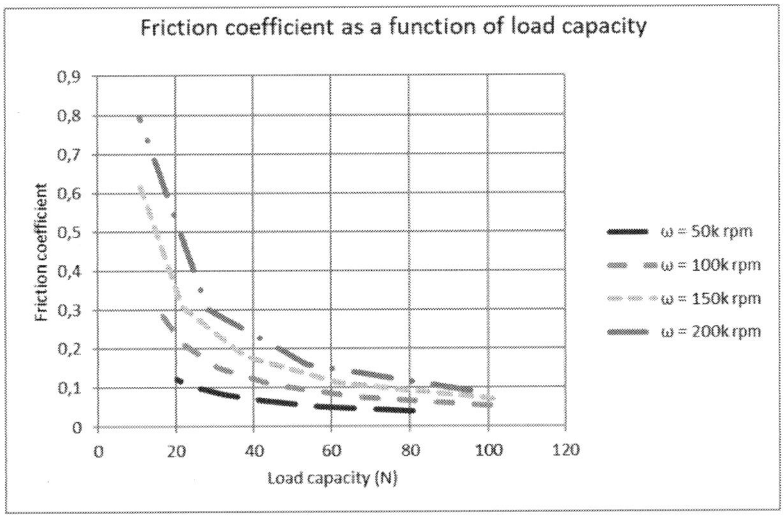

Figure 5. Friction coefficient as a function of load capacity.

According to figure 5, friction coefficient decreases with thrust, however, increases with velocity. Nevertheless, if thrust is further increased, friction coefficient is supposed to reach a total minimum. This means that bearing has optimum operating conditions regarding friction.

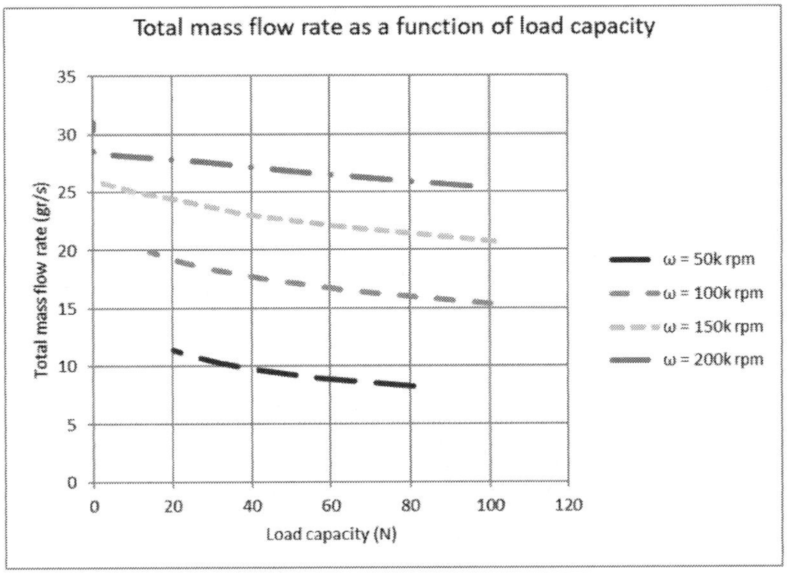

Figure 6. Total mass flow rate as a function of load carrying capacity.

Total mass flow rate decreases with thrust but increases with velocity, as clearly demonstrated in figure 6. Thrust and film thickness are inversely proportional. Thus, increase in thrust causes decrease in film thickness and eventually flow cross section. However, since runner velocity remains the same, lubricant average velocity through hydrodynamic wedge is negligibly affected. Implementing flow continuity for unchanged density, means that mass flow rate is decreased.

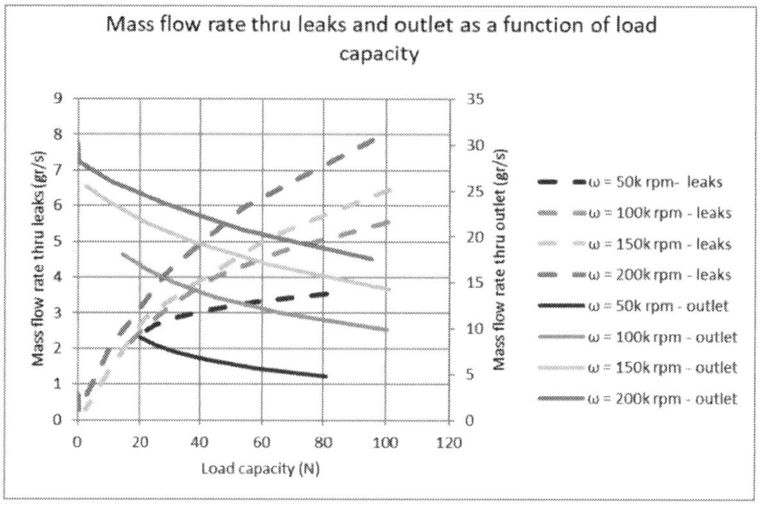

Figure 7. Mass flow rate through leaks and outlet as a function of load carrying capacity.

According to figure 7, mass flow rate through leaks rapidly increases with thrust and velocity. In order to achieve equilibrium, pressure is built into hydrodynamic wedge, counteracting external thrust. When thrust is increased, pressure gradient inside bearing is also increased and highly affects velocity profile. A lot of lubricant exits the sides of hydrodynamic wedge resulting in pressure drop inside pad bearing.

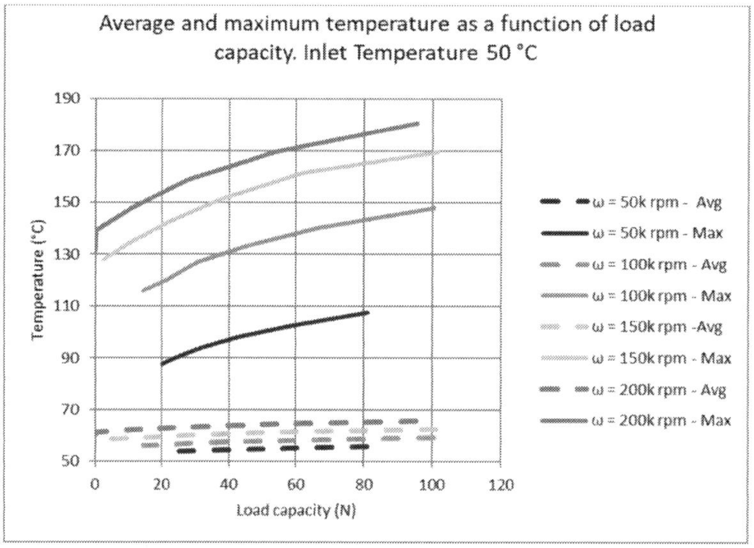

Figure 8. Average and maximum temperature as a function of load carrying capacity.

In accordance with figure 8, average and maximum temperature increase with thrust and velocity. Also, from model contours for temperature, it is obvious that maximum temperature occurs at pad trailing edge. In this area, velocity gradient or shear rate appears having a total maximum.

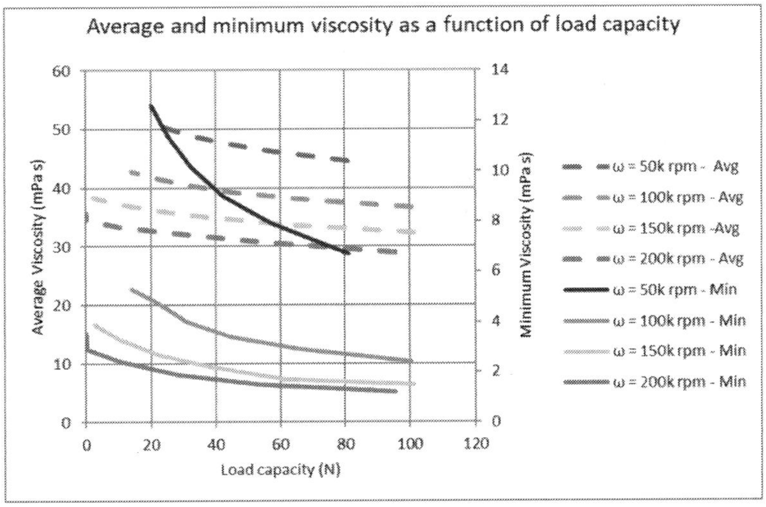

Figure 9. Average and minimum viscosity as a function of load carrying capacity

Lastly, according to figure 9, average and minimum viscosity decrease with thrust and velocity. Increase in thrust causes decrease in film thickness and increase in viscous heating. On the other hand, increased velocity leads to increased viscous heating. The outcome is decrease in viscosity due to increase in temperature for both reasons.

4. Conclusions and future work

Summarizing, in the present paper, a simulation of a thrust pad bearing lubricated by the use of a shear thinning lubricant is performed. A number of models is solved with different boundary conditions regarding to velocity and film thickness. Based on results, pseudoplastic behaviour strongly affects bearing operation due to fluctuations in viscosity, which is considered as the most important property of lubricants. In addition, it is proven that there is an optimal operating point for each bearing and it must be taken into account while designing such a system. Temperature distribution and shear rate throughout lubricant strongly impact load carrying capacity and power losses. Therefore, assumption of constant viscosity leads to precarious conclusions, which cannot be generalized. For example, by assuming constant viscosity to CFD model, load capacity changes linearly with runner rotational velocity, which is not valid in reality.

Concerning future work, it would be interesting to study the impact of shear thinning behaviour in combination with pad elastic deformation on bearing functioning parameters. Other studies prove that elastic deformation negatively affects load carrying capacity and friction coefficient [9]. However, the majority of published papers concerning pad bearings deal with phenomena occurring at steady state condition. Research has only put a little effort to study conditions during start-up of a pad bearing, where a transient model needs to be developed [10]. Studies on incomplete lubricant film and experimental verification would be of extreme importance and is to our interest for further research.

ORCID IDs
C P Salodimitris https://orcid.org/0000-0002-7805-9198
P G Nikolakopoulos https://orcid.org/0000-0002-0944-6653

5. References

[1] Jane Q J and Chung Y W 2013 *Encyclopedia of Tribology*, Springer, Boston, MA https://doi.org/10.1007/978-0-387-92897-5
[2] Azeez S and Bertola V 2020 Lubrication of journal bearings by shear thinning lubricants using different constitutive models *Proc. of the Inst. of Mech. Eng. Part J Engineering Tribology*

235 (6) 1203-10 https://doi.org/10.1177/1350650120950521

[3] Liang X, Yan X, Ouyang W and Liu Z 2019 Experimental research on tribological and vibration performance of water-lubricated hydrodynamic thrust bearings used in marine shaft-less rim driven thrusters *Wear* **426-427** (A) 778-91 https://doi.org/10.1016/j.wear.2018.11.017

[4] Peixoto T F and Cavalca K L 2020 Thrust bearing coupling effects on the lateral dynamics of turbochargers *Tribol. Int.* **145** 106166 https://doi.org/10.1016/j.triboint.2020.106166

[5] Yang S, Li S, Yiang M and Pei S 2019 A study of inlet temperature models of a large size tilting thrust bearing comparison between theory and experiment *Tribol. Int.* **140** 105881 https://doi.org/10.1016/j.triboint.2019.105881

[6] Javorova J, Radulesku A, Lovchalieva R, Nikolov N and Kostova P 2012 HD Lubrication of Journal Bearings by Pseudo-Plastic Lubricants *Tribological Journal BULTRIP* **3** 248-252 (Sofia, Bulgary)

[7] Stachoviac G W and Batchelor A W 1993 *Engineering Tribology* ISBN 978-0-12397037-3 Elsevier

[8] Gyimah G K, Guo Z, Huang P, Jiang S, Barber G C, Liu G and Liu G W 2016 Improvers of Pressure Viscosity Coefficients of two face liquid solid lubricants *International Conference on Advanced Manufacture Technology and Industrial Application (AMTIA 2016)* https://doi.org/10.12783/dtetr/amita2016/3562

[9] Chalkiopoulos M, Charitopoulos A, Fillon M and Papadopoulos C I 2020 Effects on thermal and mechanical deformations on textured thrust bearings optimally designed by a THD calculation method *Tribol. Int.* **148** 106303 https://doi.org/10.1016/j.triboint.2020.106303

[10] Sun F, Zhang X, Wei Y, Wang X and Wang D 2020 Stability analysis of rubber-supported thrust bearing in a rotor-bearing system used in marine thrusters under disturbing moments *Tribol. Int.* **151** 106356 https://doi.org/10.1016/j.triboint.2020.106356

IOP Conf. Series: Materials Science and Engineering 1235 (2022) 012060 doi:10.1088/1757-899X/1235/1/012060

Ground penetrating radar for real time surveying of soil as bioremediation precursor

N Iftimie, R Steigmann and G S Dobrescu

National Institute of Research and Development for Technical Physics, NDT Department, 47 Mangeron Boulevard, 700050, Iasi, Romania

niftimie@phys-iasi.ro

Abstract. GPR is a non-destructive method based on the emission and reception of electromagnetic waves generated by a bowtie antenna, of 400 MHz, in our case. The soil conductivity influences the propagation of electromagnetic waves, thus any inhomogeneity as water lens, chemical pollution, soil texture, buried objects modify the signal scattered on them and recorded by GPR tools. One of the potential advantages of GPR is that it is able to quickly analyses the presence in the underground of various inhomogeneities without having to come into contact with the ground and measures the time interval between pulse generation and reception, the results being presented in B-scan and linescan representation. Combining and processing the received signals, the results show as a map with emphasized watering regions. This paper presents the results of a case study where it was determined the effectiveness of GPR technology in detecting water content as well as the presence of water holes in airport runway draw. Further investigations will include the use of GPR together with electromagnetic induction method in order to make data fusion and increase the probability of detection.

1. Introduction

The soil is an inhomogeneous system of which composition, structure, and humidity, the presence of foreign matter and in-depth geological formations vary strongly, even on small distances. In order to determine changes in subsurface properties different techniques are used, such as electromagnetic (EM) and acoustic wave propagations and the electrical resistivity and magnetic properties [1,2]. The GPR technique, due to the high frequencies used, has a much lower penetration into the ground, can operate at sufficiently high speeds and has a good separation power. GPR allows knowledge of soil stratification, the existence of large porous areas, soil contamination with various hazardous chemicals, predictions of areas at risk of collapse and/or avoids construction in areas with problems in soil structure [3]. Water pollution in buried pipes is caused usually by interaction of bacteria from pipes walls or due to leaks and damages of pipelines [4]. The presence of water in the soils as well as a various contaminants is detected as the soil moisture influences the GPR signal reflection characteristics [5]. GPR applications are including archaeology, civil and environmental engineering, geophysics, cultural heritage, utilities or target detection. Within the various nondestructive testing method (NDTs) used for investigation of roads, GPR is considered as operative and substantial tool based on large domain of utility and confidence of the results, enabling determination of physical properties and dimensional conditions of the ground using short EM impulses transmitted and received in a specific frequency range. [6].

Deformations monitoring in airport runways and the surround areas are crucial, especially in cases of low-bearing capacity subgrades, such as the clayey subgrade soils. This paper presents the results recorded with a GPR tool to detect of water holes in expansion joint and touchdown zone and methods of signal processing and post – processing carried out on A-scan and B-scan data in order to obtain effective and interpretable results. The survey conditions have been first simulated by finite-difference time-domain (FDTD) [7] software for easy recognition of signals received by bowtie antenna of the GSSI SIR 3000 USA [8,9].

2. General description of the problem

The runways are central to airport infrastructure because they are dedicated to aircraft-to-aircraft take-off and landmine maneuvers. From this point of view, the runways must comply with very strict construction requirements [10] and maintenance standards [11]. It should be noted that a continuous surface area on runways is ensured throughout their life cycle and damage such as cracking, rotation, undergraduate sinking is not acceptable to compromise the safe handling of aircraft. GPR provide a high resolution, picture of a cross-section view of below the earth's surface features and reproduce an electrical properties map. The ground electrical variations can be connection to the physical variations in the scanned area such as the soil interfaces water table or contaminants, existence of buried targets, and even air voids in soil. The investigated area (Iasi airport area) is the runway area, including runways, platforms, perimeter road, partial handle road, and aircraft parking (figure 1). The incident waves penetrate the ground being propagated along the scanning line and they are scattered by water lenses leading to reflected EM pulse which are recorded as radargram B-scan to obtain 2D image of the ground (figure 2(b).

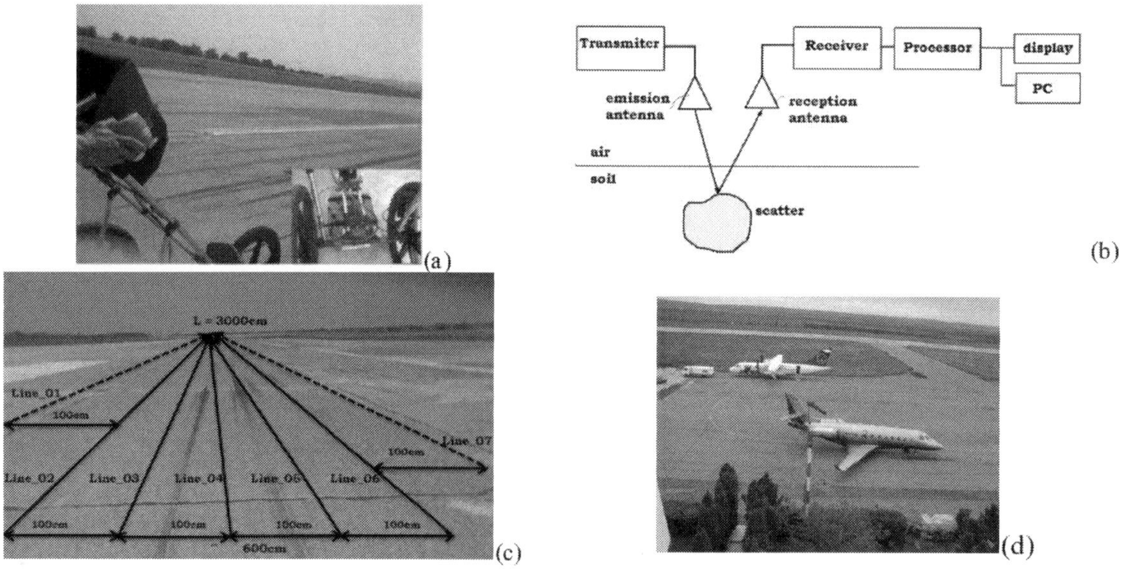

Figure 1. Experimental set-up and the investigated runway area: (a) runways (inset-GPR tool), (b) block diagram of a generic GPR tool, (c) perimeter road, and (d) aircraft parking

In the perimeter of road areas and the parking of the aircraft, an area of [3000x600] cm was scanned in order to detect water holes in the expansion joint and the contact area. The scanning has been carried out on 7 parallelly traces having 3000 cm length, from 100 cm to 100 cm, as seen in Figure 1c. Knowing the physical-mechanical and chemical properties of the terrain is very important when civil infrastructures are involved. The realization of the infrastructures for the works of art, bridges, viaducts, underground sewers suppose the exact establishment of some characteristics related to the geological, mineralogical and hydrological knowledge of the analysed lithological formations.

Previous studies have been developed on infrastructure build on the terraces of the riverbed that crosses the city and includes: upper terrace (underground various targets and drainage pipes detection), lower terrace (detection of pipes for sewer leaks [8]), and medium terrace (investigations related to spill basins and civil protection dam [9]). For the airport area, the environmental pollution risk analysis was analysed and the runway was tested by the GPR method, including the runways, platforms, perimeter road, road handling (figure 2).

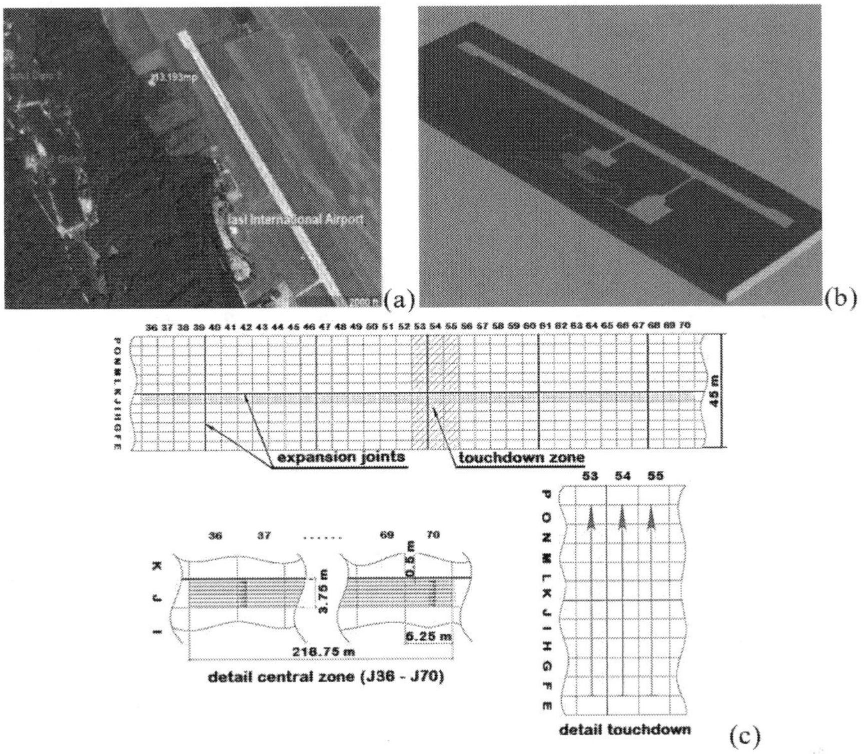

Figure 2. (a) Airport runway; (b) surveyed runway draw; (c) surveyed area

The main negative impact that will manifest itself on the environment refers to the rainwater on the runway - landing, runways, platforms, perimeter and handling roads that are collected in the rainwater network, treated in desanders - separators of petroleum products and discharged into lake in a agreement area. In general, the area where the airport is located has a system of protection of the soil and vegetation of the area. The control and characterization of the runway at airports is done by monitoring the quality of these elements of the runway structure, at given periods of time, according to the norms imposed by the Romanian Aeronautical Authority (*"regulates civil aeronautical activities and provides oversight airports, aircraft, and pilots"*) [12]. Variations in the percentage of soil moisture under the slabs as well as the wear of the slab surface led to erosion, damage and even cracking of the slabs, with serious implications in controlling the movement of aircraft on the ground.

That is why these inspections are done periodically. Destructive tests to evaluate tile thickness and material density can now be completed with non-destructive tests to assess the structural integrity of the runway, including the use of the GPR georadar. The use of GPR, as a non-destructive technique, is motivated both by the speed of the method and by the information it provides at a low cost. Following the inspection, information can be obtained both regarding the thickness of the concrete layer and regarding the presence of water or contaminants in the track structure and the existing cracks in the tiles. The GPR equipment (figure 1(a)), having a 400 MHz antenna is coupled to the inspection area through a layer of air of a few millimetres, in the regions of interest (ROI) you can lower the antenna on the track, the coupling being directly on the slab. This direct coupling reduces the inspection speed but allows the GPR signal to penetrate deeper into the ground. The propagation speed v can be defined by geometry hypothesis [1]:

$$v = \frac{2\sqrt{z^2 + \left(\frac{x}{2}\right)^2}}{t} \quad (1)$$

The scanned area consists in multi-layer targets; thus, the speed must be calibrated on layer n and the propagation speed becomes:

$$v_n = \frac{\sqrt{t_n v_n^2 - t_{n-1} v_{n-1}^2}}{t_n - t_{n-1}} \quad (2)$$

The dielectric constant of the material ε_r determines the speed of propagation of electromagnetic waves in the material, according to the relation:

$$d = \frac{c \cdot t}{2\sqrt{\varepsilon_r}} \quad (3)$$

3. Results and discussions

The GPR system permits the investigation in depth up to 4.5 m and depend of soil moisture described in [8,9] and the dielectric constant of the soil into the scanned region has been set at $\varepsilon_r = 12$ A series of regions with expansions joints and touchdown with three files each from the airport runway area has been scanned. These regions are of interest because they are areas collected in the rainwater network, treated in sanders - separators of petroleum products and discharged into lake in an agreement area. Figure 3 presents the raw data recording with GPR tool on the investigated runway area which describes the crossing over a zone where plate (detail in figure 2) has been identified. Is observed that in the case of real tests, the signal is very noisy, containing, in addition, clutters. Using GPR raw data, the simulations with synthetic data and a specific software for processing and post-processing of the signals, the obtained results provide the opportunity to estimate presence of water holes in scanned areas.

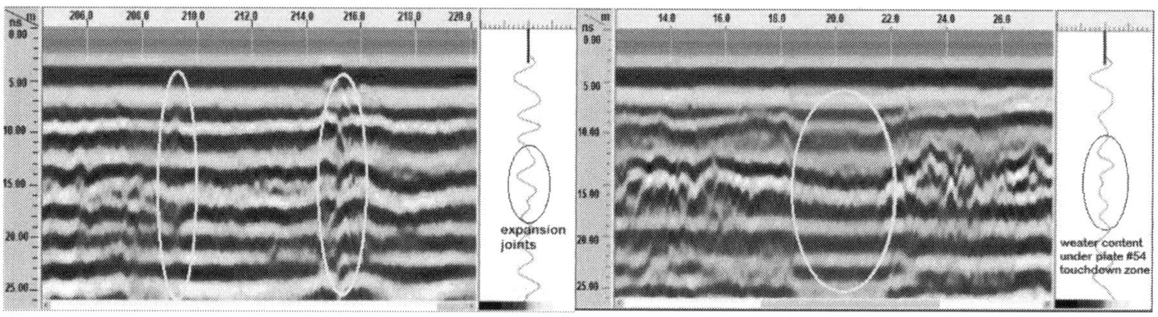

(a) (b)

Figure 3. B-scan GPR image, Linescan and Wiggle Mode interpretation: (a) zone with expansion joints and (b) touchdown zone (25.31 m).

On basis the free software which involve solved Maxwell's equation with FDTD method, the presence of water or various contaminants was been simulated. First, the reflections of the ground-air interface appear which dominate the data set, and then the hyperbola arches produced by the expansion joints, water drainage channels or water lenses under the pavement or concrete become predominant. The vertical zone represents the connection between vast sections of the water, where different and complex reflections appear. The horizontal line represents the reflections on the waste humidity on the concrete. The horizontal lines represent multiple reflections from interfaces and are given of multiple reflections on the interfaces between the soil and concrete. Figure 4 (a) presents the simulation of surveying conditions using GPRMax 2D. The presence water holes are difficult to identify in noisy data, their GPR patterns being susceptible to noises. Raw radargrams are the input for simulation, the positive segments showing the presence of water in surveyed area.

IOP Conf. Series: Materials Science and Engineering 1235 (2022) 012060 doi:10.1088/1757-899X/1235/1/012060

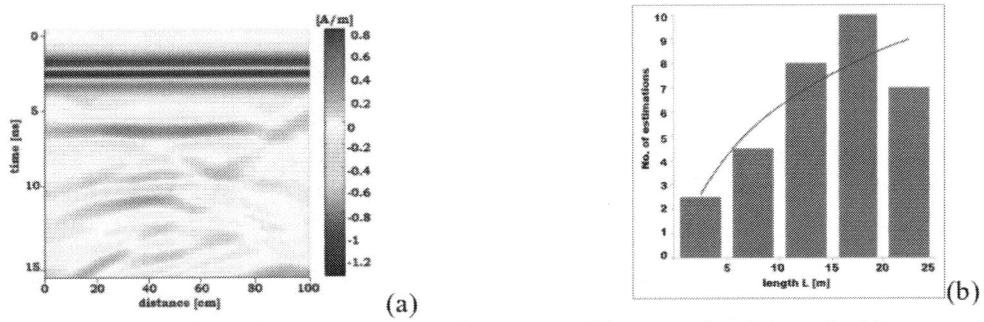

Figure 4. (a) Simulation using GPRMax 2D for a zone with expansion joints; (b) histogram of water content under plate #54 touchdown zone

The histogram of water content under plate #54 touchdown zones is presented in figure 4 (b). Among the areas selected for the examination and characterization of the soil that of the airport runway, measurements were performed over several days depending on the rainy weather. Under these conditions the probability of locating the presence of water as a "lens" under the pavement of the runway at plate #54 could be estimated.

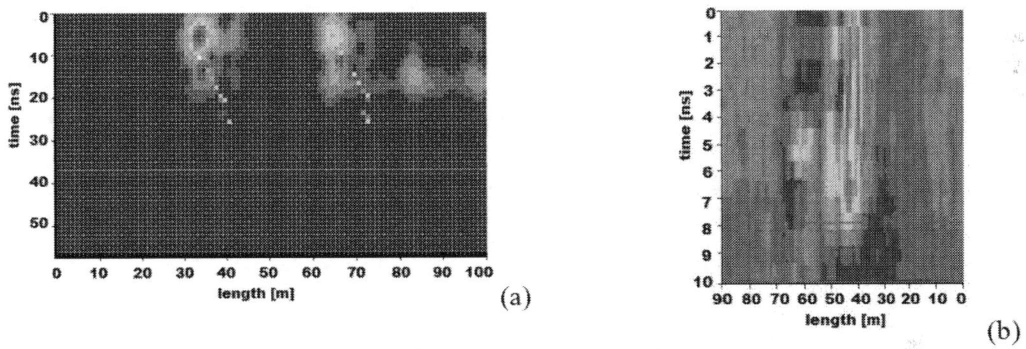

Figure 5. Ground removal and specific B-scan processing for (a) zone with expansion joints and (b) water content under touchdown zone

To each A-scan from radargram recorded, the signal processing algorithm described [8,9] is applied and therefore, background removal with window able to move smoothly along a surface with length L=20 pixels is presented in figures 5 (a) and (b). It was specific that the precision and accuracy of geophysical utility scan identification and location can vary strong because the EM waves of the ground are attenuated, the contrast at the ground and targets interface establishing the power of reflection. The GPR technique, due to the high frequencies used, has a much lower penetration into the ground, can operate at sufficiently high speeds and has a good separation power. GPR tool has is used with favourable results, in different types of applications, as the detection of soil humidity and different contaminants by soil. The GPR technique used for non-destructive soil characterization the examination provides information on soil texture and moisture as well as asphalt cracks by airport runways.

4. Conclusions
The runway is affected by long-term differential subsidence effects, which require accurate monitoring and full-depth runway rehabilitation. In order to correctly interpret the signals provided by GPR, due to the complex level of noise and in homogeneities, specific signal and image processing procedures are developed. When the terrain requires only scanning a soft area due to rain or accumulation of water holes under concrete, optimal signal and image processing algorithms are applied and the images become easier to interpret.

The B scan data raw were converted into images using Matlab, the obtained results provided the opportunity to estimate presence of water in areas scanned highlighting the zone with expansion joints and water content under touchdown zone with good accuracy.

Acknowledgments
This work was supported by the Ministry of Research, Innovation and Digitization, Romania Nucleus Program MAGNA2022, Project PN 19 28 01 02 and Bilateral Cooperation JINR under protocol 24-4-1121-2015/2021.

ORCID IDs
N Iftimie https://orcid.org/0000-0001-9614-0101
R Steigmann https://orcid.org/0000-0003-1618-7650
G S Dobrescu https://orcid.org/0000-0001-8112-2621

References

[1] Daniels D J 2004 *Ground Penetrating Radar* 2nd Edition, IEE Radar, Sonar and Navigation Series 15 (London: The Institution of Electrical Engineers) http://dx.doi.org/10.1049/pbra015e
[2] Daniels D J 2006 A review of GPR for landmine detection *Sensing Imaging* **7** 90-123 https://doi.org/10.1007/s11220-006-0024-5
[3] Ayala–Cabrera D, Herrera M, Izquierdo J, Ocana–Levario S J and Perez–Garcıa R 2013 GPR-Based Water Leak Models in Water Distribution Systems *Sensors* **13** (12) 15912-36 https://doi.org/10.3390/s131215912
[4] Cheng W, Fang H, Xu G and Chen M 2018 Using SCADA to Detect and Locate Bursts in a Long-Distance Water Pipeline *Water* **10** (12) 1727 https://doi.org/10.3390/w10121727
[5] Yu Y, Klotzsche A, Weihermüller L, Huisman J A, Vanderborght J, Vereecken H and Der Kruk J 2020 Measuring vertical soil water content profiles by combining horizontal borehole and dispersive surface ground penetrating radar data *Near Surface Geophysics* **18** (3) 275–94 https://doi.org/10.1002/nsg.12099
[6] Jol H M 2009 *Ground penetrating radar theory and applications* (Amsterdam: Elsevier) pp 3-40 https://doi.org/10.1016/B978-0-444-53348-7.X0001-4
[7] Giannopoulos A GprMax2D/3D. User's Guide 2003 Available online: http://www.gprmax.com/code/UserGuideV2.pdf (accessed on 15 January 2021).
[8] Iftimie N, Dobrescu G S, Savin A 2015 Imaging Subsurface Water Pipe using GPR and Evanescent Waves: Experimental and Simulations Data *Applied Mechanics and Material* **772** 359-64 https://doi.org/10.4028/www.scientific.net/AMM.772.359
[9] Iftimie N, Savin A, Danila N A and Dobrescu G S 2019 Radar pulses to image the subsurface using Ground Penetrating Radar (GPR) *IOP Conference Series: Materials Science and Engineering* **56** 012130 https://doi.org/10.1088/1757-899X/564/1/012130
[10] https://www.faa.gov/documentLibrary/media/Advisory_Circular/150-5370-10H.pdf
[11] https://www.faa.gov/documentLibrary/media /Advisory_Circular/150-5320-6F.pdf
[12] https://www.caa.ro/pages/pages/istoric-caa
[13] Grimberg R, Steigmann R, Iftimie N and Savin A 2011 Ground Penetrating Radar as tool for nondestructive evaluation of soil *Proceedings of the VIth workshop NDT in Progress Prague* October 10-12.
[14] Grimberg R, Udpa L, Savin A, Steigmann R, Bruma A and Udpa S S 2009 Electromagnetic evaluation of soil condition *Stud. in Applied Electromagnetics and Mechanics* **32** 363-70 https://doi.org/10.3233/978-1-60750-023-0-363

IOP Conf. Series: Materials Science and Engineering 1235 (2022) 012061 doi:10.1088/1757-899X/1235/1/012061

Optimization of risk assessment in renewable energy of Russia by applying statistical calculations of climatic characteristics and GIS technologies

L V Nefedova and Yu Yu Rafikova

Lomonosov Moscow State University, Leninskie Gory, 1, Moscow, Russia

ju.rafikova@gmail.com

Abstract. Based on the characteristics of the renewable energy sources (RES), current states and also Russia uses the analysis of emerging risks and their management methods for avoiding this problem. The analysis of methodological approaches to assessing emerging risks in the implementation of electricity generation projects based on various types of renewable energy sources is carried out. The problems of assessing the natural resource risks of using renewable energy in the Russian Federation in relation to the climatic regionalization of the territory are examined. The most representative parameters of the ecological and geographical component for assessing the degree of risk by using renewable energy are highlighted. The authors propose a methodology for assessing the degree of risk for using solar resources based on the characteristics calculated of the influx variability for solar radiation to the earth's surface, by assessing and mapping the level of natural resource risks. The levels of resource risks in solar energy development for some regions of Russia were estimated. The paper presents maps of areas with different risk levels in Russia.

1. Introduction

Solar PV accounted to the end of 2020 760 mln kW (GW) total in the world, wind power and solar PV have generated more than 20% of electricity in nine countries. China added 48.2 GW (35% of total in the World) of solar PV capacity in 2020, Asia eclipsed all other regions for new installations, even excluding China, Asia was responsible for around 23% of new capacity in 2020. The top five national markets – China, the United States, Vietnam, Japan and Germany – were responsible for almost 66% of newly installed capacity in 2020 [1]. Over the past seven years, thanks to the Government program and support grid facilities of renewable energy (RE), the process of renewable energy sources development has intensified in Russia. Domestic production of highly efficient heterojunction solar panels has been established in the country, enterprises for the production of parts for wind turbines are operating. At the beginning of 2021, photovoltaic solar power plants (SPP) - 1440.0 MW and wind power plants (WPP) - 1008.8 MW were already operating in the regions of Russia [2]. The volume of commissioning of capacities in Russia under the RE program until 2024 is planned at 5.4 GW, of which WPP - 3.4 GW, SPP - 1.9 GW.

The laboratory of renewable energy sources of Lomonosov Moscow State University is carrying out research in the field of risk factors and methods of their analysis for the development of solar energy in the regions of Russia. Risk analysis is especially important for determining the investment attractiveness of power facilities during their design and development [3,4]. It is also important to analyze territorial distribution of risk factors for the use of renewable energy sources and the dependence on their ratio - investment prospects in renewable energy. Our analysis showed that resource risks are especially important for the territory of Russia with a high spatial type of climate formation heterogeneity [5]. It is the differences in the statistical characteristics of insolation for territories located in the same region that can play a decisive role in the stability of electricity generation at photovoltaic plants.

The multifactoral nature of risk assessments and calculations in solar energy requires the use of multicriteria analysis methods. In many studies, the method of multicriteria decision analysis based on geographic information systems (GIS-MCDA) is used for creating maps that represents potential areas for mounting solar power plants in countries with different climatic conditions [6-9]. Currently, the methods of geoinformatics are actively used in the study of the solar energy potential and risk assessment for various territories [10-12]. Geographic information systems (GIS) open up opportunities for solving a number of important tasks, such as visualizing resources for calculating the values of technical and economic energy potential by types of renewable energy sources, analyzing territories by the level of resource availability and the possibility of placing solar and wind energy facilities on them in relation to different scales of territorial research. This article is devoted to our experience in optimization of natural resource risks assessment in solar energy in the regions of Russia.

2. Data and methodology

A number of the regions of Russia was selected by the authors having in view to assess the spatio temporal variability of the total solar radiation (Sr) input to the horizontal surface.

Step 1. The collection of the incident solar radiation data according to the NASA POWER database for a ten-year period (2009-2018) [13] was made by extraction NetCDF files in GIS. On the basis of specified initial materials the data arrays for calculations were formed. A space interval of 0.5°x0.5° was selected so far as it was considered sufficient.

Step 2. The numerical estimate the variability of the total solar radiation input to the horizontal surface allowed the analysis of the variation coefficient (C_v).

$$C_v^j = \frac{\left[\frac{1}{N}\sum_{i=1}^{N}\left(x_{ij}-\bar{x}_j\right)^2\right]^{\frac{1}{2}}}{\bar{x}_j} \tag{1}$$

where: x_{ij} - all sky daily insolation incident on a horizontal surface.

N – number of years in the calculation period, i = 1÷10 in our case

j = 1,...,12 – months of the year

The annual average C_v for every month were calculated and this data arrays resulted in calculation of the mean coefficient of variation for the whole year. The variability of monthly and daily mean values was assessed. The values of $Cv_{monthly}$ radiation input characterize year-to-year variability, the values of $C_{v\ daily}$ - characterize intra-annual variability.

Step 3. The authors created maps for the assessing of the territorial distribution of annual average monthly values of $Cv_{monthly}$ for ten years period (2009-2018) using GIS technologies for some regions of Russia with different natural and climatic conditions. Also, data on intra-monthly variability by calculating $C_{v\ daily}$ for the same period for the maps creating were used. The analysis of the territorial distribution of resource risks level in a number of regions of the Russian Federation was carried out. A detailed description of the methodology and results of research on the risks for solar energy systems development in the Republic of Altai were published by the authors [14]. This paper presents the results of research in two regions of the Russian Federation: Astrakhan region - on figure 1, and Orenburg region - on figure 2.

IOP Conf. Series: Materials Science and Engineering 1235 (2022) 012061 doi:10.1088/1757-899X/1235/1/012061

3. Results and discussion

For the Astrakhan region the variability of the incoming radiation and the associated risks of reducing the efficiency of solar energy is higher in the northwest area ($C_{v\ monthly}$ is more than 13%). 12 solar power plants are currently operating in the Astrakhan region and their location is presented on figure 1. The first one was the Zavodskaya SPP, commissioned by Solar Systems Group in March 2017. This Group of companies built 4 SPPs with a total capacity of 60 MW in the Astrakhan region (Zavodskaya, Promstroy, Oktyabrskaya and Peschanaya, 15 MW each). Also, 4 SPP (Volodarovka, Enotaevka, Elista Severnaya and Mikhailovskaya) 15 MW each was built by an Austrian private investment fund. However, the reported data on electricity generation of these 8 SPP is classified information and is published in limited quantities. For Zavodskaya SPP, according to the reporting data of Solar Systems Company for the period 09/01/2017-12/25/2018 the amount of produced energy was 26 484.262 MWh, and the average value of the capacity factor (CF) for 16 months was 15.26%. For Promstoroymaterialy SPP – built also by Solar System Group - according to reporting data for the period 1/05/2018-12/12/2018 the amount of produced energy was 15 283.494 MWh. According to our calculations, the average value of CF was 17.76%. Both Solar plants are located in the same power generation risk zone. Apparently, such differences in CF values are due to the fact that for the Zavodskaya SPP, a longer period of a lower insulation level (1/09/2017-30/04/2018) entered the calculated time interval.

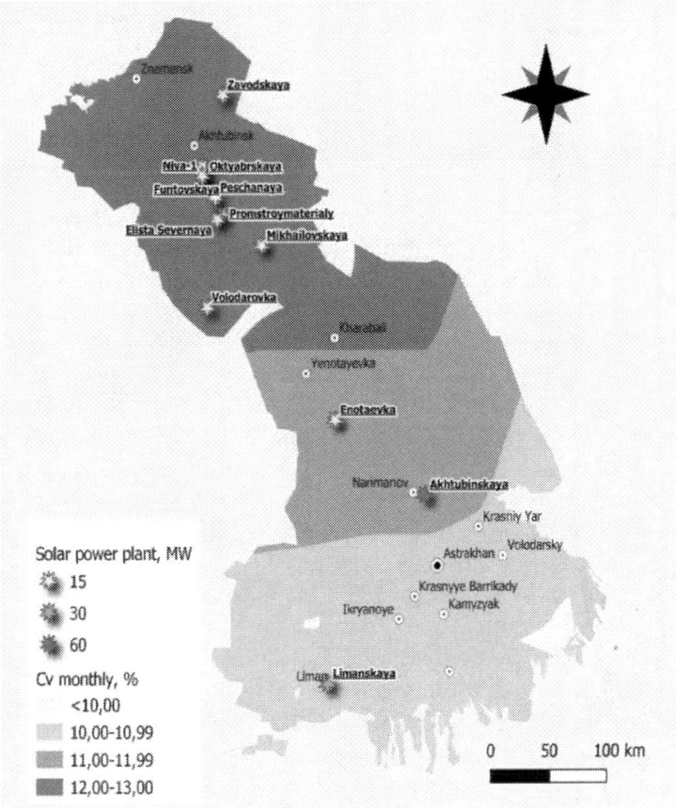

Figure 1. The map of the variation coefficient distribution (%) for the total insolation input to the horizontal surface according to the NASA POWER database in the Astrakhan region (Russia) for ten years period (2009-2018): annual average monthly values of C_v.

The necessary level of power generation stability for the solar power plants was obtained by calculating the coefficient of total solar radiation input to the horizontal surface variation We used the results of the operation of solar power plants at the Hevel Group of Companies (GC). 4 SPPs, including large SPPs of 60 MW each (Phuntovskaya and Akhtubinskaya), with a total capacity of 165 MW were built by this Group of Companies. For these purposes we compared our calculations with the values of the solar power plants operating factor in a number of regions of Russia, built by this GC using photo voltaic modules produced at the Hevel plant. The Hevel is the largest Russian solar energy GC. The plant for the production of solar modules "Hevel" is located in the city of Novocheboksarsk, Chuvash Republic and is named "Research and development center for thin-film technologies in energetics. This work centrebelongs to the Ioffe Institute in St.Petersburg and is a scientific and technical center in the field of photovoltaics. The Hevel Solar cells are manufactured using heterojunction technology and have a theoretical efficiency of 23.5%, which

IOP Conf. Series: Materials Science and Engineering 1235 (2022) 012061 doi:10.1088/1757-899X/1235/1/012061

makes it possible to provide the power of modules consisting of 60 cells up to 325 W.

The key advantages of heterojunction technology are high efficiency and stability of parameters, which allows ensuring high quality of the final product. This is achieved due to a number of technological features during production, namely according to the [15] reference as follow:
- the deposition of doped layers of amorphous silicon allows to increase the efficiency at extremely high and low temperatures, as well as in low light conditions;
- flank passivation reduces recombination (junction loss), which in it turn provides an increase in open circuit voltage and a decrease in temperature coefficient;
- the use of antireflection coatings reduces surface reflection from 30 to 10%;
- a special glass with increased permeability is used;
- the metal contacts on the surface are located close to each other; this fact allows to minimize transverse resistive losses and at the same time, very thin spaces in order to reduce shaded surface area.

Table 1. The efficiency of the SPP power generation in Astrakhan Region during 2019-2020 period

SPP's name & Installed capacity	SPP commissioning date	Capacity factor for the period (%)				
		IV-VI 2019	VII-IX 2019	X-XII 2019	I-III 2020	Average annual
Niva-1 15 MW	23/04/2018	20.32	16.93	7.56	6.89	12.92
Funtovskaya 60 MW	24/12/2018	18.29	17.79	8.01	10.67	13.69
Akhtubinskaya 60 MW	20/02/2019	21.02	19.29	9.49	10.74	15.14
Limanskaya 30 MW	28/11/2019	-	-	-	11.26	-

Source: authors' calculations based on data [2,15].

The calculations of CF in table 1 show a higher efficiency of SPP in the central and southern areas. Akhtubinskaya (CF=21.02%) is located in the zone of reduced variability of solar radiation input ($C_{v\ monthly}$ is less than 12%, $C_{v\ daily}$ is near 35%). The highest CF=11.26% in the first quarter of 2020 has the Limanskaya SPP, which is located in the zone of minimum insulation variability in the Astrakhan region. It is possible to predict in the rest of the year also the high efficiency of this SPP.

Those 12 solar power plants also were installed at the territory of the Orenburg Region by the beginning of 2021. Their total installed capacity is 330 MW. This is about 23% of all SPPs operating in Russia (1440 MW). Power plants are built from photovoltaic modules, mainly heterojunction, produced in Russia at a plant in Novocheboksarsk. T Plus Company Group has implemented 4 projects with a total capacity of 190 MW as part of his Solar System renewable energy (RE) development program. 8 SPPs were built by "Hevel" Company Group, two of them (Pleshanovskaya and Grachevskaya) at the end of 2017 were transferred to the ownership of PJSC Fortum. The author's calculations and analysis of the electricity generation at 6 SPPs, were based on data published by the Hevel [15].

Solar energy provides about 10% of electricity in the Orenburg region by 2021. The most efficient SPPs in the Orenburg region worked in the II quarter of 2021 (capacity factor is 22.73-21.36%), the minimum electricity generation fell on the IV quarter (capacity factor is 6.98-5.03%) in table 2. In the second quarter, daylight hours are already quite long, but the air temperatures are not yet as high as in July and August, the conditions for generation are optimal. In winter, the duration and intensity of solar radiation is much lower, due to snowfalls, blizzards, fogs which are frequent and reduce the generation of electricity by photovoltaic panels several times in comparison with April and June. When assessing the risks of SPP operation, it is necessary to take into account the data on the distribution of snow loads on the surface of photovoltaic panels in their most unfavourable design combinations. For predicting the performance of photovoltaic panels, the height and duration of snow cover are important characteristics. Snow covers of the working surface leads to a significant decrease in the calculated values capacity factor of the instantaneous efficiency of modules (more than 35-40%)

[16-17]. Our map of snow distribution in Republic of Altai showed a significant increase in the risks of reducing the efficiency of SPP in areas with a significant depth of snow cover [14].

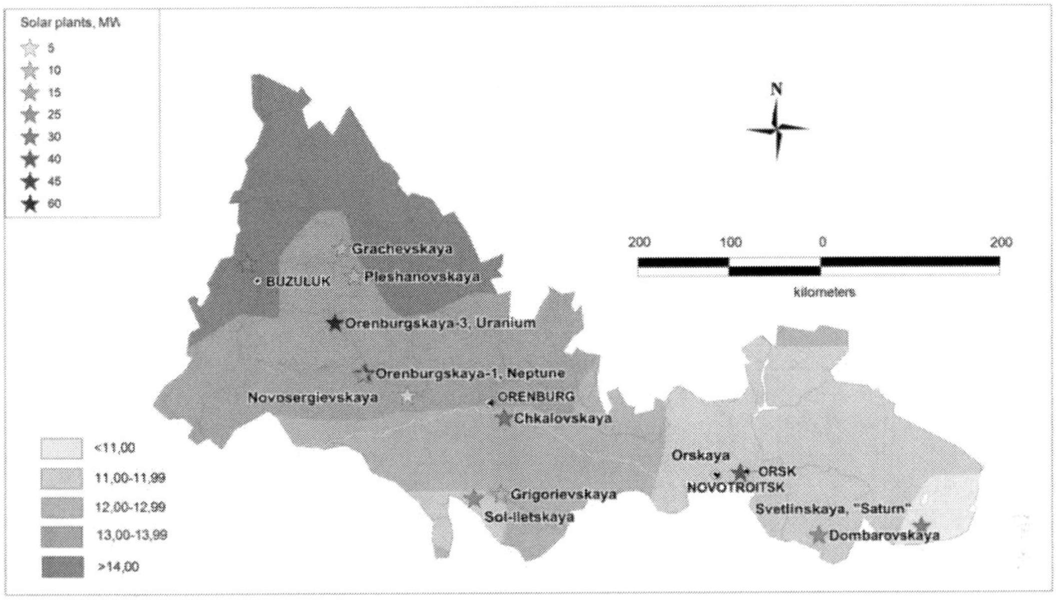

Figure 2. The map of the variation coefficient annual average monthly values distribution C_v (%) for the total insolation input to the horizontal surface according to the NASA POWER database in the Orenburg Region (Russia) for ten years period (2009-2018).

Table 2. The efficiency of the SPP power generation in Orenburg Region in 2019-2020.

SPP's name & Installed capacity, MW	SPP commissioning, date	Capacity factor for the period (%)				
		IV-VI 2019	VII-IX 2019	X-XII 2019	I-III 2020	Annual average
Perevolotsk 5 MW	22/10/2015	22.76	17.57	5.02	8.43	13.45
Sol-Iletsk 25 MW	17/04/2017	21.36	16.72	6.55	9.33	14.24
Chkalovsk 30 MW	21/02/2019	22.73	17.44	5.22	9.41	13.72
Grigorievsk 10 MW	27/05/2019	21.44	18.98	6.98	10.11	14.38
Elshansk 25 MW	27/06/2019	-	18.76	6.37	9.87	-
Dombrovsk 25 MW	28/11/2019	-	-	-	7.96	-

Source: authors' calculations based on data [2, 15].

4. Conclusion

The aim of this paper was to use the variability of solar radiation input as an indicator of the resource risks level in solar energy. The verification of the assessing variability of total insolation and data on the efficiency of operating Solar Power Plants in Regions of Russia was carried out. The insolation variability was evaluated using the method of calculating the coefficients of variation of average monthly and daily values for total insolation input. In the Orenburg region, the highest capacity Factor (CF) values are specific for the Grigorievskaya and Sol'-Iletskaya SPPs, located in the zone of minimum $C_{v\ monthly}$ values (less than 11%). Dombrovskaya SPP is located in the same variability zone, but is generating electricity since December, 2019 and its efficiency data will be obtained later. The use of GIS technologies made it possible to assess the temporal variability of the solar radiation input distribution in regions of Russia. On the territories of the studied regions the authors have identified zones with different indicators of solar radiation input variability, on the basis of which the analysis of risks in the stability of electricity generation at photovoltaic plants is carried out.

ORCID IDs
L Nefedova https://orcid.org/0000-0002-6022-0617
Y Rafikova https://orcid.org/0000-0003-3204-9135

5. References

[1] REN21. 2021Renewables 2021 Global Status Report. Paris: REN21 Secretariat, p 367
[2] *Association NP Market Council, official website 2021.* Available at: https://www.np-sr.ru/ru/market/vie/index.htm Accessed 5 July 2021
[3] Nefedova L V and Solov'ev A A 2020 Financial methods of risk management when using RES *Energy policy* **5** (147) 62–75 (in Russian)
[4] Kozlova M and Collan M 2020 Renewable energy investment attractiveness: Enabling multi-criteria cross-regional analysis from the investors' perspective *Renew. Energy* **150** 382-400 https://doi.org/10.1016/j.renene.2019.12.134
[5] Nefedova L V, Solov'ev A A, Shilova L A and Solov'ev D A 2016 Risk Factors in the Construction of Power Facilities Using Renewable Energy Sources in Russia *MGSU Bulletin* **12** 79–90 (in Russian)
[6] Shorabeh S, Firozjaei M, Nematollahi O, Firozjaei H and Jelokhani-Niaraki M 2019 A risk-based multi-criteria spatial decision analysis for solar power plant site selection in different climates: A case study in Iran *Renew. Energy* **143** 958-73 https://doi.org/10.1016/j.renene.2019.05.063
[7] Guaita-Pradas I, Marques-Perez I, Gallego A and Segura B 2019 Analysing territory for the sustainable development of solar photovoltaic power using GIS databases *Environ. Monit. Assess.* **191** 764 https://doi.org/10.1007/s10661-019-7871-8
[8] Türk S, Koç A and Şahin G 2021 Multi-criteria of PV solar site selection problem using GIS-intuitionistic fuzzy based approach in Erzurum province/Turkey *Sci. Rep.* **11** 5034 https://doi.org/10.1038/s41598-021-84257-y
[9] Rediske G, Siluk J C M, Michels L, Rigo P D, Rosa C B and Cugler G 2020 Multi-criteria decision-making model for assessment of large photovoltaic farms in Brazil *Energy* **197** 117167 https://doi.org/10.1016/j.energy.2020.117167
[10] Doorga J, Rughooputh S and Boojhawon R 2019 Multi-criteria GIS-based modelling technique for identifying potential solar farm sites: a case study in Mauritius *Renew. Energy* **133** 1201-19 https://doi.org/10.1016/j.renene.2018.08.105
[11] Choi Y, Suh J and Kim S-M 2019 GIS-Based Solar Radiation Mapping, Site Evaluation, and Potential Assessment: A Review *Appl. Sci.* **9** (9) 1960 https://doi.org/10.3390/app9091960
[12] Al Garni H, Awasthi A and Ramli M 2018 Optimal design and analysis of gridconnected photovoltaic under different tracking systems using HOMER *Energy Convers. Manag.* **155** 42-57 https://doi.org/10.1016/j.enconman.2017.10.090
[13] *NASA POWER database.* Available at: https://power.larc.nasa.gov/data-access-viewer Accessed 5 July 2021
[14] Nefedova L V, Rafikova Y Y and Soloviev D A 2021 Application of climatic characteristics statistical assessments and GIS technologies for solar energy systems development IOP Conf. Ser.: Mater. Sci. Eng. 1079 032086 https://doi.org/10.1088/1757-899X/1079/3/032086
[15] Hevel Group of Companies official website. Available at: https://www.hevelsolar.com Accessed 5 July 2021
[16] Andenæs E, Jelle B P, Ramlo K, Kolås T, Selj J and Foss S E 2018 The influence of snow and ice coverage on the energy generation from photovoltaic solar cells *Solar Energy* **159** 318–28 https://doi.org/10.1016/j.solener.2017.10.078
[17] Kolomiets Yu, Tarasenko A, Tebuev V and Suleymanov M 2018 Investigation of the influence of various pollution types on operational efficiency of solar power installations in Moscow, *Alternative Energy and Ecology (ISJAEE)* **4-6** pp 12-24, https://doi.org/10.15518/isjaee.2018.04-06.012-024

Desalination by endothermal generators

B Monteiro and M Cavique

Naval Academy, Base Naval de Lisboa, Alfeite, 2810-001 Almada, Portugal

cavique.santos@marinha.pt; miguel.cavique@cest.pt

Abstract. Many regions of the planet are under water stress that can create humanitarian disasters. Where available, seawater can undergo a desalination process to produce drinkable water. The desalination processes may use thermal and mechanical energy. Methods based on evaporation and flashing are thermal driven, whether water osmosis uses mechanical energy. An endothermal motor can generate both mechanical and thermal energy. This paper focuses on a Diesel generator able to supply an electrical power of 362 kW. It evaluates the water production of a combined system with a Multi-Effect Evaporator and a Reverse Osmosis system. The Multi-Effect Evaporator uses the thermal energy from the exhaust gases; the Reverse Osmosis uses the generator's electrical power. The generator is typical for marine applications and can be deployed in humanitarian operations. The system can produce 1,011 m^3 per day of drinkable water, enough to supply the daily basic needs of a population of about 50,500 persons.

1. Introduction

As the population grows, the demand for drinking water around the world gradually increases. However, sources are limited, and many places face problems of water scarcity [1]. The sea represents 97% of the water on the planet. Desalination processes can produce freshwater using large amounts of available seawater [2]. Around 15,906 desalination plants worldwide produce about 95 million m^3 of water per day [3].

Thermal motors and generators exist everywhere. They can be merged with desalination packages or have a second life in desalination plants. A large Diesel motor has an efficiency of about 45%. It releases 55% of the energy as thermal energy. Nearly half of the thermal energy released goes through exhaust gases. The reuse of the thermal energy of the exhaust gases depends on the temperature of the application. Exhaust gases can produce power in an organic Rankine cycle, feed an absorption refrigeration system, or be used in any thermal system that requires less than 200 °C [4].

A desalination system removes the concentrations of salts present in seawater to produce fresh water. The desalination process generates freshwater without salt concentrations and a brine flow with high salinity [1].

A portable electric generator working as an auxiliary source for a solar distiller can increase drinkable water production by 20 L [5] in a multi-effect diffusion distiller. The heat rejection can be used in thermal systems to produce potable water. The residual steam from plants can feed the multi-effect desalination plants, improving water production and reducing energy consumption [6].

The classification of desalination systems depends on the activation energy. It can use thermal energy, mechanical energy, and chemical processes. Chemical processes are seldom used. The most mature thermal processes are Multi-Stage Flashing (MSF) and Multi-Evaporator Distillation (MED). MSF evaporates the water by heating the seawater and submitting the stream to a sudden pressure

IOP Conf. Series: Materials Science and Engineering 1235 (2022) 012062 doi:10.1088/1757-899X/1235/1/012062

drop. MED evaporates water at a pressure close to the atmospheric pressure, requiring a higher temperature source than MSF. However, it does not need to depressurize the system.

Reverse osmosis (RO), electrodialysis, and nano-filtration are the most common mechanical energy systems, the most predominant being RO. In the RO, seawater is pressurized against a semipermeable membrane over the osmosis pressure. It makes water go to the other side of the membrane in equilibrium with the brine, where salt and other minerals remain.

A generator produces mechanical and thermal energy. Therefore, these engines can produce freshwater by recovering the heat from exhaust gases at a MED system and using electrical power to run the RO systems.

In this paper, it is presented the modeling of the MED and the RO systems. The first model uses the EES (Engineering Equation Solver) platform, and the WAVE package models the RO system.

The novelty of the study is the evaluation of water production of a marine generator using the well-known MED and RO technologies. It helps to account for the deployment capacity of a vessel with the mentioned capabilities using wide available seawater.

The combined system can produce 1,011 m^3/day of drinkable water, enough to warrant the basic needs of a population of 50,500 inhabitants.

The following section describes the desalination processes and mathematical models, and section 3 computes the solutions.

2. Desalination processes

This work aims to seek for a marine solution to deliver water to dry areas. It is a possible solution for supplying the needs of drinkable water in regions with scarce water availability. The solution uses a combined MED and RO system. The exhaust gas heat allows producing steam to feed the MED system. MED systems work at atmospheric pressure and are easy to implement than flashing equipment.

Moreover, the study constraints the steam pressure to a maximum of 10 gauge bar to make construction easy and reliable. Three MED stages are used due to dimension constraints of the vessel.

Moreover, the combined system uses the maximum number of well-known RO systems compatible with the available electrical power.

2.1. Multi-effect desalination (MED)

MED systems are among the most well-known thermal desalination systems that allow large amounts of water flow and accept high salinity water. MED uses many stages, usually parallel feed [7]. As depicted in Figure 1, the feedwater flow $\dot{m}_f$ comes across a water pump and is sprayed at each stage. Mass steam $\dot{m}_s$ from an external source heats the first stage. It evaporates the feed water producing desalinated water $\dot{m}_{D1}$. Stage two works at a lower pressure than stage one. The mass flow $\dot{m}_{D1}$ can produce vapor ($\dot{m}_{D2}$) at more downward pressure in stage 2, which in turn goes to the next step.

The process may have many steps. In this work, three stages are used for the simplicity of construction and space constraints. The last stage exchanges heat with $\dot{m}_C$ the feed water from the sea. Each step produces brine water with high salinity $\dot{m}_B$. The total valuable water $\dot{m}_{Dtotal}$ is the sum of all parallel desalination water flows.

MED mainly needs thermal energy from a heat source. It also needs electrical power to drive the electric motors of the pumps, control, and sensors. Electrical energy accounts for less than a quarter of thermal energy [8].

The thermodynamic model is described by three equations: mass, salinity, and thermal energy balance. Moreover, the model needs to use the heat transfer equations regarding boiling and condensation convection to describe the thermal phenomena.

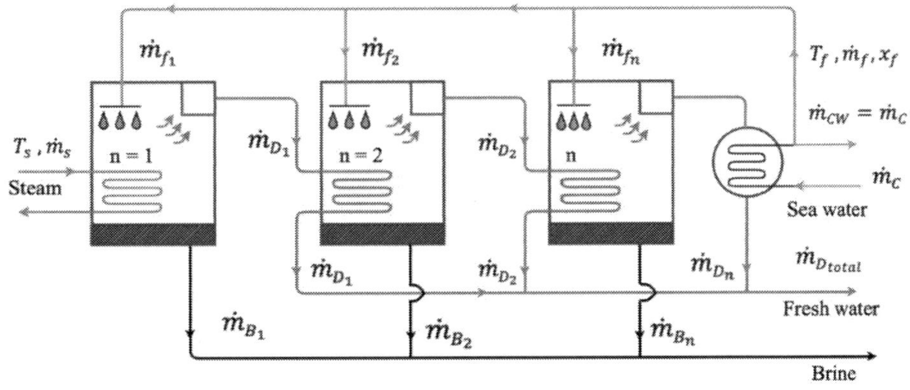

Figure 1. MED system with parallel feed

The thermal energy $\dot{Q}$ to boil ($\dot{m}$) is close to the latent enthalpy (h_{fg}) at temperature T, $h_{fg} = \lambda = \lambda(T)$ Equation 1 show the thermal needs for the isothemal and isobaric process.

$$\dot{Q} = \dot{m}\, h_{fg} \tag{1}$$

At each stage, a vacuum ejector removes noncondensable gases and helps to control pressure. The system usually works between 90 °C and 50 °C with a constant temperature drop between the n stages. Each stage receives an equal amount of salted water in the parallel configuration arrangement. This arrangement is the most convenient system to use when seawater is widely available. Figures 2 and 3 show the first and an n stage configurations.

At the first stage, the water mass and salt balance are defined according to Equations 2 and 3.

$$\dot{m}_{D_1} = \dot{m}_{f_1} - \dot{m}_{B_1} \tag{2}$$

$$x_{B_1} = \frac{\dot{m}_{f_1}\, x_f}{\dot{m}_{B_1}} \tag{3}$$

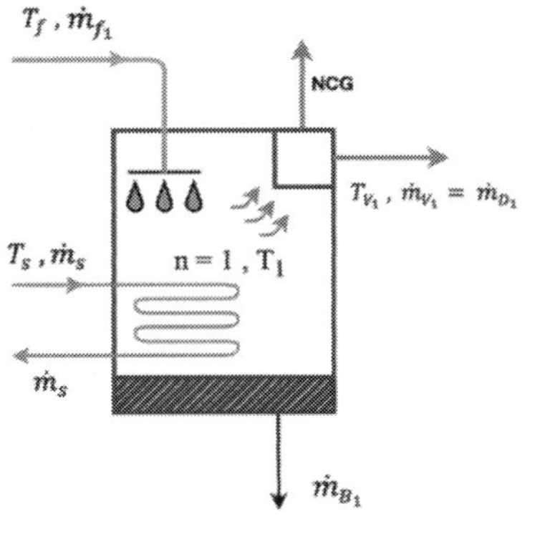

Figure 2. First MED stage

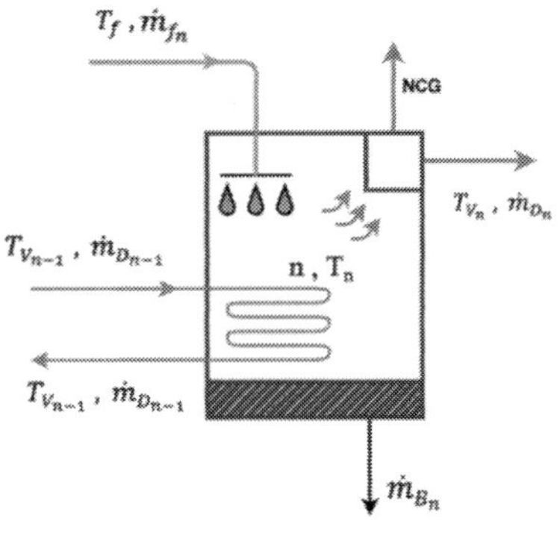

Figure 3. Last MED stage

And the energy balance is according to Equation 4

$$\dot{m}_s\, \lambda_s = \left(\dot{m}_{D_1} h_{TV1} + \dot{m}_{B_1} h_{T1}\right) - \dot{m}_{f_1} h_{T_f}$$

$$\dot{m}_{D_1} = \dot{m}_s \frac{\lambda_s}{\lambda_1} - \dot{m}_{f_1} y_1 \tag{4}$$

Where y_1 is the evaporation rate, as per Equation 5:

$$y_1 = \frac{c_P(T_1 - T_f)}{\lambda_1} \tag{5}$$

The heat exchanger uses steam to boil the seawater. The heat transfer equation is according to Equation 6, A_1, the heat exchange area and U the overall heat transfer coefficient. U mostly depends on the internal condensation coefficient. Sprays induce forced convection, resulting in a high boiling heat transfer and, therefore, a neglectable low thermal resistance from the external tube surface.

$$\dot{m}_s \lambda_s = A_1 U_1 (T_s - T_1) \tag{6}$$

All other stages i=1, 2, 3, ...n Equations 7 and 8 describe the mass and salt balance.

$$\dot{m}_{D_i} = \dot{m}_{f_i} - \dot{m}_{B_i} \tag{7}$$

$$x_{B_i} = \frac{\dot{m}_{f_i} x_{f_i}}{\dot{m}_{B_i}} \tag{8}$$

The energy balance follows equations 4, 5, and 6 similarly, described by Equations 9, 10, and 11.

$$\dot{m}_{D_n} = \dot{m}_{D_{n-1}} \frac{\lambda_{n-1}}{\lambda_n} + \dot{m}_{f_n} y_n \tag{9}$$

$$y_n = \frac{c_P(T_n - T_f)}{\lambda_n} \tag{10}$$

$$\dot{m}_{D_{n-1}} \lambda_{n-1} = U_n (T_{V_{n-1}} - T_n) \tag{11}$$

Finally, the condenser uses seawater ($\dot{m}_c$) to distilled the vapor $\dot{m}_{Dn}$. Moreover, the condenser pre-heat the feedwater $\dot{m}_f$ and discards wasted seawater $\dot{m}_{cw}$. Figure 1 aforementioned depicts the process, and Equation 12 and 13 describe the mass and energy balance at the condenser.

$$\dot{m}_c = \dot{m}_{cw} + \dot{m}_f \tag{12}$$

$$Q = \dot{m}_c A_c U_c (LMTD)_c \tag{13}$$

LMTD being the logarithmic mean temperature difference:

$$LMTD = \frac{\Delta T_1 - \Delta T_2}{ln(\frac{\Delta T_1}{\Delta T_1})} \tag{14}$$

$$\Delta T_1 = T_{V,in} - T_f \quad ; \quad \Delta T_1 = T_{V,out} - T_c$$

The most common performance coefficients are the recuperation rate, RR, and the specific heat consumption, SHC, described in Equations 15 and 16.

$$RR = \frac{\dot{m}_{D_{total}}}{\dot{m}_f} \tag{15}$$

$$SHC = \frac{\dot{m}_s \lambda_s}{\dot{m}_{D_{total}}} \tag{16}$$

2.2. Reverse Osmosis

Reverse osmosis uses pressure as the driving force to move water from seawater using a semipermeable membrane. Two reservoirs separated by a semipermeable membrane, one with distilled water and the other with salted water, experience the distilled water passing through the membrane to create an equilibrium distilled water concentration. Therefore, the column of salted water increases, displaying the so-called osmotic pressure. Applying a similar pressure over the salted column reverses the osmosis process [9]. Reverse osmosis (RO) is one of the most well-known methods for desalination. It became popular in the last decades because it can be used in small to large plants fed by different types of non-drinkable water [10].

Figure 4 shows a RO plant. The system uses a pre-treatment section built with a sand filter and a cartridge matter filter. Then a high-pressure pump, typically of 40 to 80 bar, creates the osmotic pressure over the semipermeable membrane. It establishes a drinkable water flow and a high concentration brine flow. Then, the brine flow is discarded, and the potable water enters a post-treatment section to tune ions and water pH.

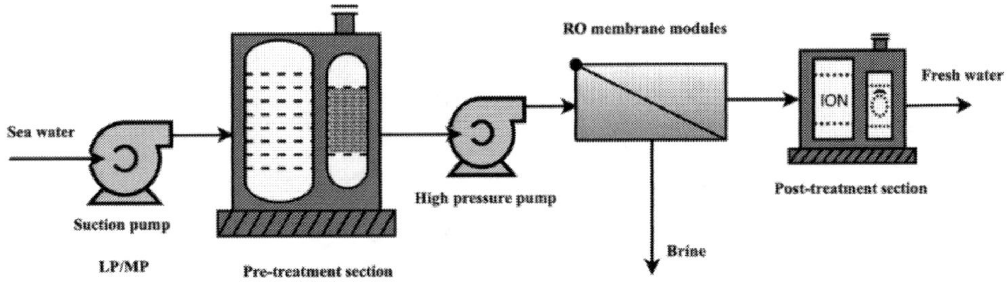

Figure 4. Reverse osmosis plant

Where seawater is widely available, the plant has many RO membrane blocks fed in parallel. Seawater flows across the RO blocks typically with a RR of 50% of produced water Q_p. Equation 17 shows the flow balance between the produced water (p), feed (f), and discharged brine (B).

$$\dot{Q}_f x_f = \dot{Q}_B x_B + \dot{Q}_p x_p \quad (17)$$

The produced water, or permeate, is the sum of solvent and solute flows. The solute depends on the difference of salt concentrations and is a small quantity compared to the solvent flow. Equation 18 shows the water flow rate per unit area of membrane (J_w). Jw depends on the membrane permeability A_w, the applied pressure feed P_f, pressure drop across treatment sections P_d, production pressure P_p and osmotic pressure $\Delta\pi$ [11]. The production pressure is close to the atmospheric pressure.

$$Jw = A_w(P_f - P_d - P_p - \Delta\pi) \quad (18)$$

The permeability (A_w) of the plant, given by Equation 19, depends on the water temperature T and plant and salt coefficients α_1, α_2.

$$A_w = A_{w0}\, exp\left(\alpha_1 \frac{T - 273}{273} - \alpha_2(P_f - P_d)\right) \quad (19)$$

Moreover, there is a solute flow J_s of the salt. The permeate flow depends almost only on J_w. The osmotic pressure $\Delta\pi$ is a function of the salt feed concentration x_f and production x_p. The production x_p increases along the RO semipermeable membrane. Sea incrustation creates a flooding factor that can decreases production by about 20%.

IOP Conf. Series: Materials Science and Engineering 1235 (2022) 012062 doi:10.1088/1757-899X/1235/1/012062

3. Computation

3.1. MED system

The generator has a maximum electrical power of 362 kW. It might work at 75% of the maximum capacity to last longer, so the calculation considers an available electrical power of 272 kW. In such conditions, the exhaust gases go out at 320 °C at a rate of $\dot{m}_g$=0.82 kg/s. A shell heat exchanger can use the energy of exhaust gases to produce steam. Figure 5 shows a scheme of the heat exchanger. Exhaust gases enter at (1) and go out (2) at 250 °C. The going-out temperature is a conservative choice not to create any damage to the generator. In these circumstances, the exhaust gases transfer a heat power of 62 kW to a water flow (3) to produce steam at (4).

An EES program simulates the overall system behavior of MED using equations of section 2.1. Equation 20 describes the heat transfer from the gases to steam production. The primary convection resistance comes from the external surface of the tube where the exhaust gases flow. It creates an overall heat transfer coefficient in the gas heat exchanger of 750 W/m^2°C.

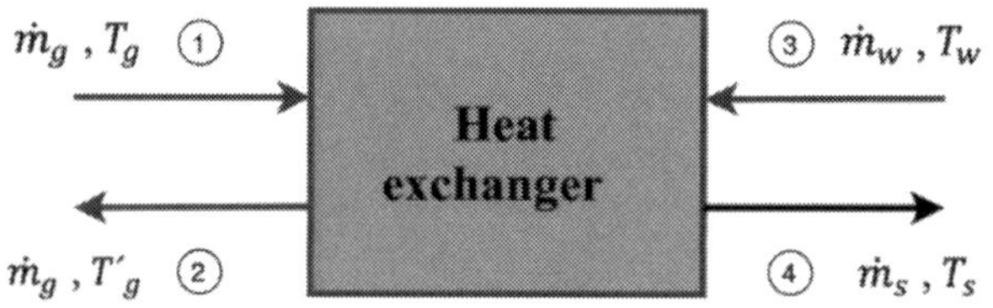

Figure 5. Heat exchanger - exhaust gases from the generator to produce steam

$$Q = \dot{m}_g c_{p,g}\left(T_g - T'_g\right) = \dot{m}_s\, h_{fg} \tag{20}$$

Thus, steam is used to create vapor from the seawater at the first stage. Maintaining the heat power, increasing steam temperature reduce the steam mass. The function is depicted in Figure 6. The steam design temperature (T_S) was defined to be 180 °C, corresponding to a gauge pressure of 9 bar. This pressure allows an easy construction. At this temperature, the steam flow is 0.99 kg/s, and the heat exchanger needs an area of 7 m^2. The seawater feed is 5 kg/s ($\dot{m}_f$) with 36000 ppm.

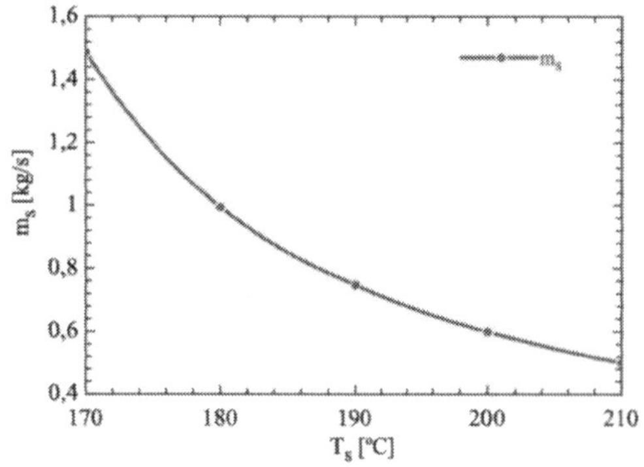

Figure 6. Steam mass flow at steam temperature.

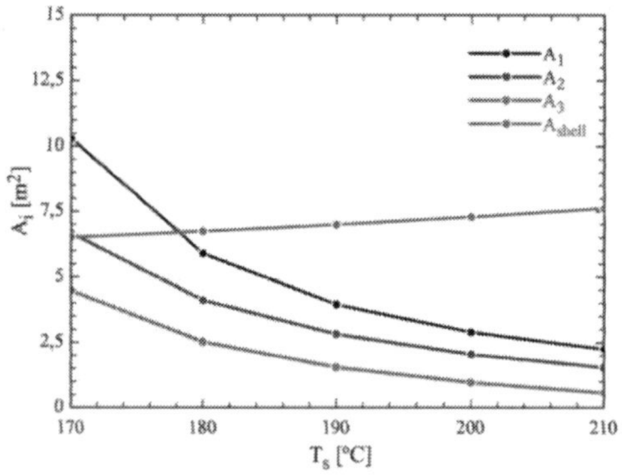

Figure 7. Heat exchange areas of the shell (exhaust gases to steam) and stage areas 1, 2, and 3 vs. steam temperature.

Increasing steam temperature increases the shell area due to a minor temperature difference between gases and steam. On the contrary, the stages area decreases due to the opposite outcome. Figure 7 shows the mentioned effects.

Moreover, it shows the decrease of the area along the sequence of stages, according to the distillate production. The result is a decrease in the heat exchange area as steam temperature increases, as depicted in Figure 8.

For a fixed number of stages, the distillate water production decreases (Figure 9). The decrease is a consequence of the decrease in the exchange area. It means that more stages would be needed for higher steam temperatures.

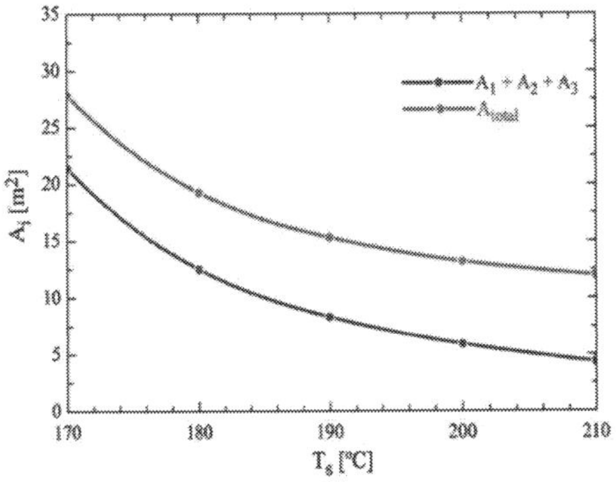

Figure 8. The total area function of the steam temperature.

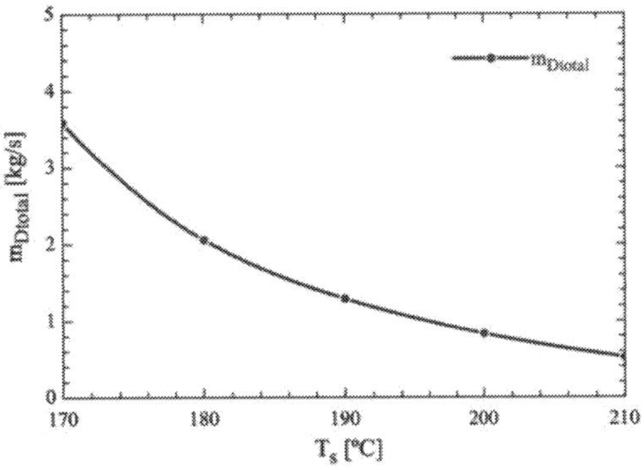

Figure 9. The distilled water of the system function of steam temperature.

Regarding the feedwater temperature T_f, when T_f increases, the water production ($\dot{m}_D$) increases and, therefore, the recovery ratio (RR) increases as well (Figures 10 and 11). The distillate production reduces from the first to the third stage. Moreover, the dependency of feed temperature increases along stages due to the reduction of steam temperature.

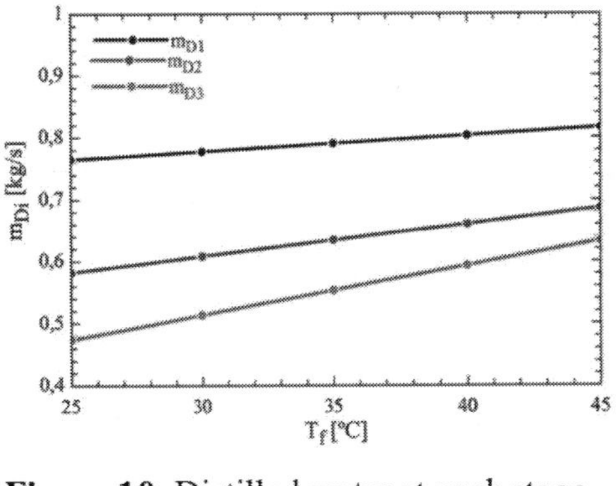

Figure 10. Distilled water at each stage, 1, 2, and 3 vs. feed temperature.

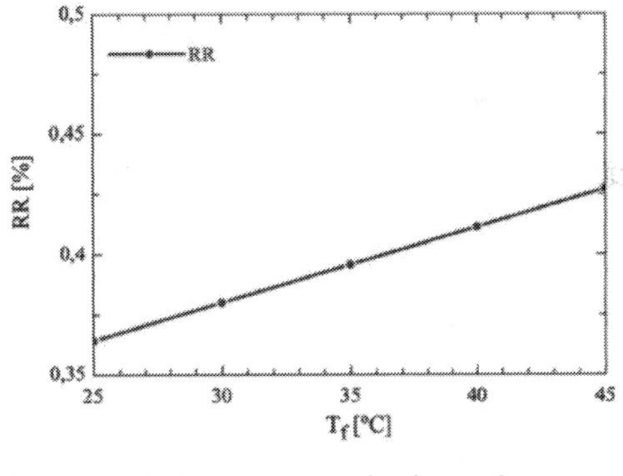

Figure 11. Recovery ratio dependency on feed temperature.

Figure 12 shows the brine salt concentration increases with T_f. Notice that the brine salt concentration and the recovery ratio (RR) decrease along the stages because the steam temperature drops. The heat exchange areas depend on the overall heat transfer coefficients U. According to empirical expressions [12], the convection coefficients depend on working temperature. The U values for the first, second, and third stages are 2.7, 3.6, and 3.7 kW/m^{2o}C for bank tubes of 30 mm. The U values are mostly dependent on the inside condensation coefficient. On the external side, the spray creates a forced-convection boiling. This type of boiling has convection coefficients higher than the computed ones of orders of magnitude. The overall heat transfer coefficients U are in the range of expected values of MED systems [8].

Figure 13 shows the heat needed to produce a kg of distilled water. For the 60 kW of available thermal load, the MED system can make 1.9 kg/s of water. According to [13], it will need an electric power of about 14 kW.

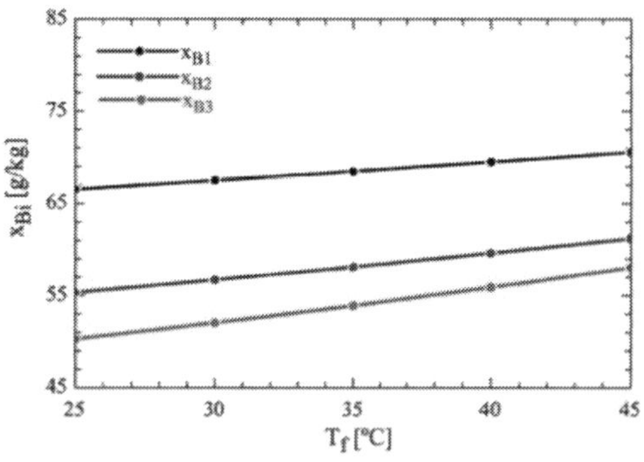

Figure 12. Brine salt concentration at each stage, 1, 2 and 3 vs. feed temperature.

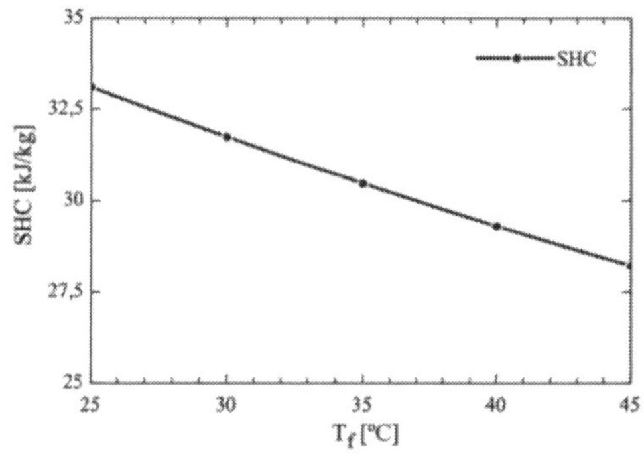

Figure 13. Specific heat consumption function of feed temperature.

3.2. Reverse osmosis

The study uses the *WAVE* software to analyze the behavior of the semipermeable membranes at high pressure (Figure 4). The computation uses a package available salt concentration of 35,500 mg/L, close to the one used in the MED system, at a temperature of 25 °C. On the pre and post-treatment sections, there is a pressure drop of 5 bar, used to compute the electrical energy needs of the RO system.

The osmotic pressure for the mentioned salt concentration is about 27 bar. Typically the osmotic pressure increases by 0.77 bar for each 1,000 mg/L of salt in the water.

Due to the high availability of seawater, the RO system designed has a single-stage module in a parallel plant. The selected semi permeate membrane is the *SW30ULE-440i* feed at 50 m^3/day of seawater.

The software allows running the system against different feed pressures (Figures 14 and 15). When the feed pressure increases from 40 to 80 bar, the recovery ratio increases (Figure 14) almost linearly at about 3% for each 5 bar increase. Consequently, the permeate flow increases as well from 12 to 28 m^3/day. Moreover, the salt concentration decreases from 360 to 250 ppm (Figure 15).

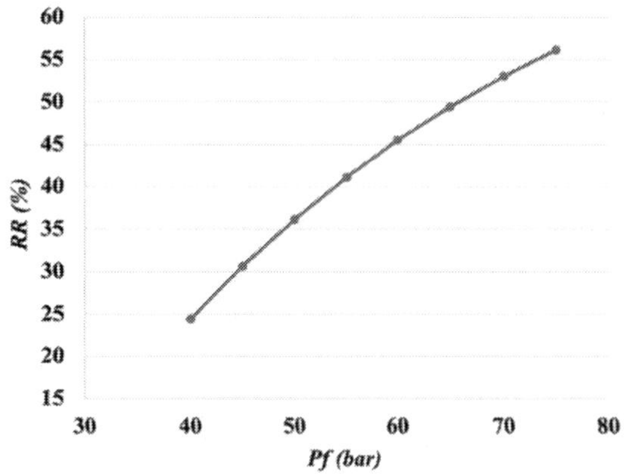

Figure 14. Recovery ratio as a function of feed pressure

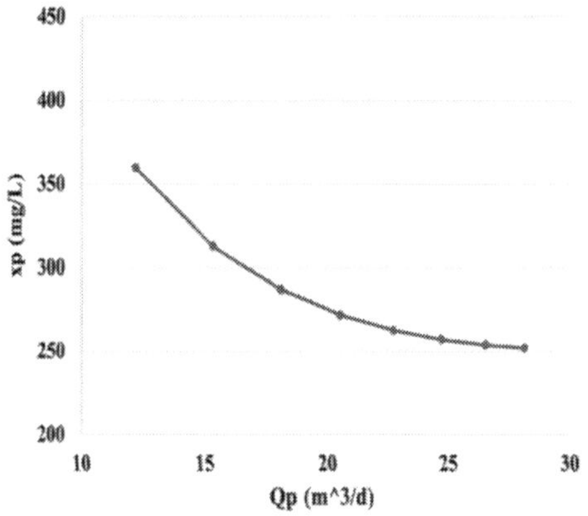

Figure 15. Salt concentration dependency by increasing permeate flow due to the pressure increase of Figure 14.

RO system is highly dependent on feed water temperature. The system was designed for 70 bar feed pressure; the RR increases by about 10% when feed temperature increases from 5 to 30 °C, as depicted in Figure 16. However, the rise in feed temperature makes the permeate flow have a higher salt content (Figure 17).

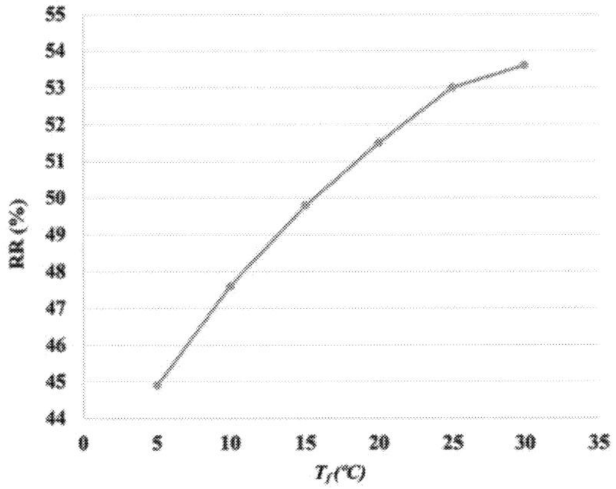

Figure 16. Recovery ratio as a function of feed water temperature.

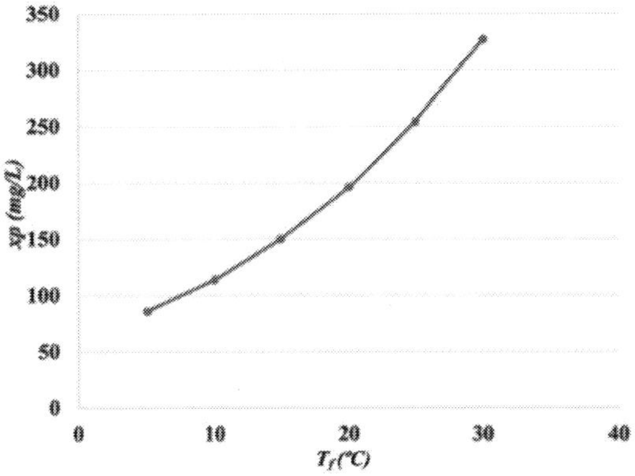

Figure 17. Salt concentration dependency of feed water temperature.

The power needed depends on feed pressure according to Equation 21, where v is the specific volume and η the overall pump efficiency to pump a mass flow $\dot{m}_p$. The operating pressure is 65 bar, which needs 4.8 kW, to pump a mass flow of 0.58 kg/s with a pump efficiency of η=70%.

$$Q = \frac{P_f \, v \, \dot{m}_p}{RR \, \eta} \tag{21}$$

4. Conclusion

This study uses a 362 kW marine electrical generator at 75% of its capacity to produce desalinated water. It feeds a reverse osmosis (RO) system and a multi-effect desalination (MED). The RO runs with 258 kW of electrical energy, and the MED system uses 60 kW of thermal power recovered from exhaust gases. Moreover, the MED system needs an extra 14 kW to run pumps and auxiliary equipment.

The unitary RO set needs 4.8 kW to produce 16 m^3/day of drinkable water. The recovery rate (RR) of the RO is 36%. Therefore, the generator can feed 53 RO systems that allow 848 m^3 of water daily production. The MED system can make 163 m^3/day. Thus, installing a MED in a generator can increase the overall capacity to produce water by about 13% if we consider its electrical energy use.

The whole system makes 1,011 m^3/day capable of giving the essential water needs to 50,500 inhabitants.

In future works, it is worth discussing the possibility of dropping the temperature of the exhaust gases under 250 °C. Moreover, it is necessary to evaluate the available space accurately to define the likely number of stages.

The deployment of vessels to deliver water to dry regions in humanitarian operations is for sure a valuable area to explore.

ORCID IDs

Miguel Cavique https://orcid.org/0000-0002-9204-7521

References

[1] Khawaji A D, Kutubkhanah I K and Wie J M 2008 Advances in seawater desalination technologies *Desalination* **221** 47-69 https://doi.org/10.1016/j.desal.2007.01.067

[2] Mutaz I S A and Wazeer I 2014 Comparative performance evaluation of conventional multi-effect evaporation desalination processes *Applied Thermal Engineering* **73** (1) 1194-203 https://doi.org/10.1016/j.applthermaleng.2014.09.025

[3] Jones E, Qadir M, Vliet MTV, Smakhtin V and Kang S 2019 The state of desalination and brine production: A global outlook *Science of the Total Environment* **657** 1343-56 https://doi.org/10.1016/j.scitotenv.2018.12.076

[4] Talbi M and B Agnew B 2002 Energy recovery from diesel engine exhaust gases for performance enhancement and air conditioning *Applied Thermal Engineering* **22** (6) 693-702 https://doi.org/10.1016/S1359-4311(01)00120-X

[5] Tanaka H and Park C 2010 Distillation utilizing waste heat from a portable electric generator *Desalination* **258** (1-3) 136-42 https://doi.org/10.1016/j.desal.2010.03.025

[6] Sommariva C 2008 Utilisation of power plant waste heat steams to enhance efficiency in thermal desalination *Desalination* **222** (1-3) 592-5 https://doi.org/10.1016/j.desal.2007.01.122

[7] Frantz C and Seifert B 2015 Thermal analysis of a multi effect distillation plant powered by a solar tower plant *Energy Procedia* **69** 1928-37 https://doi.org/10.1016/j.egypro.2015.03.190

[8] Alasfour F N 2020 Introduction to Desalination: Systems, Processes and Environmental Impacts (Kuwait: Wiley-VCH)

[9] Zhao T, Niu R, Su M and Anderson T 2011 Steady State and Dynamic Modeling of RO Desalination Modules and System using EES *IEEE International Conference on Robotics and Automation* https://doi.org/10.1109/ICRA.2011.5980587

[10] Seoudy A, Lotfy H and Saleh I 2008 Simulation of a reverse osmosis seawater desalination plant Part 1: Theoretical modeling of a real plant and its validation *13th International Conference on Applied Mechanics and Mechanical Engineering* 464-82 https://doi.org/10.21608/AMME.2008.39764

[11] Murthy Z and Gupta SK 1997 Estimation of mass transfer coefficient using a combined nonlinear membrane transport and film theory model *Desalination* **109** (1) 39-49 https://doi.org/10.1016/S0011-9164(97)00051-9

[12] Holman JP 2010 *Heat transfer* (New York: McGraw-Hill) pp 487-551

[13] Zhang H, Wang E and Fan B 2013 Heat transfer analysis of a finned-tube evaporator for engine exhaust heat recovery *Energy Conversion and Management* **65** 438-47 https://doi.org/10.1016/j.enconman.2012.09.017

[14] Maheswari K, Murugavel K K and Esakkimuthu G 2015 Thermal desalination using diesel engine exhaust waste heat - An experimental analysis *Desalination* **358** 94-100 https://doi.org/10.1016/j.desal.2014.12.023

[15] Youssef PG, AL-Dadah RK and Mahmoud SM 2014 Comparative Analysis of Desalination Technologies *Energy Procedia* **61** 2604-7 https://doi.org/10.1016/j.egypro.2014.12.258

[16] Jiang A, Ding Q, Wang J, Jiangzhou S, Cheng W and Xing C 2014 Mathematical Modeling and Simulation of SWRO Process Based on Simultaneous Method *Journal of Applied Mathematics* **2014** 908569 https://doi.org/10.1155/2014/908569

IOP Conf. Series: Materials Science and Engineering 1235 (2022) 012063 doi:10.1088/1757-899X/1235/1/012063

Biogas from poultry waste - a source of energy

A I Halmaciu, I Ionel, M R Wächter and I Vetres

Politehnica University of Timisoara, Romania, Faculty of Mechanical Engineering, Bv. M. Viteazu, Nr. 1, 300222, Timisoara, Romania

ioana.ionel@upt.ro

Abstract. The latest excessive industrial development has led to both the increase of the fossil fuel consumption and an excessive pollution. A saving solution would be considered renewable energy sources, best based on wastes, which prove energy content. The current work focuses on a case study for the biogas production, using anaerobic digestion of poultry dejections, by means of a latest generation bio-processor. These dejections are presently and normally used only as fertilizer. The experiment was performed simultaneously in 6 digesters, which were loaded daily with a quantity of material resulting from the calculated recipe, based on the characteristics of the poultry litter. Preliminary experimental analysis such as calorific value, humidity and volatile content were run. Following the results obtained, one can state that the poultry litter can be used as raw material for the anaerobic biogas production, in the future, as well.

1. Introduction

Together with the humankind evolution, the energy has become an important part of the economic development, significantly being generated by the increase of the population`s life standards [1]. This rapid economic development has come, up not only with benefits, but it also has brought a series of disadvantages. Thus, one of them demonstrates that energy constitutes a major factor in the evolution and the increase of the global economy [2].

As a result of the significant increase in electricity consumption, scientists are urging countries around the world to turn their attention to obtaining renewable energy [3].

Also, the use of renewable energy will generate lower production costs and will significantly contribute to solving the problem of global warming [4].

The climate changes the increase of the fuel prices [5], the environment regulation and irreversible changes are facts, more and more demanding [6]. The rapid increase of the energy consumption, but also, the fear of fossil fuels exhaustion, especially in the countries where the natural resources are reduced, have determined the community of the scientists to search alternative resources. Best of them are characterised by a low or neutral CO_2 emission, generally speaking a low exhaust of green gases are not affecting the climate.

An innovating solution would be considered as usage of recycling energy resources, as being most viable, cheap, ecologic, and avoidable for any state, poor in natural resources. The increase of the consumption of recycling energy by producing electrical energy and heat is at the basis of an ecological energetic transaction [7]. For the countries that are rich in biodegradable litter the designing of the biogas installation would be first option. This is a viable method to solve, not only the issue related to the energy, but also the issue of eliminating the litter, in an economic circle. In case of Romania, this provides a wide variety of biodegradable resources such as: biomass, the energy cultures [8], the litter obtained from the animal farms, the agricultural litter, the industrial litter, and the municipal solid litter.

Raising animals for slaughtering from meat processing and the large number of slaughterhouses have positive effects on the population, by creating new jobs and ensuring a fresh diet. These also lead to the generation of a large amount of waste and for the elimination of this, urgently solutions must be found [9]. A suitable solution would be to capitalize this waste and use it in a circular economy, by producing biogas, not before it has gone through a pre-treatment process (biological, physical, chemical and mechanical) [10].

The biogas is obtained through the biological decomposing of organic material in the absence of the oxygen [8]. Through this technology, the anaerobic mass digest is changed into biogas [11], while the organic substrate, resulted due to this process, is used in agriculture [12].

Concerning the methane production, the researchers are doing detailed tests to obtain bigger amounts of this biogas using as raw the whey resulted after the process of milk manufacturing, the pig litter, the cow litter [13] and the poultry litter.

Following the studies made until now, it is concluded that by mixing several types of raw materials, could be generated a huge amount of methane [14].

2. The used material and the method of testing sampling

2.1. The experimental configuration

The lab installation used for this experiment is a bioreactor simulator (BRS), type CSTR-2G (Continuous Stirred- Tank Reactors- 2 litres, G-glass). This installation was developed by Bioprocess Control Sweden AB (BPC) for the on-line monitoring and control in a continual way of the anaerobic fermentation of the experiment [15]. The installation has the capacity to calculate and to show in real time the rate of the organic charging (ORL), the hydraulic time retention (HRT) and the significant gas production (SGP) [15]. The experimental device used is equipped with 6 graded glass digesters up to 1800 ml. Although, the digester has maximum capacity of 2000 ml, but during the experiment the charging rate was to 1800 ml. Each one of those 6 reactors has 3 orifices: an orifice is intended for feeding raw material supply, an orifice used to evacuate the digestate in the reactor and an orifice used in continuous measurement of pH or temperature values.

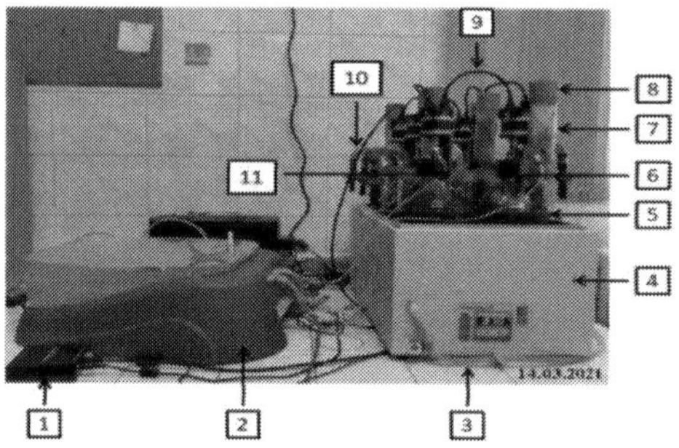

Figure 1. The experimental lab installation for the bio-gas production: 1 - device for the motor control, 2- device of measuring the gas volume, 3 - tube of water evacuation, 4 - bath of thermostatic water, 5 - six glass reactor with three holes, 6 - plastic screw cap without hole, 7 - funnel for charging, 8-silicon corks, 9 - engine cable, 10 - discharging tube, 11 - plastic clamps for tubes [16].

Before starting the experiment, water was put in the gas volume measuring device, in the thermostatic water bath and with the help of the motor cables, the digesters were connected in series. Then batch of thermostatic water was set at 37 °C, with the role of maintaining a constant temperature for the 6 digesters. This temperature offers the bacteria a mesophilic environment. Daily digesters have been charged/ according to the recipe.

The installation is a very high performance one because it offers to the user the possibility to choose for each digester the type of substrate used, as well as its concentration. The 6 digesters are running in parallel, but independent one from another, main data (such as flow rate) being automatic measured and stored, according to a selected programme. Moreover, the equipment used offers the possibility to use a preset program for loading/unloading time of the digesters. With the help of a laptop or phone the user can overview, start, interrupt or even to finish the experiment [15].

Until now There was made another experiment on this facility, using pig litter as the raw material [16]. Comparing the two experiments, can be observed that the lowest specific value of gas production recorded in pig litter was about 0.189 nl/(l·day), whereas in bird litter with value of 0.229 nl/(l·day). Analysing the highest biogas production value, should be mentioned that in pig waste the value is 0.898 nl/(l·day), and in bird waste the recorded a value of 0.812 nl/(l·day). As regards to the gas flow, the lowest value for pig litter was 0.340 nL/day, while the highest value was 1.6 nl/day. In case of chicken litter was recorded 0.412 (nl/day), which representing the lowest value, while 1.4 nl/day indicates the highest value of the gas flow [16].

2.2. Storing and pre-treating of the raw material

The poultry litter was collected from a poultry breeding farm and inoculums were provided to biogas installation. Throughout, in the whole experiment the raw material was stored in a fridge while the inoculums were incubated at a temperature of 37 °C for a week. Before being introduced in the digester, the poultry litter was grounded with the help of a mixer, being afterwards diluted with warm water. Analysis concerning humidity, volatile content and the heat value were simultaneously run, according to the recommended procedure specified by EN 14774 standard, Solid biofuels- determination of moisture content, EN 14775- Solid biofuels- determination of ash content, EN 14918- Solid biofuels- for determination of calorific value, EN 15148- Solid bio fuels-for determination of volatile matter content [17].

As a result of the preliminary tests, was concluded that for obtaining an optimal receipt were need of: 1065.7 g of inoculums, 620.79 g of warm water and 113.51 grams of poultry litter, in mixture, for each digester. During the experiment the daily rate of discharging of the 6 digesters was of 168.9 grams/digester, while the daily dose of their charging was about 26.1 grams of poultry litter and 142.8 grams of warm water/digester. In the last 13 days how long lasted the experiment, the loading rate of the digesters was doubled. During the whole experiment, the digesters were kept in a thermostatic water bath at a temperature of 37 °C. The samples used reveals specific characteristic that are turning it into a good biodegradable material, as it has pH of 6.8-7.3 and a specific proportion *C/N*= 15-20 (Carbon to Nitrogen ratio). Based on the information obtained from the farm owners, the use of antibiotics, phenols, or detergents in the practice is not usual.

To reach the anticipated results, following procedures/steps were accomplished: (a) The numbering and the weighing of six empty crucibles with a capacity of 100 ml; (b) The adding of 150 grams of wet material in every crucible; (c) The introduction of the crucibles in the oven and their keeping at a temperature of 105 °C for 24 hours; (d) After 24 hours the dishes are removed from the oven and are left to cool down at the room temperature in several desiccators; (e) Then the six crucibles were weighed and their values marked, these values representing the amount of dried substance; (f) The weighed crucibles are introduced in a calcinatory for two hours at a temperature of 550 °C; (g) After the two hours the six crucibles were removed from the calcinatory and introduced in the desiccators, to be cooled down (h). After that step the samples were weighed, and was obtained values for the amount of the burnt substance; (j) The steps were repeated, if necessary, until the final weight (mass) remained constant.

Based on the previously presented steps one could calculated the necessary amount of material to obtain an efficient receipt of the biogas production.

2.3. The supervision of the optimal environment of the microorganisms. Results.
The pH value offers significant information about the acid and the basic measure of the digested. The studies have proved the fact that the bacteria has been developed better in neutral conditions up to light alkaline, where the optimal value of pH is between 7 and 8.5 [18]. Another important factor in the developing of the bacteria consists in the temperature that has to be stable. In case of mesophilic bacteria, the temperature must be between 30-45 °C, for the thermophilic bacteria should be between 45-55 °C [19,20]. Within the whole experiment with the help of a portable pH-meter of high accuracy, the pH value of the digestate was measured. The recorded values throughout the experiment varies between 7.35 and 7.87, this thing proves the fact that the optimal environment was offered to the bacteria.

Figure 2 represents the pH variation that was recorded during the experiment of anaerobic digestion. If in the first part of the experiment the pH value was higher, this oscillating between the values 7.56- 7.87; one can notice that towards the end of the experiment the pH the registered values present a slight decrease, varying between 7.36-7.49.

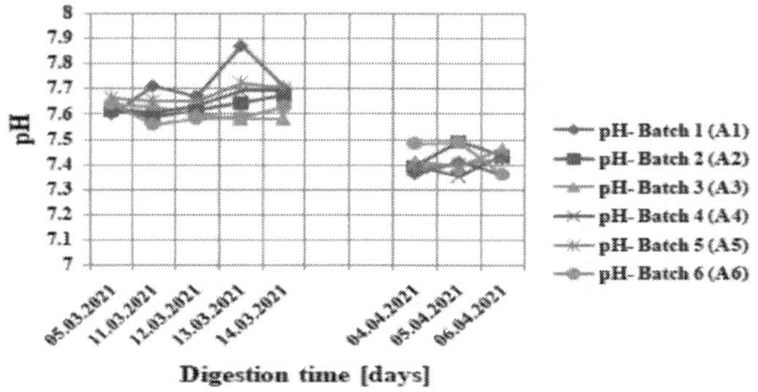

Figure 2. The pH value for the six digesters.

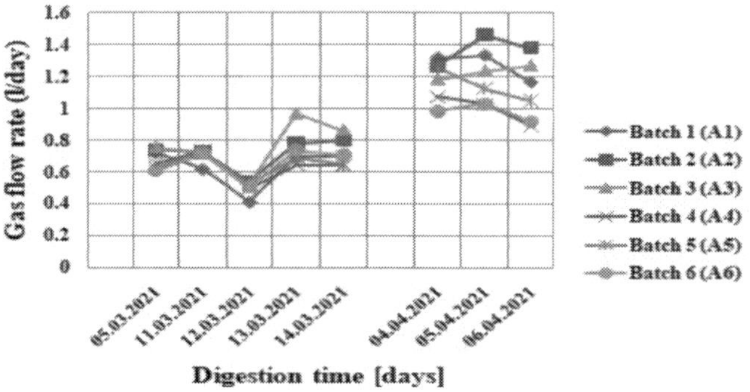

Figure 3. The days with the most relevant gas flow.

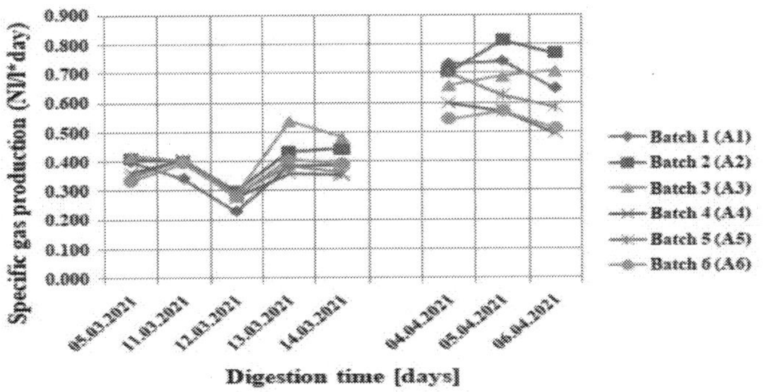

Figure 4. The specific gas production.

The values presented in figure 3 and figure 4 are measured and stored automatically by the measurement system of the control unit. Being a laboratory installation, the values generated are relatively small and the literature doesn't provide measurement results in the International Measurement System related to this kind of experiment, the used units are known as specific (traditional, dedicated) used one.

The results are covering the most relevant gas flow rate, aspect presented in the figure 3. If in the first part of the experiment, those 6 digesters were loaded with 168.9 g (water + poultry manure), during this period the lowest value of the gas flow it was recorded at line 1 (A1), while the value of the highest was recorded at line 3 (A3). Towards, the end of the experiment, the loading rate of the digesters was doubled. During this period the lowest gas flow was recorded at line 4 (A4), and the highest gas flow recorded at line 2 (A2).

Figure 4 represents the amount of biogas (daily specific gas production) obtained at the beginning and at the end of the experimental period. At the beginning of the experiment, the amount of gas generated was reduced and stable. After that, it can be observed a slight increase, followed by a sudden decrease and then an increase in the specific gas production. By analysing the final stage of the experimental period, fluctuations were as well registered, but, in comparison with the beginning of the experiment, they were not so sudden. The highest amount of gas recorded was on line (A2) and took place at the end of the experimental stage.

3. Conclusions

This research was carried out to determine the energy properties of the poultry litter in the biogas production, using as experimental method the anaerobic digestion. The experiment was accomplished with the support of an ultramodern equipment made up with 6 digesters, filled preliminary with the same amount of material and inoculums. The advantages of this installation consist in the fact that it has the capacity to store the measured data and information on a computer and that it can be supervised, in real time, from a laptop or a mobile. Within the whole experiment, the pH was measured, which offers the possibility to conclude that the bacteria provide a proper environment for developing and supporting the CH_4 generation, under anaerobic conditions, in mesophilic conditions.

Even those 6 digesters were daily filled with the same dose of under-layer, the biogas amount generated in the 6 lines was different. An explanation consists by differences on the raw material, as even one mixed very well each batch, the variety of the original poultry manure which may vary, the composition could be slightly different. Another explanation is connected to the sensibility of the process versus the pH value which must be controlled and corrected. In the case of this experiment, no corrections were needed, because during the whole experiment the pH value was maintained between 7.35 and 7.87.

Based on the results obtained, it can sustain that this new source of recyclable energy, based on poultry manure, can be easily implemented worldwide, and also in Romania, being an important source of "clean energy". Due to the great diversity of biodegradable waste which can be use, it is desired that in the future the recipes for biogas production, using various mixtures of waste should be developed, by experiments and similar research as in this article, preferable at larger scale. But still, a good start has been achieved.

Acknowledgments
This work was carried out within the research activity of the first author as PhD student, in the frame of the PhD School of the Politehnica University of Timisoara.

ORCID IDs
A I Halmaciu https://orcid.org/0000-0002-9932-7366
I Ionel https://orcid.org/0000-0002-2999-1210
M R Wächter https://orcid.org/0000-0002-0667-3818
I Vetres https://orcid.org/0000-0002-6425-9818

4. References

[1] Diagi E A, Akinyemi M L, Emetere M E, Ogunrinola I E and Ndubuisi A O 2019 Comparative analysis of biogas produced from cow dung and poultry droppings *IOP Conf. Ser.: Earth Environ. Sci.* **331** 012064 https://doi.org/10.1088/1755-1315/331/1/012064
[2] Opurum C C 2021 Kinetic study on biogas production from cabbage (Brassica oleracea) waste and its blend with animal manure using logistic function model *J. adv. microbiol.* **21** 34-43 https://doi.org/10.9734/JAMB/2021/v21i130317
[3] Qudrat-Ullah H and Nevo C M 2021 The impact of renewable energy consumption and environmental sustainability on economic growth in Africa *Energy Rep.* **7** 3877-86 https://doi.org/10.1016/j.egyr.2021.05.083
[4] Derasid N A C, Tahir I M, Musta'amal A H, Bakar Z A, Mohtaram N, Rosmin N and Ali M F 2021 Knowledge, awareness and understanding of the practice and support policies on renewable energy: Exploring the perspective of in-service teachers and polytechnics lecturers *Energy Rep.* **7** 3410-27 https://doi.org/10.1016/j.egyr.2021.05.031
[5] Shinde M A, Dikshit A K, Odlare M, Thorin E and Schwede S 2021 Life cycle assessment of bio-methane and biogas-based electricity production from organic waste for utilization as a vehicle fuel *Clean Technol. Environ. Policy* **23** 1715-25 https://doi.org/10.1007/s10098-021-02054-7
[6] Pant K K and Mohanty P 2014 Biomass, Conversion Routes and Products – An Overview *Transformation of Biomass: Theory to Practice* ed A Hornung (Wiley) chapter 1 pp 1-30 https://doi.org/10.1002/9781118693643.ch1
[7] Lambrecht H, Lewerenz S, Hottenroth H, Tietze I and Viere T 2020 Ecological scarcity based impact assessment for a decentralised renewable energy system *Energies* **13** (21) 5655 https://doi.org/10.3390/en13215655
[8] Bhatia S C 2014 Biogas *Advanced Renewable Energy Systems* (Woodhead Publishing India) chapter 17 pp 426-72 https://doi.org/10.1016/B978-1-78242-269-3.50017-6
[9] Baldi G, Soglia F and Petracci M 2021 Valorization of meat by-products *Food Waste Recovery* ed C M Galanakis (San Diego: Academic Press) chapter 21 pp 419-43 https://doi.org/10.1016/B978-0-12-820563-1.00017-2
[10] Kumar P S, Saravanan A, Jayasree R and Jeevanantham S 2020 Food industry waste biorefineries *Refining biomass residues for sustainable energy and Bioproducts* ed R P Kumar, E Gnansounou, J K Raman and G Baskar (Academic Press) chapter 18 pp 407-26 https://doi.org/10.1016/B978-0-12-818996-2.00018-1
[11] Usman A, Noor H, Malik B H, Ali H W and Muzaffar I 2020 Applications of Cloud and IOT

technology for the development of agricultural sector *Int. J. Sci. Technol. Res.* **9** 52-7
[12] de Castro e Silva H L, Silva A M L, Barros R M, dos Santos I F S and de Freitas J V R 2021 Addition of iron ore tailings to increase the efficiency of anaerobic digestion of pig manure: A technical and economic analysis *Biomass Bioenergy* **148** 106013 https://doi.org/10.1016/j.biombioe.2021.106013
[13] Lee J T E, Khan M U, Dai Y, Tong Y W and Ahring B K 2021 Influence of wet oxidation pre-treatment with hydrogen peroxide and addition of clarified manure on anaerobic digestion of oil palm empty fruit bunches *Bioresour. Technol.* **332** 125033 https://doi.org/10.1016/j.biortech.2021.125033
[14] Acosta N, Kang I D, Rabaey K and De Vrieze J 2021 Cow manure stabilizes anaerobic digestion of cocoa waste *J. Waste Manage.* **126** 508-16 https://doi.org/10.1016/j.wasman.2021.02.010
[15] Operation and Maintenance Manual 2017 Bioreactor Simulator
[16] Halmaciu I A, Wächter M R and Ionel I 2021 Turning a waste producer into an energy producer *J. Phys.: Conf. Ser.* **1781** 012061 https://doi.org/10.1088/1742-6596/1781/1/012061
[17] Cioabla A E 2020 Habilitation thesis: The use of biomass through biodegradation processes to obtain clean energy. Technology an experiments, Timisoara (original in Romanian: Habilitation Thesis: Utilizarea biomasei prin procese de biodegradare pentru obtinerea de energie curata. Tehnologie si experimente)
[18] Korbag I, Omer S M S, Boghazala H and Abusasiyah M A A 2021 Recent advances of biogas production and future perspective *Biogas - Recent Advances and Integrated Approaches* ed A E-F Abomohra, M Elsayed and Z Qin (Rijeka: IntechOpen) chapter 1 https://doi.org/10.5772/intechopen.93231
[19] Darwesh M R and Ghoname M S 2021 Experimental studies on the contribution of solar energy as a source for heating biogas digestion units *Energy Rep.* **7** 1657-71 https://doi.org/10.1016/j.egyr.2021.03.014
[20] Li X, Damartzis T, Stadler Z, Moret S, Meier B, Friedl M and Maréchal F 2020 Decarbonisation in complex energy systems: A study on the feasibility of carbon neutrality for Switzerland in 2050 *Front. Energy Res.* **8** 549615 https://doi.org/10.3389/fenrg.2020.549615

IOP Conf. Series: Materials Science and Engineering 1235 (2022) 012064 doi:10.1088/1757-899X/1235/1/012064

Increasing energy efficiency of counter-rotating wind turbines by experimental modelling

G Cîrciumaru, R-A Chihaia, N Tănase, L-A El-Leathey and A Voina

National Institute for Research and Development in Electrical Engineering ICPE-CA Bucharest, Splaiul Unirii 313, district 3, Bucharest, Romania

gabriela.circiumaru@icpe-ca.ro

Abstract. With the significant advance in exploiting wind energy, there has been a shift in research from the study of conventional systems to the study of counter-rotating systems due to additional energy input of the second rotor. The paper presents the results of the research on achieving an increase in the energy efficiency of counter-rotating wind power conversion systems. For this purpose, there have been designed different sizes of wind rotors, were 3D printed and tested in an open-circuit aerodynamic tunnel for different wind velocities and axial distances between rotors. The constructive design consisting of a smaller diameter front rotor and a larger rear rotor was chosen. The individual wind rotors were used to configure two experimental models of counter-rotating wind systems. The testing results analysis and interpretation enabled the establishing of the design and operating conditions that provide the highest power extracted from wind at 10 m/s velocity. A higher efficiency of the wind turbine system is achieved for a lower ratio between the front and rear turbines. In the case of the analysed experimental models, an increase in system efficiency of 49.14% is achieved for a 0.845 diameter ratio, and of 39.02% for 0.945 diameter ratio, respectively.

1. Introduction

Given the modern economic and environmental challenges regarding the energy production and consumption, an advance in the research of innovative or improved wind energy conversion solutions has been registered. The climate change and its effects have led to formulation and approval of stringent measures, policies and regulations regarding the use of fossil fuels across the world. Consequently, clean and renewable energy sources must be harnessed more efficiently and on a larger scale. Energy efficiency has multiple benefits and is a central pillar of the economy. These benefits can be maximized through continuous development of improved or innovative solutions and their implementation, in line with the European policies. This approach can improve the quality of life and environment, and can contribute to achieving sustainable economic growth.

The wind industry registered a record in 2020. Thus, 93 GW wind power were installed (86.9 GW installed onshore), meaning a year-over-year growth of 53% [1]. This increase makes the cumulative wind power capacity reach 743 GW. The analysis made in [1] show that China and US remain the world's largest markets for new onshore installations. Still, 180 GW wind power per year should be installed to reach the net zero CO_2 emissions objectives by 2050, as set by the Intergovernmental Panel on Climate Change [2], established by the World Meteorological Organization and the United Nations Environment Programme, which aims to asses several scientific, technical and socio-economic aspects relevant for the risk of human-induced climate change.

In this context, an increased number of theoretical and experimental investigations on improving the performance of wind turbines or developing new wind energy conversion systems has been reported in literature.

In this regard, a series of studies approached the counter-rotating wind turbines (CRWT). These systems consist of two rotors, one rotating clockwise and the other counter-clockwise, placed either at the same side of the tower, either one in front of the electric generator and one in the rear of the generator. In literature there are reported studies on different configurations of the counter-rotating systems. Thus, CRWT with rotors of same diameter [3-5] or different rotor diameters [6,7] are investigated. The CRWT systems can use one electric generator for each rotor [8], one generator with a differential planetary system coupled to the rotors [6,9] or one unconventional permanent magnet generator with both rotational armatures [7,10].

Most of the research reported in literature approached the investigation of CRWT by CFD simulations. However, there are several studies that analyse these systems both theoretically and experimentally. In [11], the interaction of two rotors of same radius, *R*, placed at an axial distance d/R=0.25, 0.5, 0.75, 1 were investigated using the free-wake vortex lattice method. The solidity of the rotors was considered the same as single rotor turbine solidity. The front rotor power coefficient (C_p) increased along with the axial distance, while C_p of the rear rotor decreased, thus leading to an almost constant overall efficiency of the system. This finding was compared to the experimental results reported in [12]. There was found a good compliance between the theoretical and the experimental axial velocity behind the front rotor. Another theoretical study [13] aimed at extracting the maximum energy from wind by both front and rear rotors of a CRWT system, analysed the aerodynamic performance of the system by CFD simulation. Considering *D* the diameter of the front rotor, it resulted an optimum axial distance between rotors of 0.65*D*, leading to a maximum power increase of 9.67% compared to the single rotor case. The results were compared with the experimental data reported in [6]. Although a good correlation between results was identified, the estimates were slightly higher than the experimental values. The CRWT system analysed in [6] had a front rotor of 5.5 m diameter and a rear rotor of 11 m diameter. The rotors were mounted at several axial distances in the range of (0.125÷0.5)*D* and a differential planetary system was used to connect them to the electric generator. The best efficiency of the 30 kW CRWT system was identified at approximately 0.5*D*. A CRWT system consisted of two 0.8 m diameter rotors and separate generators was numerically and experimentally investigated in [4]. The separate testing of each rotor and of the counter-rotating system showed an increase in the maximum extracted power of 9% of the CRWT system compared to the single rotor case. The share of power provided by the front rotor, namely 74% from the overall power of the system, confirms the choice of making the front rotor smaller than the rear rotor. In [3] a CRWT system consisted of two 3-bladed rotors of same diameter (1.4 m), optimized for tip speed ratio TSR = 5, were simulated for 5 axial distances between rotors: 0.1*D*, 0.25*D*, 0.5*D*, 0.75*D*, 1*D*. The first rotor produces the most power of the system and its power coefficient is considerably higher than the second rotor's coefficient. The optimum identified case was for 1*D* axial distance. Also, the rotational speed of the electric generator has doubled its value. The second rotor reduced the power of the first one by 9.36%. Even if the rotors have been designed for TSR = 5, their optimum in operation was achieved at TSR = 6. The power of the CRWT system increased by 0.69% for 0.1*D* and by 8.01% for 1*D* compared to single rotor case. Since significant dynamic interactions have been found for axial distances lower than 0.5*D*, it was concluded that the CRWT system should operate with the rotors placed at axial distances of 0.5*D* to 1*D*. Also, given the optimum axial distance, CRWT systems with rotors placed upwind and downwind are more appropriate than the ones with the rotors at the same side of the tower.

The studies in literature analysed the influence of different parameters on CRWT system performance: blade type, solidity of the rotors, size of the rotors, diameter ratio of the two rotors, axial distance between them, and movement transmission system from the turbine's shaft to the electric generator. Since a series of parameters influence the CRWT performance, there is need for more in-depth studies to be developed for this type of energy conversion systems.

The paper presents the findings of some research projects that envisaged achieving an increase in the energy efficiency of wind power conversion systems in order to improve the capitalization of high wind density regions, to limit the expansion of the land occupied by wind farms. The novelty of the research consists in determining the effects on varying both the axial distance between the rotors and their diameter ratio. In addition, by performing an experimental analysis of the CRWT models, a more in-depth knowledge of the associated phenomena of these systems operation is enabled, which would be difficult to obtain only by CFD studies. This approach can also provide valuable conclusions to be taken into account when upscaling CRWT type systems. For this purpose, there have been designed, developed and tested in a wind tunnel small experimental models of CRWT with adjustable distance between the rotors. The models are 3-bladed and respect the geometric similarity. In order to achieve an increased efficiency of the CRWT system by experimental modelling, some requirements to be met by the rotors have been imposed in the designing phase. The experimental models were tested in different operating conditions. The analysis and interpretation of the results enabled establishing the design and operating conditions that provide the highest power extracted from wind at 10 m/s air velocity.

2. The axial distance between counter-rotating wind rotors

Since the second rotor operates in the wake of the first rotor, in the design of counter-rotating wind turbines it is required to consider the aerodynamics of the wake. This problem has been investigated in several papers [13-19]. For example, wind turbine models with 0.9 m rotor diameter have been the subject of several studies. In [16] the wake of wind turbine models placed at 3 and 5 diameters, respectively, was investigated by measuring the axial velocity component. In [18] experiments were carried out in a wind tunnel regarding the near and far velocity field in the case of a 0.9 m diameter rotor and in the case of two counter-rotating rotors of the same diameter (figure 1).

Compared to the near wake, in the case of the far wake a considerable reduction in the velocity and a non-uniform distribution was found, both for single turbine operation and for tandem operation. In the case of tandem operation there is a slight increase both in the rotation of the wake and in the area covered by it compared to the case of single turbine operation. It has also been identified that the inner part of the wake rotates faster than the outer part of it, but this cannot be generalized.

Experiments in wind tunnel on the wind turbines wake have also been reported in [19]. It was concluded that the highest velocity is found in the area corresponding to the periphery of the blades. However, it should be noted that different results can be obtained depending on the blades constructive characteristics.

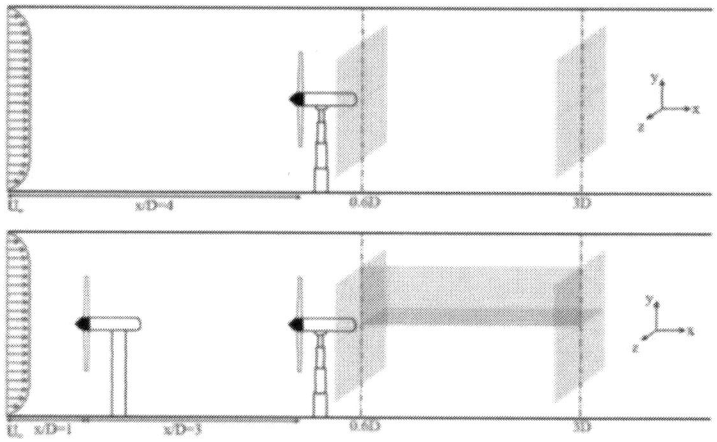

Figure 1. Sketch of the experimental setup for the case of one turbine and two turbines [18].

According to [20], in the case of wind farms, the usual distance between turbines is 6 to 10 diameters, in the predominant wind direction, even if the turbulence effects can be identified at a distance of 15 rotor diameters, downstream a turbine.

Given that the wake is the main phenomenon that leads to the kinetic energy recovery in the case of wind farms and determines the interaction between turbines, in [21] the velocity field located downstream (up to 5 diameters) of a two-blade turbine model with 0.6 m rotor diameter was investigated using PIV (Particle Image Velocimetry), in a closed-circuit wind tunnel. The performed experimental and statistical study determined that the velocity becomes negative (the vortex behind the rotor moves in the opposite direction to the rotor) to about 0.2 diameters behind the rotor for both considered TSR values: 4.8 and 6; TSR = 4.8 was obtained at a wind velocity of 3.8 m/s and TSR = 6 at a wind velocity of 5.7 m/s.

In [22] and [23] a study was carried out on models for the case of a wind farm where the individual turbines are being either in co-rotation or in counter-rotation. The distance between the two rotors (of same diameter D) of a counter-rotating wind system was less than $1D$. The study revealed that: for a distance of $0.7D$ between turbines, the power of the rear counter-rotating turbine is up to 20% higher than the power of the rear co-rotating turbine; when the distance is $5D$ and the rotors are counter-rotating, this advantage is reduced to only about 4%.

3. Design of wind turbine experimental models for counter-rotating systems

In order to achieve experimental models of counter-rotating wind turbine rotors with imposed power, whose assembly and operating conditions to be known and established following theoretical and experimental studies and analyses, the authors conducted applied research activities in the frame of the contract [24]. The research envisaged to obtain an improved energy efficiency of the CRWT systems. Thus, there have been designed, developed and tested in open-circuit wind tunnel experimental models of counter-rotating wind rotors with different diameters. In order to achieve the experimental models, there were performed the following: design of different models of wind rotors, their printing by using a 3D printer followed by testing the models in the ICPE-CA open-circuit wind tunnel and analysing the experimental results.

The research aimed to achieve two experimental models of CRWT rotors having the diameters ratio D_f / D_r = (1: 1.05; 1: 1.4) = (0.714–0.952), with D_f the front rotor diameter and D_r the rear rotor diameter. This range of the diameter ratio was established based on the results of the research reported in [7] concerning the operation and efficiency of 1 kW counter-rotating wind turbine experimental model and on the conclusions of the numerical study reported in [25]. By performing CFD simulations, the aerodynamics of CRWT systems with diameters ratio of 0.5–1.5 operating at TSR = 3 was investigated. It was found that the flow separation occurring in the hub area of the rear rotor has the tendency to fade away when the rotor diameter ratio increases. At low TSR values, for D_f / D_r = 0.5–1, the power coefficient of the rear rotor is higher than in the case of $D_f / D_r > 1$. Also, the estimated power of individual wind turbine models should be in the range of 20–45 W, at 10 m/s wind velocity. In order to achieve an assembly (a system) of counter-rotating turbines, it is necessary to have two individual wind turbines, with diameters in the range imposed by the project. This results in the need to develop at least 3 individual experimental models of wind turbine rotors. The constructive design consisting of a smaller diameter front rotor and a larger rear rotor was chosen for the CRWT system.

In the design of experimental models of CRWT rotors, it was taken into account that, in order to characterize the counter-rotating assembly, the rotors are first independently tested and then in tandem operation. It was decided to design and make a single rear rotor, with 620 mm maximum rotor diameter and to vary the size for the front rotors. Thus, two front rotors were made - of 586 mm and 524 mm maximum diameter.

For designing the experimental models of counter-rotating turbines, several types of NACA aerodynamic profiles were analysed with the help of the QBlade open access calculation and simulation program. Details regarding the simulation and selection of the profiles are given in [26].

IOP Conf. Series: Materials Science and Engineering 1235 (2022) 012064 doi:10.1088/1757-899X/1235/1/012064

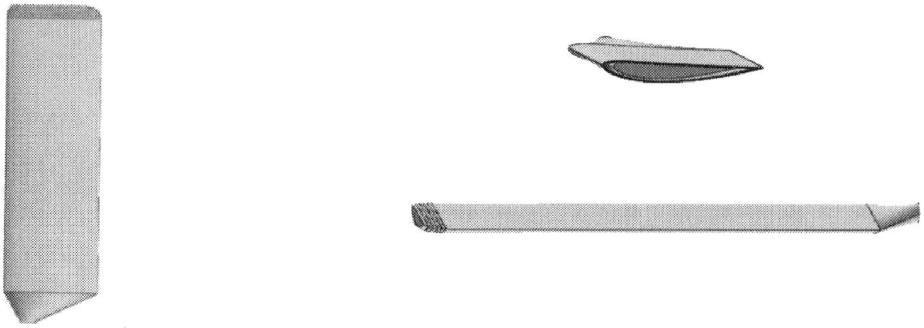

Figure 2. NACA 4412 aerodynamic profile with c=80 mm and TSR=3.

NACA 4412 aerodynamic profile was selected for the blades of the experimental models. A blade chord c of the largest rotor (rear rotor) of 80 mm was considered and the blade was optimized for TSR = 3 (figure 2). The blade has a constant chord of 80 mm, except for the periphery, where, for aerodynamic and mechanical reasons, the profile is rounded and the chord size gradually decreases.

Table 1. Geometrical characteristics of the CRWT system rotors.

	Maximum chord (mm)	Aerodynamic profile (–)	Blade size (mm)	Rotor size (mm)	CRWT system diameter ratio
Rear Rotor	80		260	620	–
Front Rotor 1	75		243	586	$D_{f1} / D_r = 0.945$
Front Rotor 2	66	NACA 4412	212	524	$D_{f2} / D_r = 0.845$

Thus, to achieve two experimental models of counter-rotating wind rotors, there have been designed 3 pairs of blades, respecting the geometric similarity (see table 1). The rotor size takes into account the rotor hub (100 mm) and the length of the blades.

Since the rear rotor must rotate in the opposite direction to the front rotor, after designing the rear blade, it was "mirrored" in order to obtain the front blades. Their length was set to values that meet the condition imposed by the research project, namely the rotors diameter ratio of the CRWT system should be in the range 0.714...0.952. The designed wind rotors can be combined in the following configurations, resulting 3 CRWT systems: 524 mm front rotor and 620 mm rear rotor, 524 mm front rotor and 586 mm rear rotor, 586 mm front rotor and 620 mm rear rotor. The blades were made of acrylonitrile butadiene styrene (ABS) by 3D printing using a dedicated printer (Fortus 360 mc large model).

Envisaging to test the wind rotors in the open-circuit wind tunnel of ICPE-CA first in separate/independent operation and then in tandem operation, the elements necessary for carrying out the tests had been also designed. Thus, in order to determine the best operating conditions leading to extracting the highest power from wind, the operation of the wind rotors at different axial mounting distances d_m was envisaged. In this regard, a system that allows adjusting the distance d_m between the two experimental models of wind turbine rotors was designed (figure 3).

Table 2. Configuration of the counter-rotating wind rotors.

Crt. no.	Rotor diameter (mm)	Position (–)	d_m (mm)	Distance relative to the diameter of the front rotor
1	620	rear	0 (fixed rotor)	–
2	586	front	650; 480; 330	1.109; 0.819; 0.563
3	524	front	650; 480; 330	1.240; 0.916; 0.629

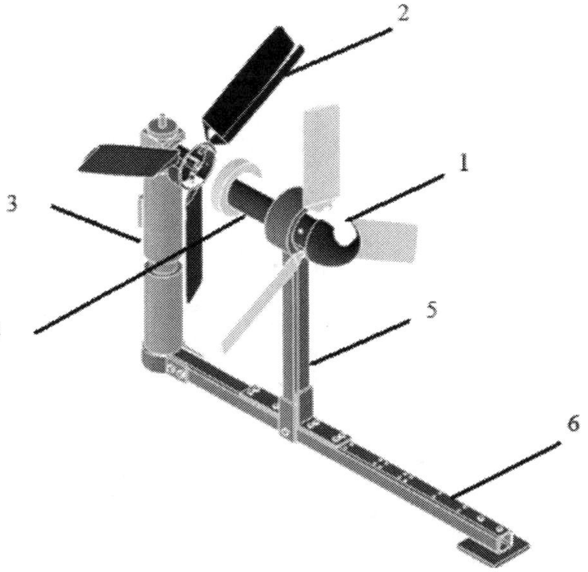

Figure 3. Counter-rotating wind turbines testing assembly: 1 – front rotor, 2 – rear rotor, 3 – rear rotor support tower with electromagnetic brake and integrated torque and speed transducers, 4 – magnetic powder brake for the front rotor, 5 – front rotor support tower, 6 – perforated horizontal system allowing to adjust the distance between the two rotors.

The envisaged d_m distances for tandem operation as well as the configuration of the counter-rotating wind system are presented in table 2.

4. Testing of the experimental models of counter-rotating wind turbine rotors

4.1. Experimental setup

The most appropriate way to test and characterize wind turbine rotors is by performing experiments in wind tunnels. Therefore, to determine the performance characteristics of the developed wind rotors, there was used the experimental setup in figure 4, consisting of an open-circuit wind tunnel provided with different measuring and control equipment. By using this aerodynamic tunnel and its related equipment, scale models of horizontal and vertical axis wind turbines can be tested for velocities ranging between 2 m/s and 30 m/s. The flow uniformity maximum deviation in the test section of the tunnel is: ± 0.3 m/s for velocities smaller than 10 m/s and ± 3% from the pre-set velocity. The tunnel is provided with a 22 kW fan and a frequency converter AVD200 type that ensures the velocity control. The control panels of the experimental setup equipment are located in the proximity of the 1x1 m test section.

The counter-rotating wind system consists of the following:

- a front turbine, located on a vertical support tower (with a specific shaft end for the turbine and with electromagnetic brake and torque and speed transducers included) and an anchoring system on a perforated horizontal rod allowing fixing the rotors at different distances. A data acquisition system composed of torque transducer and a brake based on eddy currents is integrated on the support tower. The brake, model FAT 20, is produced by Mobac Gmbh; the torque transducer, model DR 2335, is produced by Lorenz Messtechnik Gmbh. These devices measure a maximum torque of 2 N·m.

Figure 4. Open-circuit wind tunnel.

- a rear turbine, located on a vertical tower with adjustable height, provided with a subassembly with pinions for changing by 90° the rotating shaft location. The tower has embedded an eddy current brake FRAT 50 type, produced by Mobac GmbH, with a maximum torque of 5 N·m, and speed and torque transducers.

4.2. Wind turbine rotors testing methodology
The experimental research carried out on the CRWT systems aimed at identifying the constructive solution and the operating conditions that provide the highest energy efficiency and the highest energy extracted from wind, respectively.

To get the performances of the counter-rotating wind rotors, first was separately tested each of the developed rotors for wind velocities between 9 and 12 m/s; then the torque-speed and power-speed curves were drawn. The testing of the individual rotors was presented in [26]. The results for testing the rotors at 10 m/s showed the following power output: 48 W at rotational speeds $n \approx 900$ rpm for the 620 mm diameter rotor, 37 W at $n \approx 850$ rpm for the 586 mm rotor and 34 W at $n \approx 860$ rpm for the 524 mm rotor. The testing of the CRWT experimental models with adjustable distance between rotors aimed to determine the characteristic power curves P=f(n) and torque curves T=f(n) for each pair of rotors and the axial mounting distance, respectively. For the tandem operation testing, each rotor had been provided with its own braking (loading) equipment and measurement system. By acting this way, each turbine can be separately loaded and the behaviour of the experimental CRWT system can be studied at different torque values at the shafts of the two turbines. To identify the optimal mounting distance between the two wind rotors, the tests were repeated for different axial distances between the front and rear turbines.

To test the CRWT system, the following steps were followed: wind rotors were installed at various distances relevant to the test purpose, using the perforated horizontal system allowing to adjust the distance between the two rotors presented in figure 3; the CRWT systems were tested for each set distance in the range 330 mm to 650 mm; the preliminary results were evaluated and, if necessary, the test procedure was repeated for verifying the results repeatability; power curves P=f(n) and torque curves T=f(n) were plotted for each pair of rotors and axial distance; the families of curves were analysed and the axial mounting distance providing the highest energy efficiency of the wind system was determined.

IOP Conf. Series: Materials Science and Engineering 1235 (2022) 012064 doi:10.1088/1757-899X/1235/1/012064

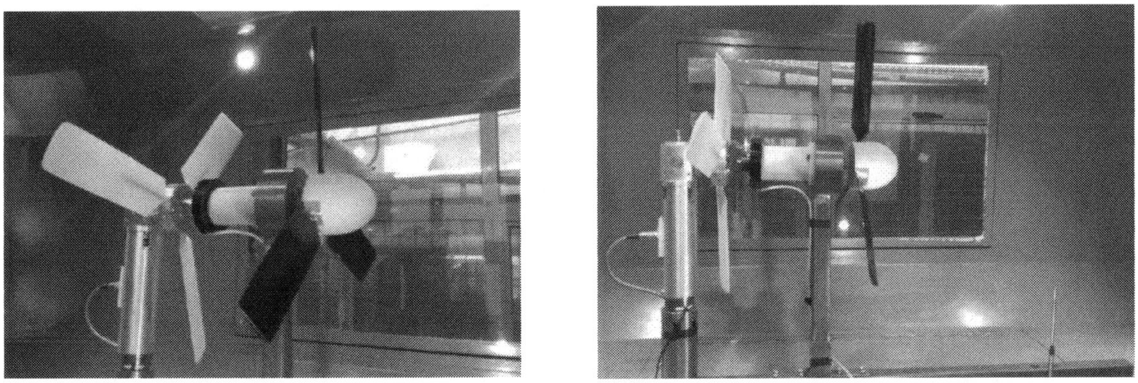

Figure 5. The CRWT system mounted in the wind tunnel test chamber.

5. Results and discussions

Given the fact that each wind rotor had already been characterized for velocities between 9 and 12 m/s, as reported in [26], the tests of the counter-rotating system were performed at 10 m/s wind velocity. The operation of the assembly was checked by: simultaneous loading of the two rotors; bringing the two rotors to a dynamic equilibrium in operation and loading only the rear turbine and, afterwards, only the front turbine. The performed experiments proved that a lower pitch angle leads to a higher starting velocity and to increased power and speed of the turbine.

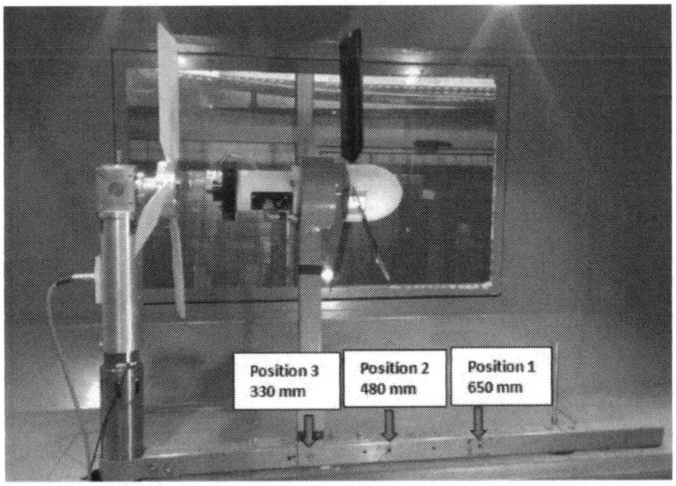

Figure 6. The 3 mounting and testing positions of the CRWT experimental models.

The best testing results were obtained in the following conditions:

- the rotors "share" the total power (when the front rotor operates at a lower C_p than the maximum power coefficient), as also demonstrated in the research project [27], which analysed the operation of a real scale CRWT system of 1 kW;
- the two rotors operate in dynamic equilibrium, each having a power lower than the maximum power, but as close as possible to it;
- after achieving the dynamic equilibrium, the rear turbine is progressively loaded, until it stops (indicating that the maximum power was reached; at this moment, the speed of the front turbine is rapidly increasing).

It is worth mentioning that the two turbines influence each other in operation; thus, not only the front turbine, which takes over a large part of the airstream kinetic energy, influences the operation of the rear turbine (theoretically benefiting of only 1/3 from the wind velocity), but also vice versa.

Thus, if the front turbine is kept braked/loaded and the rear turbine operates at or around maximum power (maximum speed) in order to achieve the dynamic equilibrium in operation, then the front turbine does not start or starts very hard. Most likely, this is due to the changes in the air tube streamlines in the immediate upstream vicinity of the downstream rotor.

Following the methodology described above, there were plotted the power and torque curves of the counter-rotating system for 10 m/s wind velocity. In figures 7 and 8 P_{tot} represents the total power of the CRWT system, $P_{tot}=P_{rear}+P_{front}$; likewise, $n_{tot}=n_{rear}+n_{front}$ represents the total speed of the CRWT experimental model. Also, d_m represents the axial mounting distance between the two rotors.

Figure 7 presents a comparison between the total power variation with total speed of the experimental model of CRWT system consisting of the rear rotor of 620 mm diameter and front rotor of 586 mm diameter, operating at different axial distances between rotors. It can be seen that the highest efficiency of the system is obtained for the rotors operation at an axial distance of 650 mm (meaning 1.109 D_f).

After performing the tests of the counter-rotating wind system consisting of the 620 mm diameter rear rotor and of the 586 mm diameter front rotor (i.e. $D_f/D_r = 0.945$), a second configuration consisted of the 620 mm rear rotor and the 524 mm front rotor (i.e. $D_f/D_r= 0.845$) was tested.

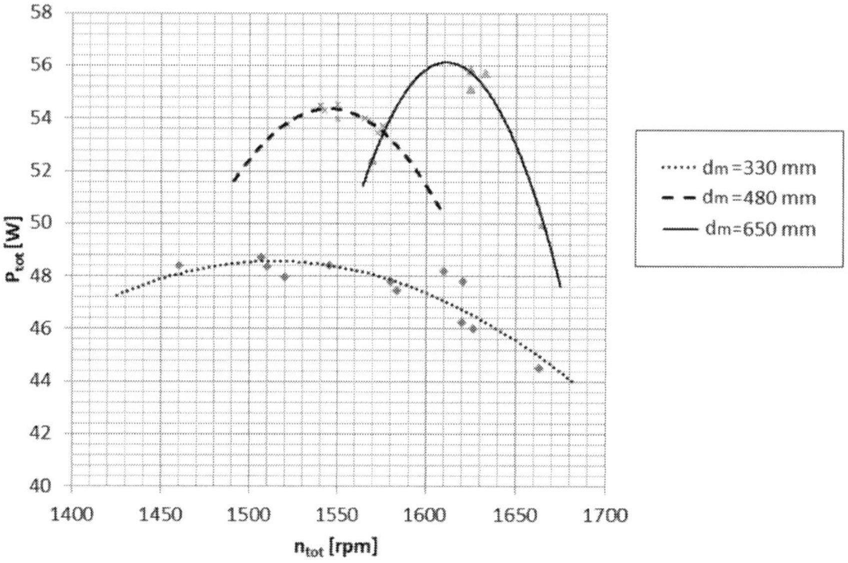

Figure 7. Power variation with speed for the CRWT model, with 620 mm rear rotor and 586 mm front rotor.

In figure 8 a comparison between the total power variation with total speed of this experimental model of CRWT system is presented. The highest efficiency of the system is obtained for an axial distance of 650 mm (meaning 1.24D_f).

After perfoming the tests of the counter-rotating system consisting of the 620 mm diameter rear rotor and of the 586 mm diameter front rotor (i.e. $D_f/D_r = 0.945$), a second configuration consisted of the 620 mm rear rotor and the 524 mm front rotor (i.e. $D_f/D_r= 0.845$) was tested.

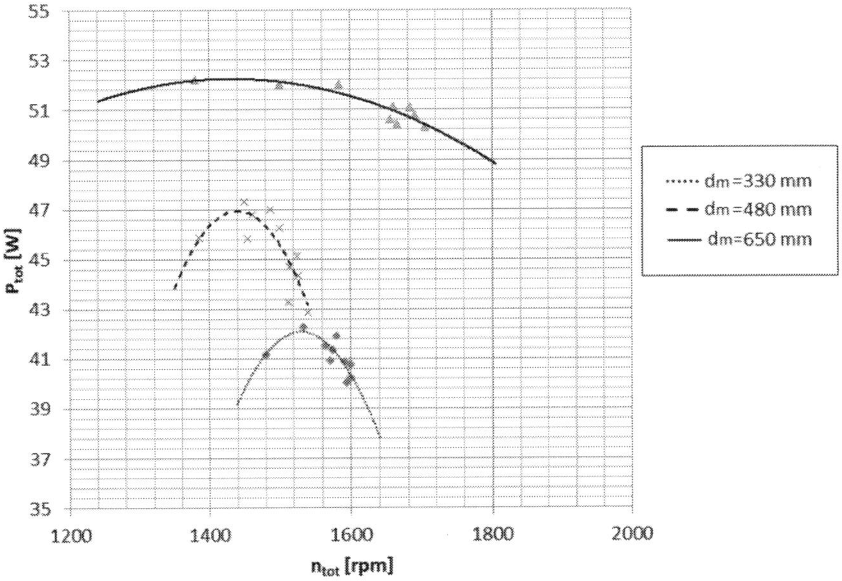

Figure 8. Power variation with speed for the CRWT model, with 620 mm rear rotor and 524 mm front rotor.

The tests performed with different configurations of counter-rotating wind turbines at three axial distances between rotors aimed at identifying the distance determining the highest conversion of available wind energy, respectively the highest extracted power. Thus, to determine the additional contribution of the second rotor, the total power of the system is calculated in relation to the power of the front rotor, P_f.

$$\Delta P=(P_{tot}-P_f)\cdot 100/P_f,\ [\%] \quad (1)$$

Table 3. CRWT system power compared to the power of a single turbine (front turbine) at 10 m/s wind velocity for different configurations and mounting distances between rotors.

Crt. no.	Counter-rotating system	d_m (mm)	d_m/D_f (–)	P_f (W)	P_{tot} (W)	ΔP (%)
1	620 mm+586 mm	330	0.563	41	48.7	18.78
2	620 mm+586 mm	480	0.819	41	54.3	32.44
3	620 mm+586 mm	650	1.109	41	57	39.02
4	620 mm+524 mm	330	0.629	35	42.6	21.71
5	620 mm+524 mm	480	0.916	35	47	34.29
6	620 mm+524 mm	650	1.240	35	52.2	49.14

Table 3 summarizes the results regarding the efficiency of the counter-rotating system in two configurations, at three different axial mounting distances between rotors.

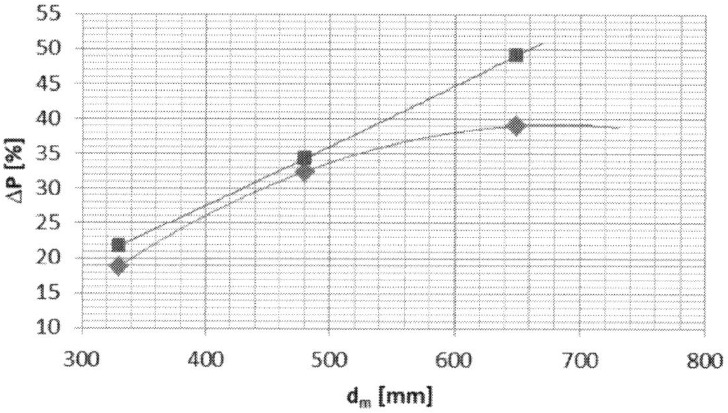

Figure 9. Power increase variation with the axial mounting distance between the rotors of the CRWT system models.

Figure 9 shows that the variation of the additional power input of the second rotor of the counter-rotating wind system differs for the two tested experimental models. Also, the analysis of the curves shows that for the case of the system consisted of the 620 mm rear rotor and 586 mm front rotor, the maximum was reached. It means that, for axial distances between the rotors greater than 720 mm, the power of the CRWT system decreases. In conclusion, the highest efficiency of the counter-rotating system in this configuration is obtained for mounting distances in the range 650–720 mm, respectively (0.814–0.9)D_f.

In the case of the CRWT wind system consisted of the 620 mm rear rotor and 524 mm front rotor, the highest efficiency of the system is obtained for an axial distance between rotors of 650 mm (1.24 D_f, respectively); however, there is no tendency of the curve to reach its maximum. This may be due to the different development of the downstream vortex, whose flow/evolution is strongly asymmetric and may be influenced by certain characteristics of the rotor, such as its size, configuration and interactions between rotor, hub and nacelle.

6. Conclusions

The testing of the experimental models of counter-rotating wind turbines with adjustable distance between rotors was performed for two configurations, with the diameter ratio D_f/D_r=(1:1.05;1:1.4) and the estimated power in the range 50–70 W for 10 m/s wind velocity. The rotors were placed at different axial distances (650 mm, 480 mm and 330 mm, respectively) and each configuration was tested for 10 m/s wind velocity; power curves P=f(n) and torque curves T=f(n) were plotted for each configuration and axial distance, respectively; the curves were analysed and the mounting distance providing the highest energy efficiency of the wind system was identified.

Following the performed tests there was concluded:

For the counter-rotating system consisted of 620 mm rear rotor and 586 mm front rotor:

- the total power of the system, P_{tot}, increases by 39.02% compared to the power of the front rotor for an axial distance between rotors d_m of 650 mm (1.109, D_f);
- P_{tot} increases by 32.44% for d_m= 480 mm = 0.819D_f;
- P_{tot} increases by 18.78% for d_m= 330 mm = 0.563D_f.

For the counter-rotating system consisted of 620 mm rear rotor and 524 mm front rotor:

- P_{tot} increases by 49.14% for d_m= 650 mm = 1.24D_f;
- P_{tot} increases by 34.29% for d_m= 480 mm = 0.916D_f;
- P_{tot} increases by 21.71% for d_m= 330 mm = 0.629D_f.

Compared to the power of the front turbine, the results showed that a higher efficiency of the system is obtained in case of a lower ratio between the diameters of the front and rear turbines. Thus, in the case of the counter-rotating experimental models developed and characterized, an increase in the system efficiency of 49.14% is obtained in the case of a D_f/D_r=0.845 ratio, and 39.02% for D_f/D_r=0.945, respectively. Compared to the results in literature, the improvement of the counter-rotating system efficiency is superior to some data reported in [27,28], and comparable to the efficiency determined in [7].

Also, it was determined that the wind system extracts maximum power under the following conditions: the rotors "divide" the total power, the pitch angle of the rear rotor blades is less than that of the front rotor blades and the speed of the second rotor is less than the speed of the first rotor.

Acknowledgments
This work was financially supported by Ministry of Research, Innovation and Digitization, NUCLEU Programme, projects no. PN16110101 (14N/2016) and PN 19310201 (49N/2019), respectively.

ORCID IDs
G Cîrciumaru https://orcid.org/0000-0002-7814-195X
R A Chihaia https://orcid.org/0000-0002-5084-3212
N Tănase https://orcid.org/0000-0002-2426-5607
L A El-Leathey https://orcid.org/0000-0003-2517-7788
A Voina https://orcid.org/0000-0002-5155-7103

References

[1] Global Wind Energy Council 2021 *Global Wind Energy Report* www.gwec.net
[2] Intergovernmental Panel on Climate Change 2020 *IPCC Special Report Emissions Scenarios, Summary for Policymakers* ISBN 92-9169-113-5
[3] Pacholczyk M, Blecharz K and Karkosinski D 2019 Paper no. ICERT 119 *Proc. of the 5th World Congress on New Technologies (Lisbon Portugal)*
[4] Herzog R, Schaffarczyk A P, Wacinski A and Zürcher O 2010 *European Wind Energy Conf. & Exh. (Warsaw Poland) 20-23April*
[5] Faisal M, Zhao X, Kang M H and You K 2020 Aerodynamic Performance and Flow Structure Investigation of Contra-rotating Wind Turbines by CFD and Experimental Methods *ATDMAE 2020 IOP Conf. Series: Materials Science and Engineering* **926** https://doi.org/10.1088/1757-899X/926/1/012017
[6] Jung S N, No T S and Ryu K W 2005 Aerodynamic performance prediction of a 30 kW counter-rotating wind turbine system *Renew. Energ.* **30** 631–44 https://doi.org/10.1016/j.renene.2004.07.005
[7] Mituleț L A, Oprina G, Chihaia R A, Nicolaie S, Nedelcu A and Popescu M 2015 Wind tunnel testing for a new experimental model of counter-rotating wind turbine *Procedia Engineering* **100** 1141-49 https://doi.org/10.1016/j.proeng.2015.01.477
[8] Debleser Y 2003 Jeumont Industrie *Wind turbine with counter rotating rotors* US Patent No. 6504260
[9] Habash R W Y, Groza V, Yang Y, Blouin C and Guillemette P 2011 Performance of a Contrarotating Small Wind energy Converter *ISRN Mech. Eng.* **2011** 1-10 https://doi.org/10.5402/2011/828739
[10] Sutikno P and Saepudin D B 2010 Design and Blade Optimization of Intelligent Wind Turbine *Most* no. December 1-15
[11] Lee S, Son, Lee S 2012 Velocity interference in the rear rotor of a counter-rotating wind turbine *Renew. Energ.* **59** 235-240 https://doi.org/10.1016/j.renene.2012.08.003
[12] Schepers J G, Boorsma K, Bon A, Kim C and Cho T 2014 *J. Phys.: Conf. Ser.* **555** 012089
[13] Smith D and Taylor G J 1991 *Proc. of the 13th BWEA Wind Energy Conf. (Swansea UK)* 325

[14] Vermeer L J, Sørensen J N and Crespo A 2003 Wind turbine wake aerodynamics *Progress Aerospace Sciences* **39** 467-510 https://doi.org/10.1016/S0376-0421(03)00078-2
[15] Maeda T, Yokota T, Shimizu Y and Adaki K 2004 Wind tunnel study of the interaction between two horizontal axis wind turbines *Wind Engineering* **28** (2) 197-212 https://doi.org/10.1260/0309524041211396
[16] Bartl J, Pierella F and Sætran L 2012 Wake measurements behind an array of two model wind turbines *Energy Procedia* **24** 305-12 https://doi.org/10.1016/j.egypro.2012.06.113
[17] Blomhoff H P 2012 *An experimental investigation of wind turbine wakes* Master thesis Norwegian University of Science and Technology
[18] Schümann H, Pierella F and Sætran L 2013 Experimental investigation of wind turbine wakes in the wind tunnel *Energy Procedia* **35** 285-96 https://doi.org/10.1016/j.egypro.2013.07.181
[19] Medici D 2005 *Experimental Studies of Wind Turbine Wakes - Power Optimisation and Meandering in Mechanics* – Technical Report from KTH Mechanics - Royal Institute of Technology KTH: Stockholm
[20] Sanderse B 2009 *Aerodynamics of wind turbine wakes* Literature review ECN-E-09-016 Wind Energy
[21] Lignarolo L E M, Ragni D, Krishnaswami C, Chen Q, Simão Ferreira C J and Van Bussel G J W 2014 Experimental analysis of the wake of a horizontal-axis wind-turbine model *Renew. Energ.* **70** 31-46 https://doi.org/10.1016/j.renene.2014.01.020
[22] http://www.computescotland.com/wind-turbine-layout-wakes-and-rotors-7183.php Accesed
[23] Yuan W, Tian W, Ozbay A and Hu H 2014 An experimental study on the effects of relative rotation direction on the wake interferences among tandem wind turbines *Sci China-PhysMechAstron* **57** 5 https://doi.org/10.1007/s11433-014-5429-x
[24] Innovative constructive solutions for wind turbines developed by small-scale modelling to increase the energy efficiency 2016, contract PN 16 11 01 01 no. 14N/2016 research theme 5101/2016
[25] Koehuan V A et al. 2019 Numerical Analysis on Aerodynamic Performance of Counter-rotating Wind Turbine through Rear Rotor Configuration *Modern Applied Science* **13** (2) 240-57 https://doi.org/10.5539/mas.v13n2p240
[26] Cîrciumaru G, Chihaia R A, Ovezea D, Chiriță I, Nicolaie S, El-Leathey L A and Nedelcu A 2020 *7th Int. Conf. on Energy Efficiency and Agricultural Engineering (EE&AE) (Ruse Bulgaria)* pp 1-4
[27] Kumar P S, Bensingh R J, Abraham A Computational analysis of 30 kW contra rotor wind turbine, *ISRN Renew. Energ,* article ID 939878 https://doi.org/10.5402/2012/939878
[28] Bramantya M A, Al Huda 2016 *6th International Annual Engineering Seminar (InAES) Yogyakarta Indonezia* 1-3 August pp 212-217

IOP Conf. Series: Materials Science and Engineering 1235 (2022) 012065 doi:10.1088/1757-899X/1235/1/012065

Theoretical and experimental research on the quality of thermoformed foils using 3D-printed plastic moulds

M A Boca[1,2], A Sover[2] and L Slătineanu[1]

[1] "Gheorghe Asachi" Technical University of Iaşi, Machines Manufacturing Technology Department, D. Mangeron Blvd 59A, 700050 Iaşi, Romania
[2] Ansbach University of Applied Sciences, Technology Departments, Residenzstraße 8, 91522 Ansbach, Germany

marius-andrei.boca@student.tuiasi.ro

Abstract. An innovative approach to additive manufacturing consists in the possibility of producing moulds that can be used for traditional manufacturing processes such as thermoforming. The quality and dimensional precision of the interior surfaces of the plastic foils obtained after thermoforming is closely related to the used mould. The influencing factors considered are the different thermal and mechanical properties offered by the wide range of available plastic materials for additive manufacturing (AM), the multitude of printing parameters, and the technological process by means of which the mould was obtained. For the production of plastic moulds, Fused Filament Fabrication (FFF) and Selective Laser Sintering (SLS) processes, as well as materials specific to each technology, are used. This paper identifies and explains issues concerning the surface quality of the thermoformed foil through a systemic analysis approach and by conducting laboratory experiments. Optical 3D scanning equipment and software are employed to determine the proper additive process and materials for thermoforming process. The best quality of the finished product was obtained in the case of a male mould made by the SLS technology and PS foil with a thickness of 0.7 mm.

1. Introduction

The global plastic packaging market is the largest end-user segment for plastic products, and it was valued at 348.08 billion $ in 2020 [1]. On the other hand, the market value of thermoforming plastics totalled 3.8 billion $ in 2019 [2], and it is expected that, in the following five years, the thermoforming market value will reach nearly 45.9 billion $ worldwide by 2024, at a Compound Annual Growth Rate (CAGR) of 5.7% between 2019 and 2024 [3].

Thermoforming is considered one of the oldest methods of forming products where the plastic articles are formed by stretching a heated foil over a mould until it is cooled enough to retain the shape of the mould [4]. However, new technologies (additive manufacturing or 3D printing) have emerged to replace subtractive ones when it comes to mould manufacturing. In this case, plastic products are obtained by adding successive layers of material or bonding agents. Unlike thermofoming, the firsts attempt to create a 3D object using photopolymers and a laser was made at the end of 1960 at Battelle Memorial Institute (Holography [5]). After that, the additive technology began to gain popularity, starting with Stereolithography SLA [6] in 1984, followed by Selective Laser Sintering SLS and Fused Deposition Modelling-FDM in 1989 [7].

Based on the above consideration, the attention of both research communities and independent users was directed to the design and production of moulds, tools and accessories for thermoforming using additive manufacturing. The moulds are vital components, and the costs and time of production directly influence the end-use product. The realisation of such a mould is a challenging technical problem due to the multitude of additive processes and their specific design rules, printing parameters, thermoforming design rules and materials that can be applied.

Researchers as I.Gajdoš, T.Jachowicz [8] and J. Serrano-Mira [9] prove in their papers that the moulds made using low-cost additive manufacturing concepts are entirely suitable for a small production of 500 lid trays or small production batches of tactile graphics for people with visual impairment. This approach is even feasible when using high-performance thermoplastic material, such as Ultem 9085. Furthermore, Sunil Bhandari [10] demonstrates that the costs of such a mould production are reduced to 91.6% compared to stainless steel moulds and 87% compared to aluminium moulds.

Motivated by the positive results of the previous research and the fact that none of those works integrates optical 3D scanning equipment and software, this paper presents a conceptual solution for choosing the proper additive process and materials for thermoforming. The developed framework was validated through experimental tests of the surface quality of the obtained thermoforming foils, and the results attained highlight the great potential of the proposed concept.

2. Methodology

Due to the industry's rapid evolution, the customer requirements become more stringent, and in consequence, the dimension and surface quality of the thermoformed foil become critical elements in fulfilling those requirements. Unfortunately, finite element analysis or the visual and classical methods of inspection of the obtained foils cannot always detect all defects or not offer an accurate visualisation of the problems that occurred during the thermoforming process. Therefore, the optical scanning systems aid and facilitate the quality control process, offering many useful features.

For the experimental study, a particular custom design mould for soap casting was used. Design requirements of the thermoforming and 3D printing processes are employed as a guideline to improve the thermofomed foils' geometry and quality. For the production of the foils, six mould concepts have been tested and manufactured through two different technologies (FFF and SLS) with two specific materials for each technology (PLA and PA12) at 1.5 mm Layer Height. In addition, two types of material (black PS and transparent APET) with different sheet thicknesses (0.3 mm, 0.6 mm and 0.7 mm) were chosen for the thermoforming. The experimental study starts from a 70 mm x 70 mm male type mould with a 2 mm height custom extruded design. After few iterations, the mould design presented in Figure 1 was obtained, based on the thermoforming and 3D printing design rules for 2D and 3D corners, draft angle, webbing, minimum features' dimensions and vacuum holes.

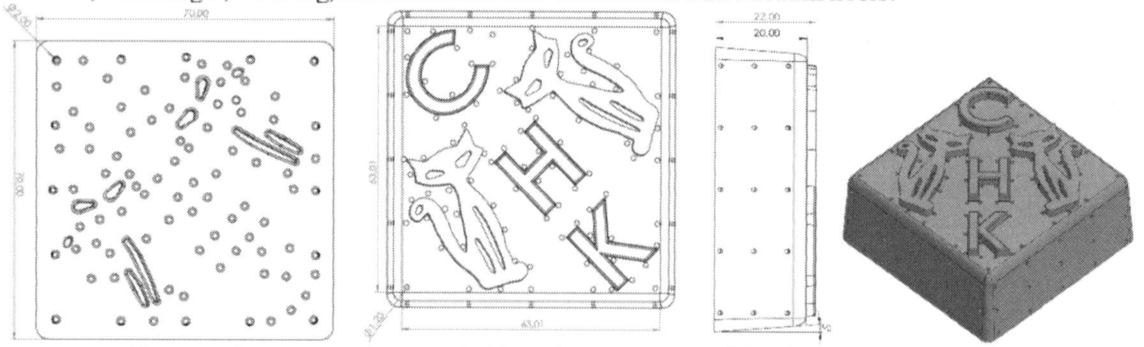

a. Bottom view b. Top view c. Side view d. Isometric view

Figure 1. Custom design mould for thermoforming

For this application, two different equipment units have been used:

- Ultimaker 2+, a reliable and robust Cartesian, single Bowden extrusion FFF 3D printer, with a 0.4 mm brass nozzle size that works with filaments of 2.85 mm in diameter;

- Sinterit Lisa PRO SLS with a 5 W IR Laser diode (λ = 808 nm) that works with 18-90 μm granulation PA12 powder.

In mould design, different approaches have been considered:

- Concept I: FFF mould printed with 0.4 mm diameter nozzle, with top and bottom layers and gradual infill (Printing time: 6 h 43 min., Figure 2. a);
- Concept II: FFF mould printed with 0.25 mm diameter nozzle, with top and bottom layers and gradual infill (printing time: 7 h 44 min., Figure 2. b);
- Concept III: FFF mould printed with no top and bottom surfaces and increased degree of gradual infill and walls (printing time: 7 h 37 min., Figure 2. c);
- Concept IV: FFF mould with no infill, no bottom surface and 3 mm wall thickness (printing time: 5 h 12 min., Figure 2. d);
- Concept V: SLS mould with top and bottom surfaces and infill at the interior (printing time: 14 h 13 min., without pre- and post-processing, Figure 2. e);
- Concept VI: SLS mould with no infill, no bottom surface and 3 mm wall thickness (printing time: 7 h 15 min., without pre- and post-processing, Figure 2. f).

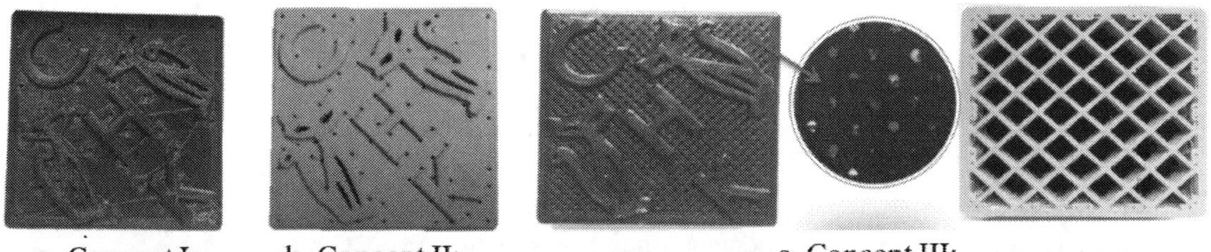

a. Concept I; b. Concept II; c. Concept III;

d. Concept IV; e. Concept V f. Concept VI;

Figure 2. Obtained 3D printed male mould.

All male moulds were successfully printed with 1.5 mm Layer Height, without warpage, shrinkage and layers' delamination but with some top surface problems and visible layers, depending on the used technology and printing orientation.

3. Results and discussion

3.1. Preliminary experiments and 3D scanning of the obtained foil

For thermoforming, a Formech Model 450 DT has been used, which has a maximum forming area of 430 mm x 280 mm and allows working with semi-finished products with a maximum thickness of 6 mm [11]. The moulds were successfully used for thermoforming tests with the 0.3 mm, 0.6 mm and 0.7 mm APET and PS foil. The obtained thermoformed sheet will be subsequently used as a mould for soap casting. Therefore, the inner surface quality is more important than the outer one. The obtained thermoformed foils have been validated using a 3D scanning process. The equipment used to scan and study the thermoformed foils is a GOM ATOS Core 200, a Blue Light Technology, non-contact, triple scan technology that measures millions of accurate points per quick scan and two software programs (GOM Inspect and GOM Scanning [12]).

During the first test, due to its lower glass transition point, between 55 °C and 60 °C, the thin wall thickness and the vacuum system of thermoforming equipment, the PLA mould starts to deform (the top surface is drawn inwards). However, this effect is not visible for the empty SLS mould due to the higher glass transition point of the PA12. Therefore, a material with a higher glass transition temperature or a higher temperature resistance is recommended for the further use of Concept IV.

Using Concept III, the infill becomes visible due to the porous top surface structure. The irregularities (bulges and depressions) are also visible with the GOM Inspect software on the surface quality inspection. The numerical values of the studied points indicate the distance in mm, from the computed theoretical surface using spline approximation and each individual studied point. Based on these considerations, the concept will be used only for female moulds or male moulds with deep cavities in future. For the 0.6 mm gauge APET foil (Figure 3, on the right side), a higher percentage of deviation values is in the range of -0.03 mm and -0.01 mm (for negative values) and between 0.01 mm and 0.035 mm (for positive values). Whereas regarding the 0.3 mm gauge APET foil (Figure 3, on the left side), for the negative area, most of the deviation values are between -0.038 m and -0.01 mm and the positive range is between 0.015 mm and 1.1 mm.

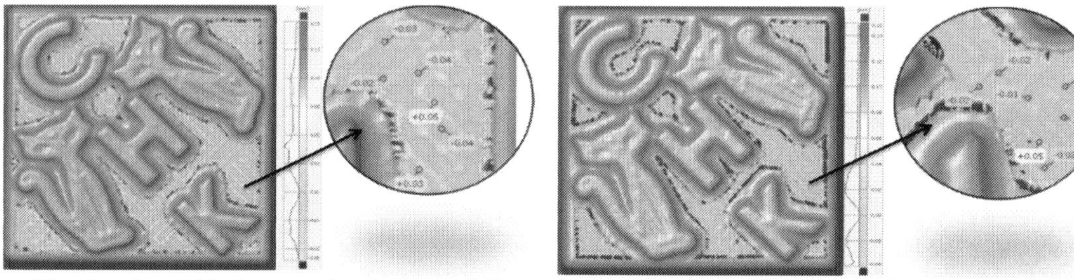

a. Obtained moulded part using 0.3 mm foil; b. Obtained moulded part using 0.6 mm foil

Figure 3. Scans of thermoformed foils obtained using Concept III.

From this preliminary test, it can be concluded that a thinner foil will facilitate the defects from the mould to the obtained thermoformed foil.

3.2. Validation of thermoformed foil using 3D scanning device and software

Besides the multitude of factors of concern such as polymer shrinkage, warpage and equipment capabilities, other part design rules must be considered when designing the mould. Below, some problems related to vacuum holes, draft angle and draw ratio have been studied.

In the aluminium mould case, a minimum diameter of 0.34 mm (#80) can be used for the drilled vacuum holes. When vacuum holes with a diameter larger than the foil thickness are used, the material will draw into the holes and specific thermoforming marks (nib and nipple) will appear on the final product. For the manufacture of the mould, 1.2 mm diameter vacuum holes have been used due to the limitations of the equipment and manufacturing processes involved, and in consequence, bulges and depressions appear on the top and side surfaces. Besides those marks, the irregularities caused by the nozzle path can be seen for the Concept I mould, obtained through the FFF process and into the obtained foil. The defects obtained can be seen in Figure 4 (in the detailed view A and B, on the right side) and appear on both PS and APET foils. These problems, specific to the FFF process, can be reduced by using a smaller nozzle diameter. In the case of the moulds obtained through SLS, the layers on the top surface (detailed view A and B from Figure 4, left side), as a consequence of the part's orientation during printing, cause visible lines to appear on the thermoformed foil.

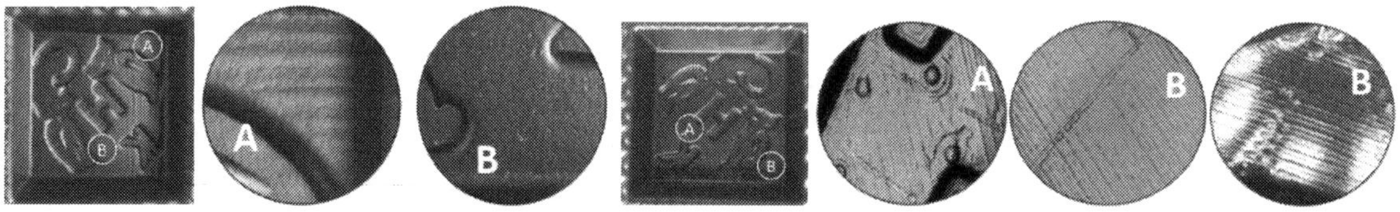

a. Obtained moulded part using Concept V (left 3 picture); b. Obtained moulded part using Concept I;

Figure 4. Surface defects of the thermoformed foils.

During thermoforming, a common problem is the thinning of the foils, especially in the corners and at the edges. Moreover, with a deeper or taller part, a larger radius or chamfer is required. Besides the retention of material thickness, those features are needed to improve the strength and aesthetics of the obtained foil. In the case of the mould obtained by the FFF process with PLA, first heat deformations appear in the corner and edge areas while for the mould manufactured by PA12 through SLS, no deformation appears for the same testing conditions.

In contrast to the previous cases, in which the thinning can be controlled by adding chamfers or radii, the sheet will thin at a much faster and uncontrollable rate in the case of tall walls or deep cavities. In this case, the thinning can be approximated by calculating the draw ratio based on formula (1) [4], which will also offer a recommendation for a minimum starting gauge of the foil by using formula (2) [4].

$$\text{Draw ratio} = \text{Surface Area of mould} / \text{Footprint of the mould} \quad (1)$$

$$\text{Draw ratio} * \text{Desired Finished Gauge} = \text{Minimum Starting Gauge} \quad (2)$$

Using the CAD software SolidWorks and the above formulas (1) and (2), the desired area has been calculated, and theoretical minimum gauges have been obtained for all used foils.

$$\text{Draw ratio} = \frac{10466.59\ mm^2}{4428.13\ mm^2} = 2.363 \quad (3)$$

$$\text{Finished Gauge (0.7 mm)} = \frac{\text{Starting Gauge}}{Draw\ ratio} = \frac{0.7\ mm}{2.363} = 0.296\ mm \quad (4)$$

$$\text{Finished Gauge (0.6 mm)} = \frac{\text{Starting Gauge}}{Draw\ ratio} = \frac{0.6\ mm}{2.363} = 0.253\ mm \quad (5)$$

$$\text{Finished Gauge (0.3 mm)} = \frac{\text{Starting Gauge}}{Draw\ ratio} = \frac{0.3\ mm}{2.363} = 0.126\ mm \quad (6)$$

The above values ignore the effect of the used chamfers and radii, but with the help of the GOM Scan software, the actual values of the sheet thickness can be seen in Figure 5, for different used foils. For the 0.7 mm PS foil obtained using Concept VI, a reduction of 48% of the original material thickness, with a mean distance of 0.6 mm, can be seen. For the 0.7 mm PS foil (Concept I), a reduction in thickness of 55%, with a mean value of 0.55 mm, has been obtained. For the 0.3 mm APET foil (Concept V) the mean value is 0.21 mm while the foil thickness has been reduced by 43%, with a min. value of the thickness of 0.13 mm (resembling to the above-calculated value). Thus, the thinning is not influenced by the type of manufacturing process and the mould material.

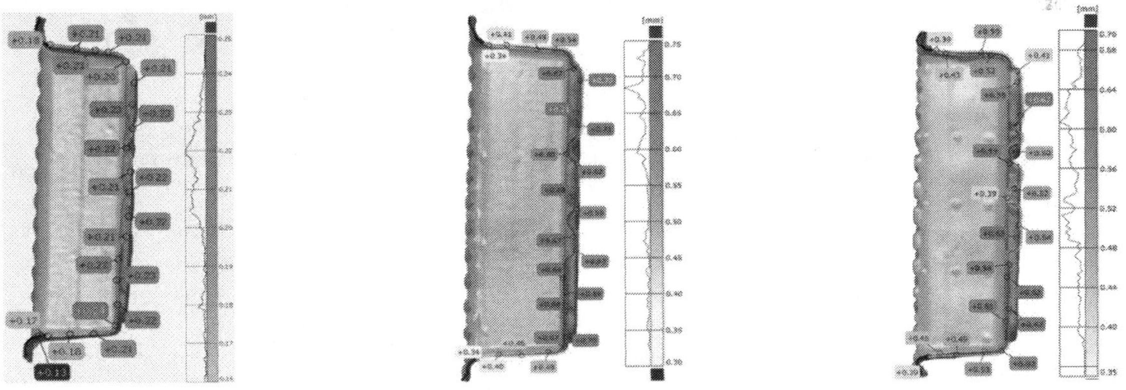

a) 0.3 APET foil (Concept V) b) 0.7 PS foil (Concept VI) c) 0.7 PS foil (Concept I)

Figure 5. Material thickness see-through GOM Inspect for different foils.

Another important feature in designing a mould for thermoforming is the draft angles that depend on the mould type (female or male). During thermoforming, the polymer sheet behaves differently depending on the type of the used mould. During cooling, due to shrinkage, the thermoformed foil increases in density and tightens on the male mould or on the positive portions of the mould, limiting the mould's release. In the case of the female type, the material will shrink away from the vertical surfaces allowing easy extraction of the thermoformed part. On a male mould, a minimum recommended draft angle is 3 degrees and can be increased depending on the height of the parts, particularly when the height exceeds the width or length. A five-degree draft angle is used for the sidewalls for this application.

4. Conclusion

This paper proposes a framework for developing thermoformed foils, subsequently used as a mould for soap casting, using a mould obtained through additive technology (FFF and SLS processes).

Some issues related to the additive technology for producing vacuum thermoforming moulds were presented and Concepts I, III and IV were rejected due to the multiple problems caused by the defects on the mould, which were further transferred to the thermoformed foil.

The advantage of this approach is that it is possible to obtain a ready-to-use custom mould within hours or days, with optimised air suction holes and cooling channels manufactured in the same process and at a low cost for equipment compared to classical technologies. In order to obtain the final soap product, a 0.7 mm PS foil and Concept V have been successfully used.

Acknowledgements

This paper was funded by "Gheorghe Asachi" Technical University of Iasi through the CNFIS-FDI-2020-0354 grant.

ORCID IDs

M-A Boca https://orcid.org/0000-0001-8560-1355
A Sover https://orcid.org/0000-0003-3408-898X
L Slătineanu https://orcid.org/0000-0002-5976-9813

5. References

[1] Grand V R 2021 https://www.grandviewresearch.com/industry-analysis/plastic-packaging-market Accessed 29.04.2021
[2] Tusei I 2021 https://www.statista.com/statistics/1110202/global-market-size-thermoforming-plastics/ Accessed 29.04.2021
[3] Markets A M 2019 https://www.marketsandmarkets.com/Market-Reports/thermoforming-plastic-market-97181143.html Accessed 29.04.2021
[4] Throne L J 2008 *Understanding Thermoforming* vol 2, (Munich/ Carl Hanser Verlag) pp 1-3
[5] Swainson W K 1971 Method of producing a three-dimensional figure by holography *Patent GB 12430044 A* Accessed 18.03.2021
[6] Hull W C 1989 Apparatus For Production Of Three-Dimensional Objects By Stereo Thography *Patent US4575330B1* Accessed 18.03.2021
[7] Freedman H D 2011 https://www.technologyreview.com/2011/12/19/20869/layer-by-layer/ Accessed 18.03.2021
[8] Gajdos I, Maňková I, Jachowicz T and Tor-Świątek A 2016 Application of Rapid Tooling approach in the process of thermoforming mold production *Bánki/ 8th International Engineering Symposium* paper 32
[9] Serrano-Mira J, Gual-Orti J, Bruscas-Bellido G and Abellán-Nebot J V 2017 Use of additive manufacturing to obtain moulds to thermoform tactile graphics for people with visual - impairment *Procedia Manufacturing* **13** 810-7 https://doi.org/10.1016/j.promfg.2017.09.113
[10] Bhandari S 2017 *Feasibility of using 3D printed molds for thermoforming thermoplastic composites* (Maine/ University of Maine- Electronic Theses and Dissertations 2655)
[11] Formech 2020 https://formech.com/wp-content/uploads/Spec-sheet-450DT.pdf Accessed 19.12.2020
[12] Atos C 2019 https://www.gom.com/metrology-systems/atos/atos-core.html Accessed 23.11.2020

IOP Conf. Series: Materials Science and Engineering 1235 (2022) 012066 doi:10.1088/1757-899X/1235/1/012066

Turning tool with tangential cutting line. Concept and constructive solution

M V Puriciuc, R Cioară and C Pisarciuc

Department of Engineering and Industrial Management, Transilvania University of Brasov, 29 Eroilor av., Brasov, Romania

cioarar@unitbv.ro

Abstract. New ways to increase the performance of metal cutting are constantly being sought. Although diverse and increasingly performing tools are designed and used, especially those with inserts, rarely new processing schema are identified in order to ensure favourable conditions to obtain a higher quality for generated surfaces. The paper presents a particular and innovative processing schema, as well as a specific tool for machining cylindrical surfaces by turning with a tangential cutting line. The active part of the latter is an insert whose cutting line is placed frontally in a vertical plane and considerably inclined in relation to the horizontal plane. In order to validate the concept, a variation of the tool has been developed and tested, one that doesn't present the option of adjusting the angle of the cutting line. The success thus achieved prompted the design of a tool that allows the cutting line to be tilted at any angle in relation to the horizontal plane. A patent application has been filed in this regard.

1. Introduction

Turning is a frequently used machining process, which justifies the interest in identifying optimal technological conditions that lead to superior performance in terms of quality of surface and dimensional accuracy. A different approach from the traditional one, from the point of view of the processing scheme, can lead to useful results for the technological aspects.

There is a wide variety of constructive solutions for inserts used in cutting tools [1], in particular lathe tools, and new and new solutions are proposed through patents [2-6]. At the same time, new materials generate a demand for adequate innovative tools [4]. Unfortunately, new machining scheme are rare, most of the time being only particular applications.

2. Turning tool with tangential cutting line. Preliminary concept

In this article is presented a new design regarding a new processing method based on a lathe tool with inclined tangential cutting line. To evaluate the concept feasibility, it was designed a new lathe tool, as presented in figure 1. This new tool was obtained starting from right hand lathe tool, ISO 6 (DIN 4980) whit active part made of carbide P30. The front of the tool was cut off and welded to another in a way that the long edge of the carbide insert is, on the one hand, contained in the front plane of the tool, perpendicular to the axis of its body, and, on the other hand, is inclined with respect to the support plane of the knife body at an angle of 70° approximately.

Experimental tested, it provided very good results, at least in terms of quality of machined surfaces. These preliminary results were the motivation for further research, an objective being the study of the influence of the edge angle on the quality and accuracy of the processed surfaces, and in general on the optimal processing conditions.

Figure 1. Turning tool, with inclined tangential cutting line in working position.

3. Turning tool with tangential inclined cutting line. Concept and construction

The experimental research carried out with the tool originally designed provided a number of useful information in order to design and make a tool with the same characteristics but with the possibility that the cutting line could be rotated at different angles. In relation to the workpiece, the cutting edge is in the front plane of the tool, a vertical plane tangent to the surface to be machined and must be able to be inclined in relation to the axis of the workpiece at any angle by turning to the left the support 3, figure 2.

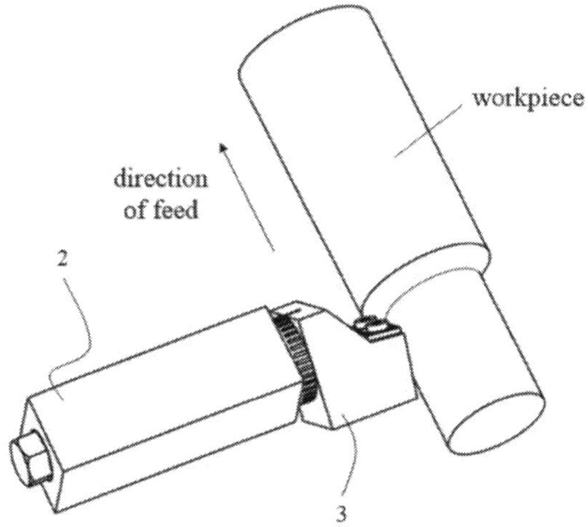

Figure 2. The relative position workpiece -tool, with the cutting line inclined at 45° toward right.

The lathe tool with inclined cutting line was designed in ProEngineer. The 3D model of the assembled tool is presented in figure 3, and in exploded view in figure 4.

The tool is composed from a body 2, for mounting the tool in the lathe holder and a holder for the insert 1 (or 7). This one can rotate relative to the body 2 thus ensuring the desired inclination of the cutting edge. The insert 1, with long useful cutting edge, it is fixed in the support 3 with the help of two screws 6. Between the insert 1 and the support 3 is placed a support plate 5, which protects the insert 1.

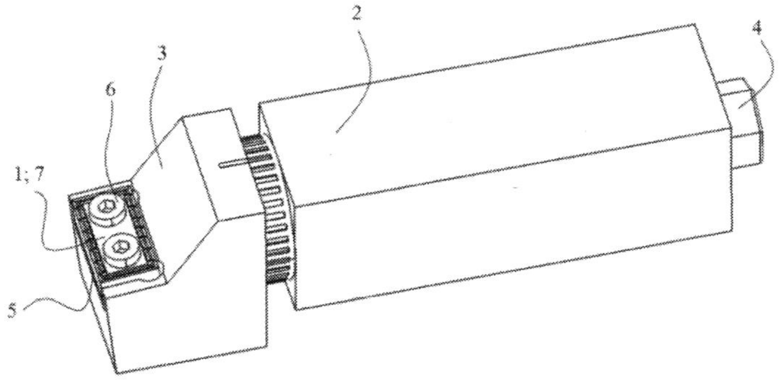

Figure 3. 3D model of the assembled tool.

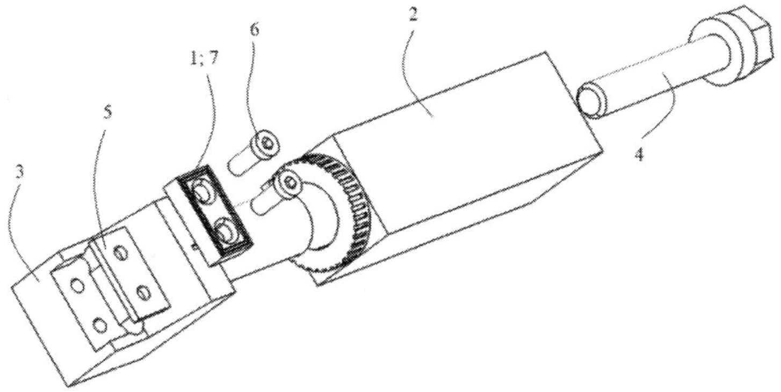

Figure 4. Exploded view of the tool.

The mounting of support 3 to the body 2 is done by friction, the necessary tightening force being developed by the screw 4. The tail z of the support 3 may be cylindrical, figure 5, in which case the required friction force is manifested mainly between the flat front part of the body 2 and the flat surface ah of the support 3 (the 3D model shown in figure 3 corresponds to such a situation), or conical, figure 6, which is preferable, in which case the contact between the front flat surface of the body 2 and the flat surface ah of the support 3, figure 7 is excluded.

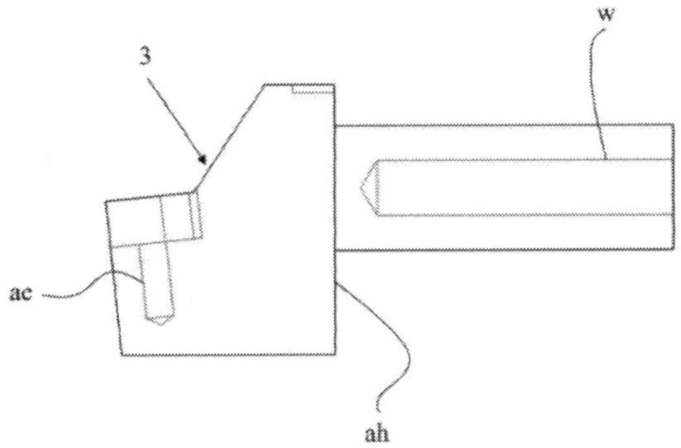

Figure 5. Support for removable insert with cylindrical tail.

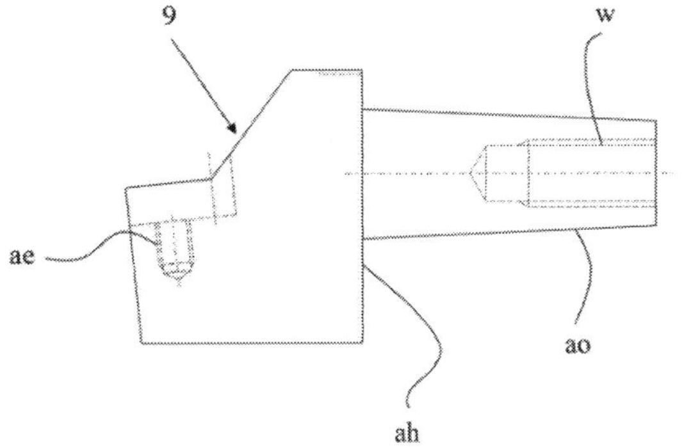

Figure 6. Support for removable insert with conical tail.

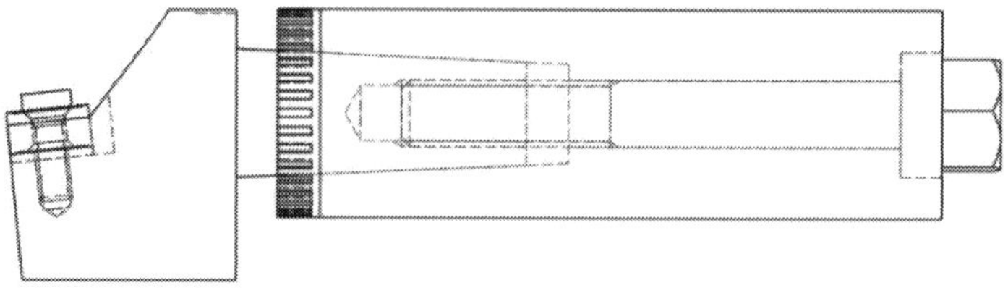

Figure 7. Side view of the turning tool

The body 2 has in its front part a cylindrical area with longitudinal grooves necessary for measuring the angle of inclination of the edge, and the support 3 has, for the same purpose, a single groove ag, figure 7 and 8.

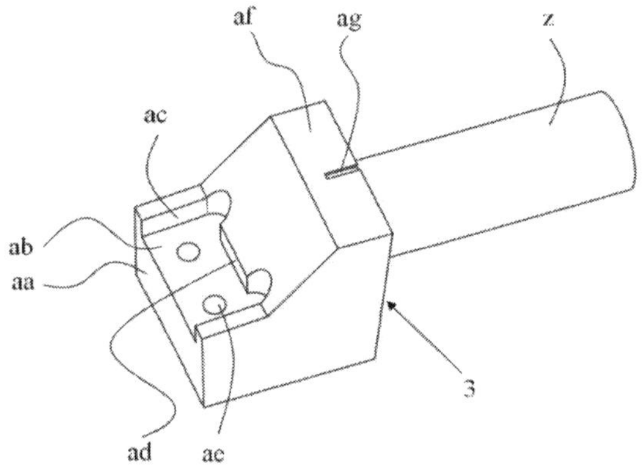

Figure 8. Axonometric view of the tool body

It is very important that the surface ab of the support 3, the surface which is in contact with the support plate 5, to be made in such a way as to obtain the desired positioning angles (rake and clearance angle). In the same time, it is very important that the axis of the tail z intersects the cutting edge of the removable insert 1. If this this requirement is satisfied, the point of the cutting edge of the insert 1 which is in contact with the machined surface of the workpiece will be invariant when changing the inclination of the cutting edge.

The minimum required length of the useful edge of the removable insert 1 is dependent on the radius of the generated cylindrical surface and the thickness of the layer of material removed in a single pass. Designed only for turning with tangential edge, removable insert 1 has a parallelepipedal shape, and has four useful edges, all four long edges of the parallelepiped that delimit it, and the short length sides g of are not useful, figure 9.

The full use of the cutting possibilities of the new removable inset means that the edges g must be provided as useful edges, figure 10, in which case the plate 7 can be mounted on left hand lathe tools, figure 11, and then could be used for turning according to traditional turning. Obviously, in this case the edge l of the parallelepiped is rounded and thus the radius at the tip necessary for the process results.

The described lathe tool is the subject of a patent application [8], the applicants being the Transilvania University of Braşov, Romania in partnership with the "Lucian Blaga" University of Sibiu, Romania.

The novelty of the concept requires its validation through experimental research. For this, a technical project was elaborated in close accordance with the documentation imposed by the patent application. The only difference was the impossibility of using a specific removable insert. Consequently, a commercially available insert with a sufficiently long edge (19 mm) was used. The geometric characteristics of the slot in support 3 for the removable insert were thus reconsidered.

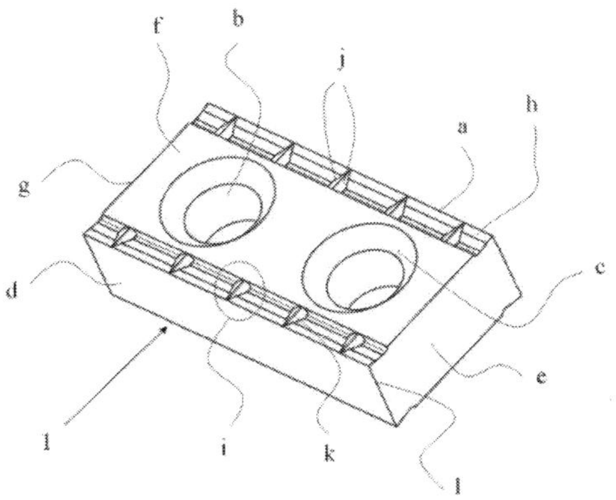

Figure 9. Removable insert for tangential turning.

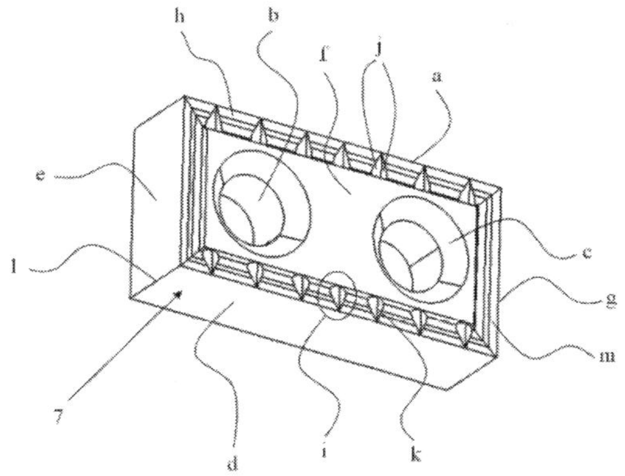

Figure 10. Removable insert for tangential and conventional turning.

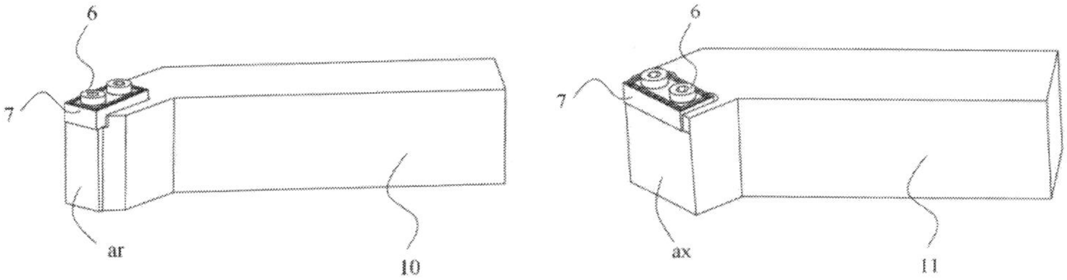

Figure 11. Left side tangential lathe tool with removable insert.

Under the conditions presented, the lathe tool was physically made and preliminary successfully tests were performed. An image of the two large parts of the tool, the body intended to be fixed in the tool holder of the lathe and the support of the removable insert, is shown in figure 12.

Figure 12. Experimental tangential lathe tool.

Figure 13 shows the tool in the working position, fixed in the tool holder of a lathe, as well as the cutting edge in contact with a workpiece. The active edge of the insert is inclined at an angle of 45°, by turning the support to the left.

Figure 13. Experimental tangential turning tool, mounted in lathe support and in working position.

4. Conclusions

Although turning is a frequently used machining process, new processing schemes rarely appear and there are intended only for particular applications. It is found that the concern of specialists is focused on removable inserts, which already exist in a wide variety and for which new solutions are proposed through patents.

A different approach from the traditional one regarding the processing method can lead to unprecedented results. The authors of this paper devised such a method - turning with a tool with an inclined tangential cutting edge - which was proved to be correct. Moreover, from the preliminary tests performed, compared to traditional turning methods, superior results were obtained in terms of the quality of the generated surfaces.

In early stages, the experimental research was performed using a turning tool whose tangential edge is not adjustable in position. The development of the research has imposed to design a lathe tool that allows the inclination of the cutting edge from various angles and implicitly the study of the influence of the inclination angle over the quality of the generated surface and on other characteristics of the machined part and the cutting process.

The concept and the constructive solution elaborated for the turning with a tool with an inclined tangential cutting edge are original, which motivated the filing of a patent application, the applicants being Transilvania University of Braşov, Romania in partnership with the "Lucian Blaga" University of Sibiu, Romania.

A lathe tool with adjustable inclined tangential cutting edge, fully consistent with the patent application, has been physically made and is being tested. At this, the inclination of the tangential edge can be done for any angle, being useful the range 5 - 85°.

ORCID IDs
M V Puriciuc https://orcid.org/0000-0001-6246-7608
R Cioară https://orcid.org/0000-0002-9756-2776
C Pisarciuc https://orcid.org/0000-0003-0595-1668

References

[1] *Metal cutting tools* (https://www.sandvik.coromant.com/en-gb/products/pages/tools.aspx)
[2] Satran A, Eizen Y and Penhas Y 2001 Tangential cutting insert *US 6238146 patent*
[3] Webb S W and Eiler J T 2003 Cutting tool inserts and methods to manufacture *US 2004/0228694 patent application*
[4] Hyatt G A, Andras L R and Massa T R 2004 Rotating cutting insert assembly and cutting insert *EP 1 660 260 patent*
[5] Jonsson M and Eriksson T 2006 A cutting insert for turning with a recess intended to give room for a cooling jet *EP 1 904 254 patent*
[6] Kazuki Y 2013 Cutting insert, cutting tool, and production method for cut workpiece *EP 3015205 patent*
[7] Asghari S 2019 *Innovative cutting tool design for machining advanced engineering materials* (PhD thesis, University of Bath, UK)
[8] Cioară R, Puriciuc M V, Ţîţu A M, Oprean C and Pisarciuc C 2020 Procedeu de strunjire cu tăiş tangenţial înclinat, cuţit de strung şi plăcuţă amovibilă pentru acesta *Patent application RO a 2020 00738*

IOP Conf. Series: Materials Science and Engineering 1235 (2022) 012067 doi:10.1088/1757-899X/1235/1/012067

A study of effect the supporting space and tool path in sheet forming, thickness evaluation, part D

K I Abass

Mechanical Engineering Department, College of Engineering, Mustansiriyah University, Baghdad, Iraq

k.i.abass@uomustansiriyah.edu.iq, khalil1969a@yahoo.com

Abstract. The main purpose of any sheet metal forming process is producing a product with high accuracy of the dimensions in three areas (planes). Incremental Forming (IF) is a process of redistributing the areas. Thickness is the main value and is an indicator of the complexity of the forming ability in depth. It is also an indicator value of the area that represents the product section. IF includes general plastic deformation, thus the explanation of the process by nonlinear boundary condition settings is more difficult. Especially when considering the interaction between surfaces and the frictional properties. The study involved six series that investigate the supporting space, the die radius and the forming tool path. The research presents the main results of a study for evaluating the thickness distributions and the thinning along the axisymmetric product section. The incremental forming has been considered under certain boundary conditions by reference to the ANSYS application. The study shows that the model of the simulation can expect the process track and predict the product profile and its thickness distribution. Additionally, the present paper illustrates the behavior of the forming tool during direct and indirect contact for evaluating the accuracy of IF product.

1. Introduction

The Incremental Forming (IF) process supports manufacturing product profiles without die sets. As a result, the process leads to reduction of the period for performance. Therefore, it can be implemented in prototyping main products for automobiles, aircraft or biomedical parts [1]. The essential process involves a set that includes fixture and tool forming fixed on the CNC milling machine, and a program to simulate the tool path for forming the product in three dimensions [2,3]. The accuracy of surface finish, wall angle and thickness distribution are the main issues [4,5], considering the complicated parts and high number of parameters: product properties (material, product profile and blank thickness), and forming tool (tool path, speed, step size, tool profile).

2. Literature review

The supporting distance or the gap is the distance between the initial forming tool position and the support die edge. Deviation of the final product profile is affected by this gap. In case the gap is less than the blank thickness, it causes the ironing influence, while a large distance causes losing the installation torque. Pohlak M. et al. [6] discovered that the optimal position of the tool for starting the forming process is at a distance equal to the initial blank thickness "the gap is forming tool-support space". The un-deformed blank material is comparatively hard; therefore, the profiles of the areas which are not in direct forming contact return to the previously forming positions.

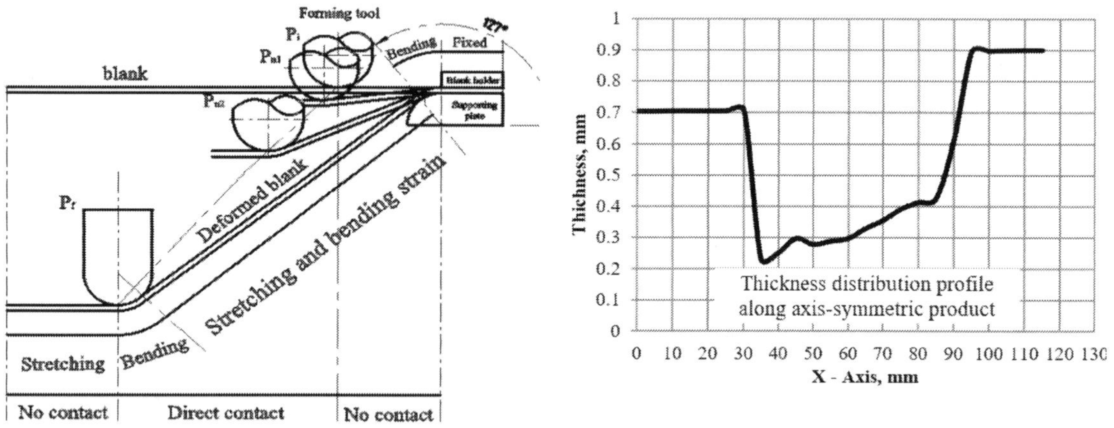

Figure 1. Toolpath profile and thickness distribution

2.1. Sheet Thickness

The main forming mode of the IF process was compared with the shear forming. Szekeres *et al.* [7] confirmed that using high blank thickness would increase the forming angle which can be formed, due to the fact that formability is in direct relation with the blank thickness. Jeswiet J. [8] expressed the sine law, Eq. (1), based on the forming angle, λ and the initial blank thickness, calculating the final product thickness.

$$t_f = t_o \sin \lambda \tag{1}$$

where t_f is the final product thickness and t_o is the initial blank thickness.

2.2. Forming Angle

The forming angle, λ is a part of the tool path profile, and the range value depends on mechanical properties and the initial thickness of the blank used. Martins *et al.* [9] expressed the maximum value of the wall angle referring to the mechanical material properties and the forming variables as in Eq. (2), where ϵt is the strain of the thickness at the maximum limit of the formability.

$$\lambda = \pi/2 - e^{\epsilon_t} \tag{2}$$

2.3. Forming Toolpath

To represent the toolpath profile in IF process, there are variables that should be considered: the horizontal contour on xy-sheet plane "translate into x and y values", and the vertical movement value in the z-axis direction "translate to step size value". Both expressions involve the angular values between any two points during the forming tool movement. Additionally, this process causes surface roughness on the product profile (physical indentation) for each step and contour [10,11].

The study involved six series that consider the supporting space, the die radius and the forming tool path [12-14]. The paper presents the data and main results of a study for evaluating the thinning and the thickness distribution along axisymmetric product profile section. Aluminum material alloy sheets, A1025 with t = 0.90 mm thickness, are used to produce the profile and the mechanical properties illustrated in Table 1.

The research goal is to determine the ratio involving the thickness distribution, the supporting space, die radius and the forming tool path. Additionally, the study of the maximum local thinning in the deformed profile section and the main conditions of the final products to crack have been evaluated. The methodology is considered on six series and is based on producing similar products (outer diameter and depth, 226.0 mm and 47.0 mm, respectively) and on using identical x - y step size, 0.1 mm, and Friction coefficient, μ = 0.05.

3. Development of FE Model

The tool sets and the blank materials are represented in the application program "Ansys". The elements selection is considering the properties for each of them, as illustrated in detail in previous studies [12-14] and Table 2. The nonlinear analysis considers the process, and the friction coefficient, step size and the fixed parts involved within the boundary conditions.

Table 1. Material properties of the blank, Al (AA1050)

Young's modulus, E	Poisson's ratio, ν	Density, ρ	Tangent modules, Eτ	Tensile Strength
75 GPa	0.3	2700 g/m³	0.5 GPa	105 - 145 MPa

Table 2. Methodology of the six series

a	Inner radius value of the supporting die, 90 mm; Forming tool position, 80 mm; and Tool path angle, 45°.
	Die edge radius, R, 1, 2.5, 5, 7.5, 10 mm
b	Die edge radius, 5 mm; Forming tool position, 80 mm; and Tool path angle, 45°.
	Inner supporting die radius, D, 80 mm with step value 2.5, 5, 7.5, 10, 12.5, 15 mm
c	Inner radius value of the supporting die, 80 mm; Forming tool position, 80 mm; and Tool path angle, 45°.
	Die edge radius, R, 2.5, 5, 7.5, 10, 12.5, 15 mm
d	Supporting space, Ss, 10 mm; Forming tool position, 80 mm; and Tool path angle, 45°.
	Die edge radius, R, 2.5, 5, 7.5, 10, 12.5, 15 mm
e	Die edge radius, 5 mm; Product base radius, P_{Base}, 33 mm.
	Supporting space, Ss, 0.98, 2.5, 5, 7.5, 10 mm
f	Die edge radius, 5 mm; Tool path angle, 45°.
	Supporting space, Ss, 0.35, 2.5, 5, 7.5, 10 mm

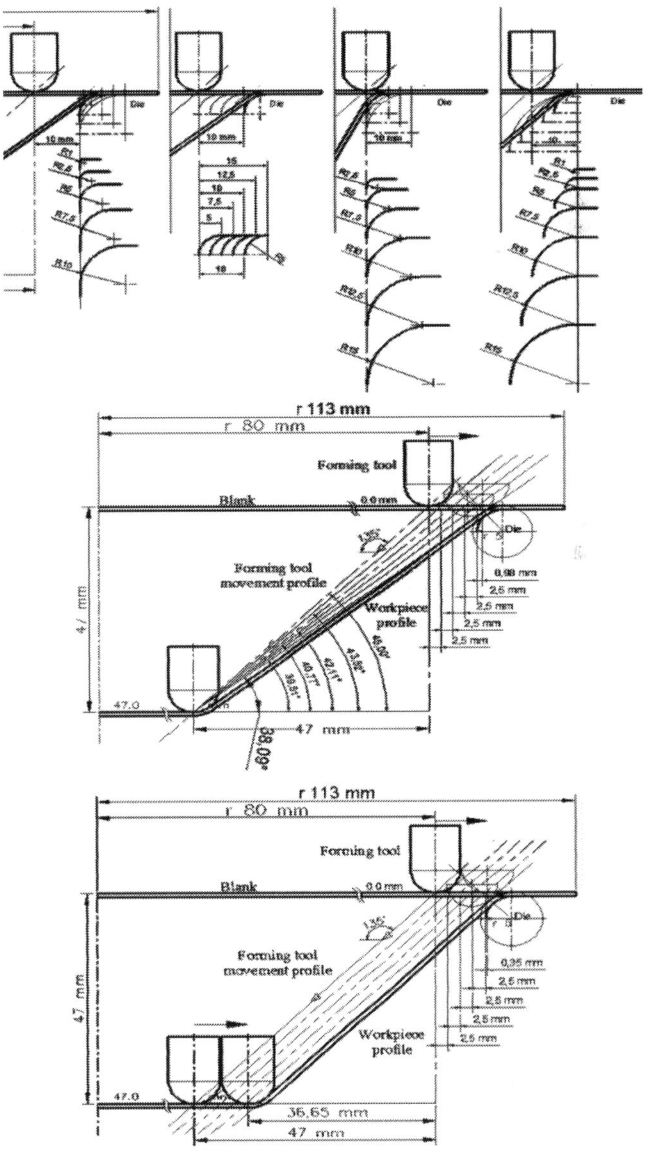

Figure 2. Profiles of six series

4. Results and Discussion

The main requirement for accepting any product that results from the sheet forming process is the final profile-thickness distribution ratio. This ratio is a limitation of the metal ability to respond to the forming process. Therefore, the evaluation of the profile-thickness ratio in order to determine the acceptance of the product depends on the following: thinning evaluation during forming steps, depth; thickness calculation along the product section; upper and lower values of thinning and the relation with the forming process variables; thinning evaluation by comparing the behaviour of the values related of each series.

Severe bending status is occurring in the case of 1 mm die edge radius, leading to high values of thinning, which promotes the neck formation. Next to this point, the part failure is obviously specified. The blank profile section is separated into four sections, as follows (blank centre is used as reference point for calculating all the dimensions):
i. fixed region (100 to 113 mm), not affected by the tool contact movement;
ii. direct forming tool effect region (33 to 90 mm), clear influences in thickness distribution;
iii. indirect contact effect regions, supporting and free spaces, (0 to 33 mm) and (90 to 100 mm), respectively. In these cases, the thinning percentage is clearly affected by decreasing the thickness, as illustrated in Figure 3.

4.1. Evaluation of the forming steps:

Thickness distributions is concentrated in the direct contact area, the influence of the location of the forming tool, which is between 30 to 100 mm is slightly decreases thickness by 1.11%; at 47 mm in depth, the highest percentage occurs. The general behaviour of changing thickness is similar across all stages, as follows:

- Gradual decrease of the thickness when increasing depth (90-100 mm) is affected by the radius of the die.
- Decrease the thickness affected by the radius of the tool (direct contact, 55-90 mm).
- Deficiency the thickness is affected by the last contact area of the tool "forming tool centre", (38-40 mm).
- Decrease the thickness affected by tool radius (42-55 mm), build-up of plate metal around the forming tool.

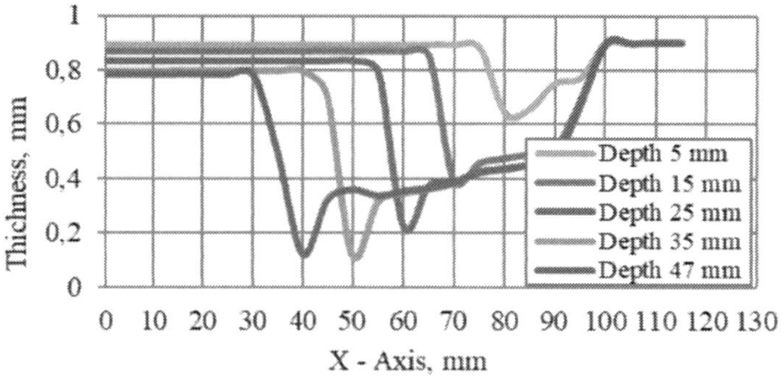

Figure 3. Profiles of thickness distributions, in depth

4.2. Evaluation of the product thickness:

a. The stability of the thickness in the installation area (100-115 mm).
b. A gradual subsidence that follows the curvature of the die surface (90-100 mm), moving away or approaching according to the value of the parameters of that area.
c. A high drop in thickness at the area of the fixation space (80-90 mm), moving away or approaching according to the value of the area variables.
d. A gradual decrease in thickness in the direct contact area (80-40 mm).

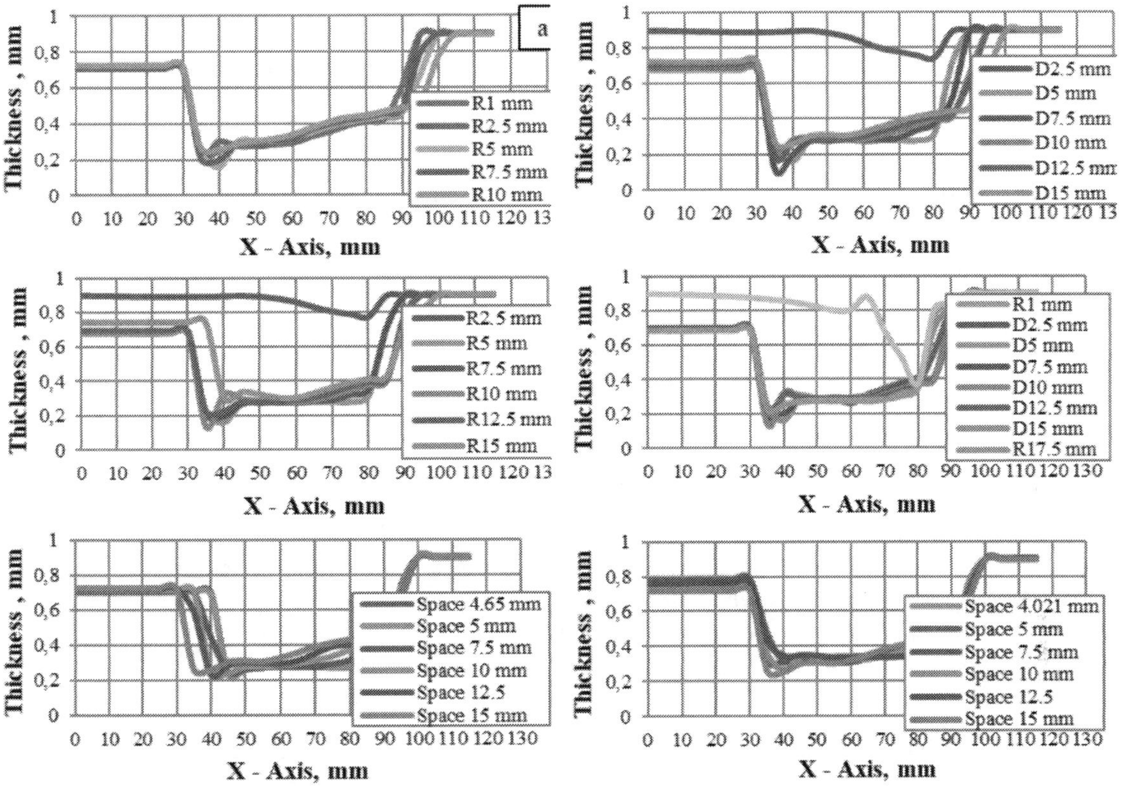

Figure 4. Thickness distributions of the series a, b, c, d, e, and f.

The thickness evaluations are illustrated in Figure 4. In the fixed region between 110 to 113 mm from the centre of the blank, the thickness distribution does not change because of the inability to move, as the blanks are fixed between the blank holder and the upper surface of the die.

The lowest value for thickness is shown in the Table 3:

a. The difference came gradually with the progress of the forming tool horizontally depending on the formation of layers and vertically with depth.

b. Systematic effect of changing the thickness of the area between the centre of the product and up to 30 mm.

It is the non-direct contact area of the tool.

Failure to form the plate occurred due to following causes:

a. The intermediate distance between the die and the tool when following the tool path reaches less than the plate thickness and thus the ironing state occurs, leading to failure, Table 4.

b. The effect of the variable change is evident in the regions indicated in the Table 5.

c. The indirect impact area of the tool contact shows the lowest and highest values of the thickness in that area, as shown in the Table 6.

Table 3. Minimum thickness, mm.

Series	a	b	c	d	e (space)	f (path)
State	R 5	D 7.5	R 10	R 10	α° 5	P_{Base} 15
Mini. thickness, mm	0.17	0.109	0.138	0.138	0.20	0.23
Max. depth, mm	40.3	45	44.9	44.9	44.10	40.30

Table 4. Maximum depth, mm.

Series	a	b	c	d	e (space)	f (path)
State	R --	D 2.5	R 2.5, 15	R 20	α° --	P_{Base} --
Max. depth, mm	----	0.80	1.2, 42	9.20	---	---

Table 5. Effect region from the blank centre, mm.

Series	a	b	c	d	e (space)	f (path)
Region	85 to 105	35 to 100	35 to 100	35 to 100	30 to 90	35 to 90

Table 6. Minimum and maximum blank thickness at the indirect contact region of the forming tool.

Series	a	b	c	d	e (space)	f (path)
Min. thickness, mm	0.70	0.67	0.67	0.68	0.69	0.71
Max. thickness, mm	0.73	0.73	0.75	0.70	0.72	0.80

4.3. Evaluation of the product thinning:

With the exception of the cases of failures shown in the Table 3, the relations that are correlated to the percentage of thinning are shown in the following Figure 5.

The series did not produce any clear direct or opposite relation for all cases, in other words:

• A positive increase in a variable within the series does not necessarily lead to an increase or decrease in thinning percentage;

• The supporting space variable, Ss is effectively interacting with the die radius of series a, c and d;

• Series c, a and d indicated less variation in thinning percentage 3%, 6% and 9% respectively, while series b, e and f indicated higher variance, respectively.

From this, it becomes clear that the change of the supporting space Ss with the change of the tool path is the most influential relationship on the thickness distribution of the product section (series c), and the change of the die radius with the stability of the supporting space Ss is the relationship least affecting the thickness distribution (series f).

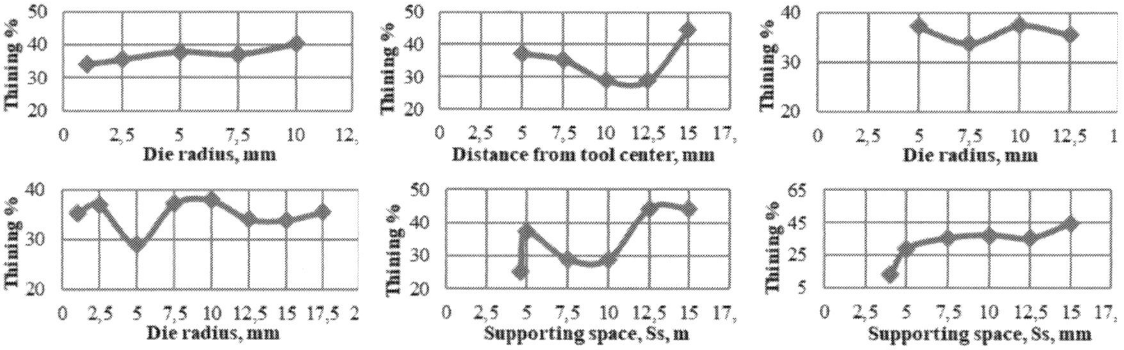

Figure 5. Max. thinning of the products

5. Conclusion

The investigation indicated the effect of the supporting space and die radius related to the tool path influences by means of the formability with thickness indicator.

• The position of the primary tool and its change during the progression of the forming process makes the momentary forces change accordingly, so the ability of the metal to stretch changes, and the change of thickness follows the change of the strength of the stretch.

• The localized formation of the tool raises the temperature locally, thus increasing the formability in the direct contact sites of the tool, and the change of thickness follows this effect.

• The increase in the voids are used in characterized the material structure, and thus the ability of the plate to absorb the forces of flexible forming is higher, "decreasing with the increasing depth of forming".
• Increasing the radius of the die impedes the formation by restricting the flow of the metal.
• Reducing the die radius increases the strain rate, the potential of tearing the blank in the forming process.
• The tool path is a reflection of the shape of the desired product, and the forming angle limits the elastic and ductile of the metal. Thus, it is a limiting factor in the thickness of the metal.

ORCID IDs

K I Abass https://orcid.org/0000-0001-6176-3894

6. References

[1] Dittrich M A, Gutowski T G, Cao J, Roth J T, Xia Z C, Kiridena V and Henning H 2012 Exergy analysis of incremental sheet forming *Prod. Eng. Res. Devel.* **6** 169-77 https://doi.org/10.1007/s11740-012-0375-9
[2] Centeno G, Silva M B, Cristino V A M, Vallellano C and Martins P A F 2012 Hole-flanging by incremental sheet forming *Int. J. Mach. Tools Manuf.* **59** 46-54 https://doi.org/10.1016/j.ijmachtools.2012.03.007
[3] Benmessaoud R 2014 *A Tool path generation method for the multi-pass incremental forming process investigation Int. J. Adv. Research in Computer Science and Software Eng.* **4** (5) 1035-44
[4] Suresh K, Khan A and Regalla S P 2013 Tool Path Definition for Numerical Simulation of Single Point Incremental Forming *Procedia Eng.* **64** 536-45 https://doi.org/10.1016/j.proeng.2013.09.128
[5] Wu S H, Reis A, Andrade Pires F M, Santos A D and da Rocha A B 2012 Study of tool trajectory in incremental forming *Adv. Mat. Res.* **472-475** 1586-91 https://doi.org/10.4028/www.scientific.net/AMR.472-475.1586
[6] Pohlak M, Majak J and Küttner R 2007 Manufacturability and limitations in incremental sheet forming *Proc. Estonian Acad. Sci. Eng.* **13** (2) 129-39
[7] Szekeres A, Ham M and Jeswiet J 2007 Force Measurement in Pyramid Shaped Parts with a Spindle Mounted Force Sensor *Key Eng. Mater.* **344** 551-8 https://doi.org/10.4028/www.scientific.net/KEM.344.551
[8] Jeswiet J 2005 Asymmetric Incremental Sheet Forming *Adv. Mat. Res.* **6-8** 35-58 https://doi.org/10.4028/www.scientific.net/AMR.6-8.35
[9] Silva M B, Skjoedt M, Atkins A G, Bay N, and Martins P A F 2008 Single-point incremental forming and formability–failure diagrams *J. Strain Anal. Eng. Des.* **43** (1) 15-35 https://doi.org/10.1243/03093247JSA340
[10] ***, What is robo forming http://www.alexjfischer.com/what-is-roboforming (accessed on 01.09.2015).
[11] Wu S H, Reis A, Andrade Pires F M, Santos A D, da Rocha A B 2012 Study of tool trajectory in incremental forming *Adv. Mat. Res.* **472-475** 1586-91 https://doi.org/10.4028/www.scientific.net/AMR.472-475.1586
[12] Abass K I 2018 Evaluation of the supporting plate edge radius and position effects on springback in incremental forming, part A *MATEC Web of Conferences* **178** 02007 https://doi.org/10.1051/matecconf/201817802007
[13] Abass K I 2018 Supporting plate edge radius and internal hole diameter relation effect on the springback in incremental forming, part B *MATEC Web of Conferences* **178** 02012 https://doi.org/10.1051/matecconf/201817802012
[14] Abass K I 2021 The forming tool, primary start position and toolpath angle movement effect on the geometry accuracy "springback", part C *IOP Conf. Ser.: Mater. Sci. Eng.* **1009** 012001 https://doi.org/10.1088/1757-899X/1009/1/012001

Study of roller-shoe contact during high-pressure injection pump normal operation

R C Iordache, N D Petrea and C Bujoreanu

"Gheorghe Asachi" Technical University of Iasi-Romania, Faculty of Mechanical Engineering, Blvd. Mangeron, No. 59A, 700050, Iasi,

cbujorea@tuiasi.ro

Abstract. Roller-shoe mechanism is very often used in diesel high pressure pumps. The most encountered issues are caused by this mechanism components which are in direct contact. Lubrication, roughness, elements dimensions and profiles, coatings, working conditions can be adjusted so that the issues are reduced. Our paper aims to identify the most critical factors which influence the roller-shoe mechanism and to calculate the contact parameters during normal engine operation. The loads applied on the components are the key factors which contribute to all parameters calculation that ensure the contact optimization. The components clearance, friction coefficients, medium contact pressure, minimum lubricant film, equilibrium temperature are some of important parameters which are the basis of a correct functioning. Analysing the obtained results, it can be observed that the lubricant film thickness can be adjusted to establish a minimum safely drive shaft rotation. Considering these results new improvements methods will be analysed for roller- shoe contact development.

1. Introduction

The continuous improvement is the one of the success factors that lead to the products development promoted by various companies. In the automotive industry, subassemblies based on purely mechanical principles represent an important majority which is the basis of the vehicles operation.

Thus, the classical mechanical principles are continuously applicable and in the same time represent an open space for permanent development. Roller-shoe mechanism is analysed in this paper.

The interaction between solid surfaces is related to contact mechanics and also tribology area. This interaction/contact can take place on many different scales, ranging from nanoscale aspects to the size of tires in contact with roads and even contact between tectonic plates [1,2].

Basically, the roller-shoe assembly can be considered as a cylinder inside another cylinder. Thus, when numerical calculations are needed, some specific key parameters will be followed. These parameters can depend of the contact surface area, contact pressure and the clearance between components. The latter, together with some components, constructive characteristics can be adjusted in order to improve the contact and to significantly reduce the wear.

The analysis of constructive solution, the operating environment and the role of this high-pressure pump part are some of the factors that can justify the need to improve the contact between roller and shoe.

IOP Conf. Series: Materials Science and Engineering 1235 (2022) 012068 doi:10.1088/1757-899X/1235/1/012068

2. Contact profile

Although the shoe is not a well-defined cylinder, it is considered a sliding bearing with a cylindrical inner shape. Thus, from a static point of view the contact surface and the pressure distribution between these two components are presented in figure 1.

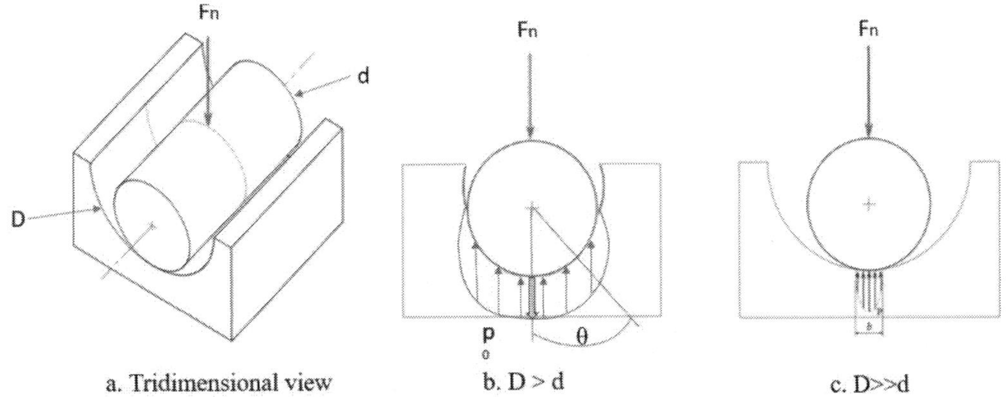

a. Tridimensional view b. D > d c. D>>d

Figure 1. Pressure distribution for roller and shoe (D - shoe diameter and d - roller diameter)

Depending of these two diameters' dimensions, D and d, we can have two cases [2,3]:

- If D >> d, the contact is considered as a Hertzian type;
- If D> d, the contact is considered as a conformal type and the pressure distribution is cosinusoidal $p = p_o \cdot \cos(\theta)$;

3. Components dimensional characteristics

The most important dimensional characteristics are highlighted on the sketch from figure 2. The measurement results are presented in table 1.

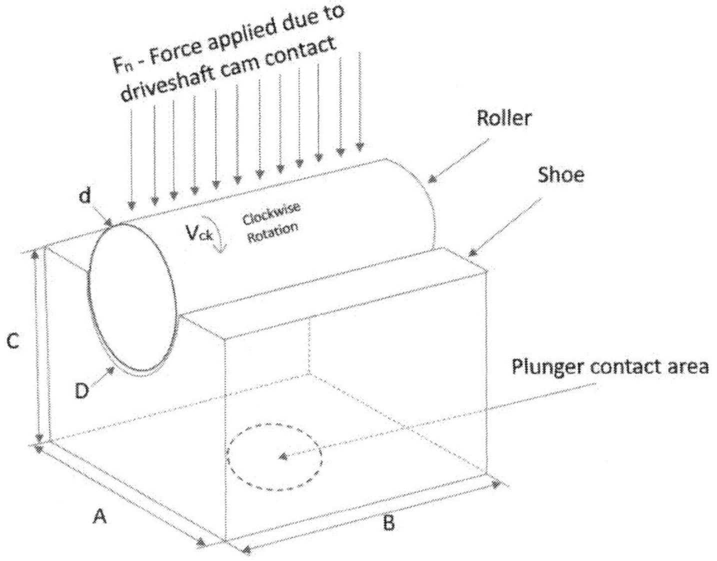

Figure 2. Roller and shoe subassembly – sketch with dimensional characteristics.

Table 1. Roller and Shoe – Basis dimensions

Dimension	Drawing Symbol	Measured Value	Measurement Unit
Roller diameter	d	12.029	
Shoe diameter	D	12.05	
Shoe length (same with roller length)	B	21	mm
Shoe width	A	17	
Shoe height	C	15	

Considering above dimensions the clearance between roller and shoe is:

$$J = D - d = 21\ \mu m \quad (1)$$

4. Roller and Shoe lubrication regime

In order to identify the lubrication regime, it is necessary to find out the minimum lubricant film thickness. The steps for this parameter calculation are described below.

- Total load applied is Fn = 709.775 N;
- Medium pressure calculation:

$$p_m = \frac{F_n}{B \cdot D} = 2{,}805 \cdot 10^6\ \frac{N}{m^2} \quad (2)$$

- Diesel fuel dynamic viscosity is η =1.352 *mPa s* (0.01352 *Pa s*) [4];
- Relative clearance:

$$\Psi = \frac{D-d}{d} = 1.746 \cdot 10^{-3} \quad (3)$$

- Roller rotation velocity calculated is $n = 114.6\ min^{-1}$;
- Sommerfeld number:

$$S = \frac{\eta * n}{p_m * \Psi^2} = 0.018 \quad (4)$$

- Minimum lubricant film thickness:

$$h_{min} = \delta \cdot \frac{\Psi \cdot D}{2} = 1.473\ \mu m \quad (5)$$

The δ coefficient can be approximate using the graph from figure 3 considering the B/D report and Sommerfeld number S.

Considering the above parameters, the δ coefficient is approximated 0.14.

The roller and shoe roughness have an important role for surfaces contact area and will be compared with film minimum thickness.

After measurements the roller roughness is Ra1= 0.096 and shoe roughness is Ra2 = 0.16.

If Rq1 and Rq2 are the square mean deviations of the roll and shoe roughness when determining the lubrication parameter λ:

$$\lambda = \frac{h_{min}}{\sqrt{Rq1^2 + Rq2^2}} \quad (6)$$

where $R_a = R_q{}^2$

- Roller Rq roughness (measured on roller surface): $R_{q1} = \sqrt{R_a} = \sqrt{0.096} = 0.31\ \mu m$;
- Shoe Rq roughness (measured on shoe surface): $R_{q2} = \sqrt{R_a} = \sqrt{0.16} = 0.4\ \mu m$;

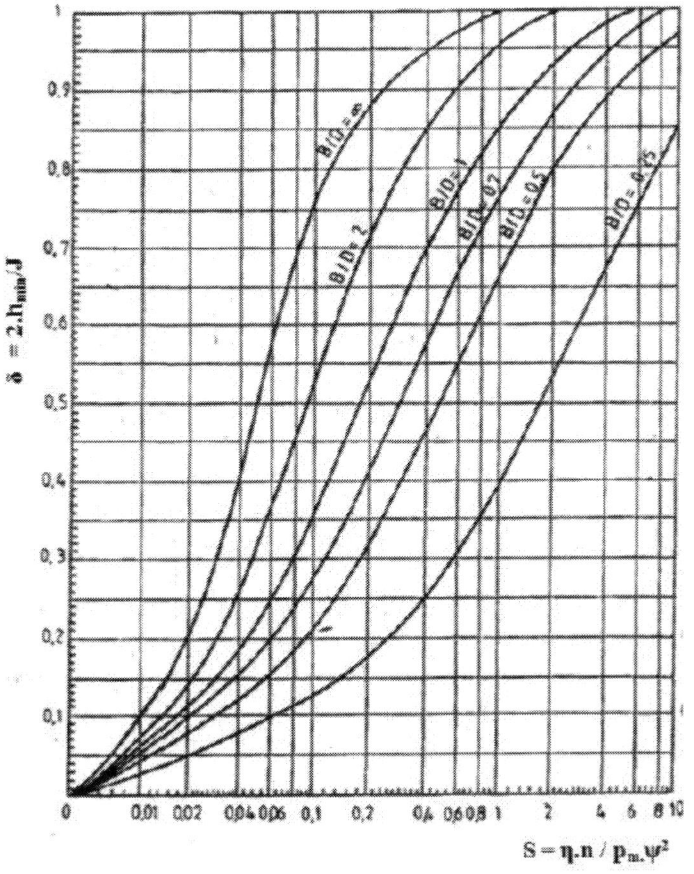

Figure 3. Variation of the relative minimum thickness δ depending of B/D (B/D = 1.74) report and S [5]

Based on lambda parameter calculation are identified two cases:

- If λ> 3 the lubrication regime is with continuous film;
- If λ <3, the lubrication regime is mixed.

The lubricant parameter value after calculation is λ = 2.91. Considering this value, it is obvious we are situated in the second situation, which means that the lubrication regime is a mixed one [6].

5. Mixed regime parameters calculation

- Friction coefficient value imposed is $\mu = 0.01$ (considering the components material);
- Pressure distribution – cosinusoidal form, where p_o is maximum pressure on Fn direction:

$$p = p_o \cdot \cos(\theta) \tag{7}$$

- Fn calculation considering the balance of forces in the horizontal direction:

$$F_n = 2 \cdot \int_0^{\frac{\pi}{2}} p_0 \cdot \frac{D}{2} \cdot B \cdot \cos(\theta)^2 \, d\alpha \tag{8}$$

$$\text{where } p_0 = \frac{4}{\pi} \cdot p_m \tag{9}$$

- Friction torque between roller and shoe:

$$M_f = 2\int_0^{\frac{\pi}{2}} \mu_3 \cdot p_0 \cdot \frac{D}{2} \cdot B \cdot \frac{D}{2} \cdot \cos(\theta)\, d_\alpha \tag{10}$$

$$\int_0^{\frac{\pi}{2}} \cos(\theta) d_\alpha = 1 \tag{11}$$

$$M_f = \frac{4}{\pi} \cdot \mu \cdot p_m \cdot B \cdot (\frac{D}{2})^2 = 54.164\ Nmm \tag{12}$$

- Tangential force required to overcome the friction torque between roller and shoe:

$$F_{fc} = 2 \cdot \frac{M_f}{D} = 9.027\ N \tag{13}$$

- Roller-cam friction coefficient:

$$\mu_1 = \frac{F_{fc}}{F_n} = 0.13 \tag{14}$$

The value of this coefficient suggests that we have sliding between the roller and the cam. We will consider a sliding of 3%: $n_r = 0.97 \cdot 114.6 = 111.162\ rot/sec$.

- Power dissipated by friction between the roller and the shoe:

$$P = M_f \cdot 10^{-3} \cdot \omega = 18.915\ W \tag{15}$$

The angular velocity of the roller is calculated using a basic equation, $\omega = \pi \cdot n_r$.

- Temperature developed in the imposed conditions, with heat dissipation through the shoe:

$$T = \frac{P}{K \cdot A_1} + T_0 = 360.081\ °C \tag{16}$$

The T_0 parameter represents the ambient temperature and the value of this is $T_0 = 20\ °C$; K is a global heat transfer coefficient and the chosen value is K=30. The range of values for K coefficient is (10...30). In our case is used a superior value due to forced air circulation around the roller and shoe mechanism.

A_1 represents the outer surface of the shoe through which the heat is exchanged and it is calculated using equation (17).

$$A_1 = (2 \cdot A \cdot B + 2 \cdot A \cdot C + 2 \cdot B \cdot C) \cdot 10^{-6} = 1.854 \cdot 10^{-3}\ m^2 \tag{17}$$

The value of this calculated temperature could be lower if we take into account the heat taken over by the lubricant. Considering this we can recalculate the temperature using the parameters and coefficients described below.

- Lubricant Specific Heat, $c_L \approx 1880 \frac{N \cdot m}{Kg \cdot {}^oC}$ [5];
- Lubricant Density, $\rho_L \approx 900\ \frac{kg}{m^3}$ [5];
- Diesel fuel inlet temperature measured, $T_i = 60°C$ [4];
- Calculated flow:

$$Q_s = C_{qs} \cdot D^2 \cdot B \cdot n \cdot \Psi = 0.4 \cdot 10^{-9} \frac{m^3}{s} \qquad [18]$$

C_{qs} is the side loss coefficient. Considering the graph from figure 4, its value is $C_{qs} = 0.65$.

- Recalculated temperature:

$$T_1 = \frac{(P_3 + A_1 \cdot K \cdot T_0 + Q_s \cdot c_L \cdot \rho_L \cdot T_i)}{A_1 \cdot K + Q_s \cdot c_L \cdot \rho_L} = 82.788\ °C \tag{19}$$

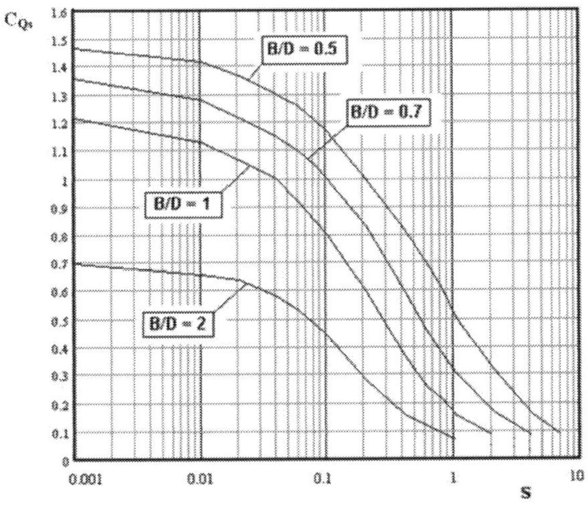

Figure 4. Variation of the C_{qs} parameter depending of B/D (B/D = 1.74) report and S [5]

6. Conclusions

Surface and lubrication parameters are on the basis for transition between lubrication regimes. The calculation models can be diverse and can be developed according to these parameters. As we observed in the previous calculations, the lambda lubrication parameter is the one that helps us to establish the lubrication regime depending of the thickness of the lubricant film and the roughness of the contact surfaces. By adjusting these parameters in the first phase we can obtain a value of the lambda lubrication parameter that corresponds to the continuous lubrication film regime. Thus, a different calculation model would be obtained, with significant improvements in the contact between the surfaces. Also, reducing the developed temperature, the friction torque or the tangential force needed to overcome this friction torque would be some consequences of this improvement. All this combined helps to reduce the risk of scuffing occurring and increase the service life of the mechanism. New calculation models will be developed for new research on this topic.

ORCID IDs

R C Iordache https://orcid.org/0000-0002-0984-810X
N D Petrea https://orcid.org/0000-0003-0299-0792
C Bujoreanu https://orcid.org/0000-0002-5680-2788

7. References

[1] Robert L J, Hamed G, Hyeon L, Amir R and Xianzhang W 2013 *Tribology for Scientists and Engineers:From Basics to Advanced Concepts* (New York: Springer) pp 93-140 https://doi.org/10.1007/978-1-4614-1945-7_3

[2] Fischer-Cripps A C 1999 The Hertzian contact surface *J. Mater. Sci.* **34** 129-37 https://doi.org/10.1023/A:1004490230078

[3] Trivedi H K and Bhatt D V 2017 Tribological Behaviour under Conformal and Non-Conformal Contact Condition of Piston Ring and Cylinder Liner Material in a Reciprocating Bench Test *IOP Conf. Ser.: Mater. Sci. Eng.* **239** 012014 https://doi.org/10.1088/1757-899X/239/1/012014

[4] Kumbar V and Skrivanek A 2015 Temperature dependence viscosity and density of different biodiesel blends *Acta Univ. Agric. et Silvic.* **63** 1147-1151 https://dx.doi.org/10.11118/actaun201563041147

[5] Stachowiak G W and Batchelor A W 1998 *Engineering Tribology* (Butterworth&Heinemann) vol 1 (Elsevier) p 871

[6] Dong Z and Qian W 2012 On the λ ratio range of mixed lubrication *P. I. Mech. Eng. J-J Eng.* **226** (12) 1010-22 https://dx.doi.org/10.1177/1350650112461867

Experimental study of the removable assemblies with screws from high pressure injection pumps

D P Bistriceanu, C Bujoreanu and S G Pal

"Gheorghe Asachi" Technical University of Iasi, Department of Mechanical Engineering, Mechatronics and Robotics, Blvd. Mangeron, No. 43, 700050, Iasi, Romania

cbujorea@tuiasi.ro

Abstract. High pressure injection pumps use removable assemblies with screws. They allow the subsequent assembly and disassembly, without destroying the components and the entire assembly, whenever is necessary. Our paper is an experimental analysis highlighting the influence of tightening parameters on components deformations. We perform the study using controller, Janome screwdriver, ZEISS measurement system and SEM (Scanning Electron Microscope) used in the metallographic analysis. Various torques applied on the screws give us results that show the interactions and their impact on the pumps components, producing deformation on the screws, gaskets and even front plate. A trial was done to establish the upper limit for the torque of the screws. After applying several tightening's at different values, we conclude that the optimal upper limit for screws tightening between front plate and housing is 45 Nm. Applying a torque bigger than 50 Nm, the screws are destroyed, shear. This can generate numerous defects on the sealing gasket area and on the front plate in the screws holes area. The circularity of the front plate is impacted by the thickness of the wall and the inner / outer diameters.

1. Introduction

The high-pressure pump was created and designed as a pump that uses detachable assemblies. Although these removable assemblies are generally more expensive than non-removable ones, they are widely used due to the many advantages they offer.

This type of pump allows its assembly and disassembly in times to evaluate the components inside the pump. The use of large-scale demountable assemblies allows their use in complex situations, throughout the industry such as pressure vessels, high-pressure pumps, tubular structures, etc. Circular flange joints are subjected to tensile forces and therefore the design of such a joint requires knowledge of several areas, knowledge of the type, material from which the flange, housing, pipe or component to be assembled is made [1].

In the specialized literature, we find numerous studies and analyzes regarding the joints made with circular flanges. A simple method used to find the forces acting on a flange is to transform them into a torque operated on the edges of the flange [2-4].

The aim of our paper is to present the interaction between the components of the pump, to establish the maximum acceptable limits and to observe their impact on the components.

2. Constructive considerations

Literature presents a general solution of a joint with a flange that is subjected to axial forces, a solution that can be applied to a variety of circular joints that use flanges. Figure 1 presents the general case of a circular flange subjected to tension loading from a tube of radius R wherein the force T, the applied tensile force is the force evenly distributed along a circumference of the radius of the flange (screw area) [5-7].

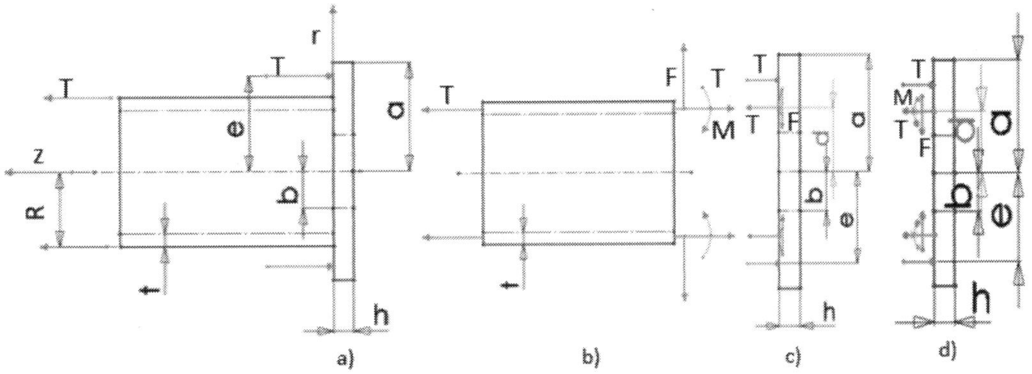

Figure 1. Modelling of a general circular flange joint [1].

The screws (figure 2) used in tightening the joint flanges are produced in different classes of properties depending on their applicability. The classes are divided by points for example 8.8, 10.9 or 12.9. The first digit indicates 1/100 of the nominal tensile strength measured in Newton per square millimeter. The second digit indicates ten times the ratio between the stress at 0.2% non - proportional elongation Rp 0.2 and the nominal tensile strength.

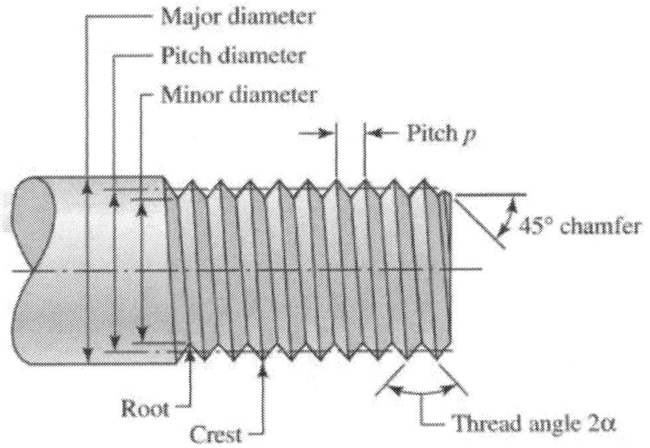

Figure 2. Thread standards.

The fixing areas are three critical points in the thread coupling zone exemplified as A, B, and C in figure 3 a). Point A is generally ignored because we have no relative movement between the screw and the nut or between the contact surface of the housing and the front plate after assembly and there is no risk of wear. Point B is critical because highlights potential ways of stripping and fatigue of first coil of the thread, as well as point C.

Figure 3 b) presents the fourth critical point that is point D which is placed at the boundary between the assembled components (housing and front plate) and must be taken into account only if the joint is subjected to shear force.

IOP Conf. Series: Materials Science and Engineering 1235 (2022) 012069 doi:10.1088/1757-899X/1235/1/012069

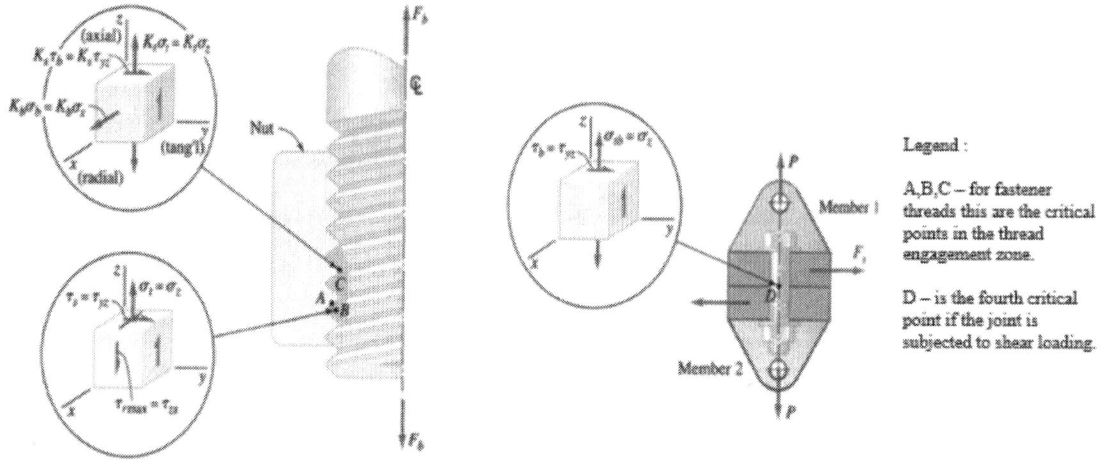

a) b)

Figure 3. Critical points of screw [1].

The screws tightening is influenced by several factors, including the tightening method. One of these methods would be to control the torque reached by the screw, in which case it is monitored and controlled. Another method would be to control the angle. The torque control method uses the torque wrench or monitored spindles depending on the friction between the components. The variation of these coefficients of friction may vary between 0.08 and 0.16 (friction coefficient class B in VDO 2230). The tightening with a specific torque can lead to a widespread of the tightening force (VDI 2230-2003) [8,9].

The screws analyzed are shown in figures 4 and 5. The screw in figure 4 is a Metric 7, class 10.9 and it's used to assembly the front plate on the housing of pump. The screw in figure 5 is used to fix the hydraulic head on the housing body.

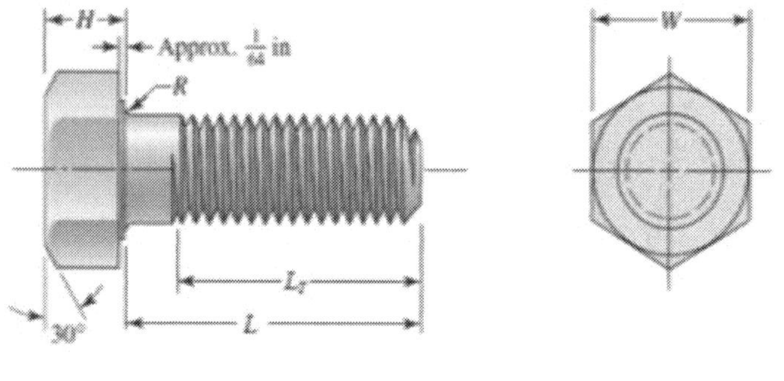

Figure 4. Screw front plate to housing.

Figure 5. Screw hydraulic head to housing.

3. Results and discussion

Using Janome screwdriver several trials were performed using different torque sequences in order to establish the upper and lower limits of screwing torque process. These screwdrivers are commonly used in different fields of activities because torque and angle values can be recorded for each screw. Also, depending on the field of activity in which they are used, different programs can be created in order to set a specific value where the screwdriver stops the screwing process. Changing the angle,this can affect the screw torque and the final preload. To have a better understanding of the impact generated by obtained values, we performed several tests to establish the admissible limits. These tests were performed on an experimental bench using specific fixtures in order to keep the pump in the same position throughout the test. Two programs have been made adapted for the Janome screwdriver to record the torque of thread, one program for hydraulic head screw to housing and the second for screw front plate to housing.

To be sure, the conformity of the parts is respected according to international standards, without cracks, we performed various measurements with SEM. The first analysis performed was to check the microstructure of the screw. The results of the analysis (figure 6) show that the microstructure of the screw is correct for this hardened steel screw and the heat treatment is similar to all the samples analyzed. There are no signs of decarburization on the screw surface [10,11].

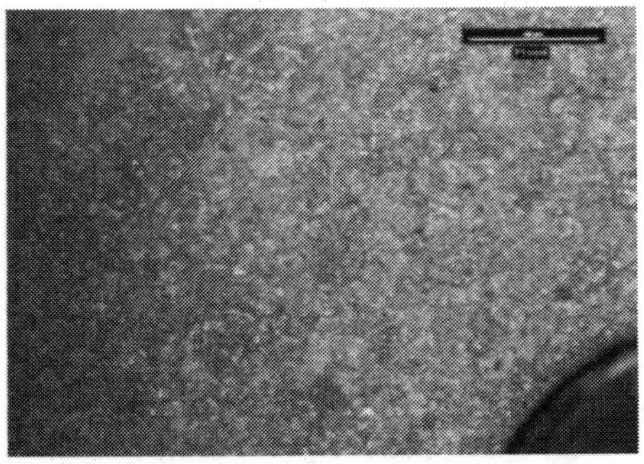

Figure 6. Screw microstructure.

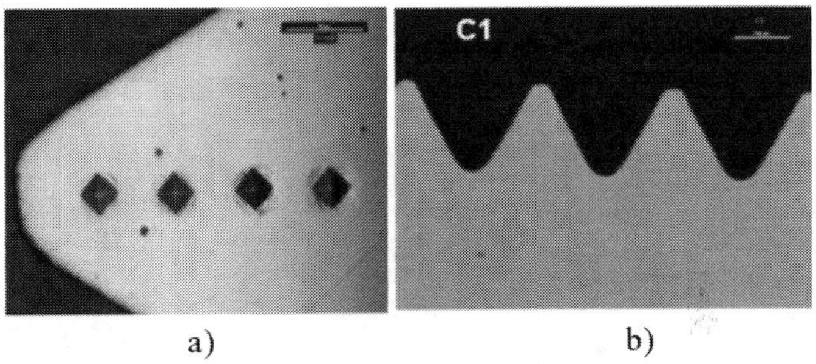

Figure 7. a) Screw microhardness; b) Polished sample.

Figure 7 shows another analysis, the microhardness of the screws, to determine the properties of metallic materials based on hardness values. The HV05 microhardness has been converted to HRC and the values obtained are within the standard requirements. In figure 7 b) it can be seen the image of a polished sample where the aspect and integrity of the material look homogenous, without any pre-existing crack or microfracture for all the analyzed samples.

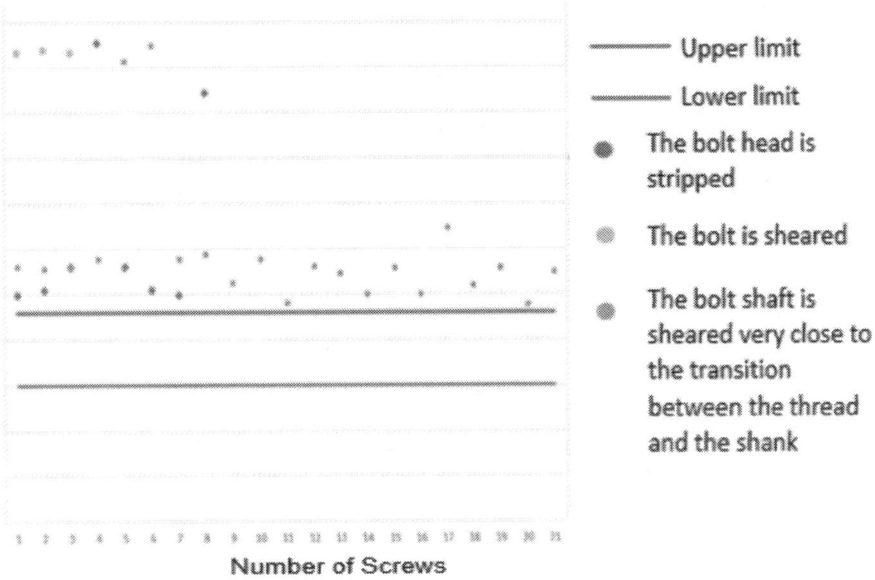

Figure 8 . Values obtained at torque for screw hydraulic head to housing.

After making sure that the screws that fix the components are in accordance with the imposed standards, we performed the first tests to determine the limits of the screw regarding the torque and the angle at which they must be tightened. The experimental results obtained after applying a higher torque have shown that there are three possible defects that we can generate to the screws.

Tests made for the hydraulic head to housing screw revealed three possible defects. One of these defects was the shearing of the screws, another defect was the shearing of the screw head and the third defect created to the screw was sheared very close to the transition between the thread and the shank [12,13]. The recorded torque for screw hydraulic head to housing can be found in figure 8, the values were achieved on twenty-one screws.

The same tests were performed for the M7 screws used between the front plate and the pump body to confirm the limits on tightening. The recorded torques can be found in figure 11.

All three types of defects exemplified in the above graph can be found in figure 9. Applying a higher torque to the screw, the bolt is stripped figure 9 a), the second defect may be the bolt head is shear figure 9 b) and third defect, figure 9 c) when the bolt shaft is sheared very close to the transition thread-shank.

a) b) c)

Figure 9. Type of failure of screws.

It has been shown that by applying a torque of more than 50 Nm the screws are destroyed and the defects can also appear on the seal area.

Figure 10 shows the defects that may appear on the front plate, in the area where the three screws are placed.

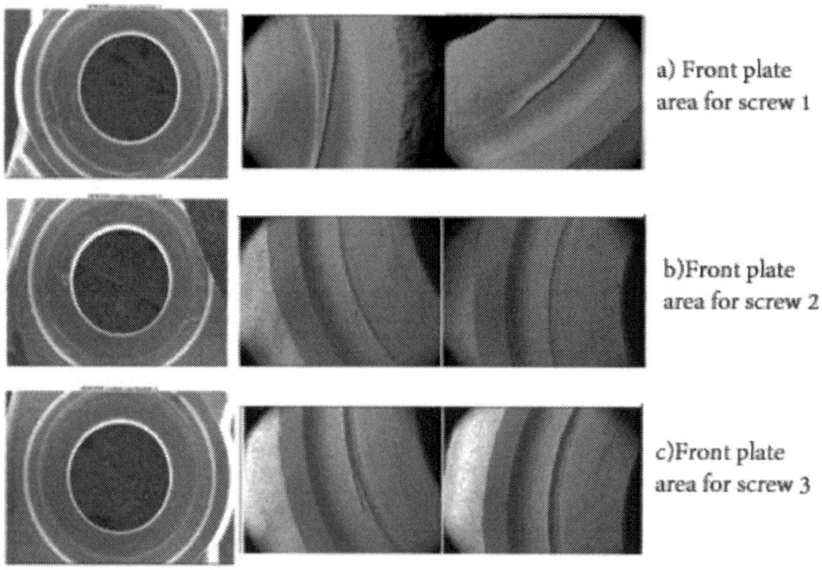

Figure 10. Type of defects on front plate.

To verify that we have selected the right limits without having a negative impact on the functionality, two pumps were assembled with these screws. Figure 11 shows the values obtained at torque: one pump has the screws on lower torque limit for the hydraulic head and for the front plate and the second pump has the screws on the upper limit and tested on the rig to simulate the conditions from the engine.

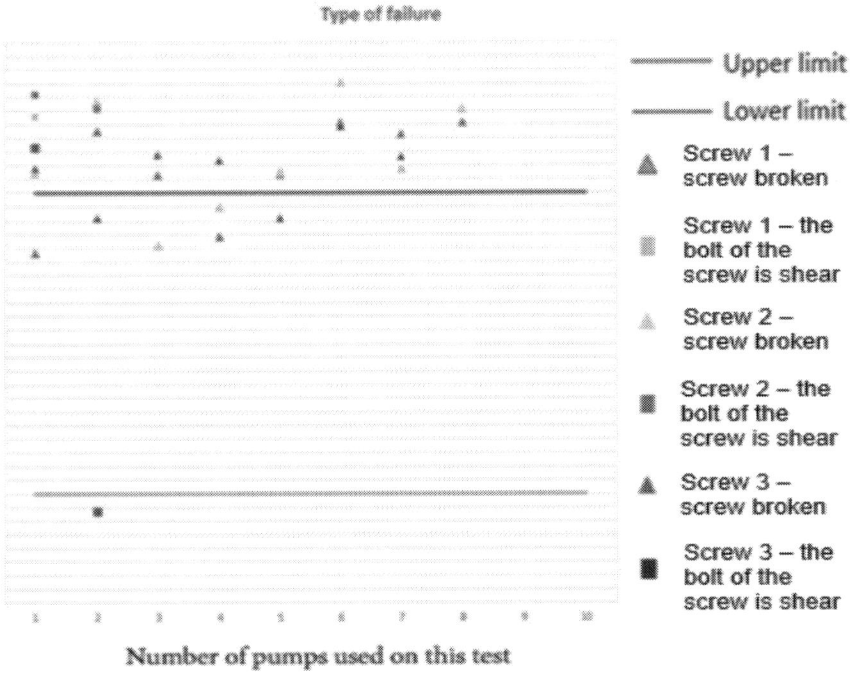

Figure 11. Values obtained at torque for screw front plate to housing.

4. Conclusions

During the assembly of the high pressure pumps, when the screws are tightened we can generate a series of defects on the components, if the tightening limits of the screws are not controled. In order to find the tightening limits, we performed a series of analyzes and tests. After we have identified the differences between two customers, several tests were performed on the pumps in order to establish the lower and the upper limit. With the limit defined, some pumps were assembled and tested during 1400 hours. At the end of the tests the pumps do not show leaks/wear between components and the screws maintained the initial toque without any impact generated by vibrations from the engine.

ORCID IDs
D P Bistriceanu https://orcid.org/0000-0001-8578-9749
C Bujoreanu https://orcid.org/0000-0002-5680-2788
S G Pal https://orcid.org/0000-0002-9934-211X

5. References

[1] Cao J and Bell A J 1993 Elastic analysis of a circular flange joint subjected to axial force *Int. J. Pres. Vessel. Pip.* **55** 435-49 https://doi.org/10.1016/0308-0161(93)90062-X
[2] Park S J, Park S D, Kim S K and Hyun Cho H J 2006 Reliability evaluation for the pump assembly using an accelerated test *Int. J. Pres. Vessel. Pip* **83** 283-6 https://doi.org/10.1016/j.ijpvp.2006.02.014

[3] Speck A J 2015 *Mechanical Fastening Joining and Assembly* (Boca Raton: CRC Press) https://doi.org/10.1201/9781351228640
[4] Walter G 1993 Kunststoffe und *Elastomere in Kraftfahrzeugen (Plastics and Elastomers in Motor Vehicles). Verlag W. Kohlhammer*, Stuttgart, ISBN 3-170-08833-5 (in German)
[5] Young W C 1989 *Roark's Formulas for Stress and Strain* 6th edition (New York: McGraw-Hill) pp. 391-486, 533-4
[6] Drago J 2008 *World Pumps* Publisher European Committee of Pump Manufacturers Elsevier 498 pp 25-27
[7] Cao J and Bell A J 1992 General solutions for a circular flat plate with a central hole loaded by transverse force or bending moment uniformly distributed along a circumference *Int. J. Pres. Vessel. Pip.* **51** (2) 155-73 https://doi.org/10.1016/0308-0161(92)90078-T
[8] Zhang Z, Wang D and Guo Y 2019 Fretting friction and wear behaviors of spiral wound gasket (SWG) sealing surface *Tribol. Int.* **133** 236-45 https://doi.org/10.1016/j.triboint.2019.01.017
[9] Aibada N, Manickam R, Gupta K K and Raichurkar P 2017 Review on various gaskets based on the materials their characteristics and applications *Int. J. Text. Eng. Process.* **3** (1) 12-8
[10] Astalosch C 2009 Comparison betwenn EPDM - and TPE-extrusion. Study for the company Fornix d.o.o., Dugi Rat/Croati Engineering and Processes 3
[11] British Standards Institution 1986 DD 133 Draft for Development. Code of Practice for Strength Assessment of Members of Lattice Towers and Masts
[12] Voitik R M 1993 Realizing Bench Test Solutions to Field Tribology Problems *Tribology: Wear Test Selection for Design and Application* pp 45-59 https://doi.org/10.1520/STP15964S
[13] Bayer R G 2002 Wear Analysis for Engineers *Ind. Lubr. Tribol.* **60** (4) https://doi.org/10.1108/ilt.2008.01860dad.004

IOP Conf. Series: Materials Science and Engineering 1235 (2022) 012070 doi:10.1088/1757-899X/1235/1/012070

Research concerning the influences related to the perforated hole diameters in long pipes

D I Poiană, I Ionel and G Drăghici

Politehnica University of Timisoara, Faculty for Mechanical Engineering, PiataVictoriei no 2, 300006, Timisoara, Romania

ioana.ionel@upt.ro

Abstract. The purpose of this paper is to highlight the influence of the dimensional characteristics and mechanical strength characteristics of the pipe material on the diameter of the perforated holes in these long pipes, to maintain the stability of the pipe shape in the perforated area. The research aimed to establish the maximum limit dimensions of holes that can be made in long round pipes with thin walls, through perforating by cold plastic deformation in air, using only a punch applied to on the outside of the pipe, for a particular case of pipe clamping. Following the research, the mathematical relations between the maximum diameters of the holes that can be obtained in round pipes, in correlation with the dimensional characteristics and the mechanical resistance of the material, are presented. The characteristics taken into account are the outer diameter of the pipes, the thickness of the pipe wall, and the permissible resistance to rupture of the part material. In the conclusions, it is specified that the determined mathematical relations allow the construction of variation diagrams and the determination of the maximum possible diameters, for other dimensional characteristics of pipes and for other mechanical strength characteristics of different materials.

1. Introduction

To make the holes in the long tubular parts, several processing processes are used [1-5]: perforation by cold plastic deformation, laser perforation, hydro-perforation, thermal erosion perforation, ultrasonic perforation.

One of the procedures used to make the holes in tubular parts is the perforation by cold plastic deformation in the air environment, with associated edges, a punch applied on to the outside of the pipe and the cutting plate inserted on the inside of the pipe. This variant of perforation by cold plastic deformation, with associated edges is applied only for the holes located at the ends of the long pipes. This variant with associated edges is avoided for making holes arranged along parts such as long pipes with thin walls, because there is no convenient solution to insert a cutting plate on the inside of the pipe in the area of the hole.

The execution of cold plastic deformation, in the air environment, of the holes arranged along the length of the pipe is used less frequently, in which case only a secure cut is used, the punch applied from the outside of the pipe, without the cutting plate on the inside of the pipe. Such holes made with only one edge, the punch, can be made in good condition only for pipes with sufficiently high rigidity, so that the pipe does not deform under the action of the punch.

In the present paper, the influence of the pipe diameter, pipe wall thickness, and the material characteristics on the maximum diameter of the holes and the rigidity of the pipe were investigated.

IOP Conf. Series: Materials Science and Engineering 1235 (2022) 012070 doi:10.1088/1757-899X/1235/1/012070

2. Research hypothesis

The study of the perforation process of long tubular parts, with a singular edge with a punch, is based on the following considerations and recommendations:

- the perforation is accomplished by cold plastic deformation in an air environment, with a singular edge of a punch type;
- a clamping device according to figure 1 is used, thus the part is placed on a plate with a flat surface, the side walls in contact with the part;
- a punch guide plate is used;
- the workpiece is a hold for a length of 80 mm, centred towards the hole to be made.

Figure 1 shows schematically, in this section, the process of making holes by punching applied from the outside of the pipe and the device proposed for use in this case.

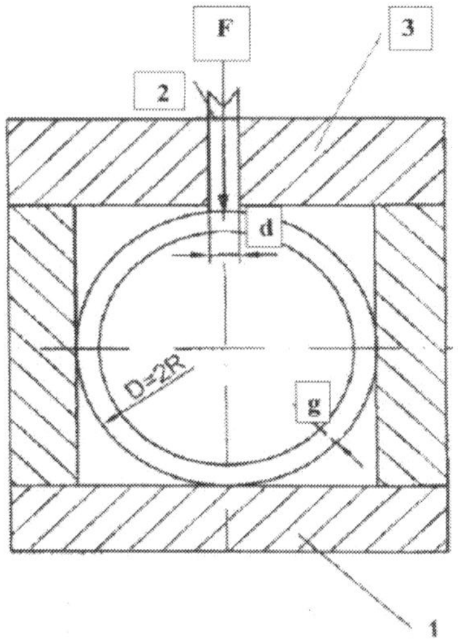

Figure 1. Device for clamping the tubular pipe in the hole area.

The notations are:
1 - the lower plate of the clamping device of the pipe;
2 - the punching punch;
3 - the upper plate of the clamping device and the punch guide;
F - punch actuation force;
D - outer diameter of the pipe;
g - the thickness of the pipe wall;
d - the hole diameter/ the punch diameter.

3. The state of stress and solving indetermination

Because the holes in the round pipe are made by a plastic deformation process, at the moment of impact of the punch on the round pipe, the tensions appear in the workpiece. When the sheer force appears, this one causes the rupture and detachment of the material subjected to the perforation process.

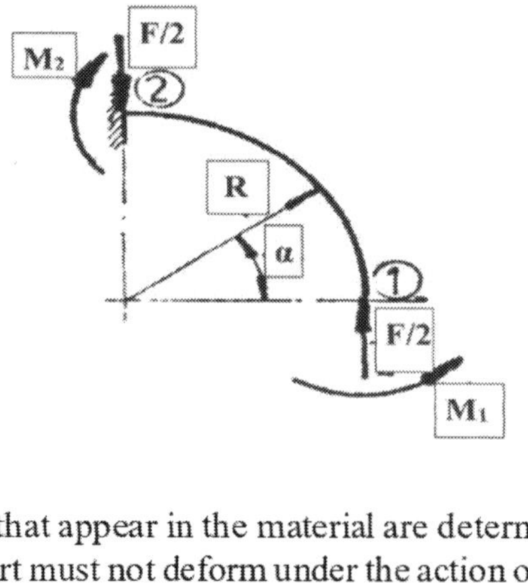

Figure 2. The balance diagram of the forces and moments that appear in the process, in a cross section through the pipe.

The tensions that appear in the material are determined at the moment of the action of the punch on the piece. The part must not deform under the action of the punch. This tension can be determined based on known issues in the strength of materials, using analytical computational relationships and constructing equilibrium schemes of the forces and moments that appear for the considered case.

Figure 2 shows the specific schemes of the strength of the materials taken into account for this case studied, with the forces and moments that appear [6].

4. Mathematical considerations. Interdependent relationships

The mathematical relations between the maximum diameters of the holes that can be obtained in round pipes, depending on the dimensional characteristics of the pipes and the mechanical resistance of the material. For the present case in the study of the perforation process for parts of the type of long pipes, we chose the Alberto Castigliano method, which is one of the energetic methods widely used in the study of cold plastic deformation processes.

The Castigliano theorem is applied to solve the indeterminacy of the diagram in figure 2. In section 1, the zero rotation is considered, and the relation is obtained

$$\phi_1 = \frac{\partial u}{\partial M_1} = \frac{1}{EI}\int M_i \frac{\partial Mi}{\partial M_1} ds = 0 \tag{1}$$

The bending moment in zone 1-2 will have the expression:

$$\mathrm{M}_{1-2} = \mathrm{M}_1 + \frac{\mathrm{F}}{2}\mathrm{R}(1-\cos\alpha) \tag{2}$$

Where, F is the punching force.

Replacing relation (2) in relation (1) results:

$$\int_0^{\frac{\pi}{2}} \left[M_1 + \frac{F}{2} R\left(1-\cos\alpha\right) \right] R \cdot d\alpha = 0 \tag{3}$$

After integration you get:

$$M_1 = -0.18F \cdot R \tag{4}$$

The expressions of the efforts in a certain cross section are:

$$\mathrm{N}_{1-2} = -\frac{\mathrm{F}}{2}\cos\alpha$$

$$\mathrm{T}_{1-2} = -\frac{\mathrm{F}}{2}\sin\alpha \tag{5}$$

$$M_{1-2} = \left(0.32 - 0.5\cos\alpha\right) F \cdot R$$

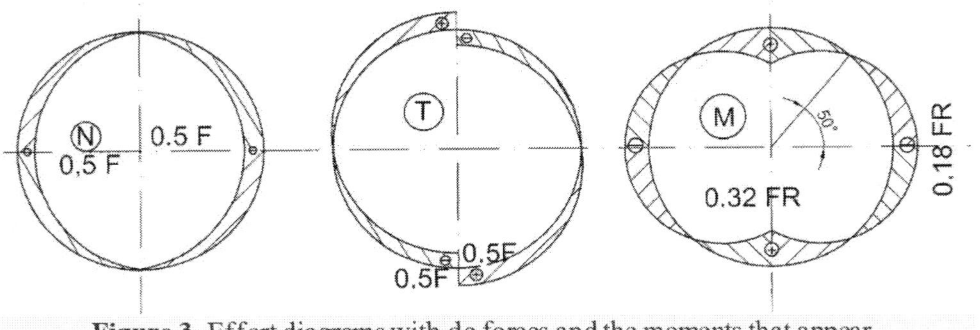

Figure 3. Effort diagrams with de forces and the moments that appear.

The stress diagrams are shown in figure 3. In critical section 2, efforts have values:

$$N_2 = 0$$

$$\mathrm{T}_2 = -\frac{\mathrm{F}}{2} \tag{6}$$

$$M_2 = 0.32F \cdot R$$

To determine the maximum value of the diameter of the holes that can be made by punching with a die without cutting plate, the condition of resistance to the compound stress of the part will be taken into account, on which the drilling force acts and the diagram in figure 3 is considered N, T and M_i.

In the critical section 2, the normal tension σ_2 will be:

$$\sigma_2 = \frac{N_2}{A} \pm \frac{M_2}{Wz} \leq \sigma_{ai} \tag{7}$$

In which:

$$A = \lambda \cdot g \tag{8}$$

$$W_z = \frac{\lambda \cdot g^2}{6} \tag{9}$$

Where, λ is the length of the pipe corresponding to the deformation zone [mm]; g is the pipe wall thickness [mm]; σ_{ai} is the permissible bending strength of the pipe material [MPa].

For the perforation force F, the known relation is considered:

$$F = \pi \cdot d \cdot g \cdot \sigma_r \tag{10}$$

Where, d is the diameter of the perforated hole and σ_r is the breaking strength of the pipe material. By replacing (8), (9) and (10) in (7), relationship (11) is obtained:

$$d \leq \frac{\lambda \cdot g}{1.91 \cdot \pi \cdot R} \cdot \frac{\sigma_{ai}}{\sigma_r}, \text{ [mm]} \tag{11}$$

The deducted relations (11) can be verified in the future by experimental research by various other methods appropriate to the case studied, such as example, the finite element method (MEF), for example, using Solid Works and/or Forge7.

5. The variation curves of the maximum diameter of the holes based on the mathematical relations

Based on the relationship (11), the values were drawn up in table 1 and figure 4. Thus, one indicates the relations existing between the diameters of the perforated holes, the dimensional and material characteristics of the perforated part, in case the piece is placed on a device with a flat surface, according to figure 1.

Table 1. Results concerning calculated values for the diameters according to different wall thickness.

Pipe wall thickness g [mm]	D=15, Maximum diameter of the perforated hole d [mm]	D=20, Maximum diameter of the perforated hole d [mm]	D=25, Maximum diameter of the perforated hole d [mm]	D=30, Maximum diameter of the perforated hole d [mm]	D=40, Maximum diameter of the perforated hole d [mm]	D=60, Maximum diameter of the perforated hole d [mm]
0.5	0.4042	0.30315	0.24252	0.2021	0.15157	0.10105
1	0.8084	0.6063	0.48504	0.4042	0.30315	0.2021
1.5	1.2126	0.90945	0.72756	0.6063	0.45472	0.30315
2	1.6168	1.2126	0.97008	0.8084	0.6063	0.4042
2.5	2.021	1.51575	1.2126	1.0105	0.75787	0.50525
3	2.4252	1.8189	1.45512	1.2126	0.90945	0.6063
3.5	2.8294	2.12205	1.69764	1.4147	1.06102	0.70735
4	3.2336	2.4252	1.94016	1.6168	1.2126	0.8084

4.5	3.6378	2.72835	2.18268	1.8189	1.36417	0.90945
5	4.042	3.0315	2.4252	2.021	1.51575	1.0105
5.5	4.4462	3.33465	2.66772	2.2231	1.66732	1.11155
6	4.8504	3.6378	2.91024	2.4252	1.8189	1.2126
6.5	5.2546	3.94095	3.15276	2.6273	1.97047	1.31365
7	5.6588	4.2441	3.39528	2,8294	2.12205	1.4147

Figure 4 shows the corresponding curves for parts of the long pipe type, with the dimensional characteristics:

- Outer diameter of the parts of: $D = 15, 20, 25, 30, 40, 60$ mm,
- The wall thickness of the pipes for each diameter considered was $g = 0.5$ - 7 mm.

For the length of the device, one selected $\lambda = 80$ mm.

The characteristics of the material taken into account for the material used in automotive industry are: $\sigma_r = 330$ MPa and $\sigma_{ai} = 150$ MPa [2,7,7,8].

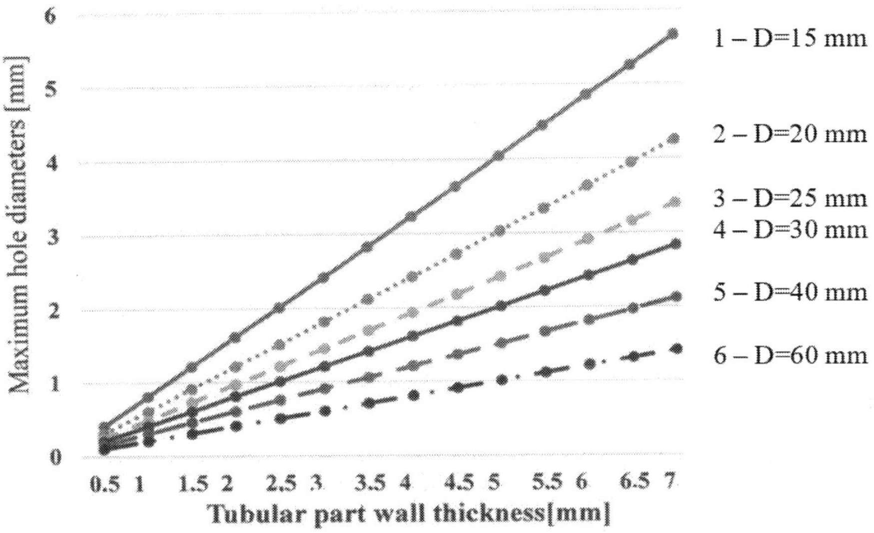

Figure 4. Variation curves of the maximum diameter of the perforated holes, depending on the dimensions of the parts, outer diameters, and the thickness of the wall, under the conditions of maintaining the rigidity of the part, in the area of the perforated hole.

By analysing the diagram in figure 4, it is observed that for the pipes with a wall thickness of 1 mm, the following maximum punch diameter results:

- ϕ 0.8 mm → for the pipes with D = 15 mm,
- ϕ 0.6 mm → for the pipes with D = 20 mm,
- ϕ 0.4 mm → for the pipes with D = 25 mm,
- ϕ 0.4 mm → for the pipes with D = 30 mm,
- ϕ 0.3 mm → for the pipes with D = 40 mm,
- ϕ 0.1 mm → for the pipes with D = 60 mm.

For the pipes with a wall thickness of 7 mm, the following maximum punch diameter results:

- ϕ 5.6 mm → for the pipes with D = 15 mm,
- ϕ 4.2 mm → for the pipes with D = 20 mm,
- ϕ 3.3 mm → for the pipes with D = 25 mm,
- ϕ 2.8 mm → for the pipes with D = 30 mm,

- ϕ 2.1 mm → for the pipes with D = 40 mm,
- ϕ 1.4 mm → for the pipes with D = 60 mm.

6. Conclusions

The research scope is confirmed, because a general relationship was established, which allows the calculation of the maximum limits of the diameters for the perforated holes in round pipes, for different types and dimensions of material.

It was applied, for the case referring to the clamping part. It proofs the interdependence between:

- the diameter of the holes, respectively, the diameter of the punch, d;
- the outer diameter of the pipe, D;
- the pipe wall thickness, g;
- the characteristics of the material, σ_r and σ_{ai}.

Based on this relation (11), specific diagrams were deducted. They allow determining the maximum diameter of the holes, depending on the thickness of the pipe wall, for certain pipe diameters.

The determined relation (11) allows the construction of similar diagrams as well, for pipes with different outer diameters and for other types of materials, with different mechanical characteristics.

Acknowledgements

The paper is accomplished in the frame of the PhD research activity of the first author, at the Politehnica University of Timisoara.

ORCID IDs

D I Poiană https://orcid.org/0000-0002-4417-2803
I Ionel https://orcid.org/0000-0002-2999-1210
G Drăghici https://orcid.org/0000-0002-3003-5015

7. References

[1] Poiana D I and Ionel I 2020 Study of modern perforation processes specific to the automotive industry IOP Conference Series: Materials Science and Engineering 898 012043 https://doi.org/10.1088/1757-899X/898/1/012043
[2] Koç M 2008 *Hydroforming for advanced manufacturing*, Woodhead Publishing and Maney Publishing on behalf of The Institute of Materials, Minerals & Mining (Boca Raton Boston New York Washington DC: CRC Press)
[3] Neugebauer R 2007 *Hydro-Umformung* (Berlin Heidelberg: Springer) https://doi.org/10.1007/978-3-540-49013-5
[4] Singh H 2003 *Fundamentals of Hydroforming* (Michigan, Society of Manufacturing Engineers) pp 219
[5] Kumar R and Hynes N R J 2019 Thermal drilling processing on sheet metals: A review *International Journal of Lightweight Materials and* Manufacture **2** (3) 193-205 https://doi.org/10.1016/j.ijlmm.2019.08.003
[6] Tripa P and Hluşcu M 2006 *Material resistance (*original in Romanian: *Rezistenta materialelor)* computerized writing (Timişoara: Mirton) pp 24-351
[7] *Pipe steels* (original in Romanian: *Oteluri pentru ţevi*, SR EN) http://www.utchimsteel.ro/date-tehnice/Calitati-de-oteluri-uzuale-pentru-tevi-de-otel-2 Accessed in 05.01.2020
[8] *Aluminum and its alloys. Types of aluminum alloys* (original in Romanian: Aluminiul si aliajelelui. Tipuri de aliaje de aluminiu) https://www.alumil.com/extrusion/ro/info/knowledgebase-in-aluminium Accessed in 05.01.2020

IOP Conf. Series: Materials Science and Engineering 1235 (2022) 012071 doi:10.1088/1757-899X/1235/1/012071

Using 3D scanning in assessing the dimensional accuracy of mechanically machined parts

Cezarina Afteni[1], Viorel Paunoiu[1], Mitica Afteni[2] and Virgil Teodor[1]
[1] "Dunarea de Jos" University, Department of Manufacturing Engineering, Domneasca Street 111, 800201 – Galati, Romania
[2] Rulmenti S.A., Barlad, Republicii Street 320, 731108 – Barlad, Romania

cezarina.afteni@ugal.ro

Abstract. In recent years, productivity and competitiveness have played a major role in manufacturing. To improve these aspects, many companies use computer-aided inspection (CAI) technologies. Inspection of parts, using 3D scanning technologies, is one the most modern and competitive ways to ensure the quality control of products. This paper presents a study on the assessment of the machined parts' dimensional accuracy. An Atos Core device for 3D scanning of the part was used. The Atos device allows a non-contact inspection of the part's entire surface, generating a very dense *point cloud,* which stores, with great accuracy, all the details of the inspected object. The machined part is compared with its numerical model, obtained by computer-aided design (CAD). The accuracy of the machining was proven to be very good, and the inspection's result proves that the dimensions of the piece fall within the tolerance limits. A complete inspection of the piece, using 3D scanning, guarantees a fast and accurate quality control, thus reducing the production times.

1. Introduction

In recent years, the elements' reconstruction, based on existing physical parts and reverse engineering, plays an important role in research and everyday life.

Since the early 1990s, productivity and competitiveness have been the main concerns of the global manufacturing industry. To meet these goals and meet the challenges, companies have adopted advanced manufacturing technologies such as *i)* computer-aided design of technological processes, *ii)* computer-aided manufacturing, and *iii)* flexible manufacturing systems, etc.

Due to the pressure of international competition and the global market in the 21st century, there is an increasing need to reduce the production costs, as well as the marketing time of a new product, without diminishing its quality and functional properties.

Quick response and flexibility, together with the new business opportunities are some of the most important factors influencing competitiveness. To meet these challenges, advanced manufacturing technologies involve computerization and automation of all stages of the product's development process. Advanced manufacturing technologies are rapid prototyping, reverse engineering, and advanced CNC processing methods. These require the direct involvement of computer-controlled systems.

3D scanning is a new technology with a lot of exciting possibilities and diverse applications. By 3D scanning process is obtaining information about the geometry of a physical object. This process is also known as *digitization.*

Digitization/3D digitization is a method that uses a contact or non-contact digitization system to inspect the surface of an object and rebuilds him in a virtual computer space through a network of points (xyz), in the form of a 3D graphic representation.

In recent years, 3D scanning technologies applications have increased approximately more than ten times. Many papers treat 3D scanning of various mechanical objects for quality assurance. Paper [1] presents a methodology for assessing the accuracy of digitization using 3D optical scanners. The author deals with practical implementation of an acceptance test for ATOS contact-less 3D scanner, designing and manufacturing its own test etalon for the determination of its nominal dimensions, by digitization. In [2], a 3D scanning methodology is used for measuring the part geometry in different stages of the laser metal deposition process. In this way, the manufacturing process is accomplished with the geometry control of the part. The result will be a better process control with favourable effect toward the part quality. The manufacturing process will be more stable and the reproduction of the part geometry, from part to part, will be assured. This will lead in saving manufacturing time and cost. In automotive industry, using 3D scanning technology helps the manufacturers to obtain accuracy for large and complex parts and dies specific for this type of industry. Two case studies are presented in [3], showing how 3D scanning technology can help in achieving the customers' requests and expectations by providing data for applications as cutting, metal forming, tools and devices and other parts with low tolerances. The first study approaches the scanning process applied in case of a re-manufacturing process of a metal cutting die damaged. In the second study the scanned point clouds data were compared with the 3D CAD data. This comparison was developed for inspection purposes. In [4], the author evaluates the effect of optical work parameters of 3D scanners on the accuracy of the geometry of the vehicle element – the rear door of the Skoda Octavia in two variants, non-deformed and deformed. The GOM ATOS optical 3D scanner was used during measurements. In [5], the authors present reverse engineering, which is focused on obtaining a digital version of an object, using 3D scanning technology. Park and Chang propose in [6] a scanning procedure using a structured light system. In order to obtain a complete model using the structured light system, it is necessary to perform multiple scans of the objects, because in structured light system only the visible areas can be measured from a specific direction. In order to identify the missing areas, it was proposed a procedure that uses multiple range images (scanning data) and try to find additional scan guidelines to fill in the missing area. In [7], the authors approach the mechanical object 3D meshes, in order to have an automatic retro-engineering process. The authors define this new method, which consists of the following three main steps: *i)* extraction of elements, *ii)* building of elements - lines, curves, and *iii)* rebuilding 3D model. Beccari *et al.* present in their paper, [8], a method of reverse engineering for fast, simple, and interactive acquisition and rebuilding of a virtual 3D model. They propose an active stereo acquisition system, based on two infrared cameras and a wireless active-pen device. To create the polygonal network, the wireless device tracks the profile of the 3D part in different sections. The system automatically produces the polygonal network, which is then processed on smaller surfaces. This solution allows the scanning of complex surfaces. From industrial perspective, a number of sixteen applications of 3D scanning were identified and discussed by Javaid *et al.* in paper [9]. In general, the technology provides possibilities to conduct reverse engineering by scanning the object and transforming it into CAD with different surface reconstruction techniques. Helle and Lemu reported in [10] a case study conducted to investigate the challenges of 3D scanning process for reverse engineering and product quality control. Krznar *et al.* shows in [11] a case study of developing the 3D CAD model of the fixture that holds parts during 3D scanning for many similar parts.

This paper presents a study on the assessment of the dimensional accuracy of mechanically machined parts using 3D scanning. The main objective of this paper is to evaluate the deviation and the dimensional accuracy of 3D scanned parts compared to the 3D CAD model. The main reasons for using 3D scanning, apart from its physical reproduction, are the accuracy and time saving, compared to manually recreating a part in software alone. An Atos Core device for 3D scanning of the part was used. This device is used in various industries, such as in automobile factories, aeronautics, reverse engineering, topography, research and education institutions, and others [9]. After obtaining the 3D model, its dimensional analysis was performed using the GOM Inspect software color map, which evaluated the dimensions and the geometric deviations. Comparing the manual measurement of the part and then design with the process of digitizing the part and then creating the CAD model using 3D scanning shows the advantage of less time consuming and also improved accuracy.

The paper is organized as follows: Section II describes the 3D scanning technology and presents the steps of 3D scanning. Section III shows the device used to scan the parts, namely ATOS Core 200. Section IV describes the 3D scanning process. This includes two main steps: *i)* scanning the part and *ii)* the dimensional evaluation and comparison of results based on the part design. Finally, section V presents the paper's conclusions.

2. Technology of 3D scanning

Technological developments have brought many benefits in many areas of activity. One of the most used and appreciated technologies is 3D scanning. 3D scanning objects to digitize them is becoming increasingly common. This is a potential tool for increasing productivity while securing quality in product development [3].

Using 3D scanning, the physical objects are transposed in a digital format based on a data point collection. These data are used to build the 3D models [12].

In the industrial field a threshold is needed for high data quality. Technical requirements in the execution drawings impose very strict conditions (deviations located on the threshold ±0.001 mm to ±0.01 mm). Although there are many manufacturers of equipment for 3D scanning today, very few of them manage to provide good data quality that can be used in industrial applications.

3D scanning is used in the manufacturing stage for applications such as:

- *Tooling design for manufacturing process*: reverse engineering of dies, fixtures, jigs and patterns, update the CAD file to reflect as-built tooling measurements, and tooling validation/Inspection.
- *Assembly and production*: virtual assembly, tool/robot path programming, and part assessment before machining, and
- *Quality control*: first inspection, part to CAD inspection, and supplier quality inspection.

In 3D scanning, processing the scanned object is transposed into digital geometric shape of *point cloud/mesh* type. The resulting files after the scanning can be used in different activities or applications. In order to obtain the digital shape of a 3D object a variety of technologies were developed. These techniques work with different types of sensors: optical, laser scanning, thermal and seismic acoustic, radar. It's already known the classification of inspection equipment's in contact and non-contact. 3D contact scanners inspect the object by physical touch, the object being positioned on a precision flat surface plate. 3D non-contact scanners can be classified into two categories as: active and passive. The active sensors emit radiation or light and detect the reflection or radiation passing through to probe an object or environment. The passive technique does not emit radiation or light, these techniques rely on detecting reflected ambient radiation [12].

In figure 1 the steps of non-contact active 3D scan are presented.

Step 1: *Hardware Setup*, ensure that the connection between the Atos Core 3D scanner and laptop is tight and the USB dongle is tightly inserted into the USB port.

Step 2: *Part Preparation*, apply some reusable adhesive to the bottom of the part and attach the part at the centre of the rotary table.

Step 3: *Launch Scanning Software*, power on the laptop, login account *user*, launch the scanning software *GOM Scan* and ensure the ATOS Core 3D scanner is ON (a blue light in the front panel of the 3D scanner appears).

Step 4: *3D Scanning*.

- *Create new project*: pull-down menu: File → New Project.
- *Cross lines alignment*: check the black and white cross lines at the Camera window and make sure that the bounding rectangles of the two crosses are overlapped and if the two crosses are not overlapped well, then adjust the alignment by relocating the tripod of the scanner.

- *Exposure adjustment*: adjust the knob (by turning the middle mouse button) as shown in the lower left-hand corner of the Camera window so that no red colour is shown in the captured image.
- *Scanning*: press the *Space* bar of the keyboard to start the scanning process, and the *Measurement Delay* will be displayed, if the *Measurement Error* dialog is displayed, then please click on the button *Use*. After the first scanning is complete, the scanned model will be displayed in the graphics screen.

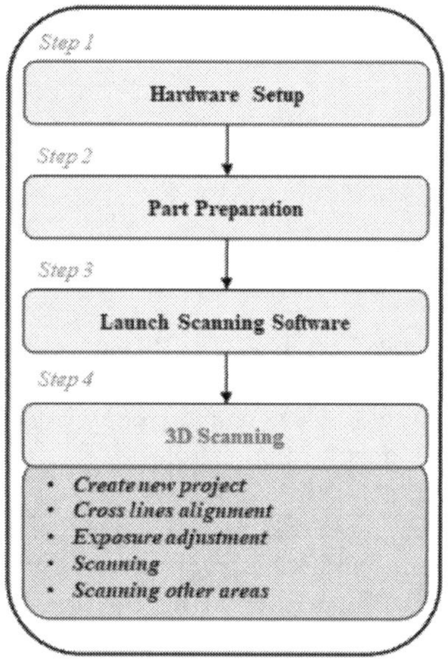

Figure 1. Steps of the 3D scan.

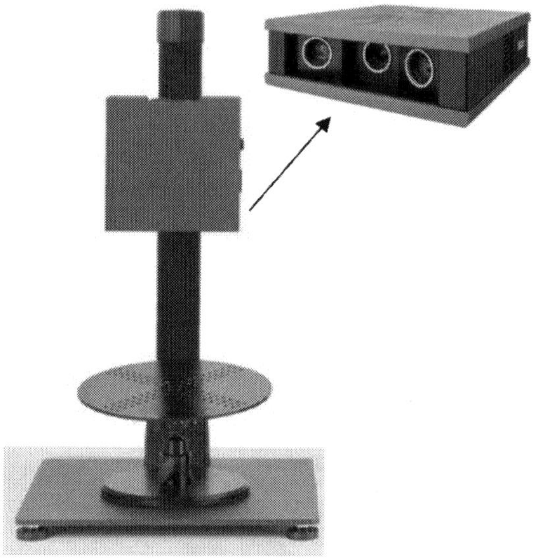

Figure 2. Atos Core - 3D scanner.

- *Scanning other areas*: turn the rotary table by an angle (say, 10° – 15°) so that the consecutive area of the object can be viewed by the camera.
- *Save scanned model file*: combined scanning results by pull-down menu: Acquisition → Measurement Series → Polygonize and Recalculate. Select the appropriate polygonization resolution, say *Standard*.
- *Save the GOM project*: pull-down menu: File → Save.
- *Select drive* for storing of project file, and then enter your desired project name.
- Export combined mesh model to *.STL* file by pull-down menu: *File → Export → Mesh → .STL*.
- Specify the *.STL* filename at Drive D by clicking the *right arrow* button of *Filename* edit box, and then enter the desired *.STL* filename and please make sure that the *Export in one file* is checked ON, click the *OK* button to create the specified *.STL* file.

3. 3D scanner – ATOS Core

ATOS Core 200 is an optical 3D scanner. This scanner is based on fringe projection, with LED Blue light technology, which delivers accurate and traceable 3D coordinates. The mentioned scanner is produced by GOM Germany, Spectromas being the main Romanian distributor, figure 2.

Using the ATOS Core scanner, the medium or small object can be three-dimensional measured.

The active working range (available measuring volume) is:

- Minimum L 180 x l 140 x h 1140 mm;
- Maximum L 220 x l 160 x H 160 mm.

The accuracy of measuring the 3D coordinates is minimum +/- 10 µm on the available measuring volume.

The measuring tasks as simple 3D scanning or fully automated measurement/inspection processes use a basic element which a sensor [13]. The obtained files resulting after the scanning can be used in different applications as: dimensional inspections, reverse engineering, 3D printing and digital archiving. The digital archived data can be then used to develop the products and processes and to optimise the lead times [16].

The main stages in 3D scanning are, according to [14]:

- *Sensor positioning and scanning*. ATOS scanner is positioned for captured the part area. This could be done in a step or in multiple steps. The result will be a 3D *point cloud*.
- *Inspection and evaluation*. The 3D point cloud is automatically transformed in a polygon mesh which represents the part surface. This surface could be compared with 3D model of the part. Also, 2D analysis of different sections of the part could be made.
- *Results and reports*. All the data could be saved in different formats, including *pdf*, and used for different application. The software assures the obtaining of detail reports of measurements, images, diagrams, etc.

4. 3D scanning process

The scanning process includes two steps:

- Step 1. Scanning the part.
- Step 2. Dimensional evaluation and comparison of the results based on the execution drawing.

4.1. Scanning the part

To prove a good efficiency of the ATOS Core 3D scanner, a sample part was chosen. This part was made by mechanical processing, figure 3.

Figure 3. The analysed part.

Before making the scanning, it was necessary to clean the part. In order not to reflect the light and to get the right point cloud, it was necessary to coat the part, using a special powder, in red colour in this case.

The reference markers were positioned on both the table of the scanner and on the part, so that they can be recognized by the scanner as reference points. Before each measurement, it is necessary to perform a calibration/verification of the ATOS Core 3D scanner.

Some reusable adhesive was applied to the bottom of the part and the object was attached at the centre of the rotary table. The part was scanned in three positions (figure 4).

(a) (b) (c)

Figure 4. Positioning the part (a) front view; (b) top view; (c) rear-view.

For the first position, front view, 13 scans were performed (figure 4.a). For the second position, top view, 11 scans (figure 4.b), and for the third position, rear-view, 7 scans were performed (figure 4.c). After each captured image, the table on which the part is positioned is rotated at a 15° angle so that the camera's focus can obtain part views from all angles. After the three scanned positions follows the process of polygonization by command *Polygonize and Recalculate*. With this function, the device calculates the network of inspected points on the surface of the part. The polygonization process merges/overlaps scanned images into a mesh.

This part of measuring process is called registration. By registration all the captured images are correlated to obtain the complete numerical model of the part.

The reference thus obtained is saved as a file with the extension. STL (Standard Tessellation Language) to allow accessing it with several programs such as Catia, Inventor, AutoCAD, etc. and including 3D printing programs. Figure 5 shows the mesh of the part obtained after scanning.

Figure 5. Mesh after scanning.

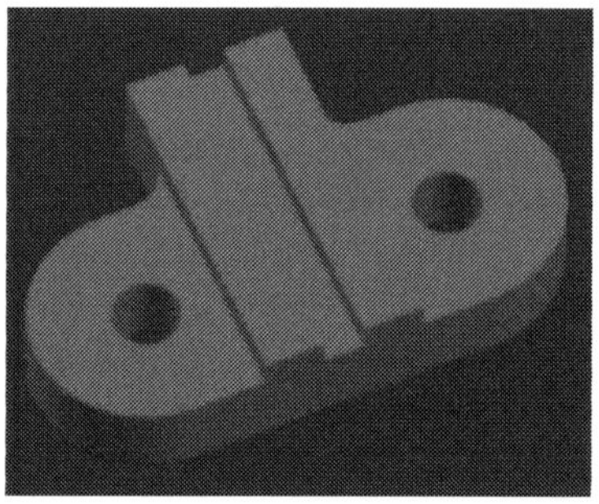

Figure 6. 3D representation of the part.

4.2. Dimensional evaluation and comparison of results based on the execution drawing

Starting from this execution drawing, the three-dimensional numerical model was designed in the Autodesk AutoCAD software (see figure 6). The drawing is saved as a file with the *.STP* extension to allow inspection in the GOM software, to obtain dimensional accuracy.

In the GOM measurement software, choose *New project*, and import the two files (*.STL* and *.STP*). In the inspection sheet they are displayed in different directions (figure 7.a). The two models merge into a single reference system, which determines their overlap (figure 7.b).

The comparison of the two models is made. It is preferable that this is done on the CAD model with the help of the following commands: choose the file *Inspection* → CAD Comparisons → Surface Comparisons on CAD → choose a maximum distance to search for the points where the part is inspected.

IOP Conf. Series: Materials Science and Engineering 1235 (2022) 012071 doi:10.1088/1757-899X/1235/1/012071

5. Conclusions

In an integrated production environment, 3D scanners are capable to measure all the necessary parameters in one stage, without errors, and render results in the same way as computer-equipped manufacturing networks.

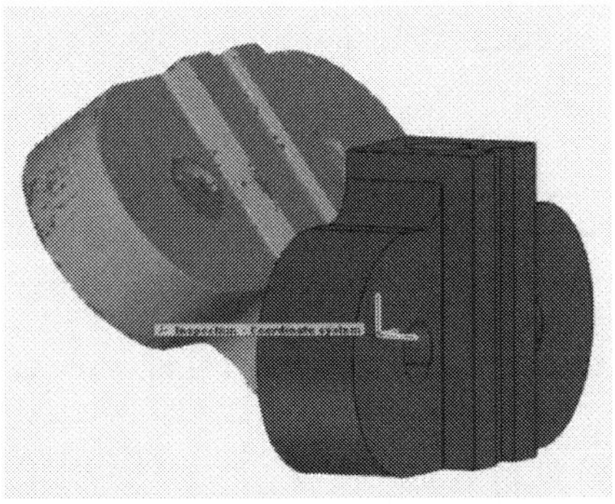
(a)

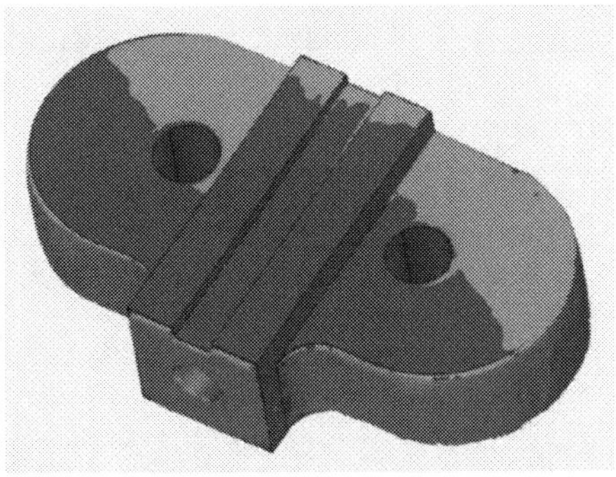
(b)

Figure 7. Models imported into GOM (a), overlapping models (b).

In this paper, the steps for a part inspection used for machining laboratory experiments were presented.

The model was scanned on its entire volume and the results obtained allowed to highlight the following aspects:

- Atos Core 3D scanners allow the entire measurement of the surfaces of prototypes, molds or parts and finished products, regardless of their size.
- The complete measurement of the surfaces, guarantees a fast inspection, thus reducing the production times.
- Atos Core 3D scanners provide accurate measurement data at high speeds, improving chain-from planning to production and subsequent maintenance.

Following the inspection, the accuracy of the part was analysed. The part, obtained by milling, in this case, has some important deviations compared with the CAD model, figure 8.

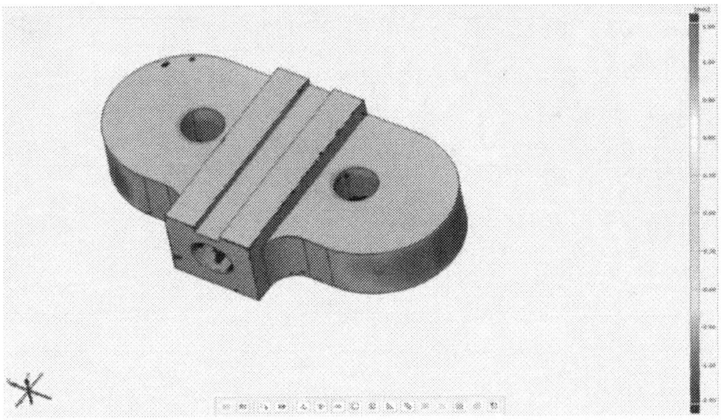

Figure 8. The deviation color map regarding the accuracy of the part, the range is in mm.

The colors signification is:

- *The green color* represents deviations of ±0.30 mm difference, this covering the part in proportion of 90%.
- *The blue color* represents deviations of -1.50 mm, which appear in the areas difficult to access the measurement.

- *The red color* represents deviations of +1.50 mm, meaning surplus of material.
- *The yellow color* represents deviation of +0.60 mm, being determined by imperfections caused by the lack of material.

This analysis demonstrates the correctness of the measurements.

Based on the experience gained during the scan of the landmark, the following observations could be made in the field of computer-assisted inspection of mechanical parts used in industry, special attention is required to avoid errors in the part scanning stage; in general, due to the specifics of this type of operation, errors can have two main sources:

- *The scanning process itself.*

The scanning process involves surface preparation, and it is known that dark, glossy and transparent surfaces cannot be scanned using this method. In the surface preparation stage, it is covered with a matte, very thin layer, using a special spray. If the coating is uneven, errors may occur in the form of gaps in the point cloud. If the layer thickness is too large, exaggerated shape deviations may occur.

During the scanning operation, the Atos Core system allows the choice of different exposure times, so that a complete image of the scanned surface can be obtained even in the conditions of the existence of areas with relatively important color differences. If the exposure times are not chosen correctly, as number or duration, overexposed or underexposed areas will appear. These are manifested as gaps in the resulting point cloud.

Positioning the object during scanning can also be a source of errors. The positioning must be done to avoid that the surfaces of the object are parallel to the incident beam. If this cannot be avoided, scanning will be done in several landmark positions. Thus, views are obtained in which the surfaces make wide enough angles with the incident beam to avoid gaps in the point cloud.

- *The shape of the scanned part.*

Although contactless scanning allows for a faster point cloud, a larger number of points, and access to narrower channels or bores than contact scanning, there are situations when landmarks cannot be reached. These surfaces are of the type of holes or narrow channels, whose deep areas are shaded by their walls.

Sometimes, it is possible to inspect them by choosing a convenient position of the part, but there are cases when they cannot be reached, being necessary the physical sectioning of the piece or giving up the inspection of the respective areas.

Acknowledgements

This work was supported by the Romanian Ministry of Research and Innovation, CCCDI –UEFISCDI, project number PD-III-P1-1.2-PCCDI-2017-0446 / *Intelligent manufacturing technologies for advanced production of parts from automobiles and aeronautics industries* (TFI PMAIAA) – 82 PCCDI/2018, within PNCDI III.

ORCID IDs

C Afteni https://orcid.org/0000-0003-2296-4751
V Paunoiu https://orcid.org/0000-0001-7846-4102
M Afteni https://orcid.org/0000-0001-8292-1863
V G Teodor https://orcid.org/0000-0002-3764-4243

References

[1] Mendricky R 2016 Determination of measurement accuracy of optical 3D scanners *MM Science Journal* **06** 1565-72 https://doi.org/10.17973/MMSJ.2016_12_2016183
[2] Garmendia I, Flores J, Madarieta M, Lamikiz A, Uriarte L G and Soriano C 2020 Geometrical control of DED processes based on 3D scanning applied to the manufacture of complex parts *Procedia CIRP* **94** 425-9 https://doi.org/10.1016/j.procir.2020.09.158

[3] Kuş A 2009 Implementation of 3D optical scanning technology for automotive applications *Sensors* **9** (3) 1967-79 https://doi.org/10.3390/s90301967
[4] Karczewski M 2019 Influence of 3D scanner parameters on accuracy evaluation of vehicle element deformation *AIP Conference Proceedings* **2078** 020012 https://doi.org/10.1063/1.5092015
[5] Paulic M, Irgolic T, Balic J, Cus F, Cupar A, Brajlih T and Drstvensek I 2014 Reverse Engineering of Parts with Optical Scanning and Additive Manufacturing *Procedia Engineering* **69** 795-803 https://doi.org/10.1016/j.proeng.2014.03.056
[6] Park S C and Chang M 2009 Reverse engineering with a structured light system *Computers & Industrial Engineering* **57** (4) 1377-84 https://doi.org/10.1016/j.cie.2009.07.005
[7] Bénière R, Subsol G, Gesquière G, Le Breton F and Puech W 2013 A comprehensive process of reverse engineering from 3D meshes to CAD models CAD *Computer-Aided Design* **45** (11) 1382-93 https://doi.org/10.1016/j.cad.2013.06.004
[8] Beccari C V, Farella E, Liverani A, Morigi S and Rucci M 2010 A fast interactive reverse-engineering system *Computer-Aided Design* **42** (10) 860-73 https://doi.org/10.1016/j.cad.2010.06.001
[9] Javaid M, Haleem A, Singh R P and Suman R 2021 Industrial perspectives of 3D scanning: Features, roles and it's analytical applications *Sensors International* **2** 100114 https://doi.org/10.1016/j.sintl.2021.100114
[10] Helle R H and Lemu H G 2021 A case study on use of 3D scanning for reverse engineering and quality control *Materials Today: Proceedings* **45** (6) 5255-62 https://doi.org/10.1016/j.matpr.2021.01.828
[11] Krznar N, Pilipović A and Šercer M 2016 Additive manufacturing of fixture for automated 3D scanning-Case study *Procedia Engineering* **149** 197-202 https://doi.org/10.1016/j.proeng.2016.06.656
[12] https://en.wikipedia.org/wiki/3D_scanning
[13] https://www.1st.com.my/atos-core
[14] https://www.atos-core.com/en/features.php
[15] www.gom.com
[16] https://scanare3d.com/services/3d-scanning-professional-3d-measurement-services/?lang=en

Holistic management of next generation manufacturing system

Gabriel Frumuşanu, Cezarina Afteni and Alexandru Epureanu
"Dunărea de Jos" University, Manufacturing Engineering Department, Domnească Street 111, 800201 – Galaţi, Romania

gabriel.frumusanu@ugal.ro

Abstract. The manufacturing system is a widely addressed subject. Many researchers have worked since a significant number of years in this field. Still, manufacturing system performance is capped by the insufficient level of accordance between manufacturing system and product. At the same time, global energy and global information networks have been created during recent years. Besides other applications, these networks can be turned in hard and soft infrastructure supporting the manufacturing system. For this reason, many specialists consider that the conceptual rebuilding of the manufacturing system is a notable challenge. This paper answers to this challenge by suggesting a holonic technique for manufacturing system management at basic, operational level. The proposed technique lays on the following hypothesis: *i)* next generation manufacturing system will be a modular cyber-physical system, *ii)* the typical management task is decision making, and *iii)* the procedure for decision making consists in options identification, followed by their comparative assessment. The approach is to manage the manufacturing system activity by holonic modelling of the system functioning, followed by the identification of a dedicated holarchy. The proposed technique aims to mitigate system performance in what concerns responsiveness capacity, investment efficiency and production costs. The technique is presented in architectural conceptual manner.

1. Introduction

In broad sense, the manufacturing system may be defined as an ensemble of facilities and assets, selected such as they are appropriate for supporting the manufacturing activity. More specific, the manufacturing system consists in a combination of manufacturing equipment and humans bound by a common material and information flow [1].

In economic approach, the manufacturing system is seen as a system in which raw materials are processed from one form into another, known as a product, gaining a higher or added value in the process and thus creating wealth in the form of a profit [2]. This is illustrated in figure 1.

In what concerns the global manufacturing trends, product complexity and variety continue to increase, product life-cycle are becoming shorter, markets have become multinational and global competition has been increasing rapidly, market conditions fluctuate widely and customers are consistently demanding high-quality, low-cost products and on-time delivery [3].

Under these circumstances, manufacturing system management, referred in the dedicated literature as *operations management*, plays an essential role in ensuring the required competitiveness on market.

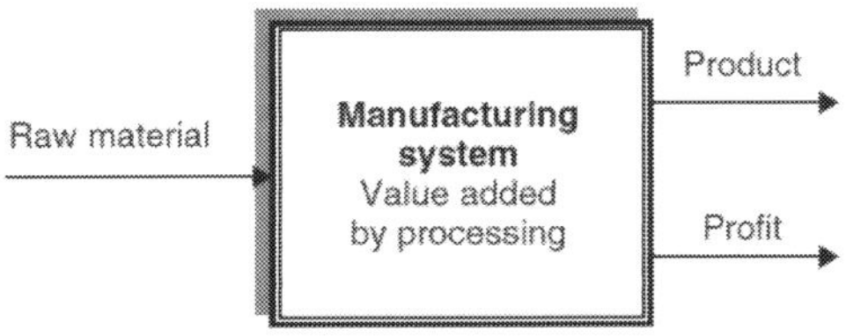

Figure 1. Basic model of manufacturing system adding value [2]

Operations management has numerous definitions – e.g. "the area of management concerned with designing and controlling the process of production and redesigning business operations in the production of goods or services" [4]. The operations management involves three main activities [5,6]:

- *Production planning.* During production planning, managers determine how goods will be produced, where production will take place, and how manufacturing facilities will be laid out.
- *Production control.* Once the production process is under way, managers must continually schedule and monitor the activities that make up that process. They must solicit and respond to feedback and make adjustments where needed. At this stage, they also oversee the purchasing of raw materials and the handling of inventories.
- *Quality control.* Finally, the operations manager is directly involved in efforts to ensure that goods are produced according to specifications and that quality standards are maintained.

Several management techniques have been developed during recent years in order to enhance operation management performances, [7]: Just in Time, Total Quality Management and Six Sigma, Business Process Reengineering, Enterprise Resource Planning. In [8] the authors propose some principles for adapting the operations management to the specificities of new-born Industry 4.0.

Despite the research interest concerning the operations management subject, the existing approaches present some notable shortcomings:

- The management is inconsistent, being often performed in a single decisional layer.
- The manufacturing process is not efficiently managed because it is addressed in its wholeness and not in its profoundness.
- There is a lack of systemic approach, actually required by the huge diversity of the manufacturing tasks.

Starting from here, the challenge is to improve the existing operations management by introducing a management new paradigm. In present paper, this paradigm is sampled through a management technique, namely the holonic management, specific to next generation manufacturing system, enabled by the evolutions regarding the manufacturing system [9,10] and on the large amount of information released by manufacturing system & process monitoring. The proposed management technique concerns the basic, operational level of manufacturing. It lays on the following hypothesis: *i)* next generation manufacturing system will be a modular cyber-physical system, *ii)* the typical management task is decision making, and *iii)* the procedure for decision making consists in options identification, followed by their comparative assessment. The approach is to manage the manufacturing system activity by holonic modelling of the system functioning, followed by the identification of a dedicated holarchy. The proposed technique aims to mitigate system performance regarding responsiveness capacity, investment efficiency and production costs. The technique is presented in architectural conceptual manner.

In what concerns paper structure, the next section briefly analyses the supply chain structure. The third section deals with holonic modelling and the architectural holarchy of the next generation manufacturing system. The fourth section introduces the proposed management technique, while the last section is for conclusion.

2. The supply chain

Supply chain is nowadays one of the most addressed topics in manufacturing management field and it has numerous diverse definitions. According to a general one, the supply chain consists of all parties involved, directly or indirectly, in fulfilling a customer request [11]. Supply chain activities involve the transformation of natural resources, raw materials, and components into a finished product that is delivered to the end customer [12]. Thus, the supply chain includes not only the manufacturer and suppliers, but also transporters, warehouses, retailers, and even customers themselves. For strategic decision-making, one of the most promising supply chain models is Supply Chain Operations Reference (SCOR) model [13]. According to it, the supply chain comprises five different sections: *Plan*, *Source*, *Make*, *Deliver* and *Return*.

The manufacturing process takes place into *Making* section, inside the manufacturing environment (figure 2). In general, the objective of *Making* section is to transform the raw material into final product. For example, in the case of a mechanical product, *Making* involves design, planning and processing. Activities such as assembling, testing and packing also happen here. Different types of resources are needed in order to perform *Making*: manufacturing modules & energy (brought from Physical space), cyber modules & information (brought from Cyber space), and qualified personnel (coming from social environment). Managerial resources are also necessary, for organizing the manufacturing environment according to a given product.

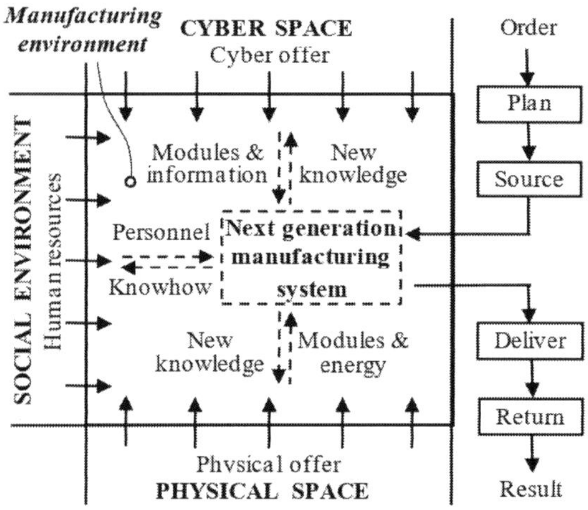

Figure 2. Making section of supply chain

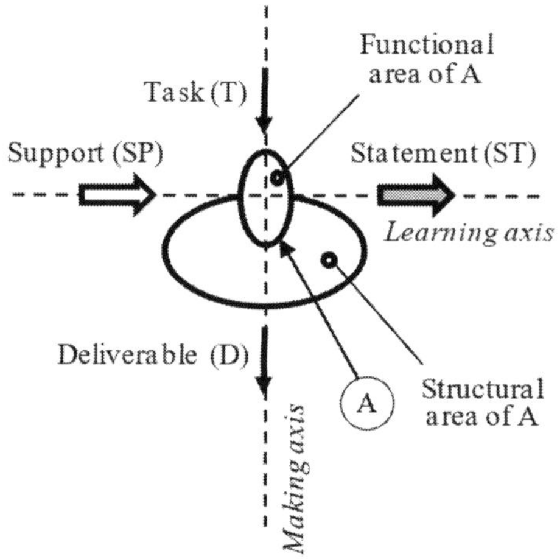

Figure 3. Holonic model of structural constituent

3. Next generation manufacturing system

3.1. System function

The manufacturing system function F is to accomplish the manufacturing task T. This means to bring the product from initial state S_i to final state S_f, through a manufacturing process, according to the diagram from below.

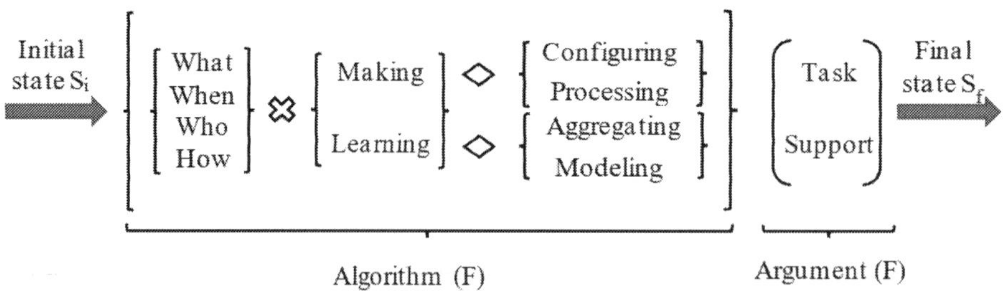

Before manufacturing function exerting, four issues must be established [9], namely: *what*, which refers to the task definition, *when* – to the precedence conditions, *who* – to the support needed for task accomplishment (humans included), and *how* – to the procedure following to be applied.

Function exerting involves two component activities, taking place concomitantly: *making* and *learning*. Making refers to actually transforming the product, while learning – to ascertaining the transformation results. *Configuring* action concerns both the task, and the support needed for *processing*. Task configuring means the repeated division/subdivision of initial manufacturing task among successive generations of manufacturing holons. Support configuring means to assemble all the manufacturing assets following to be used by each holon for its task accomplishment. The results learned about product transforming are the subject of *aggregating* action, which enables the manufacturing system *modelling*, with application in system control and management.

The manufacturing function argument comprises both the task T and the available support (meaning a selection of manufacturing assets chosen from the architectural supports pool).

3.2. System modular architecting
The term *architecting* regards the design of the manufacturing environment structure (figure 2). All resources brought here in order to build the manufacturing system of next generation type are expressed in modular manner and referred as architectural constituents.

3.3. System holonic modelling
A *holon* is something that is simultaneously a whole in and of itself, as well as a part of a larger whole [14]. In our vision, the modular structural constituents are modelled as holons. The holonic model of a structural component of the manufacturing system, referred as holon A, is presented in figure 3.

The generic holon has two components. The first one is the functional area, and means the place where the manufacturing action actually takes place. The second one is the structural area, where other structural constituents (offspring holons) may be co-opted to accomplish sub-tasks derived from initial task T.

Related to the two component activities of the manufacturing function, the holonic model of the structural constituent is bidimensional. The first dimension is represented by *Making axis*, along to which there are the task T, as input, and the deliverable D, as output. The second dimension corresponds to *Learning* axis, with support SP as input, and statement ST as output.

3.4. System architectural holarchy
A hierarchy of holons is called *holarchy*, which is defined [14] as a hierarchy of self-regulating holons that function (a) as autonomous wholes in supra-ordination to their parts; (b) as dependent parts in sub-ordination to controls on higher levels; or (c) in co-ordination with their local environment.

Because, in general, a single manufacturing holon cannot accomplish the entire task, this has to collaborate with other manufacturing holons, by task configuring/reconfiguring, and so on. This way, successive generations of holons take birth and an architectural holarchy of the system is generated (figure 4). Each holon belonging to this holarchy has assigned its specific support, which, associated with the received task, defines it. At the same time, each holon concludes it task by drawing up the corresponding statement.

3.5. System operation
The manufacturing system operation is illustrated by architectural holarchy diagram and can be explained starting from this. Let us consider the example from figure 4. The holons are denoted by capital letters A, B, C, ... M. The tasks accomplished by each holon are denoted by T_A, T_B, T_C ... T_M, respectively, while the released tasks, by T_1, T_2, T_3 ... T_{16}.

In the beginning, the holon A, of first generation, receives the task T and configures it. Thus, three new tasks result, T_A (which it assumes), T_1 and T_2 (which it releases towards holons B and C):

$$T = T_A + T_1 + T_2. \quad (1)$$

The second generation holons B and C, at their turn, configure the tasks T_1 and T_2, according to relations:

$$T_1 = T_B + T_3 + T_4, \text{ respective } T_2 = T_C + T_5 + T_6. \tag{2}$$

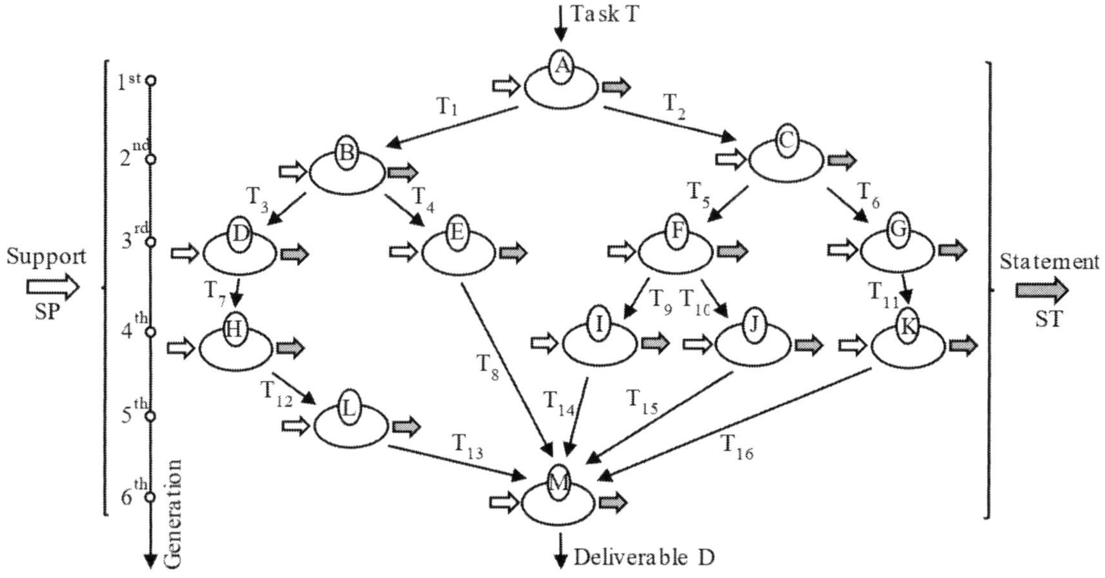

Figure 4. Architectural holarchy diagram

The things repeat at the levels of third, fourth and fifth generations of holons, until, finally, the holon M receives the tasks T_8, T_{13}, T_{14}, T_{15} and T_{16}, executes its resulted task:

$$T_M = T_8 + T_{13} + T_{14} + T_{15} + T_{16}, \tag{3}$$

and the deliverable D is obtained. The diagram closing relation is:

$$T = T_A + T_B + T_C + \cdots + T_M. \tag{4}$$

4. Holonic management

4.1. Management function

In the current approach, each of the holons involved in the architectural holarchy associated to the manufacturing of a given product (e. g. figure 4) have a management function. This concerns both the received task configuring and the support selection. More specific, each holon decides which part of received task follows to be assumed by itself and which tasks will be further released to offspring holons, and which is the most appropriate support that it will use to accomplish the task.

4.2. Management relations & management holarchy

The traditional approach of manufacturing system management, involving a single, top decisional layer proves to be, many times, inconsistent, due to the high gap existing between decider and executants.

At the opposite pole, the holonic approach of manufacturing system management supposes that each holon from the holarchy generated for manufacturing a certain product makes its managerial decisions after the tasks of the holons from previous generations are completely accomplished. This is the most profound management strategy, because the current decider holon should not make detailed prognoses for the manufacturing process, so these prognoses will be as correct as possible. However, as it can be noticed from the diagram of the holarchy depicted in figure 4, they can be many holons involved in the holarchy, hereby the functional relations among them might become complicated, as well as the information exchanges through individual statements.

The proposed method for holonic management of next generation manufacturing system aims to reduce the inconvenience of this approach by modifying the management relations inside the holarchy, on the base of a compromise between the number of holons with management function and the management performance. When configuring its task, the holon is able to find the causal relations among the tasks it deals with. If the causal relations between the task it assumes and the released tasks are weak or inexistent [15], then it does not keep any managerial prerogatives about them. On the contrary, if the causal relations are strong, then it forms from these tasks a cluster that it will manage, extending its managerial area to more than a single task. The other holons included in such managerial areas have exclusively execution function. Thus, only a part of the holons will keep their management function and become manager-holons. The making of decisions about the needed support, as well as the statement preparing are done by the manager-holon, at the level of its managerial area. The functional relations also establish only among manager-holons.

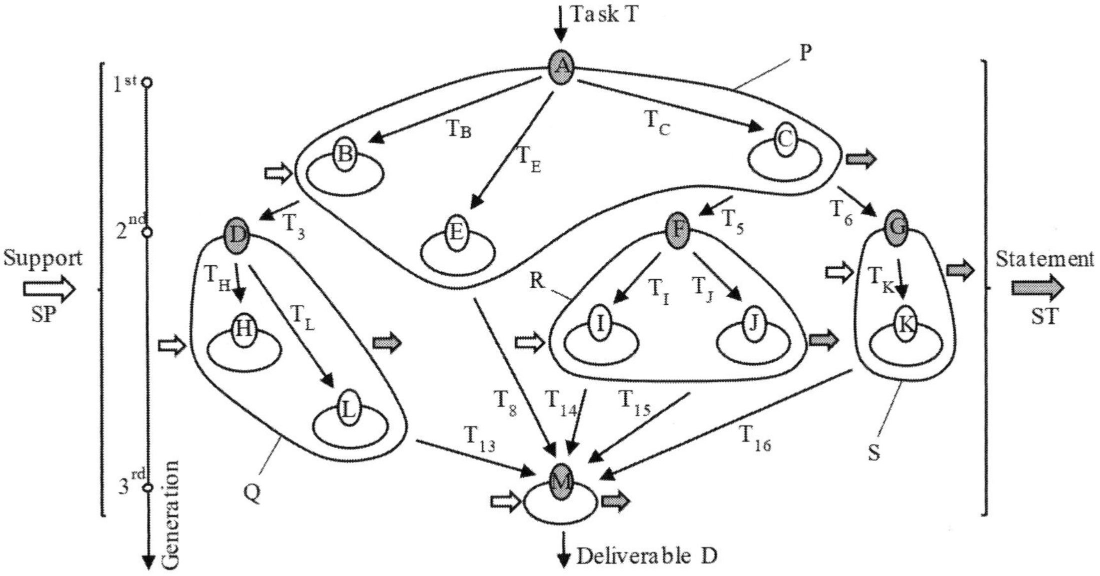

Figure 5. The management holarchy

Method application is sampled in the case of the architectural holarchy presented in figure 4, the result being the management holarchy from figure 5. The *A*, *D*, *F*, *G* and *M* holons assume themselves as manager-holons, their functional areas being shadowed in grey. Each of them receives a task, configures it and releases tasks to a number of execution-holons, which entirely accomplish the received tasks and compose the managerial area of the manager-holon. In the case of addressed example, if denoting the resulted managerial areas of manager-holons by *P*, *Q*, *R*, and *S*, the following relations can be written regarding the tasks:

$$T = T_P + (T_3 + T_8 + T_5 + T_6), \text{ where } T_P = T_A + T_B + T_E + T_C, \tag{5}$$

$$T = T_P + (T_Q + T_{13}) + T_8 + (T_R + T_{14} + T_{15}) + (T_S + T_{16}), \tag{6}$$

$$T = T_P + T_Q + T_R + T_S + T_M = T_A + T_B + T_C + \cdots + T_M. \tag{7}$$

The application of proposed method for holonic management in the case of the holarchy presented in figure 4 leads to a lower number of holons with management function (five instead of thirteen). Thus, the manufacturing system management will be leaner and more efficient. At the same time, the management holarchy will have only three generations of manager-holons (instead of six) – hence three management levels. Last but not least, the manufacturing system management at operation level will be still much better performed than in traditional approach, when decisions are made at the top level, while the management profoundness can be adjusted by setting the number of management levels.

4.3. Management loops

The management loops in the case of the proposed method are easier to be explained if considering the simplest system architecture, involving two making holons (A and B), and a learning holon C (see figure 6). The holon A receives a given task T_1. At first, it configures by assuming for itself the task T_A and the support SP_A (which consists in architectural modules, taken from the dedicated pool), and by releasing the task T_2^* to B holon. Then, it passes to processing action for accomplishing T_A and, after doing this, it draws up the statement ST_A, specifying the obtained results. The holon B also accomplishes its task T_2 and draws up the statement ST_B.

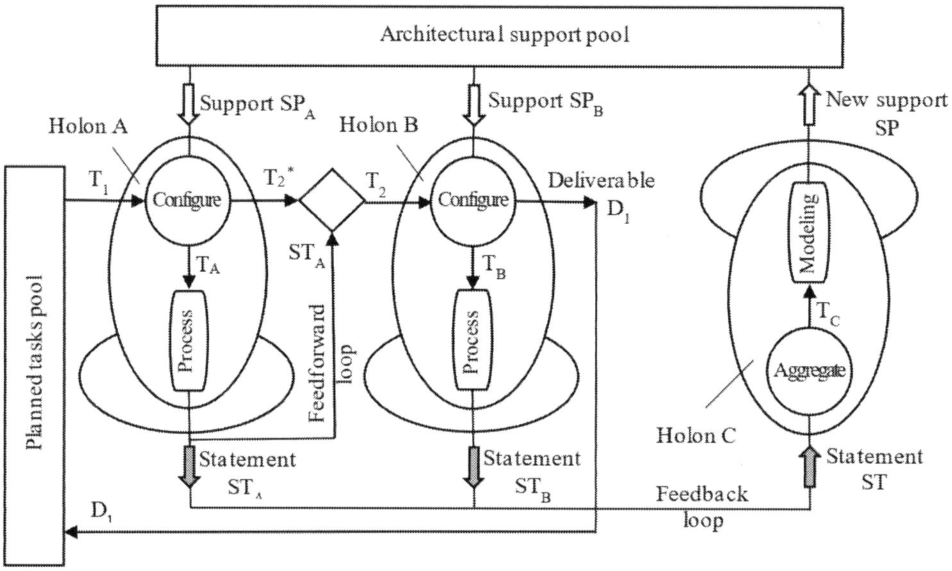

Figure 6. The management loops

The result of T_A accomplishment, according to ST_A is compared to the specified result of T_A. If differences are acknowledged, the released task T_2^* is reconfigured to T_2, such as holon B may compensate these differences, in order to obtain the expected deliverable D_1. This means the *feedforward loop* of management.

At the same time, through the *feedback loop*, the statements ST_A and ST_B are aggregated by holon C, whose task T_C is to deliver the new models of T_1, T_A, and T_B tasks accomplishment by the couple of holons A and B to the architectural tooling pool. If T_1, T_A, and T_B tasks must be repetitively accomplished, at next iteration the model will be used for improving the manufacturing performance.

5. Conclusion

This paper proposes a new management technique, grounded on a new paradigm, namely the holonic management. The proposed management technique concerns the basic, operational level of manufacturing, which is supposed to be executed by holons. The technique is based on the analysis of the causal relations among the manufacturing tasks and consists in building clusters from the tasks among which strong causal relations are found, and defining a managerial area for each such cluster. The criterion for establishing the causal relations can determine the number of tasks clusters, hence of managerial areas. Due to this specificity, the technique enables to make a tradeoff between the number of holons with management function and the management performance.

Looking from another perspective, the proposed technique introduces an important novelty, namely the correction of manufacturing system management on the base of two loops – the feedforward loop and the feedback loop, since in the exiting dedicated literature we did not find a similar approach. The feedforward loop enables a rapid correction of the deficiencies acknowledged in the manufacturing process, without needing to wait for process ending before taking reparatory measures. This management feature is very important in what concerns the global performance of the manufacturing system. The feedback loop enables, at its turn, to preserve the entire manufacturing experience of the system and to capitalise it by structural implementation, aiming its application in future processes.

Acknowledgements
The participation at conference was financially supported by the project *Internationalization for institutional development at Dunarea de Jos University of Galati – IDEI (UGAL)[21]*, project code CNFIS-FDI-2021-0349.

ORCID IDs
G Frumusanu https://orcid.org/0000-0003-3038-8758
C Afteni https://orcid.org/0000-0003-2296-4751
A Epureanu https://orcid.org/0000-0001-6939-9433

References
[1] Chryssolouris G 2005 *Manufacturing Systems: Theory and Practice* (New York: Springer Verlag) https://doi.org/10.1007/0-387-28431-1
[2] Scallan P 2003 *Process Planning* (Butterworth-Heinemann) https://doi.org/10.1016/B978-0-7506-5129-5.X5000-4
[3] Kalpakjian S and Schmid S R 2014 *Manufacturing Engineering and Technology* Prentice Hall
[4] https://en.wikipedia.org/wiki/Operations_management accessed in 22.03.2021
[5] Collins K 2012 *An Introduction to Business* Creative Commons
[6] Stevenson W 2018 *Operations management* McGraw-Hill Education
[7] Silver E 2004 Process Management Instead of Operations Management *Manufacturing & Service Operations Management* **6** (4) 273-9 https://doi.org/10.1287/msom.1040.0055
[8] Filipov V and Vasiliev P 2016 Manufacturing Operations Management – the Smart Backbone of Industry 4.0 *Scientific Proceedings of I-st International Scientific Conference "Industry 4.0"* **1** 71-6
[9] Frumuşanu G and Epureanu A 2021 Architectural Holarchy of Next Generation Manufacturing System, accepted for publication in *International Journal of Modeling and Optimization*
[10] Frumuşanu G and Epureanu A 2019 New Vision on the Manufacturing Process Control *International Journal of Modern Manufacturing Technologies* **XI** (3) 57-62
[11] Chopra S and Meindl P 2013 *Supply Chain Management. Strategy, Planning, and Operation* (Pearson Education)
[12] Kozlenkova I, Hult T M, Lund D, Mena J and Kekec P 2015 The Role of Marketing Channels in Supply Chain Management *Journal of Retailing* **91** (4) 586-609 https://doi.org/10.1016/j.jretai.2015.03.003
[13] https://www.iimu.ac.in/blog/what-are-the-five-basic-components-of-a-supply-chain-management-system/ accessed in 26.03.2021
[14] https://en.wikipedia.org/wiki/Holon_(philosophy) accessed in 30.03.2021
[15] Frumuşanu G, Afteni C and Epureanu A 2021 Data Diven Causal Modeling of the Manufacturing System, accepted for publication in *Transactions of FAMENA* **45** (1) 43-62 https://doi.org/10.21278/TOF.451020920

IOP Conf. Series: Materials Science and Engineering 1235 (2022) 012073 doi:10.1088/1757-899X/1235/1/012073

Hybrid organization of manufacturing and operations: the Ukrainian case under quarantine restrictions

T Omelianenko, K Korotkova, P Banschykov, M Sahaidak and T Sobolieva

Kyiv National Economic University named after Vadym Hetman, Management Department, 54/1 Prospect Peremogy, Kyiv, Ukraine

korotkovakse@gmail.com

Abstracts. The paper presents the dynamics of the ratio of productive and non-productive spheres in the economy of Ukraine in the pre-quarantine period. Trends in changes in the business processes of manufacturing companies caused by the introduction of strict quarantine in Ukraine due to the COVID-19 pandemic have been identified. It was analysed the official statistics on business closures. It was determined which businesses developed rapidly during the quarantine period and the starting point of this success is established - a hybrid approach to the organization of production. That is a combination of traditional and digital approaches, i.e. offline and online. It was analysed the successful cases of Ukrainian manufacturing companies, which quickly reoriented to the hybrid approach. It was summarized the results of the recommendations on the application of the hybrid approach in the reorganization of manufacturing and operations.

1. Introduction.

In modern conditions of business functioning, in order to maintain their competitiveness, organizations must take care of the strategic development of their activities. Regardless of the field of activity (industrial or non-industrial), a successful organization must be flexible and quickly respond to changes in the external environment. Often, it's the production organizations that find it most difficult to quickly restructure their operating processes in a short time, and especially those that work with the use of foreign raw materials. At the end of 2019, the whole world was swept by a pandemic, which forced absolutely everyone, from individual entrepreneurs to corporations, to restructure their production lines as soon as possible, as well as to revise their strategic plans. For many, these changes were fatal, which led to a partial or complete closure of the business. But at the same time, many companies not only remained to function but also grew several times. Scientists all over the world do their research on the impact of Covid-19 on the operations of companies. For example, the research [1] discusses various operations and supply chain perspectives to address such disruptions in the future due to the introduction of lockdowns. In this article we will analyse company cases that will help to understand the success of function in the quarantine restrictions.

2. Analytical base of the research.

First, let's dwell on the ratio of production and non-production areas of activity on the example of Ukraine (Table 1). As of 2012, there were 1,600,127 business entities in Ukraine, of which 14.58% were production entities (industry, construction, agriculture, forestry and fisheries). At the same time, already in 2019, the number of all business entities increased by 17.6% (a total of 1,941,701 business entities), but the number of the production sector has already reached 13.51%, which, in relation to 2012, decreased by 1% (a fairly large figure for the period of 7 years) [2].

It can be concluded that the non-production sphere of activity is 7 times higher than the production one, if we consider the period 2012 – 2019, and every year the percentage of the production sphere is rapidly decreasing. This phenomenon can be explained by the fact that running a business in the service sector is less expensive, there are fewer barriers to entry into activities, and the service sector is more in demand in B2C business. But despite such a demand for business in the service sector, this type of business is the most susceptible to unforeseen changes.

Table 1. Number of business entities by type of economic activity in 2012 and 2019.

	Type of activity / number of business entities	2012		2019	
		Units	%	Units	%
1	Agriculture, forestry and fisheries	68497	4.28	75450	3.89
2	Industry	114028	7.13	130324	6.71
3	Construction	50830	3.18	56855	2.93
4	Wholesale and retail trade; repair of motor vehicles and motorcycles	843075	52.69	834159	42.96
5	Transport, warehousing, postal and courier activities	87252	5.45	96235	4.96
6	Temporary accommodation and catering	44085	2.76	69689	3.59
7	Information and telecommunications	66568	4.16	206147	10.62
8	Financial and insurance activities	10342	0.65	10086	0.52
9	Real estate transactions	81726	5.11	98361	5.07
10	Professional, scientific and technical activities	87431	5.46	134800	6.94
11	Activities in the field of administrative and support services	38769	2.42	53787	2.77
12	Education	7317	0.46	15429	0.79
13	Health care and social assistance	16479	1.03	30994	1.60
14	Arts, sports, entertainment and recreation	9238	0.58	15168	0.78
15	Provision of other types of services	74490	4.66	114217	5.88
	Total	**1600127**	**100**	**1941701**	**100**

In late 2019 - early 2020, the world was swept by the Covid-19 pandemic. To prevent the massive spread of the virus, the government of most countries introduced a full lockdown, which meant the closure of all organizations that are not vital (retail, entertainment, services, etc). As much as possible, all activities and work were moved to online mode. The life of almost all Ukrainian business became paralyzed for 3 months (March - June 2020) due to the inability to work in the usual conditions. It's worth noting the fact that in Ukraine 99% of all economic entities are small businesses, and especially the non-production sphere [2]. It is this part of all business organizations that is most susceptible to drastic and unforeseen changes. As a result of the inability to quickly reorganize into a remote work format, and for many organizations, it was completely impossible to carry out their work remotely or online (in the service sector, offline retail trade, hotels and restaurants, travel agencies), small businesses were forced to close their business entities (business closure statistics are shown in Table 2).

According to the Opendatabot platform, from March 2020 to March 2021, 215.3 thousand individual entrepreneurs closed in Ukraine. This is almost 50 thousand more closed businesses than in the same months of 2019-2020. Most enterprises closed in the areas of warehousing and transport (almost 11 thousand), wholesale (5.6 thousand), television and radio broadcasting (2.69 thousand). The number of people who decided to open a sole proprietorship during quarantine was 240.9 thousand, which is 5 thousand less than in the same months before quarantine [3].

Even during the relaxation of quarantine restrictions, the situation with some types of business (especially clothing retail) didn't improve. During the quarantine conditions, Ukrainians began to buy fewer clothes through a complete transfer to teleworking from home and no need to visit the office. Retail sales for June-December 2020 couldn't be restored until the end of 2020, even to the level of January-March 2020 [4].

3. Case study

But as mentioned above, not all organizations have a lockdown as an obstacle to growth and development. We would like to dwell not on those types of business that continued to function practically unchanged during the quarantine period (logistics, medicine, food trade), but on those that were affected by general restrictions.

Table 2. Top 10 economic activities where companies closed most often in the first half of 2020 [4].

Ranking place	Name	Quantity, in thousands	Share, %
1	Retail sale of food, beverages and tobacco products in non-specialized stores	7.0	6.6
2	Other types of retail sale in non-specialized stores	5.7	5.3
3	Provision of services of hairdressing and beauty salons	4.8	4.5
4	Restaurant activities	4.6	4.3
5	Retail sale from trays and in the markets of other goods	4.2	4.0
6	Computer programming	3.9	3.7
7	Retail sale from trays and markets of clothing and footwear	3.8	3.6
8	Freight road transport	2.5	2.4
9	Retail sale of clothing in specialized stores	2.5	2.3
10	Retail sale from trays and markets of food, beverages and tobacco products	2.1	2.0

The successful example of growth during the period of quarantine restrictions was "Epicenter K" hypermarket, a chain of shopping centres in Ukraine. The main specialization of the company is retail trade in home furnishings, furniture, household chemicals, automotive goods, garden and vegetable garden goods, children's goods, and construction materials. This is the kind of business that, like many others, was forced to stop offline trading. At the same time, online sales of the "Epicenter K" network during the quarantine tripled - from 3% to 10%. This has greatly accelerated the adoption of various online services that have helped improve the shopping experience in the omnichannel ecosystem. The company is actively working on increasing its own capacity - building logistics complexes that will serve not only shopping centres, but also online commerce. According to the CEO of the company, any crisis is both a challenge and a chance. "Epicenter K" has always followed this rule and tried to develop even faster during crises. In 2019, the company invested more than UAH 9.5 billion in the Ukrainian economy - this is the construction of elevators, the purchase of equipment, the construction of new shopping centres, and the completion of a ceramic tile factory. And COVID-2019 didn't stop the business, but only partially changed priorities. The coronavirus pandemic has changed the habits of Ukrainians, so a business that wants to remain relevant in the "new reality" must very quickly adapt to the new wishes of customers and new conditions.

It should be noted that 39% of managers in Ukraine and 81% in the world believe that the pandemic has accelerated the digitalization of operations and the creation of a next-generation operating model. That is, if the business wants to continue to be profitable, it must be as flexible as possible to external circumstances [5].

The reasons for the growth of such companies all over the world were the same - an increase in demand for online purchases for a decrease in offline sales. More interesting is the experience of those

IOP Conf. Series: Materials Science and Engineering 1235 (2022) 012073 doi:10.1088/1757-899X/1235/1/012073

companies that have secured good results in their activities under lockdown conditions. Such companies should include "RITO", "Dnipro M" and others that use a hybrid approach to their operations. Below we will consider the specifics of their functioning under conditions of quarantine restrictions.

The Ukrainian company "Dnipro M" is a manufacturer of power tools, hand tools and equipment for repair and construction, gardening equipment in the household and professional segment, which sells products in its own company stores, in an online store and a partner network. The company has been present on the offline market for more than 10 years, while it should be noted that the first brand store appeared only in 2018, as of 2019 the company had 120 brand stores and 900 thousand customers, and at the end of 2020 - already 244 stores and about 1.5 million clients. That is, despite the hard lockdown and the inability to work in the usual format, the company doubled in opening new offline stores and increased the number of customers by 60%.

The company managed to achieve such indicators due to active promotion and transition to online trading. The key task of the company was to combine the promotion and brand awareness on the Internet with the rise in sales from the site. The brand name in Google search results was strengthened, trade ads were created, the "persistence" of remarketing increased, that is, the "catch-up" of users who have already visited the site, at which they regularly appear reminding users what products they have viewed, and the video project "Drift -Korch" is a series about building a racing car from an old BMW using a company tool. An important condition for success was the free delivery of large-sized equipment, which won over the users to choose exactly "Dnipro M". Thanks to such actions, the company has more than doubled during the quarantine period. But also need to consider about the messages that are delivered offline. All channels need to show the consumer the same "face" and brand benefits, which will help unite shop and website buyers into one audience [6].

4. Hybrid organization of manufacturing and operations as a guarantee of business success during the period of quarantine restrictions

This approach to organizing business can be a successful example of using a hybrid approach in your activities, that is, combining online and offline activities. The term "hybrid" means one that arose from the combination of two other species. A hybrid approach has already been used in the formation of the operational strategy in its general understanding. For example, in the article by J. Köbera and G. Heinecke, a hybrid approach to product was described [7]. At the same time, the main advantage of the hybrid approach is that it includes all the positive aspects of the other two components, while eliminating their disadvantages.

An important condition for the formation and use of an effective hybrid strategy was the presence of production units at "Dnipro M" and other similar companies. This freed these enterprises from dependence on foreign supplies, which became unstable in a pandemic. The production units in such companies had a relatively high technological level, providing them with high adaptability and maneuverability in the production of high-quality products. In addition to production units, these companies had a good network of offline stores, their own websites to promote their products. Conventionally, the organizational structure of such companies can be represented in the form of 3 elements: Production, Offline network and Internet sites. We consider the intensive exchange of management, marketing and operational information between these departments to be a manifestation of a hybrid strategy. Of course, depending on changes in conditions in the external environment, the emphasis will shift in information flows between these departments. Let's consider 2 variants of a hybrid strategy depending on the market situation.

Option 1 - no restrictions on offline sales (before COVID-19 lockdowns, or after its completion). In this case, marketing services that ensure effective sales through offline stores will have priority in providing management and operational information to the rest of the divisions (figure 1).

Double arrows 1 and 6 - marketing research; arrows 2 and 7 - orders for the production unit; arrows 3 and 8 - orders completed by the production unit; arrows 4 and 9 - sale of manufactured products; arrow 5 - transmission of marketing information.

It is these services that carry out a deep and comprehensive market research, study of demand from different groups of buyers, the actions of competitors, etc. Based on this information, the company's behaviour policy in the offline sales markets is formed (price strategy, supply volumes, promotional events, etc.). Based on marketing research and monitoring of the results of the implementation of the offer policy, the relevant information is promptly transmitted to the production units of the company. These divisions, if necessary, make changes to the production program as soon as possible (change the output structure, make constructive and technological changes to the manufactured products) and transfer the updated products to sales services. In this case, offline stores have the opportunity to satisfy the needs of their consumers most fully and in a short time. At the same time, the online sales divisions are tasked with ensuring the sale of products through their distribution channels at minimal cost. In such conditions, online sites will use data on customer preferences obtained from the offline sales marketing service, supplementing it with a more accurate portrait of customers. Such a distribution of emphasis will allow saving on marketing Internet research and maximizing the benefits from the company's offline sales.

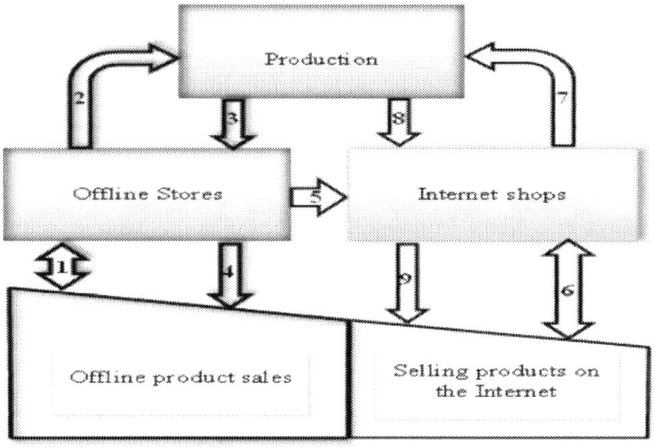

Figure 1. Interaction of the offline network, production and online stores in the absence of restrictions on sales in the offline network (priority of offline sales - bold lines).

Option 2 - restrictions on offline sales due to quarantine measures. In this case, the company's online stores begin to play a dominant role in marketing research and sales volumes (figure 2).

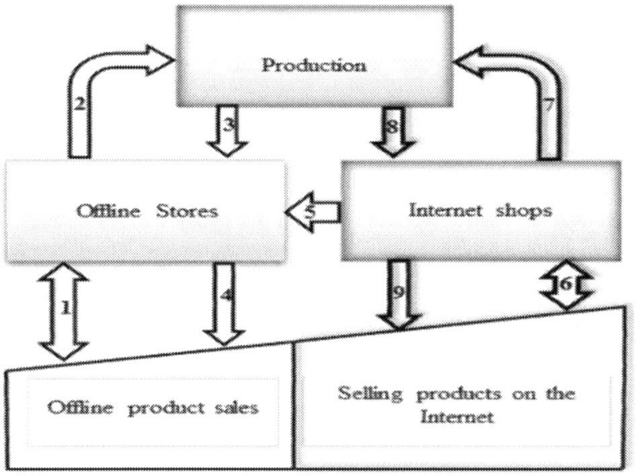

Figure 2. Interaction of the offline network, production and online stores in the context of the prohibition of offline sales (priority is given to online stores - bold lines).

They use all Internet marketing tools available to them, form target audiences of buyers, determine the characteristics of their consumer preferences in relation to the company's products and quickly transmit this information in two directions. First of all, to the production department and to the offline sales department. As in the 1st variant, the production department fulfils orders and transfers them to the online store. It is he who becomes the main distribution channel in the company. However, offline sites do not completely cease their activities. With market research from online sales teams, offline stores know the consumer preferences of their customers. Therefore, they begin to offer high-demand products to their customers using the offsite trade. This can be done especially well when a company sells clothes. So, the company RITO here needs a link to the site RITO during a hard lockdown, offered buyers to select 3-4 models from the catalog of their products and sent them by courier to the buyer's address [8]. Of course, the company attributed this form of sales to concern for the health of buyers and the desire to reduce dangerous contacts. In the case of corporate clients, it is possible to sell goods from the company's warehouse. In addition, some of the staff of offline stores can be temporarily transferred to the online sales service. For example, by increasing the number of departments associated with taking orders, completing them and sending them to consumers. This will increase the speed of order fulfilment. In addition, some of the staff of offline stores can be used as couriers to deliver goods to customers. This will bring additional benefits to both the company and the consumer. It is known that during the quarantine restrictions of offline sales, delivery services could raise prices for their services. By not using a third-party shipping service, you can reduce costs and provide better deals for your customers.

Rapidly shifting focus from one sales channel to another ensures a successful hybrid strategy. As noted above, the presence of a high-tech, adaptive production unit is of great importance in using a hybrid approach. Its absence can be replaced by reliable suppliers who are willing to work in the face of rapidly changing consumer preferences. But in a pandemic, there is a high probability that these suppliers will not be able to promptly supply products in the required quantity, range and required time frame. That is why a high production capacity is required to implement a hybrid strategy within, and not outside the country or even a separate region.

5. Conclusion

In our case, the hybrid approach is based on combining online and offline sales, which allows sales to be carried out regardless of the influence of external factors (the introduction of a lockdown, the absence of an offline store in the village, etc).

Taking a hybrid approach to operational strategy formulation using online and offline activities is a significant advantage in today's business environment. This approach allows to continue activities in any conditions of remote work and quarantine restrictions without endangering business. There are more and more companies that could effectively use hybrid strategies. It should also be seen as a signal for the formation of long-term development strategies of companies, which include the creation of adaptive and high-tech production units. It is they who provide the very opportunity for the successful implementation of hybrid strategies in operations. The provided version of the hybrid operating strategy is an effective method in the conditions of the VUCA-world. In this case, a hybrid strategy can be based not only on a combination of online and offline activities, but also on other types of activities, which is a very promising area for further research.

ORCID iDs

T Omelianenko https://orcid.org/0000-0001-6218-0455
K Korotkova https://orcid.org/0000-0002-2670-8643
P Banschykov https://orcid.org/0000-0003-3720-0578
M Sahaidak https://orcid.org/0000-0001-6526-1170
T Sobolieva http://orcid.org/0000-0002-7818-145X

References

[1] Kumar A, Luthra S, Mangla S and Kazançoğlu Y 2020 COVID-19 impact on sustainable production and operations management *Sustainable Operations and Computers* **1** 1-7 https://doi.org/10.1016/j.susoc.2020.06.001
[2] State Statistics Service http://www.ukrstat.gov.ua/ Accessed 18.02.2021

[3] Vinokurov J 2021 https://www.epravda.com.ua/rus/publications/2021/03/24/672241/_God_–_pljus_FLP – minus_:_kak_malyj_biznes_perezhil_pervyj_god_karantina_Jekonomicheskaja Pravda Accessed 22.02.2021
[4] Sereda E 2020 https://zn.ua/macrolevel/armija-chepeshnikov-kto-oni.html_Melkij_biznes_Ukrainy_v_cifrah_i_faktah Accessed 16.01.2021
[5] KPMG Research The view of business leaders in Ukraine 2020 Special issue: COVID-19 https://assets.kpmg/content/dam/kpmg/ua/pdf/2020/10/CEO-Outlook_2020_Ukraine.pdf
[6] Kondratenko E and Evtushenko S 2019 https://rau.ua/ru/news/dnipro-m/_Kak_uvelichit_onlajn_-_prodazhi_na_375_%_v_nishe_instrumentov-—_kejs_Dnipro_M Accessed 13.10.2020
[7] Köber J and Heinecke G 2012 Hybrid Production Strategy between Make-to-Order and Make-to-Stock – A Case Study at a Manufacturer of Agricultural Machinery with Volatile and Seasonal Demand *Procedia CIRP* **3** 453-8 https://doi.org/10.1016/j.procir.2012.07.078
[8] RITO https://rito.ua/index.php?route=common/home Accessed 04.03.2021

IOP Conf. Series: Materials Science and Engineering 1235 (2022) 012074 doi:10.1088/1757-899X/1235/1/012074

Industry management transformation in digital age

Olena Shatilova, Tetiana Sobolieva, Liudmyla Batenko, Mykhailo Sahaidak and Tetiana Omelianenko

Kyiv National Economic University named after Vadym Hetman, Management Department, 54/1 Prospect Peremogy, Kyiv, Ukraine

tetiana.sobolieva@kneu.ua

Abstract. The article is devoted to the topical problems of adapting company management to the modern conditions of the digitization and Industry 4.0 formation. The essence of digital transformation is characterized and the main spheres of its impact are highlighted. There are the main characteristics of the modern companies that operate under the conditions of Industry 4.0, problems, prospects and key factors of ensuring the competitiveness of companies in the context of digital transformation. The ways of the company management development under the influence of digital transformation have been investigated.

1. Introduction

The modern world is impossible to imagine without the use of digital devices, digital social networks that have radically changed the nature and manner of human communication. In the context of the COVID-19 pandemic, the processes of digital transformation will only accelerate, changing the established way of life of people and the rules of doing business around the world and adapting them to the new reality. The use of digital tools will allow business organizations of various activities to ensure their competitiveness, in particular by meeting such tangible and intangible needs of their customers as safe communication and product selection. Digital transformation is an inevitable process and the focus for business management should be on how to find its place in the new global digital economy

2. Digital transformation and Industry 4.0

Digitalization, as a process of penetration of digital technologies, automation and IT technologies into all spheres of life and economy, began at the end of the twentieth century and was named "Technological order 3.0". According to Harvard Business Review survey there are only 23% of the companies that doesn't require intensive digitalization [1].

At the same time, in recent years, under the influence of horizontal and vertical integration of IT technologies, the creation of the latest cyber systems and artificial intelligence, there has been a complete rethinking of the ways of doing business and a change in business models. Nevertheless just 44% of the companies are ready to such digital transformation even taking into account that 87% of the companies expect the disruption of their industries by IT-technologies [2].

Digital era requires digital transformations in all spheres, including industry management. In today's world, digital technology has become an integral part of everyday life. One of the global challenges of our time is digital transformation, which manifests in different aspects of human life at different levels. In addition, the modern realities of digitalization of the world are changing the nature and manner of socio-economic relations in almost all areas of business [3]. In a broad sense, digital transformation is the emergence of new or modification of existing processes using digital devices and digital technologies. Digitization is rapidly capturing almost all aspects of people's daily and professional lives; in particular, by influencing the way they disseminate information and changing habits and preferences.

It should also be noted that not only private business organizations are affected by the digital transformation, but also digitalization has affected state organizations. In particular, the Society 5.0 initiative was launched in Japan [4]. That is, we are talking about the fact that digitization covers almost all the reality around us. Not only is the internal environment of business organizations digitizing, also so is the environment in which they operate. This, in turn, affects the objective transformation of the management process of business organizations. Because management activities synthesize various aspects of the business organization and is very sensitive to changes in factors, both external and internal, the issue of flexibility of the management system of the business organization becomes relevant.

Digital transformation manifests itself at different levels and in different ways. For a more comprehensive understanding of the processes of digital transformation, we propose to consider their features at three levels: at the level of society, at the level of the business environment and at the level of the management system of the business organization (figure 1).

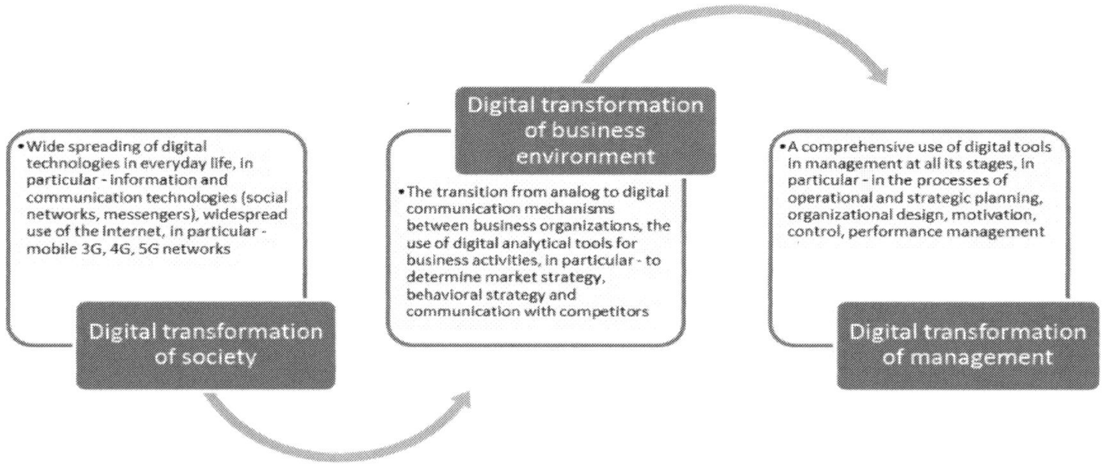

Figure 1. Areas of digital transformation influence [5].

The main vectors of digitalization of business organizations are digitalization of business organization personnel, digitalization of the transformation process within the business organization, in particular, the transformation of internal business processes, as well as digitization of operational activities in general and individual operations in particular [6].

Today, to characterize the digital transformation in the economy (digitalization) of mass consumption, the term "Industry 4.0" has been acquired [7].

Among the main characteristics of the modern organization of the era of Industry 4.0, the following should be highlighted [7,8]:

• development of interaction between people and machines through the use of the Internet of things;

• accumulation of comprehensive information about all processes by creating digital copies of real objects, systems and functions, which ensures their transparency;

• full support for the implementation and tracking in real time of most production processes, which may consist in the complete replacement of a person with a machine;

• automation of management activities due to the detailing of management decisions and their delegation to cyber-physical systems;

• automation of management activities due to the detailing of management decisions and their delegation to cyber-physical systems;

• cheaper and faster vertical and horizontal integration to ensure full control and improve the efficiency of the organization.

Digital transformation is a pervasive trend in the modern world. Companies use it for ensuring potential growth on the market (51%) or getting additional competitive advantages (41%) as well for meeting legislation requirements (for example, GDPR) – 38% [9].

Technological progress rapidly changes not only the ways of communication, work, study, life in general, but also leads to a change in the conditions for the development and competition of modern organizations. The emergence of the Internet changed not only the way people connect, but also led to the phenomenon of the "Internet of Things", which allows machines and equipment to be networked both in private life and in organizations and enterprises of various sectors of the economy. By IDC estimation [10], not less than 70% of the companies will integrate digital technologies to transform business processes in the near future. At the same time there are high expectations of the future growth of the Internet of Things (IoT), AR/VR technologies [11].

The big accelerator of digital transformation is the pandemic. According to some survey, in order to overcome the consequences of the pandemic, significant efforts of the CEO were aimed at implementing projects to ensure remote work of employees (about 37%) and transfer of the company's processes to a digital format (one third of respondents) [12].

Digital transformation statistics shows such influence of the COVID-19 pandemic on digitalization processes in the companies [13] as increasing significantly (two times) digitization rate in financial services and infrastructure, accelerating digital changes by 20-25 times, shortening time for come IT-solutions dramatically – from one year to 11 days etc.

Fortune states that 77% of Chief executive officers expect an acceleration of business digitalization [14]. If we talk about the driving forces accelerating business digitization in a broader context it is primarily the strengthening of the digital environment of the companies. This is important both for enhancing competitive advantages and for harnessing the power of digital technologies, for example, to improve the quality of customer service, and to force partners to adapt to digital business processes. Digital transformation can also be the result of a competitive defence strategy. At the same time, the era of Industry 4.0 not only opens up new perspectives, but also poses certain challenges for the development and functioning of modern organizations such as resistance to changes, lack of initiative, needs for additional skills and technical and organizational support.

Some survey show that companies have certain obstacles in digital transformation because on the one hand it requires additional cost (28%) but on the other hand it is difficult to estimate ROI (29%). Juridical and compliance aspects (26%) with staff resistance (26%) also slow down digital transformation processes [9]. Among interviewed retailers 76% underline that the complication of digitization is high or very high, 75% stress on necessity of adequate human resources and additional cost, 75% concern because of time to profit. 74% of respondents pay attention on security and privacy issues [15].

Ensuring competitiveness of modern companies in the context of digital transformation is primarily associated with global production chains and minimizing costs, optimizing product mix based on synergy and emergence, increasing productivity and restructuring activities through outsourcing, asset consolidation and the use of new digital technologies.

3. Industry management transformation

In the modern conditions of digital transformation, high degree of uncertainty and variability of the functioning environment, the problem of rethinking traditional management concepts and business models is becoming increasingly important. Business management is directly affected by digitization, as the directions and pace of digitization significantly depend on the level and vector of management development. Digitalization determines the following directions in the management transformation:

3.1 Necessity of digital strategy development. It was mentioned that proactive digital strategy development and implementation increase market capitalization by 10% [16]. Latest tendencies show that 70-90 of the companies already have or just develop digital transformation strategies [17,18].

3.2 Changes in entire business model or in its elements, first of all in business processes. The interview of executives brightened the positive influence of digital business model on operational efficiency (40% of respondents), shortening time to market (36%), and better respond to the customer needs (34%) [19]. It also vital to use data analysis (Big Data) for supporting decision making processes [20], keep both strategic and operational objectives of digital transformation in focus at the same time [21].

3.3 New approaches to organizational structures. Different locations of businesses, culturally specific features create some difficulties in term of communications and coordination. It requires developing appropriate communication procedures and channels with adequate technical support and cyber security measures from relevant departments.

Management transformation in industries in the era of digitalization is carried out by digitization of business processes and usage of digital tools for the managers' functions and tasks execution. Digitalization of industries is also exacerbated by the use of new digital technologies. Companies use the Internet and websites to increase the number of consumers, facilitate interaction and improve the experience with target audiences, promote products, increase sales and strength positive images and brands. According to [21], near 77% of EU enterprises had a website in 2019. The share of such companies was increased by one percentage point compared to 2017. There are significant fluctuations in this indicator in different countries. More than 90% of companies used the website in Denmark and the Netherlands. Sweden, Austria, Germany and Belgium were close to 90%. Accordingly, less than 60% of enterprises that had website were observed in Greece, Portugal, Bulgaria and Romania.

Digital technology and software can improve management efficiency by increasing the amount of information available for analysis and decision-making as well as speeding up its processing. Software application such as enterprise resource planning (ERP) allows combining internal and external information to manage several subsystems of companies: personnel, marketing, purchases, sales, planning, finance and customer relationship. The share of EU companies used ERP was 36% in 2019, as well as in 2017. This low level of ERP usage is explained by big number of small businesses, that didn't use it, while among large companies there were much more ERP users. The largest share of companies that use ERP is in the information and communication sector (figure 2) [21]. More than 40% of the ERP users are in manufacturing (47%), in electricity, gas, steam and air conditioning (43%) and in wholesale and retail trade (43%). Less than 30% of enterprises used ERP are in construction (22%), in accommodation, transport and storage (27% each).

Customer relationship management (CRM) application is an effective digital tool in marketing management. The share of EU companies used CRM was 33% in 2019. Almost 60% of enterprises in the information and communication sector used operational CRM. The same shares (40%) of enterprises used operational CRM were in wholesale and retail trade, in accommodation, in professional, scientific and technical activities, and in real estate business.

According to survey [22], technologies, data and Internet of Things (IoT) are digital sources of enterprise competitiveness, flexibility and resilience. Examples of such smart companies are Infineon (production of electronics and semiconductors), Schneider Electric and Siemens Electronics Works (industrial automation), Bonfiglioli (production of power transmission components), Moderna (pharmaceutical industry).

Digital transformation elevates important task for industrial management to develop appropriate ecosystems by involving and connecting different organizations, their staff with unique competencies and technologies in order to develop and implement common projects, create business, social and nature – friendly environment, get achieved benefits based on synergy [23].

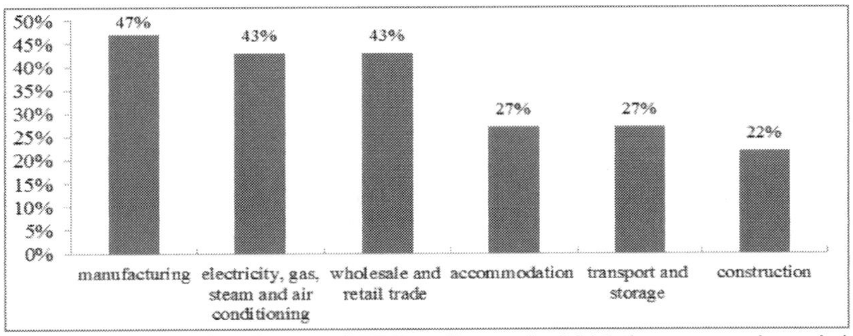

Figure 2. EU companies that used ERP software in 2019, by economic activities [21].

Digitization requires crucial changes in personnel staff, their education, knowledges and skills. For these purposes necessary to provide different training programs. The transformation of organizational culture plays an important role for this. In 2019, companies spent $1.2 trillion on digital transformation but only 13% of companies' leaders say that their organizations are ready for an era of digital transformation, almost 90% of the companies are confronted with the need to re-qualify employees [24].

Companies need not only to overcome understandable resistance, moreover, its' vital to create "digital-friendly" internal environment, correlated organizational culture in order to lead and succeed in life-long digital transformation way.

4. Conclusion

Summarizing, we can generalize that in modern realities, digitalization has different vectors and levels of influence on the nature of socio-economic processes, in particular, changes society, business environment and management of organizations. Digital transformation is a pervasive trend in the modern world. The big accelerator of digital transformation is the pandemic. The era of digitization and Industry 4.0 not only opens up new perspectives, but also poses certain challenges for the development and functioning of modern organizations. Digitalization determines the following directions in the management transformation: digitization strategy development, business model and processes transformation, additional internal and external services need to be provided by appropriate (may be just created) departments, adequate conditions for staff security, remote work, training and support should be provided. Indicators of digital integration in enterprise management of various industries are Internet visibility through a website, the use of software applications and platforms to enhance the ability of companies to analyse data, flexibility and speed of management decisions. Smart factories are represented in many industries such as pharmaceuticals and electronics and semiconductor manufacturing. Such software applications as CRM and ERP are most actively used by EU enterprises in the information and communication sector.

ORCID iDs
O Shatilova http://orcid.org/0000-0003-3883-5567
T Sobolieva http://orcid.org/0000-0002-7818-145X
L Batenko https://orcid.org/0000-0003-0126-609X
M Sahaidak https://orcid.org/0000-0001-6526-1170
T Omelianenko https://orcid.org/0000-0001-6218-0455

References

[1] Competing in 2020: Winners and Losers in the Digital Economy 2020 A Harvard Business Review Analytics Services Report *Harward Business Review* 20 p
[2] Hunt G 2016 Is your business ready for inevitable digital disruption? (infographic) https://www.siliconrepublic.com/business/deloitte-leadership-digital-disruption-infographic Accessed 01.02.2021

[3] Salesforce 2018 What Is Digital Transformation? https://www.salesforce.com/products/platform/what-is-digital-transformation/ Accessed 12.10.2020

[4] I-Scoop 2020 https://www.i-scoop.eu/digital-transformation/ _Digital_transformation_:_online_guide_to-digital_business_transformation Accessed 22.11.2020

[5] Shatilova O 2020 Digital Technologies Influence on Business Organization Management. *Management: strategic imperatives and trends of transformation* Monograph Edited by Sahaidak M and Sobolieva T Kyiv National Economic University named after Vadym Hetman pp 34-49

[6] Strohmeier D 2014 HRM in the digital age – digital changes and challenges of the HR profession *Employee Relations* **36** (4) https://doi.org/10.1108/ER-03-2014-0032

[7] Yurchak O 2016 https://appau.org.ua/publications/industriya-4-0-shho-tse-take-ta-navishho-tse-ukrayini Accessed 12.01.2021

[8] Delo 2017 https://delo.ua/business/chetverta-promislova-revoljucija-chogo-nam-ochikuvati-334676 Accessed 23.03.2021

[9] Solis B 2019 https://insights.prophet.com/the-state-of-digital-transformation-2018-2019_Digital_is_an_enterprise_-_wide_strategic_priority_—_but_there's_work_to_be_done Accessed 19.07.2021

[10] IDC 2020 https://www.idc.com/getdoc.jsp?containerId=prUS46967420_IDC_Reveals_2021_Worldwide_Digital_Transformation_Predictions Accessed 19.07.2021

[11] Digital Transformation Market by Technology (IoT, Cloud, Big Data, AI), Process Transformation (Customer, Operation, Product, Workforce), End-use Industry (Retail, Healthcare, Manufacturing, Insurance), Industry Size - Global Forecast to 2025 2020 Meticulous Market Research Pvt. Ltd. 254 p

[12] Liu Sh 2020 https://www.statista.com/statistics/1131695/top-3-priorities-for-cios/_CEOs'_top_priorities_for_CIOs_to_preserve_business_amid_COVID_-_19_2020 Accessed 19.07.2021

[13] Choudhuri A 2021 https://www.probegroup.com.au/blog/the-ultimate-list-of-digital-transformation-statistics The_Ultimate_List_of_Digital_Transformation_Statistics Accessed 15.07.2021

[14] Murray A and Meyer D 2020 https://fortune.com/2020/05/14/ceo-survey-economic-recovery-ceo-daily_CEOs_don't_see-a_full_economic_rebound_happening_for_years Accessed 16.07.2021

[15] DataDriven 2020 *Global Retail Industry Digital Transformation Trends* 74 p

[16] BCG 2021 https://www.bcg.com/capabilities/digital-technology-data/digital-strategy-roadmap_Digital_Strategy_Roadmap Accessed 15.07.2021

[17] IDG 2018 https://resources.idg.com/download/white-paper/2018-digital-business_Understand_How_Organizations_Evolve_to_a_Digital_Business_Model Accessed 19.07.2021

[18] Wachsman M 2018 https://www.zdnet.com/article/survey-despite-steady-growth-in-digital-transformation-initiatives-companies-face-budget-and-buy-in/ Accessed 19.07.2021

[19] CorporateLeaders 2018 Digital Transformation Survey *Creating Products and Services in a Digital World* 19 p

[20] Nexus Integra 2018 https://nexusintegra.io/digital-transformation-industry/_The_importance_of_digital_transformation_for_the_industry Accessed 19.07.2021

[21] Eurostat 2020 E-business integration https://ec.europa.eu/eurostat/statistics-explained/index.php?title=E-business_integration#Enterprise_resource_planning_.28ERP.29 Accessed 05.10.2021

[22] Romanov R Otchet 80 umnyh zavodov ispol'zuyut peredovye tekhnologii Interneta veshchej https://internetua.com/otcset-80-umnyh-zavodov-ispolzuuat-peredovye-tehnologii-interneta-vesxei Accessed 01.10.2021

[23] World Economic Forum 2020 Reimagining Digital Identity: A Strategic Imperative 21 p

[24] Manju G and Nokuthula L 2020 World Economic Forum https://www.weforum.org/agenda/2020/07/3-ways-digital-business-can-enable-the-great-reset/ Accessed 15.04.2021

IOP Conf. Series: Materials Science and Engineering 1235 (2022) 012075 doi:10.1088/1757-899X/1235/1/012075

Self-cleaning filtration with magnetically controlled particles

I A Shorstkii

Advanced technologies and new materials laboratory, Kuban State Technological University, 2 Moskovskaya st., Krasnodar, Russian Federation

i-shorstky@mail.ru

Abstract. In this work, a new concept of dynamic granular bed was proposed based on magnetically controlled particles. Compared with conventional moving granular bed, a rotating granular layer was formed on dynamic granular bed, which can capture solid particulates. The dust cake can be kept at a certain amount of thickness even after long-term operation. Self-cleaning function of dynamic granular bed installation provides a long-term operation. Hence, the backwash process is not required in self-cleaning installation. Rotating magnetic field was used for granular bed formation with a predetermined porosity in this work. Mechanism of a dynamic granular bed formation described using experimental and visual surface layer data. Periodic working moving granular bed is then changed to continuous and steady working dynamic granular bed. Numerical data of the main filtration process showed a good agreement with experiment. The experimental results show that there exist optimal parameters of filtration superficial velocity and mass flow rate of filter granules for dynamic granular bed. Best efficiency obtained at a filtration superficial velocity of 0.03 m/s and a mass flow rate of 10 l/h for dynamic granular bed in the experiments.

1. Introduction

Filtration is a widespread cleaning process of various kinds of fluids from suspended particles in clarification technologies, pasteurization or sterilization of juices [1], wines [2], beers [3], dairy products, and in water treatment processes.

In the most commonly used filtering process using semipermeable membranes, the components of the solution passed through the capillaries of the membrane are removed from the pressure channel, while the delayed components are collected on the surface of the membranes, which leads to concentration polarization. In such a system, due to the increase in osmotic pressure and the hydraulic resistance of the sediment, the concentration polarization, and the sediment layer reduce the driving force of the process, which negatively affects the efficiency of membrane processes.

In the development of modern filtration plants, the main efforts are aimed at the following tasks:

- reduction of sediment resistance;
- the adsorption of particles on the surface of the filter element, as well as the interaction of particles with each other in the boundary layer, should be minimal;
- reduction of concentration polarization and deposition of particles on the filter element due to the influence of external factors on the filter element.

Based on these tasks, self-cleaning filters appeared in the history of the development of filtration equipment, which has been used in industry for several decades. Such filtration plants are mainly used in industrial closed cycle processes, oil production platforms or submarine systems. The process of self-cleaning or reducing the concentration polarization and thickness of the sediment layer in such filters is performed by backwashing or using mechanical scanners-scanners [4,5]. Self-cleaning filters are also referred to as hydrodynamic filters with a fixed filter element such as a "cylinder in a cylinder" [6] and filters using external physical forces [7]. As external physical effects may be mechanical oscillations [8], directed ultrasonic vibration on the filter element, aeration and other effects [9].

One of the problems of microfiltration noted in [10] is the fixed pore size. The reason for the low cleaning efficiency is that the porous structure does not change during filtration and cleaning. As a result, the sediment that clogs the pores is almost impossible to remove [11]. Based on the described problems and problems, a filtration unit of a self-cleaning filter with a dynamic capillary structure of a filter element based on magnetically controlled particles is proposed.

The term magnetically controlled particles have become widely used in the field of transportation of medicinal products. Such particles, as a rule, are made of composite materials based on ferromagnetic nano powders [12-14], easily supplied to manipulations of external magnetic fields. The rotating magnetic field (RMF) in this paper is a system of permanent magnets rotating around a common axis. The effect of a rotating magnetic field on individual chains of ferromagnetic particles has been extensively studied [15-17]. However, data on the effect on a bulk array of particles have not been studied before. In current paper, a hypothesis was constructed on the possibility of forming volumetric dynamic arrays of particles of the same size with a given stacking structure. The purpose of the work is to describe the mechanism for the formation of a dynamic structure of a filter element based on the developed design of a self-cleaning filter and an analysis of the main characteristics of the proposed filtration plant with an assessment of the quality of food media after filtration.

2. Experimental Section

Structurally, the installation of the filter (Figure 1) consists of a platform, a cylindrical tube, a nozzle with permanent magnets and two stepper motors and a control point. Detailed information on the construction is presented in [17].

As permanent magnets, neodymium magnets are used, which are fixed in the nozzle and closely contact the end surface of the pipe. The magnitude of the distribution of the magnetic field of permanent magnets was measured with the help of a Gaussmeter (Lake Shore 450) and is shown in Figure 1, indicating the initial value of the magnetic induction $B_0 = 0.41$ T. The induction value is minimal in the region of the central axis of the vessel. The maximum possible pressure exerted on the filter element was determined experimentally and was 60 kPa.

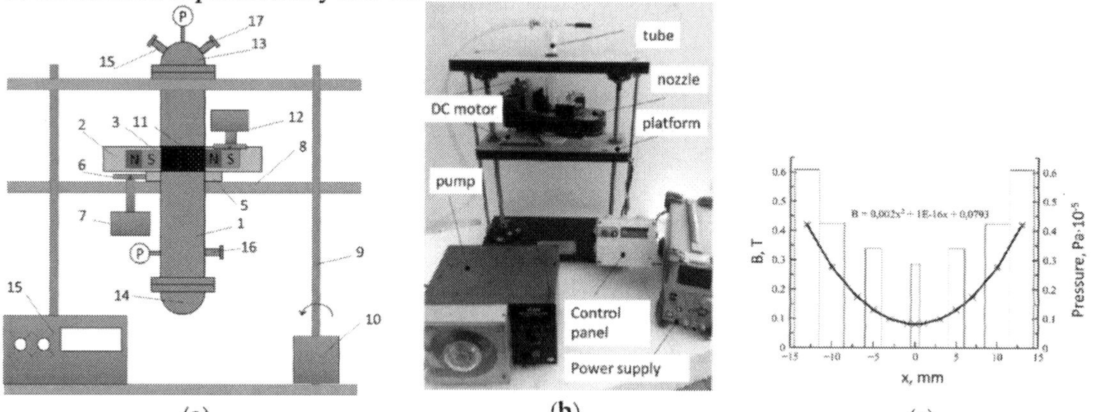

(a) (b) (c)

Figure 1. Setup scheme (a), lab scale filtration setup (b) [18] and the distribution of magnetic induction (points and line) and maximum pressure (vertical lines) along the axial line of the permanent magnets through the centre of the main vessel (c).

Before the installation starts, the magnetic particles 11 are poured into the pipe and trapped by the magnetic field of the permanent magnets 3. Due to the superposition of the rotation factor on the

IOP Conf. Series: Materials Science and Engineering 1235 (2022) 012075 doi:10.1088/1757-899X/1235/1/012075

cylindrical nozzle 2, the filter element particles are rearranged from the fibrous dispersed into a dense package, forming a capillary structure. The filtration unit works as follows: a crude substance with suspended particles under pressure through a branch pipe 15 is fed into the pipe 1 where it passes through a layer of particles held by a constant magnetic field and freed from particles of solid contaminants whose size is larger than the size of the capillaries of the filter cell. The purified substance is then removed from the filter through the discharge port 16.

The presence of a mechanical switch 12 in the form of a stepper motor for reversing the polarity of one of the permanent magnets by rotation provides a self-cleaning or sludge-relief function, which is as follows. In the filtration process, solid contaminants clog pores, increasing, in turn, the hydraulic resistance of the filter element. When the critical value of the pressure difference at the inlet and outlet of the filter element is reached, the operation of the mechanical switch 12 starts. The rotation of the nozzle stops, the polarity of one of the magnets is reversed by the electric motor by rotating a permanent magnet mounted on the bearing assembly. There is a change in polarity, while the filter element is continued to be retained by magnetic force along the pipe wall, destroying only the structure of the particle packing, as shown in Figure 2. Cleaning begins by washing by injecting a cleaning liquid in the same direction as the filtered substance. This factor avoids the use of a countercurrent cleaning system and the engineering communications that accompany it. After the end of the washing, the polarity of the magnet returns to its original position, the structure of the particle packing is restored and the filter resumes the filtration process.

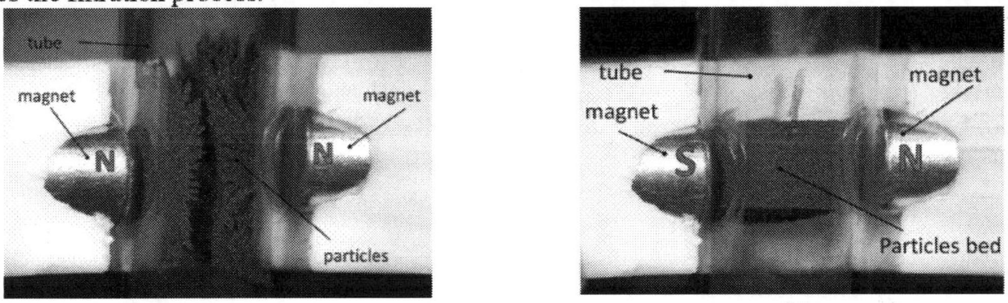

(a) (b)

Figure 2. Magnets polarity change from S-N (a) to N-N (b) around magnetically controlled particles.

As micro spherical particles of the filter element, ferromagnetic particles of a spherical shape of the "core-shell" format with an average size of 60 ± 10 μm were used (Figure 3). X-ray phase analysis (XRD) was used to determine the phase composition of the samples using a Shimadzu XRD-7000 S diffractometer (Japan). CuKα radiation was used (λ = 1.541 Å). The range of angles 2θ was from 5 to 80 degrees. The voltage on the tube is 40 kV, the current is 0.03 A. The scanning step is 0.02 degrees. The phase identification was carried out using the PDF-2 database. The mathematical processing of the diffractograms was carried out using the Shimadzu software.

(a) (b)

Figure 3. Scanning electron microscope image of microspheroidal particles (a) and spherical particle size distribution function (b).

When analysing the X-ray diffraction pattern of the investigated particle samples, it was found that a significant part of the substance is in the x-ray-morphic state. According to the results of XRF, it can be concluded that the samples under study contain phases of metallic iron coated with an inert polymer material. When compared to the PDF-2 and PDF-4 database, the material was identified as a core-shell Iron Oxide (Fe_3O_4). The use of microspherical particles of a filter element with a size of 10 μm makes it possible to form a capillary structure with a capillary diameter of 1 μm through which the filtered substance will pass.

As a tool for visualizing the motion of particles with frontal and surface fixation, the digital camera Canon EOS 70D was used with a shooting speed of 7 frames per second. For photo fixing the change in the structure of the filter element, the photo-series mode was used with a resolution of 5472 by 3648 pixels and a time interval of 500 ms. For video recording of the transition of the structure of the fibrous array into a tightly packed one, the video mode of the front and the surface camera was used with a resolution of 1920 by 1080. For the front shooting, the sector was cut from the cylindrical nozzle at an angle of 120 degrees. The resulting images and video materials of particle motion were recorded on a computer, where they were later processed in a photo and video editor (Vegas Pro 13.0, Sony).

Scanning electron microscopy was used to analyse the stacking structure of the filter element particles. The images were obtained on an SEM (JSM 6300, Jeol, IMRE, Singapore) and TEM (National University of Singapore) electron scanning microscope.

To evaluate the interaction of the filter element material with the filtered product, a study was made of the iron content for samples of distilled water and wine (Taman, Krasnodar). The iron content of Fe (μg/l) in the samples was measured using an inductively coupled plasma atomic spectrometer iCAP 6500 DUO (Thermo, USA) using the PNDF 14.1:2:4.135-98 method for measuring the mass concentration of elements in drinking samples, natural, sewage and atmospheric precipitation by atomic-emission spectrometry with inductively coupled plasma.

3. Results and discussion

Table 1 shows the dynamics of the change in the stacking pattern of the filter element particles in one turn of the nozzle. The process of change is fixed by the surface and frontal chambers. In the absence of rotational motion of the nozzle, the particles line up along the lines of force of the magnetic field, forming a dispersed structure in the form of fibres (Table 1, angle of rotation 0°).

When the process of rotation of the nozzle with permanent magnets around the cylindrical tube occurs, the spherical particles are densified with a uniform distribution of the particle density in the volume of the filter element, and the surface of the filter element acquires an even structure. One turn of the nozzle is sufficient to form the structure of the filter element particles. It should be noted that due to the rotation of the nozzle there is a constant change in the structure of the filter element pores, which positively affects the efficiency of the filtration process.

Table 1. Dynamics of changes in the structure of an array of microspherical particles in the MRF. Adopted from [19].

Rotation angel	Magnetic particles lines	Surface camera	Frontal camera
0°	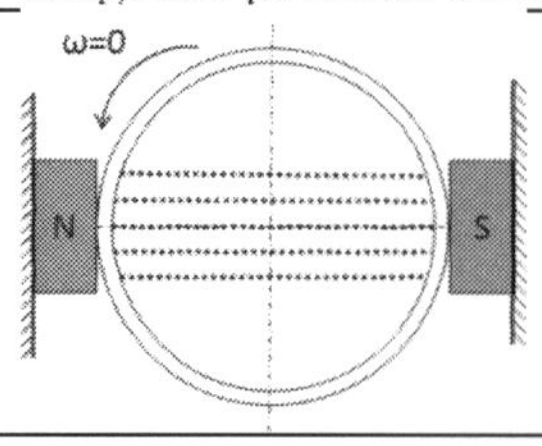	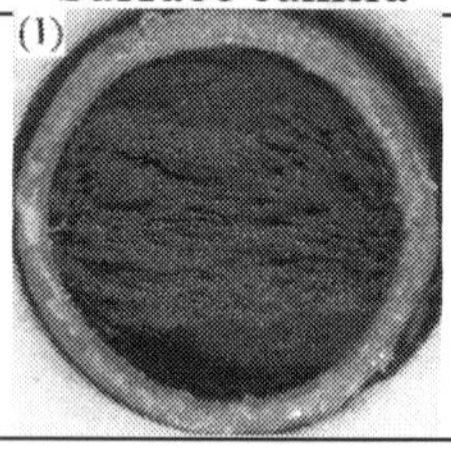	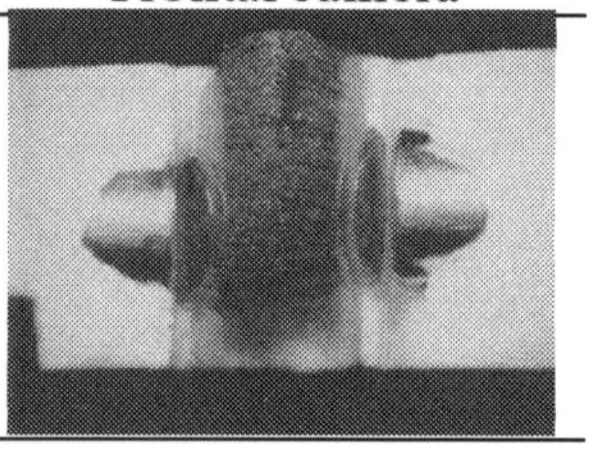

30°	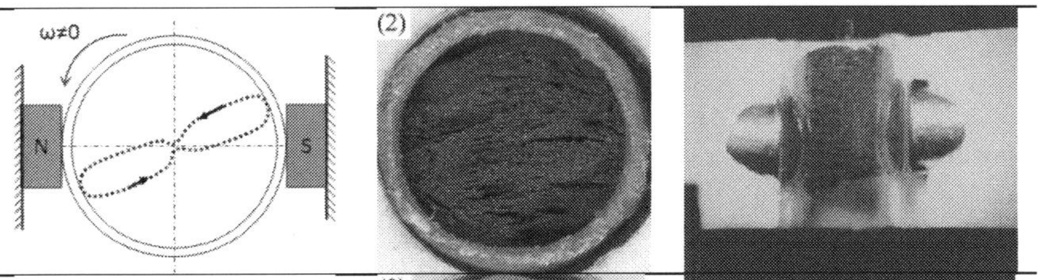
60°	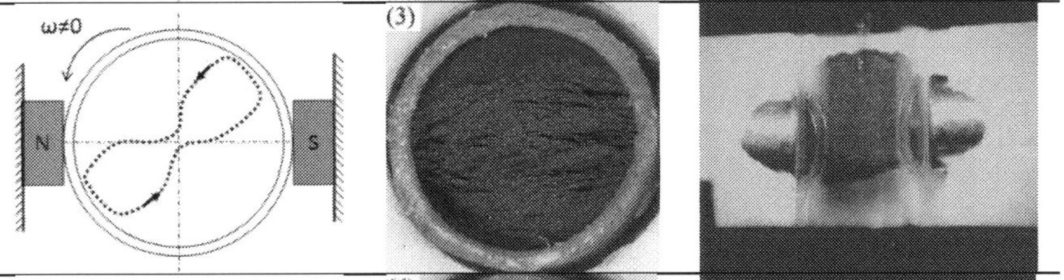
120°	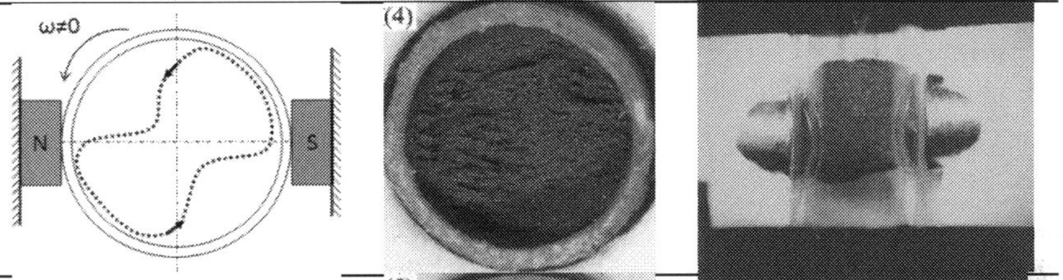
180°	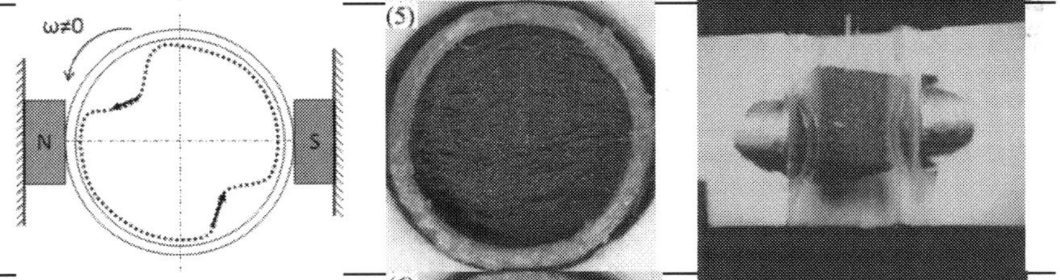
360°	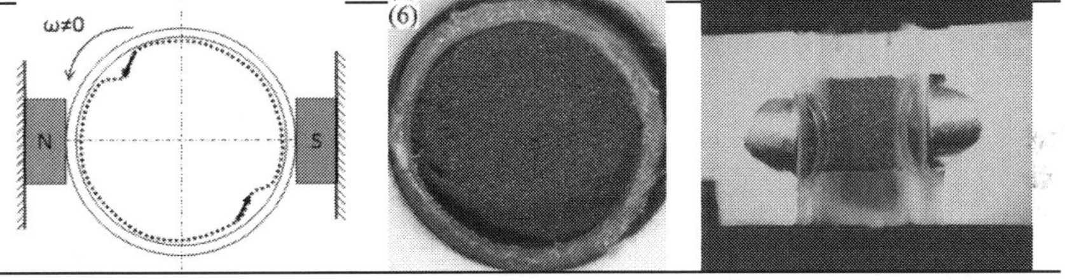

The transition from the dispersed fibrous structure of the filter element particles to the dense packing due to superposition of the VMP is also fixed according to the data of the front camera. The table displays the changes in the height of the filter element layer. In the initial state, the height of the particle layer is maximal and corresponds to two vessel diameters, where the fibrous structure of the filter element particles predominates (Table 1, 0° rotation angle) along the lines of force of the magnetic field. As the nozzle rotates, the height of the particle layer decreases, and the particles become denser. Thus, the resulting filter element of the proposed installation is a dynamic porous structure, which is an important component of the filtration process.

The structure of the filter element is an array of microspheres with porosity (volume fraction of "voids") $\varepsilon = 0.3$. The obtained value of porosity is consistent with numerous experimental data on the porosity of a freely poured layer [20].

When determining the calculated region of a layer of spherical particles, when the particles touch the wall but are not cut, it is possible to estimate the size of the region necessary to eliminate wall effects. The layer under consideration can be represented as a system that consists of two regions: the backfill core (with porosity $\varepsilon_1 = 0.259$, close to the porosity of the face-centred packing) and the boundary region (with a porosity $\varepsilon_2 = 0.476$ equal to the porosity of a simple cubic packing).

Using Darcy's law, we can calculate its effective permeability:

$$k_{ef} = \frac{F_1 \cdot k_1 + F_2 \cdot k_2}{F_1 + F_2}$$

where F_1 and k_1 are the area and permeability of the cross-section of the backfill core; F_2 and k_2 are the area and permeability of the cross-section of the boundary region.

Cross-sectional areas $F_1 = \pi a^2$ и $F_2 = \pi\,(a + d_p)^2$. The values of the permeability of the regions and, accordingly, can be calculated using a formula of the form:

$$k_i = \frac{1}{f_v} d^2 \frac{\varepsilon_i^2}{(1 - \varepsilon_i)^2}$$

The flow rate of the liquid Q to be cleaned through the grain filter element is described by the Kozeny-Karman equation [20]:

$$Q = p \cdot S_f \cdot u \tag{1}$$

where p is the separation coefficient, Pa; S_f is the area of the filter layer, m^2.

The linear velocity u of the flow (cm / s) with a pressure difference on both sides of the filtering membrane ΔP and the viscosity of the fluid η obeys the Poiseuille law [20]:

$$u = \frac{a^2}{8 \cdot q \cdot h \cdot \eta} \cdot \Delta P \tag{2}$$

where a is the diameter of the capillary, m; q - coefficient of curvature; h - filter height, m; η is the viscosity of the fluid to be passed and the pressure difference at the inlet and after the filter Pa·s.

The experimental calculation of the filtration rate was carried out according to the formula:

$$Q_{exp} = \frac{V}{t} \tag{3}$$

where V is the volume of the substance, m3, passed through the filtration barrier in time t. Substituting the obtained values of $Q_{exp} = 0.012$ m3/h.

A comparison of the main characteristics of the theoretical and experimental data of the filtration plant is presented in Table 2. The theoretical and experimental data showed good convergence.

Table 2. Comparison of the theoretical and experimental data of the characteristics of the filtration unit.

Parameter	Filtration velocity, m/h	Capacity, l/h	Porosity coefficient	Maximal pressure, kPa
Theoretical	10.8	12	0.31	90
Experiment	9	11	0.46	70

The iron content in the investigated wine and distilled water samples are shown in Table 3. The amount of iron passed through the filter element increased for Taman wine images from 1954 µg/l to 5.227 µg/l for the first and fifth cycles respectively. However, for samples of distilled water with an appropriate index of pH = 7, the iron content in the samples after the first cycle and the tenth was not detected. The obtained results of the iron content can be related to the effect of the pH factor on the efficiency of its release into the sample under study.

Table 3. Comparison of the theoretical and experimental data of filter characteristics.

Product	Fe, µg/l	pH
Distilled water (DW)	not detected	7
DW after 1 cycle	not detected	7
DW after 10 cycles	not detected	7
Wine after 1 cycle	3152	3.5
Wine after 5 cycles	5227	3.5

4. Conclusions

The possibility of controlling and manipulating the ordered arrays of filter element particles upon application of the RMF is of interest in the field of filtration of liquid alkaline media or water treatment processes. The presented self-cleaning function of the filter has the following main design advantages: supply of cleaning liquid in the direction of the main cleaned flow, absence of additional engineering communications for the filtration process in comparison with existing types of filters, possibility of changing the filtering range from 1-100 microns without changing the basis of the installation. The resulting filter element of the proposed installation is a dynamic porous structure, with the constantly changing geometry of the capillaries due to the use of a rotating magnetic field of permanent magnets, which contributes to a better filtering process. The data of the main technical characteristics showed good convergence of the experimental and theoretical data. Qualitative analysis of the medium during passage through the filter element showed the presence of iron in Taman wine, however, in presence of distilled water, the presence of iron was not detected. Along with the advantages, the installation has a number of disadvantages: high hydraulic resistance of the filter element and limited capacity of the installation.

The further direction of research of this type of filtration equipment is represented in modelling the filtration process and scaling the plant for use in the field of water treatment, filtration of food and non-food environments.

Funding
"The innovation project was carried out with the financial support of the Kuban Science Foundation in the framework of the Commercilizable scientific and innovation projects competition № НИП-20.1-63/20".

ORCID IDs
I A Shorstkii https://orcid.org/0000-0001-5804-7950

References

[1] Vodyakov V N, Kuvshinova O A, Kuznetsov V V and Paksevatkin E N 2013 Analysis of the non-stationary process of flowing microfiltration of juices *Storage and processing of agricultural raw materials* **5** 11-4 (in Russian)
[2] El Rayess Y, Albasi C, Bacchin P, Taillandier P, Mietton-Peuchot M and Devatine A 2012 Analysis of membrane fouling during cross-flow microfiltration of wine *Innovative Food Science & Emerging Technologies* **16** 398-408 https://doi.org/10.1016/j.ifset.2012.09.002
[3] Kretov I T, Shakhov S V, Popov E S and Potapov A I 2010 Investigation of the process of beer separation from a suspension of excess yeast in a plant with tangential-flow microfiltration *Bulletin of Voronezh State University of Engineering Technologies* **1** 38–41 (in Russian)
[4] Vieira A S, Weeber M and Ghisi E 2013 Self-cleaning filtration: A novel concept for rainwater harvesting systems *Resources, Conservation and Recycling* **78** 67-73 https://doi.org/10.1016/j.resconrec.2013.06.008
[5] Yao L and He J 2014 Recent progress in antireflection and self-cleaning technology–From surface engineering to functional surfaces *Progress in Materials Science* **61** 94–143 https://doi.org/10.1016/j.pmatsci.2013.12.003
[6] Meireles M, Prat M and Estachy G 2015 Analytical modeling of steady-state filtration process in an automatic self-cleaning filter *Chemical engineering research and design* **100** 15-26 https://doi.org/10.1016/j.cherd.2015.04.030
[7] Sutherland K S and Chase G 2011 *Filters and filtration handbook* (Oxford: Elsevier) 520 p. https://doi.org/10.1016/B978-1-85617-464-0.X0001-6

[8] Maksimov N P and Mazko A I 2000 Theoretical basis for calculating the throughput through the liquid phase of the filter element of the thickener vibration filter *Mining Informational and Analytical Bulletin* **2** 198–200

[9] Kim H S, Wright K, Cho D J and Cho Y I 2015 Self-cleaning filtration with spark discharge in produced water *International Journal of Heat and Mass Transfer* **88** 527-37 https://doi.org/10.1016/j.ijheatmasstransfer.2015.04.094

[10] Museev T S and Soldatov K V 2017 Construction and the principle of operation of microfilters with variable structure of pores *Actual problems of the humanities and natural sciences* **1** 77–82

[11] Xu T, Zhu Q, Chen X, Li W 2008 Equivalent cake filtration model *Chinese Journal of Chemical Engineering* **16** (2) 214-7 https://doi.org/10.1016/S1004-9541(08)60065-8

[12] Galanov A I, Yurmazova T A, Savel'ev G G, Buldakov M A, Rudyk Yu V, Litvyakov N V, Nechaev K A, Tuzikov S A, Cherdyntseva N V and Yavorovskii N A 2008 Development of a magnetically controlled system for the delivery of chemotherapy drugs based on nanoscale iron particles *Siberian Oncological Journal* **3** 50-7

[13] Shorstkii I and Yakovlev N 2019 Synthesis of magnetically controlled Fe_3O_4 composites and their enhanced microwave absorption properties *Materials Research Express* **6** (4) 046104 https://doi.org/10.1088/2053-1591/aafc58

[14] Shorstkii I and Yakovlev N 2020 Method of absorbing material formation based on magnetically controlled Fe_3O_4 particles *Inorganic Materials: Applied Research* 11 1236-43 https://doi.org/10.1134/S2075113320050317

[15] Gao Y, Hulsen M A and den Toonder J M J 2010 Dynamics of rotating magnetic micro-bead chains. *14th International Conference on Miniaturized Systems for Chemistry and Life Science*. Groningen; Netherlands; October **2010**, 2, pp. 1055–1057.

[16] Li Y H, Chen C Y, Sheu S T and Pai J M 2012 Dynamics of a microchain of superparamagnetic beads in an oscillating field *Microfluidics and nanofluidics* **4** 579-88 https://doi.org/10.1007/s10404-012-0974-y

[17] Shorstkii I A and Savenkov K Yu 2017 Self-cleaning filter based on magnetically controlled particles *Scientific Journal NRU ITMO Processes and Food Production Equipment* **4** 20-6 https://doi.org/10.17586/2310-1164-2017-10-4-20-26

[18] Shorstkii I A 2013 Dynamic filtering device *Patent RF 2544695*

[19] Shorstkii I and Sosnin M 2021 Microwave Absorption Properties of Fe_3O_4 Particles Coated with Al via Rotating Magnetic Field Method *Coatings* **11** (6) 621 https://doi.org/10.3390/coatings11060621

[20] Minko K B, Artemov V I and Yan'kov G G 2011 Hydraulic resistance and effective thermal conductivity of backfill from spherical particles *Bulletin of MEI* **4** 47-55

Industry 4.0: challenges and opportunities in packaging logistics

Á Bányai

Institute of Logistics, University of Miskolc, 3515 Miskolc-Egyetemváros, Hungary

altagota@uni-miskolc.hu

Abstract. The fourth industrial revolution opened new opportunities for the development of industrial processes, as it created the opportunity for companies to transform their conventional systems into cyber-physical systems to meet the increasingly dynamic and diversified customers' demands. Although the primary goal of the development of these cyber-physical systems is to improve the efficiency of operations and increase the quality, but energy efficiency, environmental impact and sustainability also should be taken into consideration. Packaging is a special area of production and service, where due to the significant amount of logistics activity, the design and operation of sustainable processes with low environmental impact are emphasized. Within the frame of this article, the author provides an overview of the opportunities offered by Industry 4.0 technologies in the development of packaging processes of manufacturing and service companies and outlines the challenges to be faced in the implementation of these developments.

1. Introduction

The dynamically changed demands of customers lead to the design and operation of complex supply chain solutions, where the purchasing, production, distribution and reverse logistics have a great impact on the performance of value making chains. The packaging plays an important role in supply chain solutions and the complexity of the supply chain is especially complex in the case of returnable packaging, where the reverse processes represent an important part of the logistics related costs. There are different types of supply chain solutions for returnable packaging, like switch pool or return logistics. These solutions are based on various operation strategies but the essence of these systems is to increase the cost-efficiency, availability and sustainability. These parameters can be increased through the application of new technologies and tools.

The size of the industrial packaging market represents a dynamically changing field of the industry. As the report of the Allied Market Research shows [1], the size of the global market in the field of industry was valued at $53,753 million in 2016 and it is expected, that it will be increased up to $69,787 million until the end of 2023. The forecasted compound annual growth rate is 3.9% between 2016 and 2023. This significant change in the volume makes it necessary to improve the processes of packaging logistics. Packaging logistics includes a wide range of logistic operations, like transportation, loading, unloading, warehousing, cleaning and maintaining of packaging. The Fourth Industrial Revolution offers new opportunities for the packaging industry to improve their processes. The Industry 4.0 technologies, like digital, twin, big data analytics, cloud computing, fog computing and edge computing, augmented and virtual reality, RFID technologies, automation, human-robot interaction, autonomous vehicles and e-mobility makes it possible to make revolutionary changes in the processes of packaging supply chain and transform the conventional operations into a cyber-physical processes. Within the frame of this article the author focuses on the challenges and opportunities of Industry 4.0 technologies and summarizes the main advantages of their application. As the literature review shows, the majority of research papers in the field of packaging logistics are focusing on the optimization of conventional supply chain solutions of packaging and only a few of them describes the transformation of conventional packaging logistics into a cyber-physical system, where hyper-connected value making chains are working together. The application of Industry 4.0 technologies can increase the efficiency, availability and sustainability of packaging supply chain solutions. This paper is organized as follows. Section 2 describes a short systematic literature review focusing on the research results in the field of packaging logistics in Industry 4.0 era. Section 3 describes the model framework of a cyber-physical supply chain model of returnable containers using various industry 4.0 technologies. Section 4 presents a scenario of a multi-pool packaging logistics solution, where the different optimization potentials are demonstrated. Conclusions, results and future research directions are discussed in the last section.

2. Literature review

Within the frame of the short systematic literature review I would like to identify the main research gaps including both descriptive and content analysis. I have used the following keywords to search in the Scopus database: TOPIC: ("packaging" and "Industry 4.0"). Initially, 72 articles were identified. The search was conducted in March 2021; therefore, new articles may have been published since then. The descriptive analysis shows, that packaging logistics in Industry 4.0 era has been researched in the past eight years. The first article regarding the role of the fourth industrial revolution in packaging industry was published in 2013 focusing on the automotive industry [2]. The number of published papers (Figure 1) has been increased in the last years and this fact shows the importance of this research topic. The classification of authors by nationality shows (Figure 2), that packaging logistics is a widely researched topic and the potential in fourth industrial revolution regarding packaging supply chain knows no boundaries.

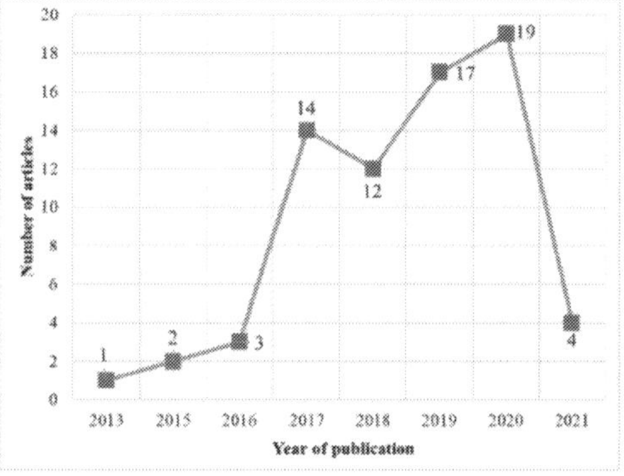

Figure 1. Classification of articles by year of publication based on search in Scopus.

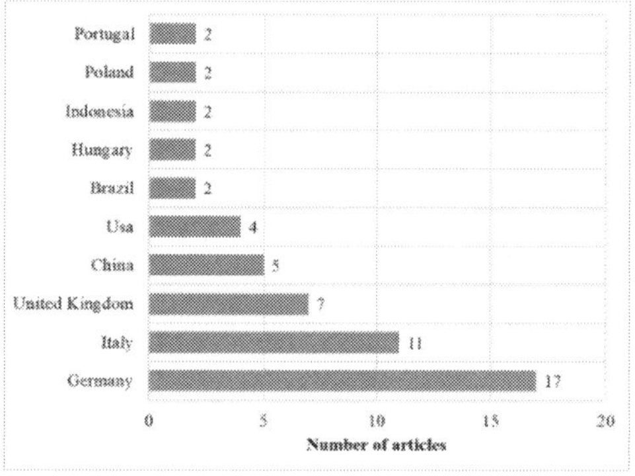

Figure 2. Classification of articles by nationality of authors based on search in Scopus.

Regarding the published articles, we can define ten important keywords (Figure 3). These classification shows the majority of manufacturing, food and automotive industry. The keywords show the importance of Industry 4.0 technologies in packaging industry and highlight the future role of learning systems, decision making methods and artificial intelligence. The classification of articles by scientific field shows the multidisciplinary approach of packaging supply chain problems including engineering, computer science, mathematics and business.

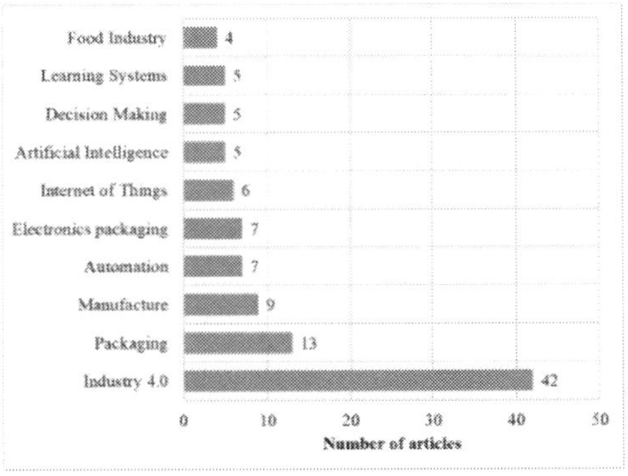

Figure 3. Classification of articles by keywords based on search in Scopus.

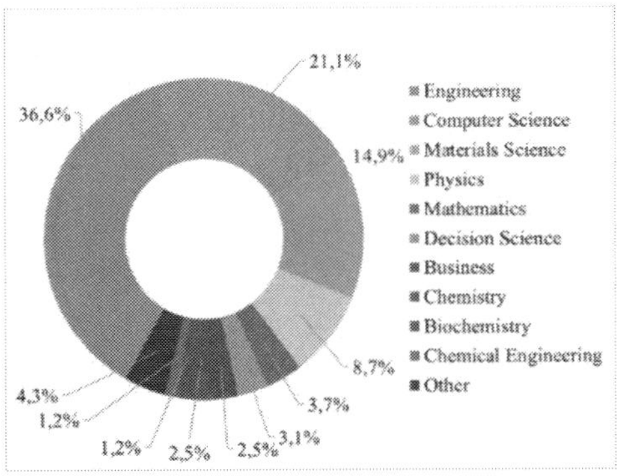

Figure 4. Classification of articles by science field based on search in Scopus.

The digital transformation of conventional food packaging processes is permanently growing and a wide range of techniques is used to improve the safety, quality, availability and productivity of production lines. The diagnostics of the air-tightness of food packaging in a flow production environment can be supported with a prototype of a self-learning software and hardware vision system [3]. The use of RFID digital technologies can also improve the performance of production lines and the safety of related supply chain through increased traceability and visibility. There are several types of applications, tools and solutions that implement RFID technologies not only in packaging but also in manufacturing, storage, assembly and testing operations [4]. The introduction of collaborative robots in production lines for the product packaging step demonstrates, that agile methodology can be used in project management and execution for the introduction of new technologies and prove the effectiveness and applicability of Industry 4.0 technologies [5]. The dynamically increased and diversified customers' demands require the personalisation of both product and packaging. Computer aided decision making processes and methods can support this personalisation trends, as the application of thing-edge-cloud collaborative computing decision-making method proved in the field of personalized customization production and packaging [6]. The performance of packaging processes depends on a wide range of parameters. One of them is the utilisation of containers. Different types of metaheuristics can be used to optimise the packing problem to minimize the empty space in the containers and maximize the number of objects in the containers [7]. The packaging processes are influenced by a wide range of uncertainties, therefore forecasting have a great impact on the performance of packaging processes. The development and the application of a deep learning model for smart manufacturing using convolutional LSTM neural network autoencoders showed, that deep learning methods can improve the reliability of decision making in uncertain environment, as shown in the case of a metal packaging company [8]. Researches highlights, that the application of state-of-the-art technologies, like imaging technology, spectroscopy, multi-sensors, electronic nose, radio frequency identification, printed sensors, acoustic impulse response can improve the quality of products during the packaging, storage, and transportation process and digital twin solutions can increase the impact of optimisation results on the physical system [9]. The Industry 4.0 technologies can be used in special sectors of packaging. An Industry 4.0 pilot project proved the impact of the application of Industry 4.0 technologies on the performance of a semiconductor packaging and test company [10]. More than 70% of the articles were published in the last four years. This result indicates the scientific potential of the transformation of conventional packaging processes into cyber-physical systems. The articles that addressed the development of packaging logistics are focusing on conventional packaging solutions and only a few of them describes the transformation potentials. According to that, the main focus of this research is to describe the challenges and opportunities offered by Industry 4.0 technologies. As a consequence, the main contributions of this article are the followings: (1) general model of packaging logistics focusing on reusable packaging supply chain including Industry 4.0 technologies; (2) mathematical description and optimisation potential in a split-pool cyber-physical packaging supply chain; (3) computational results of the described model.

3. Cyber-physical supply chain model of returnable containers

The supply chain process of returnable packaging includes a wide range of objects and sub-processes. The packaging includes primary, secondary and tertiary packaging. The sender is represented by the supplier, while the recipient is the manufacturing company receiving packaged components from the supplier. The agency organizes the process of supply chain, while the carrier is responsible for the transportation of packaged components from the sender to the receiver and the transportation of empty containers from the recipients back to the sender as a reverse logistics process. The cleaning and maintenance are an important part of the process, because the quality of containers must be checked after each usage, and they must be cleaned before each usage. The end-of-life containers and other packaging materials must be recycled or reused, therefore end-of-life containers must be transported from the end-of-life location to the recycling plant. It is possible to use one or more logistics centres as intermediate storages, depending on the type of the operation strategy (transfer system, depot system with book or deposit option) [13].

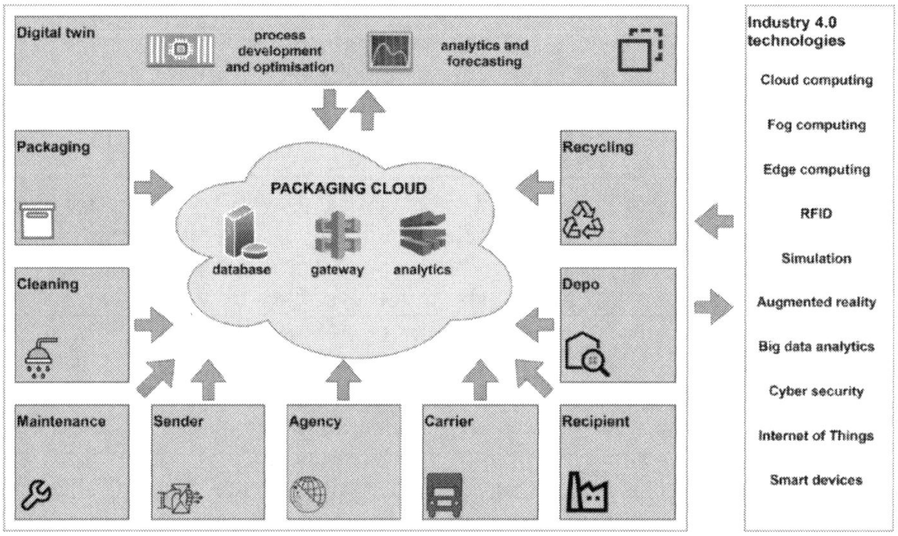

Figure 5. Structure of supply chain of reusable packaging and related Industry 4.0 technologies.

The digital twin can optimize the scheduling, assignment and routing processes regarding the whole system. Scheduling problems include the design of collection and distribution processes of reusable packages, assignment problems include the distributions of available logistics and human resources, while routing problems covers the organization of collection and distribution processes. This digital twin optimization is based on the connected physical system's parameters, which are collected with different sensors used in the system. All data born in the cyber-physical system is collected in the packaging cloud, where analytics prepares input data for the digital twin for optimization. Big data problems can be solved with an integrated cloud, fog and edge computing. The edge computing reduces the response time for optimization using smart sensors. The next session shows a scenario for the optimization of single-pool-based reusable packaging logistics.

4. Computational results

Within the frame of this scenario, we examine how to characterize each cycle of a split-pool supply of reusable packaging. In the first cycle of the examined time horizon, the quantity to be replaced by the carrier can be determined by relation (1). According to (1), if the number of packages delivered with the goods is greater than the quantity of empty reusable packaging available at the receiving party, the carrier shall replace it with a quantity s_1, otherwise it is not necessary to replace the quantity from the carrier pool.

$$b_1 \geq u_1 \rightarrow s_1 = b_1 - u_1 \text{ and } b_1 < u_1 \rightarrow s_1 = 0 \quad (1)$$

where b_1 is the quantity of reusable packages transported in the first cycle, u_1 is the quantity of available empty reusable packages at the recipient. In the first cycle, the quantity of empty packaging available at the receiving party in the following cycle can be determined as follows (2).

$$b_1 \geq u_1 \rightarrow u_2 = b_1 \cdot \vartheta \text{ and } b_1 < u_1 \rightarrow u_2 = u_1 - b_1 + b_1 \cdot \vartheta = u_1 - b_1(1-\vartheta) \quad (2)$$

where ϑ is a parameter specifying the unpacking time of packaged goods, which defines the percentage of packed goods delivered in a given cycle that can be emptied by the next cycle so that it can be returned to the sender. From the second cycle onwards, the above parameters can be determined (3-4). In a given cycle, the amount of packaging to be replaced by the carrier from its own pool can be calculated or the amount of empty reusable packaging available in the next cycle can be determined at the recipient (Table 1).

$$b_i \geq u_i \rightarrow u_{i+1} = b_2 \cdot \vartheta + b_{i-1}(1-\vartheta) \text{ and } s_i = b_i - u_i \quad (3)$$

$$b_i < u_i \rightarrow u_{i+1} = u_i - b_i + b_i \cdot \vartheta + b_1(1-\vartheta) \text{ and } s_i = 0 \quad (4)$$

Table 1. Computational results in the case of switch-pool supply chain[a].

Cycle	**1**	**2**	**3**	**4**	**5**	**6**	**7**	**8**	**9**	**10**
Sent [b]	20	30	20	80	90	70	46	20	22	70
Empty [c]	10	8	24	30	44	84	96	110	126	124
Replaced [d]	10	22	0	50	46	0	0	0	0	0
Cycle	**11**	**12**	**13**	**14**	**15**	**16**	**17**	**18**	**19**	**20**
Sent	20	10	80	90	120	39	50	125	66	50
Empty	96	126	132	90	84	102	150	144	99	134
Replaced	0	0	0	0	0	0	0	0	0	0

[a] ϑ=0,9. [b] Quantity of reusable packaging used for packaged goods shipped in a given cycle. [c] Quantity of empty reusable packaging available at the recipient's site in the current cycle. [d] Quantity of available empty reusable packaging provided by the carrier from its own pool in a given cycle.

The changing value of ϑ does not significantly affect the amount of empty packaging to be replaced from the shipping pool. Peak quantities to be replaced from the carrier's pool only occur when the amount of empty reusable packaging available at the recipient significantly exceeded by the amount of packaging associated with the delivered goods (Figure 6).

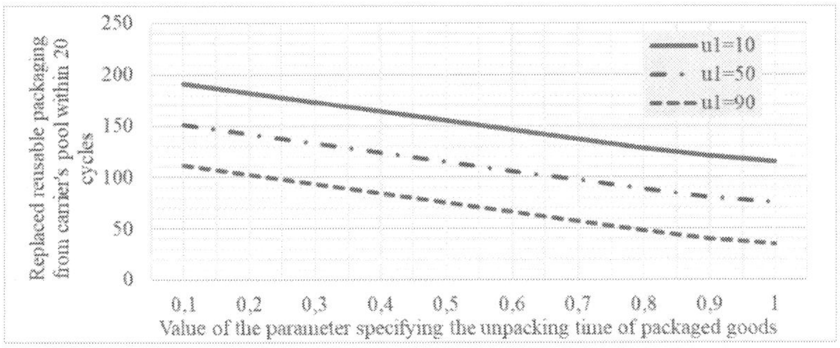

Figure 6. The amount of empty reusable packaging available at the recipient is significantly exceeded by the amount of packaging attached to the delivered goods, but even in this case the process is "damped" during the time horizon assigned to the ϑ parameter.

Applying Industry 4.0 technologies, it is possible to increase the value of ϑ, which significantly reduces the required size of the carrier's pool and at the same time makes processes more predictable, while the availability of the system is increased. The packaging cloud makes it possible to optimize the processes of the reusable packaging supply chain as a complex cyber-physical system, where the solution of above-mentioned optimization tasks leads to the global optimum solution of the supply chain. The digital twin-based optimization, prediction and forecasting offers an increased performance.

IOP Conf. Series: Materials Science and Engineering 1235 (2022) 012076 doi:10.1088/1757-899X/1235/1/012076

5. Results and conclusions

The packaging industry represents a special sector of economy, where the dynamically changed and diversified demands lead to the application of new technologies, tools and solutions to increase the performance, availability, quality, efficiency, reliability and productivity. Industry 4.0 technologies makes it possible to improve the processes of packaging supply chain. Within the frame of this article the author summarizes the available Industry 4.0 technologies from the improvement of reusable packaging supply chain point of view. The digital twin technology makes it possible to improve the potential of conventional simulation-based process development [12] and to support prediction and optimization performance of complex systems [13]. The described simple scenario shows, that the operation parameters of packaging logistics can be improved applying Industry 4.0 technologies.

ORCID IDs

Á Bányai https://orcid.org/0000-0002-2537-7301

6. References

[1] Ashmita R 2017 Industrial Packaging Market Outlook-2023. https://www.alliedmarketresearch.com/industrial-packaging-market
[2] Gruber F E 2013 *Industry 4.0: A best practice project of the automotive industry* Digital Product and Process Development Systems. NEW PROLAMAT 2013. IFIP Advances in Information and Communication Technology (Berlin: Springer) **411** 36-40 https://doi.org/10.1007/978-3-642-41329-2_5
[3] Polyakov R K and Gordeeva E A 2020 Industrial enterprises digital transformation in the context of "industry 4.0" growth: Integration features of the vision systems for diagnostics of the food packaging sealing under the conditions of a production line *Advances in Intelligent Systems and Computing* **908** 590-608 https://doi.org/10.1007/978-3-030-11367-4_58
[4] Centea D, Singh I and Boer J 2020 RFID in manufacturing: An implementation case in the SEPT learning factory *Procedia Manuf.* **51** 543-48 https://doi.org/10.1016/j.promfg.2020.10.076
[5] Franco W F and Behrens F H 2021 Agile project implementation methodology for introducing new technologies in Industry 4.0 *Smart Innovation, Systems and Technologies* **201** 469-77 https://doi.org/10.1007/978-3-030-57548-9_44
[6] Jiang C and Wan J 2021 A thing-edge-cloud collaborative computing decision-making method for personalized customization production *IEEE Access* **9** 10962-73 https://doi.org/10.1109/ACCESS.2021.3050238
[7] Martinez-Rios F, Marmolejo-Saucedo J A, García-Jacas C and Murillo-Suarez A 2020 μθ-EGF: A new multi-thread and nature-inspired algorithm for the packing problem *Mob. Netw. Appl.* **25** (6) 2105-17 https://doi.org/10.1007/s11036-020-01558-8
[8] Essien A and Giannetti C 2020 A deep learning model for smart manufacturing using convolutional LSTM neural network autoencoders *IEEE Trans. Industr. Inform.* **16** (9) 6069-78 https://doi.org/10.1109/TII.2020.2967556
[9] Onwude D I, Chen G, Eke-Emezie N, Kabutey A, Khaled A Y and Sturm B 2020 Recent advances in reducing food losses in the supply chain of fresh agricultural produce *Processes* **8** (11) 1431 https://doi.org/10.3390/pr8111431
[10] Silveira G do N, Viana R F, Lima M J, Kuhn H C, Crovato C D P, Ferreira S B, Pesenti G, Storck E and Righi R da R 2020 I4.0 pilot project on a semiconductor industry: Implementation and lessons learned *Sensors* **20** (20) 5752 https://doi.org/10.3390/s20205752
[11] Guirong Z and Zongjian Z 2012 Green Packaging Management of Logistics Enterprises *Phys. Procedia* **24** (B) 900-5 https://doi.org/10.1016/j.phpro.2012.02.135
[12] Tamás P 2020 Simulation investigational method for collection and distribution systems *J. Prod. Eng.* **23** (1) 79-82
[13] Veres P, Illés B and Bányai T 2019 Integrated assignment and facility location approach based on black hole optimization *Acad. J. Manuf. Eng.* **17** (2) 66-71

IOP Conf. Series: Materials Science and Engineering 1235 (2022) 012077 doi:10.1088/1757-899X/1235/1/012077

Impact of digital twin technology on the real-time optimisation of material handling processes

T Bányai

Institute of Logistics, University of Miskolc, 3515 Miskolc-Egyetemváros, Hungary

alttamas@uni-miskolc.hu

Abstract. The application of Internet of Things (IoT) technologies makes it possible to improve the performance of technological and logistic processes. These processes can include a wide range of IoT technologies and solutions, like cloud, fog or edge computing, intelligent tools and gentelligent products, mobile robots, 3D printing, virtual reality or last but not at least digital twin solutions. Within the frame of this article the author describes the potential of digital twin technology to improve the performance of real-time optimisation in complex logistics systems. After a short literature review, this paper introduces the conception of digital twin solutions. The model of digital twin supported real-time optimisation of an in-plant logistics system is described including the structure, the mathematical model, the solution algorithm, and the computational results. The scenario analysis of the described system validates the model and the efficiency of digital twin solutions.

1. Introduction

Today's economy has undergone major changes in recent years that have significantly changed the expectations of companies operating in the manufacturing and service sectors. The dynamic characteristics of customers' needs and the diversified products have a great impact on the challenges of production and related logistics and supply chain processes.

Industry 4.0 technologies makes it possible to improve the performance of production and service processes. These technologies including cloud, fog and edge computing, digital twin, big data, intelligent tools and gentelligent products, automation, RFID-based monitoring, real-time data management and optimization can be applied both in external and internal logistic processes. External logistic processes include purchasing, distribution and reverse sub-processes, while internal processes cover in-plant material handling solutions, like transportation, loading, unloading, packaging, warehousing. Digital twin solutions represent potential tools to improve the performance of optimization of connected processes and systems. The performance of processes can be measured with availability, cost- and energy-efficiency, quality, utilization of resources, and environmental impact.

This paper studies the impact of digital twin technology on in-plant supply optimization. As the literature review section will show, the majority of the articles in the field of Industry 4.0 technologies are focusing on conventional optimization methods and only a few of them describe the aspects of digital twin supported optimization. This paper is organized as follows. Section 2 presents a short literature review. Section 3 describes the concept of digital twin solutions. Section 4 presents a digital twin-based optimization of logistic operations, where models and methods are described. Conclusions and future research directions are discussed in Section 5.

2. Literature review

Within the frame of this chapter a short literature review is presented. The analytical analysis of the Web of Science-based search shows that 44 articles were published in the field of real-time optimization of logistics processes. As Figure 1 shows, the majority of the articles were published in the last five years; this fact shows that the mentioned topic is extensively discussed.

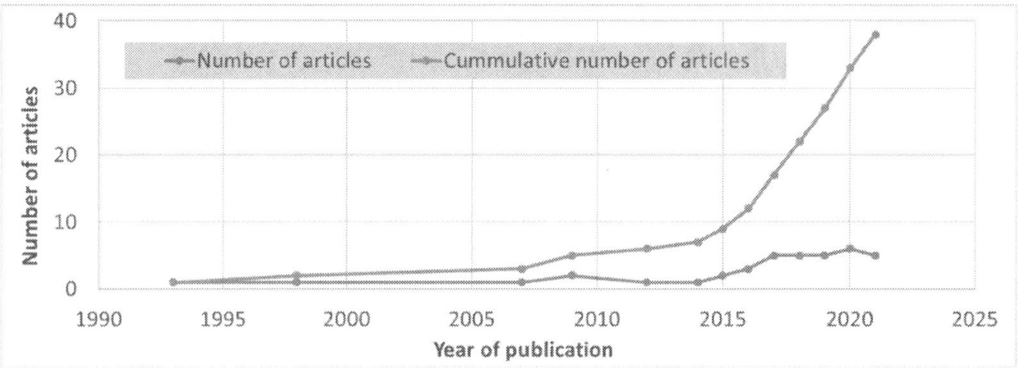

Figure 1. Distribution of publication years and their cumulative value based on search in WoS.

The distribution of research area of analyzed articles based on the search results in Web of Science shows the majority of engineering sciences, while the application of operation research and computer sciences is unavoidable (Figure 2).

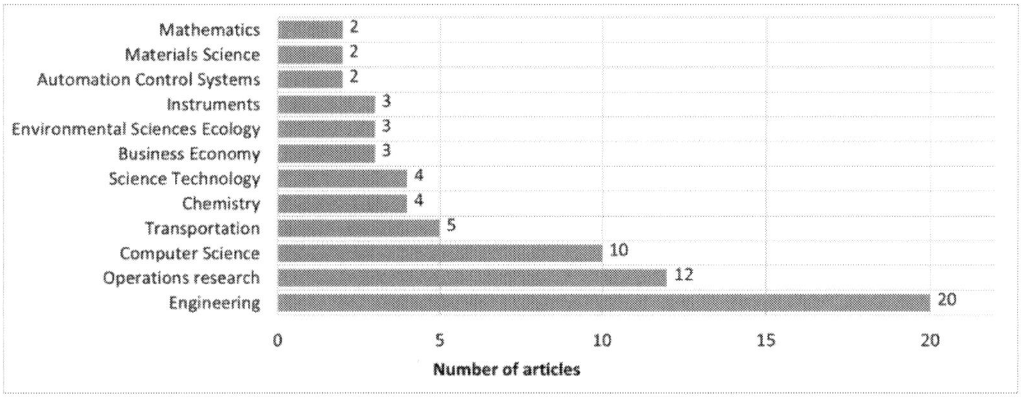

Figure 2. Distribution of research areas of analysed articles based on the search results in WoS.

It is possible to be argued that Industry 4.0 technologies can pave the way for up-to-date in-plant material handling solutions. The literature introduces a wide range of areas and methods used to analyze, solve, and optimize real-time problems. This chapter gives a short overview about the research results of real-time optimization problems. The application potential of FPGA processor real-time system and machine learning is demonstrated in [1], where support travel demand prediction and dynamic route planning in last mile is used in shared subway shuttle bus services. A cloud-based framework shows a novel approach in an integrated product family –supply chain design environment, which demonstrates how both the product and the supply chain can simultaneously be designed based on real-time data consideration [2]. Real-time data management can also be used for intermodal freight transportation. A novel real-time decision support system is proposed by [3]. The decision support system is based on a hybrid simulation-optimization approach, which combines the offline planning and the online re-planning of intermodal transportation tasks. Edge computing, fog computing and cloud computing enabled digital twin solution can support the real-time monitor, decision and control of synchronized production logistics system [4]. The scheduling represents a special type of production optimization problems, where the optimization of NP-hard problems can be solved with information theory inspired real-time self-adaptive methods including real-time task triggering, collaborative acquisition of real-time information and adaptive scheduling [5]. Real-time monitoring and optimization based indoor positioning systems in production and logistics processes significantly improve the efficiency and availability of production systems [6]. The complexity of time-critical freight logistics systems needs real-time optimization [7]. The proposed approach includes order consolidation, defines transshipment locations and assigns shipments to potential consolidation points dynamically and in real time. The cloud manufacturing platforms represents complex manufacturing systems, where tasks must be allocated to resources scattered in various locations for processing. In this complex system artificial neural networks make it possible to schedule tasks in real-time [8] and these solutions can be added to smart manufacturing solutions because of its high response performance. The real-time optimization has a wide range of application areas. A case study shows the potentials of real-time design in the case of precast concrete component transportation and storage [9]. The case study shows the importance of the integrated Industry 4.0 technologies, global navigation satellite system and radio frequency identification, especially from real-time design point of view. The integration of satellite systems and radio frequency identification can support ocean logistics, where the quality and safety of fish catches and supply chain can be improved. Vehicle and vessel tracking, parameter monitoring (temperature tags) and data management improve the safety and efficiency of the ocean supply chain process [10]. The real-time optimization is especially important in the case of first-mile and last-mile logistics, where the demands are dynamically changing. There are different types of solutions for this optimization problem including assignment, scheduling and routing. It is possible use different types of metaheuristic solution, like black hole algorithm, flower pollination heuristics [11], savings-based algorithm [12], tabu search [13], simulated annealing [14], decision diagrams combined with branch-and-bound [15,16] or simulation [17]. The main findings of the above-mentioned review can be summarized as follows:

- The articles that addressed the real-time optimization of logistics systems are focusing on conventional optimization methods without using digital twin technology to improve performance.
- Few of the articles have aimed to include Industry 4.0 technologies to provide additional input for the design and optimization process.
- The number of articles regarding real-time optimization of logistics operations is dramatically increased in the last few years.
- Real-time optimization becomes more and more important in the field of both production and services, while the potential solutions of Industry 4.0 technologies are also taken into consideration.

The main contribution of this article includes: (1) a short overview of research results focusing on real-time optimization of logistic problems in external and internal systems; (2) model and optimization method of digital twin-based real-time material handling optimization; (3) computational results to validate the impact of digital twin on the performance of real-time optimization.

3. Concept of digital twin solutions

The concept of digital twin was anticipated by David Gelernter [18]. Digital twin solutions are Internet of Things platforms that enables to create a digital representation of objects and processes of the real-world. There are four different types of digital twin solutions: parts twin focuses on the digital representation of components, product twin focuses on the interoperability and connection of components, system twin is limited for physical objects while process twin includes workflows [19]. Other classification of digital twin technologies includes digital prototype, digital instance and aggregate, and these types are working in a digital environment. In a digital twin solution, smart sensors are connected to the objects and processes of the real-world and they collect data and send it back to the digital twin. The optimization of the real-world system is based on this interaction.

4. Digital twin based optimisation of logistic operations – model and method

Within the frame of this section a short summary of a digital twin based optimization concept is described. The optimization process can be divided into two main parts. The first part is an extended optimization, where the ERP-based operations are optimized including routing, scheduling, resource assignment and inventory planning. The second part of the optimization includes the optimization of in-plant material handling operations in real-time. The digital twin represents the objects and processes of the manufacturing system and its related material handling system. The digital twin collects data from the physical system with smart sensors and uses the collected data to support prediction and optimization performance (Figure 3).

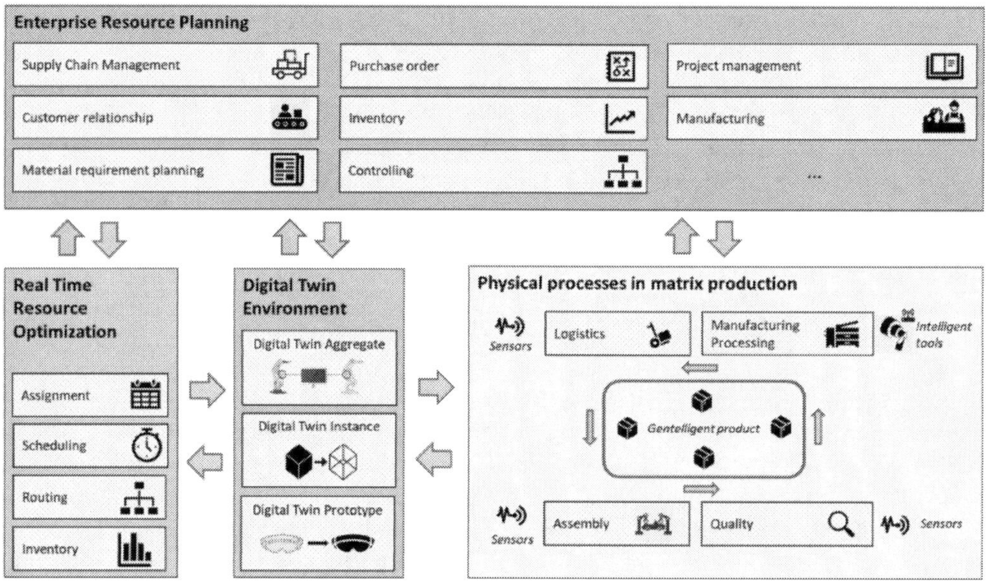

Figure 3. Digital twin-based real-time optimization of in-plant material handling process [20].

The extended optimization has a time-based and an energy consumption-based objective function, while time- and capacity-based constraints has to be taken into consideration. In the second phase of the optimization (real-time optimization) the objective function is the minimization of energy consumption, while capacity- and time-based constraints limit the solution. For the solution of this multi-phase optimization a metaheuristic optimization was used integrating a black hole and a flower pollination algorithm. The optimization process is shown in Figure 4.

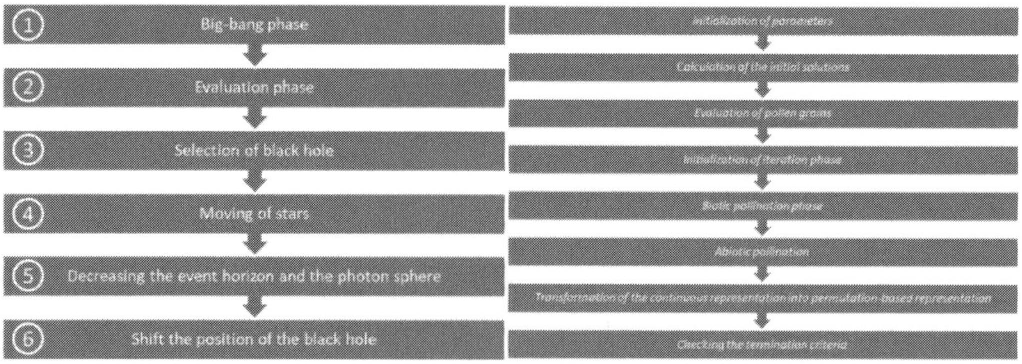

Figure 4. Process of extended a real-time optimization including black hole and flower pollination heuristics.

5. Results and conclusions

The black hole algorithm based clustering of material handling operations and the flower pollination based routing and scheduling makes it possible to find an optimal solution for the extended optimization based on the supply demands of the ERP. Figure 5. shows the input parameters of a simple scenario, where 12 production cells are arranged into a matrix grid and the material handling operations are performed with AGVs. Within the frame of the scenario three in-plant transportation must be performed among matrix production cells (brown arrows). There are supply demands from the supermarket and from the tool storage to the production cells, and some final product must be transported from production cells to the warehouse.

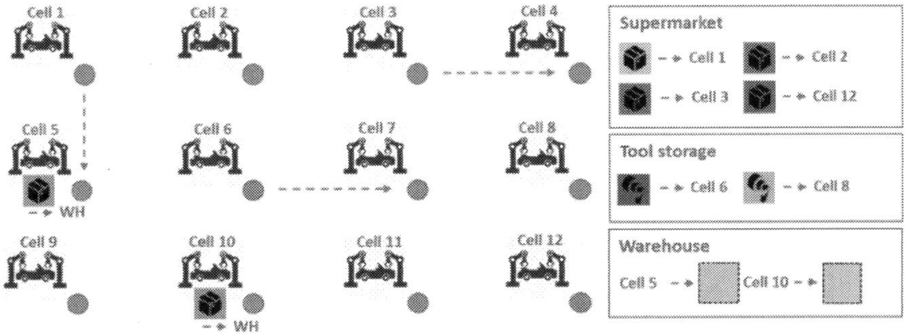

Figure 5. Process of extended a real-time optimization including black hole and flower pollination heuristics.

Figure 6 demonstrates the result of the extended optimization. In this scenario one machine failure was detected by the digital twin for cell 9, therefore the optimal route of the extended design (green arrows) must be changed so, that the deleted operation must be rescheduled to another production cell and the required material handling demand (red arrows) must be performed within a predefined timeframe.

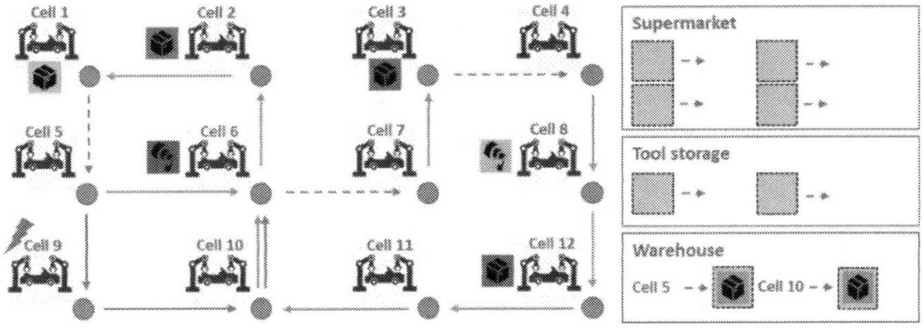

Figure 6. Process of extended a real-time optimization including black hole and flower pollination heuristics.

As the above-mentioned scenario shows, the digital twin technology makes it possible to make predictions based on smart sensor data and simulation. Based on these predictions we can modify the scheduled material handling operations in the production system in real-time. The real time optimization requires fast algorithms, which are able to compute the optimal solution in real-time. The added value of this paper is a short description of potentials of digital twin-based real-time optimization, which makes it possible to insert new supply demands caused by failures in technology or logistics. The results can be generalized for different types of production environment and for global supply chain solutions

[21]. The described methodology makes it possible to support managerial decisions, because the real-time optimization has a great impact on in-plant material handling strategy.

ORCID IDs
T Bányai https://orcid.org/0000-0002-0229-4781

6. References

[1] Dai X Z, Chen M Q and Zhou Y N 2021 *Microprocess. Microsyst.* **80** 103621 https://doi.org/10.1016/j.micpro.2020.103621
[2] Ali S I, Ali A, AlKilabi M and Christie M 2021 *Comput. Oper. Res.* **126** 105112 https://doi.org/10.1016/j.cor.2020.105112
[3] Hrusovsky M, Demir E, Jammernegg W and Van Woensel T 2021 *J. Clean Prod.* **280** (2) 124826 https://doi.org/10.1016/j.jclepro.2020.124826
[4] Pan Y H, Qu T, Wu N Q, Khalgui M and Huang G Q 2021 *J. Manuf. Syst.* **58** (B) 246-60 https://doi.org/10.1016/j.jmsy.2020.10.015
[5] Yang W, Li W, Cao Y, Luo Y and He L 2020 *Sensors* **20** (24) 7007 https://doi.org/10.3390/s20247007
[6] Racz-Szabo A, Ruppert T, Bántay L, Löcklin A, Jakab L and Abonyi J 2020 *Sensors* **20** (23) 6766 http://dx.doi.org/10.3390/s20236766
[7] Salhi S, Gutierrez B, Wassan N, Wu S, Kaya R 2020 *Expert Syst. Appl.* **158** 113574 https://doi.org/10.1016/j.eswa.2020.113574
[8] Chen S, Fang S and Tang R 2020 *Applied Sciences* **10** (7) 2491 https://doi.org/10.3390/app10072491
[9] Liu D, Xin L, Chen J and Jin R 2020 *Adv. Civ. Eng.* **2020** 5714910 https://doi.org/10.1155/2020/5714910
[10] Hao Q, Wang Z F and Qin L-L 2019 *J. Coast. Res.* **94** 204-7 https://doi.org/10.2112/SI94-043.1
[11] Banyai T 2018 *Energies* **11** (7) 1833 https://doi.org/10.3390/en11071833
[12] Li H, Liu Y, Jian X and Lu Y 2018 *Transp. Res. Pt. B-Methodol.* **110** 239-60 https://doi.org/10.1016/j.trb.2018.02.015
[13] Yuan B and Jiang Z 2017 *Sustainability* **9** (12) 2178 http://dx.doi.org/10.3390/su9122178
[14] Heilig L, Lalla-Ruiz E and Voß S 2017 *Flex. Serv. Manuf. J.* **29** 504-34 https://doi.org/10.1007/s10696-017-9280-z
[15] O'Neil R J and Hoffman K 2019 *Oper. Res. Lett.* **47** (3) 197-201 https://doi.org/10.1016/j.orl.2019.03.008
[16] van Lon R R S, Branke J and Holvoet T 2018 *Genet. Program. Evol. Mach.* **19** 93-120 https://doi.org/10.1007/s10710-017-9300-5
[17] Umang N, Bierlaire M and Erera A L 2017 *J. Sched.* **20** 67-83 https://doi.org/10.1007/s10951-016-0480-2
[18] Gelernter D 1991 *Mirror Worlds: or the day software puts the universe in a shoebox—How it will happen and what it will mean* (Oxford New York: Oxford University Press) https://doi.org/10.1093/oso/9780195068122.001.0001
[19] Parks M 2021 Digital twinning: types of digital twins. https://eu.mouser.com/applications/digital-twinning-types/ Accessed 05.05.2021
[20] Bányai Á, Illés B, Glistau E, Machado N I C, Tamás P, Manzoor F and Bányai T 2019 *Appl. Sci.-Basel* **9** (7) 1287 http://dx.doi.org/10.3390/app9071287
[21] Borodavko B, Illés B, Bányai Á 2021 Role of artificial intelligence in supply chain *Acad. J. Manuf. Eng.* **19** (1) 75-9

Development of a "Virtual Learning Factory" for learning and applying specific Lean manufacturing methods

G C Neacsu, I G Pascu, E L Nitu, L Ionescu, A C Gavriluta and N Belu

University of Pitesti, Faculty of Mechanics and Technology, Manufacturing and Industrial Management Department, Targul din Vale Street No.1, Romania.

georgiana.neacsu@upit.ro

Abstract. The processes and systems of production and training, influenced by the digital era, are forced to evolve at an accelerated pace to satisfy the need of innovation and digitization. Thus, the development of virtual learning factories within universities, which can adapt to the requirements of innovation and digitalization, is a solution. In this context, the paper presents how to development a virtual learning factory, flexible-reconfigurable to allow easier learning and experimentation of methods, techniques and tools used in Lean manufacturing. To achieve this, specific Lean manufacturing methods will be digitized for integration into the virtual factory, through programming in the Unity platform, using its specific C # language. Thus, at the end of the paper we want to obtain a virtual, mobile and flexible learning factory, which will help the user to understand, learn and experiment easily and in an attractive way the methods and techniques used in Lean manufacturing.

1. Introduction

To keep up with the needs of today's industry, learning and training processes in the field industrial production need to be developed and updated using digital tools. Thus, in recent years, new learning technologies are applied more and more frequently in education and training [1]. The concept of learning factory meets the current requirements of learning and training [2]. By definition, a learning factory is a small-scale factory which closely mimics a real factory in which participants can learn by doing [3]. Thus, due to the limitations of physical learning factories, in recent years, virtual learning factories have been developed, which combine the concept of learning factory with VR technologies [4]. Virtual learning factories are digital models of learning factories, where training is not determined by location or time, and the number of participants is not limited [5].

The objective of the paper is the development of a virtual learning factory, which would allow the learning and application of specific Lean manufacturing methods. This thing consists in the realization of different virtual spaces associated with production systems (workstation, production line, etc.) in which it is possible to apply different specific Lean manufacturing methods (Dojo, 5S, VSM, Kanban, etc.). The purpose of developing this learning factory is to be able to be used in the university environment, for training students' skills in the field of Lean manufacturing and the use of digital techniques, but also for researching industrial processes and collaborations with the socio-economic environment for the training of specialists in the field of Lean manufacturing. The original idea of this paper is first of all the digitization of specific Lean manufacturing methods and then the development of a virtual learning factory that would allow the learning and application of these methods. For the

IOP Conf. Series: Materials Science and Engineering 1235 (2022) 012078 doi:10.1088/1757-899X/1235/1/012078

fulfilment of the main objective the following steps have been performed: virtual learning factory design, development of the necessary virtual production space, digitization of Lean manufacturing methods, adapting specific LM methods and techniques to the requirements of the virtual learning factory; realization of the learning factory, integration of specific Lean manufacturing methods and techniques in the virtual factory, testing the virtual learning factory, implementation of LM methods and techniques in the virtual learning factory. Thus, a virtual learning factory was developed within the Laboratory of Production Systems Engineering & Lean Manufacturing within the University of Piteşti, for the beginning, in the form of a Dojo workshop, for training operators in the assembly process. The model of this virtual learning factory is the real learning factory [6] existing in the laboratory. Among the advantages of developing this virtual learning factory are: higher simulation speed, lack of space limitations due to the virtual environment, performing simulations of several different scenarios, reducing safety and risk issues, and the number of participants is not limited. The virtual learning factory will be used for training and will have an interface that will allow modelling the equipment, interacting with them and moving the operator in the virtual environment.

2. Presentation of the real learning factory

To meet the requirements of innovation and digitization of production and training processes and systems, a virtual learning factory was developed in the laboratory of Production Systems Engineering & Lean Manufacturing of the University of Pitesti. The model of this virtual learning factory is the real learning factory [6] existing in the laboratory.

Within the learning factory there are the following components:

- modular-flexible assembly line, consisting of three modular-flexible workstations, modular-flexible supply and transfer systems and a digitized workstation (figure 1a);
- support elements for analysis and Lean manufacturing research platforms (figure 1b);
- two Oculus Rift S devices (Oculus Rift, VR headset and joysticks - figure 1c);
- Desktop system and Laptop Gaming
- Unity platform.

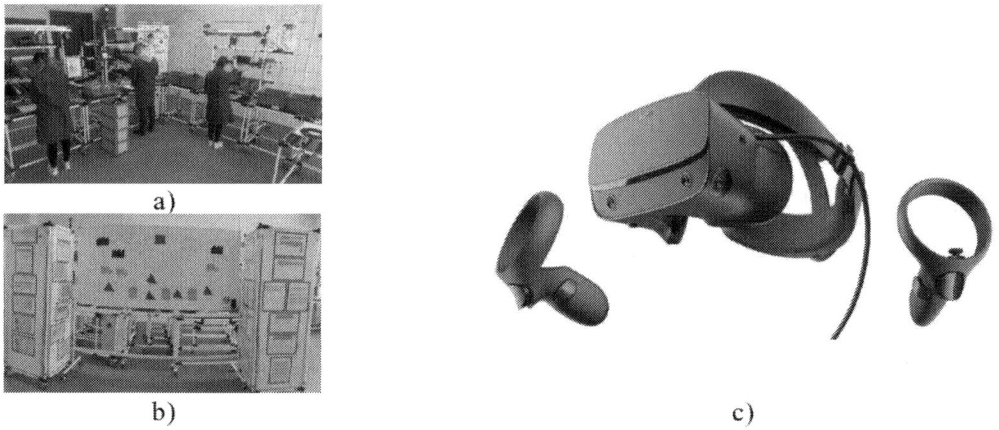

a) b) c)

Figure 1. Real learning factory: (a) modular-flexible assembly line; (b) support elements for analysis and Lean manufacturing research platforms; (c) Oculus Rift S device

In the learning factory, a steering wheel [7], figure 2, is used as an experimental product, a product with a wide variety of components and whose assembly technology includes manual activities and which generate specific production flows / similar to those in the automotive industry. The process of assembling it was chosen as a research and experimentation process. The assembly process of the steering wheel product consists of 4 operations divided into several phases: positioning, threading, assembly and detachment. Most of these will be done manually, and the screwing ones with mechanized means (electric screwdriver).

The experimentation of the methods used in Lean Manufacturing was performed in the experimental laboratory, applying the methods for different situations: VSM, Kaizen, Poka Yoke, 5S, Dojo, Kanban, to later be adapted for experimentation in the virtual learning factory in order to compare the two environments

Figure 2. Experimental product - steering wheel

3. Presentation of the virtual learning factory

For the design of the virtual learning factory, more precisely of the virtual production space, the Unity platform for creating and operating real-time interactive 3D content (RT3D) developed by Unity Technologies was used.

In the construction phase of the virtual learning factory, the following were done:

- designing 3D models, within the Unity platform, of the "fixed" components of the virtual factory (workstations, equipment, etc.) and their integration, as close as possible to the real ones;
- design of 3D models in specialized CAD environments (CATIA V5) of the "mobile" components of the virtual factory: parts and tools and their integration within the Unity platform;

3.1. Design and integration of 3D models of "fixed" components in the virtual factory

The Unity platform was used to design 3D models of "fixed" components (figure 3), workstations and equipment. The 3D models were built using simple geometric shapes of cylinder and cube type, from the collection of 3D objects of the platform, to which different textures were applied. The dimensions used to design the 3D models of the "fixed" components are: 101x56x194 mm for the workstation, 101x75x164 mm for the workstation shelf and 71x51x86 mm for the conveyor. The work area of the workstation and the conveyor, as well as the shelves are made of steel rails with platinum rollers, to facilitate the transfer of components. The "fixed" components were made according to the model of workstations and equipment in the real learning factory, to help the operator to adapt to the real work environment.

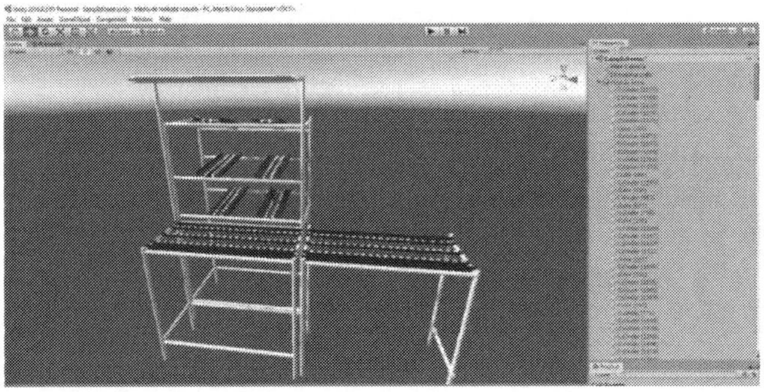

Figure 3. 3D model of the "fixed" components of the virtual factory

3.2. Design and integration of 3D models of "mobile" components in the virtual factory

The design of the 3D models of the "mobile" components: upper body steering wheel, lower steering wheel body, switches, screws, threading and storage box, was made in CATIA V5 software (figure 4) after which they were saved in *.stp format and converted into *.obj format to be imported into the Unity project. After the components were imported, they were placed on the scene (figure 5).

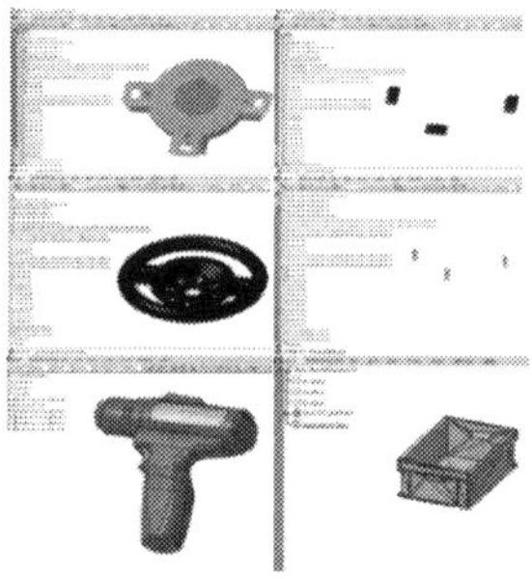

Figure 4. "Mobile" components made in CATIA

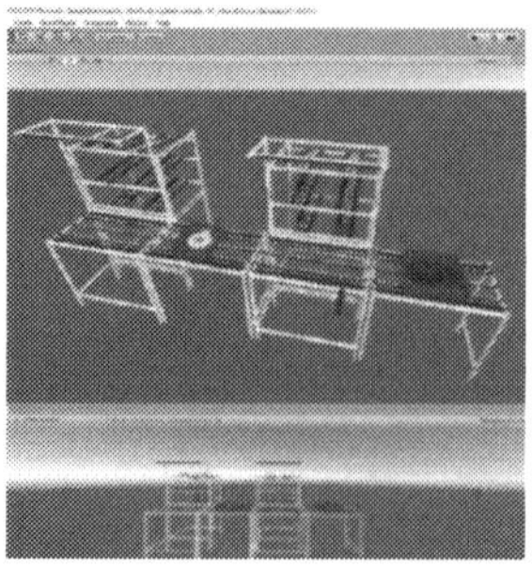

Figure 5. Placing components on stage

The "mobile" components: upper steering body, lower steering body, switches and screws will be handled by the operator to carry out the process of assembling the final product. The thread is used by the operator to fix the screws and the storage boxes are used for the intermediate stock of components. The characteristics of the "mobile" components are specified in Table 1.

Table 1. The characteristics of the "mobile" components

Component name	Material	No pieces / product	Overall size [mm]
Upper body steering wheel	Resin	1	160x160x10
Lower steering wheel body	Resin	1	280x280x15
Switches	ABS	3	21x15x4
Screws	Steel	3	M5x10
Storage box	ABS	4	500 x 350 / 200 x 150

4. Digitization of the learning factory

The digitalization of the learning factory refers to the implementation of the actions of movement, grasping and manipulation. These actions have been achieved using the Unity graphical interface and the C # programming language and are necessary for the most realistic virtualization of specific Lean Manufacturing methods.

4.1. Implementation of movement actions in the virtual factory

In the virtual factory, movement actions were implemented in the form of two types of movements: physical movement and teleportation movement. Thus, for teleportation it is necessary to use the controls to determine the place of movement, and by operating one of them (figure 6), the teleportation is done in that place. The teleportation action is done by building a script (figure 7) in the C # programming language of the Unity platform.

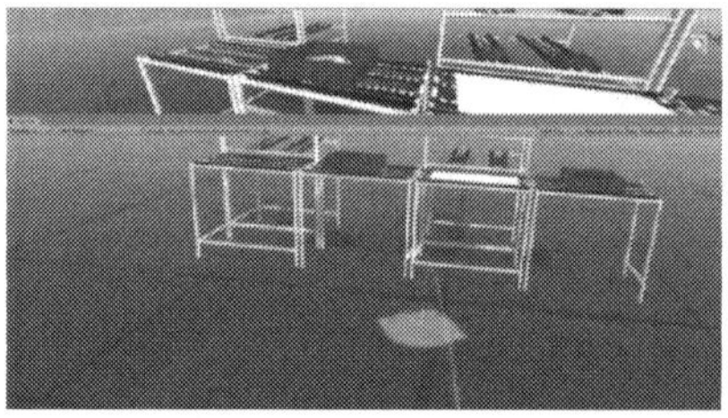

Figure 6. Teleportation to the virtual factory

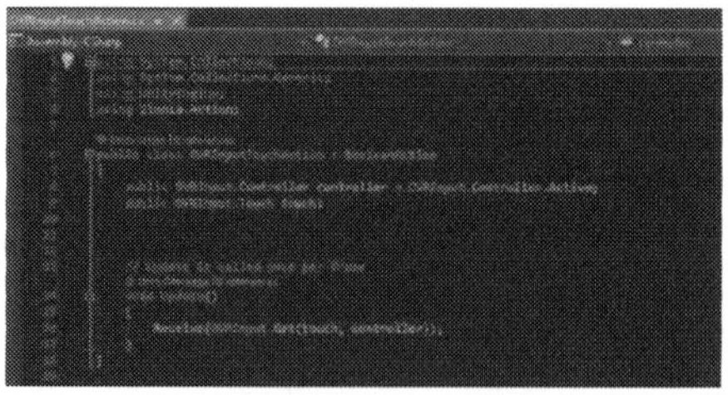

Figure 7. Script for teleportation in the C # programming language

4.2. Implementation of grasping and manipulation actions in the virtual factory

The grabbing and manipulating actions (figure 8) to be done in the virtual factory are grabbing and manipulating the 3D models of the "mobile" components (virtual objects) by moving or pressing the controller. These actions are done by building a script (figure 9) in the C # programming language of the Unity platform.

Figure 8. Manipulating objects in the virtual factory

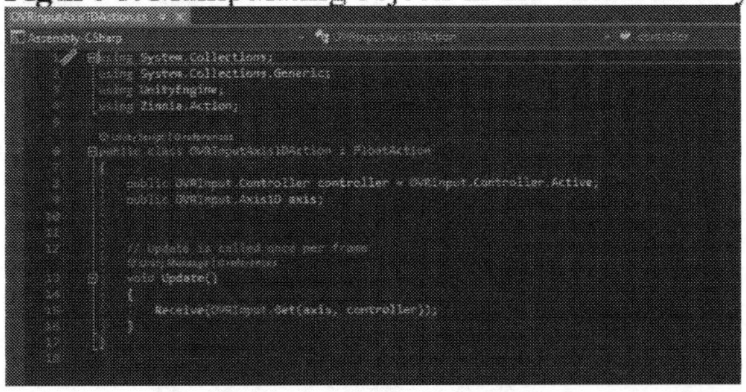

Figure 9. Script for grabbing and manipulating in the C # programming language

4.3. Implementation of the pick to light system

The role of the pick to light system is to guide the operator in the assembly process of the final product, by indicating the order of choice of components and their positioning. For this, two sphere-type objects were created (figure 10). The principle of operation of the pick to light system was made by building a script in the C # programming language of the Unity platform.

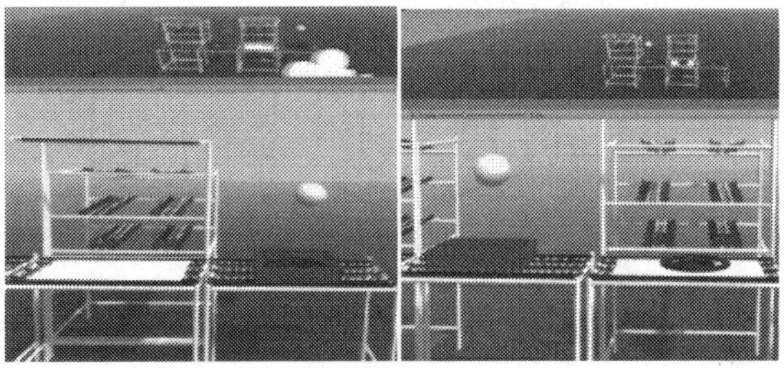

Figure 10. Pick to light system

5. Testing the virtual learning factory

The process of experimenting with the virtual learning factory refers to experiencing its functionality. Thus, the stages of the experimentation process in the virtual learning factory are represented by the movement, grasping, manipulation and assembly of 3D models of "mobile" components helped by the pick to light system and the movement of the operator from one workstation to another. Pressing the Play button on the toolbar leads to the appearance of 3D models of the "mobile" components in the box, as well as the indicator specific to the pick to light system (sphere) which indicates the first step to be perfomed in the assembly process. Thus, the operator is placed to the workstation (figure 11a), from where he has to move (physically or by teleportation) to the first component indicated by the pick to light system (figure 11b). After the operator has taken the component, it must move (physically or by teleportation) to the work area (figure 11c) where he placed it (figure 11d). The next step is for the operator to track the pick to light system (figure 11e) and move to where it needs to pick up the next component (figure 11f). After the operator has taken the component, he must move (physically or by teleportation) to the work area (figure 11g) where he placed it over the first component (figure 11h), in order to carry out the assembly (figure 11i). After the assembly has been carried out by pressing the S key, the whole process is resumed. Figure 11 shows the stages of the assembly process made in the virtual learning factory. After assembly, pressing the S key resumes the whole process.

IOP Conf. Series: Materials Science and Engineering 1235 (2022) 012078 doi:10.1088/1757-899X/1235/1/012078

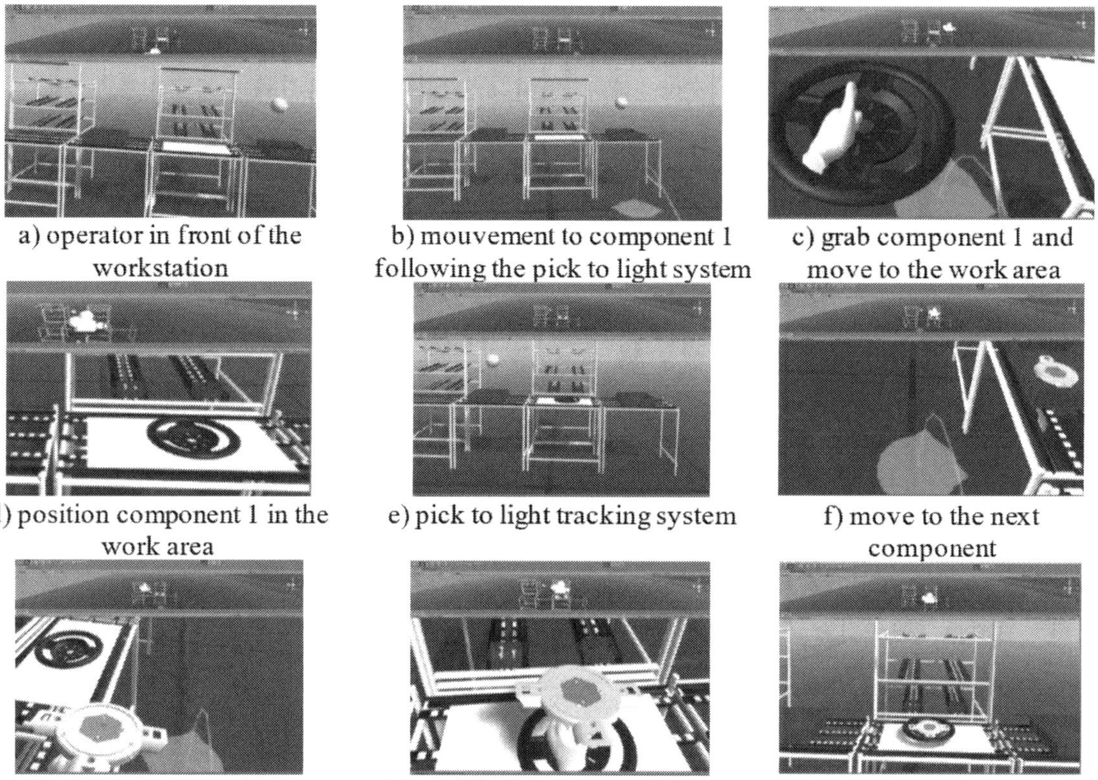

a) operator in front of the workstation
b) mouvement to component 1 following the pick to light system
c) grab component 1 and move to the work area
d) position component 1 in the work area
e) pick to light tracking system
f) move to the next component
g) grab component 2 and move to the work area
h) positioning component 2 on component 1 (assembly)
i) final product (steering wheel)

Figure 11. Testing the virtual learning factory

6. Conclusions

The paper shows how an existing learning factory began to be virtualized. The application within the virtual learning factory was developed in the form of a Dojo workshop, for training operators in the assembly process. Thus, after completing all the stages of development of the virtual learning factory and experimenting with its operation, it was found that the application is functional and can be used as a Dojo workshop. The paper addressed issues related to the design and integration of 3D models of components in the virtual factory, implementation of basic actions of VR applications: moving, grasping and handling, digitization of methods by programming in C # and integration and use of a pick to light system. Research will continue by experimenting with the Dojo application and comparing it with experimentation in the real learning factory. The aim is to adapt the virtual learning factory for learning and experimentation and other methods, techniques and tools used in Lean Manufacturing, as well as to develop the interaction of several users in the virtual environment. It is also hoped that this virtual learning factory can be used for teaching and research and development, as well as for training, by conducting training courses.

Acknowledgments
This work was done within the project Development of a "Virtual Learning Factory" for learning and applying specific Lean manufacturing methods, no. CIPCS-2020-04, funded by the University of Pitesti.

ORCID IDs
G C Neacsu https://orcid.org/0000-0002-4131-0991
I G Pascu https://orcid.org/0000-0001-8992-8232

E L Nitu https://orcid.org/0000-0003-4571-3488
L Ionescu https://orcid.org/0000-0002-4774-9524
A C Gavriluta https://orcid.org/0000-0001-9121-8475
N Belu https://orcid.org/0000-0002-5961-8783

7. References

[1] Riemann T, Kreß A, Roth L, Klipfel S, Metternich J and Grell P 2020 Agile implementation of virtual reality in learning factories *Procedia Manufacturing* **45** 1-6 https://doi.org/10.1016/j.promfg.2020.04.029

[2] Hellriegel J and Čubela D 2018 Das Potenzial von Virtual Reality für den schulischen Unterricht – Eine konstruktivistische Sicht *MedienPädagogik: Zeitschrift für Theorie Und Praxis Der Medienbildung* pp 58-80 https://doi.org/10.21240/mpaed/00/2018.12.11.X

[3] Abele E, Metternich J, Tisch M, Chryssolouris G, Sihn W, ElMaraghy H, Hummel V and Ranz F 2015 Learning Factories for research, education and training *Procedia CIRP* **32** 1-6 https://doi.org/10.1016/j.procir.2015.02.187

[4] Haghighi A Shariatzadeh N and Gunilla S 2014 http://conferences.chalmers.se/index.php/SPS/SPS14/paper/view/1729 Accessed: 04/12/2020

[5] Yildiz E, Møller C and Bilberg A 2020 Virtual Factory: Digital Twin Based Integrated Factory Simulations *Procedia CIRP* **93** 216-21 https://doi.org/10.1016/j.procir.2020.04.043

[6] Nițu E L and Gavriluță A C 2019 Lean Learning Factory at the University of Pitesti *IOP Conf. Series: Materials Science and Engineering* 591 012095 https://doi.org/10.1088/1757-899X/591/1/012095

[7] Gavriluță A, Nițu E L, Gavriluță A, Anghel D C, Stănescu N D, Radu M C, Crețu Gh, Biriș C M and Păunoiu V 2018 The development of a laboratory system to experiment methods to improve the production flows *Proceedings in Manufacturing Systems* **13** (3) 127-32

IOP Conf. Series: Materials Science and Engineering 1235 (2022) 012079 doi:10.1088/1757-899X/1235/1/012079

Experimental tests to validate the usage domain for personalised formworks

Raluca Diaconu and Paul Dan Brindasu

University "Lucian Blaga" of Sibiu, Faculty of Engineering, Emil Cioran 4, Sibiu, Romania

raluca.diaconu@ulbsibiu.ro

Abstract. Analysing the nowadays tendency for customised products, the personalised formwork is a solution that is worth taking into account. Not only that the concrete mould (due to the composite membrane) can take various shapes, function of the designer's requirements, but it can be reusable and remodelled at each usage. By using the Product Lifecycle Management (referred as PLM) principles, two directions were analysed: a pushing system without any auxiliary reinforcement of the formwork and a pulling system with additional reinforcement under the formwork membrane. The composite membrane is subjected to multiple tests with different variants. By analysing the resulting values regarding the final shape of the product (concrete element), the proposed solutions were validated by defining the usage domain for each type of personalised formwork.

Introduction

In the construction domain, a known fact is the significant influence of formwork in the product's cost. The year 2021 is the year with significant price escalation for wood (up to 250%) and metallic structure (100%). Also, an important factor for pouring a concrete elements is the selection of the correct formwork. Hence, the cost will be risen from labour expenses, rentals or even recasting [1].

By achieving a complex experiment regarding types of personalized formwork products, one can determine when and where to use them, in a correct and productive manner.

1. Obtained technical solutions

By passing through each PLM step [2], two theoretical solutions were selected in order to be subjected to experimental testing (see Fig. 1,2): a pushing system without any auxiliary reinforcement of the formwork (solution 1) and a pulling system with additional reinforcement under the formwork membrane (solution 2) [3].

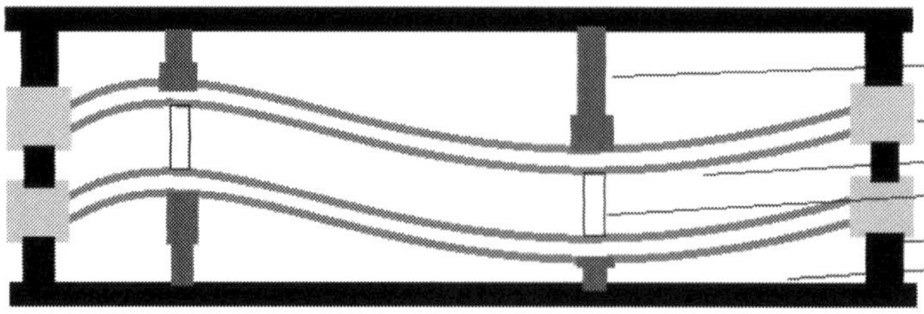

Figure 1. Solution 1 for personalised formwork - sketch

IOP Conf. Series: Materials Science and Engineering 1235 (2022) 012079 doi:10.1088/1757-899X/1235/1/012079

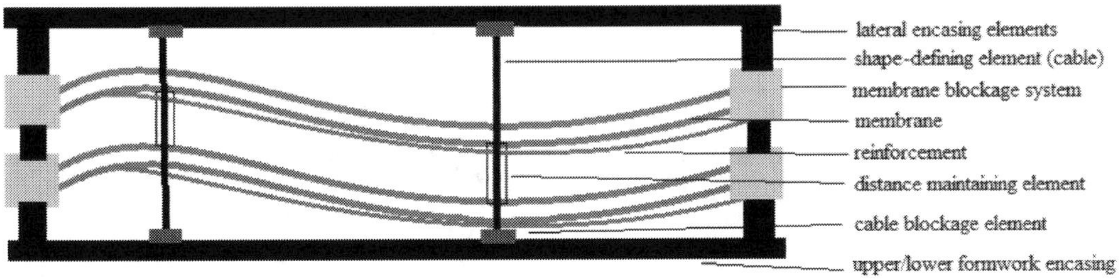

Figure 2. Solution 2 for personalised formwork – sketch

2. Input experimental data

In order to have a complex and enlarged view of the proposed personalised equipment, three test series were analysed for each of the two solutions: a single curvature element (meaning the equipment has only a single shape-defining element), double curvature products (with two shape-defining elements) and a multiple curvature element (with more than three modelling elements). The input data for the experiment are listed in the table below (Table 1).

Table 1. Experiment variables.

Variable type	Variable name	Variable values		
	Material thickness		SBR 1 mm, SBR 3 mm with fiber insertion, SBR 3 mm, NBR 1 mm, NBR 2 mm with fiber insertion, EPDM 1 mm	
Independent variables	Shape-defining element height	h_{st} = 10 mm	h_{st} = 10 mm h_{st} = 15 mm	h_{st} = 5 mm h_{st} = 10 mm h_{st} = 15 mm
Dependent variables	2D Shape-defining element position: the xOy system	St 1 (100,100)	St 1 (100,100) St 2 (150,150)	St 1 (50,100) St 2 (170,80) St 3 (150,150)
	Deformation height (the height between two shape-defining elements)	$h_{inter-st}$ – according to final measurements		
Control variables	Concrete characteristics	$\gamma_{concrete}$ = 2500 kg/m^3 $g_{concrete}$ = L dist = 50 mm		
	Shape and type of shape-defining element	$diam_{cable}$ = 3 mm/$diam_{st}$ = 8 mm		
	Membrane dimensions	L x l: 250 mm x 250 mm		
	Concrete pressure	P_{beton} = 2.584 kN/m^2 (computed for a concrete thickness of 50 mm)		
	Experiment environment (pouring concrete)	T (°C) constant, low constant humidity, fixed testing base		

Note: the shape-defining elements were chosen stochastic to have a larger diversity of the results.

For comparison purpose, the results are listed in parallel for each curvature type, analysing the pushing non reinforced system and the reinforced proposed solution.

3. Single curvature concrete element

For the first test, a single shape-defining element was positioned on each membrane type (1-6) (Fig. 3, 4)

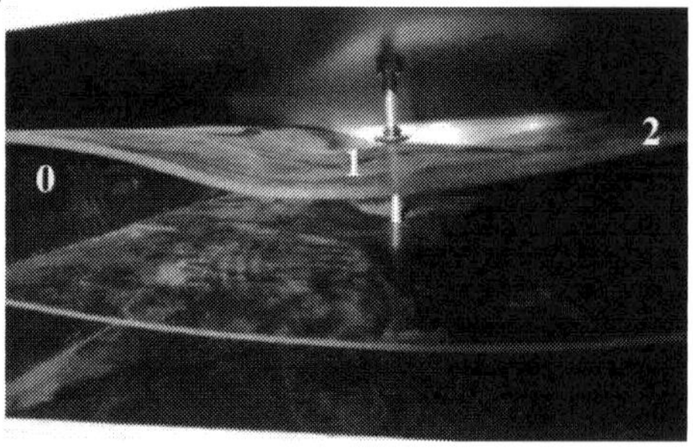

Figure 3. Single pushing shape-defining element

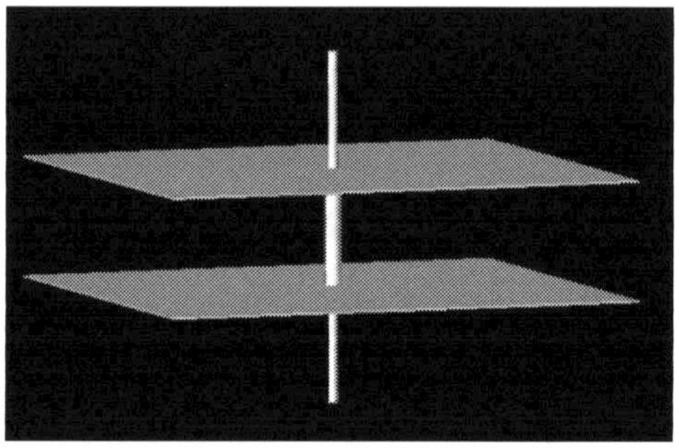

Figure 4. Single pulling element

Measurements were made for the $h_{inter\text{-}st}$ (in the median zone) taking into account the imposed initial length and the final dimension (Table 2). The final measurements were made after concrete pouring, when the element reached its concrete strength for dismantle.

Table 2. Experiment variables.

Membrane type	H shaping element [mm]	$h_{inter\text{-}st}$	Initial dimension [mm]	Final dimension sol. 1 [mm]	Final dimension sol. 2 [mm]
	10	0-1	5.5	4.7	5.51
1	10	1-2	5.5	4.8	5.505
2	10	0-1	5.5	5.0	5.51
2	10	1-2	5.5	4.9	5.51
3	10	0-1	5.5	5.0	5.5
3	10	1-2	5.5	5.0	5.49
4	10	0-1	5.5	5.1	5.5
4	10	1-2	5.5	5.2	5.51
5	10	0-1	5.5	5.2	5.5
5	10	1-2	5.5	5.2	5.5
6	10	0-1	5.5	5.2	5.5
6	10	1-2	5.5	5.3	5.5

Note: 0-1-2 (from $h_{inter\text{-}st}$) are the starting point (0), the shape defining point (1) and the end point of the membrane (the last point- in this case 2).

By using ANOVA [4] and introduction into the program the difference between the initial height and its final dimension ($h_{st\,init} - h_{st\,fin}$), we obtained the following summarised data (Table 3,4):

For solution 1:

Table 3. ANOVA – solution 1

Source of Variation	SS	df	MS	F	P-value	F crit
Sample	0.175	5	0.035	21	1.4869 E-05	3.105875
Columns	547.215	1	547.215	328329	5.376 E-28	4.747225
Interaction	0.175	5	0.035	21	1.4869 E-05	3.105875
Within	0.02	12	0.001667			
Total	547.585	23				

For solution 2:

Table 4. ANOVA –solution 2

Source of Variation	SS	df	MS	F	P-value	F crit
Sample	0.000155	5	3.1E-05	3.311111	0.041428	3.105875
Columns	600.3501	1	600.3501	64037339	9.77 E-42	4.747225
Interaction	0.000155	5	3.1E-05	3.311111	0.041428	3.105875
Within	0.000112	12	9.37E-06			
Total	600.3505	23				

From all the data obtained, the one we need to focus on is the P-value. Hence, for coefficient P-value in the sample variation < 0.05 signifies that the membrane type is important, for coefficient P-value in the column variation < 0.05 means that the height difference (h_{st}) has a significant influence regarding the deflected shape of the membrane and the P-value in the interaction variation < 0.05 means that the results are statistically relevant (have interaction).

In other words, by looking at the final dimensions of two consecutive defining points, the deflections of each type of membrane are greater than in the case of the reinforced solution. This fact is also seen in the P sample value for solution 2 where it tends to reach 0.05, hence, the membrane type being unimportant because regarding which type of membrane we subject to the test, the reinforcement is the one maintaining the imposed formwork shape.

4. Double curvature concrete element

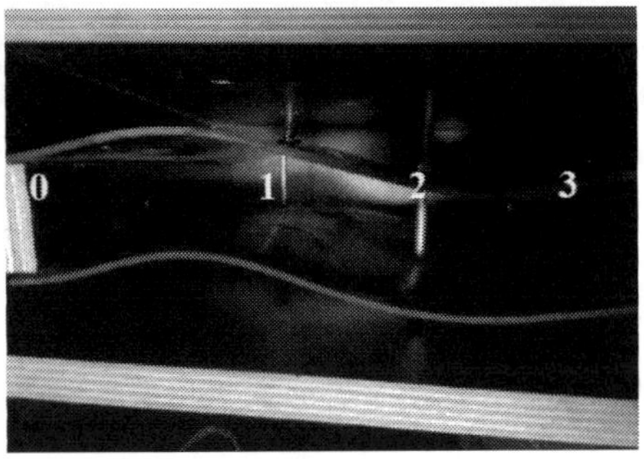

Figure 5. Two pushing shape-defining elements

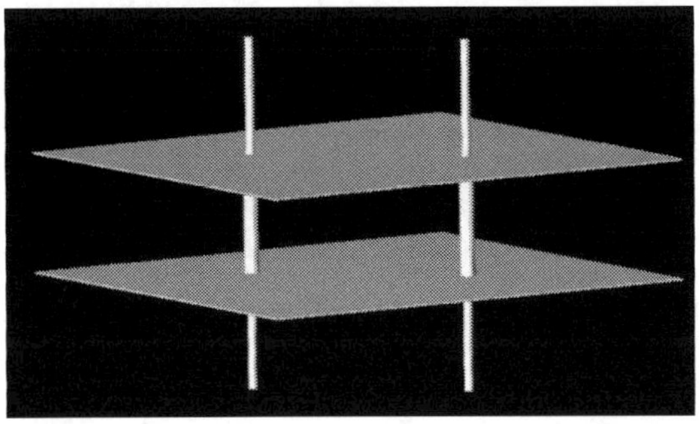

Figure 6. Two pushing shape-defining elements

As in the case for a single curvature element, the initial dimensions for $h_{inter-st}$ were measured (from point 0 to point 1= 5.5 mm, point 1to point 2 = 2.6 m and from point 2 to point 3 = 8.1 mm) and compared with the final results (for both solutions). After using the ANOVA analysis (the same input data principle as stated before), we obtained the following summarised concluding data (Table 5,6):

For solution 1:

Table 5. ANOVA – solution 1

Source of Variation	SS	df	MS	F	P-value	F crit
Sample	0.294722	5	0.058944	0.004715	0.999995	2.620654
Columns	824.6469	1	824.6469	65.9571	2.41 E-08	4.259677
Interaction	0.294722	5	0.058944	0.004715	0.999995	2.620654
Within	300.0667	24	12.50278			
Total	1125.303	35				

For solution 2:

Table 6. ANOVA –solution 2

Source of Variation	SS	df	MS	F	P-value	F crit
Sample	0.001353	5	0.000271	2.17E-05	1	2.620654
Columns	897.2521	1	897.2521	71.77818	1.13 E-08	4.259677
Interaction	0.001353	5	0.000271	2.17E-05	1	2.620654
Within	300.0083	24	12.50035			
Total	1197.263	35				

When increasing the number of shape-defining elements, the membrane type/thickness does not influence the final deflection/deformation (P-value in the sample variation > 0.05). Also, the independent variables do not influence each other (no interaction detected: P-value in the interaction variation > 0.05).

5. Multiple curvature concrete element

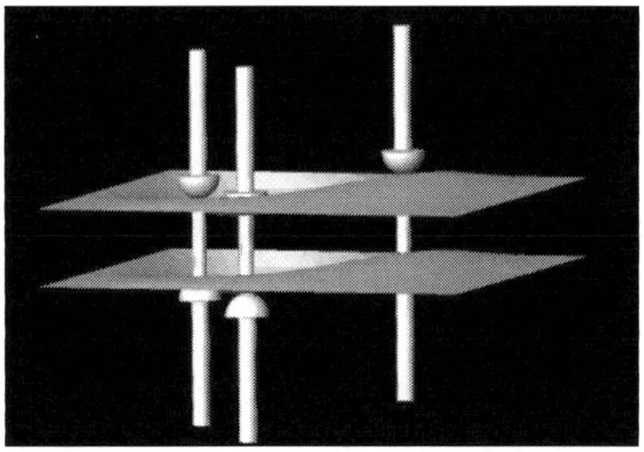

Figure 7. Three pushing shape-defining elements

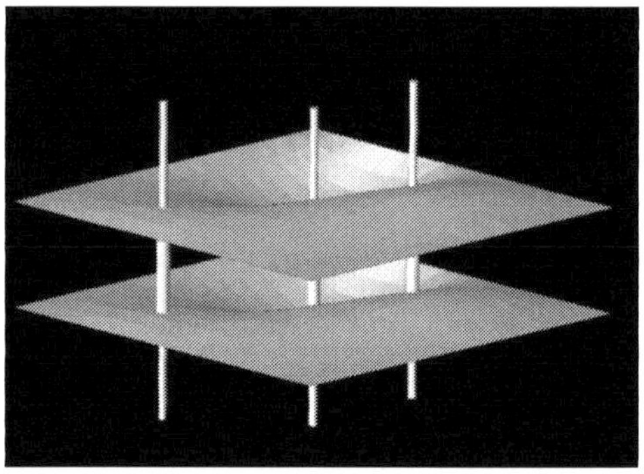

Figure 8. Three pulling shape-defining elements

The ANOVA concluding data are (Table 7,8):

For solution 1:

Table 7. ANOVA – solution 1

Source of Variation	SS	df	MS	F	P-value	F crit
Sample	0.3615	5	0.0723	0.007226	0.999986	2.408514

Columns	877.8375	1	877.8375	87.73038	2.06 E-12	4.042652
Interaction	0.3615	5	0.0723	0.007226	0.999986	2.408514
Within	480.292	48	10.00608			
Total	1358.853	59				

For solution 2:

Table 8. ANOVA –solution 2

Source of Variation	SS	df	MS	F	P-value	F crit
Sample	0.0045	5	0.0009	9E-05	1	2.408514
Columns	952.8135	1	952.8135	95.27857	5.56 E-13	4.042652
Interaction	0.0045	5	0.0009	9E-05	1	2.408514
Within	480.014	48	10.00029			
Total	1432.837	59				

The conclusions are similar to the ones from the double curvature elements, stating that: membrane type is not influencing the results and there is no interaction between the independent variables.

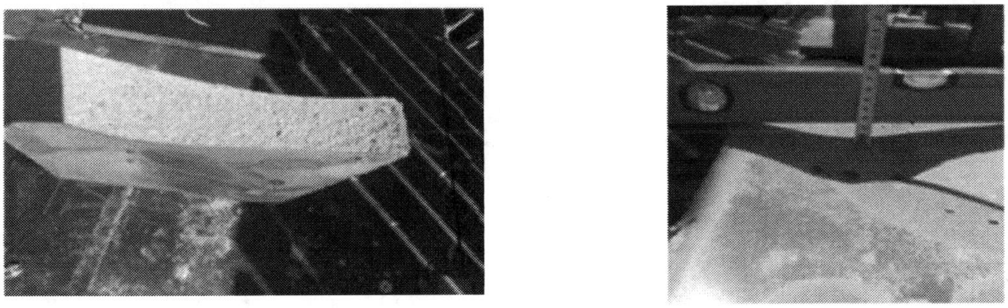

Figure 9. Final shape of the concrete element and measurement details

General conclusions

By taking into account the final shape of each poured element, the obtained measurements (Fig. 9) and the presented data, the system with no reinforcement can be use when multiple shape-defining elements are present (the shaping elements play also the fixing role). Otherwise, this experiment shows no restrictions for the multi –curvature elements.

ORCID IDs

R Diaconu https://orcid.org/0000-0003-2497-8276

References:

[1] Agenda Constructiilor – Utilaje si echipamente, 19.05.2021 https://www.agendaconstructiilor.ro /files/utilaje-echipamente/hunnebeck-cum-poate-impacta-alegerea-cofrajului-costul-unui-proiect-de-constructii.html

[2] Diaconu R and Brindasu P D 2016 Personalised formwork design by using PLM *Academic Journal of Manufacturing Engineering* **14** (3) 41-9

[3] Diaconu R and Brindasu P D 2019 Personalised Formwork – scientific approach for new solution variants *MATEC Web Conf.* 290 (1) 04005 https://doi.org/10.1051/matecconf/201929004005

[4] Henson R N 2015 Analysis of Variance (ANOVA) *Reference Module in Neuroscience and Biobehavioral Psychology - Brain Mapping - An Encyclopedic Reference* vol 1, p 477-481 ed Academic Press

AUTHOR INDEX